BAYESIAN ANALYSIS IN ECONOMETRICS AND STATISTICS

ECONOMISTS OF THE TWENTIETH CENTURY

General Editors: Mark Perlman, *University Professor of Economics, Emeritus, University of Pittsburgh* and Mark Blaug, *Professor Emeritus, University of London, Professor Emeritus, University of Buckingham and Visiting Professor, University of Exeter*

This innovative series comprises specially invited collections of articles and papers by economists whose work has made an important contribution to economics in the late twentieth century.

The proliferation of new journals and the ever-increasing number of new articles make it difficult for even the most assiduous economist to keep track of all the important recent advances. By focusing on those economists whose work is generally recognized to be at the forefront of the discipline, the series will be an essential reference point for the different specialisms included. Wherever possible, the articles in these volumes have been reproduced as originally published using facsimile reproduction, inclusive of footnotes and pagination to facilitate ease of reference.

A list of published and future titles in this series is printed at the end of this volume.

Bayesian Analysis in Econometrics and Statistics

The Zellner View and Papers

Arnold Zellner

H.G.B. Alexander Distinguished Service Professor Emeritus of Economics and Statistics, University of Chicago, US

ECONOMISTS OF THE TWENTIETH CENTURY

Edward Elgar

Cheltenham, UK • Lyme, US

Published by
Edward Elgar Publishing Limited
8 Lansdown Place
Cheltenham
Glos GL50 2HU
UK

Edward Elgar Publishing Company
1 Pinnacle Hill Road
Lyme
NH 03768
UK

A catalogue record for this book is available from the British Library

Library of Congress Cataloging-in-Publication Data

Zellner, Arnold.
Bayesian analysis in econometrics and statistics : the Zellner view and papers / Arnold Zellner.
(Economists of the twentieth century series)
Includes bibliographical references.
1. Econometrics. 2. Bayesian statistical decision theory.
I. Title. II. Series.
HB139.Z443 1997
330'.01'5195—dc21 96–48222
CIP

ISBN 1 85898 220 0

Printed and bound in Great Britain by
Biddles Limited, Guildford and King's Lynn

Contents

PART IV BAYESIAN METHODS

Section A Estimation

Section B Testing and Model Selection

Acknowledgements

The author and publisher wish to thank the following who have kindly given permission for the use of copyright material.

Academic Press for article: 'Sensitivity of Control to Uncertainty and Form of the Criterion Function' with Martin S. Geisel, *The Future of Statistics*, 1968, 269–89.

American Statistical Association for articles: 'Statistics, Science and Public Policy', *Journal of the American Statistical Association*, **87** (417), March 1992, 1–6; 'Optimal Information Processing and Bayes's Theorem' (with discussion), *The American Statistician*, **42** (4), November 1988, 278–84; 'Bayesian and Non-Bayesian Analysis of the Regression Model with Multivariate Student-*t* Error Terms', *Journal of the American Statistical Association*, **71** (354), June 1976, 400–5; 'Bayesian Estimation and Prediction Using Asymmetric Loss Functions', *Journal of the American Statistical Association*, **81** (394), June 1986, 446–51; 'Minimum Expected Loss (MELO) Estimators for Functions of Parameters and Structural Coefficients of Econometric Models' with Soo-Bin Park, *Journal of the American Statistical Association*, **74** (365), March 1979, 185–93; 'Prediction and Decision Problems in Regression Models from the Bayesian Point of View' with V. Karuppan Chetty, *Journal of the American Statistical Association*, 1965, 608–16.

Cambridge University Press for articles: 'The ET Interview: Professor Arnold Zellner', interviewed by Peter E. Rossi, *Econometric Theory*, **5**, 1989, 287–317; 'Bayesian Analysis, Model Selection and Prediction' with Chung-ki Min in *Physics and Probability: Essays in Honor of Edwin T. Jaynes*, 1993, 195–206.

Econometric Society, for article: 'Bayesian Econometrics', *Econometrica*, **53** (2), March 1985, 253–69.

Elsevier Science Publishers B.V. for articles: 'Bayesian Analysis in Econometrics', *Journal of Econometrics*, **37**, 1988, 27–50; 'Bayesian and non-Bayesian Approaches to Statistical Inference and Decision-making' in *Journal of Computational and Applied Mathematics*, **64**, 1995, 3–10; 'Maximal Data Information Prior Distributions' in *New Developments in the Applications of Bayesian Methods*, A. Aycak and C. Brumat (eds), 1977, North-Holland Publishing Co., 211–32; 'Bayesian Analysis of Dichotomous Quantal Response Models' with Peter E. Rossi, *Journal of Econometrics*, **25**, 1984, 365–93; 'Estimation of Functions of Population Means Expected Loss (MELO) Approach', including Structural Coefficients: A Minimum Expected Loss (MELO) Approach', *Journal of Econometrics*, **8**, 1978, 127–58; 'Jeffreys-Bayes Posterior Odds Ratio and the Akaike Information Criterion for Discriminating between Models', *Economics Letters*, **1**, 1978, 337–42; 'Forecasting International Growth Rates Using Bayesian Shrinkage and Other Procedures' with Chansik Hong, *Journal of Econometrics*, **40**, 1989, 183–202; 'Forecasting Turning Points in International Output Growth Rates Using Bayesian Exponentially Weighted Autoregression, Time-varying Parameter, and Pooling Techniques' with Chansik Hong and Chung-ki Min, *Journal of Econometrics*, **49**, 1991, 275–304; 'Bayesian and non-Bayesian Methods for Combining Models and Forecasts with Applications to Forecasting International Growth Rates' with Chung-ki Min, *Journal of Econometrics*, **56**, 1993, 89–118.

Institute of Mathematical Statistics for article: 'Sir Harold Jeffreys, F.R.S.: 1881–1989', *Institute of Mathematical Statistics Bulletin*, **18** (3), 336–7.

International Economic Review for articles 'Estimation of Regression Relationships Containing Unobservable Independent Variables', **11** (3), October 1970, 441–54; 'An Error-Components Procedure (ECP) for Introducing Prior Information about Covariance Matrices and Analysis of Multivariate Regression Models', **20** (3), October 1979, 679–92.

Kluwer Academic Publishers for article: 'Bayesian Methods and Entropy in Economics and Econometrics' in *Maximum Entropy and Bayesian Methods*, W.T. Grandy, Jr and L. H. Schick (eds), 1991, 17–31.

Macmillan Ltd for article: 'The Quality of Quantitative Economic Policy-making When Targets and Costs of Change are Mis-specified' in *Selected Readings in Econometrics and Economic Theory: Essays in Honour of Jan Tinbergen Part 2*, 1973, 147–64.

Springer-Verlag for paper: 'Bayesian Method of Moments (BMOM) Analysis of Mean and Regression Models' in *Prediction and Modelling Honoring Seymour Geisser*, J.C. Lee, A. Zellner and W.O. Johnson (eds), forthcoming 1996.

University of Chicago Press for article: 'Posterior Odds Ratios for Regression Hypotheses: General Considerations and Some Specific Results' in *Basic Issues in Econometrics*, 1984, 275–305.

University Press for article: 'Posterior Odds Ratios for Selected Regression Hypotheses' with A. Siow in *Bayesian Statistics: Proceedings of the First International Meeting Held in Valencia (Spain), May 28 to June 2, 1979*, Valencia, 1980, 585–603.

John Wiley & Sons, Ltd for article: 'To Combine or not to Combine? Issues of Combining Forecasts' with Franz C. Palm, *Journal of Forecasting*, **11**, 1992, 687–701.

Every effort has been made to trace all the copyright holders but if any have been inadvertently overlooked the publishers will be pleased to make the necessary arrangements at the first opportunity.

Preface

'Why and when did you become a Bayesian?' To provide something in the way of an answer to this frequently asked question, I have included the text of an interview, conducted by Peter E. Rossi, a former student of mine and current colleague at the University of Chicago. It provides an answer to the question and additional information about my life and Bayesian adventures.

It was from a reading of Sir Harold Jeffreys's book, *Theory of Probability*[1] in the early 1960s that I determined to commence a programme of research designed to compare Bayesian and non-Bayesian approaches to inference and decision-making and determine how well they work in solving theoretical and practical problems. After more than 30 years of research, I have concluded that the Bayesian approach has won this 'horse race', as is evident from views expressed in my American Statistical Association Presidential Address included in this volume. This address provides a report to the 'public' on Statistics, Science and Public Policy in which the important role of Bayesian analysis is described.

Indeed, it appears that a Bayesian era has arrived with an upswell in the volume of Bayesian research since the 1960s and the creation in 1992 of both the International Society for Bayesian Analysis and the American Statistical Association's Section on Bayesian Statistical Science. These organizations have already sponsored several very successful research meetings and publications and have ambitious plans for the future.

While some have suggested Bayes-non-Bayes 'compromises', it is a fact that much past and recent Bayesian research has broken new ground and is addressing many important inference and decision problems from new perspectives. New developments have occurred in information processing, model formulation and selection, estimation, testing, forecasting, decision-making, computational techniques, etc. These exciting new developments make compromises with former paradigms difficult. As I see it, our task is to develop, test and apply new, fruitful methods that incorporate, when appropriate, techniques and results from the past.

Since Harold Jeffreys was a valued friend and a world famous scientist, it was a great honour to be invited to write an obituary for him. As indicated in my essay, Jeffreys produced many fundamental philosophical, theoretical and applied Bayesian results. Also, that a natural scientist would recommend and use a subjective 'degree of reasonable belief' definition of probability surprises many. Using his subjective definition, he made many key contributions. That he made both theoretical and applied contributions to Bayesian analysis and provided an axiom system for scientific learning from data and experience that are of great value in all the sciences is a remarkable achievement.

The next three papers on Bayesian analysis in econometrics and statistics provide an introduction to various basic issues and controversial points. To sharpen the discussion, in my paper for the International Conference on the Foundations of

Statistics, I solved several estimation, prediction and control problems using the Bayesian approach and challenged non-Bayesians and others to provide better solutions. As yet, I have not heard of any. Also, to sharpen the qualitative discussion, several explicit challenges were stated that relate to fundamental Bayes-non-Bayes issues. After all these years, I still enjoy talking about these challenges and attempts to meet them.

Similarly, in my invited Fisher-Schultz Lecture for the Econometric Society, I used specific examples to make the general point that both non-Bayesians and Bayesians use considerable prior information in building and analysing models. Some of the negative points about complicated, multi-equation models have since become commonplace, particularly in view of the poor performance of many complicated macroeconometric models in recent years.

The last paper in this section, an invited talk to the ASA Harrisburg Chapter's Annual Meeting on Applications of Bayesian Analysis presents a summary description and discussion of Bayes-non-Bayes 'operating characteristics' of standard statistical practice. By making such simple, explicit comparisons of Bayes-non-Bayes procedures, many saw and understood the fundamental points without getting tears in their eyes from reading pages and pages of heavy text. Also, I learned from comments of participants.

In Part III, the first paper tackles a problem that has been facing Bayesians for literally centuries, namely the satisfactory formulation of prior densities. As mentioned in the paper, maximal data information priors (MDIPs) came on the scene in the summer of 1968 when I commenced work on the problem to get a reproducible, satisfactory procedure for formulating both 'non-informative' and 'informative' priors. I commenced by trying to derive priors that maximize the 'information provided by an experiment'. After weeks of fruitless labour, I decided that something was wrong with the criterion functional and changed it to the average information in the likelihood function or data density minus the information in the prior. Lo and behold, use of this criterion function led to many useful results, some of them detailed in the paper. When Ed Jaynes saw the MDIP procedure and its results for the first time, he remarked, 'Anything that simple must be right'. I did not argue with him.

Later, I discovered that the information in a joint density for the observations and parameters, as measured by its negative entropy relative to uniform measure, breaks up into the sum of two parts. The first is the average information in the data density or likelihood function and the second is the information in the prior density. Thus the criterion functional, denoted by G, that I intuitively adopted to produce MDIPs is just the difference of these two components of the total information in the joint density as explained in the second paper in this section. More recently, it dawned on me that G represents the 'total information provided by an experiment' and that earlier measures of the 'information provided by an experiment' do not include all of the information as explained in the papers in this Part. Also, many operational MDIPs and MDIP odds are presented in these papers.

The third paper in this Part takes up the problem of providing additional results on the formulation of models, prior densities and information processing procedures. After discussing the difficult problem of model formulation, it is pointed out that a reproducible, dependable procedure for producing models would be invaluable. Since maxent has worked well in physics and chemistry to produce important models, it was decided to derive some of the basic statistical models employed in statistics, econometrics and other sciences by maxent with many results provided in the paper.

Given a model produced by maxent, one fundamental input is available for use in processing sample information. A second component is a suitable prior density for the model's parameters. Subject matter knowledge, or lack of it and degree of invariance are relevant in formulating priors. As mentioned above, maximal data information priors (MDIPs) are optimal, explicit, reproducible priors that can incorporate subject matter information and invariance restrictions. Thus, two inputs needed for use of Bayes's Theorem or other information processing rules (IPRs) are available.

While Bayes's Theorem has been rationalized in various ways, in the paper, 'Optimal Information Processing and Bayes's Theorem', it is shown that Bayes's Theorem is an optimal information processing rule (IPR) in the sense that it optimizes an information theoretic criterion functional. That is, it minimizes the difference between measures of the output information (information in a 'post data' parameter density and a marginal data density) and the input information (information in the prior and data densities). Further, when the optimal Bayes' IPR is employed, it is the case that output information = input information, that is that no information is lost or added in processing the input information to obtain the output information, an important information conservation principle. Thus, Bayes's Theorem is a 100 per cent efficient IPR.

When one considers inference to be information processing or learning (note that Jeffreys regards learning from experience to be a fundamental objective of science), it is conceivable that there may be alternative learning models as some have already recognized. In the paper, 'Bayesian Methods and Entropy in Economics and Econometrics', one such alternative model is explicitly derived as an optimal IPR rule that permits the quality of sample and prior information to be different. The optimal IPR is derived and found to be in a form different from that of Bayes's rule. Also, with respect to temporal, sequential information processing with costs associated with obtaining new information and adjusting to it, a dynamic optimal IPR will probably differ in some respects from Bayes's rule. Such new IPRs can be evaluated in practice and if found to be useful will be an important new dimension to Bayesian analysis.

In Part IV, estimation, testing, model selection, prediction, control and forecasting problems are analysed using Bayesian methods and results are compared to results produced by non-Bayesian methods.

First, in the area of estimation, the problem of using 'unobservable' or proxy variables in regression analysis is considered. Bayesian and non-Bayesian methods are presented and compared with the finding that finite-sample Bayesian results are available and thus there is no need to rely on approximate asymptotic Bayesian or non-Bayesian procedures.

The next paper in Part IV deals with Bayesian and non-Bayesian analyses of a regression model with non-normal error terms, an important topic since not everything in the world is normal. While the multivariate Student-*t* error term assumption is broader than the usual normal assumption, it is remarkable that certain results under assumed normality still hold. In the paper on an error-components procedure for introducing prior information about covariance matrices' parameters, it is shown how to introduce inequality constraints on variances and covariances quite simply by use of error components, a procedure that generally leads to a reduction in the number of free parameters in an error term covariance matrix and that imposes the above inequality constraints. The effects on precision of estimation of using such constraints are analysed.

In a joint paper with Peter Rossi, Bayesian analyses of dichotomous response models, including logit and probit models are carried forward using importance function Monte Carlo integration techniques. For many years, I had been curious about the accuracy of asymptotic approximate results for this class of models. Calculations presented in the paper provide evidence that for small sample sizes asymptotic approximate results are not very accurate even for small models. The operational numerical integration procedures used in the paper make it unnecessary to rely on large sample approximate Bayesian or non-Bayesian results.

While many have used symmetric loss functions, it has long been recognized that in many problems, say bridge construction, medicine, real estate assessment, etc. that errors of the same magnitude but of different algebraic sign can have very different consequences and thus the need for asymmetric loss functions. In 1971 on a pleasant visit to the University of California at Berkeley, I brought this point to the attention of Hal Varian, then a graduate student who was working on real estate tax assessment problems. In a very short time, he formulated what has become to be the famous LINEX asymmetric loss function and used it in a path-breaking paper that he published. In my paper in the present volume on 'Bayesian Estimation and Prediction Using Asymmetric Loss Functions', further properties of optimal Bayesian estimators and predictors are derived. Surprisingly, it is found that these Bayesian estimators and predictors uniformly dominate the usual least squares estimators and predictors, often by a large margin. Currently, many have come to understand the importance of using estimates and predictions appropriate for asymmetric loss functions in solving many practical problems.[2]

With respect to the two papers on minimum expected loss (MELO) estimation, it is shown how functions of parameters, e.g. reciprocals and ratios of regression parameters and structural coefficients of usual econometric simultaneous equations models can be optimally estimated relative to sensible, simple generalized quadratic loss functions. With diffuse priors and normal likelihood functions, the resulting optimal MELO estimates are in simple forms that can be easily computed and compared to maximum likelihood (ML) and other estimates. While ML estimators are known not to possess finite moments, it is shown that the Bayesian MELO estimators generally possess finite moments. Also, the MELO estimate for coefficients of simultaneous equations models can be very close to or equal to the classical least squares estimate in small samples but not in large samples. See the applied estimation results in the Zellner-Park paper, included in this volume, to see how MELO estimates compare with others for parameters of two well known models. Further results on MELO estimates' and estimators' properties relative to a 'goodness of fit' and to combined 'goodness-of-fit' and 'precision-of-estimation' loss functions are presented in Chapter 21. Many are happy that after all these years, Bayesian methods have produced easily computable optimal point estimates, as well as complete posterior densities for parameters of the important simultaneous equations model.

In the last two papers of this section, a new Bayesian approach, the Bayesian method of moments, instrumental variable (BMOM/IV) approach is explained and applied in analyses of mean and regression models. It is designed to produce moments and probability density functions for parameters, realized error terms and future observations from a given set of observations in situations in which the form of likelihood function is unknown. With simple assumptions, it is shown how to compute post data moments and maximum entropy distributions for parameters and future observations as well as optimal point estimates of structural equations' parameters.

Bayesian testing and model selection are considered in the four papers included in Part IV, Section B. After a discussion of general issues, such as 'to test or not to test?' and empirical information on the scandalous manner in which hypotheses are being tested in the literature, Bayesian methods for testing hypotheses, some originally developed by Jeffreys, are described and applied, including posterior odds for hypotheses relating to regression coefficients. With a certain class of Cauchy priors, it is possible to relate Bayesian posterior odds directly to usual sampling theory t and F statistics. R.A. Fisher's P-values and other commonly used measures. These posterior odds have been used in the paper by Min and Zellner to evaluate 256 alternative forecasting equations. Corrections for 'selection effects', as Jeffreys calls them or for 'data-mining' are discussed and is a topic that deserves much more attention. Alternative model selection techniques are applied and discussed, a topic pursued in the third paper in this section. In it, a detailed comparison of the Akaike Information Criterion (AIC) and the Bayesian Posterior Odds Criterion (POC) is presented with the finding that the AIC contains some but not all the elements incorporated in the POC.

The papers on prediction and control, included in Part IV, Section C, treat central problems that previously had been treated rather qualitatively or approximately. Explicit Bayesian solutions to control problems are compared to approximate 'certainty equivalence' solutions with the result that the latter solutions are found to be suboptimal. Also, sensitivity of control solutions to the form of loss functions is investigated in two of the papers in this section. It is found that misspecification of a loss function can result in sub-optimal solutions. However, in certain cases, surprisingly errors can be off-setting. For example, if a policy-maker overstates a target value for the dependent variable in a regression and at the same time overstates the cost of changing the independent, control variable, the second error helps prevent the control variable from being changed so as to realize the overstated target value. Conditions are given such that the errors offset and result in an optimal solution. However, overstating target values and understating costs of changing control variables are reinforcing errors that lead to very suboptimal control solutions. When one thinks in terms of 'conservatives' and 'liberals', and the errors that they make, are the former more prone to make offsetting errors than the latter?

Several papers are presented that illustrate how Bayesian techniques have been developed and used in forecasting in Part V. Many Bayesian techniques have been developed, applied and evaluated analytically and by use of 'out-of-sample' forecasting experiments. These include various Bayesian shrinkage forecasts, combined forecasts, turning point forecasts, etc. That use of these Bayesian methods produces improved forecasts is a most important finding. Further, that variants of a relatively simple autoregressive, leading indicator (ARLI) model forecasts reasonably well, is a welcome result, one that raises a challenge, namely, can it be rationalized and/or produced by economic theory? Answers to this challenge have appeared in past work of Chansik Hong, Chung-ki Min, Luis Manas-Anton, and Arnold Zellner who have shown that it is possible to derive the ARLI models from simple macroeconomic theoretical models. Perhaps these are the first steps in a process that will produce macroeconometric models that work well in explaining the past, in predicting future outcomes and in helping to solve policy problems.

My Bayesian research has been made possible by the freedom and support provided by the University of Chicago, the University of Wisconsin and the National Science Foundation. In addition, colleagues and students, too numerous to list by name, have been most helpful and supportive for which I am most appreciative. I am also most grateful to Pamela Eckert, my secretary for many years, for her extremely valuable assistance.

Endnote

1. H. Jeffreys, *Theory of Probability*, London: Oxford University Press, first ed., 1939 and revised editions, 1948, 1961, 1967 and 1988.
2. See several papers on asymmetric loss functions in D.A. Berry, K.M. Chaloner and J.F. Geweke (eds), *Bayesian Analysis in Statistics and Econometrics: Essays in Honor of Arnold Zellner*, Wiley, 1996.

PART I

INTRODUCTION

[1]

Econometric Theory, 5, 1989, 287–317. Printed in the United States of America.

THE ET INTERVIEW: PROFESSOR ARNOLD ZELLNER

Interviewed by Peter E. Rossi

Arnold Zellner stands out among econometricians as distinctive in the breadth of his contributions to many different areas of econometrics. His pioneering work in systems of equations, Bayesian statistics and econometrics, or time series analysis would each have earned him worldwide recognition. In spite of a prodigious volume of theoretical work, Zellner retains a strong interest in applications which has fostered applied work in fisheries conservation, production theory, forecasting, and many other fields. His record of service to both the econometrics and the statistics profession includes such impressive achievements as the founding of two major journals, organizing two NBER/NSF seminar series, and the supervision of over thirty Ph.D. dissertations in economics, finance, econometrics, and statistics.

As a newly minted Ph.D. in economics in the late 1950s, Zellner set to work on the important problems of estimation and hypothesis testing in systems of equations. He soon produced fundamental work on the seemingly unrelated regressions (SUR) model and three-stage least squares. His 1962 article on SURs is one of the most cited articles in econometrics (Arnold often jokes that if he had a nickel for every SUR fitted, he would be a very wealthy man). His work on systems of equations made him aware of the limitations of large-sample theory. This interest in finite-sample results led to a close study of the multivariate analysis literature and contributed to Zellner's attraction to Bayesian analysis in which exact results are possible.

In the early 1960s, Zellner embarked upon a research program to evaluate the usefulness of the Bayesian approach. Since Bayesian statistics was not

 0266-4666/89 $5.00 + .00

very well-developed and Bayesian econometrics was practically nonexistent, Zellner developed the field by tackling, one by one, most of the important econometric models. His classic 1971 book, modestly entitled *An Introduction to Bayesian Econometrics*, provides Bayesian solutions for many important estimation problems in econometrics. In addition, the book provides important guidance in the philosophy of science and new approaches to hypothesis testing. More recently, Zellner has added to this line of research by providing exact sample results for the simultaneous equations model through what he calls the "direct Monte Carlo simulation approach," Bayesian and non-Bayesian treatment of regression models with nonnormal errors, and a MELO estimator for the simultaneous equations model which has impressive sampling properties.

Participation in a time series seminar as a visitor to Yale University in the late fifties fueled Zellner's interest in time series analysis. Along with Franz Palm, Zellner established the close link between multivariate time series models and dynamic structural equations models. This link provides new methods of estimation and diagnostics for structural econometric models. It also anticipates much of the linear-quadratic rational expectations literature in which economic models impose nonlinear cross-equation restrictions on vector autoregressions. A keen interest in the problems of seasonal adjustment led Zellner to organize two influential conferences on seasonal adjustment of time series for the Bureau of the Census. Recent projects on forecasting international output series and forecasting turning points continue Zellner's time series research.

Zellner's commitment and service to the econometrics profession is well-known. What is perhaps less well-known is his commitment to his colleagues and students. Colleagues can expect that their papers will be read carefully and that they will receive many insightful and useful comments. Students visit Zellner's office in Rosenwald 205D and return brimming over with many ideas, any one of which, if pursued vigorously, could develop into a Ph.D. thesis. Once a paper or idea impresses Zellner, he will urge his colleagues to learn about it so that the progress of econometrics research may continue.

The hallmark of Zellner's career is a commitment to scholarship, service, and the belief that complicated problems can be solved by the application of a few powerful, simplifying concepts.

It is my privilege to report this interview to you. The interview took place in Arnold Zellner's office at the Graduate School of Business, University of Chicago on May 31, 1988. I have organized the questions into four major categories: 1. Professional Career; 2. Simultaneous Equations Models; 3. Philosophy of Science and Statistical Methodology; and 4. Bayesian Methods.

1. PROFESSIONAL CAREER

How did you get interested in econometrics? What teachers and/or papers were influential in attracting your interest to econometrics?

A "few" years ago when I was a graduate student in physics at Berkeley, I became acquainted with some of my brother Norman's graduate student friends majoring in economics and agricultural economics. From discussions with them, I became convinced that there was a great opportunity to apply mathematical and quantitative techniques in the analysis of economic problems. Norman, who did a Ph.D. in agricultural economics at Berkeley, was very helpful in orienting me with respect to the quantitative study of economics, a subject which was not very well-treated in the four undergraduate economics courses which I took. I've always had a strong interest in economics and business and perhaps if my undergraduate courses had been more quantitatively oriented I might have switched from physics to economics much earlier. But I did make the change later at Berkeley in the 1950s. George M. Kuznets, my thesis chairman, and Ivan M. Lee of Berkeley's Department of Agricultural Economics, and Norman Buchanan, Robert Dorfman, Robert Gordon, and Walter Galenson of Berkeley's Economics Department were influential in stimulating and guiding my interest in economics and econometrics.

The switch that you refer to is between your undergraduate major in physics and graduate program in economics or did it occur after your graduate study in physics? How did that work?

I had had about a year and a half of graduate study in physics at Berkeley before switching into economics. Also, when I was in the army, I worked for a year and a half in biophysics and then came back to finish my graduate studies in economics.

The switch from physics to economics at Berkeley wasn't too hard because I'd had four economics courses as an undergraduate and a considerable amount of math which enabled me to enroll in graduate economics courses. At that time not many students in economics took mathematics or statistics. Anyone who knew a little math, I think, had a decided advantage.

What did you take in statistics as an undergraduate? Did you have statistics courses as part of your physics major?

Physicists were, in fact still are perhaps, babes in the woods when it comes to statistics. In physics you weren't required to take any statistics courses at all. I did get into statistical thermodynamics and statistical mechanics where you do direct probability in maximizing entropy and things of that sort. But when it came to analyzing data, we had no training at all except for a few laboratory sessions introducing us to probable error which few fully understood. Least squares was for me a calculus problem essentially. We had no formal statistical inference training at all. I picked that up in graduate econometrics and statistics courses. These courses seemed easy in one respect, namely, that most everything was assumed normally distributed without much controversy. In physics, theory is used to derive the forms of distributions.

How would you contrast the working atmosphere at Washington, Wisconsin, and Chicago (the three schools at which you have held appointments)?

These three institutions are impressive universities and all of them fortunately provided me with complete freedom to do the research that I wanted to do for which I'm most grateful. The most significant difference I think among the three institutions is that both Washington and Wisconsin had very large undergraduate programs, whereas Chicago is mainly a graduate institution. That I think is the biggest difference. At the University of Chicago, I am very fortunate to have had the opportunity to interact with many colleagues in economics and business since most are directly involved in research and don't have heavy undergraduate teaching loads. You get more of a research atmosphere in economics here at Chicago perhaps than at other places. What I like about Chicago particularly, and it's true too at Washington and Wisconsin, is that there's a strong interaction between theory and application.

Very few econometricians find their home in a business school. What particular advantages or disadvantages has this afforded you?

It's hard for me to generalize since I have had a position in just one business school since 1966. I think our school is unique in emphasizing the discipline approach to business and economic problems and we have, as you know, various disciplines very strongly represented in the school—economics, econometrics, statistics, management science, and behavioral science. Having these various disciplines represented in the school has had a very beneficial broadening effect on me. I think that's rather unique with our School of Business. Then another feature of Chicago's School of Business which appeals to me very much is its interactive nature. Since being here, I've had very close interaction with the Department of Economics, particularly in the Doctoral Econometrics Program. We've had a joint doctoral program in econometrics for many years and many of my students come from the Economics Department since all my courses are cross listed. At Chicago's School of Business, I think it's almost like having an appointment in a University within the University. It's a very exciting environment. We have financial economists, labor economists, microeconomists, marketing researchers, accountants, and others. My interaction on econometric topics with colleagues in different areas has been very stimulating.

Is there any one place or time in your career which you single out as the most productive or stimulating?

I've been fortunate in the sense that almost always I've had a number of projects underway about which I've been very enthusiastic. Some people call me an optimist because I'm very optimistic about getting results on almost any project. However, there are a few projects which stand out that were really exciting for me. One of them was the two-year study that I did at the University of Washington with Jimmy Crutchfield in the 1950s to evaluate the world's most famous fishery conservation program, the International Pacific Halibut Fishery Conservation Program. That was a very exciting and useful project. Then in the early 1960s at Wisconsin and at Chicago over the years, I've worked on a comparative evaluation of the Bayesian and other approaches to econometrics and statistics. That has been a very exciting experience involving interaction with a number of colleagues and students. Also, while at the University of Washington, I got the idea of writing systems in single-equation form and that gave me the seed for the idea of the seemingly unrelated regression paper that I did later and the three-stage least-squares paper that I did later jointly with Hans Theil. I had these ideas expressed in a paper in the IER in 1961 and took a year's leave as a Fulbright professor in the Econometric Institute and the Netherlands School of Economics in Rotterdam for a year. During that year, I had the good fortune to complete the work on the seemingly unrelated regression problem and on three-stage least squares. These were several very exciting experiences. Another one was in the construction of the Susquehannah River Basin Economy Model in the 1960s done in consulting work with the Battelle Memorial Institute. And then a very stimulating research experience was the joint work I did with Franz Palm in the early 1970s in the time series area. We emphasized and analyzed the link between statistical multiple time series models and dynamic structural econometric models and suggested a synthesis of the two approaches. I still find it a very exciting topic and I'm currently working on it. These are some of the items that come to mind but there are many other projects. Current topics involve work on a new derivation of Bayes' Theorem and its interpretation, information-processing, and our international forecasting and modeling project in which the SEMSTA approach that Palm and I developed is being used along with Bayesian shrinkage and turning-point forecasting techniques.

It is very rare that one person should have been involved with the founding of two major journals in econometrics/statistics, the *JBES* and the *JE*. Why did you co-found the *JE*? Why did you found the *JBES*? In particular, how is the mission of the *JBES* different from that of *JE*? Has the focus of *JE* remained true to your original intent?

As regards the *Journal of Econometrics*, Dennis Aigner and I founded it in the early 1970s mainly because we thought that a journal devoted to econometric statistics and econometric statistical methodology and applications was needed. Some journals in the econometric area had come to stress mathematical economics at the expense of econometrics. We thought there was an opportunity to develop a new journal to reflect work in that area—a growing amount of work, I should add. Currently, we are very happy with the journal. The journal has grown in size. We now have three volumes a year including special annals issues that treat specific topics in depth.

As regards the *Journal of Business and Economic Statistics*, the idea to found this journal emerged from two very successful applied time series conferences, one on seasonal analysis and adjustment and the other an extension of the first. These were held under the auspices of the Bureau of the Census, the American Statistical Association, and the National Bureau of Economic Research and brought together leading researchers from academia, industry, and government. These fruitful conferences and the wonderful papers presented at them suggested to me that this sort of interaction would be useful to foster, particularly in a journal devoted to a combination of theory and application. That was the way the *Journal of Business and Economic Statistics* got started. It took a couple of years to convince the ASA to found the journal and since it's been founded it's been successful. I take great pleasure and enjoyment from innovating, in general, and in setting up these two journals. It involved a lot of hard work with a good deal of satisfaction in seeing them started and quite successful.

2. SIMULTANEOUS EQUATIONS MODELS

It would be fair to say that your work on the SUR model is one of the most frequently cited and used techniques in econometrics today. How did you get the idea for SUR?

The origins essentially arose from my efforts to understand multivariate analysis; in particular, the multivariate regression model and other multivariate systems. As is usual when I try to seek understanding, I try to simplify problems. On a rainy night in Seattle in about 1956 or 1957, I somehow got the idea of algebraically writing a multivariate regression model in single-equation form. When I figured out how to do that, everything fell into place because then many univariate results could be carried over to apply to the multivariate system and the analysis of the multivariate system is much simplified, notationally, algebraically, and conceptually. I first applied the idea to the analysis of autocorrelated errors in multivariate regression in a 1961 IER paper, and then I noticed in analyses of panel data, in contrast to the traditional regression model, that different matrices of independent variables appeared in each equation. On applying generalized least squares to the system, it turned out that the generalized least-squares estimator differed from the equation-by-equation least-squares estimator. That was the tip-off that you could get some added precision in analyzing the equations jointly rather than one by one. The only fly in the ointment was that the generalized least-squares estimator for the system involved the unknown covariance matrix of the error terms. However, by substituting a consistent estimate, that provided a large-sample solution to the problem. Currently, Bayesian methods provide finite-sample solutions to SUR estimation, prediction, and testing problems.

On the whole, it is fair to say that the SEM has attracted a great deal of econometric research following your early work on 3SLS. However, linear SEMs are not very widely used in applied economic research. What accounts for the declining use of SEMs?

For many years, considerable attention has been given to the linear simultaneous equation model, theoretically and in applications. And at present, linear supply and demand systems are used to analyze markets quite broadly and there are many other problems that come up in the linear simultaneous equation form. Also, as you know, there have been some major developments in the analysis of the SEM system from the Bayesian point of view. Now through the use of direct Monte Carlo simulation and Monte Carlo numerical integration, we can provide finite-sample analyses for SEM systems. So there's been a fair amount of continuing work on those systems. Also what you call a linear simultaneous equation model is linear in the structural coefficients and error terms but it's not linear in the reduced form coefficients. You have the unrestricted reduced form subject to some nonlinear restrictions involving the structural coefficients and the reduced form coefficients. When we look at this simultaneous equation model in that form, it is in the form of certain rational expectations models. Some rational expectations models are in fact forms of the linear simultaneous equation model. And then there's an overlap between the linear simultaneous equation model and errors in the variables models. Thus, I don't think it's entirely accurate to say that interest has declined in the general SEM system but that new variants of the SEM have come to the fore.

Do rational expectations models pose unique econometric problems? Or are these models simply special cases of nonlinear regression models?

As I pointed out before, there are certain rational expectations models that are special cases of nonlinear regression models. In particular, you can view some of these models as unrestricted reduced form equations subject to nonlinear restrictions and that's exactly the form of the linear simultaneous equation problem. So in part, I agree with you that there are cases in which that can happen. Do rational expectations models pose unique econometric problems? I would say definitely yes—more on formulation of the models than on the statistical implementation, though on both scores. For example, many rational expectations models assume that everyone holds the same expectations. Then too, often it is assumed that everyone is in agreement about what is the true model. However, these assumptions have been relaxed in certain analyses. Some have allowed for dispersion of beliefs. Others have allowed for the possibility that several models are entertained. If you go ahead with the problem allowing for dispersion of beliefs and the possibility that several models are entertained rather than just one, I think one will have to consider the use of prior distributions and posterior probabilities on models and then combine results by averaging over beliefs and over models. That will involve new econometric methodology which will draw heavily on the Bayesian model selection literature.

3. PHILOSOPHY OF SCIENCE AND STATISTICAL METHODOLOGY

You have often stressed the importance of simple models in scientific inference. How did you come to form this view? After all, you have been involved in several large-scale modeling projects.

To be more accurate, I have emphasized the importance of "sophisticatedly simple" models. Some simple models are just absurd. Thus, I refer to models that are "sophisticatedly simple." I have come to the view that sophisticatedly simple models are to be preferred because they seem to work better in explanation and prediction. I can list a number of relatively simple models that have performed well in explanation and prediction but I find it hard to find any complicated models that work well in explanation and prediction in any area of science. Let me give you a few sophisticatedly simple models that seem to work well. From physics, $s = 1/2gt^2$. That works like a charm. Another one is $E = mc^2$. That is very simple and works very well too. I can mention many more simple equations from physics that work very, very well. Now from economics, we have supply and demand models that are relatively simple. If it weren't for them, I think economists would be having a really hard time. Similarly, some consumption theories, for example Friedman's consumption theory, are very simple and work fairly well. Arbitrage relations are relatively simple and very important. When I look at economics and other sciences, I see a number of sophisticatedly simple models that work fairly well. I have yet to see one complicated model that works well in explanation and prediction. You mentioned my work on the Susquehannah River Basin economy. That was a big model because we had a lot of detail to take into account. We had to subregionalize the Susquehannah River Basin economy. We had many industries, a water sector, an employment sector, and a demographic sector. But we could boil that model down into nine equations. The workings of the model were very simple. We had much detail because we disaggregated by age group, by region, by industry, and the water sector was part of the model, and so forth. Even though we had a lot of detail, we tried to keep everything as simple as possible. The other project I worked on with a large model was one of the macroeconometric models of the U.S. economy. I urged the model builders to simplify their model and, to make the point very dramatically, Steve Peck and I did a paper in which we simulated the model. The simulation experiments produced a number of surprises. The model builders did not know that they had built into their model certain features that showed up in the simulation experiments. So there I tried to urge them to simplify their model and see what they gained or lost by complicating their model. Some years later, I was told that workers at the Federal Reserve Board developed a core model. At the time their large model had 179 equations. Their core model was much, much smaller and attempted to simplify the big, complicated model—179 equations, many of them nonlinear stochastic difference equations—a very complicated model.

Some statisticians regard it as fundamentally impossible to test causal theories without the use of experimental data. For this reason, they regard empirical work in economics as irrelevant to the testing of competing economic theories. Do you agree that it is impossible to test economic theories using *one sample* of passively collected economic data?

I agree that if you have good experimental data and a lot of it, it makes testing theories much easier. I'm all for experimental data. There are a number of experiments that can be performed to shed light on the validity of competing theories, but as you say most of our data are passively generated by the system and it takes a good deal of skill to test alternative economic theories using one sample of passively collected economic data. Now if you just have one sample, I think the real task is to quantify accurately the uncertainty regarding competing theories. Given the information in the data, to what extent do the prior odds change as you analyze the data, and transform the prior odds to the posterior odds? For example, if you start one-to-one on two theories and the data move you to 1.5 to 1, you have a slight preference for one theory vis-à-vis the other. On the other hand, there is often no need to limit analyses to one sample. There's always the possibility of getting more samples. For example, in our current forecasting work, we have data for 18 countries and in the IMF-IFS data base they have data for 100 or more countries. So I think sometimes in testing theories we all tend to be lazy. We don't collect enough data. There is much more data out there many times than one would think. It takes a lot more work to collect the data and there's usually more than one sample of data relevant for testing important economic theories.

Many statisticians scoff at hypothesis testing as a means of testing model adequacy and advocate the increased use of graphic methods. Economists, on the other hand, persist in the parametric restriction approach to testing of economic theories. How do you stand on this issue?

First off, some of my best friends are statisticians and they often scoff at many things. The fact that they scoff at many things doesn't mean that what they scoff at is necessarily unsatisfactory. Some have suggested that instead of hypothesis testing that we all compute confidence intervals rather than test sharp nulls or precise nulls. Now you know the old joke about the statistician. He has his head in the refrigerator and his feet in the oven and says "on average I'm comfortable." Then he computes a confidence interval and says "I'm uncertain." Years ago, Jeffreys in his *Theory of Probability* discussed hypothesis testing at length. More recently, Berger and others in *JASA* (1987) and *Statistical Science* (1987) reexamine this issue. Many scientists, including physicists and economists, want to test sharp null hypotheses. Is the income elasticity of permanent consumption with respect to permanent income equal to one? Others want to test hypotheses such as that a variable has no effect on the dependent variable in a regression equation. Some people say these hypotheses are not realistic but that's really not a very sensible reaction. If a person wants to test such a hypothesis, he should have good methods for doing it. As you know, there are good methods for analyzing sharp nulls. Now if you have no reason for entertaining sharp nulls, then you don't want to be testing sharp nulls. You want to pursue an estimation approach and get the point estimate and standard error and measure the size of the effect. If the theory suggests a sharp null, then you should be able to test that very well.

I'm sympathetic with graphic methods but at present, graphic methods are really pretty much an art. What we're trying to do in econometric and statistic methodology is to turn art into science. So there's really great room for the development of scientific graphical methods—a theory of graphics. I think Persi Diaconis has attempted a contribution in that direction. There's a great opportunity to make graphic methods scientific, scientific in the sense that two people given the same data and priors will come up with approximately the same conclusions using satisfactory graphic methods.

You have always emphasized the close link between economic theory and econometric models. Doesn't the increased specialization of econometrics preclude this important mixture?

From Adam Smith's writings, I've had great respect for the effectiveness of specialization. There's no question about it that specialization can lead to enhanced output and production, but the question asks whether we want to forego a close link with economic theory and I guess that's what you mean by econometric models—applied econometric work. Is that what you mean?

Yes. I mean that the econometric specification should be closely tied to the economic theory. The economic theory should be aimed at generating an econometric specification, not just some sort of a loose suggestion of what variables would belong in the equation.

We have to distinguish a couple of situations here. In some analyses of data there is no satisfactory economic theory. We have this problem currently in certain macro areas. We go at the problem empirically. We do what you might call exploratory data analysis—to define models that forecast well. Forecasting and prediction are very important activities. If you can do those things well, it is quite an achievement. So we take it as an exploratory data analysis problem to find out what relationships work well in prediction using heuristic information about economics and so forth. Then when we have something that works well, we back up and try to get the economic theory to explain why it works well. So that's one situation where you don't have well-formulated and precise economic theory. You have a lot of background information about the subject but you don't have a well-formulated mathematical economic theory. And then exploratory data analysis of various types can dredge up all kinds of interesting facts and relationships that may work well in forecasting that theorists can then attempt to explain. This is emphasized in the structural econometric modeling time series analysis approach that Palm and I put forward. If you do have a well-formulated mathematical economic theory, it would be stupid not to take account of that theory in approaching the data, just to see if it works well would be of great interest. It really is a tremendous achievement to have a mathematical economic theory that works well in explanation and prediction. So its a very different situation to have a full blown, well-developed mathematical economic theory that works well in explanation and prediction. It would be very exciting to implement a theory like that. Now I think Friedman, to a certain extent in his book, *A Theory of the Consumption Function*, had a fairly well-developed, rather simple, model of consumer behavior and then implemented it with data throughout his book and made a number of predictions (eight or ten predictions near the end of his book) and told others what data to gather to test further implications of his theory. It's the only work in economics, I think, in which an author has made a list of predictions and told others how to perform the analyses and predicted the outcomes. So a lot depends upon the quality of the theory, the quality of the data as to what you do in applied econometric work, and one has to be very flexible. If the theory isn't there you can't insist on using theory. So there an exploratory data analysis approach could be very fruitful. But I do think, as I indicated, that a close interaction between applied workers and theoreticians is very fruitful in promoting the work of both. That's not to say that everybody should be involved in such work but perhaps a larger fraction of the profession should be involved in the close interaction between applied work and theory—of course not everybody.

4. BAYESIAN METHODS

How did you become a Bayesian? Are there any particular milestones associated with your "conversion" to the Bayesian philosophy?

As I mentioned earlier, in the early 1960s I embarked on a program of research to evaluate the Bayesian approach to inference and decision making relative to other approaches. So I never got converted. I started work on the problem and if the Bayesian approach had produced silly answers or poor answers, I would have dropped it like a hot potato. Over the years (about 26 years, I guess), I found that the Bayesian approach has been very fruitful and useful. It's still not perfect but no approach to inference is perfect. You can always develop additional systems that may be better. But as it stands now, I think the Bayesian approach comes off very well relative to current competitors. As regards milestones in the development of my interest in the Bayesian approach, in the early 1960s I participated in a University of Wisconsin Statistics Department seminar devoted to a reading of Harold Jeffreys' book, *Theory of Probability.* As you know, he's a leading natural scientist who along with Laplace and some others viewed the Bayesian approach as appropriate for statistical work in all areas of science and produced Bayesian solutions to many statistical problems. This caused me to rethink my own approach to econometrics. Then in 1962, I stumbled onto the *JRSS* paper by Barnard, Winston, and Jenkins on likelihood inference in time series analysis which emphasized that a weighted likelihood approach yields exact finite-sample inferences for time series problems. I was very worried about asymptotic approximations even way back then in 1962, so seeing that Barnard, Winston, and Jenkins had produced this weighted likelihood approach had a great impact on me. Their weighting functions could be interpreted as prior distributions and thus you had a complete finite-sample analysis of time series problems. I thought it would be worthwhile to do more research to make explicit comparisons between Bayesian and non-Bayesian inference and decision techniques which I've done over the years. Pursuing this comparative approach has made my research exciting and interesting to many different parties, including myself.

Currently, the Bayesian learning model is one model of learning. The fact that the Bayesian approach has a learning model embedded in it is quite unique among systems of inference. But there is a possibility that one can improve on the Bayesian learning model and I'm currently trying to get some results in this area by considering optimal information processing, in which we have an information criterion functional and optimize it subject to side conditions. The solution produced using calculus of variations techniques is an optimal information processing rule. Under certain conditions, it turns out that Bayes' Theorem is an optimal information processing rule, but that's under static conditions. If you change the conditions of the problem to make the problem more dynamic, it is conceivable that other solutions may emerge and they may be improved learning models. So solutions can change just as they change in firm problems. Static optimization leads to certain rules for the firm. If you use a dynamic model with costs of adjustment built in, you get other rules for profit maximization. Thus, going from the static case to the dynamic case you might expect different solutions to emerge. I've got some results along these lines where I allow for differing quality of the sample and prior information used in inference problems, and I'm currently working on other problems in this area. So my interest in the Bayesian approach hasn't been a religious interest. It's been a scientific interest and certainly motivated a lot by my curiosity to see which of several different approaches works better or best in solving econometric problems—a very pragmatic interest as well as an intellectual interest has motivated my work in this area.

What prompted you to write your classic book on Bayesian econometrics? Why did you find it necessary to write a book instead of articles?

At the time, Bayesian articles were published in many leading journals but scattered around the literature, and I felt they weren't having the impact that a book would have. Also, I felt that bringing together the results that we had at that time would be very useful for others so they could see and view the book as a report of our progress in the comparative evaluation of alternative approaches in econometrics. The book was in the nature of a progress report to the profession. I've always taken the responsibility to the profession very seriously. Since society is making funds available to support us while we're doing our research, it's very important to be accountable. So the book was an effort to be a little bit accountable—not too much, but a little bit.

What has changed in Bayesian econometrics since the publication of your book?

Of course we've solved many more problems from the Bayesian point of view, and the Bayesian solutions have compared very favorably with non-Bayesian solutions. Also there are many more Bayesian econometricians and statisticians now, including the Chairman of Harvard's Statistics Department, Don Rubin, the Director of the University of Minnesota's School of Statistics, Seymour Geisser, and the President of Carnegie-Mellon University, Dick Cyert. Thus, we have many, many more Bayesians now. Really in the 1960s, the number of Bayesians was very limited as well as the number of Bayesian publications. So that's a radical difference now as compared with 1971. The number of Bayesians has increased, the number of Bayesian publications has increased, the number of problems that are solved by Bayesian methods has increased, etc. It's a very different situation now than it was in the 1960s when Bayesians were very few and far between.

We now have more powerful computer techniques. That's perhaps currently the most important factor. We can compute posterior distributions for whatever problem we want by numerical integration or direct Monte Carlo simulation approaches and so forth. So progress on the computational side is very impressive and important. Earlier, we often had to use asymptotic approximations because we couldn't compute posterior distributions. Now for many, many problems, we can compute finite-sample posterior distributions for parameters, predictive densities for future observations, optimal point estimates, posterior odds for evaluating models, hypotheses, etc. That's a big difference. On the conceptual side, I think people are much, much more sophisticated now than they were back in the 1960s about Bayesian analysis. At that time, you might say the field was rather thin and anyone with a decided viewpoint had an inordinate effect on the market—the thin market. By now we've digested a lot of the controversial issues about uninformative priors, about utility theory and the expected utility hypothesis, Jeffreys' improper priors, Jeffreys' and Laplace's approaches to hypothesis testing, and other topics. For example, "Is Jeffreys a Necessarist?" That was argued until people were blue in the face years ago. But now I think people are much more sophisticated about these issues.

Let me also add along those lines that your work on the MELO estimation approach in structural equation models was not yet developed in the book and I think it really fills an important gap that was not covered.

It is obvious that the work of Sir Harold Jeffreys has greatly influenced you. How did you discover Jeffreys' work and why do you find it so much more persuasive than the work of Savage and de Finetti?

As I mentioned earlier, there was a seminar in the Department of Statistics at Wisconsin devoted to a reading of Jeffreys' work with George Box, Irwin Guttman, George Tiao, Norman Draper, Mervyn Stone, and some others involved in the reading of chapters in the book. We spent a whole semester reading that book. I should add that I was assigned to report on the most difficult chapter of the book, the chapter on hypothesis testing, and I came away from that seminar very impressed with Jeffreys as a philosopher of science, an applied mathematician, a statistician, a mathematician, a geophysicist, an astronomer—a man of all parts, you might say. It was very impressive to me that a person could cover so many different areas so knowingly. You know Good has reviewed that book and stated that Jeffreys has more of value on the philosophy of science as it relates to probability theory than all the professional philosophers have written about the subject lumped together. This is in a review that appeared some years ago. So it's a very impressive book on the philosophical side, on the theoretical inference side, and then also on the applied side. He had what are now called M estimators in his *Theory of Probability* years ago. He had nonparametric Bayesian analysis in his book *Theory of Probability*—how to make inferences about the median of a population given independent observations from a population and his contributions and applications in a number of different areas have been very impressive. I also find very impressive Jeffreys' many contributions to different areas of science and his keen interest in applications along with the philosophy of science. As regards Savage and de Finetti, they came at the problem in the way that Ramsey came at the problem—through the expected utility hypothesis. They started with utility theory and developed inference based on utility theory. Jeffreys considered the Ramsey-Bayes approach and then in a few pages decided that, while you could develop the subject that way, he preferred to develop it in the way that he did with probability, a fundamental concept and then the axioms of probability theory giving the rules for manipulating probabilities and leading to Bayes' Theorem. Then at the end if you want to introduce a utility function you can do that and maximize expected utility. But Jeffreys and I currently find utility theory somewhat controversial—let's put it that way. Mark Machina and others have written about the current status of the expected utility hypothesis and it's not as secure I think, as Savage and de Finetti believed earlier. Jeffreys avoided this problem by introducing probability as a fundamental concept on which to build and avoided reliance on the expected utility hypothesis that's central in Savage's and de Finetti's work.

The most common objection to Bayesian methods is that some sort of prior is required and some regard this as a diversion from the task of data analysis. How do you answer these objections?

The task of data analysis is a pretty loose term. I don't know exactly what you mean by that. I've kidded some data analysts about their poking through data without theory as really going back to the stone ages. I mentioned the data analytic approach earlier. That is in need of theory really. Data analysis—Do you mean by that descriptive statistical analysis? Descriptive statistics?

I don't subscribe to this view but I think if I could paraphrase it, they would say that you have to introspect and produce this prior distribution or density and that time spent on introspection may be more profitably spent on plotting the data or fitting models to the data or something like that which is what I think they mean by data analysis.

Well Bayesians of course plot data or look at the data. They don't spend all their time thinking of priors. Also, I should add that non-Bayesians sit around thinking about restrictions on simultaneous equation models. That's prior information. Others think about what to assume about the error terms properties. That's many times prior information. Others sit around thinking about how to formulate a model for the observations. That involves a tremendous amount of prior information. Others worry about how to choose a significance level and the power of their tests which involves a lot of prior information. So the thought that non-Bayesians are not using prior information I find hard to believe. The statements of non-Bayesians, John Tukey, David Freedman, Eric Lehmann, and others indicate that non-Bayesians use a lot of prior information—judgment in the choice of functional forms for relationships, judgment in accepting "reasonable estimates," judgment in the assumptions about error terms' properties, judgment about the total form of the model, etc. And then let's take the non-Bayesians who work with time-varying parameter models or random coefficient models. They put a distribution on the parameters just as Bayesians put a prior distribution on them. Is the distribution of the random parameters a prior or is it part of the model? Well if you like time-varying parameter models and you put an ARMA process on the time-varying parameters, you're involved I think in formulating your prior beliefs about parameter variation. In multivariate regression, you may have regression coefficient vectors that you assume normally distributed around some mean vector. Is that part of the likelihood function or the prior information? Non-Bayesians worry about whether to assume the vectors normally distributed, independently distributed, or identically distributed, just as Bayesians worry about a prior distribution. In my opinion, the amount of time spent on prior information is probably not far different for Bayesians and non-Bayesians.

Recent papers by Diaconis and Freedman and Sims have convinced some that Bayesian methods have some difficulty when applied to nonparametric problems or problems with an infinite number of parameters. Do you agree with this conclusion?

Which conclusion? That they have convinced some?

No, I think we can agree with that.

Well the people they've convinced I don't think know much about nonparametrics and models with an infinite number of parameters. First off, tell me about one infinite parameter model that works well in explanation and prediction. Can you name one? I can't.

Well, people use histograms quite a bit.

For explanation and prediction?

Not in prediction.

I said both, explanation and prediction. Because those are two vital activities in science—explanation and prediction. Prediction is involved in definitions of causality. Prediction is involved in definitions of induction. And explanation is what we're trying to do—learn about how things work. I, frankly, can't think of any infinite parameter models that work well in prediction and explanation, in economics or any field of science. Now when some are faced with these models with an infinite number of parameters, the natural inclination is to put a hyper-distribution on the parameters and reduce the problem to a model with a finite number of parameters. As regards the implication of these articles for Bayes and non-Bayes controversies, it carries on in a tradition. If we go back a few years, we had the marginalization paradox. If we go back further, we had some other mathematical quirk. Over the years, it's been one mathematical argument after another purporting to show that the Bayesian approach is defective. It goes back to R.A. Fisher too, and others. And yet the Bayesian approach continues on. These arguments are, I think in part, contrived. Certainly, the marginalization argument as answered by Ed Jaynes is a contrived criticism of the Bayesian approach. Now let me take an example to illustrate what I mean. Let's take an infinite parameter model; the simplest model I can think of: $y_i = \theta_i + u_i$, where u_i is an error term, θ_i is the mean of y_i, and i runs from 1 to n. As n increases, the number of θ's increases. So when we have n observations, we have n θ's, with $n + 1$ observations, we have $n + 1$ θ's, etc. If we think the error terms are independently normally distributed with zero means and say constant variance, 1 or σ^2, the maximum likelihood estimate for each θ_i is y_i. It is also a diffuse prior Bayes' estimate of θ_i. Now this estimate, $\theta_i = y_i$, has the property that is inconsistent. That's the nature of these arguments that these people are putting forward. A procedure leads to inconsistent estimates, and, therefore, it's defective. In this case, maximum likelihood and Bayes lead to inconsistent estimates. I would say that the fact that they lead to inconsistent estimates is a virtue and not a defect in both approaches because you have just one observation per parameter. To have the distribution of the estimator become degenerate in the parameter value in a situation like this would be all wrong. Here the estimator for θ_i is y_i. You have just one y_i and clearly the estimator can't have a distribution degenerate into θ_i. It shouldn't, given this. So you can see that maximum likelihood and Bayes with diffuse or proper priors for each individual θ_i. leading to inconsistent estimates is a virtue rather than a vice and really leads to my greater confidence in those approaches.

Let's think about what I said before about putting a hyper-distribution on the parameters. Suppose y_i is estimated total income for an economy in a particular year, the ith year estimated by blowing up a survey estimate. θ_i is the true population income for the country that year, and u_i is the error term of the ith year, the year of the survey. θ_i is a parameter to the survey statistician, total income. What do we do in economics? What do macroeconomists do? They model the θ_i's. That's macroeconomics. So therefore they put hyper-distributions on the θ_i's (I hope successfully). So it's really very traditional to put a distribution on the θ_i's and explain the variation of the θ_i's through time. And that gets around the infinite number of parameters problem because then you can introduce a finite number of hyperparameters and make inferences about them. So as I say, I can't think of any models with an infinite number of parameters that have been effective in economics or any other science. Maybe you can think of some. I should add that Jeffreys discusses this problem in one of the sections of his book—adding a parameter for each observation. And he says that's not satisfactory because the law changes and with each observation you get a new parameter or if you go at it in terms of polynomial expansions, if you keep increas-

ing the degree of a polynomial as the sample size increases, the nature of your law can change tremendously from one polynomial to another. So unless you can convince me that there are virtues in applying infinite parameter models in explanation and prediction, I don't think I'll be tempted to use them.

Recent surveys suggest that the influence of Bayesian ideas has been increasing in both statistics and econometrics. However, in econometrics only a small fraction of applied work uses Bayesian methods. What accounts for the slow diffusion of Bayesian ideas into empirical work?

I'm glad to hear you say that theoretical and methodological Bayesian work has infiltrated econometrics to that extent. On the applied side, keep in mind that many of the middle-aged applied econometricians working today (middle-aged and above) were trained in a period when they rarely were introduced to Bayesian methods. That's one fact, that a large number of applied workers have never been exposed to Bayesian methods. Some of the younger people like Geweke, Litterman, and others are very prominent in applied Bayesian work. Sims has become quite Bayesian. He's doing applied Bayesian work. Leamer, for example, got exposed to Bayesian methods at Michigan and at Harvard. He studied I think with Bruce Hill at Michigan and at Harvard with colleagues there, Dempster, and others. So the younger econometricians have been exposed to Bayesian methods. Years ago, the textbooks had very little Bayesian material in them. Then another factor impeding some of the applied work is the lack of computer programs. Bayesian computer programs are really very much needed. In the last five or ten years, some rather good Bayesian computer programs have come on the scene. Litterman's RATS program has some Bayesian capabilities and is available on both mainframes and PCs. There's some limited Bayesian capabilities in the SAS programs. The time series group at Warwick has a Bayesian analysis in time series program, BATS. We've just finished a PC version of our BRAP program. Up until recently, the computing situation has not been favorable for wide-scale application of Bayesian methods. But in recent years, these programs have come upon the scene and many more people seem to be experimenting with Bayesian methods at the present, including the Federal Reserve Bank of Minneapolis which operates a Bayesian vector autoregression forecasting model due to Litterman. Now even in government circles, Bayesian methods are coming into use—at the Federal Reserve Board, at the Federal Reserve Bank of Minneapolis, the Department of Agriculture, and other places. The Japanese under Akaike have a Bayesian seasonal adjustment computer program. Thus, progress is being made.

Computing costs continue to decline at an exponential rate for the third decade in a row. What implications does this have for the future of Bayesian econometrics and econometrics in general?

For Bayesian econometrics, I think this will certainly speed up the adoption of Monte Carlo numerical integration techniques, direct Monte Carlo simulation techniques for computing posterior distributions and predictive densities, and will allow more interactive computing where users can experiment more with different models for the observations and different priors to go along with the model and be able to characterize model uncertainty better. I think improvements on the computing side, as I mentioned two or three times earlier, are key for promoting more use of Bayesian methods. As regards econometrics in general, the improved computing and also not just computing but new techniques for information storage, transfer and processing will have a great impact on the production of data and its utilization. You know years ago data were transcribed by hand and stored in books and written records. They were very inaccessible in warehouses, etc. Over the years the data storage, data retrieval, and the data gathering systems have all been computerized and that's going to have more of an impact on econometrics than the developments I mentioned earlier. Getting into data more easily, having more extensive data sets, data sets that can be stored readily and readily available, and having the capabilities to do on-line surveys or on-line field experiments will, I think, make econometrics much more effective. Getting further data generated that way and ready for use and able to be processed—that will be extremely important for the future of econometrics in general. I really think getting more and plentiful data of high quality is a high-priority item. As you know, I've worked in the last couple of years to try to promote improvements in data and accessibility of data. In the past, I think econometrics has really been hamstrung by the fact that we worked with very limited data. As you mentioned before in one of your questions, one sample of data oftentimes is all we have and the data may be highly aggregated, seasonally adjusted by some mysterious process with computed values in the series that you don't understand and subject to all kinds of systematic errors and from that series we're supposed to make good, precise inferences about whether theory A or theory B explains the data. Well, with improvements on the data side, we'll get much cleaner data and data with much more information. That I think will lead us to discover new facts, new features of economic behavior, and feedback on economic theory and give us a better understanding of economic behavior. So I really think improvements on the data front are really required and they will have a first-order impact on progress in econometrics as well as, of course, on methodological and theoretical improvements, too.

5. MISCELLANEOUS

Over the years, you have had a very large number of doctoral students in econometrics, economics, finance, and statistics. What has been your philosophy of doctoral education and why have you been so successful in stimulating large numbers of students?

I think the basic element of my philosophy is the importance of the doctoral programs in our system. Young people it seems to me, about the doctoral age, are just ready to make their most basic contributions to research and thus providing a setting in which they can work effectively on their research problems is extremely important. Doctoral programs should, in my opinion, enable a person to realize his aspirations in research. Here at Chicago, particularly, we're training research scholars and the doctoral program here should function to allow young people to realize themselves in research and if they plan to go on in teaching, in teaching. Now to do that you can't have a doctoral program with too many restrictions, as you know. Here in Chicago, we're very flexible. We try to keep the number of restrictions down to a minimum, prodded along by Harry Roberts who would like to eliminate all restrictions. Over the years, that's been a distinguishing feature of our program. It has very few requirements. We've been very flexible to allow students to do what they want to do.

Usually, students have general ideas and I think the faculty's responsibility is to give them guidance in trying to get these general ideas more precise and formulated in such a way that they can answer specific, important problems relating to their general area of interest. So as a teacher, as a thesis advisor, this is what I try to do with the students who work with me—try to find out what their interests are, try to lead them to the literature so they can see what others have done in that area, and then to help them formulate their research in such a way that it produces new, useful, and important results. What is often very important in this situation is confidence in what one's doing. It is very important to try to generate confidence in the student who has no experience or limited experience in research. They can't do the impossible (sometimes solving research problems does appear to be impossible) and keeping one's confidence up is very important. It's extremely important to have an optimistic outlook—I can solve any problem. I try to steer students to simple important problems—the problems that are important and simplest to solve. If you have an important problem that's very hard to solve, maybe if you have an equally important problem that's easier to solve, it would be good to start with that one and then work toward solving the other problem. The experience in solving one problem may be helpful in solving the second, equally important problem which is more difficult. Oftentimes, I suggest to students that they form a matrix of problems ordered by degree of importance and by degree of complexity with the hope there's a simple important problem they can solve and then I urge them to work on that. If you have an important problem that's easy to solve, you might as well work on it first.

Are there any problems in econometrics which you regret not working on?

Well, I've worked on a broad range of econometric topics and about the only one I thought was perhaps too hard or too difficult is the econometrics of world economies and that's always fascinated me, the relative growth of economies, some develop, others decline. Working on that range of problems has always had an attraction for me. I've thought those problems were very difficult and I really thought that my efforts were better directed toward simpler problems, ones that are easier to solve. But currently we are working on the international forecasting problem. We may get into some other problems that involve the world economy. I'm very excited about what we've done already in the international forecasting area—forecasting rates of growth of real GNP for 18 countries and forecasting turning points in the rates of growth of these 18 countries. And I expect, within the next few years, we'll push that work further into other areas. So maybe after all these years I'll get into the area in which I've had an interest since graduate school days. I did take courses in economic development with Norman Buchanan and just thought the problems were too hard at that time.

What recent articles or books are likely to emerge as especially useful or insightful in the future?

That's a hard thing to predict. So take these answers with a few grains of salt, perhaps many grains of salt. First, I really think that the work on Bayesian time-varying parameter state-space modeling techniques described for example in the West/Harrison/Mignon paper in JASA, March 1985, has already been important and probably will be more so in the years ahead. As you know parameters are not always constant in life. Aggregation effects, Lucas effects, and other things can cause parameters to change. At the macro level and even at the micro level, when we did our fishery conservation work, the conservation regulations caused the seasonal pattern of fish migration to change. Some of the parameters of the fishes' migration behavior changed. You can get changing parameters at the macro or micro level and having good models for coping with time-varying parameters, I think will be very important. Richard Highfield in his thesis work at Chicago used some time-varying parameter models in building a quarterly forecasting model for the U.S. economy. Then, second, I think that there will be some steps forward in improving our capability of learning from data. Currently, we do it mostly, particularly non-Bayesians, in an informal way. Bayesians use Bayes' Theorem as a learning model. Well, Robin Hogarth and the late Hilly Einhorn, our colleagues in psychology here in the School, and David Grether, Tversky, and Kahneman and others are studying how people learn from data. I fully believe that we'll get better learning models that will help us a good deal in econometrics and statistics. With better learning models, we'll be better able to learn more effectively from our data.

Then testing in econometrics is currently in terrible shape, I think. We don't test hypotheses very well. And I think the work of Berger and his students, Jeffreys' work in his *Theory of Probability*, and some of your work in the production area where you're testing alternative models for production—that sort of work will come to the fore in the future to be reflected in changed techniques for evaluating alternative models. Then too, the work on model selection that you've done and others have done which involves getting posterior probabilities on alternative models and carrying along several models rather than just choosing "the best" model is very important because as Draper and Hodges and some others have been emphasizing, just relying on one model when there's model uncertainty doesn't give you a correct measure of the uncertainty of your inferences. It's very important to allow for model uncertainty in measuring the precision of forecasts and estimating parameters. You get different estimates with different models. That uncertainty should be reflected in your overall standard error or measure of precision. I think that will be important. Let me get back to my favorite author, Harold Jeffreys. I have promises from more and more people, for example, Arthur Goldberger and his colleagues, that they're going to read the *Theory of Probability*. That more people will read the *Theory of Probability* and *Scientific Inference*, I think will really change the way they look at science in general and applied statistics and econometrics in particular. I think that Jeffreys' *Theory of Probability* and *Scientific Inference* will be among the most influential books in the future.

How can we improve the teaching of introductory econometrics and statistics so as to attract more interest in these fields?

With respect to introductory econometrics and statistics, I think most people have a keen interest in learning about the economy and economics in general if the subjects are presented in a reasonable way. With respect to econometrics and statistics, I think having really good data bases available for undergraduate students to plot and analyze would be very helpful and then too, the relationship of the data to economics is sometimes not well taught at the undergraduate level. Students are given theory they can't relate to data and data they can't relate to theory. If you can't bring the two together, you're not going to have effective econometrics at the undergraduate level. Bringing the two together really takes some work and I think would generate a lot of enthusiasm for both econometrics and economic statistics.

Now there is a question about how econometrics and statistics are taught at the undergraduate level and some claim, in particular, John Hey who has a new textbook out entitled *Data in Doubt: An Introduction to Bayesian Statistical Inference for Economists*, that teaching statistics and econometrics from the non-Bayesian point of view is "unnatural" to quote his preface. In his teaching experience over the years, he found the results very unsatisfactory until a few years ago when in desperation, he changed to the Bayesian approach. He claims that the Bayesian approach cleared up most of his problems. The students found it natural and took to it with great success. Whether that's specific in his teaching experience or general is a question in my mind. I really would like to do some experiments to check whether his experience is unique to him or is more generally encountered. I haven't seen those experiments done so I'll just leave it to your own prior views as to whether you want to extrapolate from his experience to the general population or not. Now this is a case in which you have to use a lot of judgment. But it is interesting to look at his book and to see how he's developed mainly a diffuse prior Bayesian approach to the standard problems of making inferences about means, proportions, simple regression, multiple regression, and introductory simultaneous equation models. He's taught that in upper division courses for some years now and the book is into its second edition, so apparently he's convinced some people that it works. I don't know if the conclusion holds in general or not, but it's interesting. Then getting the students onto PCs, I think will change the situation a good deal. That will enable them to plot data, do analyses more readily, and get into the subject much more easily. Extensive use of PCs will be a decided positive factor in undergraduate econometrics and statistics courses, in my opinion.

THE PUBLICATIONS OF ARNOLD ZELLNER

BOOKS

1. *Economic aspects of the Pacific halibut fishery* (with J.A. Crutchfield), U.S. Department of the Interior, Government Printing Office, Washington, DC, 1963, iv + 173 pp.
2. *Readings in economic statistics and econometrics* (editor), Boston: Little, Brown, and Company, 1968, xiv + 718 pp.
3. *Systems simulation for regional analysis: an application to river-basin planning* (with H.R. Hamilton, E. Roberts, A.J. Pugh, J. Milliman, and S. Goldstone), Cambridge: MIT Press, 1969.
4. *Estimating the parameters of the Markov probability model from aggregate time series data* (with T.C. Lee and G.G. Judge), Amsterdam: North-Holland Publishing Company, 1970, 254 pp. [Russian translation, 1977] Second Edition, 1977.
5. *An introduction to Bayesian inference in econometrics*, New York: J. Wiley and Sons, Inc., 1971, xv + 431 pp. [Russian translation, 1980; Japanese translation] [Reprint: Robert E. Krieger Publishing Co., Malibar, Florida, 1987].
6. *Studies in Bayesian econometrics and statistics in honor of Leonard J. Savage* (co-editor with S.E. Fienberg), Amsterdam: North-Holland Publishing Company, 1975, ix + 676 pp., Two-Volume Paperback Edition, 1987.
7. *Seasonal analysis of economic time series* (editor), Proceedings of the NBER-CENSUS Conference on Seasonal Analysis of Economic Time Series, Sept. 9–10, 1976, Washington, DC, published by the U.S. Bureau of the Census, Government Printing Office, Washington, DC, 1978, xiii + 485 pp.
8. *Bayesian analysis of econometrics and statistics: essays in honor of Harold Jeffreys* (editor), Amsterdam: North-Holland Publishing Company, 1980.
9. *Seasonal adjustment of the monetary aggregates*, (with G.H. Moore, G.E.P. Box, H.B. Kaitz, D.A. Pierce, and J.A. Stephenson), Board of Governors of the Federal Reserve System, Washington, DC, 1981, 55 pp.
10. *Applied time series analysis of economic data*, (editor), Proceedings of the Conference on Applied Time Series Analysis of Economic Data, Oct. 13–15, 1981, Arlington, VA, published by the U.S. Bureau of the Census, Government Printing Office, Washington, DC, 1983, xiii + 399 pp.
11. *Basic issues in econometrics*, Chicago: University of Chicago Press, 1984, Paperback Edition, 1987, xxi + 334 pp.
12. *Bayesian inference and decision techniques: essays in honor of Bruno de Finetti* (co-editor with P.K. Goel), Amsterdam: North-Holland Publishing Company, 1986, x + 496 pp.
13. *Bayesian and likelihood inference: essays in honor of George A. Barnard* (co-editor with S. Geisser, J. Hodges, and S.J. Press), in progress.

ARTICLES

1951

1. An interesting general form for a production function, *Econometrica* 19: 188–189.

1956

2. International comparison of unemployment rates (with Walter Galenson), The National Bureau of Economic Research volume, *The measurement and behavior of unemployment*, Princeton University Press, pp. 439–581.

3. The electrophoretic properties of red blood cells (with J.B. Bateman), *Archives of Biochemistry and Biophysics* 44–51.
4. The electrophoretic properties of red blood cells after reaction with influenza virus hemagglutinin (with J.B. Bateman, M.S. Davis, and P.A. McCaffree), *Archives of Biochemistry and Biophysics* 384–391.

1957

5. Consumption and the consumption function in the 1948–1949 recession, *Review of Economics and Statistics* XXXIX: 303–311.
6. The short-run consumption function, *Econometrica* 25: 552–567.

1958

7. A statistical analysis of provisional estimates of gross national product and its components, of selected national income components, and of personal saving, *Journal of the American Statistical Association* 53: 54–65.
8. On parameter estimates provided by various methods, *Metroeconomica* 106–107.
9. Statistical techniques for the analysis of cost and earnings data, *Report of the Technical Meeting on Costs and Earnings of Fishing Enterprises*, UN FAO, London, pp. 109–113.
10. The corporate income tax in the long run: a comment, *Journal of Political Economy* LXVI: 444–446, and "Rejoinder," in same issue, p. 448.
11. The foreign-trade and balanced-budget multipliers (with F.D. Holzman), *American Economic Review* XLVIII: 73–91.

1959

12. Comment on the 1959 conference on consumption and saving: basic elements of a general model of consumer behavior, *Proceedings of the Wharton School Conference on Consumption and Savings*, (I. Friend and R. Jones eds.), Vol. II, pp. 488–498.
13. Review of M.H. Quenouille's *The analysis of multiple time series* (New York: Hafner Publishing Co., 1957), *Journal of Farm Economics* 682–684.
14. Sequential growth, the labor-safety-valve-doctrine and the development of American unionism (with G.G.S. Murphy), *Journal of Economic History* 402–421. [Reprinted in *Turner and the sociology of the frontier*, Basic Books, New York, 1968].

1960

15. Tests of some basic propositions in the theory of consumption, *AEA Proceedings* 565–573.

1961

16. Econometric estimation with temporally dependent disturbance terms, *International Economic Review* 2: 164–178.
17. The error of forecast for multivariate regression models (with John Hooper), *Econometrica* 29: 544–555.

1962

18. An efficient method of estimating seemingly unrelated regressions and tests for aggregation bias, *Journal of the American Statistical Association* 57: 348–368.
19. Application of mathematical programming techniques to a resource management problem, *Economic Effects of Fishery Regulation*, FAO, Rome, pp. 515–534.
20. Estimation of cross-section relations: analysis of a common specification error, *Metroeconomica* XIV: 111–117.

21. Further properties of efficient estimators for seemingly unrelated regression equations (with D.S. Huang), *International Economic Review* 3: 300–313.
22. Three-stage least squares: simultaneous estimation of simultaneous equations (with H. Theil), *Econometrica* 30: 54–78.
23. War and peace: a phantasy in game theory? *Journal of Conflict Resolution* 39–41.

1963

24. Decision rules for economic forecasting, *Econometrica* 31: 111–130.
25. Estimators for seemingly unrelated regression equations: some exact finite-sample results, *Journal of the American Statistical Association* 58: 977–992.

1964

26. Bayes' theorem and the use of prior knowledge in regression analysis (with G.C. Tiao), *Biometrika* 51: 219–230.
27. Estimation of parameters in simulation models of social systems, *Computer Methods in the Analysis of Large-Scale Social Systems*, Proceedings of a Conference sponsored by the Joint Center for Urban Studies, MIT and Harvard University, (J.M. Beshers, ed.), pp. 97–116.
28. On the Bayesian estimation of multivariate regression (with G.C. Tiao), *Journal of the Royal Statistical Society* 26: 277–285.

1965

29. Bayesian analysis of a class of distributed lag models (with C.J. Park), *Indian Economic Journal* 13 (3): 432-444.
30. Further analysis of the short-run consumption function with emphasis on the role of liquid assets (with D.S. Huang and L.C. Chau), *Econometrica* 33: 571–581.
31. Joint estimation of relationships involving discrete random variables (with T.H. Lee), *Econometrica* 33: 382–394.
32. Prediction and decision problems in regression models from the Bayesian point of view (with V.K. Chetty), *Journal of the American Statistical Association* 60: 608–616.

1966

33. Computational accuracy and estimation of simultaneous equation econometric models (with H. Thornber), *Econometrica* 34: 727–729.
34. On the analysis of first-order autoregressive models with incomplete data, *International Economic Review* 7: 72–76.
35. Specification and estimation of Cobb-Douglas production function models (with J. Kmenta and J. Drèze), *Econometrica* 34: 784–795.

1968

36. Maximum likelihood and Bayesian estimation of transition probabilities (with T.C. Lee and G.G. Judge), *Journal of the American Statistical Association* 63: 1162–1179.

1969

37. Generalized production functions (with N.S. Revankar), *Review of Economic Studies* 36: 241–250.
38. On the aggregation problem: a new approach to an old problem, In K. Fox et al. (eds.), *Economic Models, Estimation, and Risk Programming: Essays in Honor of Gerhard Tintner*, pp. 365–374. Berlin: Springer Verlag.
39. Sensitivity of control to uncertainty and form of the criterion function (with M. Geisel), In D.G. Watts (ed.), *The Future of Statistics*, pp. 269–289. New York: Academic Press.

1970

40. Analysis of distributed lag models with applications to consumption function estimation (with M.S. Geisel), *Econometrica* 38: 865–888.
41. Estimation of regression relationships containing unobservable independent variables, *International Economic Review* 11: 441–454. [Reprinted in *Latent variables in socio-economic models* (A. Scott, ed.), Amsterdam: North-Holland Publishing Co., 1977, Chapter 5].
42. Management of marine resources: some key problems requiring additional analysis, *Economics of Fisheries Management: A Symposium*, (A. Scott, ed.), University of British Columbia, Institute of Animal Resource Ecology, Vancouver, pp. 109–115.
43. The care and feeding of econometric models, University of Chicago, Graduate School of Business, Selected Paper No. 35, Chicago: University of Chicago Press, 18 pp.

1971

44. A study of some aspects of temporal aggregation problems in econometric analyses (with C. Montmarquette), *Review of Economics and Statistics* 53: 335–342.
45. Bayesian and alternative approaches in econometrics, *Frontiers in Quantitative Economics*, (M.D. Intriligator, ed.), Amsterdam: North-Holland Publishing Co., pp. 178–210. [Reprinted in *Studies in Econometrics and Statistics* (S.E. Fienberg and A. Zellner, eds.), Amsterdam: North-Holland Publishing Co., 1975, pp. 39–54.]

1972

46. On constraints often overlooked in analyses of simultaneous equation models, *Econometrica* 40: 849–853.

1973

47. Bayesian analysis of the Federal Reserve-MIT-Penn model's Almon lag consumption function (with A.D. Williams), *Journal of Econometrics* 1: 267–300.
48. Comment on "Marginalization Paradoxes in Bayesian and Structural Inference" by A.P. Dawid, M. Stone, and J.V. Zidek, *Journal of the Royal Statistical Society B*.
49. Real balances and the demand for money: comment (with D.S. Huang and L.C. Chau), *Journal of Political Economy* 81: 485–487.
50. Simulation experiments with a quarterly macroeconometric model of the U.S. economy (with S. Peck), In A.A. Powell and R.A. Williams (eds.), *Econometric Studies of Macro and Monetary Relations*, pp. 149–168. Amsterdam: North-Holland Publishing Co.
51. The quality of quantitative economic policy-making when targets and costs of change are misspecified, In W. Sellekart (ed.), *Selected Readings in Econometrics and Economic Theory: Essays in Honour of Jan Tinbergen*, Part II, pp. 147–164. London: Macmillan.
52. Use of prior information in the analysis and estimation of Cobb–Douglas production function models (with J.F. Richard), *International Economic Review* 14: 107–119.

1974

53. Review of G.B. Hickman (ed.), *Econometric models of cyclical behavior*, Vols. 1 and 2, NBER, Columbia University Press, 1972, *Journal of Business* 47: 286–290.
54. Time series analysis and simultaneous equation models (with F. Palm), *Journal of Econometrics* 2: 17–54.

1975

55. Bayes–Stein estimators for k-means, regression, and simultaneous equation models (with W. Vandaele), In S.E. Fienberg and A. Zellner (eds.), *Studies in Bayesian Econometrics*

and Statistics in Honor of Leonard J. Savage, pp. 627–653, Amsterdam: North-Holland Publishing Co.

56. Bayesian analysis of regression error terms, *Journal of the American Statistical Association* 70: 138–144.
57. Comments on "Stopover monetarism: supply and demand factors in the 1972–74 inflation" by Edmund S. Phelps, In D.I. Meiselman and A.B. Laffer (eds.), *The Phenomenon of Worldwide Inflation*, Washington, DC: American Enterprise Institute.
58. Time series analysis and econometric model construction, In R.P. Gupta (ed.), *Applied Statistics*, pp. 373–398. Amsterdam: North-Holland Publishing Co.
59. Time series analysis of structural monetary models of the U.S. economy (with F. Palm), *Sankhya C* 37: 12–56.

1976

60. Bayesian and non-Bayesian analysis of the regression model with multivariate student-t error terms, *Journal of the American Statistical Association* 71: 400–405.

1977

61. Evaluation of econometric research on the income tax and charitable giving, *Research Papers*, Commission on Private Philanthropy and Public Needs, Vol. III, Treasury Department, Washington, DC.
62. Maximal data information prior distributions, In A. Aykac and C. Brumat (eds.), *New Developments in the Applications of Bayesian Methods*, pp. 211–232. Amsterdam: North Holland Publishing Co.
63. Comments on "Time series analysis and causal concepts in business cycle research," In C.A. Sims (ed.), *New Methods in Business Cycle Research*, pp. 167–173, Federal Reserve Bank of Minneapolis.

1978

64. Bayesian analysis and computing, In A.R. Gallant and T.M. Gerig (eds.), *Proceedings of Symposium on Computer Science and Statistics*, pp. 195–201. Institute of Statistics, North Carolina State University.
65. Estimation of functions of population means and regression coefficients including structural coefficients: a minimum expected loss (MELO) approach, *Journal of Econometrics* 8: 127–158.
66. Folklore vs. fact in forecasting with econometric models, *Journal of Business* 51: 587–593.
67. Jeffreys-Bayes posterior odds ratio and the Akaike information criterion for discriminating between models, *Economics Letters* 1: 337–342.
68. Posterior distribution for the multiple correlation coefficient with fixed regressors (with S.J. Press), *Journal of Econometrics* 8: 307–321.
69. Retrospect and prospect, In A. Zellner (ed.), *Seasonal Analysis of Economic Time Series*, pp. 451–456. U.S. Government Printing Office, Washington, DC.

1979

70. An error-components procedure (ECP) for introducing prior information about covariance matrices and analysis of multivariate regression models, *International Economic Review* 20: 679–692.
71. Causality and econometrics, In K. Brunner and A.H. Meltzer (eds.), *Three Aspects of Policy and Policymaking*, Vol. 10, pp. 9–54. Carnegie-Rochester Conference Series, North-Holland.
72. Minimum expected loss (MELO) estimators for functions of parameters and structural coef-

ficients of econometric models (with S.B. Park), *Journal of the American Statistical Association* 74: 185–193.

73. Statistical analysis of econometric models, *Journal of the American Statistical Association* 74: 628–651.

1980

74. A note on the relationship of minimum expected loss (MELO) and other structural coefficient estimates, *Review of Economics and Statistics* 62: 482–484.
75. Comments on "Information sources for modeling the national economy" by Jay W. Forrester, *Journal of the American Statistical Association* 75: 567–569.
76. Discussion of "Advances in model-based seasonal adjustment" by W.P. Cleveland, A.P. Dempster, and J. Stith, *1980 Proceedings of the Business and Economic Statistics Section of the American Statistical Association*, p. 37.
77. Invited contribution to the discussion of G.E.P. Box's paper, "Sampling and Bayes—Inference in scientific modeling and robustness," *Journal of the Royal Statistical Society A* 143: 424–425.
78. Large-sample estimation and testing procedures for dynamic equation systems (with F. Palm), *Journal of Econometrics* 12: 251–283.
79. Posterior odds ratios for selected regression hypotheses (with A. Siow). In J.M. Bernardo, M.H. DeGroot, D.V. Lindley, and A.F.M. Smith (eds.), *Bayesian Statistics: Proceedings of the First International Meeting held in Valencia (Spain), May 28 to June 2, 1979*, pp. 585–603. Valencia, Spain: University Press.
80. Statistical analysis of hypotheses in economics and econometrics, In *1980 Proceedings of the Business and Economic Statistics Section of the American Statistical Association*, pp. 199–203.

1981

81. Bayesian inference, In D. Greenwald (ed.), *Encyclopedia of Economics*, pp. 65–69. New York: McGraw Hill Book Co.
82. Large-sample estimation and testing procedures for dynamic equation systems: a Rejoinder (with F. Palm), *Journal of Econometrics* 17: 131–138.
83. Philosophy and objectives of econometrics, In D. Currie, R. Nobay, and D. Peel (eds.), *Macroeconomic Analysis: Essays in Macroeconomics and Econometrics*, pp. 24–34. England: Croom Helm Publishing Co.

1982

84. Basic issues in econometrics: past and present, *The American Economist* 26: 5–10.
85. Invited discussion of John Geweke's paper "The measurement of linear dependence and feedback between multiple time series." *Journal of the American Statistical Association* 77: 313-314.
86. Is Jeffreys a "Necessarist?" *The American Statistician* 36: 28–30.
87. Reply to a comment on "Is Jeffreys a "Necessarist"?" *The American Statistician* 36: 392–393.

1983

88. Applications of Bayesian analysis in econometrics, *The Statistician* 32: 23–34.
89. Bayesian analysis of a simple multinomial logit model, *Economics Letters* 11: 133–136.
90. Canonical representation of linear structural econometric models, rank tests for identification and existence of estimators' moments, In S. Karlin, T. Amemiya, and L.A. Goodman (eds.), *Studies in Econometrics, Time Series and Multivariate Statistics*, pp. 227–240. New York: Academic Press.

91. Comment on Eric Sowey's "University teaching of econometrics: a personal view, *Econometric Reviews* 2: 323-327.
92. Discussion on the paper "Exogeneity, causality, and structural invariance in econometric modeling" by R.F. Engle et al. In G.C. Chow and P. Corsi (eds.), *Evaluating the Reliability of Macro-Economic Models*, pp. 112-118. New York: John Wiley & Sons.
93. Overview of the ASA-Census-NBER conference on applied time series analysis of economic data, In A. Zellner (ed.), *Applied Time Series Analysis of Economic Data*, pp. 351-359. Washington, DC: Government Printing Office.
94. Statistical theory and econometrics, In Z. Griliches and M.D. Intriligator (eds.), *Handbook of Econometrics*, Vol. I, Chapter 2, pp. 67-178. Amsterdam: North-Holland Publishing Co.

1984

95. Bayesian analysis of dichotomous quantal response models (with P.E. Rossi), *Journal of Econometrics* 25: 365-393.
96. Comment on "On prior distributions for binary trials" by Seymour Geisser, *The American Statistician* 38: 249-250.
97. Modeling a competitive industry with entry: implications for demand and supply analysis (with W. Veloce), *Economics Letters* 16: 71-75.
98. Posterior odds ratios for regression hypotheses: general considerations and some specific results, In A. Zellner, *Basic Issues in Econometrics*, pp. 275-305. University of Chicago.
99. The current state of Bayesian econometrics, In T.D. Dwivedi (ed.), *Topics in Applied Statistics*, New York: Marcel-Dekker.

1985

100. Applications in Bayesian analysis, In *SUGI 10: A New Decade Begins*, pp. 923-929. Cary, NC: SAS Institute.
101. Bayesian statistics in econometrics, In J.M. Bernardo, M.H. DeGroot, D.V. Lindley, and A.F. Smith (eds.), *Bayesian Statistics 2*, pp. 571-586. North-Holland Publishing Co.
102. Entry and empirical demand and supply analysis for competitive industries (with W. Veloce), *Journal of Econometrics* 30: 459-471.
103. Bayesian regression diagnostics with applications to international consumption and income data (with Brent R. Moulton), *Journal of Econometrics* 53: 187-211.
104. Bayesian econometrics, *Econometrica* 53: 253-269.
105. Estimating gross labor force flows (with J. Abowd), *Journal of Business and Economic Statistics* 3: 254-283.

1986

106. A tale of forecasting 1001 series: the Bayesian knight strikes again, *International Journal of Forecasting* 92: 491-494.
107. Bayesian estimation and prediction using asymmetric loss functions, *Journal of the American Statistical Association* 81: 446-451.
108. Bayesian solutions to a problem posed by Efron, *The American Statistician* 40(4): 330-331.
109. Biased predictors, rationality, and the evaluation of forecasts, *Economics Letters* 21: 45-48.
110. Further results on Bayesian minimum expected loss (MELO) estimates and posterior distributions for structural coefficients, *Advances in Econometrics* 5: 171-182.
111. On assessing prior distributions and Bayesian regression analysis with g-prior distributions, In P.K. Goel and A. Zellner (eds.), *Bayesian Inference and Decision Techniques: Essays in Honor of Bruno de Finetti*, pp. 233-243. Amsterdam: North-Holland Publishing Co.

1987

112. Bayesian prediction with random regressors (with S.B. Park), In David Giles and Maxwell King (eds.), *Specification Analysis in the Linear Model*, pp. 234–249. England: Routledge & Kegan Paul PLC.
113. Evaluating the methodology of social experiments (with P.E. Rossi), In Joe Peckman (ed.), *The Income Maintenance Experiments*, Federal Reserve Bank of Boston, Boston.
114. Invited discussion of "Testing precise hypotheses" by J.O. Berger and M. Delampady, *Statistical Science* 2: 339–341.
115. Macroeconomic forecasting using pooled international data (with A. Garcia-Ferrer, R.A. Highfield, and F. Palm), *Journal of Business and Economic Statistics* 5: 53–67.
116. Macroeconomics, econometrics, and time series analysis, *Revista Espanola de Economia* 4(1): 3–9.
117. Science, economics, and public policy, *The American Economist* XXXI(2): 3–7.

1988

118. Bayesian analysis in econometrics, *Journal of Econometrics* 37: 27–50.
119. Bayesian inference, In John Eatwell (ed.), *The New Palgrave*, London: Macmillan Press, Ltd.
120. Calculation of maximum entropy distributions and approximation of marginal posterior distributions (with R.A. Highfield), *Journal of Econometrics* 37: 195–209.
121. A Bayesian era, In J.M. Bernardo, M.H. DeGroot, D.V. Lindley, and A.F.M. Smith (eds.), *Bayesian Statistics 3*, Oxford University Press 509–516.
122. Bayesian methods for forecasting turning points in economic time series: sensitivity of forecasts to asymmetry of loss structures (with C. Hong), In K. Lahiri and G. Moore (eds.), *Leading Economic Indicators: New Approaches and Forecasting Records*, England: Cambridge University Press (in press).
123. Bayesian specification analysis and estimation of simultaneous equation models using Monte Carlo methods (with L. Bauwens and H.K. van Dijk), *Journal of Econometrics, Annals* 38: 39–72.
124. Causality and causal laws in economics, *Journal of Econometrics* 39: 7–21.
125. Discussion of "Risk and the economy: a finance perspective" by K.C. Chan and R.M. Stulz, In C.C. Stone (ed.), *Financial Risk: Theory, Evidence, and Implications*, pp. 125–137. Massachussets: Martinus Nijhoff Publishing Co.
126. Forecasting international growth rates using Bayesian shrinkage and other procedures (with C. Hong), *Journal of Econometrics, Annals* (P. Schmidt, ed.), *Issues in Econometric Forecasting* 40: 183–202.
127. Optimal information processing and Bayes' theorem (with discussion), *The American Statistician* 42: 278–284.
128. Turning points in economic time series, loss structures, and Bayesian forecasting (with C. Hong and G.M. Gulati), In S. Geisser, J. Hodges, S.J. Press, and A. Zellner (eds.), *Bayesian and Likelihood Methods in Statistics and Econometrics: Essays in Honor of George A. Barnard*, Amsterdam: North-Holland Publishing Co. (in press).

[2]

The IMS Bulletin
Vol.18, No.3, 1989, 336-337.

Sir Harold Jeffreys, F.R.S. : 1891-1989

Sir Harold Jeffreys, F.R.S., was born on 22 April 1891 in Fatfield, County Durham, England, and died on 18 March 1989 in Cambridge, England, at the age of 97. His wife, Lady Bertha Jeffreys survives him.

Sir Harold was educated at Armstrong College, Newcastle upon Tyne, and at St. John's College, Cambridge University. After achieving a brilliant scholastic record, he became a Fellow of St. John's College in 1914. From 1946 to 1958 he was Plumian Professor of Astronomy and Experimental Philosophy at Cambridge University. In 1925, he was elected a Fellow of the Royal Society and later became its Bakerian Lecturer and Royal Medalist. In 1960, he received the Copley Medal.

According to the obituary in *The Times* of London, 23 March 1989, Jeffreys was knighted in 1953, President of the Royal Astronomical Society, 1955-57, and President of the International Association of Seismology and the Physics of the Earth's Interior, 1957-60. In 1962, the Harold Jeffreys Lectureship was created by the Royal Astronomical Society and Jeffreys presented the inaugural lecture in 1963. *The Times* also notes that Jeffreys's "famous treatise, *The Earth: Its Origin, History and Physical Constitution*, which with re-editing and improvements, went through seven editions between 1924, when it first appeared, and 1976, did perhaps more than any other book to make the geophysics of the solid earth a coherent study. His researches in seismology, in particular, established his reputation as one of the leading authorities on that subject".

His early research dealt with topics in astronomical and meteorological dynamics. He was an accomplished mathematician and authored two monographs entitled *Cartesian Tensors* and *Operational Methods in Mathematical Physics*. He and his wife, Dr. Bertha Jeffreys, wrote *Methods of Mathematical Physics*, which was published in several editions by Cambridge University Press and became a highly respected and widely cited classic. In a review of the work in *Nature*, it is stated that, "The authors, distinguished for their knowledge of mathematical physics and cosmogony, and by their researches in these fields, possess also a remarkably comprehensive and deep understanding of pure mathematics. Their purpose has been to synthesize this mathematical knowledge for the service of physics ... The work is a benefaction to the cause of progress in natural philosophy".

The *Collected Papers of Sir Harold Jeffreys on Geophysics and Other Sciences* was published in 1971 by Gordon & Breach, London, in six volumes. Volume 6 is entitled, *Mathematics, Probability and Miscellaneous Other Sciences*. As is evident, probability theory was just one area in which Jeffreys worked. This work reflected his broad-ranging experience in other areas of science. Since he began research on probability theory after World War I in a series of papers written jointly with Dr. Dorothy Wrinch, his contributions have been of fundamental importance to the philosophy of science, scientific method and theoretical and applied statistics. His major books in this area are *Theory of Probability*, 1939, and *Scientific Inference*, 1931. Later editions of both books have been published. The general nature of this research has been stated in the preface to the second edition of *Scientific Inference* as follows: "The general standpoint, that scientific method can be understood if and only if a theory of epistemological probability is provided, remains unaltered. Consequently, I maintain that much that passes for theory of scientific method is either obscure, useless or actually misleading." In his book, *Theory of Probability*, Jeffreys presents a theory of epistemological probability, discusses its relevance for work in all areas of science and applies it to solve many practical, important statistical problems. Irving J. Good, who reviewed the 1961 edition of *Theory of Probability*, wrote that Jeffreys's book "is of greater importance for the philosophy of science, and obviously of greater immediate practical importance, than nearly all the books on probability written by professional philosophers *lumped together*." See also Seymour Geisser (1980), Irving J. Good (1980), and Dennis V. Lindley (1980) wherein Jeffreys's research contributions are discussed. Edwin T. Jaynes (1983, 1984) regards Jeffreys's work to be a direct continuation of Laplace's research on Bayesian theory, applications and philosophy.

As mentioned above, Jeffreys provided an axiom system for probability theory which was designed to be useful to scientists in their efforts to learn from their data and experience. For Jeffreys, a numerical probability is a measure of degree of reasonable belief in a proposition. His axiom system provides procedures for combining and changing such probabilities as new data are obtained, an ingenious inductive calculus useful in all areas of science. He does not link his definition of probability to utility or betting considerations as Bayes, Ramsey, Savage and others have done because, as discussed in the *Theory of Probability*, Jeffreys has some reservations about the adequacy of utility theory. However, he does note that his probabilities and the operations which he has produced from his axiom system can be employed to compute the mathematical expectation of utility functions and to obtain expected utility maximizing solutions. Further, his critiques of frequency, axiomatic, and other definitions of probability in Chapter 7 of his book are ingenious and devastating.

In addition to producing an inductive statistical calculus for science, Jeffreys made many other outstanding contributions to statistical science. He was proudest of his theory of significance testing based on the use of posterior odds which he regarded as much more sensible than testing based on "p-values" à la Fisher, his colleague at Cambridge, or on Neyman-Pearson testing techniques. Distribution theory, asymptotic theory, asymptotic expansions, robust estimation, invariance theory via differential geometry, integral divergence measures (later called Kullback-Leibler measures), randomization, mixture distributions, contingency tables and a key nonparametric problem are just some of the topics treated in his book. In addition, as Edwin T. Jaynes (1984) has noted, "Indeed the variety of [Bayesian] applications demonstrated by Jeffreys included all those for which 'Student', Fisher, Neyman, and Pearson had developed 'orthodox' methods." Further, Jeffreys provided a key modification of Laplace's Rule of Succession – see Geisser (1980) for discussion of this result. He also presented many practical, important applications of his methods to scientific problems.

From the 1920s, Jeffreys pressed ahead with his research on Bayesian philosophy, methods and applications in the face of great opposition from orthodox statisticians. That he had the courage and determination to complete his important work under such adverse conditions is an indication of his strength of character. We are all deeply indebted to Sir Harold Jeffreys for his fundamental contributions to the philosophy of science, mathematics, and many sciences including statistical science.

SELECTED REFERENCES

Jeffreys, Harold (1931), *Scientific Inference*, Cambridge University Press. [Reissued with Addenda, 1937; Second edition 1957; Third edition 1973.]

Jeffreys, Harold (1939), *Theory of Probability*, Oxford University Press. [Second edition, 1948; Third edition, 1961; Third revised edition, 1967, reprinted 1988.]

Jeffreys, Harold (1971), *Collected Papers of Sir Harold Jeffreys on Geophysical and other Sciences*, Vols. 1-6, Gordon & Breach, London.

Geisser, Seymour (1980), The contributions of Sir Harold Jeffreys to Bayesian inference, in Zellner (1980), pp. 13-20.

Good, Irving J. (1980), The contributions of Jeffreys to Bayesian statistics, in Zellner (1980), pp. 21-34.

Jaynes, Edwin T. (1983), *Papers on Probability, Statistics and Statistical Physics* (R.D. Rosenkrantz, *ed.*), D. Reidel, Dordrecht.

Jaynes, Edwin T. (1984), The intuitive inadequacy of classical statistics, *Epistemologia*, VII (Special Issue on Probability, Statistics and Inductive Logic), 43-74.

Lindley, Dennis V. (1980), Jeffreys's contribution to modern statistical thought, in Zellner (1980), pp. 35-39.

Zellner, Arnold, *ed.* (1980). *Bayesian Analysis in Econometrics and Statistics: Essays in Honor of Harold Jeffreys*, North-Holland, Amsterdam. [Reprinted by Krieger, Malabar, Florida, 1989.]

Arnold Zellner
University of Chicago
Chicago, Illinois

PART II

OVERVIEW OF BAYESIAN ANALYSIS

[3]

Statistics, Science and Public Policy

ARNOLD ZELLNER*

For many years, more than I would like to count, I have worked on problems involving statistics, science, and policy. Since understanding each of these subjects and their interrelations is quite difficult, some have suggested that we undertake a crash program in research to reach definitive conclusions useful, for example, to D. Allan Bromley and his client, President Bush. However, as you probably know, crash programs in research are rarely successful even though they are a lot of fun for everyone except the one paying the bills.

In connection with statistics, it's relevant to recognize the substantial growth in the number of statisticians and members and chapters of ASA during the decades of this century as shown in Figures 1 and 2. Given ASA's new constitution, the beginning of its 1991 membership drive and use of the freehand method with its optimal properties, I project that ASA membership, chapters, and sections will continue this amazing growth for many decades in the future.

Along with this growth in statistics, there has been similar growth in the "metric" fields which utilize statistics extensively, namely, econometrics, cliometrics, sociometrics, psychometrics, chemometrics, biometrics, technometrics, and so on. What has prompted this pronounced growth and what does the future hold for us? In part, this growth has occurred because, in many of the fields mentioned, there has been an effort to turn art into science using the methods of statistics, namely, experimental and survey design, estimation, testing, prediction, and other methods. To attempt to answer these questions more fully, let me review some of my experiences with statistics, science, and policy and present some of the lessons which have emerged from these experiences.

In my undergraduate and graduate study in physics years ago, we were told that in science measurements or data are fundamentally important. We did many experiments to generate data. However, we didn't receive any formal instruction in statistics to show us how to analyze and learn from our data in an effective manner. On many occasions, I used the well-known, optimal freehand method, which, as you know, produces the "right" answer all the time. For example, in analyzing my data from the Millikan oil drop experiment, it was easy to use the freehand method to fit a regression and from it the "right" electronic charge. However, many years later, my next door neighbor Jack Halpern, who has a chair in chemistry at Chicago, informed me that there were some low points in Millikan's data, about the only evidence at the time for the existence of subelectronic particles, namely, quarks. I couldn't resist saying that I had many "quarks" in my experiment. Also, I wonder what would have resulted had Millikan and I employed robust regression techniques in analyzing our data. I'm almost willing to bet that we would have gotten the "wrong" answer.

Later in research in biophysics while in the army, I had the good fortune to work in a lab which had a consulting group of statisticians, including Don Marquardt, who were kind enough to help me with analyses of my data which were subsequently published. From this and later experiences, I learned that many physical and other scientists are not well trained in statistics and on many occasions require the services of statisticians to produce and analyze their data satisfactorily. For example, in the late 1950s, I was surprised to learn that many economists were unable to perform regression analyses, let alone interpret regression results in a meaningful fashion because they were not well trained in statistics. Literally, they required help which I have provided over the years with great returns to me in the form of joint works and new ideas which stimulated my work on statistical methods. For example, it was my efforts to produce more satisfactory analyses of panel data and to simplify multivariate regression analysis that led to my multivariate "seemingly unrelated regression" estimation and testing techniques.

Many of the physical and other scientists with whom I

* Arnold Zellner is H. G. B. Alexander Distinguished Service Professor of Economics and Statistics, Graduate School of Business, University of Chicago, IL 60637. This article was presented as the presidential address at the 1991 annual meeting of the American Statistical Association in Atlanta, Georgia. Research financed by the National Science Foundation and by income from the H. G. B. Alexander Endowment Fund.

In the Public Domain
Journal of the American Statistical Association
March 1992, Vol. 87, No. 417, Presidential Address

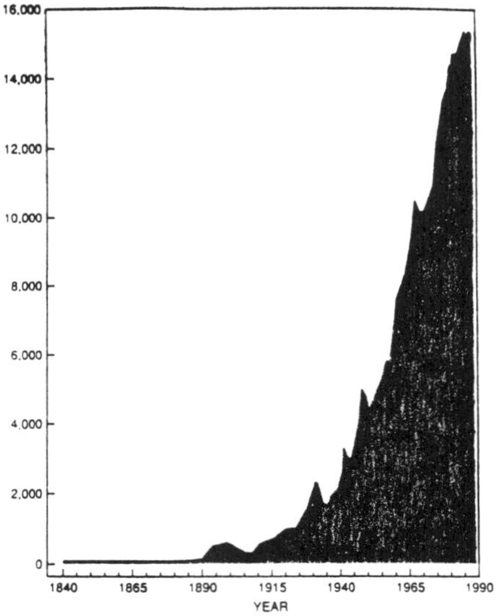

Figure 1. ASA Membership by Year. Source: Robert L. Mason, "A Golden Era of Statistics in America," in Proceedings of the American Statistical Society: Sesquicentennial Invited Paper Sessions, August 6–10, 1989 and August 22–25, 1988, 486–495.

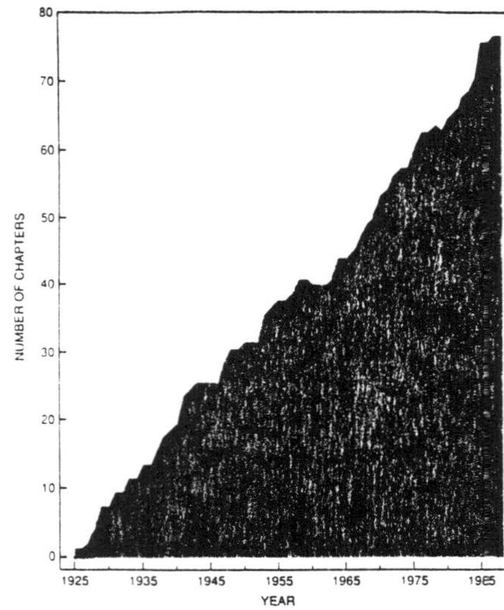

Figure 2. Number of ASA Chapters by Year. Source: Robert L. Mason, "A Golden Era of Statistics in America," in Proceedings of the American Statistical Society: Sesquicentennial Invited Paper Sessions, August 6–10, 1989 and August 22–25, 1988, 486–495.

worked over the years claimed that they deal with hard, cold, objective data. Hogwash! In a number of instances, there are problems associated with data in science as revealed by the cold fusion episode, by Millikan's notebooks, by Census undercounts, and by studies of those who report what they want to believe. As regards the cold fusion episode, the activities of Kevin Wolf, who is referred to in *Science*, October 1989, as "a hard-headed physicist [who] is skeptical about cold fusion," are described as follows:

> Wolf, among others, is looking to get good, statistically valid data on what nuclear products may be produced in the fusion cells. "That's what is lacking in this whole business—statistics," he says. Because the Pons/Fleischmann effect, if it exists, is so hard to reproduce, researchers have generally spent most of their time and money just trying to get cells that work. They generally have not set up control cells . . .

The article goes on to explain how Wolf's group and another group have done experiments ". . . with a proper blank control—cells with regular water in place of heavy water so that no . . . fusion is possible," as a check on the palladium possibly being contaminated with tritium. It's clear that in this instance producing uncontaminated data is far from a routine process.

In many other cases and fields producing good data is not usually a routine, objective, reproducible process. Thus, always, it is wise to heed the advice of Alan Enthoven, a former Undersecretary of Defense, who strongly advised us in a Chicago workshop meeting to GET THE FACTS! He made his point vividly by pointing out that in the 1950s it was estimated and believed that the Russians had twelve million military personnel in the field. During these cold war years, our Secretary of State threatened to drop the bomb on the Russians while the Europeans were singing better red than dead. At the end of the decade, someone asked how much of Russian GNP it would take to support their twelve million military personnel. After a little arithmetic and investigation, he debunked the twelve million figure by pointing out that many of the Russian divisions were reserve divisions and their division size was about seventy-five hundred rather than eleven thousand, as had been assumed. With these and other corrections, the estimated number was brought down to three to four million, however not before almost a decade of a cold war that could have turned into a hot, nuclear war at any time. Thus, then as now, the advice GET THE FACTS has great relevance for statistics, science and policy, and, fortunately, many statisticians are in the forefront in producing reliable facts. Perhaps this cold war episode points to a need for a new ASA section which would develop methods for measuring and monitoring the accuracy of news reported in the media. Certainly schools of journalism need good statistics courses.

In this connection, a 1989 cartoon in *Pravda* showed the Soviet public roaring in laughter on viewing the official Soviet economic statistics indicating strong growth and little inflation. Also, according to Chuck Waite of our Census Bureau, when Soviet government workers were told about U.S. procedures for releasing sensitive economic statistics, one of them asked, "How can the president change a number if he doesn't like it?"

With our own official government statistics, it's hard to know what the U.S. saving rate is—see, for example, Judge (1989). Based on Department of Commerce personal savings data, the saving rate is in the vicinity of 5%. On the other hand, individual savings as reported by the Federal Reserve Board yields a much higher savings rate, in the vicinity of 8 to 9%. With other reasonable adjustments, some estimate the savings rate to be in the vicinity of 15%. This is but one example of many others pointed out by Michael Boskin and Donald Ratajczak, our excellent 1991 B&E Section Forecast Luncheon Speaker, of weak spots in our national data base that make analysis, forecasting and policy-making extremely difficult. Even though our data are much, much better than the Soviets' data, we must work hard to improve them.

As another example illustrating the importance of getting the facts, some years ago an economist did a theoretical analysis of a famous marine conservation program, the International Pacific Halibut Conservation Program set up under a treaty between Canada and the United States and studied for several years by James Crutchfield and myself (1963). In his analysis, he assumed a constant equilibrium fish population and criticized the program for not charging fishermen a fee for fishing for halibut in the North Pacific. While his theoretical argument is probably valid under some conditions, this bright analyst forgot to GET THE FACTS. The facts, carefully collected over the years by marine biologists and the statistician Douglas Chapman, clearly revealed that the halibut population was declining for many decades and not equal to a constant equilibrium value. Here indeed is an example of what Thomas Huxley has called a tragedy of science, the destruction of a beautiful hypothesis or analysis by an ugly fact.

I have become convinced over the years that we should work harder to produce more ugly facts. One way to do this is to push current theories, hypotheses, models, beliefs, et cetera in any area of science, to extremes, deduce their implications and predictions, and then do the dirty work by producing the ugly facts to show that they are outlandish. The coup de grace can be applied most effectively using statistical expertise and finesse.

In addition to thinking of statistical and other procedures to produce ugly facts rather than just dull, run-of-the-mill facts, note that the great mathematician Hadamard (1945) surveyed his fellow mathematicians and found from their responses that recognition of unusual facts played a key role in producing major breakthroughs in mathematics. While Hadamard's description of this process is fascinating, I shall just remark that recognition of a new, unusual fact prompts many minds to work to produce explanations of it, that is, new hypotheses and theories. There are many well-known examples of this process in all fields of science. Thus a statistical problem is how to design investigations to produce unusual facts rather than humdrum, old, dull facts. Some years ago, I suggested a number of rather obvious ways to produce unusual facts, namely, study unusual historical periods, survey unusual groups, look for regularities in data, collect data on incorrect predictions and forecasts of scientists and their models, study equations and properties of large-scale econometric, global warming, macromolecular and other complicated models (almost every equation of such econometric models contains at least one unusual fact), and so on—see Zellner (1984, pp. 8–10). Having produced an unusual fact, a statistician may think the job is done. If so, a glowing opportunity is being missed, namely, the creation of an explanation or hypothesis to explain the unusual fact, which, if supported by much additional data, could lead to a Nobel Prize, if only there were a Nobel Prize in statistics. In my opinion, the Nobel Prize Committee needs educating on the importance of statistics in science and policy-making and how to distinguish between loose mathematicians and fine statisticians.

Speaking of Nobel Prizes, I must mention the great empirical economic statistical work of the Nobel Prize winner Simon Kuznets. He is credited with the basic research that led to our national income and product accounts, data which are invaluable to researchers, forecasters, and private and public policy-makers. He and others recognized the importance of measurement and description in science and policy. Thus, as a graduate studuent in economics I was shocked to learn that Kuznets's, Burns's, Mitchell's, and others' work on measurement of key economic variables was damned as "measurement without theory" by a leading, theoretically oriented econometrician. Later, I coined the phrase *theory without measurement* to denigrate the irrelevant work of some theorists and also to emphasize that in all fields both theorists and empirical workers have produced important research contributions. Instead of knocking heads much of the time, it would be better for theorists and empirical workers to interact more fruitfully in statistics and other sciences.

The usefulness of such interaction is evidenced by the research of Nobel Prize winners Milton Friedman, Merton Miller, Theodore Schultz, and George Stigler, my colleagues in economics and business at the University of Chicago. All of them have done outstanding work in establishing what are the facts in their fields, analyzing the facts using modern statistical methods and relating their results to economic theory and policy. Their work was enriched by close contact with econometricians and statisticians in our Business School and University Departments of Statistics and Economics in research workshop meetings, private consultations, and Chicago luncheon discussions. With a large statistics and econometrics group in our university, fruitful interactions between members of this group and applied empirical workers and subject matter theorists take place frequently and benefit all concerned. Applied workers improve their analyses, theorists learn how to relate their theories to data, and exposure to new problems often causes statisticians and econometricians to produce new statistical and computational methods.

Further, fine work in the quality management and improvement area in our Business School involving the statisticians George Easton, Edward George, William Golomski, Rob McCulloch, Harry Roberts, George Tiao, Ruey Tsay, and others has benefited from interactions with our

School's management scientists, economists, behavioral scientists, and other specialists. In this area, the facts have been assembled and analyzed using modern time series and other techniques with important results that have, along with results of many other researchers, had a significant impact on U.S. industry. Coupled with Deming's, Taguchi's, Box's, and others' methods and philosophy, this work has resulted in a growing appreciation of statistics and statisticians. I believe that these examples of the key role of statistics and statisticians in university research are not specific to Chicago and should be recognized generally by universities, especially those without departments of statistics. In this day and age, it's my view that a university without a department of statistics is like an automobile without a motor or a soldier without a rifle.

To return to the facts, usual, unusual, ugly, et cetera that are inputs to science, what are some of the objectives of science? I have come to agree with Karl Pearson (1938) and Harold Jeffreys (1988) that a major objective of science is learning from our data and experience and another is solving practical problems. Central in the learning process and in solving practical problems are the methods of statistics, which are at the heart of the unity of science. That is, similar statistical methods are not only used in many different fields of science but also use of good statistical methods for learning from data and experience is necessary for a field to be termed scientific. This central role of statistics in science involving data production, description, design of experiments and surveys, estimation, testing, prediction, risk analysis, and decision making has come to be widely appreciated and thus, in large part, the growth of statistics, econometrics, and all the other "metrics" fields in the decades of this century.

When it is stated that learning from our data and experience is a fundamental objective of science, one may ask, "How do you learn?" Some learn informally, while Bayesians use Bayes's theorem as a formal learning model. See, for example, Jeffreys (1957, 1988), Berger (1985), Box and Tiao (1973), DeGroot (1970), Hill (1974), Press (1989), and Zellner (1971). For years, many, including members of our NSF-NBER Bayesian Seminar group, have compared results of Bayesian and non-Bayesian analyses of specific problems. I have concluded from this and other work that the Bayesian learning model is effective for use in many, frequently encountered problems and yields rather good results. This is not to say that the Bayesian model is perfect—no model in science is perfect—but just that it has been shown to be quite useful in analyzing facts with given models and in making decisions in many different fields of science. Perhaps Good's (1991) good Bayes/non-Bayes compromise put forward for many years and reviewed in an invited paper presented at this meeting will be successful in uniting Bayesians and non-Bayesians. For such an accomplishment, Good deserves the Nobel Peace Prize.

At this point, I cannot resist saying a few words about use of Balanced Loss Functions (BLFs) which will, in my opinion, create more balance between Bayesians and non-Bayesians. Consider the n-means problem, $y_i = \theta_i + \varepsilon_i$, $i = 1, 2, \ldots, n$, where θ_i is the mean of y_i, or in vector notation, $y = \theta + \varepsilon$. A BLF for this problem is $L(\theta, \hat{\theta}) = w(y - \hat{\theta})'(y - \hat{\theta}) + (1 - w)(\hat{\theta} - \theta)'(\hat{\theta} - \theta)$, where $\hat{\theta}$ is some estimate of θ and w is a given weight, $0 \le w \le 1$. The first term of this BLF reflects goodness of fit and the second, precision of estimation. Given that a posterior density for θ is available, it can be used to obtain posterior expected loss, $E_\theta L(\theta, \hat{\theta})$, which can be minimized with respect to $\hat{\theta}$ to obtain the minimum expected loss (MELO) estimate, $\hat{\theta}^* = wy + (1 - w)\bar{\theta}$, where $\bar{\theta}$ is the posterior mean of θ. Currently, many non-Bayesians use $w = 1$ while most Bayesians use $w = 0$. Why not be balanced and use the MELO estimate based on a reasonable value of w which balances goodness of fit and unbiasedness and precision of estimation? For posterior expected losses, risk functions and Bayes risks of $\hat{\theta}^*$, y and $\bar{\theta}$, see my recent paper (1991), which also considers estimation of regression coefficients.

After facts or data are assembled and described, a fundamental problem is present. What model shall we utilize in analyzing the data? In this connection, I have become enamored with the precept KISS, *k*eep *i*t *s*ophisticatedly *s*imple, not as they say in industry, *k*eep *i*t *s*imple *s*tupid because some simple models are stupid. I can think of many sophisticatedly simple models in science that work well in explanation and prediction and have challenged many audiences to tell me about a single complicated model in any field of science that works well in explanation and prediction. As yet, I have not heard of any but have discovered a number of religious believers in mind-boggling, complex models and methods. These people need saving!

Usually, given the theoretical and judgmental information that we have, an initial, tentative sophisticatedly simple model can be formulated and analyzed. When little information is available, the initial model or benchmark model might be a random walk model or some such simple model as suggested by Harold Jeffreys, Milton Friedman, and others. If there is a need to capture much detail, a model may have to be large but with care it can be sophisticatedly simple. Many large econometric and other models on the scene involve hundreds of nonlinear stochastic difference equations. It's not clear whether these models have one or many solutions and the dynamic properties of these black-box monstrosities are not well understood. Years of effort and much money have been spent to build these Edsels and more is being spent to try to salvage them. Since they are dangerous to use in investigating scientific questions and making policy, I recommend that we export them to Saddam Hussein and hope that he uses them.

The problem of how to formulate models to explain past data, to predict new data successfully, and to be useful in policy-making is a first-order problem in science. From my experience in research, consulting on several industry and governmental model-building projects, and innumerable doctoral dissertation projects, I believe that we need a good *How to Build a Model* manual. Currently much multivariate model-building is still an art, not a science. If the manual could show us how to build multivariate models as successful as the Model T Ford, it would be a best seller. However, until it appears, the advice KISS is relevant and has been successfully utilized in many fields of science.

An ingenious approach to model-building is the maximum entropy (maxent) approach used in physics and linked to statistics by Shannon, Jaynes, and others. Jaynes's (1991) invited presentation at this ASA meeting provides a number of successful applications of the approach to statistical problems. In the maxent approach a model for the observations which maximizes entropy or minimizes cross-entropy subject to side conditions reflecting given initial information is recommended since it is a conservative choice of model which incorporates the initial, available information. In addition, it has been used to produce diffuse and informative prior densities for models' parameters. Also, in working on the problem of how to produce the Pearson system of densities by maxent several years ago, it occurred to me that perhaps Bayes's Theorem could be produced as a maxent solution. With a little effort, it turned out the Bayesian learning model, that is, Bayes's Theorem, could indeed be produced as the solution to an entropy optimization problem—see Zellner (1988). Thus maxent can produce models for observations, prior densities and a learning model with its associated posterior and predictive densities. This appears to me to be a fruitful, formal framework for model-building, learning from data, and making predictions. I hasten to add that using a model to make predictions and checking their accuracy is critical in appraising the quality of models and their associated prior information, as Geisser (1980, 1991) has emphasized. This is a direct way of evaluating the quality and usefulness of different prior information embedded in alternative models and their associated prior densities as emphasized by serious model-builders and statisticians. That is, the proof of the pudding is in the eating and it is noteworthy that more and more delicious pudding has been produced in recent decades.

On policy uses of statistics, producing data to describe and monitor problem areas such as human rights violations, discrimination, the environment, business cycles, cancer incidence rates, fish populations, et cetera can help considerably to characterize problems, further understanding of them, and appraise the costs and possible outcomes of alternative policies designed to achieve what are considered to be desirable objectives. In the case of the International Pacific Halibut Conservation Program, which I mentioned earlier, this is exactly what happened with the institution of a joint Canadian–U.S. conservation program in 1923. By establishing a sensible catch limit, based on much information and a good understanding of the halibut fishery and industry, the halibut population in the North Pacific grew considerably. By the 1960s, the annual catch was about 50% larger than in the 1920s, all achieved with minimal interference in industry operations. This relatively successful conservation program (note that the Atlantic halibut population was decimated) is a good example of how initial, extensive statistical information and subsequent statistical monitoring of a program's operation, coupled with a good, relatively simple behavioral model and sensible implementation policies, can lead to reasonably successful outcomes. Perhaps some of our educational reformers who are dealing with other kinds of fish would benefit from study of the halibut conservation program.

On communication of results by statisticians to the public and policy-makers, I believe that often there are serious problems. For example, consider the following exchange between a statistician and the President of the United States:

Statistician: President, my forecast of the rate of growth of real GNP for next year is 3.0% with a 1.0 percentage point margin of error.

President: What the hell is a 1.0 percentage point margin of error?

Statistician: It means that a 90% prediction interval stretches from 2.5 to 3.5 percent growth.

President: Oh, you mean that the probability is .90 that next year's growth rate is between 2.5 and 3.5.

Statistician: Well, a Bayesian would say it that way but I mean that an interval computed over and over again from repeated samples of data would have a probability of .90 of including the random variable, next year's growth rate.

President: You must be joking. There's only one sample of data. I'll stick with the .90 figure that it's between 2.5 and 3.5 even though I'm not sure what you mean by probability. Why don't you guys talk so that I can understand you?

This example involves a conditional probability statement, the one that the President opted for and a probability statement about the operating characteristics of a procedure for computing prediction intervals. Knowing when to use one or the other of these probability statements in connection with various types of intervals, point estimates and forecasts, test statistics, et cetera has caused problems for students and practitioners for years. Further, see Jeffreys (1988, Ch. 7) for a brilliant discussion of alternative definitions of probability. Certainly, if we statisticians can't be understood, few will listen to us.

To conclude, statistical data, reasoning, and methods have come to be employed widely in all sciences and in policy-making. The number of theoretical and applied statisticians has grown markedly over the years, and many of them have assumed leading roles in academia, government, and industry. To have these favorable trends continue in the future, we must work hard to improve the quality of statistical data and methods for communicating, storing, and displaying them. There has to be a greater emphasis on the production of ugly and unusual facts and explanations of them. More work has to be done on methods for learning from data and on building models which work well in explanation and prediction and are useful in policy-making. In all these efforts and in applied work on data production, forecasting, marketing analyses, censuses and surveys, quality management and improvement, drug testing, legal statistics, and so on, if more statisticians want to assume leadership roles, I believe that there has to be more investment not only in the study of statistics but also in relevant subject matter fields, for example, medicine, biology, engineering, business, law, et cetera. Along with many others, I view the future as being bright for statisticians and particularly bright for those who have solid training in statistics and a subject matter area. Finally, the American Statistical Association has to be given a good deal of credit for its constructive role in fostering the successful development of statistics in the past and present. Let us all join in a toast to ASA and in an effort to make it even more successful in the future.

REFERENCES

Berger, J. (1985), *Statistical Decision Theory and Bayesian Analysis* (2nd ed.), New York: Springer-Verlag.

Box, G. E. P. and Tiao, G. C. (1973), *Bayesian Inference in Statistical Analysis*, Reading, MA: Addison-Wesley.

Crutchfield, J. and Zellner, A. (1963), *Economic Aspects of the Pacific Halibut Fishery*, Washington: U.S. Government Printing Office.

DeGroot, M. H. (1970), *Optimal Statistical Decisions*, New York: McGraw-Hill.

Geisser, S. (1980), "A Predictivistic Primer," in *Bayesian Analysis in Econometrics and Statistics: Essays in Honor of Harold Jeffreys*, ed. A. Zellner, Amsterdam: North-Holland, pp. 363–382.

——— (1991), "Some Statistical Issues in Medicine and Forensics," President's Invited Address, ASA Meeting, Atlanta, Georgia.

Good, I. J. (1991), "The Bayes/Non-Bayes Compromise: A Review," invited paper presented at the General Methodology Session of ASA, Statistics in the Natural Sciences, Atlanta, Georgia.

Hadamard, J. (1945), *The Psychology of Invention in the Mathematical Field*, New York: Dover Publications.

Hill, B. M. (1974), "On Coherence, Inadmissibility and Inference About Many Parameters in the Theory of Least Squares," in *Studies in Bayesian Econometrics and Statistics in Honor of L. J. Savage*, eds. S. E. Fienberg and A. Zellner, Amsterdam: North-Holland, pp. 555–584.

Jaynes, E. T. (1991), "Probability Theory in the Natural Sciences," invited paper presented at the General Methodology Session of ASA, Statistics in the Natural Sciences, Atlanta, Georgia.

Jeffreys, H. (1957), *Scientific Inference* (2nd ed.), Cambridge: Cambridge University Press.

——— (1988), *Theory of Probability* (reprinted 3rd ed.), Oxford: Oxford University Press.

Judge, C. S. (1989), "Problem Isn't Rate of U.S. Saving, but Where the Money Goes," *Wall Street Journal*, March 23, p. A18.

Pearson, K. (1938), *The Grammar of Science*, London: Everyman Edition.

Press, S. J. (1989), *Bayesian Statistics: Principles, Models and Applications*, New York: John Wiley.

Zellner, A. (1971), *An Introduction to Bayesian Inference in Econometrics*, New York: John Wiley.

——— (1984), *Basic Issues in Econometrics*, Chicago: University of Chicago Press.

——— (1988), "Optimal Information Processing and Bayes's Theorem," *The American Statistician*, 42, 278–284.

——— (1991), "Bayesian and Non-Bayesian Estimation Using Balanced Loss Functions," Technical Report, H.G.B. Alexander Research Foundation, Graduate School of Business, University of Chicago.

Journal of Econometrics 37 (1988) 27–50. North-Holland

BAYESIAN ANALYSIS IN ECONOMETRICS*

Arnold ZELLNER

University of Chicago, Chicago, IL 60637, USA

After briefly reviewing some aspects of the history of Bayesian Analysis in Econometrics, five basic propositions regarding econometrics are put forward and discussed. Challenges relating to these propositions are issued. Then Bayesian estimation, prediction, control and decision procedures are discussed and a number of canonical econometric problems are described and analyzed to illustrate the power of the Bayesian approach in econometrics and other areas of science.

1. Introduction

Since the early 1960s there has been a considerable growth in the volume of Bayesian econometric research. This research has been focussed on the development and application of Bayesian methods to solve econometric problems. These include general problems of science, namely, describing, understanding and modeling economic phenomena, testing hypotheses suggested by economic theory, using econometric models and methods for prediction and policy analysis, and solving a variety of practical economic and business decision problems. Before the early 1960s, almost all econometricians employed non-Bayesian methods in their work in spite of the fact that Edgeworth, an eminent economist and statistician, and Jeffreys, a leading natural scientist, had produced many useful and important Bayesian methods by building on the earlier work of Laplace and others – see, e.g., Bowley (1928), Stigler (1978), Jeffreys (1967 / 1st ed. 1939 and 1973 / 1st ed. 1931), Geisser (1980), Good (1980), and Lindley (1980). It appears that the works of Savage (1954), Friedman and Savage (1948, 1952), Raiffa and Schlaifer (1961) and a rediscovery of Jeffreys's work stimulated several econometricians, including Albert Ando, Jacques Drèze, Walter Fisher, Tom Rothenberg, Arnold Zellner, and several others to commence research on Bayesian econometrics in the early 1960s. In Zellner (1984) further information is provided about these Bayesian researchers and their publications.

During the 1960s, Bayesian econometric research progressed rapidly and in 1970 the first meeting of the NBER–NSF Seminar on Bayesian Econometrics

*Research financed in part by the National Science Foundation and in part by the H.G.B. Alexander Endowment Fund, Graduate School of Business, University of Chicago.

was held at the University of Chicago that brought together leading Bayesian econometricians and statisticians. Since 1970, semi-annual meetings of the Seminar have been important in fostering much research in Bayesian econometrics and statistics, some of which has been published in Fienberg and Zellner (1975), Zellner (1980) and Goel and Zellner (1986). Also, since the mid-1960s a number of other Bayesian works have been published including Lindley (1965, 1971), DeGroot (1970), Lempers (1971), Morales (1971), Zellner (1971a), Press (1972), Box and Tiao (1973), Aykac and Brumat (1977), Leamer (1978), Bernardo et al. (1980, 1985), Berger (1980), Drèze and Richard (1983), Bauwens (1984), Kadane (1984), Boyer and Kihlstrom (1984), and Broemeling (1985). In addition, current and recent general econometrics textbooks include varying amounts of Bayesian material – see, e.g., Judge et al. (1985), Malinvaud (1980), Maddala (1978), Intriligator (1978), and Theil (1971). These works attest to the remarkable progress that has been made in Bayesian statistics and econometrics. It is accurate to state that the Bayesian approach to inference has had a significant impact on econometrics and that its influence is growing rapidly. Zellner (1984, 1985a) provides additional references to theoretical and applied Bayesian econometric studies.

Since this Conference is concerned with issues in the foundations of statistical inference and their impact on applied statistical work in econometrics and other fields, I shall review some elements of the foundations of Bayesian inference that I have found to be particularly important in econometric applications. Further, some comparisons of particular Bayesian and non-Bayesian solutions to specific estimation, prediction, control and other problems will be provided, in part an extension of the considerations presented in Rothenberg (1975) and Zellner (1975). Finally, I shall state some general propositions that I believe are of key importance for applied work in econometrics and other sciences and issue some challenges relating to them.

2. Foundational issues and applied econometrics

In considering alternative systems of inference for use in econometrics and other sciences, researchers have devoted considerable attention to issues in the philosophy of science. While I shall not attempt to review this vast literature, I shall present some general propositions that I believe have great relevance for applied econometric work. Many references will be made to Jeffreys's work because as Good (1980, p. 32) has written:

> 'In summary, Jeffreys's pioneering work on neo-Bayesian methods, at a time when such methods were very unpopular, was stimulated by his interest in philosophy, mathematics, and physics, and has had a large permanent influence on statistical logic and techniques. In my review

> (Good (1962b)) I said that Jeffreys's book on probability 'is of greater importance for the philosophy of science, and obviously of greater immediate practical importance, than nearly all the books on probability written by professional philosophers *lumped together*'.'

Thus, I and many others believe that there is much to be learned about foundational matters in Jeffreys's work that Jaynes (1983, 1984) regards as a direct continuation and extension of Laplace's work on Bayesian philosophy, inference and applications. Note that Laplace, Jeffreys and Jaynes have their roots in the physical sciences which have been on the whole very successful. It is quite probable that foundational propositions that these Bayesians regard as central are worth considering seriously by econometricians, other social scientists and statisticians. Some of these propositions will be stated and briefly discussed below. To sharpen the discussion and to provoke further considerations, the propositions will be followed by specific challenges.

Proposition 1. The Unity of Science Principle (USP) is valid.

Karl Pearson (1938), Jeffreys (1967) and many others subscribe to the USP that Pearson (1938, p. 16) states as follows: 'The unity of science consists alone in its method, not in its material ... It is not the facts themselves which form science, but the methods by which they are dealt with.' Also, Jeffreys (1967, p. 7) remarks: 'No matter what the subject-matter, the fundamental principles of the method must be the same. There must be a uniform standard of validity for all hypotheses, irrespective of the subject. Different laws may hold in different subjects, but they must be tested by the same criteria; otherwise we have no guarantee that our decisions will be those warranted by the data and not merely the result of inadequate analysis or of believing what we want to believe.'

Thus the USP implies that the same inductive methods are to be employed in making inferences in the natural, biological, and social sciences and in experimental and non-experimental sciences. That is, the principles of estimation, testing, prediction, etc. that are employed in learning from data in different areas of science should be the same according to the USP.

Challenge I. Demonstrate that fundamentally different inference principles for learning from data, that is fundamentally different principles of testing, estimation, prediction, etc. are required and are fruitful in different areas of science, say in the social sciences as contrasted with the physical and biological sciences.

A second proposition to which many physical scientists, statisticians and some econometricians subscribe is the belief that sophisticatedly simple hypotheses or models will probably work well in explaining facts and in prediction. This belief in the efficacy of sophisticatedly simple hypotheses and models is sometimes supported by an appeal to Ockham's Razor, the Principle of Parsimony or the Jeffreys–Wrinch Simplicity Postulate. Jeffreys (1967, p. 47) explains the Simplicity Postulate as follows:

> 'Precise statement of the prior probabilities of the laws in accordance with the condition of convergence requires that they should actually be put in an order of decreasing prior probability. But this corresponds to actual scientific procedure. A physicist would test first whether the whole variation is random as against the existence of a linear trend; then a linear law against a quadratic one, then proceeding in order of increasing complexity. All we have to say is that the simpler laws have the greater prior probabilities. This is what Wrinch and I called the simplicity postulate.'

As Jeffreys (1973, pp. 38–39) further explains, scientists cannot consider all possible models at once because they are all not known. What is done is to adopt a simple model that fits the data within a range of error. As new observations are obtained, the model is modified, usually implying greater complexity. However, sometimes this is not the case because a new simple conceptualization is put forward that explains the past data and makes good predictions. Einstein's general theory and quantum theory are cited as examples. With respect to these new theories or laws, Jeffreys (1973, p. 39) writes: '... the very fact that the new laws were simpler than the old ones was widely and influentially claimed to be a reason for accepting them. This amounts to saying that in the absence of observational evidence, the simpler law is the more probable...'

In view of the above considerations, the following proposition is put forward.

Proposition 2. The Jeffreys–Wrinch Simplicity Postulate (SP) is a very useful working principle in econometrics and other sciences.

Many natural scientists and statisticians will probably regard Proposition 2 to be 'obvious'. For example, Jaynes (1985, p. 334) writes: 'We keep our model as simple as possible so as not to obscure the point to be made; and also to heed Arnold Zellner's wise advice about 'sophisticatedly simple' models.' Also, some leading quantitative economists, including the Nobel Prize winners Jan Tinbergen, Ragnar Frisch, Milton Friedman, Kenneth Arrow, George Stigler, James Tobin, and Theodore Schultz all appear to appreciate the value of sophisticated simplicity in their work. The same can be said for many leading statisticians including George Box, Morris DeGroot, Seymour

Geisser, Bruce Hill, George Tiao, and others. However, there are a number of econometricians and economists who believe strongly that complicated models and methods are required and will work better in explaining and predicting economic phenomena than sophisticatedly simple models. Since I have not found this to be the case in my own econometric work on methods and applications, I have issued the following challenge on many occasions:

Challenge II. Demonstrate that a very complicated model in any area of science has performed well in promoting understanding of past data and in predicting as yet unobserved data.

With respect to very complicated macroeconometric models containing hundreds of non-linear stochastic difference equations and thousands of parameters, the eminent econometrician Christ (1975, p. 59) has written: '...they [the models] disagree so strongly about the effects of important monetary and fiscal policies that they cannot be considered reliable guides to such policy effects, until it can be determined which of them are wrong in this respect and which (if any) are right.' Also Litterman (1980, 1985) and McNees (1985) have compared the forecasting performance of several of these large, complicated macroeconometric models and concluded that it is not superior to that of a relatively simple Bayesian vector autoregressive model developed by Litterman (1980, 1985). It is the author's view that a prior preference for complicated methods and models, not only in macroeconomics but also in other areas, has significantly impeded the progress of econometrics and possibly of other social sciences.

A third proposition which most scientists support is the Prediction Principle (PP), namely that prediction is central in science. Indeed, in a review of philosophical concepts of causality Feigl (1953, p. 408) summarizes philosophers' views by defining causality as predictability according to a law or set of laws. In Zellner (1984), this concept of causality is compared with others and it is concluded that there is no need for a special definition of causality in econometrics in accord with the USP. Further Jeffreys (1967, p. 8) defines induction as follows: '...part [of our knowledge] consists of making inferences from past experience to predict future experience. This part may be called generalization or induction.' Thus prediction is central in science as explained above and leads to the following proposition:

Proposition 3. The Prediction Principle (PP) that predictive performance is central in evaluating hypotheses and models is of key importance in econometrics and other sciences.

While the content of Proposition 3 has been emphasized by many, particularly by Friedman (1953) in economics and by Geisser (1980) in statistics, it is a fact that many econometricians and other social scientists apparently are not aware of its importance or are openly opposed to it. As Geisser might put it, they are overly concerned with unobservables, e.g., estimating parameters, and thus neglect and/or fail to appreciate the all important role of prediction. Further, some of those who work with very complicated models find it difficult or perhaps unwise to implement predictive testing of their models. For these and others who do not stress the importance of prediction, the following challenge is put forward:

Challenge III. Demonstrate that a principle other than the Prediction Principle has been effective in producing sound results in econometrics and other sciences.

Above in the statement of the Jeffreys–Wrinch Simplicity Postulate, probabilities were associated with hypotheses and models and such probabilities are not 'objective' or 'frequency-based'. Bayesians interpret such probabilities as measures of degrees of belief. Jeffreys (1967, p. 15) views a numerical probability as measuring 'the degree of confidence that we may reasonably have in a proposition, even though we may not be able to give either a deductive proof or disproof of it.' Savage (1962, p. 163) writes: 'Personal probability is a certain kind of numerical measure of the opinions of somebody about something.' These are subjective concepts of probability that Bayesians find most useful in applied work. On the other hand, many econometricians and other workers claim that they use 'objective' or 'frequency' concepts of probability in their work, a claim that Jeffreys (1967, p. 369) disputes in the following words: 'In practice no statistician ever uses a frequency definition, but ... all use the notion of degree of reasonable belief, usually without even noticing that they are using it and that by using it they are contradicting the principles they have laid down at the outset.'

With respect to the classical or axiomatic, the Venn limiting frequency and the Fisher hypothetical infinite population definitions of probability, Jeffreys (1967, ch. 7) provides devastating critiques of these definitions that have been unanswered, perhaps because as some have remarked, they are unanswerable.

In view of the above consideration, I put forward the following proposition:

Proposition 4. A subjective concept of probability is more useful in research and applied work in econometrics and other sciences than are other concepts of probability.

Challenges related to Proposition 4 are:

Challenge IV.a. Demonstrate that Jeffreys's criticisms of axiomatic, limiting frequency, and hypothetical infinite population concepts of probability are invalid.

Challenge IV.b. Demonstrate that subjective probability concepts are not widely used in econometric and other applied areas of research by non-Bayesians.

In connection with Challenge IV.b, the following remarks of several leading non-Bayesians are illuminating. Tukey (1978, p. 52) writes:

> 'It is my impression that rather generally, not just in econometrics, it is considered decent to use judgment in choosing a functional form, but indecent to use judgment in choosing a coefficient. If judgment about important things is quite all right, why should it not be used for less important ones as well? Perhaps the real purpose of Bayesian techniques is to let us do the indecent thing while modestly concealed behind a formal apparatus. If so, this would not be a precedent. When Fisher introduced the formalities of the analysis of variance in the early 1920's, its most important function was to conceal the fact that the data was being adjusted for block means, an important step forward which if openly visible would have been considered by too many wiseacres of the time to be 'cooking the data'. If so, let us hope that day will soon come when the role of decent concealment can be freely admitted. ...The coefficient may be better estimated from one source or another, or, even best, estimated by economic judgment...
>
> It seems to me a breach of the statistician's trust not to use judgment when that appears to be enough better than using data.'

Freedman (1986, p. 127) expresses his views as follows:

> 'When drawing inferences from data, even the most hardbitten objectivist usually has to introduce assumptions and use prior information. The serious question is how to integrate that information into the inferential process and how to test the assumptions underlying the analysis.'

Also, Lehmann (1959, p. 62) writes:

> 'Another consideration that frequently enters into the specification of a significance level is the attitude toward the hypothesis before the experiment is performed. If one firmly believes the hypothesis to be true, extremely convincing evidence will be required before one is willing to give up this belief, and the significance level will accordingly be set very low.'

These quotations reveal that leading non-Bayesians use non-frequency, subjective, judgmental information in making inferences in so-called 'objective' or 'frequency-based' approaches.

Finally, Bayesians who use Bayes's Theorem as a learning model in their work find it extremely useful. Cox (1961), Jaynes (1974, 1983, 1984) and others have provided fundamental analysis rationalizing Bayes's Theorem as a coherent model of inductive reasoning. In Boyer and Kihlstrom (1984), many uses of Bayes's Theorem as a learning model in mathematical economics are described. In estimation, testing, prediction, control, design, and other applied problems, the Bayesian learning model has been found to be very useful. This range of considerations leads to the following proposition:

Proposition 5. The learning model embedded in Bayes's Theorem is very useful in econometrics and other sciences.

Associated with Proposition 5 is the following challenge:

Challenge V. Demonstrate that some other learning model is more general and useful than the Bayesian learning model in econometrics and other areas of science.

These then are some basic propositions and challenges that I put forward that relate to foundational issues that appear to me to be of fundamental importance in applied econometrics. In my own applied econometric work, I have found it fruitful to abide by the USP. In the applied projects on which I have worked, I have found that the best solutions have involved simple formulations in accord with the SP. 'Reality' may appear to be complex, but this seems to me to be no more than saying that understanding is lacking. Appropriate, sophisticatedly simple concepts and formulations have been found valuable in accord with the SP. Similarly, the PP has played a vital role in appraising alternative formulations and models. Finally, in many applied studies, use of subjective probabilities and Bayes's Theorem has been very useful. Since it would take too long to describe these specific studies in great detail, in the next section I shall just describe selected 'canonical' statistical problems that arise in applied econometrics, provide Bayesian solutions and challenge non-Bayesians to provide better solutions.

3. Specific problems and Bayesian solutions

In this section, Bayesian solutions to some statistical problems that arise in applied econometrics will be reviewed. Before presenting these problems and solutions, it is relevant to point out that most non-Bayesian inference results in econometrics can be produced by Bayesian methods under special assumptions. For example, with a large sample assumption, posterior means that are optimal point estimates relative to a quadratic loss function in general are approximately equal to maximum likelihood estimates that are widely used by econometricians and others. Also, many non-Bayesian estimates such as seemingly unrelated regression coefficient estimates, two- and three-stage least squares coefficient estimates for linear structural econometric models, Cochrane–Orcutt regression coefficient estimates that take account of autocorrelated errors, etc., have been shown to be means of conditional posterior pdfs based on diffuse prior pdfs, that is conditional posterior pdfs in which 'nuisance' parameters are set equal to sample estimates [see Zellner (1971a) for analyses illustrating these points]. Also sampling theory likelihood ratio test statistics have been related to posterior odds calculated using particular prior distributions – see, e.g., Jeffreys (1967), Leamer (1978), Schwarz (1978), Bernardo (1980), Zellner and Siow (1980), Klein and Brown (1984), and Zellner (1984). Also, many non-Bayesian point predictions can be produced by Bayesian methods. Thus, it is accurate to state that the Bayesian approach encompasses non-Bayesian approaches in the sense that it can produce many important non-Bayesian results as well as other useful results that are difficult to obtain by non-Bayesian methods. The problems and Bayesian solutions to them, presented below, will illustrate this latter point. This situation appears to parallel that in statistics which Jaynes (1984, p. 44) described as follows:

> 'Indeed, the variety of [Bayesian] applications demonstrated by Jeffreys [1967] included all those for which 'Student', Fisher, Neyman and Pearson had developed 'orthodox' methods.'

3.1. Selected point estimation problems

Here we shall present Bayesian solutions to some point estimation problems that arise in applied econometrics. For all of these problems, complete posterior distributions are available. However, since point estimation is emphasized, probably over-emphasized, in the econometric literature, Bayesian solutions to various central point estimation problems will be presented.

It is well known that when quadratic loss functions are appropriate, means of posterior pdfs are optimal, when zero–one loss functions are appropriate, posterior modal values are optimal and when absolute error loss functions are appropriate, posterior medians are optimal. These are very general useful results that relate to a wide range of point estimation problems encountered in applied econometrics. In what follows, general results like these for other types of loss functions and basic point estimation problems are presented.

Problem 1: Estimation of reciprocals of parameters

Let the parameter of interest be $\theta = 1/\beta$ and assume that a posterior pdf for β is available that possesses a finite second moment. This posterior distribution may be based on a diffuse or informative prior distribution for β.

Further, assume that a relative squared error loss function is appropriate, that is

$$L = (\theta - d)^2/\theta^2 = (1 - d\beta)^2, \tag{1}$$

where d is an estimate of θ. Using the posterior pdf for β, posterior expected loss is $\mathrm{E}L = 1 - 2d\mathrm{E}\beta + d^2\mathrm{E}\beta^2$ and the optimal point estimate, d^*, is given by

$$d^* = \frac{\mathrm{E}\beta}{\mathrm{E}\beta^2} = \frac{1}{\mathrm{E}\beta} \cdot \frac{1}{1 + \mathrm{var}(\beta)/(\mathrm{E}\beta)^2}, \tag{2}$$

where $\mathrm{E}\beta$ and $\mathrm{var}(\beta)$ are the posterior mean and variance of β, respectively. The simple result in (2) can be applied to all kinds of 'reciprocal estimation' problems, including estimation of reciprocals of population means, regression coefficients, variances, standard deviations, population proportions, etc. To illustrate in the case of the reciprocal of a population mean, assume that the posterior pdf for β is Student-t with mean $\bar{y}$, a sample mean and variance, $\bar{s}^2/n$, where

$$\bar{s}^2 = \sum_{i=1}^{n} (y_i - \bar{y})^2/(n-3),$$

as would be the case for a normal problem with n iid observations analyzed with a diffuse prior, $p(\beta, \sigma) \propto 1/\sigma$. Then the optimal point estimate is $d^* = (1/\bar{y})1/(1 + \bar{s}^2/n\bar{y}^2)$. See Zellner (1978), where it is proved that this estimator is admissible and where other prior assumptions are considered as is also done in Zaman (1981a). Note that the maximum likelihood estimator for θ, $\hat{\theta} = 1/\bar{y}$, does not possess any moments and thus has unbounded risk relative to the loss function in (1). The simplicity, generality and flexibility of the result, d^*, in (2) are noteworthy and it appears difficult, if not impossible, to produce as useful a result in any other system of inference.

Problem 2: Estimation of ratios of parameters

Suppose that the parameter of interest is $\theta = \beta_1/\beta_2$ and that a posterior pdf for β_1 and β_2 is available that possesses finite second moments. Let the loss function be a generalized quadratic loss function,

$$L = \beta_2^2(\theta - d)^2 = (\beta_1 - d\beta_2)^2, \tag{3}$$

where d is an estimate of θ. From the second line of (3), posterior expected loss is $EL = E\beta_1^2 - 2dE\beta_1\beta_2 + d^2E\beta_2^2$ and thus the optimal value of d, d^*, is

$$d^* = \frac{E\beta_1\beta_2}{E\beta_2^2} = \frac{E\beta_1}{E\beta_2}\left(\frac{1+\mathrm{cov}(\beta_1,\beta_2)/E\beta_1E\beta_2}{1+\mathrm{var}(\beta_2)/(E\beta_2)^2}\right), \tag{4}$$

where $\mathrm{cov}(\beta_1,\beta_2)$ and $\mathrm{var}(\beta_2)$ denote the posterior covariance of β_1 and β_2 and variance of β_2, respectively. See Zellner (1978), Zellner and Park (1979) and Zellner and Rossi (1984) for applications of (4) to problems of estimating ratios of population means, ratios of regression coefficients, ratios of variances, and lethal dosage rates in bioassay models (or qualitative dichotomous response models as they are called in the econometric literature). Also, (4) is relevant for point estimation in the calibration problem. Maximum likelihood estimators for many of these problems, e.g., $\hat{\theta} = \bar{y}_1/\bar{y}_2$ for the ratio of two normal populations' means, where $\bar{y}_1$ and $\bar{y}_2$ are sample means, do not possess finite moments and hence are inadmissible, whereas it has been established that Bayesian estimators for this problem have finite risk relative to the loss function in (3) and a number of others. Further, it should be appreciated that informative prior pdfs for β_1 and β_2, for example a prior pdf limiting the range of β_2 to 0 to ∞, can be utilized in computing d^* in (4). Thus a wide range of ratio estimation problems relative to the loss function in (3), or generalizations of it, e.g., $L = g(\beta_2^2)(\theta - d)^2$, where $g(\beta_2^2) \propto \beta_2^{2m}$, $m \geq 1$, can be readily solved by Bayesian methods. In addition, the complete posterior pdf of $\theta = \beta_1/\beta_2$ is available.

Problem 3: Estimation of Stein's n-means

Let y be an $n \times 1$ vector of observations assumed generated by $y = \xi + \varepsilon$, where ξ is an $n \times 1$ vector of means and ε is an $n \times 1$ vector of errors drawn from a population with a zero mean. Assume that a posterior pdf for ξ is available and that the appropriate loss function is

$$L = (\xi - d)'(\xi - d)/(\xi'\xi)^m, \tag{5}$$

where d is an estimate of ξ and $m \geq 1$. Then posterior expected loss is given by $EL = E(\xi'\xi)^{1-m} - 2d'E\xi(\xi'\xi)^{-m} + d'dE(\xi'\xi)^{-m}$ and the optimal point estimate, d^*, is

$$d^* = E\xi(\xi'\xi)^{-m}/E(\xi'\xi)^{-m}, \tag{6}$$

provided, of course, that the relevant moments exist. This is a very general solution to the Stein problem since many different models for the error vector ε and many different prior pdfs for ξ can be employed – see Zellner and Vandaele (1975) for such results for a simple case. If the integrals in (6) cannot be evaluated analytically, it will be possible in many cases to evaluate them numerically. A Monte Carlo simulation approach would involve making independent draws from the posterior pdf for $\boldsymbol{\xi}$ and using them to evaluate averages of the quantities in the numerator and denominator of (6). In the context of Friedman's (1957) economic theory, for individual i, y_i is the measured income, ξ_i is 'permanent income', ε_i is 'transitory income', and (6) provides estimates of individuals' permanent incomes.

Problem 4: Estimation of structural parameters

The first problem considered is the estimation of a structural coefficient in an errors in the variables problem. Let $\boldsymbol{y}$ and $\boldsymbol{x}$ be two $n \times 1$ vectors of measurements generated by $\boldsymbol{y} = \boldsymbol{\eta} + \boldsymbol{u}$ and $\boldsymbol{x} = \boldsymbol{\xi} + \boldsymbol{v}$, where $\boldsymbol{\eta}$ and $\boldsymbol{\xi}$ are mean vectors and $\boldsymbol{u}$ and $\boldsymbol{v}$ are zero mean error vectors. Assume that posterior distributions for $\boldsymbol{\eta}$ and $\boldsymbol{\xi}$ are available, that interest centers on the structural coefficient β in $\boldsymbol{\eta} = \boldsymbol{\xi}\beta$, and that the appropriate loss function is the following generalized quadratic loss function:

$$L = (\beta - d)^2 \boldsymbol{\xi}'\boldsymbol{\xi} = (\boldsymbol{\eta} - \boldsymbol{\xi} d)'(\boldsymbol{\eta} - \boldsymbol{\xi} d), \tag{7}$$

where d is an estimate of β. Taking the posterior expectation of the loss function yields $\mathrm{E}L = \mathrm{E}\boldsymbol{\eta}'\boldsymbol{\eta} - 2d\mathrm{E}\boldsymbol{\xi}'\boldsymbol{\eta} + d^2\mathrm{E}\boldsymbol{\xi}'\boldsymbol{\xi}$ and thus the optimal value of d, d^* is

$$d^* = \frac{\mathrm{E}\boldsymbol{\xi}'\boldsymbol{\eta}}{\mathrm{E}\boldsymbol{\xi}'\boldsymbol{\xi}} = \frac{\bar{\boldsymbol{\xi}}'\bar{\boldsymbol{\eta}}}{\bar{\boldsymbol{\xi}}'\bar{\boldsymbol{\xi}}}\left[\frac{1 + \mathrm{E}(\boldsymbol{\xi} - \bar{\boldsymbol{\xi}})'(\boldsymbol{\eta} - \bar{\boldsymbol{\eta}})/\bar{\boldsymbol{\xi}}'\bar{\boldsymbol{\eta}}}{1 + \mathrm{E}(\boldsymbol{\xi} - \bar{\boldsymbol{\xi}})'(\boldsymbol{\xi} - \bar{\boldsymbol{\xi}})/\bar{\boldsymbol{\xi}}'\bar{\boldsymbol{\xi}})}\right], \tag{8}$$

where $\bar{\boldsymbol{\xi}}$ and $\bar{\boldsymbol{\eta}}$ are the posterior means of $\boldsymbol{\xi}$ and $\boldsymbol{\eta}$, respectively. Thus given the generalized quadratic loss function in (7) and a posterior pdf for $\boldsymbol{\eta}$ and $\boldsymbol{\xi}$, d^* in (8) can be readily computed as can also the posterior pdf for $\beta = \boldsymbol{\xi}'\boldsymbol{\eta}/\boldsymbol{\xi}'\boldsymbol{\xi}$, either analytically or by a Monte Carlo simulation approach. Note that this analysis applies to many possible likelihood functions, including those based on normality and a variety of possible prior distributions.

The second problem involving structural parameters is a structural econometric modeling problem in which we have a reduced form system (or multivariate regression model),

$$\Big(\underset{n\times 1}{\boldsymbol{y}} \,\vdots\, \underset{n\times m}{Y}\Big) = \underset{n\times k}{X}\Big(\underset{k\times 1}{\boldsymbol{\pi}} \,\vdots\, \underset{k\times m}{\Pi}\Big) + \Big(\underset{n\times 1}{\boldsymbol{v}} \,\vdots\, \underset{n\times m}{V}\Big), \tag{9}$$

subject to the restrictions

$$X\boldsymbol{\pi} = X\Pi\boldsymbol{\gamma} + X_1\boldsymbol{\beta} = Z\boldsymbol{\delta}, \tag{10}$$

where $\boldsymbol{\delta}' = (\boldsymbol{\gamma}' \vdots \boldsymbol{\beta}')$ is a vector of structural coefficients, X_1 is an $n \times k_1$ submatrix of X, that is $x = (X_1 \vdots X_0)$, and $Z = (X\Pi \vdots X_1)$. In (10), we assume that the matrix Z is of full column rank. The problem is to estimate $\boldsymbol{\delta}$ given the data $(y_1 \vdots Y_1)$, X and a posterior distribution for $(\boldsymbol{\pi} \vdots \Pi)$ that possesses finite second moments. Assume that the following generalized quadratic loss function is appropriate:

$$L = (\boldsymbol{\delta} - \boldsymbol{d})'Z'Z(\boldsymbol{\delta} - \boldsymbol{d}) = (X\boldsymbol{\pi} - Z\boldsymbol{d})'(X\boldsymbol{\pi} - Z\boldsymbol{d}), \tag{11}$$

where $\boldsymbol{d}$ is an estimate of $\boldsymbol{\delta}$ and in going from the first right-hand side to the second $Z\boldsymbol{\delta} = X\boldsymbol{\pi}$ has been employed. From the second right-hand side of (11), posterior expected loss is $\mathrm{E}L = \mathrm{E}\boldsymbol{\pi}'X'X\boldsymbol{\pi} - 2\boldsymbol{d}'\mathrm{E}Z'X\boldsymbol{\pi} + \boldsymbol{d}'\mathrm{E}Z'Z\boldsymbol{d}$ and thus the optimal value of $\boldsymbol{d}$ is

$$\boldsymbol{d}^* = (\mathrm{E}Z'Z)^{-1}\mathrm{E}Z'X\boldsymbol{\pi}. \tag{12}$$

The point estimate $\boldsymbol{d}^*$ in (12) is in a simple form depending on just second posterior moments of $(\boldsymbol{\pi} \vdots \Pi)$. If, for example, the rows of $(\boldsymbol{v} \vdots V)$ in (9) have been independently drawn from a zero mean multivariate normal distribution with positive definite symmetric covariance matrix Ω and if the standard diffuse prior pdf for $(\boldsymbol{\pi} \vdots \Pi)$ and Ω is employed, then it is well known that the marginal posterior pdf for the regression coefficients is a matrix Student-t posterior pdf. Using the second-order moments for this pdf, $\boldsymbol{d}^*$ was evaluated in Zellner (1978) and found to be in the form of a 'K-class estimate', with a value of K less than one in finite samples. Thus a form of a particular ad hoc non-Bayesian estimate is a special case of (12) – see Zellner and Park (1979) for computed examples involving the use of $\boldsymbol{d}^*$ in (12) and computation of its approximate risk. In Park (1982), Monte Carlo experiments were performed to evaluate this estimator that has been shown analytically to possess at least finite second-order sampling moments under general conditions in contrast to several other estimators, including the ML estimator that do not possess finite moments either in general or in important specific cases and are thus inadmissible relative to many loss functions, including that shown in (11). Also note that the optimal estimate in (12) can be computed using normal and non-normal likelihood functions, diffuse and informative prior pdfs and for 'undersized' samples – see Swamy (1980) and Swamy and Mehta (1983).

Third, the loss function in (11) can be elaborated to take account of goodness-of-fit considerations as follows:

$$L = (X\boldsymbol{\pi} - Z\boldsymbol{d})'(X\boldsymbol{\pi} - Z\boldsymbol{d}) + g(y - Z\boldsymbol{d})'(y - Z\boldsymbol{d}), \tag{13}$$

where g is a non-negative, given constant. The first term on the right-hand side of (13) reflects how well the restrictions in (10) are satisfied, while the second term reflects how well the model fits. On taking the posterior expectation of the loss function in (13) and minimizing it with respect to $\boldsymbol{d}$, the resulting optimal value for $\boldsymbol{d}$ is

$$\boldsymbol{d}^{+} = (\boldsymbol{d}^{*} + g\boldsymbol{d}_1^{*})/(1+g), \tag{14}$$

where $\boldsymbol{d}^{*}$ is given in (12) and $\boldsymbol{d}_1^{*} = (\mathrm{E}Z'Z)^{-1}\mathrm{E}Z'\boldsymbol{y}$ is the quantity minimizing the posterior expectation of $(\boldsymbol{y} - Z\boldsymbol{d})'(\boldsymbol{y} - Z\boldsymbol{d})$. Thus the optimal estimate in (14) is an average of $\boldsymbol{d}^{*}$ in (12) and d_1^{*}. When $\boldsymbol{d}^{+}$ was evaluated using the matrix Student-t posterior distribution for $(\pi \vdots \Pi)$, mentioned above, it was found that $\boldsymbol{d}^{+}$ assumed the form of a 'double-K-class estimate', another ad hoc non-Bayesian estimate – see Zellner (1985d) for details. Also, in this last reference, $\boldsymbol{d}^{+}$ was explicitly derived employing an informative natural conjugate prior distribution. Van Dijk and Kloek (1980) compute complete posterior distributions for structural coefficients using Monte Carlo numerical integration techniques.

Problem 5: Estimation using an asymmetric loss function

Let θ be a parameter to be estimated and d an estimate and assume that the following asymmetric LINEX loss function, due to Varian (1975), is appropriate:

$$L = b\left[\mathrm{e}^{a(d-\theta)} - a(d-\theta) - 1\right], \qquad b>0, \quad a \neq 0. \tag{15}$$

Given that we have a posterior pdf for θ, posterior expected loss is $\mathrm{E}L = b[\mathrm{e}^{ad}\mathrm{E}\mathrm{e}^{-a\theta} - a(d - \mathrm{E}\theta) - 1]$ and the value of d that minimizes $\mathrm{E}L$ is, from Zellner (1985b),

$$d^{*} = (-1/a)\log(\mathrm{E}\mathrm{e}^{-a\theta}). \tag{16}$$

In (16), note that $\mathrm{E}\mathrm{e}^{-a\theta}$ is the moment generating function for the posterior pdf for θ. In the special case in which θ has a normal posterior pdf with mean $\bar{y}$, a sample mean, and variance σ^2/n with σ^2 assumed known,

$$d^{*} = \bar{y} - a\sigma^2/2n. \tag{17}$$

With $\bar{y}$ normally distributed with mean θ and variance σ^2/n, the risk of d^{*} in (17) relative to the loss function in (15) is uniformly lower than the risk of $\bar{y}$, as shown in Zellner (1985b), and thus the sample mean $\bar{y}$ is inadmissible relative to a LINEX loss function. The general solution in (16) can be applied to a number of problems. Also, an extended form of the LINEX loss function involving a vector of parameters has been formulated and used in Zellner

(1985b). The ability to analyze problems conveniently with asymmetric loss functions is important since symmetric loss functions are not always appropriate – see Varian (1975) for an applied study using the asymmetric LINEX loss function.

3.2. Some prediction problems

As mentioned above, prediction is central in science and thus it is fortunate that operational Bayesian prediction procedures are available. These utilize predictive pdfs that are defined as follows:

$$p(y|D) = \int_{\Theta} f(y|\boldsymbol{\theta})\pi(\boldsymbol{\theta}|D)\,\mathrm{d}\boldsymbol{\theta}, \tag{18}$$

where $f(y|\boldsymbol{\theta})$ is the pdf for the as yet unobserved vector of observations, y, $\boldsymbol{\theta}$ is a vector of parameters contained in the parameter space Θ and $\pi(\boldsymbol{\theta}|D)$ is the posterior pdf with D denoting past sample and prior information. Thus (18) yields the complete pdf for the as yet unobserved observations, y. If a point prediction is desired relative to a specific convex predictive loss function, $L(y, \hat{y})$, where $\hat{y}$ is a point prediction, an optimal value for $\hat{y}$ can be obtained, as is well known, by solving the following minimization problem:

$$\min_{\hat{y}} \int L(y, \hat{y})p(y|D)\,\mathrm{d}y. \tag{19}$$

The solution to this minimization problem, say $\hat{y}^*$, is the optimal Bayesian point prediction. If $L(y, \hat{y}) = (y-\hat{y})'Q(y-\hat{y})$ with Q a given positive definite symmetric matrix, the optimal value of $\hat{y}$ is the mean of the predictive pdf. For absolute loss functions, $L = |y-\hat{y}|$, the median of the predictive pdf for y is optimal while for zero–one loss functions the modal value is optimal. With diffuse prior pdfs for parameters, some Bayesian point predictions are the same as certain non-Bayesian point predictions. However, this will not be true when informative prior pdfs are employed and in other circumstances. See Litterman (1980, 1985) for use of predictive pdfs in forecasting macroeconomic variables and Monahan (1983) for their use in connection with Box–Jenkins ARMA time series models. Monahan's computed predictive pdfs are noteworthy in that they reflect restrictions on the parameter space associated with stationarity and invertibility assumptions as well as other information regarding parameters' probable values. Aitchison and Dunsmore (1975) also present a number of valuable Bayesian prediction procedures. In what follows several point prediction problems will be solved using Bayesian methods that appear difficult to solve using non-Bayesian techniques.

Problem 6: Prediction using asymmetric loss functions

Let y be a future value of a variable with a given prediction pdf and assume that the following class of asymmetric LINEX loss functions is appropriate:

$$L = b\left[e^{a(\hat{y}-y)} - a(\hat{y}-y) - 1\right], \qquad b>0, \quad a \neq 0, \tag{20}$$

where $\hat{y}$ is some point prediction. Using the predictive pdf for y to compute expected loss and minimizing it with respect to $\hat{y}$ yields the following optimal point prediction – see Zellner (1985b):

$$\hat{y}^* = (-1/a)\log \mathrm{E}e^{-ay}, \tag{21}$$

a general solution applicable to a range of different problems. For example, if y has a normal predictive pdf with mean m and variance v, $\hat{y}^* = m - av/2$, a result used by Varian (1975) in a regression study with informative prior pdfs to predict house prices from observed house characteristics for purposes of tax assessment. Over-assessment is often more costly than under-assessment and thus the asymmetric loss function in (20) with particular values of $a>0$ and $b>0$ was deemed more appropriate than a symmetric loss function. It appears difficult to obtain a result like (21) in a non-Bayesian approach.

Problem 7: Prediction using relative squared error loss

This prediction problem involves predicting y when $z = \log y$ has a normal predictive pdf with mean m and variance v and the following relative squared error loss function is employed:

$$L = (y-\hat{y})^2/y^2 = (1-\hat{y}/y)^2, \tag{22}$$

where $\hat{y}$ is a point prediction. The expectation of this loss function is $\mathrm{E}L = 1 - 2\hat{y}\mathrm{E}(1/y) + \hat{y}^2\mathrm{E}(1/y^2)$ and the optimal value of $\hat{y}$ is

$$\hat{y}^* = \mathrm{E}(1/y)/\mathrm{E}(1/y^2) = e^{m-3v/2}, \tag{23}$$

since $\mathrm{E}(1/y) = \mathrm{E}e^{-z} = e^{-m+v/2}$ and $\mathrm{E}(1/y^2) = \mathrm{E}e^{-2z} = e^{-2m+2v}$. The simple result in (23) can be computed readily for predictive pdfs based on diffuse or informative prior pdfs and is very useful when logarithmic transformations and relative squared error loss functions are employed. Note that $\hat{y}^*$ differs from the mean of the log-normal predictive pdf for y, namely $\mathrm{E}y = \mathrm{E}e^{z} = e^{m+v/2}$ that is optimal for a squared error loss function and the posterior median of y, e^m, that is optimal for an absolute error loss function, results that are well known and very useful when these loss functions are appropriate. See Zellner (1971b, 1985c) for further consideration of these problems.

3.3. Control and decision problems

Bayesian methods are particularly useful in solving many control and decision problems that arise in econometrics because they permit incorporation of both sample and judgmental prior information and yield good solutions that are difficult to obtain by non-Bayesian methods – see, e.g., Bawa, Brown and Klein (1979), Fisher (1962), Bowman and Laporte (1975), Harkema (1975), Wright (1983), Zellner and Geisel (1968), and Zellner (1971a, ch. 11; 1973). Below a few problems are analyzed that illustrate the power of the Bayesian approach in this area.

Problem 8: Controlling regression processes

Let $y = x\beta + u$ be a simple regression model for the as yet unobserved dependent variable y with x a variable under control, β a regression coefficient with unknown value and u an unobserved random error with zero mean and variance σ^2. Assume that a posterior pdf for β and σ^2 is available based on past data and prior information. Of course if no past data are available, as is the case in some important control and decision problems, the pdf for β and σ^2 would be a prior pdf. Assume further that the problem is to choose a value for x so as to keep y close to a given target value, denoted by a, and that the following squared error loss function is appropriate:

$$L = (y - a)^2 = (x\beta - a + u)^2. \tag{24}$$

With β and u assumed independent, expected loss is given by $\mathrm{E}L = \mathrm{E}[x(\beta - b) + xb - a + u]^2$, or

$$\mathrm{E}L = x^2\mathrm{E}(\beta - b)^2 + (xb - a)^2 + \sigma_p^2, \tag{25}$$

where b and σ_p^2 are the posterior means of β and σ^2, respectively. On minimizing $\mathrm{E}L$ with respect to x, the optimal setting for x, x^*, is

$$x^* = \frac{a}{b}\left[\frac{1}{1 + \mathrm{var}(\beta)/b^2}\right], \tag{26}$$

with $\mathrm{var}(\beta) = \mathrm{E}(\beta - b)^2$, the posterior of β. Note that uncertainty about the value of β, as measured by $\mathrm{var}(\beta)$, has an impact on the value of x^* in contrast to the sub-optimal certainty-equivalence solution, $x^{\mathrm{ce}} = a/b$ in which no such dependence is present. x^{ce} minimizes just the second term on the right-hand side of (25) but fails to take account of the first term that involves x and hence is sub-optimal – see Zellner (1971a, ch. 11), Harkema (1975) and Prescott (1975) for further analysis of the sub-optimality of the certainty-equivalence 'solution' that was widely used in econometrics. Further, it is direct to solve the above one-period control problem with the loss function elaborated to take account of costs of changing the control variable, that is $L=(y-a)^2+c(x-x_0)^2$ with $c>0$ and x_0 the given initial value of x. See Zellner (1973) for the effects of using mistaken values for a and c in solving this problem, an aspect of robustness with respect to errors in formulating loss functions.

Above only one control variable was considered. If $y=\boldsymbol{x}'\boldsymbol{\beta}+u$ is the regression model generating y, where $\boldsymbol{x}$ is a $k\times 1$ vector of control variables, $\boldsymbol{\beta}$ is a $k\times 1$ vector of regression coefficients with unknown values and u is a zero mean error term with variance σ^2, then, with a given posterior distribution for $\boldsymbol{\beta}$ and σ^2 and the quadratic loss function in (24), an optimal value of the vector $\boldsymbol{x}$ can be obtained by minimizing

$$\mathrm{E}L=\mathrm{E}[\boldsymbol{x}'(\boldsymbol{\beta}-\boldsymbol{b})+\boldsymbol{x}'\boldsymbol{b}-a+u]^2=\boldsymbol{x}'V\boldsymbol{x}+(\boldsymbol{x}'\boldsymbol{b}-a)^2+\sigma_p^2, \qquad (27)$$

where $\boldsymbol{b}$ and $V=\mathrm{E}(\boldsymbol{\beta}-\boldsymbol{b})(\boldsymbol{\beta}-\boldsymbol{b})'$ are the posterior mean and covariance matrix for $\boldsymbol{\beta}$, respectively. The value of $\boldsymbol{x}$ that minimizes $\mathrm{E}L$ in (27) is

$$\boldsymbol{x}^*=(V+\boldsymbol{b}\boldsymbol{b}')^{-1}\boldsymbol{b}^2=\left[\frac{1}{1+\boldsymbol{b}'V^{-1}\boldsymbol{b}}\right]V^{-1}\boldsymbol{b}a, \qquad (28)$$

using $(V+\boldsymbol{b}\boldsymbol{b}')^{-1}=V^{-1}-V^{-1}\boldsymbol{b}\boldsymbol{b}'V^{-1}/(1+\boldsymbol{b}'V^{-1}\boldsymbol{b})$. $\boldsymbol{x}^*$ can be readily computed for a variety of regression assumptions and prior pdfs. Further, when $\boldsymbol{x}^*$ is substituted in (27), the result is

$$\mathrm{E}(L|\boldsymbol{x}=\boldsymbol{x}^*)=a^2/(1+\boldsymbol{b}'V^{-1}\boldsymbol{b})+\sigma_p^2, \qquad (29)$$

a rather simple expression that can be employed to compute average or Bayes risk. If Bayes risk is finite, as it usually is when proper prior pdfs are employed, then the decision $\boldsymbol{x}=\boldsymbol{x}^*$ minimizes Bayes risk and is admissible. See Stein and Zaman (1980) and Zaman (1981b) for admissibility considerations when a diffuse, improper prior, $p(\boldsymbol{\beta})\propto$ const. is employed with σ^2 known and a normal regression likelihood function. For this problem they find $\boldsymbol{x}^*$ is admissible when it contains four or fewer elements but not for five or more. However, they do not consider at all the use of proper prior pdfs that most practitioners would use and that lead to admissible solutions. Also, with a compact sample space, it is possible that 'diffuse prior' solutions may be conditionally admissible in the sense of Villegas (1980, p. 54).

Problem 9: Portfolio selections

Portfolio selection problems are central in both theoretical and applied economics and econometrics. Earlier attempts to solve portfolio selection problems were plagued by the appearance of parameters with unknown values in 'solutions'. In Zellner and Chetty (1965), it was indicated how Bayesian methods can be employed to overcome this problem and since then Bayesian methods have been used in the work of Brown (1976, 1978), Bawa, Brown and Klein (1979), Jorion (1983, 1985), and others.

In a simple one-period portfolio problem the investor allocates his initial wealth, W_0, among n assets, the ith of which has a random, unknown return per dollar invested, y_i. The amount to be invested in the ith asset is w_i with $\sum_{i=1}^{n} w_i = W_0$. Subject to this constraint, the investor must choose the w_i's values so as to maximize expected utility. Let the $n \times 1$ vector of future returns, $\boldsymbol{y}$, be generated by

$$\boldsymbol{y} = \boldsymbol{\theta} + \boldsymbol{u}, \tag{30}$$

where $\boldsymbol{\theta}$ is an $n \times 1$ mean vector with unknown value and $\boldsymbol{u}$ is a zero mean error vector with an $n \times n$ positive definite symmetric covariance matrix Σ. Assume that a posterior pdf for $\boldsymbol{\theta}$ and Σ is available. Further, for any selection of the w_i's, the return on the portfolio is $R = \boldsymbol{w}'\boldsymbol{y}$, where $\boldsymbol{w}' = (w_1, w_2, \ldots, w_n)$. The utility of R, the random return, is given by a utility function, $U(R)$. The problem then is to maximize $\mathrm{E}U(R)$ with respect to $\boldsymbol{w}$ subject to $\boldsymbol{\iota}'\boldsymbol{w} = W_0$, where the $1 \times n$ vector $\boldsymbol{\iota}' = (1, 1, \ldots, 1)$ has all elements equal to one. If we use a quadratic utility function

$$U(R) = c_1 R - c_2 R^2, \qquad c_1, c_2 > 0, \tag{31}$$

since $R = \boldsymbol{w}'\boldsymbol{y} = \boldsymbol{w}'\boldsymbol{\theta} + \boldsymbol{w}'\boldsymbol{u}$, (31) can be expressed as $U = c_1(\boldsymbol{w}'\boldsymbol{\theta} + \boldsymbol{w}'\boldsymbol{u}) - c_2(\boldsymbol{w}'\boldsymbol{\theta} + \boldsymbol{w}'\boldsymbol{u})^2$ and expected utility, $\mathrm{E}U$, can be evaluated using the posterior pdf for $\boldsymbol{\theta}$ and Σ and minimized with respect to $\boldsymbol{w}$ subject to $\boldsymbol{w}'\boldsymbol{\iota} = W_0$ to yield the optimal value of $\boldsymbol{w}$, say $\boldsymbol{w}^*$, that depends just on the past data and prior information underlying the posterior pdf for $\boldsymbol{\theta}$ and Σ. Since the solution is direct, it is left as an exercise for the reader.

In the literature, portfolio problems have been solved using normal and Student-t data processes, diffuse and informative priors and various utility functions. In Brown (1976, 1978) and Jorion (1983, 1985) such Bayesian solutions have been compared with certainty-equivalence solutions with the general finding that Bayesian solutions are better. Also Jorion (1983, 1985) finds that prior assumptions underlying a Bayes–Stein approach to the analysis of the means problem in (30) usually produce better results than use of a diffuse prior pdf for $\boldsymbol{\theta}$ and Σ.

Finally, in solving these decision and control problems, we have expressed loss or utility in terms of the parameters of the problems and used parameters' posterior pdfs to compute posterior expectations. As is well known, these problems can also be solved using predictive pdfs for as yet unobserved variables, e.g., the vector y in (30) – see Zellner (1971, ch. 11) for examples.

4. Summary and concluding remarks

In section 2 several key propositions were put forward that have important implications for the inductive process in econometrics and for foundational considerations in statistics. The present writer finds them compelling and urges others to consider them carefully. In essence, they lead to emphasis on the fruitfulness of the unity of methods in science, sophisticatedly simple models and methods, the predictive criterion, subjective probability and the Bayesian learning model in making inferences and decisions. Then, in section 3, several problems were described and solved using Bayesian methods rather readily even though most of the problems are difficult to solve satisfactorily using non-Bayesian approaches. In the way of an overall conclusion, it is suggested that more wide-spread adherence to the propositions described in section 2 and further use of Bayesian methods will do much to promote more rapid progress in econometrics and other sciences.

References

Aitchison, J. and I.R. Dunsmore, 1975, Statistical prediction analysis (Cambridge University Press, Cambridge).

Aykac, A. and C. Brumat, eds., 1977, New developments in the applications of Bayesian methods (North-Holland, Amsterdam).

Bauwens, L., 1984, Bayesian full information analysis of simultaneous equation models using integration by Monte Carlo (Springer-Verlag, Berlin).

Bawa, V.S., S.J. Brown and R.W. Klein, 1979, Estimation risk and optimal portfolio choice (North-Holland, Amsterdam).

Berger, J.O., 1980, Statistical decision theory (Springer-Verlag, Berlin); 2nd ed. in 1985.

Bernardo, J.M., 1980, A Bayesian analysis of classical hypothesis testing, in: J.M. Bernardo et al., eds., Bayesian statistics: Proceedings of the first international meeting held in Valencia (Spain), May 28 to June 2, 1979 (University Press, Valencia, Spain) 605–618.

Bernardo, J.M., M.H. DeGroot, D.V. Lindley and A.F.M. Smith, eds., 1980, Bayesian statistics: Proceedings of the first international meeting held in Valencia (Spain), May 28 to June 2, 1979 (University Press, Valencia, Spain).

Bowley, A.L., 1928, F.Y. Edgeworth's contributions to mathematical statistics (Royal Statistical Society, London); reprinted in 1972 (Augustus M. Kelley Publishers, Clifton, NJ).

Bowman, H.W. and A.M. Laporte, 1975, Stochastic optimization in recursive equation systems with random parameters with an application to control of the money supply, in: S.E. Fienberg and A. Zellner, eds., Studies in Bayesian econometrics and statistics in honor of Leonard J. Savage (North-Holland, Amsterdam) 441–462.

Box, G.E.P. and G.C. Tiao, 1973, Bayesian inference in statistical analysis (Addison-Wesley, Reading, MA).

Boyer, M. and R.E. Kihlstrom, eds., 1984, Bayesian models in economic theory (North-Holland, Amsterdam).
Broemeling, L.D., 1985, Bayesian analysis of linear models (Marcel Dekker, New York).
Brown, S.J., 1976, Optimal choice under uncertainty, Unpublished doctoral dissertation (Graduate School of Business, University of Chicago, Chicago, IL).
Brown, S.J., 1978, The portfolio choice problem: Comparison of certainty equivalence and optimal Bayes portfolios, Communications in Statistics B, 321–334.
Christ, C.F., 1975, Judging the performance of econometric models of the U.S. Economy, International Economic Review 16, 54–74.
Cox, R.T., 1961, The algebra of probable inference (Johns Hopkins University Press, Baltimore, MD).
DeGroot, M.H., 1970, Optimal statistical decisions (McGraw-Hill, New York).
Drèze, J.H. and J.F. Richard, 1983, Bayesian analysis of simultaneous equation systems, in: Z. Griliches and M.D. Intriligator, eds., Handbook of Econometrics, Vol. 1 (North-Holland, Amsterdam) 517–598.
Feigl, H., 1953, Notes on casuality, in: H. Feigl and M. Brodbeck, eds., Readings in the Philosophy of Science (Appelton–Century–Crofts, New York) 408–418.
Fienberg, S.E. and A. Zellner, eds., 1975, Studies in Bayesian econometrics and statistics in honor of Leonard J. Savage (North-Holland, Amsterdam).
Fisher, W.D., 1962, Estimation in the linear decision model, International Economic Review 3, 1–29.
Freedman, D.A., 1986, Reply, Journal of Business and Economic Statistics 4, 126–127.
Friedman, M., 1953, Essays in positive economics (University of Chicago Press, Chicago, IL).
Friedman, M., 1957, A theory of the consumption function (Princeton University Press, Princeton, NJ).
Friedman, M. and L.J. Savage, 1948, The utility analysis of choices involving risk, Journal of Political Economy 56, 279–304.
Friedman, M. and L.J. Savage, 1952, The expected-utility hypothesis and the measurability of utility, Journal of Political Economy 60, 463–474.
Geisser, S., 1980a, The contributions of Sir Harold Jeffreys to Bayesian inference, in: A. Zellner, ed., Bayesian analysis in econometrics and statistics: Essays in honor of Harold Jeffreys (North-Holland, Amsterdam) 13–34.
Geisser, S., 1980b, A predictivistic primer, in: A. Zellner, ed., Bayesian analysis in econometrics and statistics: Essays in honor of Harold Jeffreys (North-Holland, Amsterdam) 363–382.
Goel, P.K. and A. Zellner, 1986, eds., Bayesian inference and decision techniques: Essays in honor of Bruno de Finetti (North-Holland, Amsterdam).
Good, I.J., 1962, Review of Harold Jeffreys' 'Theory of probability (3rd ed.)', Geophysical Journal of the Royal Astronomical Society 6, 555–558, and Journal of the Royal Statistical Society A 125, 487–489.
Good, I.J., 1980, The contributions of Jeffreys to Bayesian statistics, in: A. Zellner, ed., Bayesian analysis in econometrics and statistics: Essays in honor of Harold Jeffreys (North-Holland, Amsterdam) 21–34.
Harkema, R., 1975, An analytical comparison of certainty equivalence and sequential updating, Journal of the American Statistical Association 70, 348–350.
Intriligator, M.D., 1978, Econometric models, techniques, and applications (Prentice-Hall, Englewood Cliffs, NJ).
Jaynes, E.T., 1974, Probability theory with applications in science and engineering: A series of informal lectures, Manuscript (Department of Physics, Washington University, St. Louis, MO).
Jaynes, E.T., 1983, Papers on probability, statistics and statistical physics, edited by R.D. Rosenkrantz (D. Reidel, Dordrecht, Holland).
Jaynes, E.T., 1984, The intuitive inadequacy of classical statistics, Epistemologia VII (special issue on probability, statistics and inductive logic), 43–74.
Jaynes, E.T., 1985, Highly informative priors, in: J.M. Bernardo et al., eds., Bayesian statistics 2 (North-Holland, Amsterdam) 329–352.
Jeffreys, H., 1967, Theory of probability, 3rd ed. (Oxford University Press, London); 1st ed. in 1939.

Jeffreys, H., 1973, Scientific inference, 3rd ed. (Cambridge University Press, Cambridge); 1st ed. in 1931.
Jorion, P., 1983, Portfolio analysis of international equity investments, Unpublished doctoral dissertation (Graduate School of Business, University of Chicago, Chicago, IL).
Jorion, P., 1985, International portfolio diversification with estimation risk, Journal of Business 58, 259–278.
Judge, G.G., W.E. Griffiths, R.C. Hill, H. Lütkepohl and T.-C. Lee, 1985, The theory and practice of econometrics, 2nd ed. (Wiley, New York).
Kadane, J.B., ed., 1984, Robustness of Bayesian analysis (North-Holland, Amsterdam).
Klein, R.W. and S.J. Brown, 1984, Model selection when there is 'minimal' prior information, Econometrica 52, 1291–1312.
Leamer, E.E., 1978, Specification searches (Wiley, New York).
Lehmann, E., 1959, Testing statistical hypotheses (Wiley, New York).
Lempers, F.B., 1971, Posterior probabilities of alternative linear models (Rotterdam University Press, Rotterdam).
Lindley, D.V., 1965, Introduction to probability and statistics from a Bayesian viewpoint, 2 vols. (Cambridge University Press, Cambridge).
Lindley, D.V., 1971, Bayesian statistics: A review (Society for Industrial and Applied Mathematics, Philadelphia, PA).
Lindley, D.V., 1980, Jeffreys's contribution to modern statistical thought, in: A. Zellner, ed., Bayesian analysis in econometrics and statistics: Essays in honor of Harold Jeffreys (North-Holland, Amsterdam) 35–39.
Litterman, R.B., 1980, A Bayesian procedure for forecasting with Bayesian vector autoregressions (Department of Economics, MIT, Cambridge, MA).
Litterman, R.B., 1985, Forecasting with Bayesian vector autoregressions: Five years of experience, Technical report (Federal Reserve Bank of Minneapolis, Minneapolis, MN); also in: Journal of Business and Economic Statistics 4, 25–38.
Maddala, G.S., 1978, Econometrics (McGraw-Hill, New York).
Malinvaud, E., 1980, Statistical methods of econometrics, 3rd rev. ed. (North-Holland, Amsterdam).
McNees, S.K., 1985, Forecasting accuracy of alternative techniques: A comparison of U.S. macroeconomic forecasts, Invited paper presented at the ASA meetings, Las Vegas, Nevada, August 1985, and with invited discussion in: Journal of Business and Economic Statistics 4, 5–23.
Monahan, J.F., 1983, Fully Bayesian analysis of ARMA time series models, Journal of Econometrics 21, 307–331.
Morales, J.A., 1971, Bayesian full information structural analysis (Springer-Verlag, Berlin).
Park, S.B., 1982, Some sampling properties of minimum expected loss (MELO) estimators of structural coefficients, Journal of Econometrics 18, 295–311.
Pearson, K., 1938, The grammar of science (Everyman Edition, London).
Prescott, E.C., 1975, Adaptive decision rules of macroeconomic planning, in: S.E. Fienberg and A. Zellner, eds., Studies in Bayesian econometrics and statistics in honor of Leonard J. Savage (North-Holland, Amsterdam) 427–440.
Press, S.J., 1972, Applied multivariate analysis including Bayesian techniques (Holt, Rinehart, and Winston, New York).
Raiffa, H. and R. Schlaifer, 1961, Applied statistical decision theory (Graduate School of Business Administration, Harvard University, Boston, MA).
Richard, J.F., 1973, Posterior and predictive densities for simultaneous equation models (Springer-Verlag, Berlin).
Rothenberg, T.J., 1975, The Bayesian approach and alternatives in econometrics, in: S.E. Fienberg and A. Zellner, eds., Studies in Bayesian econometrics and statistics in honor of Leonard J. Savage (North-Holland, Amsterdam) 55–68.
Savage, L.J., 1954, The foundation of statistics (Wiley, New York).
Savage, L.J., 1962, Bayesian statistics, in: R.F. Machol and P. Gray, eds., Recent developments in information and decision theory (Macmillan, New York) 161–194; reprinted in 1981 in: The writings of Leonard Jimmie Savage – A memorial selection (American Statistical Association and Institute of Mathematical Statistics, Washington, DC).

Schwarz, G., 1978, Estimating the dimension of a model, Annals of Statistics 6, 461–464.
Stein, C. and A. Zaman, 1980, Admissibility of the Bayes procedure corresponding to the uniform prior distribution for the control problem in four dimensions but not in five, Technical report no. 324 (Department of Statistics, Stanford University, Stanford, CA).
Stigler, S.M., 1978, Francis Ysidro Edgeworth, statistician, Journal of the Royal Statistical Society A 141, 287–313.
Swamy, P.A.V.B., 1980, A comparison of estimators for undersized samples, Journal of Econometrics 14, 161–181.
Swamy, P.A.V.B. and J.S. Mehta, 1983, Further results on Zellner's minimum expected loss and full information maximum likelihood estimators for undersized samples, Journal of Business and Economic Statistics 1, 154–162.
Theil, H., 1971, Principles of econometrics (Wiley, New York).
Tukey, J.W., 1978, Discussion of Granger on seasonality, in: A. Zellner, ed., Seasonal analysis of economic time series (U.S. Government Printing Office, Washington, DC) 50–53.
Van Dijk, H.K. and T. Kloek, 1980, Further experience in Bayesian analysis using Monte Carlo integration, Journal of Econometrics 14, 307–328.
Varian, H.R., 1975, A Bayesian approach to real estate assessment, in: S.E. Fienberg and A. Zellner, eds., Studies in Bayesian econometrics and statistics in honor of Leonard J. Savage (North-Holland, Amsterdam) 195–208.
Villegas, C., 1980, comment on 'On some statistical paradoxes and non-conglomerability' by Bruce M. Hill, in: J.M. Bernardo et al., eds., Bayesian statistics: Proceedings of the first international meeting held in Valencia (Spain), May 28 to June 2, 1979 (University Press, Valencia, Spain) 53–54.
Wright, R., 1983, Measuring the precision of statistical cost allocations, Journal of Business and Economic Statistics 1, 93–100.
Zaman, A., 1981a, Estimators without moments: The case of the reciprocal of a normal mean, Journal of Econometrics 15, 289–298.
Zaman, A., 1981b, A complete class theorem for the control problem and further results on admissibility and inadmissibility, Annals of Statistics 9, 812–821.
Zellner, A., 1971a, An introduction to Bayesian inference in econometrics (Wiley, New York).
Zellner, A., 1971b, Bayesian and non-Bayesian analysis of the log-normal distribution and log-normal regression, Journal of the American Statistical Association 66, 327–330.
Zellner, A., 1973, The quality of quantitative economic policymaking when targets and costs of change are misspecified, in: W. Sellekart, ed., Selected readings in econometrics and economic theory: Essays in honour of Jan Tinbergen, Part II (Macmillan, London) 147–164; reprinted in 1984 in: A. Zellner, Basic issues in econometrics (University of Chicago Press, Chicago, IL) 169–183.
Zellner, A., 1975, The Bayesian approach and alternatives in econometrics, in: S.E. Fienberg and A. Zellner, eds., Studies in Bayesian econometrics and statistics in honor of Leonard J. Savage (North-Holland, Amsterdam) 39–54.
Zellner, A., 1978, Estimation of functions of population means and regression coefficients including structural coefficients: A minimum expected loss (MELO) approach, Journal of Econometrics 8, 125–158.
Zellner, A., 1980, ed., Bayesian analysis in econometrics and statistics: Essays in honor of Harold Jeffreys (North-Holland, Amsterdam).
Zellner, A., 1984, Basic issues in econometrics (University of Chicago Press, Chicago, IL).
Zellner, A., 1985a, Bayesian econometrics, Econometrica 53, 253–270 (a written version of my invited Fisher–Schultz lecture, Econometric society Meeting, Pisa, Italy, 1983).
Zellner, A., 1985b, Bayesian estimation and prediction using asymmetric loss functions, Technical report (H.G.B. Alexander Research Foundation, Graduate School of Business, University of Chicago, Chicago, IL); also in: Journal of the American Statistical Association 81, 446–451.
Zellner, A., 1985c, A tale of forecasting 1001 series: The Bayesian knight strikes again, Technical report (H.G.B. Alexander Research Foundation, Graduate School of Business, University of Chicago, Chicago, IL); also in: Journal of Forecasting 2, 491–494.
Zellner, A., 1985d, Further results on Bayesian minimum expected loss (MELO) estimates and posterior distributions for structural coefficients, Technical report (H.G.B. Alexander Research Foundation, Graduate School of Business, University of Chicago, Chicago, IL); also in:

Advances in Econometrics 5, 171–182.
Zellner, A. and V.K. Chetty, 1965, Prediction and decision problems in regression models from the Bayesian point of view, Journal of the American Statistical Association 60, 608–616.
Zellner, A. and M.S. Geisel, 1968, Sensitivity of control to uncertainty and form of the criterion function, in: D.G. Watts, ed., The future of statistics (Academic Press, New York) 269–289.
Zellner, A. and S.B. Park, 1979, Minimum expected loss (MELO) estimators for functions of parameters and structural coefficients of econometric models, Journal of the American Statistical Association 74, 185–193.
Zellner, A. and P.E. Rossi, 1984, Bayesian analysis of dichotomous quantal response models, Journal of Econometrics 25, 365–393.
Zellner, A. and A. Siow, 1980, Posterior odds ratios for selected regression hypotheses, in J.M. Bernardo et al., eds., Bayesian statistics: Proceedings of the first international meeting held in Valencia (Spain), May 28 to June 2, 1979 (University Press, Valencia, Spain) 585–603.
Zellner, A. and W. Vandaele, 1975, Bayes–Stein estimators for *k*-means, regression and simultaneous equation models, in: S.E. Fienberg and A. Zellner, eds., Studies in Bayesian econometrics and statistics in honor of Leonard J. Savage (North-Holland, Amsterdam) 317–343.

ECONOMETRICA

VOLUME 53 MARCH 1985 NUMBER 2

BAYESIAN ECONOMETRICS

BY ARNOLD ZELLNER[1]

The widespread use of prior information in formulating, estimating, and using econometric models is reviewed. Attempts to avoid the use of prior information by formulating multivariate statistical VAR and ARMA time series models for economic time series data have resulted in heavily over-parametrized models. A simple demand, supply, and entry model is presented to contrast models utilizing prior information provided by economic theory and other sources with multivariate statistical time series models. Formal Bayesian methods for incorporating prior information in econometric estimation, testing, and prediction are presented. A number of published applied Bayesian studies are cited in which Bayesian methods have proved to be effective. It is concluded that wise use of the Bayesian approach will produce improved econometric results.

1. INTRODUCTION

I AM GRATEFUL to have this opportunity to share some of my thoughts on Bayesian econometrics with you. Before doing this, I would like to say that I have great admiration and respect for the work of Irving Fisher and Henry Schultz. Henry Schultz, who spent many years at the University of Chicago, made many significant research contributions. Similarly, Irving Fisher's research has had a profound effect on economics and econometrics. While I could spend the entire lecture attempting to summarize their research, I shall just emphasize that both of them produced key results relating to relatively simple models that have endured over the years. For example, Fisher put forward the famous Fisher equation that relates the nominal interest rate to the anticipated real rate and the anticipated rate of inflation. Often, when I am asked, "Are there any laws in economics?" I point to the Fisher equation as an example. Schultz worked on the laws of supply and demand, relatively simple relationships that are additional examples of sophisticatedly simple laws in economics. I shall discuss the role of "simplicity" in model-building later in my lecture.

Since the early 1960's, Bayesian econometrics has developed rapidly with considerable interaction among econometricians and statisticians. After the publication of Bayes' essay in 1763, many statisticians and scientists including Laplace, Edgeworth, Jeffreys, de Finetti, Box, Savage, Lindley, Raiffa, Jaynes, Schlaifer, Pratt, Good, Geisser, DeGroot, Tiao, and others have made important contributions to the development of Bayesian statistics. As regards modern Bayesian econometrics,[2] it commenced in the early 1960's with papers by Jacques Drèze, Tom Rothenberg, Walter Fisher, Albert Ando, Gordon Kaufman, myself, and a few others. Since then, there has been an upswell of work in Bayesian econometrics

[1] This is a written version of my Fischer-Schultz Lecture, presented at the European Econometric Society Meetings, Pisa, Italy, August 29-September 2, 1983. My research was financed in part by the National Science Foundation and by income from the H. G. B. Alexander Endowment Fund, Graduate School of Business, University of Chicago.

[2] See Zellner [61] for a review of the development of modern Bayesian econometrics.

with a good deal of interaction among econometricians and statisticians.[3] Now we have many published Bayesian monographs and papers. Also, recent and current econometrics texts include varying amounts of Bayesian material and Bayesian computer programs are available. Then too, the Bayesian learning model has been utilized in many works in economic theory.[4]

During the last twenty years, I have found econometricians and economists to be very interested in Bayesian ideas and methods. Contrary to some theories of how science progresses, it was the case that many were quite interested in learning about Bayesian ideas and methods as long as they were given evidence that the new methods work well in practice. This pragmatic attitude on the part of the profession is very commendable. I always like to learn about new ideas and approaches but the bottom line is how well they work in practice. This pragmatic dictum gets a lot of weight in the thinking of many econometricians and economists.

In view of the importance of the pragmatic dictum, mentioned above, I shall not describe in detail the various axiom systems that have been put forward to justify the Bayesian approach. Harold Jeffreys [**26**], a leading natural scientist, has an axiom system in his book, *Theory of Probability*, that leads him to conclude that the only sound approach to inference in science is the Bayesian approach based on a "reasonable degree of belief" definition of probability. In addition, Savage's [**49**] axiom system, put forward in his *Foundations of Statistics*, implies that procedures for acting rationally in making decisions under uncertainty are Bayesian procedures. While these axiom systems are not perfect, they do exist and may be of interest to some of you. As an aside, Savage and Milton Friedman were very good friends and both admirers of Bruno de Finetti's work. Savage and Friedman co-authored two well known papers on choice under uncertainty.[5] These papers brought to the fore the importance of economic utility theory in statistics by emphasizing the central role of maximization of expected utility or equivalently minimization of expected loss in making choices under uncertainty, for example in choosing estimates, predictions, and models. Their work helped to introduce economists and statisticians to the intimate connection between utility theory and statistical theory. There is thus considerable axiomatic support for Bayesian inference and decision procedures that have been applied in analyzing many problems in econometrics, for example estimation, prediction, testing, control, and other decision problems with very good results.

[3] Since 1970, semi-annual meetings of the NBER-NSF Seminar on Bayesian Inference in Econometrics have been held at which Bayesian research papers have been discussed by econometricians and statisticians. Some of these papers have been published in Fienberg and Zellner [**13**] and Zellner [**62**].

[4] See, e.g., papers and references in Fienberg and Zellner [**13**] and Zellner [**62**]. Also see the book edited by M. Boyer and R. E. Kihlstrom, *Bayesian Models in Economic Theory*, published in the North-Holland Publishing Company's *Series on Bayesian Econometrics and Statistics*, 1983.

[5] These papers, "The Utility Analysis of Choices Involving Risk" and "The Expected-Utility Hypothesis and the Measurability of Utility" have been reprinted in *The Writings of Leonard Jimmie Savage*, American Statistical Association: Washington, D.C., 1981.

Rather than dwell on axiomatic matters, I shall consider more down-to-earth matters. Why are we interested in the Bayesian approach? My point of view is that in econometrics and in other areas of science a good deal of prior information is employed, usually informally. The Bayesian approach provides formal methods for handling background or prior information. Some of you may say that this is subjective and not objective. I believe that the words, "subjective" and "objective" are loaded words fraught with difficulty. I refer to the philosopher Bishop Berkeley who was a philosophical idealist. He claimed, rightly in my opinion, that it is impossible to prove that the "real" world exists. On the other hand, there are philosophical realists who believe, without proof, that the "real" world exists and that we as scientists can learn about its properties. Finally, there are solipsists who claim that only I exist and nothing else exists. Since it's impossible to settle these issues, who is "objective" and who is "subjective"? My position is that the best we can do at present is to build models and determine how well they perform in explaining past data, predicting as yet unobserved data, and analyzing policy problems. The point that I make to you, very strongly, is that non-Bayesians, as well as Bayesians, use a good deal of prior information in building and using models. Further, since the use of prior information is unavoidable, it should be used carefully and more formally than has been done in the past. Finally, it appears to me that most econometricians usually do not use a frequency concept of probability in model-building, estimation, testing, and prediction, but rather a degree of reasonable belief concept of probability.[6]

To illustrate some of these points, let us consider modeling of time series data as, for example, in demand and supply modeling or macroeconometric modeling. How do we generally approach the modeling problem? Some take the Jeffreys-Wrinch Simplicity Postulate, Ockham's Razor, and the Principle of Parsimony very seriously and formulate relatively simple models with the *belief* that they will probably work well. If there is a need to capture much detail, these workers would formulate a large, relatively simple model and investigate the extent to which it is adequate and accomplishes the objectives of a modeling project. Those who advocate sophisticated simplicity in model-building include Harold Jeffreys, Ronald Fisher, Milton Friedman, George Box, myself, and many others. Note that there are many examples of sophisticatedly simple models that work reasonably well, for example, Newton's laws, the gas laws, Einstein's laws, Schrödinger's equation, laws of supply and demand, the Fisher equation, Friedman's consumption model, the quantity theory of money, and so on. On the other hand, there are others who *believe* that complicated models will probably work well in practice. Witness the large number of macroeconometric models that have hundreds of nonlinear stochastic difference equations containing thousands of parameters. It is not clear whether such models have unique or many solutions. I have challenged many audiences to give me examples of large, complicated models that work well in any area of science and have not yet heard of any. A main point to be appreciated in this discussion is that *prior beliefs* about the

[6] This point has also been made with respect to physical scientists in Jeffreys [26].

simplicity/complexity issue have a vital impact on the way in which econometric models are formulated.

To pursue this last point further and more explicitly, let me review several types of linear time series models that are currently being studied and used. The first is the multiple time series model or multivariate autoregressive moving average (MVARMA) model put forward by Quenouille [48] and others. The MVARMA model for a mean-corrected $p \times 1$ vector of variables, z_t, is given by

$$H(L)z_t = F(L)e_t \qquad (t = 1, 2, \ldots, T), \tag{1}$$

where $H(L)$ and $F(L)$ are matrix lag operators given by

$$H(L) = H_0 + H_1 L + H_2 L^2 + \cdots + H_r L^r,$$
$$F(L) = I + F_1 L + F_2 L^2 + \cdots + F_q L^q,$$

with L the lag operator ($L^i z_t = z_{t-i}$) and e_t is a $p \times 1$ non-serially correlated error vector with zero mean and $p \times p$ covariance matrix Σ.

The issue is how successful will we probably be in using a MVARMA model given the types of samples that we often have in econometrics, say 20 years of quarterly data? One relevant consideration is the number of parameters in the model. This number, denoted by N is given by

$$N = (r+1)p^2 + qp^2 + p + p(p+1)/2,$$

that is, $(r+1)p^2$ parameters in the H_i matrices, qp^2 in the F_i matrices, p means, and $p(p+1)/2$ distinct parameters in the error covariance matrix. Clearly, the value of N, the number of parameters will depend on the values assigned to p, r, and q. Some may consider a model with 30 variables ($p = 30$) to be a small model but for illustration I shall assume that just six variables ($p = 6$) appear in the model, $r = 3$, and $q = 4$. With these assumed values for p, q, and r, there are 395 parameters in the model. Suppose that we have 20 years of quarterly data on the six variables, that is a total of 480 observations. The observation/parameter ratio is 480/395, not even two observations per parameter which is hardly satisfactory because with so few observations, parameter estimates and predictions will generally be imprecise. Also, if tests are performed to determine whether elements of the coefficient matrices are equal to zero, they probably will not be very powerful and will have an impact on subsequent inferences, the problem of pre-testing. These then are some of the consequences of entertaining a relatively complicated model. With nonlinear versions of the above model, there would be even greater difficulties. There is a need for restrictions or prior information, perhaps derived from economic theory, that can effectively reduce the number of parameters.

Some in economics and statistics have not employed MVARMA processes but have utilized vector autoregressions. Of course, if the matrix lag operator, $F(L)$ in the MVARMA model is invertible, the MVARMA process has an infinite autoregressive representation, $F^{-1}(L)H(L)z_t = e_t$, or

$$(I - G_1 L - G_2 L^2 - \cdots - G_n L^n - \cdots)z_t = e_t.$$

If we cut off the autoregressive operator at the term involving L^n, that is we consider an nth order vector autoregression to represent our data, note that each of the G matrices is $p \times p$ and we have n of them. Thus there are np^2 parameters in the G matrices, $p(p+1)/2$ in the error term covariance matrix, and p means. If again we assume that we have 6 variables and that $n = 10$, perhaps a reasonable value for quarterly, seasonally unadjusted data, then the system contains 387 parameters. If we have 20 years of quarterly data on the 6 variables, there are 480 observations and the observation/parameter ratio, 480/387 is abysmally low. Further, consider a single equation of the above vector autoregression, say that for z_{it}. There are $np = 10 \times 6 = 60$ coefficients in this relation. Fitting by least squares is like fitting a regression with 60 independent variables that are usually highly correlated. The precision of estimation and prediction is not going to be very high. Indeed Litterman [35] has fit such relations with U.S. quarterly macroeconomic data and found that estimates and predictions are not very precise, the latter relative to predictions yielded by Box–Jenkins models, univariate autoregressions, and several macroeconometric models. Thus an attempt to determine relations solely from the data, this time by use of unrestricted vector autoregressions is foiled by the appearance of an inordinate number of parameters in the system. Again there is a need for subject matter information, or prior information, that will help to reduce the number of free parameters.

Sometimes economic theory or prior information suggests that some of the variables in a MVARMA process are exogenously determined. If we partition the z_t vector as follows, $z_t' = (y_t', x_t')$, we can write the MVARMA model in the following form:

$$\begin{pmatrix} H_{11}(L) & H_{12}(L) \\ H_{21}(L) & H_{22}(L) \end{pmatrix} \begin{pmatrix} y_t \\ x_t \end{pmatrix} = \begin{pmatrix} F_{11}(L) & F_{12}(L) \\ F_{21}(L) & F_{22}(L) \end{pmatrix} \begin{pmatrix} e_{1t} \\ e_{2t} \end{pmatrix}. \tag{2}$$

The assumption that the variables in x_t are exogenous implies, as pointed out in Zellner and Palm [64], that

$$H_{21}(L) \equiv 0, \quad F_{12}(L) \equiv 0, \quad \text{and} \quad F_{21}(L) \equiv 0.$$

Thus 3 matrix lag polynomials are identically zero, given that x_t is exogenous. With these restrictions imposed, the above system becomes

$$H_{11}(L)y_t + H_{12}(L)x_t = F_{11}(L)e_{1t}, \tag{3}$$

$$H_{22}(L)x_t = F_{22}(L)e_{2t}. \tag{4}$$

The first set of relations in (3) is in the form of a linear dynamic structural econometric model with y_t endogenous and x_t exogenous. The second set of relations in (4) is a model for the random exogenous variables that is in the MVARMA form. The question has arisen as to whether a model for x_t is needed. For certain purposes it is not needed. For example in estimation of the structural equations, provided that they are identified, one can proceed conditional upon the observed values of the x_t's. However, if the system is to be used in forecasting, the future values of x_t are needed. Currently in forecasting with structural econometric models, it is often not clear how the future values of exogenous variables are obtained. In many cases workers *guess* the future values of the exogenous variables, a very important, informal use of prior information that usually has a vital impact on forecasts and forecast intervals. If the exogenous variables are generated by a MVARMA process, as in (4), it can be used to forecast their future values.

If in the above system, 2 of the 6 variables are assumed exogenous and if we use the same assumptions as above, namely $r = 3$ and $q = 4$, the total number of free parameters is 183. While this is an improvement relative to the unrestricted MVARMA process, the observation/parameter ratio, 480/183 is still quite low. However, we generally go further by assuming that particular variables do not appear in some of the equations of (3), assumptions that lead to zero restrictions which are important in achieving identification and representing our economic knowledge about the forms of structural relations. Such restrictions, as well as the assumption that some variables are exogenous, have come under vigorous attack in the recent literature as being "unrealistic." I do not believe that this is a reasonable criticism of these systems because in many applications these restrictions are reasonable and if there is some doubt about them, many of them can be tested.

To make this last point clearly, let's come down from these abstract algebraic systems to a very simple demand and supply system, the kind that Henry Schultz and many others have used. In this connection, I shall emphasize another area, that of model formulation in which prior information is very important. In formulating an initial variant of a model, say a demand and supply model, we all use much prior information in sorting through combinations of ideas. In this process, the role of unusual facts has been emphasized by Jacques Hadamard [**21**] in his book, *The Psychology of Invention in the Mathematical Field*. From a survey of his fellow mathematicians, he found that most of them made their major research breakthroughs by discovering unusual facts and providing explanations of them. In economics, econometrics and other sciences there are a number of examples of this process. Given that unusual facts are so important, in a recent paper (Zellner [**63**]) I suggested 8 ways to produce unusual facts, one of which is to look at the equations of current macroeconometric models. Almost every equation contains one or more unusual facts. I have had several students consider such unusual facts and investigate them using micro and industry data with very illuminating results—see Chau [**8**], Laub [**29, 30**], Levedahl [**33**], Montmarquette [**38, 39**], and Peck [**44, 45**]. Another way to produce unusual facts is to push current theories to extremes. For example, with respect to the proportionality hypothesis of Friedman's [**16**] consumption theory, consider what happens as permanent income approaches zero. The ratio of permanent consumption to permanent income probably rises toward one as I suggested years ago (Zellner [**57**]) and recently work with low income Indian consumers has supported this view—see Bhalla [**4**]. Also, at very high permanent incomes, the proportion of permanent income consumed probably falls. Thus, while proportionality is a pretty good approximation over the mid-range of permanent incomes, at very low and very high levels of permanent incomes, it probably breaks down. Discovering such unusual facts and rationalizing them can lead to significant new theories of consumption and saving—see, e.g., Gersovitch [**18**].

Recently, with respect to theoretical and applied demand and supply studies, I noticed some very unusual facts. In Muth [**41**] and some other papers on rational expectations models of competitive industries, the number of firms is assumed to be a constant. On the other hand, in most applied econometric demand-supply analyses, the number of firms variable does not appear explicitly in estimated models. As is well known, Alfred Marshall and many economists have emphasized the important role of entry and exit of firms in producing long run equilibrium in a competitive industry. Also, simple aggregation considerations indicate that the number of firms in operation is a variable that should appear in a competitive industry's supply function. With these somewhat unusual facts in mind, William Veloce, a current student of mine, and I have formulated a three equation demand and supply model (Veloce and Zellner [**55**]) containing demand, supply, and net entry equations for the following endogenous variables:

$$y_{1t} = (1-L)\log Q_t, \qquad y_{2t} = (1-L)\log P_t, \qquad \text{and}$$

$$y_{3t} = (1-L)\log N_t,$$

where Q_t, P_t, and N_t, are quantity, real price, and the number of firms in operation, respectively in period t. The three equations of our demand, supply, and entry (DSE) model of a competitive industry are:

(5) Demand: $y_{1t} = a_1 y_{2t} + \boldsymbol{b}_1' \boldsymbol{x}_{1t} + u_{1t}$,

(6) Supply: $y_{1t} = a_2 y_{2t} + y_{3t} + \boldsymbol{b}_2' \boldsymbol{x}_{2t} + u_{2t}$,

(7) Net Entry: $(1-cL) y_{3t} = \boldsymbol{b}_3' \boldsymbol{x}_{3t} + u_{3t}, \quad 0 < |c| < 1$,

where $\boldsymbol{x}_{1t}$ and $\boldsymbol{x}_{2t}$ are vectors of exogenous variables appearing in the demand and supply equations, respectively, x_{3t} is a vector of predetermined variables appearing in the net entry equation, the a's, b's, and c are structural parameters, and u_{1t}, u_{2t}, and u_{3t} are zero-mean, serially uncorrelated disturbance terms.

From the net entry relation, we can solve for y_{3t}, that is

(8) $y_{3t} = (\boldsymbol{b}_3' \boldsymbol{x}_{3t} + u_{3t})/(1-cL)$

and substitute this expression in the supply equation (6) to obtain the following "reduced" two equation system:

(9) $y_{1t} = a_1 y_{2t} + \boldsymbol{b}_1' \boldsymbol{x}_{1t} + u_{1t}$,

(10) $(1-cL) y_{1t} = a_2 (1-cL) y_{2t} + \boldsymbol{b}_3' \boldsymbol{x}_{3t} + \boldsymbol{b}_2' (1-cL) \boldsymbol{x}_{2t} + u_{3t} + (1-cL) u_{2t}$.

This reduced two equation system is similar to those that have been estimated in applied econometric studies. Note that equation (9) is exactly the demand equation (5) of the three equation DSE system. However, the second equation (10) of the two equation reduced system is a mixture of the supply and net entry equations, (6) and (7) of the three equation DSE model. It contains lagged endogenous variables, autocorrelated disturbances, and predetermined variables and coefficients from the supply and net entry equations. In empirical work, lags and autocorrelated disturbances in an equation like (10) have been rationalized in terms of partial supply adjustment and/or adaptive expectations effects—see, e.g., Nerlove [**42**]. Here the lags and autocorrelation appear as a consequence of considering the three equation DSE model to be an appropriate model. In current work with annual data for a Canadian industry, Veloce has found that a three equation DSE model, like the one above, provides a good interpretation of the data.

Let us consider the number of variables in the three equation system, (5)-(7). There are three endogenous variables and the variables in three x vectors. If each of these vectors contains three exogenous variables, then there are 12 variables in the system. Consider modeling these 12 variables in a vector autoregressive approach. Each matrix in the autoregressive system is 12×12, that is contains 144 parameters. Further, the 12×12 disturbance covariance matrix contains 78 distinct elements. It is clear that with about 30 annual observations for each variable that a vector autoregressive approach, as well as an unrestricted MVARMA approach, will probably be very unsatisfactory. If we omit some of the relevant variables, we shall be analyzing marginal processes, like the two equation system (9)-(10), and it will be very difficult to interpret estimation results in behavioral terms. These then are some of the practical consequences of employing approaches that fail to take account of relevant, available economic information, that is prior information.

What is the economic information that has been employed to produce the restrictions on the three equation model above? First, for example, it is assumed that the number of firms variable does not appear in the demand equation, certainly a reasonable restriction. Also, variables such as the real wage rate and the price of capital services that might be included as exogenous variables in the supply equation probably do not belong in the demand equation. These restrictions and others like them have been embedded in the three equation model which contains about 20 parameters. If some of these restrictions seem doubtful, they can be tested. In these testing problems as well as in formulating the three equation model, it is apparent that much prior information is used. It is clearly important to use such prior information carefully and to have good methods for incorporating it flexibly in estimation, testing, forecasting, and policy analysis procedures.

To provide another example of a situation in which much prior information is used and has a vital impact on the results of analyses, consider the topic of seasonal adjustment. In much work on seasonal adjustment (see, e.g., Zellner [**59**]) an observed variable, y_t, is assumed to be represented as the sum of unobservable random trend-cycle, seasonal, and noise components, all assumed mutually uncorrelated. There does not seem to be any economic theory that suggests that the random trend-cycle component is uncorrelated with the random seasonal component. Yet this a priori assumption is employed to produce seasonally adjusted variables that are used in econometric analyses. See

Havenner and Swamy [**24**] for an analysis that casts doubt on the adequacy of this assumption.

The discussion presented above indicates that much prior information is needed and used in work with econometric models. Thus good procedures for incorporating prior information in analyses are needed. Fortunately, the Bayesian approach is one in which prior information, as little or as much as an investigator desires, can be flexibly and formally incorporated in estimation, prediction, testing, model selection, and policy analysis procedures. Let me briefly review these procedures and comment on some of their salient features.[7]

First, with respect to estimation, Bayes' Theorem yields the general result that a posterior probability density function (pdf) for the parameters of a model is proportional to a prior pdf times the likelihood function; that is,

$$\text{Posterior pdf} \propto (\text{Prior pdf}) \times (\text{Likelihood Function})$$

or

$$p(\boldsymbol{\theta}|D) = cp(\boldsymbol{\theta}|I_0)p(y|\boldsymbol{\theta}), \quad \boldsymbol{\theta} \subset \Theta, \quad y \subset R_y, \tag{11}$$

where $\boldsymbol{\theta}$ is a vector of parameters, Θ the parameter space, y a vector of observations, R_y the sample space, I_0 the prior information, $D=(y, I_0)$, c a normalizing constant given by $c^{-1} = \int_\Theta p(\boldsymbol{\theta}|I_0)p(y|\boldsymbol{\theta})\, d\boldsymbol{\theta}$, $p(\boldsymbol{\theta}|I_0)$ the prior pdf, and $p(y|\boldsymbol{\theta})$ the likelihood function. Thus for any estimation problem, given a prior pdf and a likelihood function, Bayes' Theorem provides an *exact, finite sample* posterior pdf for the parameters.[8]

The prior pdf, $p(\boldsymbol{\theta}|I_0)$, in (11) can be chosen to represent available prior information about possible values of $\boldsymbol{\theta}$, a formal, probabilistic representation. Note that the ranges of the elements of $\boldsymbol{\theta}$ can be restricted to particular intervals, say zero to one if this is suggested by economic theory or other considerations. Further, (11) can be viewed as a transformation of the prior pdf into a posterior pdf that incorporates both prior and sample information. In addition, assuming a pdf for $\boldsymbol{\theta}$ is closely related to assuming a pdf for parameters in non-Bayesian analyses of random parameter models such as random regression coefficient models, the structural form of the errors in the variables model, etc. The adequacy of the prior pdf for $\boldsymbol{\theta}$ can be assessed by use of various diagnostic checks including posterior odds, described below, for alternative priors, and most importantly by determining the quality of predictions yielded by a model incorporating the information in the prior pdf as was done in Litterman [**35**]. Of course, in structural econometric models, prior information must be introduced to identify structural parameters. On many occasions, Drèze [**11, 12**] and others have emphasized that this needed prior information can be flexibly and satisfactorily represented by use of a prior pdf.

[7] Some works that provide detailed presentations of Bayesian methods are Box and Tiao [**6**], DeGroot [**10**], Jeffreys [**26**], Leamer [**31**], Lindley [**34**], and Zellner [**58**].

[8] In Baxter [**3**], a posterior pdf for the mean rate of growth of the money supply is computed and employed in an imaginative way to provide a measure of the public's degree of confidence in a monetary reform.

Analytical or numerical integration procedures are available for analyzing properties of the posterior pdf in (11). For example, if $\boldsymbol{\theta}' = (\boldsymbol{\theta}_1', \boldsymbol{\theta}_2')$, the marginal posterior pdf for $\boldsymbol{\theta}_1$ is

$$(12) \qquad p(\boldsymbol{\theta}_1|D) = \int_{\Theta_2} p(\boldsymbol{\theta}_1, \boldsymbol{\theta}_2|D)\, d\boldsymbol{\theta}_2 = \int_{\Theta_2} p(\boldsymbol{\theta}_1|\boldsymbol{\theta}_2)p(\boldsymbol{\theta}_2|D)\, d\boldsymbol{\theta}_2$$

where $\boldsymbol{\theta}_2 \subset \Theta_2$.[9] The integration in (12) is a very convenient means of getting rid of nuisance parameters, for example autocorrelation coefficients, disturbance variances and covariances, etc., a problem that is often difficult to solve in non-Bayesian approaches. The marginal posterior pdf for $\boldsymbol{\theta}_1$ in (12) can be used to make posterior probability statements about the elements of $\boldsymbol{\theta}_1$, to compute moments of the elements of $\boldsymbol{\theta}_1$, etc.

If a point estimate for $\boldsymbol{\theta}$ is desired, given that we have a convex loss function, $L(\boldsymbol{\theta}, \hat{\boldsymbol{\theta}})$, where $\hat{\boldsymbol{\theta}}$ denotes an estimate for $\boldsymbol{\theta}$, the optimal estimate is obtained by minimizing the posterior expectation of the loss function with respect to $\hat{\boldsymbol{\theta}}$, that is by solving the following problem:

$$(13) \qquad \min_{\hat{\boldsymbol{\theta}}} \int_{\Theta} L(\boldsymbol{\theta}, \hat{\boldsymbol{\theta}})p(\boldsymbol{\theta}|D)\, d\boldsymbol{\theta}.$$

The solution to (13), say $\hat{\boldsymbol{\theta}}^*$ is the Bayesian estimate that is tailored to be optimal relative to the particular loss function that is deemed reasonable. For example, for the quadratic loss function, $L(\boldsymbol{\theta}, \hat{\boldsymbol{\theta}}) = (\boldsymbol{\theta} - \hat{\boldsymbol{\theta}})'A(\boldsymbol{\theta} - \hat{\boldsymbol{\theta}})$, with A a given, positive definite symmetric matrix, $\hat{\boldsymbol{\theta}}^*$ = the posterior mean of $\boldsymbol{\theta}$, i.e., $\hat{\boldsymbol{\theta}}^* = \int_{\Theta} \boldsymbol{\theta} p(\boldsymbol{\theta}|D)\, d\boldsymbol{\theta}$. For an absolute error loss function, the median of the posterior pdf is optimal. If other loss functions are employed, the minimization problem in (13) can be solved analytically or numerically to obtain an optimal point estimate. Generally, Bayesian estimates have been shown to have very good sampling properties in both small and large samples. A sufficient condition for Bayesian estimators to be admissible is

$$(14) \qquad \text{Average Risk} = \int_{\Theta} r(\boldsymbol{\theta})p(\boldsymbol{\theta}|I_0)\, d\boldsymbol{\theta} < \infty$$

where $r(\boldsymbol{\theta}) = \int_{R_y} L(\boldsymbol{\theta}, \hat{\boldsymbol{\theta}})p(y|\boldsymbol{\theta})\, dy$, the risk function for $\hat{\boldsymbol{\theta}}$ relative to the loss function $L(\boldsymbol{\theta}, \hat{\boldsymbol{\theta}})$. When (14) is satisfied, as it usually is when $p(\boldsymbol{\theta}|I_0)$ is a proper prior pdf, the Bayesian estimator, $\hat{\boldsymbol{\theta}}^*$, minimizes average risk given in (14). Thus Bayesian estimators have rather good sampling properties as shown analytically and in Monte Carlo studies.[10]

[9] The conditional posterior pdf, $p(\boldsymbol{\theta}_1|\boldsymbol{\theta}_2, D)$ has been used in many analyses to explore the effects of various assumptions about $\boldsymbol{\theta}_2$'s value on inferences about $\boldsymbol{\theta}_1$. See, e.g., Box and Tiao [6] and Zellner [58] for examples.

[10] See, e.g., Chetty [9], Fomby and Guilkey [15], Gibbons [19], Griffiths and Dao [20], Lee, Judge, and Zellner [32], Park [43], Surekha and Griffiths [50], Thornber [53], and Zellner [58].

As regards prediction procedures, let y_f denote a vector of as yet unobserved variables with pdf, $p(y_f|\theta)$. Since θ's value is unknown, it is hard to use this pdf in practice. However, the joint pdf for y_f and θ is $p(y_f|\theta) \times p(\theta|D)$, where $p(\theta|D)$ is the posterior pdf for θ given in (11). Thus the marginal or predictive pdf for y_f, $p(y_f|D)$, is readily obtained by integration, that is,

$$(15) \qquad p(y_f|D) = \int_{\Theta} p(y_f|\theta)p(\theta|D)\, d\theta$$

which can be used to make probability statements about the elements of y_f, to compute their moments, etc. Further, if $L(y_f, \hat{y}_f)$ is a convex loss function, where $\hat{y}_f$ is a point prediction, an optimal value of $\hat{y}_f$ can be obtained by minimizing expected loss with respect to $\hat{y}_f$, that is by solving the following problem:

$$(16) \qquad \min_{\hat{y}_f} \int_{R_y} L(y_f, \hat{y}_f)p(y_f|D)\, dy_f.$$

The solution to (16), say $\hat{y}_f^*$, is the Bayesian point prediction.[11] If, for example, the loss function is quadratic, then $\hat{y}_f^*$ is the mean of the predictive pdf, $p(y_f|D)$. As with Bayesian estimates, the Bayesian predictor has very good sampling properties in small and large samples.

With respect to large sample properties of Bayesian estimation procedures, it has been shown in the literature that the posterior pdf $p(\theta|D)$ in (11) approaches a multivariate normal pdf as the sample size grows with mean equal to the maximum likelihood estimate, $\hat{\theta}_{MLE}$, and covariance matrix equal to the inverse of the estimated Fisher information matrix, $\widehat{Inf}^{-1}$, that is $p(\theta|D) \to N(\hat{\theta}_{MLE}, \widehat{Inf}^{-1})$ as the sample size gets large. Heyde and Johnstone [**25**] and Hartigan [**23**] show that when the observations are i.i.d., the conditions needed for this result are identical to those needed for proving the asymptotic normality of maximum likelihood estimators. However, when the observations are stochastically dependent, as in time series analyses, Heyde and Johnstone [**25**] show that the conditions needed for the asymptotic normality of the posterior pdf are simpler and more robust than those needed to establish the asymptotic normality of the maximum likelihood estimator. While these results are interesting, it should be appreciated that Bayes' Theorem in (11) provides operational, exact finite sample results and thus there is no need in general to rely on approximate, asymptotic results that are often poor approximations in small sample cases.[12]

Bayesian posterior odds ratios are available for comparing and testing hypotheses and for solving model selection problems. For example, consider two mutually exclusive hypotheses, H_1 and H_2 and assume that the prior odds for these two hypotheses is Π_1/Π_2, where Π_i, $i = 1, 2$, is the prior probability associated with H_i. Then, from Bayes' Theorem, the posterior odds associated with H_1 versus

[11] See Varian [**54**] for a very interesting computation of $\hat{y}_f^*$ for an asymmetric LINEX loss function. Also Litterman [**35**] computes $\hat{y}_f^*$ in his work with Bayesian VAR models.

[12] Zellner and Rossi [**66**] provide analysis to illustrate this point with respect to logit models.

H_2, denoted by K_{12} is given by

$$K_{12} = (\text{prior odds}) \times (\text{Bayes' factor}) \tag{17}$$

$$= (\Pi_1/\Pi_2)p_1(y|H_1)/p_2(y|H_2)$$

where y is an observation vector and $p_i(y|H_i)$ is the pdf for y given H_i, $i = 1, 2$. If H_1 and H_2 are both simple hypotheses, the Bayes' factor, $p_1(y|H_1)/p_2(y|H_2)$ is identical to the likelihood ratio. On the other hand, if H_1 and H_2 are composite hypotheses involving free parameters, $\boldsymbol{\theta}_1$ and $\boldsymbol{\theta}_2$, respectively, then the Bayes' factor is given by

$$p_1(y|H_1)/p_2(y|H_2) \tag{18}$$

$$= \int_{\Theta_1} p_1(y|\boldsymbol{\theta}_1, H_1)p_1(\boldsymbol{\theta}_1|I_1)\, d\boldsymbol{\theta}_1 \bigg/ \int_{\Theta_2} p_1(y|\boldsymbol{\theta}_2, H_2)p_2(\boldsymbol{\theta}_2|I_2)\, d\boldsymbol{\theta}_2 \,,$$

where $p_i(y|\boldsymbol{\theta}_i, H_i)$ is the pdf for y given $\boldsymbol{\theta}_i$ and H_i and $p_1(\boldsymbol{\theta}_i|I_i)$ is the prior pdf for $\boldsymbol{\theta}_i \subset \Theta_i$ given prior information I_i, $i = 1, 2$. Since $p_i(y|H_i)$ can be computed before observing y, the Bayes' factor in (18) is in the form of a ratio of predictive pdfs. Bayes' factors have been computed for many hypotheses relating to means, variances, regression coefficients, coefficients of ARIMA models, etc.,[13] and in a number of cases related to sampling theory test statistics and P-values. Also, analyses of more than two hypotheses can be performed and asymptotic properties of posterior odds have been derived. While these asymptotic results are interesting, it should be noted that K_{12} in (17) is an exact finite sample measure of the relative degree of confidence associated with H_1 and H_2. Since non-Bayesian testing procedures do not involve the concept of the probability of a hypothesis, there is no non-Bayesian analogue of the posterior odds, K_{12}.

If the two hypotheses H_1 and H_2 are mutually exclusive and exhaustive, then $K_{12} = P/(1-P)$, where P is the posterior probability associated with H_1. Suppose, for example, that H_1 is $\boldsymbol{\theta} = \boldsymbol{\theta}_0$, with $\boldsymbol{\theta}_0$ a given value of $\boldsymbol{\theta}$, and H_2 is $\boldsymbol{\theta} \neq \boldsymbol{\theta}_0$. From (17), the value of $K_{12} = P/(1-P)$ can be computed and from it a value of P can be obtained. If our loss function is $L(\boldsymbol{\theta}, \hat{\boldsymbol{\theta}}) = (\boldsymbol{\theta} - \hat{\boldsymbol{\theta}})'A(\boldsymbol{\theta} - \hat{\boldsymbol{\theta}})$, with A a given pds matrix, then expected loss is given by

$$EL(\boldsymbol{\theta}, \hat{\boldsymbol{\theta}}) = PL(\boldsymbol{\theta}_0, \hat{\boldsymbol{\theta}}) + (1-P)EL(\boldsymbol{\theta}, \hat{\boldsymbol{\theta}})|H_2$$

where E on the right-hand side denotes the expectation based on the posterior pdf for $\boldsymbol{\theta}$ under H_2. On inserting the quadratic loss function and minimizing expected loss with respect to $\hat{\boldsymbol{\theta}}$, the optimal point estimate, $\hat{\boldsymbol{\theta}}^*$ is

$$\hat{\boldsymbol{\theta}}^* = P\boldsymbol{\theta}_0 + (1-P)\bar{\boldsymbol{\theta}}_2 \tag{19}$$

$$= \boldsymbol{\theta}_0 + \frac{1}{1+K_{12}}(\bar{\boldsymbol{\theta}}_2 - \boldsymbol{\theta}_0)$$

[13] See e.g., Jeffreys [26], Geisel [17], Leamer [31], Monahan [37], and Zellner [58].

where $\bar{\theta}_2$ is the posterior mean of θ under H_2. $\hat{\theta}^*$ is a Bayesian "pre-test" estimate.[14] The second line of (19) involves the "shrinkage" factor $1/(1+K_{12})$ and is in Stein-like form. The analysis can be easily modified to apply to other types of hypotheses and to combining predictions from alternative models.[15] That these procedures are relatively simple and permit flexible incorporation of prior information through the prior odds and prior pdfs for parameters is noteworthy.

With respect to control and other policy analysis problems, the Bayesian approach yields solutions that are operational, reflect uncertainty about parameters' values, and incorporate allowance for sequential learning. To illustrate, let $y_f = x\beta + u_f$ where y_f is a future value of a policy variable, x is a control variable, β is a regression coefficient, and u_f is a future error term with zero mean and variance σ^2. If the loss function is $L(y_f, y_*) = (y_f - y_*)^2$, where y_* is a given target value for y_f, expected loss is given by $E(x\beta + u_f - y_*)^2$, or

$$E(y_f - y_*)^2 = x^2 E\beta^2 - 2y_* x E\beta + Eu_f^2 + y_*^2$$

under the assumption that u_f and β are independent. The value of x that minimizes expected loss is

$$x^* = y_* E\beta / E\beta^2 = (y_*/E\beta)\frac{1}{1+\operatorname{var}\beta/(E\beta)^2} \tag{20}$$

where $E\beta$ and $\operatorname{var}\beta$ are the posterior mean and variance of β, respectively.[16] It is seen that the expression for x^* involves not only $E\beta$ but also $\operatorname{var}\beta$, the posterior variance of β, a measure of uncertainty about the value of β. In the literature, more elaborate one-period and multi-period control problems have been analyzed using Bayesian methods. In multiperiod problems, learning and design considerations enter. Comparisons of Bayesian and non-Bayesian control solutions have generally indicated that Bayesian solutions are better.[17]

Above, some Bayesian procedures and methods have been briefly reviewed. It is relevant to question whether these methods work well in applications. In this regard, it is important to realize that many, if not all, non-Bayesian results can be produced by Bayesian methods under special assumptions. For example, with a large sample assumption, it was mentioned earlier that a posterior mean that is optimal relative to quadratic loss is approximately equal to the maximum likelihood estimate. Also many non-Bayesian estimates such as SUR regression coefficient estimates, 2SLS and 3SLS estimates, Cochrane-Orcutt estimates, etc., have been shown to be means of *conditional* posterior pdfs, based on diffuse priors, that is conditional posterior pdfs in which nuisance parameters are set equal to sample estimates. In *large samples* such approximate results may be

[14] This estimate was presented in Zellner and Vandaele [**65**] and a similar approach was employed in Griffiths and Dao [**20**].

[15] Geisel [**17**] used this approach in an application to combine predictions from two non-nested models.

[16] See Fisher [**14**], Harkema [**22**], Prescott [**46**], and Zellner [**58**] for further analysis of this and other control problems.

[17] See, e.g., Harkema [**22**], Prescott [**46**], and Zellner [**58**].

satisfactory; however in small to moderate sized samples these conditional posterior pdfs are poor approximations to the relevant marginal posterior pdfs and thus such approximations lead to poor estimates and other inferences.[18]

Bayesian methods have been used successfully in a number of applied studies. I shall just mention a few examples. Litterman [**35**] has used Bayesian vector autoregressions in forecasting seven important U.S. quarterly macroeconomic variables including real GNP, the implicit deflator for GNP, the unemployment rate, etc. In his work, Litterman used and is using a cleverly formulated informative prior pdf for the many parameters of his seven variable VAR which effectively reduces the number of free parameters. His forecasts compare very favorably with those obtained from unrestricted VARs, univariate AR models, Box-Jenkins' models, and two large structural econometric models. Varian [**54**] has applied Bayesian regression techniques to the problem of real estate assessment employing an asymmetric LINEX loss function. Akaike and Ishiguro [**1**] and their colleagues have developed a Bayesian approach and computer program, BAYSEA, for seasonal adjustment. Comparative study of the output of BAYSEA with the results of other seasonal adjustment procedures indicate that the former results compare favorably with other results. Morris [**40**] has discussed and referenced many applications of empirical Bayes' procedures. Prescott [**46**], Bowman and Laporte [**5**], Harkema [**22**], and Zellner [**58**, Ch. 11] have compared Bayesian and non-Bayesian solutions to policy control problems with results indicating that the former are superior in general. Swamy [**51**] and Swamy and Mehta [**52**] have compared Bayesian and non-Bayesian solutions to the undersized sample problem with the finding that the former are as good or better than non-Bayesian solutions. Monahan [**37**] has developed and applied Bayesian procedures for Box-Jenkins' models that include very good model selection, estimation, and prediction methods. Wright [**56**] reports a very interesting application of Bayesian methods in solving an overhead cost allocation problem. Miller [**36**] reviews Bayesian applications in actuarial statistics. Many applications of Bayesian analysis are reported in Kanjii [**28**]. Fomby and Guilkey [**15**] and Griffiths and Dao [**20**] have shown in Monte Carlo experiments that Bayesian estimators have lower mean-squared error (MSE) than non-Bayesian estimators for a regression coefficient in a simple regression model with autocorrelated disturbances. Surekha's and Griffiths' [**50**] Monte Carlo results for estimation of regression coefficients when disturbance terms are heteroscedastic indicate that Bayesian estimators outperform leading non-Bayesian estimators according to a MSE criterion. Park's [**43**] Monte Carlo experimental results show that a Bayesian minimum expected loss (MELO) estimator performs somewhat better than OLS and 2SLS estimators according to several criteria including MSE and mean absolute deviation. Brown's [**7**], Bawa, Brown, and Klein's [**2**], and Jorion's [**27**] Bayesian procedures for portfolio choice perform better than certainty equivalence procedures.[19]

[18] For examples, see Box and Tiao [**6**], Leamer [**31**], and Zellner [**58**, **60**].

[19] For further references, see the many papers listed under the key words "Bayes" and "Bayesian" in annual issues of the ASA/IMS *Current Index to Statistics*, Washington, DC: American Statistical Association.

In summary, I have emphasized that substantial prior information is being used in econometric analyses, usually informally and sometimes incoherently.[20] The Bayesian approach permits investigators to use prior information formally in a flexible, coherent framework to solve estimation, testing, prediction, model selection, design, and policy analysis problems. Some studies have been cited to demonstrate that the Bayesian approach not only has good theoretical properties but has also been successfully applied in practice. It is to be expected that many more Bayesian applications will be reported in the future given that a growing number of Bayesian computer programs are becoming available—see, e.g., Press [47], and that many more econometricians are becoming familiar with Bayesian methods. Since wise use of Bayesian techniques can lead to better solutions to econometric problems, Bayesian methods can contribute importantly to the progress of econometric science.

University of Chicago

Manuscript received January, 1984; revision received May, 1984.

REFERENCES

[1] AKAIKE, H., AND M. ISHIGURO: "Comparative Study of the X-11 and BAYSEA Procedures of Seasonal Adjustment," in *Applied Time Series Analysis of Economic Data*, ed. by A. Zellner. U.S. Government Printing Office: Washington, DC., 1983 (presented at the ASA-CENSUS-NBER Conference on Applied Time Series Analysis of Economic Data, Washington, DC, 1981).

[2] BAWA, V. S., S. J. BROWN, AND R. W. KLEIN: *Estimation Risk and Optimal Portfolio Choice.* Amsterdam: North-Holland, 1979.

[3] BAXTER, M.: "The Role of Expectations in Stabilization Policy," presented at the Econometric Society meeting, Pisa, Italy, 1983.

[4] BHALLA, S. S.: "Measurement Errors and the Permanent Income Hypothesis: Evidence from Rural India," *American Economic Review*, 69(1979), 295-307.

[5] BOWMAN, H. W., AND A. M. LAPORTE: "Stochastic Optimization in Recursive Equation Systems with Random Parameters with an Application to Control of the Money Supply," in *Studies in Bayesian Econometrics and Statistics in Honor of Leonard J. Savage*, ed. by S. E. Fienberg and A. Zellner. Amsterdam: North-Holland, 1975, pp. 427-440.

[6] BOX, G. E. P., AND G. C. TIAO: *Bayesian Inference in Statistical Analysis.* Reading, Massachusetts: Addison-Wesley, 1973.

[7] BROWN, S. J.: "Optimal Portfolio Choice Under Uncertainty," Ph.D. Thesis, Graduate School of Business, U. of Chicago, 1976.

[8] CHAU, L. C.: "Stock Adjustment, Liquidity, and Normal Income in the Determination of Household Expenditures on Durable Goods," Ph.D. Thesis, Department of Economics, University of Wisconsin at Madison, 1967.

[9] CHETTY, V. K.: "*Bayesian Analysis of Some Simultaneous Equation Econometric Models*," Ph.D. Thesis, Department of Economics, University of Wisconsin at Madison, 1966.

[10] DEGROOT, M. H.: *Optimal Statistical Decisions.* New York: McGraw Hill Book Co., 1970.

[11] DRÈZE, J. H.: "Econometrics and Decision Theory," *Econometrica*, 40(1972), 1-17.

[12] ——— "Bayesian Theory of Identification in Simultaneous Equation Models," in *Studies in Bayesian Econometrics and Statistics in Honor of Leonard J. Savage*, ed. by S. E. Fienberg and A. Zellner. Amsterdam: North-Holland, 1975, pp. 159-174.

[13] FIENBERG, S. E., AND A. ZELLNER (EDS.): *Studies in Bayesian Econometrics and Statistics in Honor of Leonard J. Savage.* Amsterdam: North-Holland, 1975.

[14] FISHER, W. D.: "Estimation in the Linear Decision Model," *International Economic Review*, 3(1962), 1-29.

[20] See Swamy [51] for some striking examples of incoherent use of prior information in some leading non-Bayesian "solutions" to the "undersized sample" problem.

[15] FOMBY, T. B., AND D. K. GUILKEY: "On Choosing the Optimal Level of Significance for the Durbin-Watson Test and the Bayesian Alternatives," *Journal of Econometrics*, 8(1978), 203-213.
[16] FRIEDMAN, M.: *A Theory of the Consumption Function*. Princeton, New Jersey: Princeton University Press, 1957.
[17] GEISEL, M. S.: "Bayesian Comparisons of Simple Macroeconomic Models," in *Studies in Bayesian Econometrics and Statistics in Honor of Leonard J. Savage*, ed. by S. E. Fienberg and A. Zellner. Amsterdam: North-Holland, 1975, pp. 227-256.
[18] GERSOVITCH, M.: "Savings and Nutrition at Low Incomes," *Journal of Political Economy*, 91(1983), 841-855.
[19] GIBBONS, M. R.: "Estimating the Parameters of the Capital Asset Pricing Model (CAPM): A Minimum Expected Loss (MELO) Approach," Technical Report, Graduate School of Business, Stanford University, 1980.
[20] GRIFFITHS, W. E., AND D. DAO: "A Note on the Bayesian Estimator in an Autocorrelated Error Model," *Journal of Econometrics*, 12(1980), 389-392.
[21] HADAMARD, J.: *The Psychology of Invention in the Mathematical Field*. New York: Dover Publications, Inc., 1945.
[22] HARKEMA, R.: "An Analytical Comparison of Certainty Equivalence and Sequential Updating," *Journal of the American Statistical Association*, 70(1975), 348-350.
[23] HARTIGAN, J. A.: *Bayes Theory*. New York: Springer-Verlag, 1983.
[24] HAVENNER, A., AND P. A. V. B. SWAMY: "A Random Coefficient Approach to Seasonal Adjustment of Economic Time Series," *Journal of Econometrics*, 15(1981), 177-209.
[25] HEYDE, C. C., AND I. M. JOHNSTONE: "On Asymptotic Posterior Normality for Stochastic Processes," *Journal of the Royal Statistical Society, Series B*, 41(1979), 184-189.
[26] JEFFREYS, H.: *Theory of Probability*. London: Oxford University Press, 1967.
[27] JORION, P.: "Portfolio Analysis of International Equity Investments: Bayes-Stein Estimators for the Mean and Implications for Risk Reduction," Ph.D. Thesis, Graduate School of Business, University of Chicago, 1983.
[28] KANJII, G. K. (ED.): "Proceedings of the Annual Conference on Practical Bayesian Statistics" Cambridge University, 1982, published in *Journal of the Institute of Statisticians*, 32(1983), 1-278.
[29] LAUB, P. M.: "The Dividend-Earning Relationship: A Study of Corporate Quarterly Panel Data, 1945-65," Ph.D. Thesis, Graduate School of Business, University of Chicago, 1971.
[30] LAUB, P. M.: "Some Aspects of the Aggregation Problem in the Dividend-Earning Relationship," *Journal of the American Statistical Association*, 67 (1972), 552-559.
[31] LEAMER, E. E.: *Specification Searches*. New York: John Wiley, 1978.
[32] LEE, T. C., G. G. JUDGE, AND A. ZELLNER: *Estimating the Parameters of the Markov Probability Model from Aggregate Time Series Data*. Amsterdam: North-Holland, 1977.
[33] LEVEDAHL, J. W.: "The Effects of Households' Permanent and Transitory Income on Both Their Timing of Automobile Purchases and Their Net Outlay for Automobile Services," Ph.D. Thesis, Department of Economics, University of Chicago, 1976.
[34] LINDLEY, D. V.: *Introduction to Probability and Statistics from a Bayesian Viewpoint* (2 vols.). Cambridge: Cambridge University Press, 1965.
[35] LITTERMAN, R.: "A Bayesian Procedure for Forecasting with Vector Autoregressions," unpublished manuscript, Department of Economics, Massachusetts Institute of Technology, 1980.
[36] MILLER, R.: "Actuarial Applications of Bayesian Statistics," in *Bayesian Analysis in Econometrics and Statistics: Essays in Honor of Harold Jeffreys*, ed. by A. Zellner. Amsterdam: North-Holland, 1980, pp. 197-212.
[37] MONAHAN, J. F.: "Fully Bayesian Analysis of ARIMA Time Series Models," *Journal of Econometrics*, 21(1983), 307-332.
[38] MONTMARQUETTE, C.: "A Model of Inventory Holdings with Empirical Application to Canadian Manufacturing Industries" Ph.D. Thesis, Department of Economics, University of Chicago, 1973.
[39] ———: "Random Walk Behavior of Finished Goods Inventory Investment," *Révue Economique*, 28(1977), 352-375.
[40] MORRIS, C. N.: "Parametric Empirical Bayes Inference: Theory and Applications," *Journal of the American Statistical Association*, 78(1983), 47-54.
[41] MUTH, J.: "Rational Expectations and the Theory of Price Movements," *Econometrica*, 29(1961), 315-335.
[42] NERLOVE, M.: "Lags in Economic Behavior," *Econometrica*, 40(1972), 221-251.
[43] PARK, S. B.: "Some Sampling Properties of Minimum Expected Loss (MELO) Estimators of Structural Coefficients;" *Journal of Econometrics*, 18 (1982), 295-312.

[44] PECK, S. C.: "A Test of Alternative Theories of Investment Using Data from the Electric Utilities Industry," Ph.D. Thesis, Graduate School of Business, University of Chicago, 1973.
[45] PECK, S. C.: "Alternative Investment Models for Firms in the Electric Utilities Industry," *Bell Journal of Economics and Management Science*, 5(1974), 420-458.
[46] PRESCOTT, E. C.: "Adaptive Decision Rules of Macroeconomic Planning," in *Studies in Bayesian Econometrics and Statistics in Honor of Leonard J. Savage*, ed. by S. E. Fienberg and A. Zellner. Amsterdam: North-Holland, 1975, pp. 427-440.
[47] PRESS, S. J.: "Bayesian Computer Programs," in *Bayesian Analysis in Econometrics and Statistics: Essays in Honor of Harold Jeffreys*, ed. by A. Zellner. Amsterdam: North-Holland, 1980, pp. 429-442.
[48] QUENOUILLE, M. H.: *The Analysis of Multiple Time-Series*. New York: Hafner Publishing Co., 1957.
[49] SAVAGE, L. J.: *The Foundations of Statistics*. New York: John Wiley, 1954.
[50] SUREKHA, K., AND W. E. GRIFFITHS: "A Monte Carlo Comparison of Some Bayesian and Sampling Theory Estimators in Two Heteroscedastic Error Models," manuscript, Indian Statistical Institute, New Delhi, and University of New England, Armidale, Australia, 1983.
[51] SWAMY, P. A. V. B.: "A Comparison of Estimators for Undersized Samples," *Journal of Econometrics*, 14(1980), 161-181.
[52] SWAMY, P. A. V. B., AND J. S. MEHTA: "Further Results on Zellner's Minimum Expected Loss and Full Information Maximum Likelihood Estimators for Undersized Samples," *Journal of Business and Economic Statistics*, 1(1983), 154-162.
[53] THORNBER, H.: "Finite Sample Monte Carlo Studies: An Autoregressive Illustration," *Journal of the American Statistical Association*, 62(1967), 801-808.
[54] VARIAN, H. R.: "A Bayesian Approach to Real Estate Assessment," in *Studies in Bayesian Econometrics and Statistics in Honor of Leonard J. Savage*, ed. by S.E. Fienberg and A. Zellner. Amsterdam: North-Holland, 1975, pp. 195-208.
[55] VELOCE, W., AND A. ZELLNER: "Modelling a Competitive Industry with Entry: Implications for Demand and Supply Analysis," *Economics Letters*, 16 (1984), 71-75.
[56] WRIGHT, R. L.: "Measuring the Precision of Statistical Cost Allocations," *Journal of Business and Economic Statistics*, 1(1983), 93-100.
[57] ZELLNER, A.: "Tests of Some Basic Propositions in the Theory of Consumption," *American Economic Review*, 50(1960), 565-573.
[58] ———: *An Introduction to Bayesian Inference in Econometrics*. New York: John Wiley, 1971.
[59] ———: *Seasonal Analysis of Economic Time Series*, U.S. Bureau of the Census, Washington, DC: U.S. Government Printing Office, 1978.
[60] ———: "Statistical Analysis of Econometric Models," *Journal of the American Statistical Association*, 74(1979), 628-641.
[61] ———: "The Current State of Bayesian Econometrics," *Basic Issues in Econometrics*, ed. by A. Bellner, Chicago: University of Chicago Press, 1984, 306-321.
[62] ———: *Bayesian Analysis in Econometrics and Statistics: Essays in Honor of Harold Jeffreys*. Amsterdam: North-Holland, 1980.
[63] ———: "Philosophy and Objectives of Econometrics," in *Macroeconomic Analysis: Essays in Macroeconomics and Econometrics*, ed. by D. Currie, R. Nobay and D. Peel. England: Croom Helm, 1981.
[64] ZELLNER, A., AND F. PALM: "Time Series Analysis and Simultaneous Equation Econometric Models," *Journal of Econometrics*, 2(1974), 17-54.
[65] ZELLNER, A., AND W. VANDAELE: "Bayes-Stein Estimators for *k*-Means, Regression and Simultaneous Equation Models," in *Studies in Bayesian Econometrics and Statistics in Honor of Leonard J. Savage*, ed. by S.E. Fienberg and A. Zellner. Amsterdam: North-Holland, 1975, pp. 627-653.
[66] ZELLNER, A., AND P. E. ROSSI: "Bayesian Analysis of Dichotomous Quantal Response Models," *Journal of Econometrics*, 25 (1984), 365-393.

ELSEVIER Journal of Computational and Applied Mathematics 64 (1995) 3–10

JOURNAL OF COMPUTATIONAL AND APPLIED MATHEMATICS

Bayesian and non-Bayesian approaches to statistical inference and decision-making[1]

Arnold Zellner*

Graduate School of Business, University of Chicago, 1101 East 58th Street, Chicago, IL 60637, United States

Received 8 May 1993

Abstract

Bayesian and non-Bayesian approaches to statistical inference and decision-making are discussed and compared. A pragmatic criterion, success in practice, as well as logical consistency are emphasized in comparing alternative approaches. Attention is given to learning models, probability concepts, prior information, axiom systems and selected technical aspects of Bayesian and non-Bayesian approaches.

Keywords: Bayesian inference; Bayesian decision-making; Bayes–non-Bayes foundations

Since there are so many varieties of Bayesians (B's) and non-Bayesians (NB's), it would be presumptuous to claim that the issues between B's and NB's as I see them are one-to-one with those perceived by my colleagues in Statistics, Econometrics and many other fields which employ statistical methods. Thus what I shall present are my own personal views which began to take shape in the early 1960's and developed subsequently in connection with my program of research to evaluate the comparative merits of B and NB statistical approaches, not merely philosophically but also in actual statistical applications. In this connection, the theoretical and applied work of Jeffreys [14, 15] has had a great impact on my thinking and work as will become evident in what follows — see articles by S. Geisser, I.J. Good and D.V. Lindley in [33], in which Jeffreys' contributions to Statistics are discussed. Many others, including participants in our semiannual meetings of the NBER-NSF Seminar on Bayesian Inference in Econometrics and Statistics since 1970 and my students have shaped my thinking on B and NB issues. Also, what I have written below has benefited from listening to the fine Bayesian papers presented at this year's Annual

* E-mail: fac_azellner@gsbacd.uchicago.edu.

[1] Invited address presented at the 1992 Annual Conference of the American Statistical Association Harrisburg Chapter, 6 April 1992. Research financed in part by the National Science Foundation and by income from the H.G.B. Alexander Endowment Fund, Graduate School of Business, University of Chicago.

SSDI 0377-0427(95)00002-X

Conference of the ASA Harrisburg Chapter, entitled Applications of Bayesian Methods and from discussions with authors of these papers and other participants.

To begin with, everyone probably recognizes that part of what is done in Statistics, as well as in any other field, is art and part is science. Some examples of art in Statistics include statistical graphics, exploratory data analysis, multivariate model formulation, etc. Also, years ago, and even today, some use the free hand method, a very artful method, to fit straight lines and curves to data. Further, the production of new statistical methods is more an art than a science. On the other hand, much in Statistics is science, that is logically sound, operational procedures that yield reproducible, dependable results. Artful work in Statistics and many other fields is hard to reproduce. For example, an artful forecaster may state that next year's growth rate of the US economy will be 5%. If asked how he arrived at his 5% forecast, he may reply, "Trust my judgment". A second forecaster who forecasts a 2% growth rate may be able to indicate explicitly how he computed his forecast and an interval for it; thus anyone using this forecaster's model, data and other information can reproduce his 2% forecast, a necessary condition, in my opinion, for it to be termed a scientific forecast. Of course, it is impossible to prove deductively which of these two forecasts is better. However, in many fields scientific, reproducible procedures have been shown to be superior in explanation, prediction and problem-solving to informal, artful procedures in a large number of cases and for a wide range of problems and hence our confidence in science and its methods. Further, as Jeffreys [15] emphasizes, there is an unavoidable uncertainty associated with procedures and propositions. It is impossible to prove deductively that a procedure will work well in practice or that a proposition about the world is 100% true. This revelation has great implications for those in Statistics who claim that a particular approach is absolutely correct. What is needed, in my opinion, is evidence that a particular approach is free from logical error and is more effective in applications and treats a broader range of problems successfully than do competing approaches. I consider this pragmatic approach to evaluating alternative statistical philosophies to be fruitful and have pursued it for many years.

When we talk about the science of Statistics or Statistical Science, what do we mean? I take these terms to mean the study, creation and use of methods for producing and employing data for description, measurement, explanation, prediction, control and decision-making. These methods include design of experiments and surveys, model-building, estimation, testing, prediction, graphical analysis, data analysis, and so on. Now different statistical approaches are based on different concepts, procedures and justifications. Clearly, it is important to compare different approaches conceptually and in practice. To let the cat out of the bag, it is my view that the B approach can produce almost all reasonable NB results and others that are difficult, if not impossible to produce using NB methods. As will be seen, it is also the case that B's can use a much broader range of probability statements than is the case for NB's and this capability is most valuable in applications. Finally, B's can justify procedures not only with respect to a single application with given data but also in terms of their "operating characteristics" in relevant repeated trials.

Let me consider a number of specific B and NB issues in relation to the above view of Statistics. First there is the following issue: In B and NB approaches is a formal learning model employed? The answer is that B's use Bayes' theorem as a learning model while NB's appear to learn informally without the explicit use of a learning model. While it is true that the Bayesian learning model, Bayes' theorem has produced fine results in many problems, as with all models there is the possibility that it can be generalized. After all, in the auto industry the Model T Ford was replaced by the Model A, the Ford V8 and later improved models. It is a fact that some researchers are currently working to extend and generalize the Bayesian learning model, Bayes' theorem — see, e.g., [2, 9, 37]. And indeed Jeffreys [15, p. 25] indicates that he has attempted to generalize the product rule of probability. Thus Bayesians have a formal, reproducible learning procedure that combines prior information with new sample information to yield posterior information whereas NB's learn informally in an artful fashion.

A second issue has to do with whether one associates probabilities with hypotheses, models, and propositions. Concepts of probability used by B's are such that it is meaningful to associate probabilities with hypotheses, models, and propositions and make it possible to use Bayes' Theorem to update such probabilities to reflect new sample information and to test hypotheses in a well-defined decision theoretic framework — see, e.g., [15], a pioneering work, and [1, 10, 16, 21, 22, 31, 34, 38] for many examples. NB's who utilize a long-run frequency or other definitions of probability (see [15, Ch. 7] for a critique of various concepts of probability) do not or cannot formally associate probabilities with models, hypotheses and probabilities with models, hypotheses and propositions even though they often do so informally. Further, some NB practitioners associate p-values with degrees of belief in null hypotheses, an error that is often encountered. However, as indicated in [34, Ch. 3.7], for many testing problems with a given sample, the posterior odds expression for a null hypothesis versus a composite alternative hypothesis is a monotonic function of the usual NB p-value which proves a B interpretation of a NB nefarious practice. Since so many users of Statistics put probabilities on all kinds of propositions, hypotheses and models, it is indeed fortunate that the Reverend Thomas Bayes and others have given us formal methods for using and updating such probabilities in a formal, reproducible scientific manner. This is a very important range of considerations because policy-makers often want to know what is the probability that a measurable effect or a future outcome is between a and b, two given numbers. Or scientists and others may wish to have measures of confidence associated with alternative models which can enable them to take account of model uncertainty in making decisions — see, e.g., [13]. That B's who associate probabilities with hypotheses, models and propositions can deal with these important problems in an operational, reproducible manner is indeed fortunate. Also, with such probabilities available, it is possible to average results over alternative models, hypotheses and forecasters — see, e.g., [19, 23, 29, 30, 40].

Third, another issue that arises in B and NB comparisons is whether it is meaningful to introduce a prior distribution for parameters in a statistical analysis. Generally B's do so but NB's generally claim that they do not. Here as Drèze [6], Drèze and Richard [7], Good [11], Zellner [34] and other have pointed out, NB's use much prior information in identifying parameters in such models as the errors-in-the-variables model (EVM), structural econometric models, etc. For example, assuming that the ratio of error term variances is known in the EVM can be regarded as a dogmatic, degenerate prior density for the ratio of variances. Also, as Swamy [25] and Good [11] have emphasized, use of random parameter models by NB's involves the use of an assumed distribution for parameters which is quite like a prior density. Similarly, in a filtering problem, putting a distribution on an initial state is much like the introduction of a prior density. The same can be said for distributions placed on random effects and unobserved trend, cyclical and seasonal components of time series variables. Indeed this controversy about whether to put distributions on parameters resembles the 19th century controversy about whether to introduce unobserved error terms with assumed properties as vividly described in [24]. This range of considerations is for

Good [11] a possible basis for a B/NB compromise. However, there are still the above issues regarding an appropriate concept of probability and whether to associate probabilities with hypotheses, models and propositions.

Fourth, as regards the issue as to whether B's are subjective while NB frequentists are not, it is relevant to point out, as Jeffreys [15, Ch. 7] does, that frequentists cannot prove that long-run frequencies exist; they have to make subjective assumptions which allow them to prove certain convergence theorems which they cannot check empirically since it is impossible to observe an infinite sequence of outcomes. Further, by introducing assumptions about the real world, it is possible that elements of circularity appear in axiom systems designed to guide researchers in their efforts to learn about real world phenomena, as Jeffreys has indicated. Thus it is accurate to state that both B's and NB's use many subjective elements in sorting out hypotheses, formulating models, choosing distributions for models' error terms and random coefficients, providing identifying assumptions and in defining probabilities. And further, in evaluating models, B's consider the quality of their predictions in many different cases. Nothing in the B approach precludes evaluations in terms of the relative frequency of successful predictions as well as other criteria.

Fifth, on defining probabilities, there is an interesting issue involving B's. Probability as a measure of the degree of confidence one has in a proposition can be introduced as a axiom, as Jeffreys [15] did, or as Bayes, Ramsey, Savage and others have done through betting and utility considerations. Although Jeffreys [15, pp. 30–33] was aware of the Bayes–Ramsey betting/utility approach, he had misgivings about it and decided to introduce his definition of probability as an axiom and add other axioms and theorems to give him the laws of probability. Then he recognized that his probabilities could of course be employed in betting and utility considerations. In view of recent developments in utility theory, see e.g., [5, 18], it appears to me that Jeffreys made a wise decision in this matter.

Sixth, to return to B and NB issues, there are some formal axiomatic systems which underlie the B approach, namely those of Jeffreys, Savage, de Finetti and others. In my opinion, none of these axiom systems is perfect in the sense of meeting all of our needs in the context of learning from our data and making decisions. There is room for improvement. On the NB side, there are axiom systems for probability theory but none, so far as I know, for learning from data and making decisions. Wald's [28] work, which started out as an NB approach to decision-making turned out to yield the B approach as optimal. As regards these axiom systems, while they are impressive intellectual achievements, as stated above, they are not perfect. Thus, I do not tell others that they are incoherent because they do not follow one or another of the available axiom systems. As recognized above, researchers can behave as artists and sometimes get rather good results without rigorously following a particular axiom system. On the other hand, formal, reproducible results are to be desired and that they can be produced by the B approach for many problem is noteworthy.

Seventh, another important issue that is critical in appraising alternative approaches to inference and decision-making is what I call the conditional/unconditional probability issue. That is, in the B approach; one can make probability statements about values of parameters, future values of random variables, and alternative hypotheses or models conditional on given sample and prior information. For example, for most estimation problems, it is easy to evaluate the probability that a parameter's value lies between two given values, as Bayes showed more than two centuries ago in connection with a binomial parameter. Also, conditional on the data, it is usually straightforward to compute the probability that a future outcome will lie in a particular given interval or region.

These conditional probability statements are extremely useful in practice to scientists and decision-makers and yet they cannot be made in NB approaches based on fixed parameter models. It should be noted however that R.A. Fisher's fiducial inference procedures were designed to produce such conditional statements. However, as Savage put it, Fisher tried to make Bayesian omelettes without the eggs (prior distributions).

Eighth, B's can make "unconditional" probability statements about the operating characteristics of B procedures. For example, B's can show that certain 95% Bayesian intervals have 95% coverage in repeated trials. In addition, as is well known the operating characteristics of B point estimates and predictions have been studied in terms of risk functions and admissibility considerations, concepts which NB's also utilize extensively. However, B's go one step further to show that B estimators and predictors are those that minimize average or Bayes risk. NB's do not use the concept of average or Bayes risk since they generally do not use prior densities which are critical in defining average or Bayes risk.

Ninth, the treatment of nuisance parameters, that is, parameters not of special interest to an investigator has been emphasized as a point of difference between B and NB approaches. In the B approach nuisance parameters can be readily integrated out of a joint posterior density to obtain the marginal posterior density of the parameter or parameters of interest — see [3, 31] for many examples of this procedure. In NB approaches, it is not easy to deal with the nuisance parameter problem. For example, when an "optimal estimator", say a weighted least squares estimator, depends on parameters with unknown values, investigators often insert estimates of these parameters in the formula defining the "optimal estimator" and give the resulting estimator an asymptotic, approximate justification. Many times such an approximate estimate can be produced as the mean of a conditional posterior density — see [31] for illustrations of this point in several central estimation problems and for marginal posterior means for these problems which are optimal estimates relative to quadratic loss functions.

Tenth, in the B approach it is generally possible to produce finite sample inference results for time series problems, nonlinear problems, and many others either analytically or by use of numerical integration and other computational procedures. That is, complete posterior densities for parameters and predictive densities for future observations are generally available for use. In NB approaches, such posterior densities and predictive densities are not utilized. Usually, asymptotic probability statements about the operating characteristics of estimators, intervals and predictors are employed. In the B approach it is generally possible to show that with proper prior densities B estimators and predictors are admissible and minimize Bayes risk. If asymptotic properties of B procedures are desired, they are available for standard iid and stochastically dependent observations — see e.g., [4, 12].

Further, B's can derive finite sample estimates, intervals and point predictions which are optimal relative to a wide range of loss functions. See e.g., [1, 27, 31, 36] for discussion of B results for a variety of symmetric and asymmetric loss functions. That estimates, intervals and point predictions can be chosen so as to minimize expected loss or maximize expected utility is appealing to many on theoretical and applied grounds. NB approaches have some estimators that have optimal operating characteristics, e.g., minimal variance, unbiasedness, etc. which can often be produced by B's using diffuse prior densities. In the case of maximum likelihood (ML) estimation, the ML estimate has a conditional optimality property, namely it is the estimate that makes the probability density function of the observations as high as possible. This property is many times also shared by the modal value of a posterior density based on a diffuse prior density. Also, while modal values of posterior densities which are often equivalent to ML estimates have good operating characteristics vis-a-vis a zero–one loss function, in finite samples the ML estimator can often be suboptimal with respect to other loss functions, for example squared error loss. Particularly interesting cases are encountered in estimation of reciprocals and ratios of population means and regression coefficients in which ML estimators fail to possess finite moments and can have bimodal distributions in small samples. Also in these cases posterior distributions for these reciprocals and ratios based on diffuse or natural conjugate priors do not possess finite posterior moments and can be bimodal. See [32, 35, 39] for Bayesian analyses of these problems which provide complete posterior densities and optimal point estimates relative to certain generalized quadratic loss functions.

From what has been presented above, it is clear that introduction of a prior density to represent information about the possible values of parameters is a distinctive feature of the B approach. Given that its introduction permits use of Bayes' Theorem to obtain exact finite sample posterior densities, predictive densities, posterior probabilities associated with alternative hypotheses and models, etc., it appears that there are large net benefits associated with the introduction of prior densities in theoretical and applied statistical inference and decision-making procedures.

On the basic issue of the introduction of subjective prior information in analyses, the following quotations from the writings of several leading NB's are illuminating. Freedman [8] has written as follows:

"When drawing inferences from data, even the most hardbitten objectivist has to introduce assumptions and use prior information. The serious question is how to integrate that information into the inferential process and how to test the assumptions underlying the analysis" (p. 127).

Thus Freedman recognizes the fundamental problem but does not offer a solution to it. Without a solution, NB Statistics, as viewed by Freedman, can hardly be said to be "objective". Rather, it is somewhat of an art, as discussed above. Some NB's have come to appreciate that B methods deal with prior information in a formal and reproducible manner and have been used to produce solutions to many statistical problems.

Tukey [26] in commenting on the above issues has written as follows:

"It is my impression that rather generally, not just in econometrics, it is considered decent to use judgment in choosing a functional form, but indecent to use judgment in choosing a coefficient. If judgment about important things is quite all right, why should it not be used for less important ones as well? Perhaps the real purpose of Bayesian techniques is to let us do the indecent thing while modestly concealed behind a formal apparatus. If so, this would not be a precedent. When Fisher introduced the formalities of the analysis of variance in the early 1920's, its most important function was to conceal the fact that the data was being adjusted for block means, an important step forward which if openly visible would have been considered by too many wiseacres of the time to be "cooking the data". If so, let us hope that the day will soon come when the role of decent concealment can be freely admitted .... The coefficient may be better estimated from one source or another, or, even best, estimated by economic judgment. It seems to me a breach of the statistician's trust not to use judgment when that appears to be enough better than using data" (p. 52).

Along with Tukey, I hope that the day will soon come when "decent concealment can be freely admitted" because that will help to make statistical analyses more easily reproduced. Also, it is safe to say that the wise use of judgment will lead to better analyses in many cases.

The following quotation from Lehman [17] provides evidence that NB's are hardly "objective" in testing hypotheses:

"Another consideration that frequently enters into the specification of a significance level is the attitude toward the hypothesis before the experiment is performed. If one firmly believes the hypothesis to be true, extremely convincing evidence will be required before one is willing to give up this belief and the significance level will accordingly be set very low" (p. 62).

While Lehman's advice is probably sound, it is not what most would call an "objective" approach to hypothesis testing.

It is clear that these leading NB's see the need for judgment and prior information in statistical inference and decision-making. The Bayesian paradigm is one that allows such judgment and prior information to be introduced formally in a reproducible, unconcealed manner. That the B approach is being more widely developed and used — see, e.g., [20] for a report on Bayesian publications in leading journals over the years, is probably an indication that a growing number of "customers" are satisfied with its results.

In conclusion, I hope that I have provided you with a useful summary of B and NB approaches to statistical inference and decision-making that will help you in your deliberations on the relative merits of B and NB approaches.

References

[1] J.O. Berger, *Statistical Decision Theory* (Springer, New York, 2nd ed., 1985).

[2] R.F. Bordley, A Bayesian extension of Bayes' rule, manuscript, Operating Sciences Department, General Motors Research Labs, Warren, MI, 1991.

[3] G.E.P. Box and G.C. Tiao, *Bayesian Inference in Statistical Analysis* (Addison-Wesley, Reading, MA, 1973).

[4] C.-F. Chen, On asymptotic normality of limiting density functions with Bayesian implications, *J. Royal Statist. Soc. Ser. B* **47** (1985) 540–546.

[5] M. Cohen and J.Y. Jaffray, Rational behavior under complete ignorance, *Econometrica* **48** (1980) 1281–1300.

[6] J. Drèze, The Bayesian approach to simultaneous equations systems, ONR Research Monograph No. 67, Technological Research Institute, Northwestern Univ., Evanston, IL, 1962.

[7] J. Drèze and J.-F. Richard, Bayesian analysis of simultaneous equation systems, in: Z. Griliches and M.D. Intriligator, Eds., *Handbook of Econometrics, Vol.* 1 (North-Holland, Amsterdam, 1983) 517–598.

[8] D.A. Freedman, Reply, *J. Business Econom. Statist.* **4** (1986) 126–127.

[9] M. Goldstein, Separating beliefs, in: P.K. Goel and A. Zellner, Eds., *Bayesian Inference and Decision Techniques: Essays in Honor of Bruno de Finetti* (North-Holland, Amsterdam, 1986) 197–216.

[10] I.J. Good, *Good Thinking: The Foundations of Probability and Its Applications* (Univ. of Minnesota Press, Minneapolis, MN, 1983).

[11] I.J. Good, The Bayes/non-Bayes compromise: a review, paper presented at the American Statistical Society Meetings, Atlanta, GA, August, 1991.

[12] C.C. Heyde and I.M. Johnstone, On asymptotic posterior normality for stochastic processes, *J. Roy. Statist. Soc. Ser. B* **41** (1978) 184–189.

[13] J.S. Hodges, Uncertainty, policy analysis and statistics, *Statist. Sci.* **2** (1987) 259–275.

[14] H. Jeffreys, *Scientific Inference* (Cambridge Univ. Press, Cambridge, 2nd ed., 1957).

[15] H. Jeffreys, *Theory of Probability* (Oxford Univ. Press, Oxford, 3nd reprinted ed. 1988; 1st ed., 1939).

[16] G. Koop, 'Objective' unit root tests, *J. Appl. Econometrics* **7** (1992) 65–82.

[17] E. Lehman, *Testing Statistical Hypotheses* (Wiley, New York, 1959).

[18] M.J. Machina, Choice under uncertainty: problems solved and unsolved, *J. Econom. Perspectives* **1** (1987) 121–155.

[19] F.C. Palm and A. Zellner, To combine or not to combine? issues of combining forecasts, *J. Forecasting* **11** (1992) 687–701.
[20] D.J. Poirier, A report form the battlefront, *J. Business Econom. Statist.* **7** (1989) 137–140.
[21] D.J. Poirier, Ed., *Bayesian Empirical Studies in Economics and Finance, J. Econometrics* (Annals Issue) **49** (1991).
[22] S.J. Press, *Bayesian Statistics* (Wiley, New York; 1989).
[23] R.A. Reynolds, Posterior odds for the hypothesis of independence between stochastic regressors and disturbances, *Internat. Econom. Rev.* **23** (1982) 479–490.
[24] S.M. Stigler, *The History of Statistics* (Harvard Univ. Press, Cambridge, MA, 1986).
[25] P.A.V.B. Swamy, *Statistical Inference in Random Coefficient Regression Models* (Springer, New York, 1971).
[26] J.W. Tukey, Discussion of granger on seasonality, in: A. Zellner, Ed., *Seasonal Analysis of Economic Time Series*, US Department of Commerce, Bureau of the Census (US Government Printing Office Washington, DC, 1978) 50–53.
[27] H.R. Varian, A Bayesian approach to real estate assessment, in: S.E. Fienberg and A. Zellner, Eds., *Studies in Bayesian Econometrics and Statistics in Honor of Leonard J. Savage* (North-Holland, Amsterdam, 1975) 195–208.
[28] A. Wald, *Statistical Decision Functions* (Wiley, New York, 1950).
[29] R.L. Winkler, Combining probability distributions from dependent information sources, *Management Sci.* **27** (1981) 479–488.
[30] R.L. Winkler, Combining forecasts; a philosophical basis and some current issues, *Internat. J. Forecasting* **5** (1989) 605–609.
[31] A. Zellner, *An Introduction to Bayesian Inference in Econometrics* (Wiley, New York, 1971; reprinted by Krieger, Malabar, FL, 1987).
[32] A. Zellner, Estimation of functions of population means and regression coefficients including structural coefficients: a minimum expected loss (MELO) approach, *J. Econometrics* **8** (1978) 127–158.
[33] A. Zellner, Ed., *Bayesian Analysis in Econometrics and Statistics: Essays in Honor of Harold Jeffreys* (North-Holland, Amsterdam, 1980).
[34] A. Zellner, *Basic Issues in Econometrics* (Univ. of Chicago Press, Chicago, IL, 1984).
[35] A. Zellner, Further results on Bayesian minimum expected loss (MELO) estimates and posterior distributions for structural coefficients, *Adv. in Econometrics* **5** (1985) 171–182.
[36] A. Zellner, Bayesian estimation and prediction using asymmetric loss functions, *J. Amer. Statist. Assoc.* **81** (1986) 446–451.
[37] A. Zellner, Bayesian methods and entropy in economics and econometrics, in: W.T. Grandy Jr. and L.H. Schick, Eds., *Maximum Entropy and Bayesian Methods* (Kluwer Academic Publishers, Dordrecht, 1990) 17–31.
[38] A. Zellner, Time series analysis, forcasting and econometric modeling: the structural econometric modeling, time series analysis (SEMTSA) approach, *J. Forecasting* **13** (1994) 215–233.
[39] A. Zellner and S.B. Park, Minimum expected loss (MELO) estimators for functions of parameters and structural coefficients of econometrics models, *J. Amer. Statist. Assoc.* **74** (1979) 185–193.
[40] A. Zellner and W. Vandaele, Bayes–Stein estimators for k-means, regression and simultaneous equation models, in: S.E. Fienberg and A. Zellner, Eds., *Studies in Bayesian Econometrics and Statistics in Honor of Leonard J. Savage* (North-Holland, Amsterdam, 1975) 317–343.

PART III

BAYESIAN PRIORS, MODELS AND INFORMATION PROCESSING

[7]

Models, prior information and Bayesian analysis*

1. Introduction

In this paper, attention is given to the important role of prior information in the formulation of models for observations and prior densities for their parameters. It is widely recognized by Bayesians and non-Bayesians that much prior information is used and misused in model formulation, that is in formulating likelihood functions and hyperdistributions for parameters. Further, in many cases models are formulated heuristically in an informal, often non-reproducible manner. In this connection, Good (1963) has remarked cogently, '... hypotheses are usually assumed to be formulated before statistical theory is invoked. This is a weakness in statistical theory, regarded as part of scientific method ...' (p. 912). To emphasize the importance of this observation, several real-life examples of model formulation that have involved careless use of prior information will be reviewed to stress the obvious fact that the best statistical methods applied to poorly formulated models can produce worthless results. To help remedy this situation, it is shown how a maxent approach, employed earlier in many different contexts, can yield central econometric and statistical models and variants of them quite readily. That models be derived in a reproducible manner is viewed as desirable by many.

Just as with the derivation of models, it would be desirable to have a dependable, reproducible method for formulating diffuse and informative prior densities. Indeed, Jaynes (1980) has called this need '... the top priority research problem of Bayesian theory' (p. 618). In addition, Akaike (1978) has written,

> In spite of the apparent utility of Bayesian models in statistical inference no systematic analysis of the role played by the prior distributions has been developed to the point that it can give practising statisticians perfect confidence in using their own prior distributions ... the problem of probability assessments, essentially of the prior distribution, is apparently dismissed as a matter more for a psychologist than a mathematician. (p. 54)

Below, we shall review the rationale of the maximal data information prior (MDIP) approach which provides a general mathematical approach for deriving and justifying prior densities, review some past results yielded by it and present new results. A direct

* Research for this paper was financed in part by the National Science Foundation and income from the H.G.B. Alexander Research Endowment Fund, Graduate School of Business, University of Chicago. Some of this material was presented at the Workshop of the Americas on Statistics and Econometrics, Caracas, Venezuela, December 1992 and at the IASI–ASA Statistical Meeting, University Sao Paulo, Brazil, January 1993, and the Conference on Informational Aspects of Bayesian Statistics in Honor of H. Akaike, Fuji Conference Center, Japan, December 1993. I would like to thank the referees and the editor, Herman van Dijk for helpful comments. Reprinted from Annals Issue of the *Journals of Econometrics* 75 (1), 1996, 51–68, edited by L. Bawens, W. Polasek and H.K. van Dijk with permission of North-Holland Publishing Co.

connection between MDIPs, Jeffreys's priors and entropic priors, described in Rodríguez (1991a, b), will be derived. Further, specific priors for hyperparameters of hierarchical models, common parameters of different likelihood functions and parameters of trinomial and multinomial models will be presented. Also, the MDIP approach is applied to produce prior odds for alternative models or hypotheses.

The plan of the paper is as follows. In Section 2 a discussion of model formulation and new examples of model formulation by maxent are provided. Section 3 is devoted to a brief review of the rationale of the MDIP approach for producing prior densities, to the production of new MDIPs and demonstrating how the MDIP approach can be utilized to yield Jeffreys's priors, entropic priors and other interesting and useful prior densities. A summary and some concluding remarks are presented in Section 4.

2. Model formulation in practice and by maximum entropy

As stated above, stochastic model formulation is often done heuristically and informally in econometrics and statistics. This is not to say that such models are always unsatisfactory but rather to point to the desirability of having a general approach that produces models in a simple, explicit reproducible manner. After reviewing some examples in which prior information was used carelessly in formulating models, the maxent approach will be utilized to produce several central econometric and statistical stochastic models that have not as yet been produced by those who have used the maxent approach to produce econometric and statistical models. Some of those who have produced models by maxent (excluding references to much work in the natural sciences) are Davis (1941), Shannon (1948), Jaynes (1957, 1982a), Kullbach (1959), Good (1963), Rao (1965), Theil (1967), Kagan et al. (1973), Burg (1975), Gokhale (1975), Zellner and Highfield (1982), Golan (1988, 1989), Maasoumi (1990), Ryu (1990, 1993), Cover and Thomas (1991), Kapur and Kesavan (1992) and Soofi (1992a, b). For axiomatic support for and discussion of the general properties of the maxent approach, see Jaynes (1982a, b), Shore and Johnson (1980), Skilling (1989), Cover and Thomas (1991), and Kapur and Kesavan (1992).

2.1 Prior information and model formulation

Below examples are presented that reveal that careless use of prior information in model formulation can and has unfortunately occurred. Also, with an improperly formulated model, even use of the best statistical methods will probably lead to unsatisfactory results.

Example A. Money demand In many studies, real money, m, is related to an interest rate, r, as follows: $\ell n m = a + br + \ldots$. With b negative, this formulation implies that the elasticity of demand for money is more elastic at high than at low interest rates, contrary to what Keynes and others thought reasonable. Revising the function to $\ell n m = a' + b'/r + \ldots$ with b' positive, provides the behavior of the elasticity suggested by Keynes and others. Use of $(r^{\lambda}-1)/\lambda$ and estimation of λ and other parameters may settle the issue empirically.

Example B. Murder and execution Let the murder rate, MR, and the execution rate, XR, for states in a particular year be represented by $\ell n MR = a + b \ell n XR + \ldots$ If b is negative, as the *XR* approaches zero, the *MR* approaches infinity. However, some states that have abolished capital punishment ($XR = 0$) have finite murder rates. Perhaps a better functional form is $\ell n MR = a' + b'XR + \ldots$

Example C. Transformation to induce normality With $y > 0$, the well known transformation, $z = (y^{\lambda} - 1)/\lambda$ has been suggested as one which will induce normality. Many years ago, James B. Ramsey, a student in my course, pointed out that for $\lambda \neq 0$, z has a truncated range and therefore cannot be exactly normally distributed; see Zellner (1971, p. 163, fn. 2). Among others, the transformation $\ell n\, y + ay$, with $a > 0$, does not have a truncation problem.

Example D. Permanent income and permanent consumption With C and Y denoting permanent consumption and income, respectively, the following model was employed to test the hypothesis that $a = 0$ using a t-test: $C = a + bY + u$. Since $0 < C < Y$, the dependent variable is truncated, particularly for low values of Y, and thus assumptions of the t-test are not satisfied. Further, with $a > 0$, the alternative hypothesis, in this model it's the case that the expectation of C can be larger than Y, permanent income! Use of the following function alleviates these problems: $\ell n[(C/Y)/(1 - C/Y)] = c + d\, \ell n\, Y + e$. The dependent variable has a doubly infinite range and $d = 0$ can be tested. Further, if $d < 0$, e.g. $d = -0.2$, as Y approaches 0, C/Y approaches 1 and as Y grows, C/Y falls gradually.

Example E. Production functions Use of a Cobb–Douglas production function, $\ell n Y = a + b\, \ell n\, L + c\, \ell n\, K + u$ or any function with returns to scale equal to a constant implies that the associated long-run average cost functions are not 'U-shaped', an implication that is often unsatisfactory. Among other functions, $\ell n\, Y + gY = a + b\, \ell n L + c\, \ell n\, K + e$, with $g > 0$ can have an associated U- or L- shaped cost function.

Example F. Multi-equation non-linear models In many fields, complicated non-linear multi-equation stochastic models are employed. Such models have many implications that are not well understood, an unsatisfactory state of affairs. For example, it's not clear that they have a unique solution. Examples of simulation studies that have unearthed peculiar properties of such models are Adelman and Adelman (1959), Hickman (1972) and Zellner and Peck (1973). Thus it may be said that many such complex models incorporate inappropriate assumptions, be they explicit or implicit.

Example G. Measurement and behavioral models In seasonal adjustment, measurement models are often utilized to produce seasonally adjusted data under the assumption that trend, cyclical and seasonal components are orthogonal or independent. Such data are then employed in models to explain the nature of various cyclical properties of variables that incorporate prior information about components that is in conflict with that used in seasonal adjustment. Similarly, it has been pointed out that measurement models used to impute missing values in survey data are probably incompatible with behavioral models used to explain variation in the survey data. In these cases, careful specification of a measurement model for imputation that is logically compatible with the behavioral model used to analyse the data is desirable.

These seven examples, and many more that could be cited, illustrate that model formulation involves an important use of prior information, not just to identify parameters, important as that is in errors in the variables models, structural econometric models, seasonal component models, factor models, etc., but also in many other aspects of model formulation. This prior information may be even more important than that used in prior densities for parameters as implied by Tukey's (1978) remarks:

> It is my impression that rather generally, not just in econometrics, it is considered decent to use judgment in choosing a functional form but indecent to use judgment in choosing a coefficient. If judgment about important things is quite all right, why should it not be used for less important ones as well? Perhaps the real purpose of Bayesian techniques is to let us do the indecent thing while modestly concealed behind a formal apparatus. (p. 52).

2.2 *Producing models formally*

In many fields deterministic models are produced by use of maximization, minimization, conservation and other 'first principles'. For example, energy conservation, utility maximization, profit maximization, cost minimization, etc. have been used to produce deterministic models in many instances. While many of these deterministic models have been very useful, e.g. Newton's laws, Einstein's laws, laws of demand and supply, etc., there is generally a need to accommodate stochastic elements, for example measurement errors, errors in equations, and probabilistic assignments and hence the widespread use of stochastic models in econometrics, statistics and many other sciences. Thus, we shall concentrate attention on the production of stochastic models for observations.

Density functions for observations have been derived as solutions to differential equations as for example in the Pearson system, the Schrodinger equation, etc. A question arises as to why certain differential equations are employed. In the case of Schrodinger's partial differential equation, Jeffreys (1973) explains that it was the simplest equation capable of explaining the empirical phenomena that had to be explained. And of course, it was rationalized by mathematical physical theory. Herein, we shall indicate how choosing models to maximize entropy subject to side conditions can lead to a wide range of important, widely used stochastic models, in essence an extension of related work by Burg (1975), Cover and Thomas (1991), Ryu (1990, 1993), Zellner and Min (1993) and others.[1]

We begin our derivation of dynamic and other models by maxent with consideration of random walk models that have served as 'benchmark' models in econometrics and other areas for many years.

Example 2.1. Producing random walk models by maxent Let $y_t = Y_t - Y_{t-1}$, where Y_t might be an exchange rate or the logarithm of a stock price in period t. We require a density for y_t that incorporates what we know about the process and yet is most conservative in the sense that it is such that it maximizes entropy, denoted by H and given by:[2]

$$H = -\int_a^b f(y_t)\, \ell n\, f(y_t) dy_t. \tag{2.1}$$

Note that information is introduced by choice of the values of a and b which define the range of y_t, $b - a$. If we utilize just the side condition that $f(y_t)$ is proper, $\int f(y_t)dy = 1$, then it is well known that the entropy-maximizing density is the uniform density, $f(y_t) = 1/(b - a)$. By adding additional side conditions giving the first and second moments, Shannon (1948) showed that the entropy-maximizing density is a normal density given a doubly infinite range for y_t. On the other hand if the range is finite, the result is a truncated normal density in this case. If just the side conditions on the zero'th and first moments are utilized, the maxent density is an exponential density while if the zero'th and first four moments are utilized, the density is an exponential quartic; see, e.g., Cover and Thomas (1991), Kapur and Kesavan (1992), and Zellner and Highfield (1982).

So far we have considered just y_t, one observation. To get a random walk process, we have to consider a set of observations, and get the maxent density for say $\mathbf{y}' = (y_1, y_2, \ldots, y_t, \ldots y_T)$. Here we could follow Burg (1975) and Cover and Thomas (1991, pp. 274–5) by writing the entropy for the joint density and maximizing it subject to the conditions that the density be proper, have given means, and autocovariances and thereby derive a multivariate normal process for $\mathbf{y}$. This can give us a Gaussian random walk process. Here however, we shall pursue a different approach by writing the joint density as $f(\mathbf{y}) = \prod_{t=1}^{T} f_t(y_t)$. Here we are assuming that the y_t's are independently but not necessarily identically distributed (e.g., they may have differing means and variances). The joint entropy is given by $H = -\int f(\mathbf{y})\, \ell n\, f(\mathbf{y})dy$ and on maximizing it with respect to the f's subject to their being proper and having given means and variances, not all equal, what emerges is a normal random walk model with a drift parameter and error term variance that vary from period to period, that is, $y_t = \theta_t + e_t$ where e_t has zero mean and variance σ_t^2. Of course if the means and variances are assumed equal for all t, the process for Y_t will be a normal or truncated normal random walk process with a constant drift parameter. On the other hand, if the zero'th and first four moments are specified to be equal for all $f_t(y_t)$, the resulting maxent density for each y_t will be exponential quartic which can display asymmetry, bimodality, and other interesting features. Further, it is possible to use other side conditions to get other non-normal densities, say the y_t's independently distributed, each with a Student-t density. See, e.g. Soofi (1992b) for side conditions needed to produce Student-t densities; Zellner and Highfield, (1982) for those needed to produce members of the Cobb–Koopstein–Chen class of densities and Don (1986) for maxent generation of some error densities.

Example 2.2. Producing regression and autoregression models In producing regression and autoregression models, we depart from procedures utilized in the literature by, for example, Burg (1975), Ryu (1990, 1993), Cover and Thomas (1991) and others by using conditional density functions as a point of departure as in Zellner and Min (1993) where a third-order autoregression with lagged input variables was derived by maxent.

We begin by considering $f(y_t \mid \mathbf{x}_t, \boldsymbol{\theta})$, the conditional density of y_t given a vector of input variables denoted by $\mathbf{x}_t$ and a vector of parameters $\boldsymbol{\theta}$. We shall first consider the vector $\mathbf{x}_t$ to be a vector of independent variables as in a regression problem. Then the entropy associated with $f(y_t \mid \mathbf{x}_t, \boldsymbol{\theta})$ is given by

$$H = -\int f(y_t \mid \mathbf{x}_t, \boldsymbol{\theta})\, \ell n\, f(y_t \mid \mathbf{x}_t, \boldsymbol{\theta}) dy_t$$

which can be maximized with respect to choice of f subject to f's being proper and having the following conditional moments: (i) $\int y_t f(y_t \mid \mathbf{x}_t, \boldsymbol{\theta})\, dy_t = \mathbf{x}'_t\boldsymbol{\beta}$ and (ii) $\int (y_t - \mathbf{x}'_t\boldsymbol{\beta})^2 f(y_t \mid x_t, \boldsymbol{\theta}) dy_t = \sigma^2$.[3]. The solution to this problem, as pointed out in Zellner and Min (1993), is a conditional normal density for y_t with mean $\mathbf{x}'_t\boldsymbol{\beta}$ and variance σ^2. Further, if the joint conditional density for $\mathbf{y}' = (y_1, y_2, \ldots y_T)$ given $\mathbf{x}_1, \mathbf{x}_2, \ldots, \mathbf{x}_T$ and the parameters can be represented as follows, $\prod_{t=1}^{T} f(y_t \mid \mathbf{x}_t, \boldsymbol{\theta})$, then from the entropy of this joint density, the maxent solution using the above side conditions is that the y_t's given the $\mathbf{x}_t$'s are independently normally distributed with $f(y_t \mid \mathbf{x}_t, \boldsymbol{\theta}) \sim N(\mathbf{x}'_t\boldsymbol{\beta}, \sigma^2)$, $t = 1, 2, \ldots, T$. Of course if the conditional variances were not all assumed to be equal, for example they could be σ_t^2 rather than σ^2, then heteroskedasticity would result. Further, if the conditional means were $g(\mathbf{x}_t, \boldsymbol{\beta})$ rather than $\mathbf{x}'_t\boldsymbol{\beta}$, the maxent solution would be a non-linear regression model. Finally, if the $\mathbf{x}_t$ vector includes lagged values of y_t as well as independent variables, then this approach provides an autoregressive model with input variables, that is a particular transfer function model.

The above approach differs from that in the literature by working with the conditional density $f(y_t \mid \mathbf{x}_t, \boldsymbol{\theta})$ rather than with the joint density, say $p(y_t, \mathbf{x}_t \mid \boldsymbol{\phi})$. The latter approach involves deriving the maxent joint density and then deriving the associated regression function. If interest centers on the joint density, it can be obtained by considering the entropy of $f(y_t \mid \mathbf{x}_t, \boldsymbol{\theta}) g(\mathbf{x}_t \mid \boldsymbol{\alpha})$, assuming that f and g are proper and maximizing the following expression:

$$\begin{aligned} H &= -\int f(y \mid x) g(x) \ell n\, f(y \mid x) g(x) dy dx \\ &= -\int \left[\int f(y \mid x) \ell n\, f(y \mid x) dy\right] g(x) dx - \int g(x) \ell n\, g(x) dx. \end{aligned}$$

On choosing $f(y \mid x)$ to maximize its entropy and letting $H(f^*)$ denote the maximized entropy, the maximizing value for $g(x)$, is $g^* \propto \exp[H(f^*)]$ if only the condition that $g(x)$ be proper is utilized. Of course other side conditions on $g(x)$ can be utilized as well. Thus it is possible to obtain the maxent joint density for y and x which takes account of side conditions on the conditional density for y given x and on the marginal density of x. Last, as above, with conditions on the joint density of the y_t's given the x_t's and of that of the marginal density of the x_t's, as above the maxent density can take the form:

$$\prod_{t=1}^{T} f^*(y_t \mid x_t) g^*(\mathbf{x}_t \mid \boldsymbol{\alpha}).$$

By varying the side conditions, a broad range of linear and non-linear regression and autoregression models can be derived.

Example 2.3. Obtaining hierarchical models by maxent Here we consider $f(y \mid \boldsymbol{\theta})$ and $g(\boldsymbol{\theta} \mid \boldsymbol{\phi})$, a hyperdensity for $\boldsymbol{\theta}$ given the parameter(s) $\boldsymbol{\phi}$. The entropy of the joint density is:

$$H = -\iint f(y \mid \theta) g(\theta \mid \phi) \ln f(y \mid \theta) g(\theta \mid \phi) dy d\theta$$
$$= \int H(f) g(\theta \mid \phi) d\theta + H(g).$$

On maximizing $H(f)$ subject to side conditions, the conditionally maximized entropy is $H_c = \int H(f^*) g(\theta \mid \phi) d\theta + H(g)$ and then maximizing this subject to g's being proper yields the maxent solution for g, namely $g^* \propto \exp[H(f^*)]$. Of course other side conditions relating to g can be introduced. Finally, with various assumptions regarding the elements of the vector $\mathbf{y}' = (y_1, y_2, \ldots y_T)$, maxent joint densities can be obtained.

Example 2.4. Time-varying parameter state space models Consider the densities $f(y_t \mid \beta_t)$ and $g(\beta_t)$. Then,

$$H = -\iint f(y_t \mid \beta_t) g(\beta_t) \ln f(y_t \mid \beta_t) g(\beta_t) dy_t d\beta_t$$
$$= \int H[f(y_t \mid \beta_t)] g(\beta_t)] d\beta_t - \int g(\beta_t) \ln g(\beta_t) d\beta_t$$

given that $f(y_t \mid \beta_t)$ is a proper density. As above, we can choose $f(y_t \mid \beta_t)$ subject to side conditions to max $H[f(y_t \mid \beta_t)]$. For example the side conditions might be that the mean and variance of y_t are β_t and σ^2, respectively. Then $f^*(y_t \mid \beta_t) \sim N(\beta_t, \sigma^2)$. On inserting $f^*(y_t \mid \beta_t)$ in the expression for H above and maximizing with respect to $g(\beta_t)$ subject to side conditions, the optimal form of $g(\beta_t)$ can be found, say $g^*(\beta_t)$. If it is just assumed that $g(\beta_t)$ is proper, $g^*(\beta_t) \propto \exp\{H[f^*(y \mid \beta_t)]$. If $f^*(y_t \mid \beta_t)$ is $N(\beta_t, \sigma^2)$, $g^*(\beta_t)$ will be a uniform density. On specifying the mean and variance of $g(\beta_t)$ to be β_{t-1} and τ^2, respectively $g^*(\beta_t)$ will be $N(\beta_{t-1}, \tau^2)$ if $\sigma^2 = 1$. That is, the model under the above conditions on the first two moments of $f(y_t \mid \beta_t)$ and $g(\beta_t)$ is $y_t = \beta_t + \varepsilon_t$ and $\beta_t = \beta_{t-1} + v_t$ where ε_t is $N(0, 1)$ and v_t is $N(0, \tau^2)$. By altering forms of the conditional means and variances, a wide range of models can be generated. Further by making conditional independence assumptions, a joint maxent density for $\mathbf{y}' = (y_t, y_2, \ldots y_T)$ and $\boldsymbol{\beta}' = (\beta_1, \beta_2, \ldots \beta_T)$ can be derived given initial conditions, e.g. β_0 and β_{-1}, etc. Finally, the approach above can be generalized to vector processes.

Example 2.5. Seemingly unrelated regression models (SURs) Let $y_{it} = \mathbf{x}'_{it}\beta_i + u_{it}$, $i = 1, 2, \ldots, N$ be a set of N regression relations that we represent as

$$\mathbf{y}_t = Z_t \beta + \mathbf{u}_t$$

where $\mathbf{y}'_t = (y_{1t}, y_2 t, \ldots, y_{mt})$, Z_t is a block diagonal matrix with $\mathbf{x}'_{it}$ in the i-i'th position, $\boldsymbol{\beta}' = (\boldsymbol{\beta}'_1, \boldsymbol{\beta}'_2, \ldots \boldsymbol{\beta}'_m)$ and $\mathbf{u}'_t = (u_{1t}, u_{2t} \ldots, u_{mt})$.

Now given that $f(\mathbf{y}_t \mid Z_t\beta, \Sigma)$ is proper with $Z_t\beta$ and Σ as the mean vector and covariance matrix of this conditional density, respectively, it is the case that the maximum entropy density is an m-dimensional normal density. Altering the side conditions can of course alter the form of the maxent density. Further, if $E(y_{it} \mid Z_{it}, \beta_i) = g(\mathbf{x}_{it}, \beta_i)$, a non-linear expression, the maxent density will be in the form of a non-linear SUR system.

In summary, many central models have been derived as maxent models given certain side conditions. On changing these side conditions, variants of central models are easily obtained. Use of conditional and marginal densities has permitted these results to be obtained quite directly. With these tasks accomplished, we now turn to deriving prior densities for the parameters of models.

3. Maximal data information prior densities (MDIPs)

As explained earlier, see Zellner (1977, 1991) and Zellner and Min (1993), MDIPs are derived by maximizing a criterion functional which is the prior average information in the data density minus the information in the prior density. Thus the prior is chosen so as to maximize the average information in the data density relative to that in the prior. A second interpretation of MDIPs, mentioned in the above references, is that it is the prior which maximizes the expectation of the logarithm of the ratio of the likelihood function to the prior density. From these interpretations, it is clear that use of an MDIP leads to an emphasis on the information in the data density or likelihood function.

Let the joint density of an observation y and parameter θ, scalar or vector, be $p(y, \theta) = \pi(\theta)f(y \mid \theta)$, where $\pi(\theta)$ is a prior density and $f(y \mid \theta)$ is a proper data density given θ, say derived by maxent. Then the negative entropy of $p(y, \theta)$ relative to uniform measure,[4] a measure of the information in $p(y, \theta)$ is

$$\begin{aligned} -H &= \iint p(y, \theta)\ell n\, p(y, \theta)dyd\theta \\ &= \int [f(y \mid \theta)\ell n\, f(y \mid \theta)dy]\pi(\theta)d\theta + \int \pi(\theta)\ell n\, \pi(\theta) \\ &= \int I(\theta)\pi(\theta)d\theta + \int \pi(\theta)\ell n\, \pi(\theta)d\theta \end{aligned}$$

where

$$I(\theta) = \int f(y \mid \theta)\ell n\, f(y \mid \theta)dy$$

the negative entropy of $f(y \mid \theta)$, a measure of the information in $f(y \mid \theta)$. Thus the negative entropy of the joint density breaks up into two parts, as shown above, the average information in the data density plus the information in the prior density. Since in science there is usually a desire to emphasize the information in the data density or likelihood function, the following criterion functional is employed in the MDIP approach:

$$G[\pi(\theta)] = \int I(\theta)\pi(\theta)d\theta - \int \pi(\theta)\ell n \pi(\theta)d\theta$$

which is to be maximized by selection of $\pi(\theta)$ subject to side conditions. Before deriving MDIPs, note that by use of $p(y, \theta) = g(\theta \mid y)h(y)$, a second form of $G[\pi(\theta)]$ has been obtained in Zellner (1977, 1991):

$$G[\pi(\theta)] = \int \{\int g(\theta \mid y)\ell n[\ell(\theta \mid y)/\pi(\theta)]d\theta\}h(y)dy$$

where $\ell(\theta \mid y) \equiv f(y \mid \theta)$ is the likelihood function. Thus by maximizing $G[\pi(\theta)]$, the posterior expectation of $\ell n[\ell(\theta \mid y)/\pi(\theta)]$ is made as great as possible. For other interpretations of $G[\pi(\theta)]$ and its relation to other criterion functionals for generating priors see Soofi (1992b) and Zellner (1991). Independently Zhang and Yao (1991) and Venegas-Martinez (1990) have shown that MDIP solutions are unique.

$G[\pi(\theta)]$ above can be maximized subject to $\int \pi(\theta)d\theta = 1$. The solution is then $\pi^*(\theta) \propto e^{I(\theta)}$ and $G[\pi^*(\theta)] - G[\pi(\theta)] = \int \pi(\theta)\ell n[\pi(\theta)/\pi^*(\theta)]d\theta$, the cross-entropy of $\pi(\theta)$ relative to $\pi^*(\theta)$. In Zellner (1977) it is shown that for general location data densities, $f(y - \theta)$, the MDIP is a uniform density, for scale data densities, $\sigma^{-1}f(y/\theta)$, it is $\pi^*(\sigma) \propto 1/\sigma$, and for location-scale densities, $\sigma^{-1}f[(y - \theta)/\sigma]$, it is $\pi^*(\theta, \sigma) \propto 1/\sigma$, $a < \theta < b$ and $0 < c < \sigma < d$. Also, MDIPs have been shown to be invariant to linear transformations of the data and/or parameters. See Zellner (1971, 1977, 1991) and Zellner and Min (1993) for MDIPs for parameters of normal mean, regression, autoregression, exponential, uniform, and other models. Also, in Zellner (1991) side conditions reflecting invariance restrictions have been utilized to produce MDIPs that are invariant to specific non-linear, one-to-one continuous transformations and some comparisons with results provided by other methods for producing priors are given in Zellner (1977, 1991).

With this said about the MDIP approach and some results yielded by it in previous work, we now turn to consider new problems.

Example 3.1. MDIP for common parameters in different data densities The old problem of obtaining a prior density for $\boldsymbol{\theta}$ when $\boldsymbol{\theta}$ is in two (or more) data densities or experiments, $f_1(y \mid \boldsymbol{\theta})$ and $f_2(y \mid \boldsymbol{\theta})$ was brought to my attention by Skilling (1992). It is well known that Jeffreys's (1967), the MDIP and many other approaches for generating priors will yield different priors for $\boldsymbol{\theta}$ on consideration of $f_1(y \mid \boldsymbol{\theta})$ and $f_2(y \mid \boldsymbol{\theta})$. For example, Jeffreys's priors are $|F_1(\boldsymbol{\theta})|^{1/2}$ and $|F_2(\boldsymbol{\theta})|^{1/2}$, where $F_1(\boldsymbol{\theta})$ and $F_2(\boldsymbol{\theta})$ are the Fisher information matrices associated with $f_1(y \mid \boldsymbol{\theta})$ and $f_2(y \mid \boldsymbol{\theta})$, respectively. Similarly the MDIP approach subject just to the condition that priors be proper yields $\pi_1(\boldsymbol{\theta}) \propto e^{I_1(\boldsymbol{\theta})}$ and $\pi_2(\boldsymbol{\theta}) \propto e^{I_2(\boldsymbol{\theta})}$. If $|F_1(\boldsymbol{\theta})| \neq |F_2(\boldsymbol{\theta})|$ and $I_1(\boldsymbol{\theta}_1) \neq I_2(\boldsymbol{\theta}_2)$, these approaches will produce different priors for the same parameter(s) appearing in different data densities, e.g. $f_1(y \mid \boldsymbol{\theta})$ and $f_2(y \mid \boldsymbol{\theta})$.

To get a single prior for $\boldsymbol{\theta}$, we seek the prior that makes the average information in both densities or experiments as large as possible relative to the information in the prior. That is our criterion functional is from Zellner (1992):

$$G_2[\pi(\theta)] = \sum_{i=1}^{2}\left[\int I_i(\theta)\pi(\theta)d\theta - \int \pi(\theta)\ell n\, \pi(\theta)d\theta\right]$$

where $I_i(\boldsymbol{\theta}) = \int f_i(y \mid \boldsymbol{\theta})\ell n f_i(y \mid \boldsymbol{\theta})dy$, $i = 1, 2$. On maximizing $G_2[\pi(\boldsymbol{\theta})]$ subject to $\int \pi(\boldsymbol{\theta})d\boldsymbol{\theta} = 1$, the result is:

$$\pi_2^*(\theta) \propto e^{\bar{I}(\theta)} = [e^{I_1(\theta)}e^{I_2(\theta)}]^{1/2}$$

where $\bar{I}(\boldsymbol{\theta}) = [I_1(\boldsymbol{\theta}) + I_2(\boldsymbol{\theta})]/2$. Thus $\pi_2^*(\boldsymbol{\theta})$ is equal to the geometric mean of the individual MDIPs, $e^{I_1(\boldsymbol{\theta})}$ and $e^{I_2(\boldsymbol{\theta})}$. If more than two data densities or experiments are involved, say m, the above criterion functional can be defined over the m experiments to yield $G_m[\pi(\boldsymbol{\theta})]$. On maximizing $G_m[\pi(\boldsymbol{\theta})]$ s.t. $\int \pi(\boldsymbol{\theta})d\boldsymbol{\theta} = 1$, the result is

$$\pi_m^*(\theta) \propto \left(\prod_{i=1}^{m} e^{I_i(\theta)} \right)^{1/m},$$

again the geometric mean of the individual MDIPs.

In considering two data densities or experiments, background information may lead an investigator to assign different weights or degrees of confidence to the information provided by them. Let $w_1 = w$ and $w_2 = 1 - w$, with $0 < w < 1$, be the weights or probabilities assigned to the two data densities or experiments and consider

$$G_{2w}[\pi(\theta)] = \sum_{i=1}^{2} \left[\int w_i I_i(\theta)\pi(\theta)d\theta - \int \pi(\theta)\ell n\, \pi(\theta)d\theta \right]$$

On maximizing $G_{2w}[\pi(\theta)]$ w.r.t. $\pi(\theta)$ s.t. $\int \pi(\boldsymbol{\theta})d\boldsymbol{\theta} = 1$, the result is $\pi_{2w}^*(\boldsymbol{\theta}) \propto \exp\{[w_1 I(\boldsymbol{\theta}) + w_2 I(\boldsymbol{\theta})]/2\}$ the geometric mean of $\exp\{w_1 I_1(\boldsymbol{\theta})\}$ and $\exp\{w_2 I_2(\boldsymbol{\theta})\}$.

From what has been presented above, it is indeed possible to derive a MDIP that permits the information in several data densities or experiments to be featured as strongly as possible.

Example 3.2. MDIPs for many observations Above, MDIPs have been derived so as to maximize the prior average information in a data density for one observation (scalar or vector) relative to the information in a prior density function. Given this MDIP prior, it can be used to form a posterior density given the first observation. This posterior density can then be used as a prior in the analysis of the second observation and so on. However in many cases we have to consider n observations and obtain a prior that features the information in these n observations. To accomplish this, assume that the n observations are independently and identically distributed (i.i.d.), $f(y_i \mid \boldsymbol{\theta})$. Then

$$G_n[\pi(\boldsymbol{\theta})] = \iint f(\mathbf{y} \mid \boldsymbol{\theta})\ell n\, f(\mathbf{y} \mid \boldsymbol{\theta})d\mathbf{y}\, \pi(\boldsymbol{\theta})d\boldsymbol{\theta}/n - \int \pi(\boldsymbol{\theta})\ell n\, \pi(\boldsymbol{\theta})d\boldsymbol{\theta}$$

specializes to $G_n[\pi(\boldsymbol{\theta})] = \int I(\boldsymbol{\theta})\pi(\boldsymbol{\theta})d\boldsymbol{\theta} - \int \pi(\boldsymbol{\theta})\ell n\, \pi(\boldsymbol{\theta})d\boldsymbol{\theta}$ which is identical to the criterion for a single observation. Thus with the use of $G_n[\pi(\boldsymbol{\theta})]$ and an i.i.d. assumption, a MDIP for any one observation is the same as that obtained from use of $G_n[\pi(\boldsymbol{\theta})]$.

If the observations are not i.i.d. and the joint density is given by $\prod_{i=1}^{n} f_i(y_i \mid \boldsymbol{\theta})$, on inserting this in the expression for $G_n[\pi(\boldsymbol{\theta})]$ above, the result is

$$G_n[\pi(\boldsymbol{\theta})] = \int \bar{I}_n \pi(\boldsymbol{\theta})d\boldsymbol{\theta} - \int \pi(\boldsymbol{\theta})\ell n\, \pi(\boldsymbol{\theta})d\boldsymbol{\theta}$$

where $\bar{I}_n = \sum_{i=1}^{n} I_i(\boldsymbol{\theta})/n$ with $I_i(\boldsymbol{\theta}) = \int f_i(y_i \mid \boldsymbol{\theta})\ell n\, f_i((y_i \mid \boldsymbol{\theta})dy_i$. Then maximizing $G_n[\pi(\boldsymbol{\theta})]$ subject to $\pi(\boldsymbol{\theta})$ being proper yields $\pi_n^*(\boldsymbol{\theta}) \propto e^{\bar{I}_n}$, the geometric mean of the MDIP densities for the individual observations, $\pi_i^*(\theta) \propto e^{I_i(\theta)}$.

Example 3.3. Prior for an exchangeable sequence Assume that instead of i.i.d. observations we have two exchangeable observations, that is $f(y_1, y_2 \mid \boldsymbol{\theta}) = f(y_2, y_1 \mid \boldsymbol{\theta})$. Since the negative entropies of these two densities are the same, the MDIP prior densities based on them will be the same given the same side conditions imposed on the priors. If it is just assumed that the priors are proper, using the criterion functional G_2 in Example 3.1, the resulting MDIPs are $\pi_1(\boldsymbol{\theta}) = \pi_2(\boldsymbol{\theta}) = e^{I(\theta)}$ where $I(\boldsymbol{\theta}) = \iint f(y_1, y_2 \mid \boldsymbol{\theta})\ell n f(y_1, y_2 \mid \boldsymbol{\theta})dy_1 dy_2 = \iint f(y_2, y_1 \mid \boldsymbol{\theta})\ell n f(y_2, y_1 \mid \boldsymbol{\theta})dy_2 dy_1$. For n exchangeable observations, the proper MDIP density is $\pi(\boldsymbol{\theta}) \propto e^{I_e(\theta)}$, where $I_e(\boldsymbol{\theta})$ is the negative entropy of the n-dimensional exchangeable data density, $f(y_1, y_2, \ldots, y_n \mid \boldsymbol{\theta})$.

Example 3.4. MDIPs for hierarchical models' hyperparameters Let $f(y \mid \boldsymbol{\theta})$ be a density for y given $\boldsymbol{\theta}$ and $g(\boldsymbol{\theta} \mid \boldsymbol{\phi})$ a hyperdistribution for $\boldsymbol{\theta}$ given $\boldsymbol{\phi}$, the hyperparameters. To derive a MDIP for $\boldsymbol{\phi}$, consider

$$h(y \mid \boldsymbol{\phi}) = \int f(y \mid \boldsymbol{\phi}) g(\boldsymbol{\theta} \mid \boldsymbol{\phi}) d\boldsymbol{\theta},$$

the marginal distribution of y given $\boldsymbol{\phi}$. Then an MDIP for $\boldsymbol{\phi}$ can be obtained by maximizing the following criterion functional subject to side conditions:

$$G[\pi(\boldsymbol{\phi})] = \int I(\boldsymbol{\phi})\pi(\boldsymbol{\phi})d\boldsymbol{\phi} - \int \pi(\boldsymbol{\phi})\ell\mathrm{n}\, \pi(\boldsymbol{\phi})d\boldsymbol{\phi}$$

where $I(\boldsymbol{\phi}) = \int h(\mathbf{y} \mid \boldsymbol{\phi})\ell n\, h(\mathbf{y} \mid \boldsymbol{\phi})dy$, the information in $h(y \mid \boldsymbol{\phi})$. If the side condition is $\int \pi(\boldsymbol{\phi})d\boldsymbol{\phi} = 1$, the solution to this problem is $\pi^*(\boldsymbol{\phi}) \propto e^{I(\phi)}$. As mentioned earlier, additional side conditions can be added to this optimization problem and the MDIP will reflect the information contained in them.

Example 3.5. Deriving Jeffreys's prior and entropic priors as MDIPs Let $\mid F(\boldsymbol{\theta}) \mid^{1/2}$ be the square root of the determinant of the Fisher information matrix, a measure of information in the associated data density, $f(y \mid \boldsymbol{\theta})$. Then consider the following criterion functional:

$$G_J[\pi(\boldsymbol{\theta})] = \int \ell n \mid F(\boldsymbol{\theta}) \mid^{1/2} \pi(\boldsymbol{\theta})d\boldsymbol{\theta} - \int \pi(\boldsymbol{\theta})\ell n\, \pi(\boldsymbol{\theta})d\boldsymbol{\theta}$$

which when maximized subject to $\int \pi(\boldsymbol{\theta})d\boldsymbol{\theta} = 1$, yields as a solution, $\pi_J^*(\boldsymbol{\theta}) \propto \mid F(\boldsymbol{\theta}) \mid^{1/2}$. This prior has the form of Jeffreys's prior but is proper in contrast to Jeffreys's priors which are many times improper. Further, if other side conditions are added, say $\int \theta^i \pi(\theta)d\theta = \mu_i$, $i = 1, 2$, in the scalar case, the solution is $\pi^*(\theta) \propto \mid F(\theta) \mid^{1/2} e^{\lambda 1\theta + \lambda 2\theta 2}$ where λ_1 and λ_2 are Lagrange multipliers. Side conditions on moments can also be introduced for a vector of parameters.

To obtain entropic priors discussed by Rodríguez (1991a, b), utilize the usual MDIP criterion functional, $G(\pi) = \int \pi(\theta)I(\theta)d\theta - \int \pi(\theta)\ell n \pi(\theta)d\theta$ subject to $\pi(\theta)$ being proper, $\int \pi(\theta)\ell n \mid F(\theta) \mid^{1/2} d\theta = c_1$ and $\int \pi(\theta)I(\theta)d\theta = c_2$ to obtain

$$\pi^*(\boldsymbol{\theta}) \propto \mid F(\boldsymbol{\theta}) \mid^{\lambda_1/2} e^{\lambda_2 I(\boldsymbol{\theta})},$$

where λ_1 and λ_2 are Lagrange multipliers. For particular values of λ_1 and λ_2, $\pi^*(\theta)$ assumes the form of an entropic prior. See Rodríguez (1991a, b) for further discussion of entropic priors which have the same invariance properties that Jeffreys's priors have.

As the above examples indicate, there is an open issue regarding the information measure to use, $I(\boldsymbol{\theta})$ and/or $\ell n\ |F(\boldsymbol{\theta})|^{1/2}$. Involved in this choice is the degree of invariance desired – see Zellner (1991, pp. 24–5) for further discussion of this issue. It is indicated therein that it seems appropriate to have prior densities invariant to *relevant* transformations, just as Jeffreys recognizes in his choices of prior densities for the mean, θ, and standard deviation, σ, of a normal distribution and for normal regression models' parameters, $\boldsymbol{\beta}$ and σ for which he recommends $\pi(\boldsymbol{\theta}, \sigma) \propto 1/\sigma$, and $\pi(\boldsymbol{\beta}, \sigma) \propto 1/\sigma$, priors in the form of MDIPs for these models.

Example 3.6. MDIPs for binomial, trinomial and multinomial parameters Let $y = 1$ with probability θ and $y = 0$ with probability $1 - \theta$. Then the probability mass function for a single observation y is $\theta^y(1-\theta)^{1-y}$ and

$$\begin{aligned} I(\theta) &= Ey\ \ell n\ \theta + E(1-y)\ell n(1-\theta) \\ &= \theta\ \ell n\ \theta + (1-\theta)\ell n(1-\theta) \end{aligned}$$

is the negative entropy for a single observation. Then the proper MDIP is $\pi^*(\theta) = ce^{I(\theta)} = c\theta^\theta(1-\theta)^{1-\theta}$ with $c = 1.6186$, the normalizing constant evaluated by numerical integration – see Zellner (1977, p. 207) for a plot of this proper density that has a minimum at $\theta = 1/2$ and rises symmetrically to 1.6186 at $\theta = 1$ and $\theta = 0$. Thus $\pi(\theta)$ is between the Bayes–Laplace uniform prior $\pi(\theta) = 1$ and the Jeffreys's prior $\pi_J(\theta) = 1/B(1/2,1/2)\ \theta^{-1/2}(1-\theta)^{-1/2}$, a proper beta density with $0 \le \theta \le 1$. Jeffreys (1967, pp. 123–5) and Akaike (1978) have considered the uniform prior and Jeffreys's prior and find the former not entirely satisfactory because it doesn't rise at the end-points 0 and 1 whereas Jeffreys's prior (and Haldane's prior $\propto 1/\theta(1-\theta)$) shoots off to ∞ at $\theta = 0$ and $\theta = 1$. That the MDIP, $\pi(\theta) \propto \theta^\theta(1-\theta)^{1-\theta}$ assumes finite values at $\theta = 0$ and $\theta = 1$ may be an advantage of it for certain problems.

Since analytical integration of $\theta^\theta(1-\theta)^{1-\theta}$ is difficult, quadratic and quartic approximations have been obtained by expanding this function about its minimizing value $\theta = 1/2$, in a Taylor series. These approximations are:

$$\pi_1(\theta) = {}^{12}\!/_7[{}^1\!/_2 + (\theta - {}^1\!/_2)^2] \quad 0 \le \theta \le 1$$

and

$$\pi_2(\theta) = {}^5\!/_3[{}^1\!/_2 + (\theta - {}^1\!/_2)^2 + {}^4\!/_3(\theta - {}^1\!/_2)^4] \quad 0 \le \theta \le 1$$

and have been normalized to integrate to one. While additional terms can be added to improve these approximations, it is a fact that such 'quadratic' priors are easy to integrate analytically. Also, while beyond the scope of this paper, Kemp (1975) has shown that maximizing Rényi's (1961) second-order entropy, $-\log_2 \underset{\pi}{E}\ \pi(\theta)$ with respect to $\pi(\theta)$, for $0 \le \theta \le 1$ subject to $E\theta = a$ and $E\theta^2 = a^2 + b^2$ yields $\pi_*(\theta) = c_0 +$

$c_1\theta + c_2\theta^2$, a quadratic prior. On completing the square on θ, $\pi^*(\theta)$ can be expressed as follows: $\pi_*(\theta) = c_2\,(\theta + 1/2\, c_1/c_2)^2 + c_0 - c_1^2/4c_2$ which has a minimum at $-c_1/2c_2$ for $c_2 > 0$ and $c_1 < 0$. If $-c_1 = c_2$, this minimum occurs at $\theta = 1/2$ and the density is symmetric about $\theta = 1/2$. Further work is needed to relate second-order entropy to the MDIP approach – see Rényi (1961) and Maasoumi (1990) for discussions of generalized entropy measures.

For a trinomial process, with probability θ_i, $i = 1, 2, 3$ where $\theta_1 + \theta_2 + \theta_3 = 1$, the probability mass function is $\theta_1^{y_1}\,\theta_2^{y_2}\,(1 - \theta_1 - \theta_2)^{y_3}$, $0 \le \theta_1, \theta_2 \le 1$ and the negative entropy of this probability mass function is: $I_3(\theta_1, \theta_2) = \theta_1\,\ell n\,\theta_1 + \theta_2\,\ell n\,\theta_2 + (1-\theta_1-\theta_2)\ell n(1-\theta_1-\theta_2)$. Then maximizing

$$G_3[\pi(\theta_1, \theta_2] = \iint I(\theta_1, \theta_2)\pi(\theta_1, \theta_2)d\theta_1 d\theta_2 - \int \pi(\theta_1, \theta_2)\ell n\,\pi(\theta_1, \theta_2)d\theta_1 d\theta_2$$

subject to $\pi(\theta_1, \theta_2)$ being proper yields $\pi^*(\theta_1, \theta_2) \propto \theta^{\theta_1}\theta^{\theta_2}\,(1 - \theta_1 - \theta_2)^{(1-\theta_1-\theta_2)}$, a MDIP density that has a minimum at $\theta_1 = \theta_2 = 1/3$ and has volume over the triangular region $0 \le \theta_1 \le 1$, $0 \le \theta_2 \le 1$ and $0 \le \theta_1 + \theta_2 \le 1$. The surface is bowl-shaped and reaches finite values at the boundaries. Finally, for the multinomial, similar considerations lead to a proper MDIP $\pi^*(\theta_1, \theta_2, \ldots, \theta_{m-1}) \propto \theta_1{}^{\theta_1}\theta_2{}^{\theta_2} \ldots \theta^{\theta_{m-1}}_{m-1}\,(1 - s_{m-1})^{1-s_{m-1}}$ where $s_{m-1} = \sum_{i=1}^{m-1}\theta_i$ for $0 \le \theta_i \le 1$ and $0 \le s_{m-1} \le 1$

Example 3.7. MDIP odds Let $h_i(y) = \int f_i(y \mid \theta_i)\pi(\theta_i)d\theta_i$ and $I_i = \iint h_i(y)\ell n h_i(y)dy$, $i = 1, 2$ and consider $G = P_1 I_1 + (1 - P_1)I_2 - P_1 \ell n P_1 - (1 - P_1)\ell n(1 - P_1)$ where P_1 = probability of model 1, $f_1(y \mid \theta_1)$. On maximizing G with respect to P_1 the result is $\ell n P^*_1/(1 - P^*_1) = I_1 - I_2$. If $I_1 = I_2$, $P^*_1 = 1/2$. If $I_1 \gtrless I_2$, $P^*_1 \gtrless 1/2$. Extension to more than two hypothesized models is direct and will be considered in another paper.

In summary the MDIP has been shown to yield a wide range of diffuse and informative prior densities by maximizing a sensible criterion functional subject to side conditions that reflect investigators' available information and invariance conditions. See Zellner (1977,1991) for results on invariance and comparisons of alternative approaches for producing prior densities.

4. Summary and Conclusions

From what has been presented it is clear that many stochastic models can be produced formally and reproducibly by the maxent approach. Further, the MDIP approach has been shown to yield reasonable diffuse and informative prior densities for their parameters. Thus two components for inference are available, a model for the observations and a prior density for its parameters. The issue of how to combine these components to form a good information processing rule (IPR) has been analysed by Bayes, Laplace, Jeffreys, Savage, Akaike, and many others. Several different approaches have yielded Bayes's theorem as an explicitly derived optimal information processing model – see Zellner (1988) and the references therein. In Zellner (1991), an optimal IPR is derived that takes account of the possibly differing quality of sample and prior information. Further research is needed to produce other static and dynamic IPRs.

Notes

1. It is recognized also that entropy approaches have been utilized by many to produce income distributions and firm size distributions – see e.g. Davis (1941), Golan (1988, 1989), and Maasoumi (1990). Davis (1941) used an entropy approach to produce a model of consumer behavior while in Cover and Thomas (1991) and Kapur and Kesavan (1992) other economic applications of the maxent approach are reported. Good, Savage and others have associated utility with entropy in deriving various results that are justified in terms of maximization of expected utility. Grandy (1991) has used maxent to produce return distributions for investors in connection with portfolio and other problems. Zellner (1988) has emphasized a distinction between information and utility in connection with maxent analyses. See Judge and Golan (1992) for analysis of interesting inverse problems using maxent.
2. Here we have defined the entropy of $f(y_t)$ relative to a uniform measure or a benchmark uniform density. If another benchmark density is relevant, say $m(y_t)$, we can maximize Hm shown below (or minimize $-Hm$, the cross entropy) by choice of $f(y_t)$ subject to side conditions:

$$Hm = -\int f(y_t)\ell n\,[f(y_t) / m\,(y_t)]\,dy_t. \tag{2.1}$$

3. As emphasized in Section 2.1, the form of the conditional mean, $\mathbf{x}_t'\boldsymbol{\beta}$ must be specified with great care and thought.
4. See Zellner (1991) for use of an arbitrary measure, $m(y, \theta)$ rather than a uniform measure in the expression for $-H$.

References

Adelman, I. and F. L. Adelman (1959), 'The dynamic properties of the Klein–Goldberger model', *Econometrica*, **27**, 569–625.

Akaike, H. (1978), 'A new look at the Bayes procedure', *Biometrika*, **65**(1), 53–9.

Burg, J.P. (1975), 'Maximum entropy spectral analysis', doctoral dissertation, Dept. of Geophysics, Stanford University, Stanford, CA.

Cover, T. M. and J. A. Thomas (1991), *Elements of information theory*, New York: John Wiley.

Davis, H. T. (1941), *The theory of econometrics*, Bloomington, IN: Principia.

Don, F. (1986), 'The specification of least informative error distributions', *Journal of Econometrics*, **31**, 81–92.

Gokhale, D. V. (1975), 'Maximum entropy characterisations of some distributions', in G. P. Patil et al. (eds), *Statistical distributions in scientific work*, Vol. 3, Dordrecht, Netherlands: Reidel, pp. 299–304.

Golan, A. (1988), 'A discrete stochastic model of economic production and a model of fluctuations in production-theory and empirical evidence', doctoral dissertation, Dept. of Agricultural Economics, University of California, Berkeley, CA.

Golan, A. (1989), 'A discrete, stochastic theory of size distribution of firms', ms, Dept. of Economics, University of Haifa, Haifa, Israel.

Good, I. J. (1963), 'Maximum entropy for hypothesis formulation, especially for multidimensional contingency tables', *Annals of Mathematical Statistics*, **34**(3), 911–34.

Grandy, C. (1991), 'The principle of maximum entropy and the difference between risk and uncertainty, in Grandy and Schick (1991), pp. 39–47.

Grandy, W. T. and L. H. Schick (1991), *Maximum entropy and Bayesian methods*, Dordrecht, Netherlands: Kluwer.

Hickman, B. G. (ed.) (1972), *Econometric models of cyclical behavior*, New York, Columbia University Press.

Jaynes, E. T. (1957), 'Information theory and statistical mechanics I', *Physical Review*, **106**, 620–30.

Jaynes, E. T. (1980), 'Discussion', in J. M. Bernardo, M. H. DeGroot, D. V. Lindley and A. F. M. Smith (eds), *Bayesian Statistics*, Valencia, Spain: University Press, pp. 618–29.

Jaynes, E. T. (1982a), *Papers on probability, statistics and statistical physics*, Dordrecht, Netherlands: Reidel.

Jaynes, E. T. (1982b), 'On the rationale of maximum-entropy methods', *Proc. of the IEEE 70*, 939–52.

Jeffreys, H. (1967), *Theory of probability*, 3rd rev. edn, London: Oxford University Press.

Jeffreys, H. (1973), *Scientific inference*, 3rd edn, Cambridge: Cambridge University Press.

Judge, G. and A. Golan (1992), 'Recovering information in the case of ill-posed inverse problems with noise', ms, University of California, Berkeley, CA.

Kagan, A. M., Y. V. Linnik and C. R. Rao (1973), *Characterization problems in mathematical statistics*, New York: Wiley.

Kapur, J. N. and H. K. Kesavan (1992), *Entropy optimization principles with applications*, San Diego, CA: Academic Press.

Kemp, A. W. (1975), 'Characterizations based on second-order entropy', in G. P. Patil et al. *Statistical distributions in scientific work*, Dordrecht, Netherlands: Reidel.
Kullbach, S. (1959), *Information theory and statistics*, New York: John Wiley.
Maasoumi, E. (1990), 'Information theory', in J. Eatwell, M. Millgate and P. Newman (eds), *The New Palgrave Econometrics*, New York: W. W. Norton, pp. 101–12.
Rao, C. R. (1965), *Linear statistical inference and its applications*, New York: Wiley.
Rényi, A. (1961), 'On measures of entropy and information', *Proc. of 4th Berkeley Symposium on Mathematical Statistics and Probability*, Vol. 1, 547–61.
Rodríguez, C. C. (1991a), 'Entropic priors', ms, Dept. of Mathematics and Statistics, State University of New York, Albany, NY.
Rodríguez, C. C. (1991b), 'From Euclid to entropy', in Grandy and Schick (1991), pp. 343–8.
Ryu, H. K. (1990), 'Orthogonal basis and maximum entropy estimation of probability density and regression functions', doctoral dissertation, Dept. of Economics, University of Chicago.
Ryu, H. K. (1993), 'Maximum entropy estimation of density and regression functions', *Journal of Econometrics*, **56**, 397–440.
Shannon, C. E. (1948), 'The mathematical theory of communication', *Bell System Technical Journal*, July–October, reprinted in C. E. Shannon and W. Weaver, (1949), *The mathematical theory of communication*, University of Illinois Press, pp. 3–91.
Shore, J. E. and R. W. Johnson (1980), 'Axiomatic derivation of the principle of maximum entropy and the principle of cross-entropy', *IEEE Transactions on Information Theory*, IT-26, 26–37.
Skilling, J. (1989), *Maximum entropy and Bayesian methods*, Dordrecht, Netherlands: Kluwer.
Skilling, J. (1992.), Personal communication.
Soofi, E. S. (1992a), 'A generalizable formulation of conditional logit with diagnostics', *Journal of the American Statistical Association*, **87**, 419, 812–16.
Soofi, E. S. (1992b), 'Information theory and Bayesian statistics', ms, School of Business Administration, University of Wisconsin, Milwaukee, WI, published in D. A. Berry, K. M. Chaluner and J. K. Geweke (eds), *Bayesian Analysis in Statistics and Econometrics: Essays in Honor of Arnold Zellner*, (1996), New York: Wiley, 179–89.
Theil, H. (1967), *Economics and information theory*, Amsterdam: North-Holland.
Tukey, J. W. (1978), Discussion of Granger on seasonality, in A. Zellner (ed.), *Seasonal analysis of economic time series*, Washington, DC: US Government Printing Office, pp. 50–53.
Venegas-Martinez, F. (1990), 'On regularity and optimality conditions for maximum entropy priors', *The Brazilian Journal of Probability and Statistics*, **4**, 105–36.
Zellner, A. (1971), *An introduction to Bayesian inference in econometrics*, New York: John Wiley, reprinted in 1987 by Krieger Publishing Co., Malabar, Florida and in 1996 by Wiley.
Zellner, A. (1977), 'Maximal data information prior distributions', in A. Aykac and C. Brumat (eds), *New developments in the applications of Bayesian methods*, Amsterdam: North Holland, pp. 211–32, reprinted in Zellner (1984).
Zellner, A. (1984), *Basic issues in econometrics*, Chicago: University of Chicago Press.
Zellner, A. (1988), 'Optimal information processing and Bayes's theorem', *The American Statistician*, **42**(4), 278–84.
Zellner, A. (1991), 'Bayesian methods and entropy in economics and econometrics', in Grandy and Schick (1991), pp. 17–31.
Zellner, A. (1992), Personal communication to J. Skilling.
Zellner, A. and R. A. Highfield (1982), 'Calculation of maximum entropy distributions and approximation of marginal posterior distributions', presented to NBER–NSF Seminar on Bayesian Inference, October 1982 and published in *Journal of Econometrics*, **37**, (1988), 195–210.
Zellner, A. and C-K. Min (1993), 'Bayesian analysis, model selection and prediction', in W.T. Grandy, Jr and P. W. Milonni (eds), *Physics and probability: Essays in honor of Edwin T. Jaynes*, Cambridge: Cambridge University Press, pp. 195–206.
Zellner, A. and S. Peck (1973), 'Simulation experiments with a quarterly macroeconometric model of the U.S. economy', in A. A. Powell and R. A. Williams (eds), *Econometric Studies of Macro and Monetary Relations*, Amsterdam: North-Holland, pp. 149–68, reprinted in A. Zellner, *Basic Issues in Econometrics*, Chicago: University of Chicago Press, 1984.
Zhang, Y. T. and Q. W. Yao (1991), 'Some maximal information and generalized maximal entropy priors', ms, Dept. of Statistics, Wuhan University, China, published in *Chinese Journal of Applied Probability and Statistics*, **7** (1991), 192–200.

[8]

3.2 Maximal Data Information Prior Distributions

1 Introduction

In Bayesian analyses, it is often desirable to have a posterior distribution reflect mainly the information in a given sample of data. To achieve this objective, it is necessary to employ a prior distribution that adds little information to the sample information. Much work has been done to provide procedures for formulating such prior distributions, as, for example, Jeffreys (1967), Lindley (1956, 1961), Hartigan (1964), Savage (1961), Novick and Hall (1965), Box and Tiao (1973), Jaynes (1968), and Zellner (1971). In the present paper, a procedure described in Zellner (1971) is analyzed further and applied to a number of problems. Several comparisons with results produced by other procedures are reported.

The plan of the paper is as follows. In section 2, the general method and some of its properties are set forth. Section 3 reports the results of applying the method of generating maximal data information priors (MDIP's) to a number of problems. Last, some concluding remarks are presented in section 4.

2 Maximal Data Information Priors (MDIP's)

The basic idea underlying MDIP's is that they provide maximal prior average data information relative to the information in the prior distribution, with information being represented by Shannon's (1948) measure.[1] To be specific, let $\tilde{y}$ be a scalar random variable with a proper, probability density function (pdf), $p(y \mid \theta)$ defined to be positive in the region R_y, where θ is a scalar parameter such that $a \leq \theta \leq b$, with a and b finite. Then the information in the data distribution $p(y \mid \theta)$ is defined to be $I(\theta)$, given by:

Reprinted from *New Developments in the Applications of Bayesian Methods*, A. Aykac and C. Brumat, eds. (Amsterdam: North-Holland Publishing Co., 1977), pp. 211–32, with the kind permission of North-Holland Publishing Co.

1. Shannon's entropy (or uncertainty) measure is $W = -\int p(x) \log p(x) dx$, where $p(x)$ is a probability density function (pdf). In our work we use $-W$ as a measure of the information associated with $p(x)$.

(1) $$I(\theta) = \int_{R_y} p(y\mid\theta)\log p(y\mid\theta)dy$$

The prior average information in the data pdf, denoted by $\bar{I}$, is:

(2) $$\bar{I} = \int_a^b I(\theta)\,p(\theta)d\theta\,,$$

where $p(\theta)$ is a proper prior pdf defined to be nonnegative for $\mathbf{a} \leq \theta \leq b$ and zero elsewhere. Then the prior average information in the data pdf, $\bar{I}$, minus the information in the prior pdf, $p(\theta)$, is

(3) $$\begin{aligned} G &= \bar{I} - \int_a^b p(\theta)\log p(\theta)d\theta \\ &= \int_a^b p(\theta)I(\theta)d\theta - \int_a^b p(\theta)\log p(\theta)d\theta\,. \end{aligned}$$

As seen from (3) G is just the difference of two information measures, the first relating to the data pdf and the second to the prior pdf. An alternative view of G can be obtained by substituting for $I(\theta)$ in (3) and writing (3) as:

(4) $$G = \int_a^b \int_{R_y} \log\{p(y\mid\theta)/p(\theta)\}\,p(y,\theta)dy\,d\theta\,,$$

where $p(y,\theta) = p(\theta)p(y\mid\theta)$ is the joint pdf for θ and y. From (4) it is seen that G can be interpreted as the average log-ratio of the ordinate of the data pdf or likelihood function to the ordinate of the prior pdf.

We now present the following definition of a maximal data information prior (MDIP) pdf:

Definition: A MDIP pdf is a proper, normalized prior pdf that maximizes G, defined in (3).

The following theorem provides a simple formula for MDIP pdf's.

Theorem 1: The normalized MDIP pdf, $p*(\theta)$ maximizing G in (3) is given by:

(5) $$p*(\theta) = c\exp\{I(\theta)\} \qquad a \leq \theta \leq b$$

where c is a normalizing constant.

Proof: Let $g(\theta)$ be an arbitrary function with $\int_b^a g(\theta)d\theta = 0$ and consider:

$$G(\epsilon) = \int_a^b\big\{I(\theta) - \log\{p(\theta) + \epsilon g(\theta)\}\big\}\{p(\theta) + \epsilon g(\theta)\}\,d\theta\,.$$

Then, $G'(\epsilon) = \int_a^b g(\theta)\big\{I(\theta) - \log\{p(\theta) + \epsilon g(\theta)\}\big\}d\theta$, where $\int_a^b g(\theta)d\theta = 0$ has been used and the necessary condition for a maximum, $G'(0) = 0$, is satisfied for $p(\theta) = p_1(\theta)$, where $\log p_1(\theta) = I(\theta)$ or $p_1(\theta) = \exp\{I(\theta)\}$. Since $G''(0) = -\int_a^b\big\{\{g(\theta)\}^2/p_1(\theta)\big\}d\theta < 0$, $p(\theta) = p_1(\theta)$ is associated with a maximum of G. Finally, $p*(\theta) = c\exp\{I(\theta)\}$ is the normalized, proper pdf that maximizes G where $1/c = \int_a^b \exp\{I(\theta)\}d\theta$.

Corollary: If $I(\theta)$ is a constant, independent of θ, then the MDIP pdf is the uniform pdf, $p*(\theta) = 1/(b - a)$, $a \leq \theta \leq b$.

The following theorem provides an invariance property for MDIP pdf's.

Theorem 2: Let $\tilde{z} = d_0 + d_1\tilde{y}$, where d_0 and d_1 are given constants and the proper, normalized pdf for $\tilde{y}$ is $p(y\,|\,\theta)$, $a \leq \theta \leq b$ and $-\infty < y < \infty$. Then the MDIP pdf for θ associated with the data pdf $p(y\,|\,\theta)$ is identical to that for the data pdf for $\tilde{z}$, $p(z\,|\,\theta)$, $-\infty < z < \infty$.
Proof: Let $I_y(\theta) = \int_{-\infty}^{\infty} p(y\,|\,\theta)\log p(y\,|\,\theta)dy$ and $I_z(\theta) = \int_{-\infty}^{\infty} p(z\,|\,\theta)\log p(z\,|\,\theta)dz$.
Then from known properties of Shannon's information measure, $I_z(\theta) = \log|d_1| + I_y(\theta)$. Thus G defined in terms of $p(z\,|\,\theta)$ differs from G defined in terms of $p(y\,|\,\theta)$ by a constant independent of θ and maximization of G in either case will yield the same MDIP pdf.
Theorem 2, of course, provides an important invariance with respect to changes in the units of measurement.
Theorem 3: Let the random variable $\tilde{y}$ have a proper, normalized pdf $p(y\,|\,\theta) = p(y-\theta)$ with $a \leq \theta \leq b$ and $-\infty < y < \infty$. Now if $\theta' = k_0 + k_1\theta$, where k_0 and k_1 are given constants, the MDIP pdf's for θ and for θ' are uniform.
Proof: For $p(y\,|\,\theta) = p(y-\theta)$, $I(\theta) = \int_{-\infty}^{\infty} p(y-\theta)\log p(y-\theta)dy =$ const independent of θ since by change of variable $z = y - \theta$ the integral is $\int_{-\infty}^{\infty} p(z)\log p(z)dz$ which does not depend on θ. Similarly, $I(\theta') =$ const independent of θ'. Thus $p*(\theta) \propto \exp\{I(\theta)\} \propto$ const and $p*(\theta') \propto \exp\{I(\theta')\} \propto$ const, the MDIP pdf's, are each uniform.
Theorem 4: Let the random variable $\tilde{y}$ have a proper, normalized pdf $p(y\,|\,\theta) = (1/\theta)f(y/\theta)$, with $0 < a \leq \theta \leq b$ and $-\infty < y < \infty$. Now if $\theta_1 = k\theta$, where k is a given positive constant, the MDIP pdf's for θ and θ_1 are given by $p*(\theta) = c/\theta$, $a \leq \theta \leq b$, and $p*(\theta_1) = c_1/\theta_1$, $ka \leq \theta_1 \leq kb$, where c and c_1 are normalizing constants.
Proof: We have

$$I(\theta) = \int_{-\infty}^{\infty} p(y\,|\,\theta)\log p(y\,|\,\theta)dy = \int_{-\infty}^{\infty}\{-\log\theta + \log f(y/\theta)\}p(y\,|\,\theta)dy$$
$$= -\log\theta + \text{const}$$

since with $\omega = y/\theta$, $\int_{-\infty}^{\infty}\{\log f(y/\theta)\}p(y\,|\,\theta)dy = \int_{-\infty}^{\infty}\{\log f(\omega)\}f(\omega)d\omega =$ const independent of θ. Similarly, $I(\theta_1) = \log\theta_1 +$ const. Thus the MDIP pdf's for θ and θ_1 have the forms $p*(\theta) = c\exp\{-\log\theta)\} = c/\theta$ and $p*(\theta_1) = c_1\exp\{-\log\theta_1\} = c_1/\theta_1$, as was to be shown.

For data pdf's involving more than one parameter, say a vector of parameters, $\boldsymbol{\theta}' = (\theta_1, \theta_2, \ldots, \theta_m)$, G is defined as follows:

$$G = \int_{R_\theta} p(\boldsymbol{\theta})I(\boldsymbol{\theta})d\boldsymbol{\theta} - \int_{R_\theta} p(\boldsymbol{\theta})\log p(\boldsymbol{\theta})d\boldsymbol{\theta}, \tag{6}$$

where R_θ is a finite region for $\boldsymbol{\theta}$ and

$$I(\boldsymbol{\theta}) = \int_{R_y} p(y\,|\,\boldsymbol{\theta})\log p(y\,|\,\boldsymbol{\theta})dy. \tag{7}$$

Proceeding as in Theorem 1, the MDIP pdf is

(8) $$p*(\boldsymbol{\theta}) = c_0 \exp\{I(\boldsymbol{\theta})\}, \qquad \boldsymbol{\theta} \subseteq R_\theta$$

where c_0 is a normalizing constant.
Last for a random vector observation, $\tilde{\mathbf{y}}' = (\tilde{y}_1, \tilde{y}_2, \ldots, \tilde{y}_p)$ with pdf $p(\mathbf{y} \mid \boldsymbol{\theta})$, where $\boldsymbol{\theta}$ is a vector of parameters, we have

(9) $$G = \int_{R_\theta} p(\boldsymbol{\theta}) I(\boldsymbol{\theta}) d\boldsymbol{\theta} - \int_{R_\theta} p(\boldsymbol{\theta}) \log p(\boldsymbol{\theta})\, d\boldsymbol{\theta},$$

with

(10) $$I(\boldsymbol{\theta}) = \int_{R_{\mathbf{y}}} p(\mathbf{y} \mid \boldsymbol{\theta}) \log p(\mathbf{y} \mid \boldsymbol{\theta}) d\mathbf{y},$$

and the MDIP pdf is given by:

$$p*(\boldsymbol{\theta}) = c_1 \exp\{I(\boldsymbol{\theta})\}, \qquad \boldsymbol{\theta} \subseteq R_\theta$$

with c_1 a normalizing constant.

3 Specific MDIP Pdf's and Their Properties

In table 1, we have listed a number of well-known univariate data pdf's, their data information measures, $I(\boldsymbol{\theta})$, and the associated MDIP pdf's that are proportional to $\exp\{I(\boldsymbol{\theta})\}$.

With respect to the normal data densities, the MDIP pdf's are generally those employed in Bayesian analyses and justified in various ways. With respect to the case in line 1*c*, it is well known that Jeffreys's method involving use of the square root of the determinant of the information matrix, $|\text{inf}|^{1/2}$, yields $p(\theta, \sigma) \propto 1/\sigma^2$, a result that he deems unreasonable. By assuming θ and σ independent and evaluating $|\text{inf}|^{1/2}$ separately for these parameters, he obtains $p(\theta, \sigma) \propto 1/\sigma$, a result that he employs in practice. Note that the normal examples in 1*a*–*c* of the table are special cases of 6*a*–*c*, respectively.

The prior pdf for the parameter θ of the rectangular pdf, $p(\theta) \propto 1/\theta$, has a form that is in accord with Jeffreys's rule for a parameter that can assume values from zero to infinity. For such a parameter, the rule is to take $\log \theta$ uniformly distributed which implies $p(\theta) \propto 1/\theta$. Also, this MDIP pdf for the parameter of the rectangular pdf is a special case of 6*b*.

The MDIP prior for the exponential pdf, $p(\theta) \propto 1/\theta$, is in accord with Jeffreys's rule mentioned in the previous paragraph.

Also, this prior pdf is a special case of 6*b*. Note that the Pareto pdf, $p(y \mid A, \beta) = (1/\beta y) \exp\{-\beta^{-1} \log(y/A)\}$, with A a given constant and $0 < A \leq y < \infty$, is a "log-exponential" pdf. Letting $x = \log y/A$, $0 \leq x < \infty$, we obtain $p(x \mid \beta) = (1/\beta) \exp\{-x/\beta\}$, the exponential pdf with parameter β for which the MDIP prior is $p(\beta) \propto 1/\beta$.

In line 4 of table 1, the MDIP prior for the location and scale parameters of the Student-*t* pdf is $p(\theta, \sigma) \propto 1/\sigma$. This is also the prior for the parameters of the Cauchy pdf, the Student-t pdf with $\nu = 1$. These results are special cases of 6c.

For the log-normal pdf in line 5 of table 1, note that θ is the median of this pdf which does not depend on σ. Since both θ and σ can range from 0 to ∞,

Table 1 Continuous Univariate Data Densities, Information Measures and Maximal Data Information Prior (MDIP) Pdf's

Data Pdf	Data Pdf Information Measure, $I(\theta)$	MDIP Pdf
1. *Normal*		
1a $p(y\|\theta) = \frac{1}{\sqrt{2\pi}} \exp\{-(y-\theta)^2/2\}$	const	$p_*(\theta) \propto$ const
1b $p(y\|\sigma) = \frac{1}{\sigma\sqrt{2\pi}} \exp\{-y^2/2\sigma^2\}$	const − log σ	$p_*(\sigma) \propto 1/\sigma$
1c $p(y\|\theta,\sigma) = \frac{1}{\sigma\sqrt{2\pi}} \exp\{-(y-\theta)^2/2\sigma^2\}$	const − log σ	$p_*(\theta, \sigma) \propto 1/\sigma$
2. *Rectangular*		
$p(y\|\theta) = 1/\theta,\ 0 < y < \theta$	const − log θ	$p_*(\theta) \propto 1/\theta$
3. *Exponential (Pareto)*		
$p(y\|\theta) = \theta^{-1} \exp\{-y/\theta\}$	const − log θ	$p_*(\theta) \propto 1/\theta$
4. *Student-*t *($\nu = 1$, Cauchy)*		
$p(y\|\theta, \sigma) = \frac{k}{\sigma}\left[\nu + \left(\frac{y-\theta}{\sigma}\right)^2\right]^{-(\nu+1)/2}$	const − log σ	$p_*(\theta, \sigma) \propto 1/\sigma$
$k = \frac{\Gamma\{(\nu+1)/2\}\,\nu^{\nu/2}}{\sqrt{\pi}\,\Gamma(\nu/2)}$		
5. *Log-normal*		
$p(y\|\theta, \sigma) = \frac{1}{\sigma y\sqrt{2\pi}} \exp\{-(\log y - \log\theta)^2/2\,\sigma^2\}$	const − log σ − log θ	$p_*(\theta, \sigma) \propto 1/\theta\sigma$
6. *General forms ($-\infty < y < \infty$)*		
6a $p(y\|\theta) = f(y-\theta)$	const	$p_*(\theta) \propto$ const
6b $p(y\|\sigma) = \frac{1}{\sigma} f(y/\sigma)$	const − log σ	$p_*(\sigma) \propto 1/\sigma$
6c $p(y\|\theta, \sigma) = \frac{1}{\sigma} f\left(\frac{y-\theta}{\sigma}\right)$	const − log σ	$p_*(\theta, \sigma) \propto 1/\sigma$
7. *Exponential Class*		
$p(y\|\theta) = \exp\{f(\theta)K(y) + S(y) + q(\theta)\}$	$f(\theta)\bar{K} + \bar{S} + q(\theta)$[a]	$p_*(\theta) \propto \exp\{f(\theta)\bar{K} + \bar{S} + q(\theta)\}$

[a]Here $\bar{K} \equiv \int K(y)p(y|\theta)\,dy$ and $\bar{S} \equiv \int S(y)p(y|\theta)\,dy$.

the form of the prior for these parameters is in accord with Jeffreys's rule. In general, if $p(y|\theta,\sigma) = (1/\sigma y)f((\log y - \log\theta)/\sigma), 0 < y < \infty$, and if $E\log y = \log\theta$, the MDIP prior for θ and σ is $p(\theta, \sigma) \propto 1/\theta\sigma$.[2]

The general forms in lines 6*a–c* of the table make the MDIP approach applicable to a very wide range of data pdf's. Since θ is a location parameter and σ a scale parameter, the forms of these priors are reasonable and in accord with usual prescriptions for "diffuse" or "noninformative" prior pdf's for such parameters. Note, for example, that in the case of 6*c*,

$$(11) \qquad I(\theta,\sigma) = \int p(y|\theta,\sigma)\log p(y|\theta,\sigma)dy$$
$$= -\log\sigma + \int\frac{1}{\sigma}f\left(\frac{y-\theta}{\sigma}\right)\log f\left(\frac{y-\theta}{\sigma}\right)dy$$

If the last integral on the r.h.s. of (11) we let $x = (y-\theta)/\sigma$, it becomes $\int f(x)\log f(x)dx$ which is independent of θ and σ.

Thus $I(\theta,\sigma) = \text{const} - \log\sigma$ and $p(\theta,\sigma) \propto \exp\{I(\theta,\sigma)\} \propto 1/\sigma$, as shown in table 1.

In line 7 of table 1, the MDIP prior pdf for the exponential class of pdf's is presented.[3] That this method of generating priors works not only for special cases but also for general cases is fortunate.

Above, we have obtained MDIP prior pdf's for parameters of data densities for continuous random variables. To illustrate application of the procedure to parameters of a distribution for a discrete random variable, we consider the binomial pdf. Let $Pr(\tilde{y} = 1) = \theta$ and $Pr(\tilde{y} = 0) = 1 - \theta$. Then the pdf for $\tilde{y}$ is $p(y|\theta) = \theta^y(1-\theta)^{1-y}$, with $0 \le \theta \le 1$. The information measure $I(\theta)$ for this data pdf is[4] $I(\theta) = \theta\log\theta + (1-\theta)\log(1-\theta)$ and the MDIP pdf is:

$$(12) \qquad p*(\theta) = c\theta^\theta(1-\theta)^{1-\theta} \qquad 0 \le \theta \le 1$$

2. Note that $\log p(y|\theta,\sigma) = -\log\sigma - \log y + \log f[(\log y - \log\theta)/\sigma]$ and

$$I(\theta,\sigma) = \int_0^\infty p(y|\theta,\sigma)\log p(y|\theta,\sigma)dy = -\log\sigma - \int_0^\infty \log y p(y|\theta,\sigma)dy$$
$$+ \int_0^\infty p(y-\theta,\sigma)\log f((\log y - \log\theta)/\sigma)dy$$
$$= -\log\sigma - \log\theta + \text{const.}$$

Note that taking $x = (\log y - \log\theta)/\sigma$ in the last integral on the r.h.s. of the expression for $I(\theta, \sigma)$ reduces it to $\int_{-\infty}^{\infty} f(x)\log f(x)dx$, which is independent of θ and σ. Then $p(\theta,\sigma) \propto \exp\{I(\theta,\sigma)\} \propto 1/\theta\sigma$.

3. For some discussion of the exponential class with examples, see, e.g., Hogg and Craig (1965), pp. 223ff.

4. Note that $\log p(y|\theta) = y\log\theta + (1-y)\log(1-\theta)$ and its average value given that $\tilde{y} = 1$ with probability θ and $\tilde{y} = 0$ with probability $1 - \theta$ is just $I(\theta) = \theta\log\theta + (1-\theta)\log\theta$.

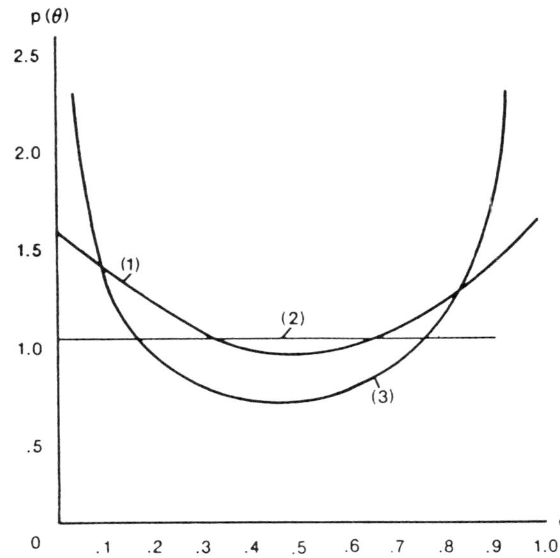

Figure 1 Prior Pdf's for Binomial Parameter θ

(1) $p_*(\theta) = 1.61856\theta^\theta(1-\theta)^{1-\theta}$

(2) $p(\theta) = 1$

(3) $p(\theta) = 0.3183\,\theta^{-1/2}(1-\theta)^{-1/2}$

where $c = 1.61856$ is the normalizing constant for this proper pdf[5] that is symmetric about 1/2, its mean. In figure 1 a plot of the pdf in (12) is shown along with plots of several other "diffuse" priors for the binomial parameter. It is seen that the plot of (12) lies "between" the uniform pdf $p(\theta) = 1$, $0 \leq \theta \leq 1$, and the proper beta prior $p(\theta) = \pi^{-1}\theta^{-1/2}(1-\theta)^{-1/2}$ that shoots off to ∞ as $\theta \to 0$ or $\theta \to 1$. This property is also possessed by the improper prior pdf $p(\theta) \propto \theta^{-1}(1-\theta)^{-1}$.

Jeffreys (1967, pp. 123–25) presents a cogent discussion of the problem of formulating a "diffuse" or "noninformative" prior for the binomial parameter in which he considers all of the prior pdf's presented in the preceding paragraph except that in (12). With respect to the uniform distribution, referred to as the Bayes-Laplace rule, he writes, "Certainly if we take the Bayes-Laplace rule right up to the extremes we are led to results that do not correspond to anybody's way of thinking. The rule $d\theta/\theta(1-\theta)$ goes too far the other way."[6] That the pdf in (12) lies above the uniform pdf at the

5. It is the case that $\lim_{\theta \to 0} \theta^\theta = 1$ and $\lim_{\theta \to 1} (1-\theta)^{1-\theta} = 1$.

6. In the quote (p. 124), Jeffreys's symbol x has been replaced by θ.

extremes, but does not shoot off to infinity, may then be a virtue associated with (12). Further, if $p(\theta) \propto 1/\theta(1 - \theta)$ is employed as a prior and if all binomial trials produce "successes," the posterior pdf will be improper whereas this problem is avoided through the use of (12).

We now consider generation of MDIP's for parameters of several continuous bivariate and multivariate data densities. Shown in table 2 are

Table 2 Continuous Bivariate and Multivariate Data Densities, Information Measures and Maximal Data Information Prior (MDIP) Pdf's

Data Pdf	Data Pdf Information Measure, $I(\theta)$	MDIP Pdf
1. *Bivariate Normal*		
1a $p(y\|\boldsymbol{\theta}) = \frac{1}{2\pi}\exp -\frac{1}{2}\{(y_1-\theta_1)^2+(y_2-\theta_2)^2\}$	const	$p_*(\boldsymbol{\theta}) \propto$ const
1b $p(y\|\sigma_1, \sigma_2, \rho) = \frac{1}{2\pi\sigma_1\sigma_2(1-\rho^2)^{1/2}}\exp\left\{-\frac{1}{2(1-\rho^2)}\left[\left(\frac{y_1}{\sigma_1}\right)^2+\left(\frac{y_2}{\sigma_2}\right)^2-2\rho\frac{y_1}{\sigma_1}\frac{y_2}{\sigma_2}\right]\right.$	const $- \log \sigma_1 \sigma_2 - (1/2) \log)(1-\rho^2)$	$p_*(\sigma_1, \sigma_2, \rho) \propto \frac{1}{\sigma_1\sigma_2}\frac{1}{(1-\rho^2)^{1/2}}$
1c $p(y\|\theta_1, \theta_2, \sigma_1, \sigma_2, \rho) = \frac{1}{2\pi\sigma_1\sigma_2(1-\rho^2)^{1/2}}\exp\left\{-\frac{1}{2(1-\rho^2)}\left[\left(\frac{y_1-\theta_1}{\sigma_1}\right)^2+\left(\frac{y_2-\theta_2}{\sigma_2}\right)^2-2\rho\left(\frac{y_1-\theta_1}{\sigma_1}\right)\left(\frac{y_2-\theta_2}{\sigma_2}\right)\right]\right\}$	const $- \log \sigma_1\sigma_2 - (1/2)\log(1-\rho^2)$	$p_*(\theta_1, \theta_2, \sigma_1, \sigma_2, \rho) \propto \frac{1}{\sigma_1\sigma_2(1-\rho^2)^{1/2}}$
2. *Multivariate Normal*		
$p(y\|\boldsymbol{\theta}, \Sigma) = \frac{\|\Sigma\|^{-1/2}}{(2\pi)^{p/2}}\exp\left\{-\frac{1}{2}(y-\boldsymbol{\theta})'\Sigma^{-1}(y-\boldsymbol{\theta})\right\}$	const $- (1/2)\log\|\Sigma\|$	$p_*(\boldsymbol{\theta}, \Sigma) \propto \|\Sigma\|^{-1/2}$
3. *General Forms*		
3a $p(y\|\theta) = f(y-\boldsymbol{\theta})$	const	$p_*(\boldsymbol{\theta}) \propto$ const
3b $p(y\|\Sigma) = \|\Sigma\|^{-1/2}f(y'\Sigma^{-1}y)$	const $- (1/2)\log\|\Sigma\|$	$p_*(\Sigma) \propto \|\Sigma\|^{-1/2}$
3c $p(y\|\boldsymbol{\theta}, \Sigma) = \|\Sigma\|^{-1/2}f\{(y-\boldsymbol{\theta})'\Sigma^{-1}(y-\boldsymbol{\theta})\}$	const $- (1/2)\log\|\Sigma\|$	$p_*(\boldsymbol{\theta}, \Sigma) \propto \|\Sigma\|^{-1/2}$

several data densities, their associated information measures, $I(\boldsymbol{\theta})$, and the MDIP's, $p*(\boldsymbol{\theta}) \propto \exp\{I(\boldsymbol{\theta})\}$. From line 1*a*, we see that the MDIP for the means θ_1 and θ_2 of a spherical bivariate normal pdf is uniform, a special case of the result in line 3*a*. In line 1*b*, the MDIP for the two standard deviations, σ_1 and σ_2, and of the correlation coefficient is $p*(\sigma_1, \sigma_2, \rho) \propto 1/\sigma_1\sigma_2(1 - \rho^2)^{1/2}$, while in line 1*c*, the MDIP for the five parameters of the bivariate normal density is given by

$$p*(\theta_1, \theta_2, \sigma_1, \sigma_2, \rho) \propto 1/\sigma_1\sigma_2(1 - \rho^2)^{1/2} \tag{13}$$

From (13), it is the case that the parameters are independent with $\log\sigma_1$ and $\log\sigma_2$ uniformly distributed. The marginal prior for ρ in (13) is $p(\rho) = c(1 - \rho)^{-1/2}(1 + \rho)^{-1/2}$, with the normalizing constant $c = 1/\pi$, a form of the beta pdf.[7] The prior for ρ, $-1 < \rho < 1$, is symmetric about $\rho = 0$ at which point the density assumes its minimal value. It is interesting to note that among other prior pdf's for ρ, Jeffreys (1967) explicitly considers $p(\rho) = 1/\pi(1 - \rho^2)^{1/2}$ that he believes is compatible with his interpretation of the correlation coefficient.[8]

With respect to the general p-dimensional nonsingular multivariate normal data pdf, considered in line 2 of table 2, its information measure is equal to $-(1/2)\log|\Sigma|$ plus a constant and thus the MDIP pdf is $p*(\boldsymbol{\theta}, \Sigma) \propto |\Sigma|^{-1/2}$, a form that implies that $\boldsymbol{\theta}$ and the distinct elements of Σ are independent, with the elements of $\boldsymbol{\theta}$ uniformly and independently distributed. The form of the marginal MDIP pdf for Σ, $p*(\Sigma) \propto |\Sigma|^{-1/2}$ is "close" to $|\Sigma|^{-1}$, a form leading to the Fisher-Cornish posterior pdf, but differs from the form often used,[9] namely, $p(\Sigma) \propto |\Sigma|^{-(p+1)/2}$. This latter form can be obtained by assuming $\boldsymbol{\theta}$ and Σ a priori independent and applying Jeffreys's rule, $|\text{Inf}|^{1/2}$, to the elements of Σ. For $p = 2$, this rule yields $p(\sigma_1, \sigma_2, \rho) \propto 1/\sigma_1\sigma_2(1 - \rho^2)^{3/2}$, a form discussed by Jeffreys,[10] who appears to favor

7. If we let $x = (1 + \rho)/2$, then $0 < x < 1$ for $-1 < \rho < 1$, and $p(x) = (1/\pi)x^{-1/2}(1 - x)^{-1/2}$ is a proper, normalized beta pdf.

8. He also discusses $p(\rho) \propto$ const and $p(\rho) \propto 1/(1-\rho^2)^{3/2}$, the latter an improper pdf produced by his $|\text{Inf}|^{1/2}$ rule.

9. See, e.g., Geisser (1965), Tiao and Zellner (1964), Savage (1961), Box and Tiao (1973), and Zellner (1971). Note, however, as pointed out by S. J. Press in conversation, that if we modify (6) to read:

$$G' = q\int p(\boldsymbol{\theta})I(\boldsymbol{\theta})d\boldsymbol{\theta} - \int p(\boldsymbol{\theta})\log p(\boldsymbol{\theta})d\boldsymbol{\theta},$$

where q is a given scalar, then the MDIP in (8) is

$$p_*(\boldsymbol{\theta}) = c_0\exp\{qI(\boldsymbol{\theta})\}.$$

For $I(\boldsymbol{\theta}) = -1/2\log|\Sigma|$, $p_*(\Sigma) = c_0|\Sigma|^{-q/2}$. By taking $q = p + 1$, the "usual" diffuse prior pdf for Σ is obtained.

10. Jeffreys (1967), pp. 187 ff.

the form shown in (13) or $p(\sigma_1, \sigma_2, \rho) \propto 1/\sigma_1\sigma_2$ even though these latter priors are not yielded by his invariant rule.

The MDIP pdf's for the general forms of data densities, given in lines 3*a–c* of table 2, are relevant for a broad range of multivariate data densities. In each case, they are derived by a straightforward evaluation of the expression for the data density information measure $I(\cdot)$ and then using $p*(\cdot) \propto \exp\{I(\cdot)\}$ to obtain the MDIP pdf. That resulting MDIP pdf's are relatively simple is indeed fortunate. Also theorems 2, 3, 4 of section 2 can be generalized to relate to multivariate pdf's.

We next consider the application of the MDIP approach to yield prior pdf's for parameters of several frequently employed models. Consider the normal multiple regression model $\tilde{y} = \mathbf{x}'\boldsymbol{\beta} + \tilde{u}$, where $\mathbf{x}'$ is a $1 \times k$ given vector, $\boldsymbol{\beta}$ is a $k \times 1$ vector of parameters and $\tilde{u}$ is a normal error with zero mean and standard deviation σ. Then the information in the pdf for $\tilde{y}$ is[11] $I(\boldsymbol{\beta}, \sigma) = \text{const} - \log\sigma$ and the MDIP pdf for $\boldsymbol{\beta}$ and σ is $p*(\boldsymbol{\beta}, \sigma) \propto \exp\{I(\boldsymbol{\beta}, \sigma)\} \propto 1/\sigma$, a "diffuse" prior pdf that is usually employed.[12]

For the multivariate normal regression model, $\tilde{\mathbf{y}}' = \mathbf{x}'B + \tilde{\mathbf{u}}'$, where $\tilde{\mathbf{y}}'$ is a $1 \times p$ random vector, $\mathbf{x}'$ is a $1 \times k$ given vector, B is a $k \times p$ matrix of regression coefficients and $\tilde{\mathbf{u}}$ is a $1 \times p$ normal error vector with zero mean vector and pds covariance matrix Σ, $I(B, \Sigma) = \text{const} - (1/2)\log|\Sigma|$. Thus the MDIP pdf for B and Σ is $p*(B, \Sigma) \propto |\Sigma|^{-1/2}$ that has B and Σ independent with the elements of B uniformly and independently distributed and the distinct elements of Σ distributed in the form $p*(\Sigma) \propto |\Sigma|^{-1/2}$.

The last process that we consider is a stationary first-order normal autoregressive process with unknown mean, that is, $\tilde{y}_t = (1 - \beta)\theta + \beta\tilde{y}_{t-1} + \tilde{u}_{t'}$, with $|\beta| < 1$ and $\tilde{u}_t$ normal with zero mean and standard deviation σ.[13] The pdf for $\tilde{y}_t$ is $p(y_t|\beta, \theta, \sigma) = (\sigma\sqrt{2\pi})^{-1}(1 - \beta^2)^{1/2}\exp\{-(1 - \beta^2)(y_t - \theta)^2/2\sigma^2\}$ and $I(\beta, \theta, \sigma) = \text{const} + 1/2\log(1 - \beta^2) - \log\sigma$. Thus the MDIP pdf is:

$$p*(\beta, \theta, \sigma) \propto \frac{(1 - \beta^2)^{1/2}}{\sigma} \qquad |\beta| < 1 \tag{14}$$

11. That is, $I(\boldsymbol{\beta}, \sigma) = \int_{-\infty}^{\infty} p(y|\boldsymbol{x}, \boldsymbol{\beta}, \sigma)\log p(y|\boldsymbol{x}, \boldsymbol{\beta}, \sigma)dy$, where

$$p(y|\boldsymbol{x}, \boldsymbol{\beta}, \sigma) = (\sigma\sqrt{2\pi})^{-1}\exp\{-\frac{1}{2\sigma^2}(y - \boldsymbol{x}'\boldsymbol{\beta})^2\}.$$

12. Jeffreys obtains this prior pdf by assuming $\boldsymbol{\beta}$ and σ a priori independent and applying his rule $|\text{Inf}|^{1/2}$ separately to $\boldsymbol{\beta}$ and to σ. If $|\text{Inf}|^{1/2}$ is applied to $\boldsymbol{\beta}$ and σ jointly, the result is $p(\boldsymbol{\beta}, \sigma) \propto 1/\sigma^{k+1}$, deemed "unsatisfactory"; see Jeffreys (1967), p. 182.

13. Note if $\tilde{y}_t = \beta_1 + \beta_2\tilde{y}_{t-1} + \tilde{u}_t$, $\tilde{y}_t = \beta_1/(1 - \beta_2) + \Sigma_{i=0}^{\infty}\beta_2^i u_{t-i}$. Given that $|\beta_2| < 1$, $E\tilde{y}_t = \beta_1/(1 - \beta_2) \equiv \theta$, the mean of the process. Thus, with $\beta_2 \equiv \beta$, we write $\tilde{y}_t = (1 - \beta)\theta + \beta\tilde{y}_{t-1} + \tilde{u}_t$ in order to parameterize the process in terms of its mean, θ.

In (14), β, θ, and σ are mutually independent with θ and $\log \sigma$ uniform. The marginal prior for β, $p_*(\beta) \propto (1 - \beta^2)^{1/2}$, $|\beta| < 1$, is symmetric about $\beta = 0$, where it reaches a maximum and falls to zero as $|\beta| \rightarrow 1$.[14]

Above we have considered the MDIP approach in terms of specific parameterizations except for theorems 3 and 4 that provide invariance properties relating to linear transformations of the parameters. The problem of dealing with nonlinear transformations of parameters is a thorny one that has received considerable attention in the literature and must be considered. Our approach is to obtain the MDIP for the particular parameterization employed. By not insisting on what Hartigan (1964) calls Ω-labeling invariance, we can achieve our objective of having a posterior distribution reflect mainly the information in a data distribution for the particular parameterization employed without being restricted by broad invariance restrictions. Given that investigators use the MDIP procedure to generate priors for any given parameterization, different investigators will obtain the same posterior pdf's.

A specific nonlinear parameter transformation that has evoked much controversy is the reciprocal transformation of a location parameter. For example, let θ be the mean of a univariate normal pdf with unit variance, $p(y|\theta) = (2\pi)^{-1/2} \exp\{-(y - \theta)^2/2\}$. For this data pdf, $I(\theta) = \text{const}$ and the MDIP pdf is $p_*(\theta) \propto \text{const}$, as pointed out above. If $\eta = 1/\theta$, $p(y|\eta) = (2\pi)^{-1/2} \exp\{-(y - 1/\eta)^2/2\}$ and $I(\eta) = \text{const}$. Thus the MDIP for η is: $p_*(\eta)$ const, for $|\eta| > 0$.[15] Thus the MDIP pdf's for θ and for $\eta = 1/\theta$ are uniform since both are pure location parameters for which $I(\cdot)$ has the same constant value. It is recognized that $p(\theta)d\theta \propto d\theta$ logically implies $p(\eta) \propto d\eta/\eta^2$.[16] However, $p(\eta) \propto 1/\eta^2$ is *not* the MDIP pdf for η, that is, one that allows the information in the data pdf to be reflected most, in the sense defined above, in a posterior pdf for η. Use of the MDIP prior pdf, $p_*(\eta) \propto$ const, achieves this latter objective.

To illustrate the analysis of the reciprocal transformation, consider the simple normal regression for income change, $\tilde{y}_t = \theta x_t + \tilde{u}_t$ where x_t is the change in autonomous spending, θ is the "Keynesian multiplier" and $\tilde{u}_t$ is a normal error term with zero mean and standard deviation σ. As explained above, the MDIP pdf for θ and σ is: $p_*(\theta, \sigma) \propto 1/\sigma$. Now let $\theta = 1/\eta$, where η is the "marginal propensity to save." The MDIP pdf for η and σ is $p(\eta, \sigma) \propto$

14. As can be seen by making the change of variable $z = (1 - \beta)/2$, $p_*(\beta) \propto (1 - \beta^2)^{1/2} \propto (1 - \beta)^{1/2} (1 + \beta)^{1/2}$ is a form of a proper beta pdf. See Zellner (1971), pp. 216–20 for discussion of other diffuse priors for β, θ, and σ.

15. Note that we exclude $\eta = 0$ since $-\infty < \theta < \infty$.

16. This is the prior pdf introduced by application of Jeffreys's Ω-labeling invariant rule, $|\text{Inf}|^{1/2}$

$1/\sigma$, that is η and $\log \sigma$ uniform and independent. On the other hand, from $\theta = 1/\eta$ and $p(\theta, \sigma) \propto 1/\sigma$, we obtain $p(\eta, \sigma) \propto 1/\eta^2\sigma$. This prior for η, defined over a finite range excluding the point $\eta = 0$, implies that small absolute values of η are more probable than large values, that is, for example, $Pr(a < \eta < a + b) > Pr(a + b < \eta < a + 2b)$ with $a, b > 0$ and the interval a to $a + 2b$ contained within the range of η. On the other hand, the MDIP pdf for η, $p*(\eta) \propto$ const, of course, implies equal prior probabilities associated with intervals of equal length for η. Thus use of the Ω-labeling invariance condition leading to priors $p(\theta) \propto$ const and $p(\eta) \propto 1/\eta^2$, both defined for a finite range, involves the implication that small absolute values of η are more probable than large values, an implication not present when MDIP pdf's for θ and η are employed. In general, if θ is a parameter of a data density for which $I(\theta) =$ const, then if $\eta = f(\theta)$ is a proper one-to-one transformation, $I(\eta) =$ const and the MDIP pdf's for θ and η are both uniform.[17]

Another important nonlinear transformation is that involving powers of a scale parameter, say the standard deviation, σ, of the normal data density, $p(y|\sigma) = (2\pi)^{-1/2} \exp\{-y^2/2\sigma\}^2$. Let $\phi_i = \sigma^i$, $i = 1, 2, 3, \ldots$. Then the MDIP for ϕ_i is given by $p*(\phi_i) \propto 1/\phi_i^{1/i}$, $i = 1, 2, 3, \ldots$. For $i = 1$, $\phi_i = \sigma$ and $p*(\phi_1) = p*(\sigma) \propto 1/\sigma$. For $i = 2$, $p*(\phi_2) = 1/(\phi_2)^{1/2}$. This latter pdf, defined for a finite range so that it is proper, is "flatter" than the Ω-labeling invariant prior, $p(\phi_2) \propto 1/\phi_2$ defined over the same finite range. As i gets large, the MDIP pdf for ϕ_i approaches a uniform pdf whereas the Ω-labeling invariant prior for ϕ_i (or the prior that is "invariant with respect to powers") $p(\phi_i) \propto 1/\phi_i$ has the same shape no matter what the value of i and is not a MDIP pdf.

In general, if $\boldsymbol{\theta}$ is a $p \times 1$ vector of parameters of a data density with information measure $I(\boldsymbol{\theta})$ and if $\theta_1 = f_1(\boldsymbol{\eta}), \theta_2 = f_2(\boldsymbol{\eta}), \ldots, \theta_p = f_p(\boldsymbol{\eta})$, is a set of proper, one-to-one transformations, denoted by $\boldsymbol{\theta} = f(\boldsymbol{\eta})$, where $\boldsymbol{\eta}$ is a $p \times 1$ vector of parameters, the forms of the MDIP priors for $\boldsymbol{\theta}$ and for $\boldsymbol{\eta}$ are given by: $p*(\boldsymbol{\theta}) \propto \exp\{I(\boldsymbol{\theta})\}$ and $p*(\boldsymbol{\eta}) \propto \exp\{I(\boldsymbol{\eta})\}$, where $I(\boldsymbol{\eta}) = I[f^{-1}(\boldsymbol{\theta})]$.

4 Concluding Remarks

In the present paper a procedure for generating MDIP pdf's has been presented and applied to several problems. The procedure utilizes an information measure that has been considered and employed by others, including Shannon (1948), Jeffreys (1967), Jaynes (1968), Lindley (1956, 1961), and perhaps others. What is distinctive about the present approach is

17. This is easily generalized to the cases involving a vector of p parameters, $\boldsymbol{\theta}$ and $\eta_i = f_i(\boldsymbol{\theta})$, $i = 1, 2, \ldots, p$. If $I(\boldsymbol{\theta}) =$ const, then $I(\boldsymbol{\eta}) =$ const and the MDIP pdf's for $\boldsymbol{\theta}$ and for $\boldsymbol{\eta}$ are each proportional to a constant.

the use of the particular function G in (3) that is used as a criterion function to be maximized by choice of a prior pdf. That use of the function G leads to simple forms for prior pdf's, some of them already in widespread use, is certainly appealing. Also, use of the MDIP approach provides a fresh view of and operational results for the problem of selecting "diffuse" prior pdf's for alternative parameterizations. That the prior pdf's yielded by the MDIP approach have been defined to be proper may appeal to some. However, if Jeffreys's (1967) arguments for the use of improper prior pdf's to represent "knowing little" or "ignorance" are considered compelling, we can take note of his remark "in the mathematical definition of an infinite integral a finite range is always considered in the first place and then allowed to tend to infinity." Thus, just as Jeffreys illustrates, we can obtain improper MDIP pdf's in many cases by allowing their ranges to tend to infinity.[18]

It is to be emphasized that it is not recommended that MDIP pdf's always be employed in Bayesian analyses. Clearly there are many circumstances in which "informative" prior pdf's will be employed. However, when an investigator knows little about parameter values or wants to proceed as if he knows little about parameter values, and wants to feature the data information in his posterior distribution,[19] it is thought that MDIP pdf's will be found useful.

Last, it is the case that prior pdf's can be found that maximize G in (3) subject not only to the condition that $\int p(\theta)d\theta = 1$ but also to one or more of the following conditions $\int \theta p(\theta)d\theta = \mu_1$, $\int \theta^2 p(\theta)d\theta = \mu_2, \ldots$, where μ_1, $\mu_2, \ldots$ are assigned values by an investigator.

Addendum (1975)

I hit on the procedure described in this paper during the summer of 1968 and included a description and some applications of it at the end of chapter 2 of Zellner (1971).[20] In recent valuable correspondence with I. J. Good, dated 8 September 1975, he drew my attention to some of his work that deals in part with related topics, in particular Good (1965, 1963). Good uses entropy measures in connection with generating null hypotheses (or prior pdf's), and writes as follows about his approach in the first reference cited above, "This method has some aesthetic appeal but is difficult to justify" (p. 75).

18. See Rényi (1970) for justification for use of Bayes's theorem with improper pdf's.

19. See Zellner and Richard (1973), where this topic is discussed in terms of a Cobb-Douglas production function estimation problem.

20. This work grew out of an investigation of a suggestion in Lindley (1956) to choose a prior to maximize the information provided by an experiment. Solutions to this latter problem proved intractable and thus I altered the criterion to the form of G in (3) or (4) that I now regard as more appropriate.

In the second reference he writes, "A possible approach, which, however, is difficult to justify, would be to maximize some linear combination of the entropy and of the log-likelihood, such as the sum." Good considers just binomial and multinomial problems for which he obtains a prior pdf in the form, $\Pi_{p_i} p_i^{-p_i}$ with $\Sigma_{i=1}^{n} p_i = 1$, a rule that differs from my (12) for the case $n = 2$.

Another point mentioned by Good and others is that MDIP pdf's forms depend on properties of likelihood functions. This is also true of Jeffreys's $|\mathrm{Inf}|^{-1/2}$ rule and the procedures put forward by Lindley (1956) and Box and Tiao (1973). Since the purpose of a MDIP pdf is to allow the information provided by an experiment to be featured, it seems natural that the form of a MDIP pdf that accomplished this objective be dependent on the design of an experiment.

In current work, the approach described in this paper has been extended by using what Good calls "invariantized" entropy, that is entropy expressions based on arbitrary measures. Also, in recent conversations with Edwin Jaynes, he pointed out another interpretation of my results, namely that MDIP pdf's can be interpreted as pdf's that maximize the entropy associated with a prior pdf subject to the side condition that the average entropy in the data pdf, $p(y \mid \theta)$, be constant. Last, further work to understand just how much or how little invariance is required for specific problems would be valuable.

References

Box, G. E. P., and G. C. Tiao (1973). *Bayesian Inference in Statistical Analysis*. Reading, Mass.: Addison-Wesley Publishing Co.

Geisser, S. (1965). "Bayesian Estimation in Multivariate Analysis," *Annals of Mathematical Statistics*, 36:150–59.

Good, I. J. (1965). *The Estimation of Probabilities: An Essay on Modern Bayesian Methods*. pp. 75–76. Cambridge, Mass.: MIT Press.

———. (1963). "Maximum Entropy for Hypothesis Formulation, Especially for Multinomial Contingency Tables." *Annals of Mathematical Statistics*, 34:911–34.

Hartigan, J. (1964). "Invariant Prior Distributions." *Annals of Mathematical Statistics*, 35:836–45.

Hogg, R. V., and A. T. Craig (1965). *Introduction to Mathematical Statistics*. 2d ed. New York: Macmillan Co.

Jaynes, E. T. (1968). "Prior Probabilities." *IEEE Transactions on Systems Science and Cybernetics*, SSC-4, pp. 227–41. New York: Institute of Electrical and Electronic Engineers.

Jeffreys, H. (1967). *Theory of Probability*. 3d rev. ed. London: Oxford University Press.

Lindley, D. V. (1956). "On a Measure of Information Provided by an Experiment," *Annals of Mathematical Statistics*, 27:986–1005.

———. (1961). "The Use of Prior Probability Distributions in Statistical Inference and Decisions." In J. Neyman, ed., *Proceedings of the Fourth Berkeley Symposium on Mathematical Statistics and Probability*, 1:453–68. Berkeley: University of California Press.

Novick, M., and W. Hall (1965). "A Bayesian Indifference Procedure." *Journal of the American Statistical Association*, 60:1104–17.

Rényi, A. (1970). *Foundations of Probability*. San Francisco: Holden-Day.

Savage, L. J. (1961). "The Subjective Basis of Statistical Practice." Department of Statistics, University of Michigan.

Shannon, C. E. (1948). "The Mathematical Theory of Communication." *Bell System Technical Journal*, July–October 1948. Reprinted in C. E. Shannon and W. Weaver, *The Mathematical Theory of Communication*, pp. 3–91. University of Illinois Press, 1949.

Tiao, G. C., and A. Zellner (1964). "On the Bayesian Estimation of Multivariate Regression." *Journal of the Royal Statistical Society*, Ser. B, 26:277–85.

Zellner, A. (1971). *An Introduction to Bayesian Inference in Econometrics*. New York: John Wiley & Sons.

Zellner, A., and J. F. Richard (1973). "Use of Prior Information in the Analysis and Estimation of Cobb-Douglas Production Function Models." *International Economic Review*, 14:107–19.

[9]

Past and recent results on maximal data information priors*

1. Introduction

The purpose of this paper is to summarize past and recent results relating to maximal data information prior (MDIP) densities and odds. The MDIP approach for producing 'non-informative' and 'informative' prior densities originated in the summer of 1968. After spending some time trying to derive priors by maximizing the information provided by an experiment, see e.g., Good (1956), Lindley (1956) and Soofi (1994), via calculus of variations techniques and getting no clean-cut analytical results, I decided that something was wrong with the criterion function and changed it to the average information in the data density or likelihood function minus the information in the prior. The objective is to obtain a prior density that maximizes the difference between these information measures subject to side conditions. A solution to this optimization problem is a MDIP density that allows the information in the data density to be featured, a usual objective in scientific work. Fortunately, solutions to calculus of variations problems using this alternative criterion functional are easy to derive. On seeing the solution procedure, Ed Jaynes remarked, 'Anything that simple must be right.' I did not argue with him.

The first MDIPs, then called minimal information priors appeared in my 1971 book. Many more MDIPs were presented in Zellner (1977) followed by other results in Sinha and Zellner (1990), Moulton (1993), Soofi (1992, 1994), Venegas-Martinez (1990), Zhang and Yao (1991), Zellner and Min (1993) and Zellner (1991, 1993). In these papers, the rationale of the MDIP approach has been developed and explained, further properties of the procedure have been established and many MDIP densities have been produced. In addition, certain invariance properties of MDIPs have been established and a procedure for deriving MDIPs that are invariant to relevant one-to-one transformations has been developed and applied. MDIPs for common parameters that appear in two or more data densities and MDIP odds for alternative models or hypotheses have been derived. Also, by use of alternative information measures and the criterion functional described above and side conditions or constraints, it has been possible to obtain Jeffreys's well known priors as well as priors similar to Rodríguez's (1991a, b) entropic priors as solutions to well defined optimization problems that employ the criterion functional, average information in the data density minus information in the prior. Finally, it has been shown that the total information in a joint density of the

* Research for this paper was financed by the National Science Foundation and by income from the H.G.B. Alexander Endowment Fund, Graduate School of Business, University of Chicago. The author gratefully acknowledges helpful comments from colleagues, especially G. A. Barnard, S. Geisser, I. J. Good, E. T. Jaynes, R. E. Kass, R. E. McCulloch, J. Skilling, E. S. Soofi and L. Wasserman.

observations and parameters can be expressed as the sum of the average information in the data density plus the information in the prior. Thus the MDIP criterion functional is just the difference of these two components of the total available information in the joint density for the observations and parameters.

The plan of the paper is as follows. In Section II, various general properties of the MDIP approach will be presented. Section III contains a summary of selected results yielded by applying the MDIP approach to a number of problems with some comparisons with results yielded by alternative approaches for producing priors. The MDIP procedure for producing prior odds for alternative hypotheses or models will be explained with examples in Section IV. Last, Section V contains a summary and some concluding remarks.

II. General properties of the MDIP approach

2.1 Informational considerations

In this section, we review some general informational features of the MDIP approach. Let us begin by considering the joint density of an observation vector **y** and a parameter vector $\boldsymbol{\theta}$, $p(y, \theta)$ with $\boldsymbol{\theta} \subset R_\theta$ and $\mathbf{y} \subset R_y$. The negative entropy of $p(y, \theta)$ relative to uniform measure, denoted by $-H(p)$, a measure of the information in $p(y, \theta)$, is given by

$$-H(p) = \int_{R_{y,\theta}} p(y,\theta)\ell n p(y,\theta) dy d\theta \tag{2.1}$$

Note that $-H(p)$ can be interpreted simply as the expectation of the logarithm of the ordinate of $p(y, \theta)$, that is, $E\ell n p(y, \theta)$. The higher on average the $1n$ joint density, the more information it contains. See, e.g., Shannon (1948), Jaynes (1982a, b), Cover and Thomas (1991), and Soofi (1994) for discussion of additional informational interpretations of entropy measures.

On using $p(y, \theta) = \pi(\theta)f(y \mid \theta)$, where $\pi(\theta)$ is the prior density for θ and $f(y \mid \theta)$ is the data density given the parameters θ, it is direct to derive (2.2), given in Zellner (1991), Zellner and Min (1993) and Soofi (1994),

$$-H(p) = \int_{R_\theta} I(\theta)\pi(\theta) d\theta + \int_{R_\theta} \pi(\theta)\ell n \pi(\theta) d\theta \tag{2.2}$$

where

$$I(\theta) = \int_{R_y} f(y \mid \theta)\ell n f(y \mid \theta) dy \tag{2.3}$$

is the information in $f(y \mid \theta)$. In (2.2) we see that the total information in $p(y, \theta)$, $-H(p)$, breaks up into two parts. The first is the prior average information in the data density,

$f(y \mid \theta)$, and the second is the information in the prior density, $\pi(\theta)$. Given that $\pi(\theta)$, the prior density, can be selected and as usual in science we wish to feature the information in the data, in the MDIP approach, the following criterion functional, denoted by $G(\pi)$, is maximized with respect to choice of a prior density $\pi(\theta)$ subject to one or more side conditions, e.g. that the prior be proper and that its mean and variance have given values.

$$G(\pi) = \int_{R_\theta} I(\theta)\pi(\theta)d\theta - \int_{R_\theta} \pi(\theta)\log \pi(\theta)d\theta \tag{2.4}$$

It is seen that $G(\pi)$ is just equal to the difference of the two terms on the right side of (2.2) that make up the total information in $p(y, \theta)$.

A second interpretation of $G(\pi)$, given in Zellner (1977), Zellner and Min (1993) and Soofi (1992), can be obtained by using $p(y, \theta) = g(\theta \mid y)h(y)$ and $g(\theta \mid y) = f(y \mid \theta)\, \pi(\theta)/h(y)$ in connection with (2.4) to obtain:

$$G(\pi) = \int_{R_y}\left[\int_{R_\theta} g(\theta \mid y)\ell n\left[\ell(\theta \mid y)/\pi(\theta)\right]d\theta\right]h(y)dy \tag{2.5}$$

where $\ell(\theta \mid y) \equiv f(y \mid \theta)$ is the likelihood function. Thus from (2.5), we see that by maximizing G by choice of $\pi(\theta)$, the average logarithm of the ratio of the likelihood to the prior will be made as large as possible. Having the likelihood function featured in this fashion is an important aspect of the MDIP approach for selecting priors.

Another general feature of the criterion functional $G(\pi)$ in (2.4) is that it can be expressed as, see Zellner (1991),

$$G(\pi) = G_1 + \int_{R_y} h(y)\ell nh(y)dy - \int_{R_\theta} \pi(\theta)\ell n\pi(\theta)d\theta \tag{2.6}$$

In (2.6), G_1 is the 'information provided by an experiment', the information in the posterior density minus the information in the prior density; see for example Good (1956), Lindley (1956) and Soofi (1994). As Soofi (1994) points out, $G(\pi)$ in (2.6) is a concave function with respect to $\pi(\theta)$ since both $G_1(\pi)$ and $-\pi(\theta)\ell n\pi(\theta)$ are concave functions. Thus the MDIP criterion functional $G(\pi)$ is the sum of the 'information in an experiment' plus the information in $h(y)$ minus the information in the prior density. While this is so, it should be appreciated that (2.6) reduces algebraically to (2.4) which has simple and reasonable interpretations.

As mentioned above, use of the 'information provided by an experiment' as a criterion functional for producing priors does not yield clear-cut analytical results. To get around this problem, some have suggested asymptotic approximations to this criterion functional and consider maximizing 'missing information' by choice of a prior; see, e.g. Lindley (1961) and Berger and Bernardo (1992). By going asymptotic, these approaches involve introducing data that we shall not have. Also by allowing samples to grow in size, the strong possibility that the model may change has not been adequately considered. This point has been made by Phillips (1983) in another context as follows:

> When a model grows in complexity with increases in the availability of data, traditional asymptotic methods of statistical analysis inevitably become self-defeating. To extend Fisher's own metaphor, by the time the cannon has rumbled into place, and been loaded and sighted to fire, the sparrow will have flown away. (p. 509, fn. omitted)

It is thus fortunate that $G(\pi)$ in (2.4) is an exact finite sample criterion functional from which optimal priors are derived.

Above, we have considered one observation y, scalar or vector, in the formulation of $G(\pi)$ in (2.4). If independently and identically distributed (i.i.d.) observations, y_1, y_2, ... y_n, are available, we can write

$$G_n(\pi) = \sum_{i=1}^{n}\left[\int I_i(\theta)\pi(\theta)d\theta - \int_{R_\theta} \pi(\theta)\ell n\pi(\theta)d\theta\right] \tag{2.7}$$

and since $I_i(\theta) = I(\theta) = \int f(y_i \mid \theta)\ell nf(y_i \mid \theta)dy_i$, $i = 1, 2, \ldots, n$, (2.7) reduces to $nG(\pi)$, with $G(\pi)$ given in (2.4). Thus maximizing $G_n(\pi)$ in (2.7) with respect to the choice of $\pi(\theta)$, subject to its being proper and possibly other side conditions, will produce the same prior for any n as pointed out in Zellner (1991). If the observations are independently but not identically distributed, then the MDIP derived from (2.7) on the basis of n observations is the geometric mean of the priors derived for the individual independently but not identically distributed observations.

2.2 *Derivation of MDIPs*

The maximization of $G(\pi)$ in (2.4) with respect to the choice of $\pi(\theta)$ subject to side conditions is a standard calculus of variations problem. At this point, we shall just consider the side condition that the prior, $\pi(\theta)$, be proper, that is

$$\int_{R_\theta} \pi(\theta)d\theta = 1 \tag{2.8}$$

where R_θ is the region containing θ. Note that R_θ is a compact region which, however, can be very large. On the other hand if, as is the case in some problems, parameters have bounded values, say limited to 0 to 1 or -1 to $+1$, R_θ in (2.8) can be defined to include such bounds on parameters values. With respect to the compactness issue, McCulloch (1992) commented:

> In the old days people wanted analytical results and given the set of mathematical tools it was actually more convenient to let the parameter space be infinite. Now most of our work is done numerically so that, in effect, we are using a compact set. (pp. 50–51)

With this said about R_θ, solution to the problem of maximizing (2.4) subject to (2.8), denoted by $\pi^*(\theta)$, is:

$$\pi^*(\theta) = \begin{cases} c\exp\{I(\theta)\} & \theta \subset R_\theta \\ 0 & \theta \not\subset R_\theta \end{cases} \tag{2.9}$$

where c is a normalizing constant, $c = 1/\int_{R_\theta} \exp[I(\theta)]d\theta$ and $I(\theta)$, given in (2.3) is the information in the data density, $f(y \mid \theta)$, as measured by its negative entropy. Since entropy expressions for many data densities, e.g. normal, beta, binomial, exponential, etc. are easily derived and have been tabulated, see, e.g., Ebrahimi and Soofi (1994) and Verdugo-Luzo and Rathie (1978), the expression in (2.9) is implemented relatively easily for many problems.

It is interesting to observe from (2.4) and (2.9), that the following result holds:

$$G(\pi^*(\theta)) - G(\pi(\theta)) = \int_{R_\theta} \pi(\theta)\ell n\left[\pi(\theta)/\pi^*(\theta)\right]d\theta \tag{2.10}$$

where $\pi(\theta)$ is any proper prior density and $\pi^*(\theta)$ is given in (2.9). The right side of (2.10) is just the Jeffreys–Kullback–Leibler information or distance measure or cross entropy relating to $\pi^*(\theta)$ and $\pi(\theta)$ and since the right side of (2.10) is non-negative, it is the case that taking $\pi(\theta) = \pi^*(\theta)$ provides a maximal value for $G(\pi(\theta))$.

As mentioned above, $G(\pi)$ in (2.4) can be maximized subject to moment and other side conditions to produce 'informative' priors. For example if the side conditions are (2.8) and the additional moment conditions, with the m_i's given,

$$m_i = \int_{R_\theta} \theta^i \pi(\theta)d\theta, \quad i = 1, 2, \ldots, q \tag{2.11}$$

the prior that maximizes $G(\pi)$ in (2.4) subject to (2.8) and (2.11) is given by

$$\pi^*(\theta) \propto \exp\{I(\theta) + \lambda_1\theta + \ldots + \lambda_q\theta^q\} \tag{2.12}$$

where the λ_i's are Lagrange multipliers. See Zellner and Highfield (1988) for a procedure for computing values of these multipliers. Further, as mentioned above, there may be restrictions on the ranges of the parameters that can be introduced without much difficulty. Finally, if the prior information relates to prior fractiles, say the median, denoted by m, such prior information can be introduced by use of delta functions, as suggested by a physicist in a meeting at the University of Wyoming several years ago,

$$\int_a^b \delta_1(\theta, m)\pi(\theta)d\theta = 0.50 \tag{2.13}$$

where $\delta(\theta, m) = 1$ for $a \leq \theta \leq m$ and $\delta(\theta, m) = 0$ for $m < \theta \leq b$. Then when $G(\pi)$ in (2.4) is maximized with respect to the choice of $\pi(\theta)$ subject to (2.8) and (2.13), the result is

$$\pi^*(\theta) = \exp\{-[\lambda_0 + \lambda_1\delta_1(\theta, m)]\} \tag{2.14}$$

where the λ_i's are Lagrange multipliers. Explicitly, $\pi^*(\theta) = 0.5/(m - a)$ for $a \leq \theta \leq m$ and $\pi^*(\theta) = 0.5/(b - m)$ for $m < \theta \leq b$. Other fractile restrictions can also be introduced via the use of delta functions.

Above, we have employed uniform measure in defining entropy in (2.1). If some other measure, denoted by $m(y, \theta)$ is employed, then the negative entropy based on this measure is

$$-H_m(p) = \int_{R_{y,\theta}} p(y,\theta)\ell n\left[p(y,\theta)/m(y,\theta)\right]dyd\theta \tag{2.15}$$

and as shown in Zellner (1991), (2.15) can be expressed as,

$$-H_m(p) = \int_{R_\theta} I_m(\theta)\pi(\theta)d\theta + \int_{R_\theta} \pi(\theta)\ell n\left[\pi(\theta)/m(\theta)\right]d\theta \tag{2.16}$$

with

$$I_m(\theta) = \int_{R_y} f(y \mid \theta)\ell n\left[\frac{f(y \mid \theta)}{m(y \mid \theta)}\right]dy \tag{2.17}$$

In (2.16), the first term on the right side is a measure of the information in the data density while the second is a measure of the information in the prior density. Then the following criterion functional, just the difference of the two terms on the right side of (2.15) can be maximized subject to side conditions to obtain MDIP priors,

$$G_m(\pi) = \int I_m(\theta)\pi(\theta)d\theta - \int \pi(\theta)\ell n[\pi(\theta)/m(\theta)]d\theta \tag{2.18}$$

If just the side condition that the prior be proper is employed, the prior that maximizes (2.18) is

$$\pi_m^*(\theta) = k \exp\{I_m(\theta)] \tag{2.19}$$

where k is a normalizing constant and $I_m(\theta)$ is given in (2.17). Also (2.18) can be maximized subject to additional side conditions, for example conditions on moments and/or fractiles.

2.3 *Invariance considerations*

In Zellner (1977), it is shown that for location-scale densities, $f(y \mid \theta,\sigma) = (c/\sigma)f[(y - \theta)/\sigma]$, MDIPs are invariant with respect to linear transformations of the parameters and observations. In the case of more general one-to-one transformations of the parameters, it is shown in Zellner (1991) that invariance conditions can be introduced as side conditions in the optimization of $G = \sum_{i=1}^{m} G_i$, where G_i is (2.4) for the i'th parameterization. In the case of m one-to-one transformations, $\eta_i = h_i(\theta)$, $i = 1, 2, \ldots m$, the resulting invariant MDIP is

$$\pi^*(\theta) = c\exp\left\{I(\theta) + \sum_{i=1}^{m} \ell n \mid h_i'(\theta) \mid /(m+1)\right\} \tag{2.20}$$

A simple application of (2.20) in Zellner (1991) is a case in which θ is a regression coefficient and $\eta = 1/\theta$. Here, with finite ranges for the parameters, (2.20) yields, $\pi^*(\theta) \propto 1/\theta$ and $\pi^*(\eta) \propto 1/\eta$, which contrasts markedly with the usual uniform prior for θ which implies the following prior on η, $\pi(\eta) \propto 1/\eta^2$. Also, (2.20) is applied to power transformations of a scale parameter σ, to yield the invariant MDIP prior, $\pi(\sigma) \propto 1/\sigma$, the 'invariant to powers' Jeffreys's prior or right Haar measure, as noted by Kass and Wasserman (1995). Note that imposition of invariance conditions changes MDIPs that do not incorporate them. For example, in the reciprocal regression problem, mentioned above, the individual MDIPs for θ and for η are both uniform whereas with invariance imposed these MDIP densities are in the forms mentioned above.

George Barnard in a letter to the author mentioned that years ago 'Student' (W.S. Gosset) solved a problem using a uniform prior on a binomial parameter. Fisher wrote to 'Student' saying that his prior does not imply a uniform prior on the binomial parameter raised to the fifth power. 'Student' replied that he has no concern about the fifth power of the parameter, an irrelevant transformation. Also, Rao (1987) discusses degrees of invariance and states, 'the choice of a metric naturally depends on a particular problem under investigation, and invariance may or may not be relevant' (p. 220). Berger (1985) comments, 'The major problem with invariance concerns the amount of invariance that can be used' (p. 423). Jeffreys (1967) also opts for limited invariance in analysing many problems, e.g. one involving a mean, μ, and a standard deviation, σ. His general invariant rule produces the prior $\pi(\mu, \sigma) \propto 1/\sigma^2$ that he deems unreasonable. Using a limited class of transformations that he deems 'reasonable' for the problem at hand, he obtains the prior, $\pi(\mu, \sigma) \propto 1/\sigma$ that is invariant with respect to the transformations that he regards as 'reasonable', and is the prior yielded by the MDIP approach. Jeffreys's example is a good illustration of the point made by Rao in the quotation above. Also, while the reciprocal of the marginal propensity to consume parameter, α, is a relevant transformation in some models, the transformation $\alpha^{5/2}$ is not. That the MDIP approach allows one to derive diffuse and informative priors that are invariant with respect to relevant transformations is indeed fortunate.

Above we have used one measure of information, the negative entropy. If we entertain other measures of information, clearly the functional $G(\pi)$ in (2.4) can be defined in terms of them or combinations of them. One example is,

$$G = \int \pi(\theta)\ell n \mid F \mid^{1/2} d\theta - \int \pi(\theta)\ell n\theta(\theta)d\theta \tag{2.21}$$

where F is the Fisher information matrix. On maximizing G with respect to choice of $\pi(\theta)$ subject to (2.8), namely that $\pi(\theta)$ be proper, the solution is

$$\pi(\theta) \propto |F|^{1/2} \qquad \theta \subset R_\theta \tag{2.22}$$

This MDIP is Jeffreys's prior defined over a finite, possibly very large, region of the parameter space. Further, moment side conditions, say $\int \theta^i \pi(\theta)d\theta = \mu_i$, $i = 1, 2$ can be employed in the optimization (2.21) to obtain $\pi^*(\theta) \propto |F(\theta)|^{1/2} e^{\lambda_1\theta+\lambda_2\theta^2}$, where λ_1 and λ_2 are Lagrange multipliers. These analyses can be extended to sets of parameters.

An extension of (2.21) is

$$G_w[\pi(\theta)] = \int [wI(\theta) + (1-w)\ell n\, |F|^{1/2}]\sigma(\theta)d\theta - \int \pi(\theta)\ell\, n\pi(\theta)d\theta \tag{2.23}$$

with $0 \leq w \leq 1$. When $G_w(\pi)$ is maximized subject to π's being proper, the result is

$$\pi_w^* \propto |F|^{(1-w)/2} e^{wI(\theta)} \tag{2.24}$$

a weighted geometric mean of Jeffreys's prior, $|F(\theta)|^{1/2}$ and of the MDIP, $e^{I(\theta)}$. Also (2.24) resembles Rodríguez's (1991a, b) entropic priors which have the same general invariance properties that Jeffreys's priors have.

While use of (2.21) and (2.23) produces interesting results, they involve alternative measures of information in the same criterion functional, e.g. $\ell n\, |F|^{1/2}$ and negative entropy, $\int \pi(\theta)\ell n\pi(\theta)d\theta$ in (2.21). Thus it seems advisable to consider maximizing $G(\pi)$ in (2.4) subject to one or both of the following constraints or side conditions:

$$\int \pi(\theta)\ell n\, |F|^{1/2} d\theta = c_1 \tag{2.25}$$

$$\int \pi(\theta)I(\theta)d\theta = c_2 \tag{2.26}$$

Use of these side conditions in optimizing (2.4) yields

$$\pi^*(\theta) \propto |F|^{\lambda 1/2} e^{\lambda_2 I(\theta)}$$

where λ_1 and λ_2 are Lagrange multipliers, a general form that includes Jeffreys's and Rodríguez's (1991a, b) entropic priors as special cases.

Thus by introducing side conditions (2.25) and/or (2.26), a variety of interesting results is obtained without altering the fundamental criterion functional, $G(\pi)$, given in (2.4).

III. MDIPs for parameters of data density and mass functions

3.1 Univariate continuous data densities

Shown in Table 1 are MDIPs for a number of univariate data densities presented in Zellner (1977). As can be seen, for normal densities, the priors for the location and scale parameters are the usual diffuse priors that are in widespread use. In the case

Table 1 Explicit examples of MDIPs

Data Pdf	Data Pdf information measure, $I(\theta)$[a]	MDIP Pdf $\pi^* \propto \exp\{I(\theta)\}$
1. Normal ($-\infty < y < \infty$)		
1a $f(y \mid \theta) = (2\pi)^{-1/2}\exp\{-(y-\theta)^2/2\}$	const.	$\pi_*(\theta) \propto$ const.
1b $f(y \mid \sigma) = (2\pi\sigma^2)^{-1/2}\exp\{-(y-\theta)^2/2\sigma^2\}$	const.$-\ell n\sigma$	$\pi_*(\sigma) \propto 1/\sigma$
1c $f(y \mid \theta, \sigma) = (2\pi\sigma^2)^{-1/2}\exp\{-(y-\theta)^2/2\sigma^2\}$	const.$-\ell n\sigma$	$\pi_*(\theta, \sigma) \propto 1/\sigma$
2. Rectangular		
$f(y \mid \theta) = 1/\theta,\ 0 < y < \theta$	const.$-\ell n\theta$	$\pi_*(\theta) \propto 1/\theta$
3. Exponential (Pareto; $0 < y < \infty$)		
$f(y \mid \theta) = \theta^{-1}\exp\{-y/\theta\}$	const.$-\ell n\theta$	$\pi_*(\theta) \propto 1/\theta$
4. Student-t ($v = 1$, Cauchy; $-\infty < y < \infty$)[b]		
$f(y \mid \theta, \sigma) = (k/\sigma)\{\nu + (y-\theta)^2/\sigma^2\}^{-((v+1)/2)}$	const.$-\ell n\sigma$	$\pi_*(\theta, \sigma) \propto 1/\sigma$
5. Log-normal		
$f(y \mid \theta, \sigma) = (2\pi\sigma^2 y^2)^{-1/2}\exp\{-(\ell ny - \ell n\theta)^2/2\sigma^2\}$	const.$-\ell n\sigma - \ell n\theta$	$\pi_*(\theta, \sigma) \propto 1/\theta$
6. General forms		
6a $f(y \mid \theta) = f(y-\theta)$	const.	$\pi_*(\theta) \propto$ const.
6b $f(y \mid \sigma) = \sigma^{-1}f(y/\sigma)$	const.$-\ell n\sigma$	$\pi_*(\sigma) \propto 1/\sigma$
6c $f(y \mid \theta, \sigma) = 1/\sigma f(y-\theta/\sigma)$	const.$-\ell n\sigma$	$\pi_*(\theta, \sigma) \propto 1/\sigma$
7. Exponential class		
$f(y \mid \theta) = \exp\{h(\theta)K(y) + S(y) + q(\theta)\}$	$h(\theta)\,\bar{K} + \bar{S} + q\,(\theta)$[c]	$\pi_*(\theta) \propto \exp\{h(\theta)\bar{K} + \bar{S} + q(\theta)\}$

[a] $I(\theta) = \int f(y \mid \theta)\ \ell nf(y \mid \theta)\ dy$
[b] Here v is assumed to have a known value.
[c] $\bar{K} \equiv \int K(y)f(y \mid \theta)dy$ and $\bar{S} \equiv \int S(y)f(y \mid \theta)dy$

of 1c, use of Jeffreys's rule, namely taking the prior proportional to the square root of the determinant of the Fisher information matrix, leads to $1/\sigma^2$ that Jeffreys deems unsatisfactory. By applying his rule separately to the location and scale parameters, he obtains $1/\sigma$, that he regards as reasonable. Note that the normal examples in 1a–c are special cases of 6a–c, respectively.

In the case of the rectangular density in line 2 of Table 1, the prior is $1/\theta$ or $\ell n\theta$ uniformly distributed, a result in accord with Jeffreys's recommendation for the form of a diffuse prior for a parameter that assumes non-zero values over the positive real line.

As regards the exponential density considered in line 3 of Table 1, the MDIP for its parameter is $1/\theta$ or $\ell n\theta$ uniformly distributed, in accord with Jeffreys's rule mentioned in the previous paragraph and a special case of 6b. Further, the Pareto density, $p(y \mid A, B) = (1/\beta y)\exp\{-(1/\beta)\ell n(y/A)\}$, with A having a known value satisfying $0 < A \leq y < \infty$, is a 'log-exponential' density. With $x = \ell n(y/A)$, $0 < x < \infty$, $p(x \mid \beta) = (1/\beta)\exp\{-x/\beta\}$ for which the MDIP is $1/\beta$. See Soofi (1992) for an MDIP for β *and* A.

For the Student-*t* density with a known value of the degrees of freedom parameter, ν, the MDIP density for the location and scale parameters is $1/\sigma$, a special case of the general result in 6b. An MDIP for the location, scale and degrees of freedom parameter has been derived by Moulton (1993) and Soofi (1992) independently.

The MDIP for the parameters of a log-normal density is given by $1/\theta\sigma$, where θ is the median and σ the standard deviation, and these priors are in accord with Jeffreys's recommendation that $\ell n\theta$ and $\ell n\sigma$ be assumed uniformly distributed.

In lines 6(a–c)–7 of Table 1, MDIPs for parameters of some general classes of data densities are presented. To illustrate the derivation of the MDIP in 6c, note that with $-\infty < y < \infty$,

$$\begin{aligned} I(\theta, \sigma) &= \int f(y \mid \theta, \sigma)\ell n f(y \mid \theta, \sigma)dy \\ &= -\ell n\sigma + \int f(x)\ell n f(x)dx = -\ell n\sigma + \text{const.} \end{aligned} \tag{3.1}$$

where $x = (y - \theta)/\sigma$. Thus $\pi^*(\theta, \sigma) \propto \exp\{I(\theta, \sigma)\} = 1/\sigma$ is the MDIP for this problem. Also, this result is in accord with the Box–Tiao (1973) 'data translation' rule both in form and in being defined as non-zero over a finite range of the parameter space and thus being a proper density. As pointed out by Kass and Wasserman (1995) this MDIP is a right Haar measure in contrast to Jeffreys's general rule which yields left Haar measure but which he sensibly modifies to right Haar measure in many cases, usually by the assumption of *a priori* independence of parameters, say location and scale parameters and applying his rule separately to these parameters.

Sinha and Zellner (1990) derived the following MDIP for parameters of the Weibull density, $f(y \mid \gamma, \theta) = (\gamma\theta)^{-1}(x/\theta)^{1/\gamma-1}\exp\{-(x/\theta)^{1/\gamma}\}$, $0 < \theta, \gamma$:

$$\pi^*(\gamma, \theta) \propto (1/\theta\gamma)\exp\{-C\gamma\} \tag{3.2}$$

with C =0.577, Euler's constant. Here the prior on $\ell n\theta$ is uniform while that for $\ell n\gamma$ is not. It is pointed out in the reference above that (3.2) is a special case of the product of two proper gamma prior densities for θ and γ. Also, the Jeffreys's prior for θ and γ is $\pi_J(\theta, \gamma) \propto 1/\theta\gamma^2$, that is θ and γ are independent *a priori* with $\ell n\theta$ uniformly distributed but with $\ell n\gamma$ not uniformly distributed. Priors for a different parameterization of the Weibull density are also presented in Sinha and Zellner (1990).

With this said about MDIPs for parameters of univariate data densities, we now turn to consider priors for parameters of certain data probability mass functions.

3.2 MDIPs for parameters of data probability mass functions

First, we consider the well known binomial process that Bayes considered many years ago. For a single observation, $y = 1$ with probability θ and $y = 0$ with probability $1–\theta$, with $0 \le \theta \le 1$. The probability mass function for a single observation is $f(y \mid \theta) = \theta^y(1–\theta)^{1–y}$ and the information in $f(y \mid \theta)$ is $I(\theta)= E\ell nf(y \mid \theta) = \theta\ell n\theta + (1–\theta)\ell n(1–\theta)$. Thus the MDIP for θ is – Zellner (1977) –

$$\pi^*(\theta) = c\theta^\theta(1–\theta)^{1–\theta} \qquad 0 \le \theta \le 1 \tag{3.3}$$

with the normalizing constant $c = 1.6186$, evaluated by numerical integration. Note that (3.3) is symmetric around 1/2, its minimal value and rises monotonically to 1.6186 at $\theta = 0$ and $\theta = 1$. This prior is proper and has finite moments of all orders. Also, it has been approximated by expanding (3.3) around $\theta = 1/2$ in a Taylor's series expansion to quadratic and quartic terms in Zellner (1993), e.g., $\pi^*(\theta) \doteq (5/3)[1/2 + (\theta - \tfrac{1}{2})^2 + (4/3)(\theta - \tfrac{1}{2})^4]$, forms that can be easily integrated analytically whereas moments, fractiles, etc. of (3.3) are available by simple univariate numerical integration.

As regards comparisons with alternatives, namely the Bayes–Laplace uniform prior, $\pi(\theta) = 1$, the Jeffreys and Bernardo–Smith (1994, p. 315) prior, $\theta^{-1/2}(1–\theta)^{-1/2}$ a proper beta density, and the Haldane improper prior, $\theta^{-1}(1–\theta)^{-1}$, it is the case that (3.3) lies above the Bayes–Laplace uniform density and below the Jeffreys and Haldane priors at $\theta = 0$ and $\theta = 1$. On this matter, Jeffreys (1967) has written, 'Certainly if we take the Bayes–Laplace rule right up to the extremes, we are led to results that do not correspond to anybody's way of thinking. The rule $d\theta/\theta(1–\theta)$ goes too far the other way'(p. 124). See also Akaike (1978) who expresses similar views. Thus that the MDIP in (3.3) does not shoot off to infinity at $\theta = 0$ and $\theta = 1$ and lies above the uniform density is in line with what Jeffreys and Akaike consider 'reasonable'. Also, use of it rather than the Bayes–Laplace uniform density will change Laplace's Law of Succession in a direction that is again in agreement with what Jeffreys considers to be reasonable; see Jeffreys (1967, pp. 127–9).

As regards the trinomial and multinomial probability mass functions, the MDIPs for their parameters are given in Zellner (1993), namely:

$$\pi^*(\theta_1, \theta_2) \propto \theta_1^{\theta_1}\, \theta_2^{\theta_2}(1 - \theta_1 - \theta_2)^{1-\theta_1-\theta_2} \tag{3.4}$$

and

$$\pi^*(\theta_1, \theta_2, \ldots, \theta_{m-1}) \propto \theta_1^{\theta_1}\, \theta_2^{\theta_2} \ldots \theta_m^{\theta_m} \tag{3.5}$$

with $\theta_m = 1–\theta_1– \ldots –\theta_{m-1}$. The bivariate MDIP density in (3.4) has a minimum at $\theta_1 = \theta_2 = 1/3$ and is bowl-shaped rising to finite values at the boundaries of the triangular region, $0 \leq \theta_1 \leq 1, \theta \leq \theta_2 \leq 1$, and $0 \leq \theta_1 + \theta_2 \leq 1$. Both (3.4) and (3.5) contrast markedly with the Jeffreys prior for multinomial parameters given and briefly discussed by Box and Tiao (1973, p. 55), namely, $\pi \propto (\theta_1\theta_2 \ldots \theta_m)^{-1/2}$, where $\theta_m = (1-\theta_1-\theta_2 \ldots -\theta_{m-1})$; see Good (1963) who also presents this prior. It is seen, just as with the binomial parameter, if individual θ_i's approach zero, this prior density grows without limit. Also as the sum of all free θ_i's approaches one, the density shoots off to infinity in contrast to the behavior of (3.4) and (3.5), which assume finite values at the boundaries of the parameter space and also are the densities which maximize $G(\pi)$ in (2.4).

3.3 MDIPs for parameters of several univariate linear models

Given a linear, normal multiple regression model, $y_i = x'_i\beta + u_i$ where the $1 \times k$ vector $\mathbf{x}'_i$ is given, β is a kx1 regression coefficient vector and u_i is an $N(0, \sigma^2)$ error term, the conditional density of y_i is $f(y_i \mid x_i, \beta, \sigma) \propto (1/\sigma)\exp\{-(y_i - x'_i\beta)^2/2\sigma^2\}$. Then $I(\beta, \sigma) = E\ell nf(y_i \mid x_i, \beta, \sigma) = -\ell n\sigma + \text{const}$. Thus the MDIP prior for β and σ is, from Zellner (1977):

$$\pi^*(\beta, \sigma) \propto e^{I(\beta, \sigma)} \propto 1/\sigma \tag{3.6}$$

In (3.6), the elements of β and $\ell n\sigma$ are independently and uniformly distributed, an assumption utilized in many Bayesian analyses of the multiple regression model and can be obtained in the Jeffreys's (1967, p. 182) approach by assuming β and σ are independent.

Next, we consider a first-order stationary normal autoregression, $y_t = (1 - \beta)\theta + \beta y_{t-1} + u_t$ with $|\beta| < 1$, $u_t \sim N(0, \sigma^2)$ and θ the mean of y_t.[1] Then from $f(y_t \mid \beta, \theta, \sigma) \propto [(1 - \beta^2)^{1/2}/\sigma]\exp\{-(1 - \beta^2)(y_t - \theta)^2/2\sigma^2\}$, $I(\beta, \theta, \sigma) = E\ell nf(y_t \mid \beta, \theta, \sigma) = -\ell n\sigma + (1/2)\ell n(1 - \beta^2)$ and the MDIP prior is

$$\pi^*(\theta, \beta, \sigma) \propto (1 - \beta^2)^{1/2}/\sigma \qquad 0 < |\beta| < 1 \tag{3.7}$$

as shown in Zellner (1977). Thus θ, β and σ are independent with θ and $\ell n\sigma$ uniformly distributed. The prior for β has a maximum at $\beta = 0$ and falls symmetrically to zero at $\beta = \pm 1$. Also $z = (1 - \beta)/2$, $0 < z < 1$, has a proper beta density. See Zellner (1971, pp. 214–19) for a discussion of other priors for this problem which differ somewhat from that in (3.7).

When $\theta = 0$, there is no intercept in the AR(1) model and the MDIP prior for β and σ is $\pi^*(\beta, \sigma) \propto (1 - \beta^2)^{1/2}/\sigma$, a form that can be compared with the Bernardo–Berger (1992) prior, $\pi(\beta, \sigma) \propto 1/\sigma(1 - \beta^2)^{1/2}$, $0 < |\beta| < 1$, a prior that has σ and β independent with $\ell n\sigma$ uniform and a marginal density for β that has a minimum at $\beta = 0$ and which shoots off to infinity as β approaches ± 1. Note that Jeffreys (1967, p. 359) discusses the stationary AR(1) model and concludes, 'The [Jeffreys] estimate rule [prior, $1/\sigma(1 - \beta^2)^{1/2}$] gives a singularity not only at $\beta = 1$, which might be tolerable, but also at $\beta = -1$, which is not.' [Jeffreys's symbol ρ has been replaced by β.] Thus Jeffreys finds the prior, $\pi(\beta, \sigma) \propto 1/\sigma(1 - \beta^2)^{1/2}$ intolerable.

If further information is available about the moments of β and/or σ, as mentioned above, MDIPs can be derived that incorporate such moment restrictions, say $E\theta = \bar{\theta} = 0.5$, $\text{Var}\theta = \sigma_\theta^2 = 0.03$, $E\beta = 0.5$, etc.

The MDIP for a stationary AR(2) process, $y_t - \theta = \rho_1(y_{t-1}-\theta) + \rho_2(y_{t-2}-\theta) + u_t$, with $u_t\ N(0, \sigma^2)$, $Ey_t = \theta$, and $\text{Var}(y_t) = \sigma^2/[1 - \rho_1^2 - \rho_2^2 - 2\rho_1^2\rho_2/(1 - \rho_2)]$, given in Zellner and Min (1993) is

$$\pi^*(\theta, \rho_1, \rho_2, \sigma) \propto [1 - \rho_1^2 - \rho_2^2 - 2\rho_1^2\rho_2/(1 - \rho_2)]^{1/2}/\sigma \tag{3.8}$$

with ρ_1 and ρ_2 restricted to lie in the region of stationarity, a triangular region in the ρ_1, ρ_2 space – see Zellner (1971, p. 196) for a plot of this region. In (3.8) θ and $\ell n\sigma$ are uniformly distributed and mutually independent of each other and of ρ_1 and ρ_2. A plot of the MDIP prior for ρ_1 and ρ_2 which has a maximum at $\rho_1 = \rho_2 = 0$ and falls to a value of zero at all points on the boundary of the triangular stationarity region is shown in Figure 1, reproduced from Zellner and Min (1993). Also in this reference, MDIPs for some non-stationary AR(1) processes are presented and discussed.

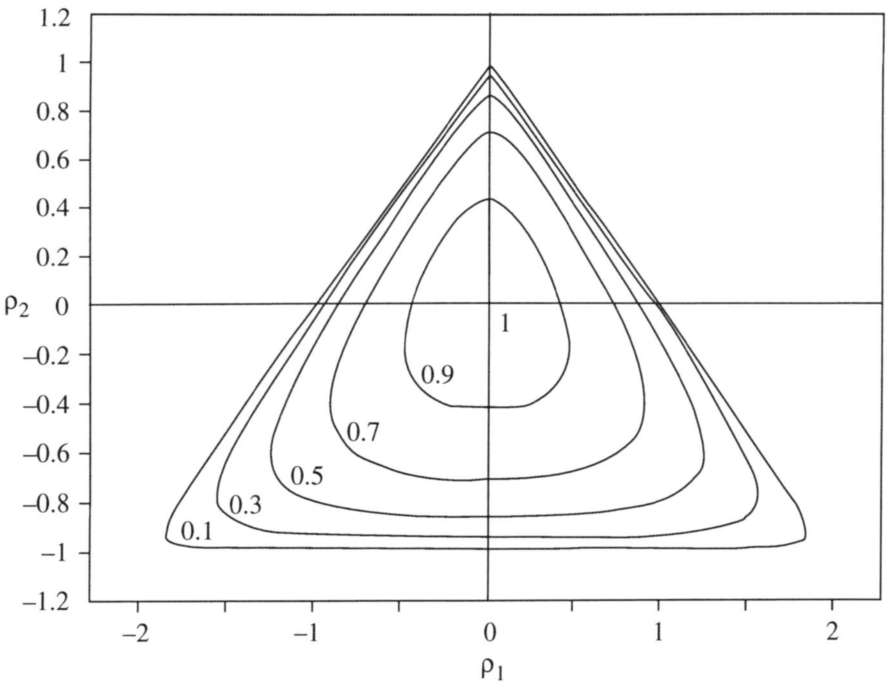

[a] The MDIP prior for ρ_1 and ρ_2, shown in the text, is

$$\pi.(\rho_1, \rho_2) \propto \left[1 - \rho_1^2 - \rho_2^2 - \frac{2\rho_1^2\rho_2}{1 - \rho_2}\right]^{1/2}$$

with $-2 < \rho_1 < 2$, $\rho_2 < \rho_1 + 1$ for $-2 < \rho_1 \leq 0$ and $\rho_2 \leq -\rho_1 + 1$ for $0 \leq \rho_1 < 2$.

Figure 1 Contours of the maximal data information prior density for parameters ρ_1 and ρ_2 of a stationary AR(2) model[a]

3.4 MDIP for Stein's n-means problem

Let ξ be an $n \times 1$ vector of means and y an $n \times 1$ vector of observable variables satisfying $y = \xi + u$ with the $n \times 1$ error vector $u \sim N(0, I_n)$. To obtain a MDIP for $\theta = \xi'\xi$, the square of the radius vector in polar coordinates, as indicated by Good and Wallace, cited in Stein (1962), it is useful to transform to polar coordinates θ and α, a vector of n–1 angles; that is

$$\xi_i = q_i(\theta, \alpha) \qquad i = 1, 2, \ldots, n \tag{3.9}$$

Then

$$f(y \mid \theta, \alpha) \propto \exp\{-(y - q)'(y - q)/2\} \tag{3.10}$$

where q is an $n \times 1$ vector with typical element $q_i(\theta, \alpha)$. Then the information in $f(y \mid \theta, \alpha)$, $I(\theta, \alpha) = E\ell n f(y \mid \theta, \alpha) =$ const. Therefore the MDIP for θ and α is $\pi^*(\theta, \alpha) \propto \exp\{I(\theta, \alpha)\}$ = const. Thus θ and the n–1 elements of α are independently and uniformly distributed. This MDIP uniform prior for θ is quite different from $\pi(\theta) \propto \theta^{(n-2)/2}$, a prior implied by the assumption that the ξ_i's are independently and uniformly distributed, $\pi(\xi) \propto$ const. This uniform prior for ξ is yielded by the invariant $|F|^{1/2}$rule. On the other hand, a uniform prior on θ and the elements of α implies the following prior on ξ, $\pi(\xi) \propto 1/(\xi'\xi)^{(n-2)/2}$, a prior that is very close to the prior, $\pi(\xi) \propto 1/\xi'\xi)^{(n-3)/2}$ that Stein (1962, p. 281) states will produce his shrinkage estimate for ξ. In addition, use of the MDIP prior $\pi^*(\theta, \alpha) \propto$ const. yields a posterior density for θ with a mean that is a reasonable estimate in contrast to the estimate obtained when the prior $\pi(\xi) \propto$ const. implying $\pi(\theta) \propto \theta^{(n-2)/2}$ is employed. See Berger (1985, p. 230) for discussion of Stein's problem of estimating θ using 'diffuse' priors.

3.5 MDIPs for hierarchical models' hyperparameters

Given $f(y \mid \theta)$, a data density for y given θ and $g(\theta \mid \phi)$, a hyperdensity for θ given ϕ, the hyperparameter(s), the MDIP for ϕ can be obtained as follows. First, obtain $h(y \mid \phi)$ given by

$$h(y \mid \phi) = \int f(y \mid \theta) g(\theta \mid \phi) d\theta \tag{3.11}$$

Then choose $\pi(\phi)$, the prior for the hyperparameter(s) ϕ to maximize the criterion functional

$$G[\pi(\phi)] = \int I(\phi)\pi(\phi)d\phi - \int \pi(\phi)\ell n\pi(\phi)d\phi \tag{3.12}$$

where

$$I(\phi) = \int h(y \mid \phi)\ell n h(y \mid \phi)dy \tag{3.13}$$

subject to $\int \pi(\phi)d\phi = 1$ and possibly other side conditions. Here, as above, it is assumed that the parameter space is compact. The unique solution to the above maximization problem subject just to the side condition that $\pi(\phi)$ be proper, given in Zellner (1993) is

$$\pi^*(\phi) \propto e^{I(\phi)} \tag{3.14}$$

with $I(\phi)$ given in (3.13).

To provide an example illustrating the above analysis, let $y = \theta + u$ and $\theta = \mu + \varepsilon$ with $u \sim N(0, 1)$ and $\varepsilon \sim N(0, \tau^2)$ independent. Then $y = \mu + u + \varepsilon$ with $Ey = \mu$ and $\text{Var } y = \sigma_y^2 = 1 + \tau^2$. Thus with $\phi = (\mu, \sigma_y)$, $h(y \mid \phi) \propto (1/\sigma_y)\exp-\{(y-\mu)^2/2\sigma_y^2\}$ and $I(\mu, \sigma_y) = E\ell nh(y \mid \mu, \sigma_y) = -\ell n\sigma_y + \text{const}$. Thus the MDIP for the hyperparameters μ and σ_y is

$$\pi^*(\mu, \sigma_y) \propto 1/\sigma_y \tag{3.15}$$

In (3.15), μ and $\ell n\sigma_y$ are independently and uniformly distributed. By adding side conditions, more informative MDIPs for hyperparameters can be obtained.

3.6 MDIPs for common parameters in different data densities

Let $f_1(y_1 \mid \theta)$ and $f_2(y_2 \mid \theta)$ be two different data densities or mass functions with the common parameter(s) θ. An example is the positive and negative binomial probability mass functions. As in Zellner (1993), a prior for θ is sought that will provide maximal data information for both $f_1(y_1 \mid \theta)$ and $f_2(y_2 \mid \theta)$. Thus the following criterion functional is used:

$$G_2[\pi(\theta)] = \sum_{i=1}^{2}\left[\int I_i(\theta)\pi(\theta)d\theta - \int \pi(\theta)\ell n(\theta)d\theta\right] \tag{3.16}$$

where

$$I_i(\theta) = \int f_i(y_i \mid \theta)\ell nf_i(y_i \mid \theta) \qquad i = 1, 2 \tag{3.17}$$

On maximizing $G_2[\pi(\theta)]$ with respect to the choice of $\pi(\theta)$ subject to $\pi(\theta)$ being proper, the solution is

$$\pi_2^*(\theta) \propto \exp\{\bar{I}(\theta)\} = \left[e^{I_1(\theta)}e^{I_2(\theta)}\right]^{1/2} \tag{3.18}$$

where $\bar{I}(\theta) = [I_1(\theta) + I_2(\theta)]/2$. Thus $\pi_1^*(\theta)$ is equal to the geometric mean of $e^{I1(\theta)}$ and $e^{I_2(\theta)}$, the MDIPs for the individual densities or experiments. With more than two densities or experiments, $f_i(y_i \mid \theta)$, $i = 1, 2, \ldots, m$, a criterion functional $G_m[\pi(\theta)]$ analogous to (3.16) but with m terms, can be employed to obtain the MDIP, $\pi_m(\theta) \propto \{\prod_{i=1}^{m}\{\exp I_i(\theta)\}\}^{1/m}$, the geometric mean of the MDIPs for the individual $f_i(y_i \mid \theta)$ densities or experiments. Also, as mentioned in Zellner (1993), instead of equal weights used in (3.16), $G_{2w} = \sum_{i=1}^{2} w_i \int I_i(\theta)\pi(\theta)d\theta - \int \pi(\theta)\ell n\pi(\theta)d\theta$ is a criterion functional that incorporates unequal weights, the w_i's that add to one. On maximizing G_{2w} subject to $\int \pi(\theta)d\theta = 1$, the solution is $\pi^*_{2w} \propto \exp\{[w_1 I_1 + w_2 I_2]/2\}$, the geometric mean of $(e^{I_i})^{w_i}$, $i = 1, 2$.

To illustrate the above analysis, let $f_1(y_1 \mid \theta) \propto \exp\{-(y_1 - \theta)^2/2\}$ and $f_2(y_2 \mid \theta) \propto 1/[1 + (y_2 - \theta)^2]$. Then $I_1(\theta) \propto$ const., $I_2(\theta) \propto$ const. and $\pi^*_2 \propto e^{(I_1+I_2)/2}$ = const. If we 'pool' the data under the assumption that y_1 and y_2 are independent, $f_1(y_1 \mid \theta)f_2(y_2 \mid \theta)$ is the joint density and the information measure for it, denoted by $I^{(\theta)}_{21}$ = constant if $I_i(\theta)$ = const. and $I_2(\theta)$ = const. Thus the MDIP prior for the pooled data is the same as that for the unpooled data.

As a second example, consider $f_1(y_1 \mid \theta) \propto (1/\theta)\exp\{-y_1 \mid \theta\}$ and $f_2(y_2 \mid \theta) \propto (1/\theta)f_2(y \mid \theta)$. For these densities, $I_1(\theta) = -\ell n\theta$ + const. and $I_2 = -\ell n\theta$ + const. and thus $\pi^*_2(\theta) \propto (e^{I_1}e^{I_2})^{1/2} \propto 1/\theta$. If the data are pooled under an independence assumption, then $\bar{I}_{12} = 1/2\iint f_1(y_1 \mid \theta)f_2(y_2 \mid \theta)\ell n f_1(y_1 \mid \theta)f_2(y \mid \theta)dy_1dy_2 = 1/2(I_1 + I_2) = -\ell n\theta$ + const. and the MDIP $\pi^*_{12}(\theta) \propto 1/\theta$ provides a maximal value for $\bar{G}_{12} = \int \bar{I}_{12}\pi_{12}(\theta)d\theta - \int \pi_{12}(\theta)\ell n\pi_{12}(\theta)d\theta$ subject to $\int \pi_{12}(\theta)d\theta = 1$.

3.7 MDIPs for parameters of multivariate densities and models

In this section, MDIPs for parameters of normal and general multivariate densities and models are presented and discussed.

Consider the bivariate normal data density, $f(y_1, y_2 \mid \theta_1, \theta_2, \sigma_1, \sigma_2, \rho) \propto [1/\sigma_1\sigma_2(1 - \rho^2)^{1/2}] \times \exp\{-[(y_1 - \theta_1)^2/\sigma_1^2 + (y_2 - \theta_2)^2/\sigma_2^2 - 2\rho(y_1 - \theta_1)(y_2 - \theta_2)]/2(1 - \rho^2)\}$. The information in this density is

$$I(\theta_1, \theta_2, \sigma_1, \sigma_2, \rho) = \text{const.} - \ell n\sigma_1\sigma_2 - 1/2\ell n(1 - \rho^2) \tag{3.19}$$

and thus the proper MDIP is from (2.9):

$$\pi^*(\theta_1, \theta_2, \sigma_1, \sigma_2, \rho) \propto 1/\sigma_1\sigma_2(1 - \rho^2)^{1/2} \tag{3.20}$$

as given in Zellner (1977).

From (3.20), it is the case that the parameters are independently distributed with the location and log scale parameters uniformly distributed. The prior for ρ, $-1 < \rho < 1$, is proper. Note that $\eta = (1 + \rho)/2$ is such that $0 < \eta < 1$ and η has a proper beta form, $f(\eta) \propto \eta^{-1/2}(1 - \eta)^{-1/2}$. Also, MDIPs can be derived for transformations of the parameters e.g. $(\sigma_1, \sigma_2, \rho) \rightarrow (\sigma_1^2, \sigma_2^2, \sigma_{12})$, etc., using results presented above. Further, if $y_3 \sim N(\theta_1, \sigma_1)$ is considered along with the bivariate normal density for y_1 and y_2 above, a MDIP prior for the parameters *vis-à-vis* the bivariate and univariate data densities can be obtained.

The prior in (3.20) is discussed in Jeffreys (1967, p. 187) who finds it compatible with his interpretation of the correlation coefficient. He also discusses a uniform prior for ρ and his invariant improper prior $\pi(\rho) \propto 1/(1-\rho^2)^{3/2}$ and seems to be ambivalent with respect to a choice among the three alternatives – see Jeffreys (1967, pp. 174, 290). The MDIP may also be compared to $\pi(\rho) \propto 1/(1 - \rho^2)$ presented in Lindley (1965, p. 219) and Bernardo and Smith (1994, p. 338). Note that if the side condition $-\int \pi(\rho)\ell n(1 - \rho^2)d\rho = c$, where c is a given constant, is introduced then the proper

MDIP that satisfies this side condition is $\pi_r^*(\rho) \propto (1-\rho^2)^{-1/2-\lambda}$ where λ is a Lagrange multiplier. For $\lambda = 0$, the unrestricted MDIP in (3.20) is obtained. For $\lambda = -1/2$, $\lambda = 1/2$, and $\lambda = 1$, the uniform, the $(1-\rho^2)^{-1}$ and $(1-\rho^2)^{-3/2}$ priors, respectively are obtained.

As regards the m-dimensional normal density, $f(y \mid \theta, \Sigma) \propto |\Sigma|^{-1/2} \exp\{-(y-\theta)'\Sigma^{-1}(y-\theta)/2\}$, $I(\theta, \Sigma) = E_y \ell n f(y \mid \theta, \Sigma) = \text{const.} - 1/2\ell n|\Sigma|$ and thus the MDIP, presented in Zellner (1977) is:

$$\pi^*(\theta, \Sigma) \propto |\Sigma|^{-1/2} \tag{3.21}$$

with the m means $\theta' = (\theta_1, \ldots, \theta_m)$ uniformly and independently distributed and independent of the distinct elements of Σ. Note also that with $f(y \mid \theta, \Sigma^{-1})$, the MDIP prior for Σ^{-1} is $\pi^*(\Sigma^{-1}) \propto |\Sigma^{-1}|^{1/2}$. Further, the invariant MDIPs for Σ and Σ^{-1} with the Jacobian of the transformation from Σ to Σ^{-1} equal to $|\Sigma|^{m+1}$ are:

$$\pi^*(\Sigma) \propto |\Sigma|^{-(m+2)/2} \text{ and } \pi^*(\Sigma^{-1}) \propto |\Sigma{-1}|^{-m/2} \tag{3.22}$$

derived using a generalized version of (2.20). Of course other relevant transformations can be introduced and MDIPs derived that are compatible with them.

The priors in (3.22) are close in form to Jeffreys's invariant priors, $\pi_J(\Sigma) \propto |\Sigma|^{-(m+1)/2}$ and $\pi_J(\Sigma^{-1}) \propto |\Sigma^{-1}|^{-(m+1)/2}$. It is interesting to note that $\pi^*(\Sigma)$ and $\pi^*(\Sigma^{-1})$ in (3.22) can be expressed as $\pi^*(\Sigma) \propto |\Sigma|^{-1/2}\pi_J(\Sigma) = |\Sigma|^{-1/2}|\Sigma|^{-(m+1)/2}$ and $\pi^*(\Sigma^{-1}) \propto |\Sigma^{-1}|^{1/2}\pi_J(\Sigma^{-1}) = |\Sigma^{-1}|^{1/2}|\Sigma^{-1}|^{-(m+1)/2}$. It is thus seen that the invariant MDIP priors for Σ and Σ^{-1} are just equal to the individual MDIP priors, $|\Sigma|^{-1/2}$ and $|\Sigma^{-1}|^{-1/2}$ times factors that ensure the required invariance condition associated with the transformation $\Sigma \leftrightarrow \Sigma^{-1}$ is satisfied.

With y and θ $m \times 1$ vectors and Σ an $m \times m$ pds matrix, we have the following MDIPs for general multivariate densities. For $f(y \mid \theta) = f(y - \theta)$, the proper MDIP is $\pi^*(\theta) \propto \exp\{I(\theta)\} \propto$ constant. For $f(y \mid \Sigma) = |\Sigma|^{-1/2} f(y'\Sigma^{-1}y)$, a class of elliptically symmetric densities, $I(\Sigma) = \text{const.} - 1/2\ell n|\Sigma|$ and $\pi^*(\Sigma) \propto |\Sigma|^{-1/2}$. Finally, for $f(y \mid \theta, \Sigma) = |\Sigma|^{-1/2} f[(y-\theta)'\Sigma^{-1}(y-\theta)]$, $I(\theta, \Sigma) = \text{const.} -(1/2)\ell n|\Sigma|$ and the proper MDIP is $\pi^*(\theta, \Sigma) \propto |\Sigma|^{-1/2}$, results given and discussed in Zellner (1977).

For the traditional multivariate regression model the $m \times 1$ vector y is related to a $k \times 1$ vector of given input variables, x, as follows, $y' = x'B + u'$ where B is a $k \times m$ matrix of regression coefficients and u' is a $1 \times m$ zero mean vector of errors with pds covariance matrix, Σ. For u with a normal or general elliptically symmetric density, $f(u \mid \Sigma) = |\Sigma|^{-1/2} f(u'\Sigma^{-1}u)$, the MDIP for B and Σ is

$$\pi^*(B, \Sigma) \propto |\Sigma|^{-1/2} \tag{3.23}$$

In (3.23) the elements of B and Σ are independently distributed with the former uniformly distributed and the latter in a form, $|\Sigma|^{-1/2}$, discussed above. Similarly for the seemingly unrelated regression model, $y_i = x'_i\beta_i + u_i$, $i = 1, 2, \ldots, m$, with $u' = (u_1, u_2, \ldots u_m)$ a $1 \times m$ vector of error terms, x'_i a $1 \times k_i$ vector of input variables and β_i a $k_i \times 1$ vector of regression coefficients, if u is MVN$(0, \Sigma)$ or elliptically symmetrically distributed, the proper MDIP for the β_i's and Σ is $\pi^*(\beta, \Sigma) \propto |\Sigma|^{-1/2}$ with $\boldsymbol{\beta}' = (\beta'_1, \ldots, \beta'_m)$. Of course other relevant side conditions can be introduced in deriving MDIPs for this model's parameters.

Having considered a range of multivariate densities and models, attention will be directed at a consideration of MDIP odds for alternative hypotheses and models in the next section.

IV. Maximal data information prior odds

In Zellner (1993), it was briefly indicated how the MDIP approach can be utilized to derive optimal prior odds relating to alternative hypotheses or models. Below, this former analysis is considered further and applied in analyses of several problems. To illustrate the approach, consider two possible densities for an observation y, scalar or vector, $f_1(y|\theta_1)$ and $f_2(y|\theta_2)$ and let $\pi_1(\theta_1)$ and $\pi_2(\theta_2)$ be prior densities, say MDIPs for θ_1 and θ_2. The steps needed to obtain prior odds relating to $f_1(y|\theta_1)$ vs $f_2(y|\theta_2)$ are as follows:

1. Compute the marginal densities of the observations

$$h_i(y) = \int f_i(y|\theta_i)\pi(\theta_i)d\theta_i \qquad i = 1, 2 \tag{4.1}$$

2. Compute the information in $h_1(y)$ and $h_2(y)$

$$I_i = \int h_i(y)\ell n h_i(y)dy \qquad i = 1, 2 \tag{4.2}$$

the negative entropy of $h_i(y)$ relative to uniform measure

3. Form the criterion functional,

$$G = \sum_{i=1}^{2} P_i I_i - \sum_{i=1}^{2} P_i \ell n P_i \tag{4.3}$$

where P_i = prior probability associated with the model $f_i(y|\theta_i)$, $i = 1, 2$ and $P_1 + P_2 = 1$, that is the models are considered to be exhaustive – see below for analysis of the case in which the entertained models do not constitute an exhaustive set.

Note that the criterion functional in (4.3) is the average information in the data densities, $\sum_{i=1}^{2} P_i I_i$, with I_i given in (4.2), minus the information in the prior probability mass function, $\sum_{i=1}^{2} P_i \ell n P_i$. Also, it is possible to express (4.3) as

$$G(P_1^*) - G(P_1) = P_1 \ell n P_1/P_1^* + (1 - P_1)\ell n(1 - P_1)/(1 - P_1^*) \tag{4.4}$$

where P_1^* is such that $\ell n P_1^*/(1 - P_1^*) = I_1 - I_2$. The expression on the right side of (4.4) is just the cross-entropy for the two probability mass functions, $\{P_1, 1 - P_1\}$ and $(P_1^*, 1 - P_1^*\}$. From the fact that the right side of (4.4) is non-negative, taking $P_1 = P_1^*$ with P_1^* such that $\ell n P_1^*/(1 - P_1^*) = I_1 - I_2$ provides a maximal value for $G(P_1)$, namely $G(P_1^*)$. Alternatively, (4.3) can be directly maximized by choice of P_1 subject to $P_1 + P_2 = 1$, to yield the following condition for a maximum of G, given in (4.3):

$$\ell n[P_1^*/(1-P_1^*)] = I_1 - I_2 \tag{4.5a}$$

with I_1 and I_2 given in (4.1) and from which

$$P_1^* = \exp\{I_1 - I_2\}/[1 + \exp\{I_1 - I_2\}] \tag{4.5b}$$

From (4.5), it is seen that:

(a) When $I_1 = I_2$, $P_1^*/(1-P_1^*) = 1$ and $P_1^* = 1/2$
(b) When $I_1 > I_2$, $P_1^*/(1-P_1^*) > 1$ and $P_1^* > 1/2$
(c) When $I_1 < I_2$, $P_1^*/(1-P_1^*) < 1$ and $P_1^* < 1/2$

The above analysis generalizes to the case of m mutually exclusive and exhaustive models. In place of (4.1), the criterion functional is

$$G = \sum_{i=1}^{m} P_i I_i - \sum_{i=1}^{m} P_i \ell n P_i \tag{4.6}$$

with $\sum_{i=1}^{m} P_i = 1$ and the values of the P_i that maximize G, subject to $\sum_{i=1}^{m} P_i = 1$, are given by

$$\ell n(P_1^*/P_j) = I_i - I_j \qquad i \neq j,\ i, j = 1, 2, \ldots, m-1 \tag{4.7}$$

To illustrate the above analysis, let $f_1(y \mid \theta_1)$ be $N(0, \sigma_1^2)$ and $f_2(y \mid \theta_2)$ be $N(\theta_2, \sigma_2^2)$ with σ_1^2 and σ_2^2 known, and assume that the prior for the mean θ_2 is $N(\mu, \tau^2)$ where μ and τ^2 are the given prior mean and variance of θ_2, respectively. Then by direct calculation, $I_1 = -(\ell n 2\pi + \sigma_1^2 + 1)/2$ and $I_2 = -[\ell n 2\pi + 1 + \ell n(\sigma_2^2 + \tau^2)]/2$ and

$$P_1^*/(1-P_1^*) = e^{I_1 - I_2} = [\tau^2/\sigma_1^2 + \sigma_2^2/\sigma_1^2]^{1/2} \tag{4.8}$$

which is the MDIP prior odds relating to the data densities $N(0, \sigma_1^2)$ and $N(\theta_2, \sigma_2^2)$ with the prior for θ_2 being $N(\mu, \tau^2)$. It is seen that when $\tau^2 \to 0$, the odds approaches 1 if $\sigma_1^2 = \sigma_2^2$. As τ^2 increases in value the prior odds grows. Thus rather than a 1:1 prior odds, the MDIP odds' value can be 1:1 under special circumstances but is not in general since the value reflects relative information contained in alternative models.

If two models considered are mutually exclusive but not exhaustive, let P_0 = probability assigned to all models not explicitly considered and assume $P_1 + P_2 + P_0 = 1$, where, as above P_1 and P_2 are probabilities associated with the explicitly given models. Then for any given P_0, it is possible to obtain values of P_1 and P_2 that maximize the criterion functional $G_0 = P_0 I_0 + P_1 I_1 + P_2 I_2 - P_1 \ell n P_1 - P_2 \ell n P_2 - P_0 \ell n P_0$, where I_0 denotes the information in models not explicitly given. The resulting MDIP odds is given by

$$\ell n[P_1^*/(1 - P_1^* - P_0)] = \ell n(P_1^*/P_2^*) = I_1 - I_2 \tag{4.9}$$

for any P_0, $0 \leq P_0 < 1$. From (4.9), it is the case that introducing P_0 does not affect the value of the odds on the alternative hypotheses relative to the exhaustive case, $P_0 = 0$. However the values of the optimal prior probabilities are affected; e.g., $P_1^* = (1 - P_0)\exp\Delta/(1 + \exp\Delta)$, where $\Delta \equiv I_1 - I_2$. Thus if $\Delta = 0$, the optimal odds value is one and $P_1^* = (1 - P_0)/2$. For $P_0 = 1/2$, $P_1^* = 1/4$ whereas when $P_0 = 0$, the exhaustive case, $P_1^* = 1/2$. Similar results are obtained in the case of m alternative models.

As a second example, let y have either an $N(0, \sigma_1^2)$ density or an $N(0, \sigma_2^2)$ density and the priors for the variances be $\pi(\sigma_i^2) = (1/a_i^2)\exp\{-\sigma_i^2/a_i^2\}$ with $0 < \sigma_i^2 < \infty$, $E\sigma_i^2 = a_i^2$ and $\text{Var}(\sigma_i^2) = a_i^4$, $i = 1, 2$. Then by direct integration, $h_i(y) = (1/\sqrt{2a_i^2})\exp\{-\sqrt{2}\,|y|/a_i\}$, $i = 1, 2$, a Laplace or double exponential density. Then $Ii = -\ell n a_i - 1/2\ell n 2 - 1$ and

$$\ell n \frac{P_1^*}{1 - P_1^*} = I_1 - I_2 = \ell n a_2 / a_1 \tag{4.10}$$

or

$$\frac{P_1^*}{1 - P_1^*} = a_2 / a_1 \tag{4.11}$$

Thus when $a_2/a_1 \gtreqless 1$, the MDIP odds, $P_1^*/(1 - P_1^*) \gtreqless 1$.

The few examples presented above illustrate the procedures that produce MDIP odds. While it has not been done above, additional side conditions can be added to the optimization problem to yield optimal restricted prior odds.

V. Summary and conclusions

Many properties of MDIP priors and odds and procedures for producing them have been presented and illustrated with central important examples. The procedures involve optimization of criterion functionals subject to side conditions and are relatively easy to apply. No asymptotics or asymptotic approximations are involved. Several comparisons with results provided by alternative procedures for producing priors indicate that MDIPs are relatively easy to produce, have reasonable properties and hopefully will be helpful to researchers and decision-makers in formulating priors and odds.

Note

1. Note if we parameterize the model as follows, $y_t = \beta_1 + \beta_2 y_{t-1} + u_t$, $y_t = \beta_1/(1-\beta_2) + \Sigma_{i=0}^{\infty} \beta_2^i u_{t-i}$. Given that $|\beta_2| < 1$, $Ey_t = \beta_1/(1-\beta_2) \equiv \theta$, the mean of the process and $\beta \equiv \beta_2$.

References

Akaike, H. (1978), 'A New Look at the Bayes Procedure', *Biometrika*, **65** (1), 53–9.

Berger, J. O. (1985), *Statistical Decision Theory and Bayesian Analysis* 2nd edn, New York: Springer-Verlag.

Berger, J. O. and J. M. Bernardo (1992), 'On the Development of Reference Priors', in J. M. Bernardo et al. (eds), *Bayesian Statistics 4*, Oxford University Press, pp. 35–49.

Bernardo, J. M. and A. F. M. Smith (1994), *Bayesian Theory*, New York: John Wiley.

Box, G. E. P. and G. C. Tiao (1973), *Bayesian Inference in Statistical Analysis*, Reading, MA: Addison-Wesley.

Cover, T. M. and J. A. Thomas (1991), *Elements of Information Theory*, New York: John Wiley.

Ebrahimi, N. and E. S. Soofi (1994), 'On the Equivalence of Entropy and Variance Orderings', ms, Division of Statistics, Northern Illinois University and School of Business Administration, University of Wisconsin, Milwaukee.

Good, I. J. (1956), 'Some Terminology and Notation in Information Theory', *Proc. of the Institute of Electrical Engineers*, C, 103, pp. 200–204.

Good, I. J. (1963), 'Maximum Entropy for Hypothesis Formulation, Especially for Multidimensional Contingency Tables', *Annals of Mathematical Statistics*, **34** (3), 911–34.

Jaynes, E. T. (1982a), *Papers on Probability, Statistics and Statistical Physics*, Dordrecht, Netherlands: Reidel.

Jaynes, E. T. (1982b), 'On the Rationale of Maximum-Entropy Methods', *Proc. of the IEEE*, 70, pp. 939–52.

Jeffreys, H. (1967), *Theory of Probability* (3rd rev. ed.), London: Oxford University Press.

Kass, R. E. and L. Wasserman (1995), 'The Selection of Prior Distributions by Formal Rules', Tech. Report no. 583, Dept. of Statistics, Carnegie Mellon University in *Journal of the American Statistical Association*, **91**, 1996, 1343–70.

Lindley, D. V. (1956), 'On a Measure of Information Provided by an Experiment', *Annals of Mathematical Statistics*, **27**, 986–1005.

Lindley, D. V. (1961), 'The Use of Prior Probability Distributions in Statistical Inference and Decisions', in J. Neyman (ed.), *Proc. Fourth Berkeley Symp. Math. Statist. and Probab.*, vol. I, 453–68.

Lindley, D. V. (1965), *Introduction to Probability and Statistics from a Bayesian Viewpoint, Part 2 Inference*, Cambridge: Cambridge University Press.

McCulloch, R. E. (1992), 'Comment,' in J. Bernardo et al. (eds), *Bayesian Statistics 4*, Oxford University Press, pp. 49-51.

Moulton, B. R. (1993), 'Bayesian Analysis of Some Generalized Error Distributions for the Linear Model', US Bureau of Labor Statistics, Washington, DC, unpublished ms.

Phillips, P. C. B. (1983), 'Exact Small Sample Theory in the Simultaneous Equations Model', in Z. Griliches and M. D. Intriligator (eds), *Handbook of Econometrics*, vol. 1, ch. 8, Amsterdam: North-Holland, 449–516.

Rao, C. R. (1987), 'Differential Metrics in Probability Spaces,' in *Differential Geometry in Statistical Inference*, vol. 10, ch. 5, Institute of Mathematical Statistics Lecture Notes – Monograph Series, Institute of Mathematical Statistics, Hayward, CA, pp. 217–40.

Rodríguez, C. C. (1991a), 'Entropic Priors', ms, Dept. of Mathematics and Statistics, State University of New York, Albany, NY.

Rodríguez, C. C. (1991b), 'From Euclid to Entropy', in W. T. Grandy and L. H. Schick, *Maximum Entropy and Bayesian Methods*, Dordrecht, Netherlands: Kluwer, pp. 343–8.

Shannon, C. E. (1948), 'The Mathematical Theory of Communication', *Bell System Technical Journal*, July–October, reprinted in C. E. Shannon and W. Weaver, *The Mathematical Theory of Communication*, University of Illinois Press, 1949, pp. 3–91.

Sinha, S. K. and A. Zellner (1990), 'A Note on the Prior Distributions of Weibull Parameters', *SCIMA*, **19** (1–2), 5–13.

Soofi, E. S. (1992), 'Information Theory and Bayesian Statistics', ms, School of Business Administration, University of Wisconsin-Milwaukee, published in D. Berry, K. M. Chaloner and J. K. Geweke (eds), *Bayesian Analysis in Statistics and Econometrics: Essays in Honor of Arnold Zellner*, New York: John Wiley, 1996, pp. 179–89.

Soofi, E. S. (1994), 'Capturing the Intangible Concept of Information', *Journal of the American Statistical Association*, **89**, (428), 1243–54.

Stein, C. M. (1962), 'Confidence Sets for the Mean of a Multivariate Normal Distribution', *Journal of the Royal Statistical Society*, Series B, **24**, 165–285.

Venegas-Martinez, F. (1990), 'On Regularity and Optimality Conditions for Maximum Entropy Priors', *The Brazilian Journal of Probability and Statistics*, **4**, 105–36.

Verdugo-Lazo, A. C. G. and P. N. Rathie (1978), 'On the Entropy of Continuous Probability Distributions', *IEEE Transactions on Information Theory*, IT-24(1), 120–22.

Zellner, A. (1971), *An Introduction to Bayesian Inference in Econometrics*, New York: Wiley, reprinted in 1987 by Krieger Publishing, Malabar, Florida and in 1996 by Wiley.

Zellner, A. (1977), 'Maximal Data Information Prior Distributions', in A. Aykac and C. Brumat (eds), *New Developments in the Applications of Bayesian Methods*, Amsterdam: North- Holland, pp. 211–32, reprinted in A. Zellner, *Basic Issues in Econometrics*, Chicago: University of Chicago Press, 1984.

Zellner, A. (1991), 'Bayesian Methods and Entropy in Economics and Econometrics', in W. T. Grandy and L. H. Schick (eds), *Maximum Entropy and Bayesian Methods*, Dordrecht, Netherlands: Kluwer, pp. 17–31.

Zellner, A. (1993), 'Models, Prior Information and Bayesian Analysis', presented at Conference on Informational Aspects of Bayesian Statistics in Honor of H. Akaike, December 1993, and to appear in L. Bauwens, W. Polasek, H. K. van Dijk (eds), special issue of *Journal of Econometrics*, November 1996, 51–68; Chapter 8 in this volume.

Zellner, A. and R. A. Highfield (1988), 'Calculation of Maximum Entropy Distributions and Approximation of Marginal Posterior Distributions', *Journal of Econometrics*, **37**, 195–210.

Zellner, A. and C-K. Min (1993), 'Bayesian Analysis, Model Selection and Prediction', invited paper presented at a Symposium in Honor of E. T. Jaynes, 15–16 May 1992 and in W. T. Grandy and P. W. Milonni (eds), *Physics and Probability: Essays in Honor of Edwin T. Jaynes*, Cambridge: Cambridge University Press, 195–206.

Zhang, Y. T. and Q. W. Yao (1991), 'Some Maximal Information and Generalized Maximal Entropy Priors', ms, Dept. of Statistics, Wuhan University, China, published in *Chinese Journal of Applied Probability and Statistics*, **7**, 192–200.

[10]

On the information provided by an experiment and maximal data information priors(MDIPs)*

Let us assume a given density function for the observation vector $y, f(y \mid \theta)$ and a given prior density for θ, $\pi(\theta)$. Then $p(y, \theta) = \pi(\theta) f(y \mid \theta)$ is the joint density of y and θ that reflects information in $\pi(\theta)$ and in $f(y \mid \theta)$. We measure the information in $p(y, \theta)$ relative to uniform measure by

$$I(p) = \iint p(y, \theta)\, \ell n p(y, \theta) dy d\theta \tag{1}$$

This is just the expectation of the ℓn-height of $p(y, \theta)$, $E\ell n\, p(y, \theta)$. The larger on average $\ell n\, p(y, \theta)$, the more informative is $p(y, \theta)$. Also $I(p)$ is the negative entropy associated with $p(y, \theta)$ or a divergence or cross-entropy distance measure for $p(y, \theta)$ relative to uniform measure. See, e.g. Jaynes (1982a, b) and Soofi (1994, 1996) for an extensive discussion of information measures such as that in (1).

Using $p(y, \theta) = \pi(\theta) f(y \mid \theta)$, it has been pointed out in Zellner (1991, 1993) that (1) can be expressed as

$$I(p) = \int \pi(\theta) I(f) d\theta + I(\pi) \tag{2a}$$

where $I(\pi) = \int \pi(\theta) \ell n \pi(\theta) d\theta$, the information in $\pi(\theta)$ and

$$I(f) = \int f(y \mid \theta) \ell n f(y \mid \theta) dy \tag{2b}$$

the information in $f(y \mid \theta)$. Note from (2) that the information in $p(y, \theta)$ breaks up into two parts, the first the average information in $f(y \mid \theta)$, $\int \pi(\theta) I(f) d\theta$ and the second the information in the prior $\pi(\theta)$, $\int \pi(\theta) \ell n\, \pi(\theta) d\theta$.

From (2) for any given choice of $\pi(\theta)$ and with a given $f(y \mid \theta)$, the *total* information provided by an experiment[1] over and above that in the prior is from (2a):

$$I(\text{Exp}) = I(p) - I(\pi) = \int \pi(\theta) I(f) d\theta \tag{3}$$

For given $f(y \mid \theta)$, I(Exp) depends on the form of the prior density, $\pi(\theta)$ and thus I(Exp)'s value cannot be determined until the form of $\pi(\theta)$ is determined. One principle for choosing the form of $\pi(\theta)$, utilized in Moulton (1993), Sinha and Zellner (1990), Soofi (1994, 1996), Venegas-Martinez (1990), Zellner (1971, 1977, 1991, 1993, 1995), and Zhang and Yao (1991) is to choose $\pi(\theta)$ to maximize the difference of the

* Research for this paper was financed in part by the National Science Foundation and by income from the H.G.B. Alexander Endowment Fund, Graduate School of Business, University of Chicago.

two terms on the right side of (2), that is the difference between the average information in the data density, $\int \pi(\theta)I(f)d\theta$, minus the information in the prior, $I(\pi)$, namely:

$$G(\pi) = \int \pi(\theta)I(f)d\theta - I(\pi) \tag{4}$$

On maximizing $G(\pi)$ subject to just the condition $\int \pi(\theta)d\theta = 1$, with $\theta \subset \Theta$, a compact parameter space, the solution is a maximal data information prior (MDIP):

$$\pi^*(\theta) = c \exp\{I(f)\} \tag{5}$$

where c is a normalizing and $I(f)$ is given in (2b). See Table 1 for some MDIP priors yielded by (5) and Kass and Wasserman (1995), Moulton (1993), Sinha and Zellner (1990), Soofi (1994, 1996), Venegas-Martinez (1990), Zellner (1971, 1977, 1991, 1993, 1995), and Zhang and Yao (1991) for others. These papers include many comparisons of MDIPs to priors produced by other procedures, e.g. those of Jeffreys, Box and Tiao, Lindley, Bernardo and Berger, and Rodríguez.

Then the information provided by an experiment when $\pi^*(\theta)$ in (5) is the prior, denoted by $I(\text{Exp}^*)$, is:

$$I(\text{Exp}^*) = \int \pi^*(\theta)I(f)d\theta \tag{6}$$

with $I(f) = \int f(y \mid \theta)\ell n\, f(y \mid \theta)dy$, the negative entropy of $f(y \mid \theta)$. Further, from (4) and (5),

$$G(\pi^*) - G(\pi) = \int \pi(\theta)\ell n\, \pi(\theta)/\pi^*(\theta)d\theta \tag{7}$$

which is just the cross entropy of $\pi(\theta)$ relative to the MDIP, $\pi^*(\theta)$. Since the expression in (7) is non-negative, taking $\pi(\theta) = \pi^*(\theta)$, given in (5) provides a maximal value for $G(\pi)$.

Note that for any prior in the form $\pi(\theta) = \phi(\theta)\pi^*(\theta)$, with $\pi^*(\theta)$ given in (5), we have from (7),

$$\begin{aligned} G(\pi^*) - G(\pi) &= \int \pi^*(\theta)\phi(\theta)\ell n\, \phi(\theta)d\theta \\ &= E_{\pi^*}[\phi(\theta)\, \ell n\, \phi(\theta)] > 0 \end{aligned} \tag{8}$$

Thus $G(\pi)$ is deflated by the amount in (8) when the prior $\phi(\theta)\pi^*(\theta)$ rather than $\pi^*(\theta)$ is employed. To maximize $G(\pi)$ with respect to choice of $\pi(\theta)$ results in having the average information provided an experiment, $I(\text{Exp}) = \int \pi(\theta)I(f)d\theta$ be as large as possible relative to the information in the prior density, $I(\pi)$.

Finally on using $p(y, \theta) = g(\theta \mid y)h(y)$, from (4) we have

$$G(\pi) = \int\int p(\theta, y)\ell n[\ell(\theta \mid y)/\pi(\theta)]d\theta dy \tag{9}$$

where $\ell(\theta \mid y) \equiv f(y \mid \theta)$ and thus choice of $\pi(\theta)$ to maximize $G(\pi)$ makes the expectation of the ℓn ratio of the likelihood function to the prior density as large as

Table 1 Explicit examples of MDIPs

Data Pdf	Data Pdf information measure, $I(f)$[a]	MDIP Pdf $\pi_* \propto \exp\{I(f)\}$
1. Normal $(-\infty < y < \infty)$		
1a $f(y \mid \theta) = (2\pi)^{-1/2}\exp\{-(y-\theta)^2/2\}$	const.	$\pi_*(\theta) \propto$ const.
1b $f(y \mid \sigma) = (2\pi\sigma^2)^{-1/2}\exp\{-(y-\theta)^2/2\sigma^2\}$	const.$-\ell n\sigma$	$\pi_*(\sigma) \propto 1/\sigma$
1c $f(y \mid \theta, \sigma) = (2\pi\sigma^2)^{-1/2}\exp\{-(y-\theta)^2/2\sigma^2\}$	const.$-\ell n\sigma$	$\pi_*(\theta, \sigma) \propto 1/\sigma$
2. Rectangular		
$f(y \mid \theta) = 1/\theta,\ 0 < y < \theta$	const.$-\ell n\theta$	$\pi_*(\theta) \propto 1/\theta$
3. Exponential (Pareto; $0 < y < \infty$)		
$f(y \mid \theta) = \theta^{-1}\exp\{-y/\theta\}$	const.$-\ell n\theta$	$\pi_*(\theta) \propto 1/\theta$
4. Student-t ($v = 1$, Cauchy; $-\infty < y < \infty$)[b]		
$f(y \mid \theta, \sigma) = (k/\sigma)\{\nu + (y-\theta)^2/\sigma^2\}^{-((v+1)/2)}$	const.$-\ell n\sigma$	$\pi_*(\theta, \sigma) \propto 1/\sigma$
5. Log-normal		
$f(y \mid \theta, \sigma) = (2\pi\sigma^2 y^2)^{-1/2}\exp\{-(\ell ny - \ell n\theta)^2/2\sigma^2\}$	const.$-\ell n\sigma - \ell n\theta$	$\pi_*(\theta, \sigma) \propto 1/\theta$
6. General forms		
6a $f(y \mid \theta) = f(y-\theta)$	const.	$\pi_*(\theta) \propto$ const.
6b $f(y \mid \sigma) = \sigma^{-1}f(y/\sigma)$	const.$-\ell n\sigma$	$\pi_*(\sigma) \propto 1/\sigma$
6c $f(y \mid \theta, \sigma) = 1/\sigma f(y-\theta/\sigma)$	const.$-\ell n\sigma$	$\pi_*(\theta, \sigma) \propto 1/\sigma$
7. Exponential class		
$f(y \mid \theta) = \exp\{h(\theta)K(y) + S(y) + q(\theta)\}$	$h(\theta)\bar{K} + \bar{S} + q(\theta)$[c]	$\pi_*(\theta) \propto \exp\{h(\theta)\bar{K} + \bar{S} + q(\theta)\}$

[a] $I(f) = \int f(y \mid \theta)\ \ell nf(y \mid \theta)\, dy$
[b] Here v is assumed to have a known value; see Moulton (1993) and Soofi (1996) for MDIPs for v.
[c] $\bar{K} \equiv \int K(y)f(y \mid \theta)dy$ and $\bar{S} \equiv \int S(y)f(y \mid \theta)dy$

possible subject to the constraints imposed, e.g. that $\pi(\theta)$ be proper and possible additional constraints on the moments of θ.

Above, the data density $f(y \mid \theta)$ was taken as given. In design problems, a data density must be selected that is satisfactory in the sense that it reflects given available information but does not incorporate additional, unfounded information. From (3), for any given prior $\pi(\theta)$, the information provided by an experiment is

$$\begin{aligned} I(\text{Exp}) &= \int \pi(\theta) I(f) d\theta \\ &= \int \left[\int f(y \mid \theta) \ell n\, f(y \mid \theta) dy\right] \pi(\theta) d\theta \end{aligned} \tag{10}$$

It is possible to choose the form of $f(y \mid \theta)$ in (10) so that it incorporates the information in side conditions, e.g. $\int y^i f(y \mid \theta) dy = \mu_i(\theta)$, $i = 0, 1, 2, \ldots, m$, given moments, where $\mu_i(\theta)$ is a given function of the elements of θ in a 'most conservative' fashion by minimizing (10) subject to the side conditions. Thus the optimal $f(y \mid \theta)$, say $f^*(y \mid \theta)$, incorporates the given information as conservatively as possible. See Jaynes (1982a, b), Verdugo Lazo and Rathie (1978), Zellner (1993), Zellner and Highfield (1988), and Zellner and Min (1993) for examples. It is also a 'maxent' solution since minimizing (10) is equivalent to maximizing the negative of (10). For example, if $Ey = \theta_1$ and var $y = \theta_2$, the solution, $f^*(y \mid \theta_1, \theta_2)$ is a normal density, as is well known.

With the form of $f(y \mid \theta)$ determined as explained above, namely $f^*(y \mid \theta)$, the MDIP prior relative to this density is $\pi^*(\theta) \propto \exp\{I(f^*)\}$ when just the side condition that $\pi(\theta)$ be proper is employed. Finally, with these elements available, the posterior density for θ, $g^*(\theta \mid y)$ is given by

$$g^*(\theta \mid y) \propto \pi^*(\theta) f^*(y \mid \theta) \tag{11}$$

Also, with $f^*(y_f \mid \theta)$, the density for the future observations y_f available, the predictive density is given by $q(y_f \mid y) = \int f^*(y_f \mid \theta) g^*(\theta \mid y) d\theta$. Note that $\pi^*(\theta)$ can be 'diffuse' or 'informative' and that $f^*(y_f \mid \theta)$ reflects what is known about various aspects of a given experiment.

Note

1. See Soofi (1994, 1996) for a review of earlier attempts to define the information provided by an experiment.

References

Jaynes, E. T. (1982a), *Papers on Probability, Statistics and Statistical Physics*, Dordrecht, Netherlands: Reidel.

Jaynes, E. T. (1982b), 'On the Rationale of Maximum-Entropy Methods', *Proc. of the IEEE*, **70**, 939–52.

Kass, R. E. and L. Wasserman (1995), 'The Selection of Prior Distributions by Formal Rules', Tech. Report no. 583, Dept. of Statistics, Carnegie Mellon University.

Moulton, B. R (1993), 'Bayesian Analysis of Some Generalized Error Distributions of the Linear Model', US Bureau of Labor Statistics, Washington, DC, unpublished ms.

Sinha, S. K and A. Zellner (1990), 'A Note on the Prior Distributions of Weibull Parameters', *SCIMA*, **19** (1–2), 5–13.

Soofi, E. (1994), 'Capturing the Intangible Concept of Information', *Journal of the American Statistical Association*, **89**, 1243–54.

Soofi, E. (1996), 'Information Theory and Bayesian Statistics', in D. A. Berry, K M. Chaloner and J. K Geweke (eds), *Bayesian Analysis in Statistics and Econometrics: Essays in Honor of Arnold Zellner*, New York: John Wiley, pp. 179–89.

Venegas-Martinez, F. (1990), 'On Regularity and Optimality Conditions for Maximum Entropy Priors', *The Brazilian Journal of Probability and Statistics*, **4**, 105–36.

Verdugo Lazo, A. C. G. and P. N. Rathie (1978), 'On the Entropy of Continuous Probability Distributions', *IEEE Transactions on Information Theory*, IT-24(1), 120–22 and in 1996 by Wiley.

Zellner, A. (1971), *An Introduction to Bayesian Inference in Econometrics*, New York: John Wiley, reprinted in 1987 by Krieger Publishing, Malabar, Florida.

Zellner, A. (1977), 'Maximal Data Information Prior Distributions', in A. Aykac and C. Brumat (eds), *New Developments in the Applications of Bayesian Methods*, Amsterdam: North-Holland, pp. 211–32, reprinted in A. Zellner, *Basic Issues in Econometrics*, Chicago: University of Chicago Press, 1984.

Zellner, A. (1991), 'Bayesian Methods and Entropy in Economics and Econometrics', in W. T. Grandy and L. H. Schick (eds), *Maximum Entropy and Bayesian Methods*, Dordrecht, Netherlands: Kluwer, pp. 17–31.

Zellner, A. (1993), 'Models, Prior Information and Bayesian Analysis', presented at Conference on Information Aspects of Bayesian Statistics in Honor of H. Akaike, December 1993, to appear in L. Bauwens, W. Polasek, H. K. van Dijk (eds), special issue of *Journal of Econometrics*, November, 1996, 51–68; and Chapter 8 in this volume.

Zellner, A. (1995), 'Past and Recent Results on Maximal Data Information Priors', presented to ASA meeting; Chapter 10 in this volume.

Zellner, A. and R A. Highfield (1988), 'Calculation of Maximum Entropy Distributions and Approximation of Marginal Posterior Distributions', *Journal of Econometrics*, **37**, 195–210.

Zellner, A. and C-K. Min (1993), 'Bayesian Analysis, Model Selection and Prediction', invited paper presented at a Symposium in Honor of E. T. Jaynes, 15–16 May 1992 and in W. T. Grandy and P. W. Milonni (eds), *Physics and Probability: Essays in Honor of Edwin T. Jaynes*, Cambridge: Cambridge University Press, pp. 195–206.

Zhang, Y. T. and Q. W. Yao (1991), 'Some Maximal Information and Generalized Maximal Entropy Priors', ms, Dept. of Statistics, Wuhan University, China, published in *Chinese Journal of Applied Probability and Statistics*, **7**, 192–200.

[11]

Optimal Information Processing and Bayes's Theorem

ARNOLD ZELLNER*

In this article statistical inference is viewed as information processing involving input information and output information. After introducing information measures for the input and output information, an information criterion functional is formulated and optimized to obtain an optimal information processing rule (IPR). For the particular information measures and criterion functional adopted, it is shown that Bayes's theorem is the optimal IPR. This optimal IPR is shown to be 100% efficient in the sense that its use leads to the output information being exactly equal to the given input information. Also, the analysis links Bayes's theorem to maximum-entropy considerations.

KEY WORDS: Information theory; Maximum entropy; Statistical inference.

1. INTRODUCTION

Bayes's theorem has been widely used as an inductive learning model to transform prior and sample information into posterior information in econometrics, statistics, physics, and other sciences. Although some works have been devoted to rationalizing Bayes's theorem as a coherent learning model (e.g., Cox 1961; Jaynes 1974, 1983, 1984; Jeffreys 1967, 1973), it does not appear that Bayes's rule has been derived as an optimal information processing rule. In what follows, this problem is addressed after describing some needed information measures and a criterion functional. This criterion functional is minimized in a calculus-of-variations approach to yield an operational, optimal information processing rule that is surprisingly identical to Bayes's rule. Further, Bayes's rule is informationally efficient in a sense defined in Section 2.

The plan of the article is as follows. In Section 2 needed concepts are introduced, the problem is explained, and its solution is presented. The derivation of the solution is given in Section 3. Some concluding remarks are provided in Section 4.

2. INFORMATION CONCEPTS AND AN OPTIMAL INFORMATION PROCESSING RULE

Let y denote the given data, and let $l(\theta|y)$ denote a likelihood function for θ—a parameter, scalar, or vector contained in the parameter space Θ. Further, let

$\pi_a(\theta|I) \equiv$ given prior or antedata probability density function (pdf) for $\theta \subset \Theta$, based on prior information I,

$\pi_a(\theta|I) \equiv$ given prior or antedata probability density function (pdf) for $\theta \subset \Theta$, based on prior information I,

$\pi_p(\theta|D) \equiv$ postdata pdf for $\theta \subset \Theta$, where $D = (y, I)$, the given sample, y, and prior information, I,

$p(y|I) \equiv$ pdf for y, given by

$$p(y|I) \equiv \int_\theta \pi_a(\theta|I) f(y|\theta) d\theta, \tag{2.1}$$

where $f(y|\theta)$ is the pdf for y given θ. [The term *postdata pdf* is employed instead of *posterior pdf* to emphasize that the optimal form of $\pi_p(\theta|D)$ is to be derived.]

Note that (2.1) is a definition that does not involve the assumption that the product rule of probability theory necessarily holds. See Jeffreys (1967, pp. 25, 52) for a discussion of assumptions needed for the product rule to be valid.

The inputs and outputs of any information processing rule (IPR) are depicted graphically in Figure 1, where $l(\theta|y)$, the likelihood function, is $f(y|\theta)$ viewed as a function of θ. Thus the information in the likelihood function $l(\theta|y)$ and the prior pdf $\pi_a(\theta|I)$ enter the IPR, whose output is the information in the postdata pdf $\pi_p(\theta|D)$ and the pdf $p(y|I)$. Different IPR's will produce different output information from the given input information. Some IPR's may be *inefficient* in the sense that the output information, measured in a suitable metric, is *less* than the input information. On the other hand, some IPR's may add *extraneous* information so that the output information is greater than the given input information, an undesirable state of affairs. A good, efficient IPR will satisfy the following principle.

Information Conservation Principle (ICP). Input information = Output information.

An IPR that satisfies the ICP is 100% efficient in the sense that the ratio of output to input information is equal to 1. An inefficient IPR has efficiency less than 100%. An IPR that *adds* extraneous information is considered unsatisfactory.

To implement these concepts, there is a need to measure the information in the input and output pdf's. (For discussion of information measures, see Kullback 1959.) The following postdata measures will be employed:

$$\text{Information in } l(\theta|y) \equiv \int_\theta \pi_p(\theta|D) \log l(\theta|y) d\theta. \tag{2.2}$$

$$\text{Information in } \pi_a(\theta|I) \equiv \int_\theta \pi_p(\theta|D) \log \pi_a(\theta|I) d\theta. \tag{2.3}$$

$$\text{Information in } \pi_p(\theta|D) \equiv \int_\theta \pi_p(\theta|D) \log \pi_p(\theta|D) d\theta. \tag{2.4}$$

$$\text{Information in } p(y|I) \equiv \int_\theta \pi_p(\theta|D) \log p(y|I) d\theta = \log p(y|I). \tag{2.5}$$

In each case, information is given as an average of a log pdf with $\pi_p(\theta|D)$ used as a "weight function."

To illustrate some of these measures, suppose $l(\theta|y) = (2\pi/n)^{-1/2} \times \exp\{-n(\bar{y} - \theta)^2/2\}$, where $\bar{y}$ = sample mean. Then $\log l(\theta|y) = -1/2 \log 2\pi/n - n(\bar{y} - \theta)^2/2$, and (2.2) yields $-1/2 \log 2\pi/n - n[\text{var}(\theta|D) + (\bar{y} - \bar{\theta})^2]/2$, where $\bar{\theta}$ and $\text{var}(\theta|D)$ are the mean and variance of

*Arnold Zellner is Professor of Economics and Statistics, Graduate School of Business, University of Chicago, Chicago, Illinois 60637. The research for this article was financed in part by the National Science Foundation and the H. G. B. Alexander Endowment Fund of the University of Chicago's Graduate School of Business.

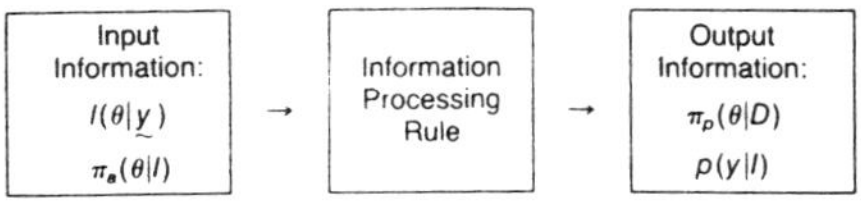

Figure 1. Inputs and Outputs of Any Information Processing Rule.

$\pi_p(\theta|D)$, respectively. The larger var$(\theta|D)$ and $(\bar{y} - \bar{\theta})^2$ are, the smaller the information in $l(\theta|y)$ is, which is reasonable. Further, if $\pi_a(\theta|I) = (2\pi)^{-1/2}\exp\{-(\theta - m)^2/2\}$, where m is a given prior mean, $\log \pi_a(\theta|I) = -1/2 \log 2\pi - (\theta - m)^2/2$ and (2.3) yields $-1/2 \log 2\pi - [\text{var}(\theta|D) + (m - \bar{\theta})^2]/2$; thus the larger var$(\theta|D)$ and $(m - \bar{\theta})^2$ are, the smaller the information in the pdf $\pi_a(\theta|I)$ is. As regards (2.4), if $\pi_p(\theta|D) = (2\pi)^{-1/2} \times \exp\{-(\theta - \bar{\theta})^2/2\}$, the information in this pdf is $-1/2 \log 2\pi - \text{var}(\theta|D)/2$, again a reasonable result because the information diminishes as var$(\theta|D)$ grows larger.

Using the information measures (2.2)–(2.5), the problem is to determine the form of $\pi_p(\theta|D)$, given the inputs $l(\theta|y)$ and $\pi_a(\theta|I)$, so as to minimize a reasonable criterion functional subject to the condition that $\pi_p(\theta|D)$ is a proper pdf. The criterion functional that will be employed is motivated by the ICP. That is, $\pi_p(\theta|D)$'s form should be such that the output information is as close as possible to the input information and, ideally, equal to it. Thus the criterion functional

$$\Delta[\pi_p(\theta|D)] = \int_\theta \pi_p(\theta|D)\log \pi_p(\theta|D)d\theta + \log p(y|I) - \int_\theta \pi_p(\theta|D)[\log l(\theta|y) + \log \pi_a(\theta|I)]d\theta \qquad (2.6)$$

will be minimized to $\int_\theta \pi_p(\theta|D)d\theta = 1$. Note from (2.2)–(2.5) that the first two terms on the right side of (2.6) represent the information in the outputs, $\pi_p(\theta|D)$ and $p(y|I)$, and from this is subtracted the information in the inputs, $l(\theta|y)$ and $\pi_a(\theta|I)$—the likelihood function and the prior pdf, respectively. Thus (2.6) represents the difference between the output and input information, and minimization with respect to the choice of $\pi_p(\theta|D)$, the postdata pdf, will make the output information as close as possible to the input information.

It is useful to note that the criterion functional in (2.6) can be expressed as

$$\Delta[\pi_p(\theta|D)] = 2\int_\theta \pi_p(\theta|D)\log\left[\frac{\pi_p(\theta|D)}{\pi_a(\theta|I)} \times \frac{p(y|I)}{f(y|\theta)}\right]^{1/2} d\theta. \qquad (2.7)$$

From (2.7), it is seen that minimizing (2.6) involves choosing $\pi_p(\theta|D)$ such that the postdata mean of the logarithm of the geometric mean of the ratios $\pi_p(\theta|D)/\pi_a(\theta|I)$ and $p(y|I)/f(y|\theta)$ will be as small as possible; in this sense the outputs, $\pi_p(\theta|D)$ and $p(y|I)$, will be as "close" as possible to the inputs, $\pi_a(\theta|I)$ and $f(y|\theta)$. Equation (2.7) can be interpreted as an information-theory divergence measure relating to the pdf's $\pi_a(\theta|y, I)p(y|I)$ and $\pi_a(\theta|I)f(y|\theta)$ and, as a referee suggested, the negative entropy of $\pi_p(\theta|D)$ relative to the measure $\pi_a(\theta|I)f(y|\theta)/p(y|I)$.

As shown in Section 3, the solution, denoted by $\pi^*(\theta|D)$, to the minimization problem is

$$\pi_p^*(\theta|D) = c\pi_a(\theta|I)l(\theta|y), \qquad (2.8)$$

with $c^{-1} = \int \pi_a(\theta|I)l(\theta|y)d\theta = p(y|I)$. From (2.8), it is seen that $\pi_p^*(\theta|D)$ is just the postdata or posterior pdf yielded by Bayes's IPR, that is, Bayes's theorem. From (2.8), $\int_\theta \pi_p^*(\theta|D)\log[\pi_p^*(\theta|D)/\pi_a(\theta|I)]d\theta = \int_\theta \pi_p^*(\theta|D)\log l(\theta|y) d\theta - \log p(y|I)$, where the quantity on the left side is the negative of the entropy of $\pi_p^*(\theta|D)$ relative to the measure $\pi_a(\theta|I)$. Thus the negative entropy of the posterior pdf can be expressed in terms of the information measures.

To check the informational efficiency of the rule in (2.8), substitute $\pi_p^*(\theta|D)$ given in (2.8), with $c = 1/p(y|I)$, into (2.6) with the result

$$\Delta[\pi_p^*(\theta|D)] = 0. \qquad (2.9)$$

From (2.9), the IPR in (2.8) is 100% efficient and therefore satisfies the ICP relative to the information measures in (2.2)–(2.5). That is, use of $\pi_p^*(\theta|D)$ as a postdata pdf makes the input information equal to the output information. No information is lost and no extraneous information is introduced by use of the Bayesian IPR in (2.8).

3. DERIVATION OF THE OPTIMAL INFORMATION PROCESSING RULE

To minimize the criterion functional $\Delta[\pi_p(\theta|D)]$ in (2.6) subject to the condition that $\pi_p(\theta|D)$ be a proper, normalized pdf, we consider the class of neighboring functions, $\bar{\pi}_p(\theta|D) = \pi_p(\theta|D) + \varepsilon\eta(\theta)$, where ε is a small quantity and $\eta(\theta)$ is an arbitrary continuous function with a value of 0 at the endpoints of the region of integration and with $\int_\theta \eta(\theta)^2 d\theta < \infty$. On substituting $\bar{\pi}_p(\theta|D)$ in (2.6) and in the side condition $\int \pi_p(\theta|D)d\theta = 1$, the Lagrangian expression, denoted by $L(\varepsilon)$, is

$$L(\varepsilon) = \int_\theta [\pi_p(\theta|D) + \varepsilon\nu(\theta)]\log[\pi_p(\theta|D) + \varepsilon\nu(\theta)]d\theta - \int[\pi_p(\theta|D) + \varepsilon\nu(\theta)]\log[\pi_a(\theta|I) + \log l(\theta|y)] + \log p(y|I) + \lambda[\int[\pi_p(\theta|D) + \varepsilon\nu(\theta)]d\theta - 1], \qquad (3.1)$$

where λ is a Lagrange multiplier. On differentiating $L(\varepsilon)$ with respect to ε and evaluating the derivative at $\varepsilon = 0$, the necessary condition for an extremum is

$$L'(0) = \int_\theta \nu(\theta)\,[\log \pi_p(\theta|D) + 1 - \log \pi_a(\theta|I) - \log l(\theta|\underset{\sim}{y}) + \lambda]d\theta = 0.$$

For $L'(0)$ to be equal to 0 for any arbitrary $\nu(\theta)$, the quantity in brackets in the integrand must be identically equal to 0, which leads to

$$\pi_p(\theta|D) = \pi_p^*(\theta|D) = c\pi_a(\theta|I)l(\theta|y), \qquad (3.2)$$

where $c = e^{-(1+\lambda)}$ is given by $c^{-1} = \int_\theta \pi_a(\theta|I)l(\theta|y)d\theta = p(y|I)$. Further, $d^2L(\varepsilon)/d\varepsilon^2$, evaluated at $\varepsilon = 0$, is given by

$$d^2L(\varepsilon)/d\varepsilon^2\big|_{\varepsilon=0} = \int_\theta \nu(\theta)^2/\pi_p^*(\theta|D)d\theta. \qquad (3.3)$$

Assuming $\pi_p^*(\theta|D) \le M$, a positive constant, (3.3) is larger than $(1/M) \times \int_\theta \nu(\theta)^2 d\theta > 0$ and the expression for $\pi_p(\theta|D)$ in (3.2) corresponds to a minimum. It is assumed that the integral in (3.3) converges as would be the case of $\lim_{\theta \to \infty}|\theta|^q[\nu(\theta)^2/\pi_p^*(\theta)] \to$ constant for $q > 1$. This requirement places a condition on the rate at which $\nu(\theta)^2 \to 0$ as $\theta \to \pm\infty$. Alternatively, if the parameter space is finite, the integral will usually converge.

Thus it is seen that (3.2), in the form of Bayes's theorem, is a solution to the constrained minimization problem. Further, when $\pi_p^*(\theta|D)$ in (3.2) is inserted in (2.6), the result is $\Delta[\pi_p^*(\theta|D)] = 0$; therefore, (3.2) is informationally 100% efficient.

4. CONCLUDING REMARKS

In this article an information processing approach has been formulated. This approach is thought to be a useful representation of the processing of information in inference situations. An optimal information processing rule was derived that is identical to Bayes's rule and is 100% informationally efficient. Further research to consider extended variants of the criterion functional used in this study as well as alternative measures of information would be valuable.

[Received October 1986. Revised July 1987.]

REFERENCES

Cox, R. T. (1961), *The Algebra of Probable Inference*, Baltimore: Johns Hopkins University Press.

Jaynes, E. T. (1974), "Probability Theory With Application in Science and Engineering: A Series of Informal Lectures," unpublished manuscript, Washington University, Dept. of Physics.

——— (1983), *Papers on Probability, Statistics and Statistical Physics*, ed. R. D. Rosenkrantz, Dordrecht, Holland: D. Reidel.

——— (1984), "The Intuitive Inadequacy of Classical Statistics," *Epistemologica*, 7, 43–74.

Jeffreys, H. (1967), *Theory of Probability* (3rd ed.), London: Oxford University Press.

——— (1973), *Scientific Inference* (3rd ed.), Cambridge, U.K.: Cambridge University Press.

Kullback, S. (1959), *Information Theory and Statistics*, New York: John Wiley.

Comment

E. T. JAYNES*

Arnold Zellner's article appears to me a potentially important one in two respects, one psychological and one theoretical. By looking at Bayes's theorem in a fresh way independent of previous arguments, it could make the use of Bayesian methods more attractive and widespread, and stimulate new developments in the general theory of inference.

In almost all real problems of scientific inference we need to take into account our total state of knowledge, only part (or sometimes none) of which consists of frequencies. For many years I have believed, and taught my students, that the fundamental justification for use of Bayes's theorem in such applications lies in the logical consistency arguments of R. T. Cox, referred to by Zellner. Cox's desiderata appeared to me more elementary—therefore, more compelling logically—than the well-known arguments of De Finetti, Jeffreys, and L. J. Savage.

Recently, however, I was taken aback in a conversation with a prominent anti-Bayesian when he opined that logical consistency is not an important desideratum at all for inference, because it gives no reason to believe that our conclusions are in any way sensible from a pragmatic standpoint.

It appears, then, that the arguments that convinced me may have little psychological force for others; perhaps this may account for the rather slow growth of Bayesian methods, in spite of their easy success in applications where sampling-theory methods would be awkward due to nuisance parameters, nonexistence of sufficient or ancillary statistics, or cogent prior information calling for an informative prior. It is surely clear to all, however, that inference is basically a procedure of information processing; some black box receives input information in the form of prior knowledge and data, and it emits output information in the form of parameter estimates, predictive distributions, and so forth.

Then a derivation of Bayes's theorem directly from desiderata of optimal information processing might have a stronger convincing power for many. An acceptable inference procedure should have the property that it neither ignores any of the input information nor injects any false information; if this requirement already determines Bayes's theorem, the issue would seem to be settled.

The logarithmic measures of information might appear arbitrary at first glance; yet as Kullback showed, this is not the case. And Bayes's theorem doubtless has more than one information-optimality property; I rather expect that, having seen this start, others different in detail and/or background conditions may be found. Indeed, the fact that many different psychological approaches point to the same actual algorithm is a major strength of "Bayesianity."

On the theoretical side, entropy has been a recognized part of probability theory since the work of Shannon 40 years ago, and the usefulness of entropy maximization as a tool in generating probability distribution is thoroughly established in numerous new applications including statistical mechanics, spectrum analysis, image reconstruction, and biological macromolecular structure determination. Grandy

*E. T. Jaynes is Wayman Crow Professor of Physics, Department of Physics, Washington University, St. Louis, Missouri 63130.

(1987), Justice (1986), Moore and Scully (1986), and Smith and Grandy (1985) provided a wealth of details. This makes it seem scandalous that the exact relation of entropy to the other principles of probability theory is still rather obscure and confused. But now we see that there is, after all, a close connection between entropy and Bayes's theorem. Having seen this start, other such connections may be found, leading to a more unified theory of inference in general. Thus in my view Zellner's work is probably not the end of an old story, but the beginning of a new one.

ADDITIONAL REFERENCES

Grandy, W. T., Jr. (1987), *Foundations of Statistical Mechanics* (Vol. 1), Dordrecht, Holland: D. Reidel.

Justice, J. H. (ed.) (1986), *Maximum Entropy and Bayesian Methods in Applied Statistics*, Cambridge, U.K.: Cambridge University Press.

Moore, G. T., and Scully, M. (eds.) (1986), *Frontiers of Nonequilibrium Statistical Mechanics*, New York: Plenum Press.

Smith, C. Ray, and Grandy, W. T., Jr. (eds.) (1985), *Maximum-Entropy and Bayesian Methods in Inverse Problems*, Dordrecht, Holland: D. Reidel.

Comment

BRUCE M. HILL*

Professor Zellner has suggested an interesting way to think about the nature of the Bayesian approach to the revision of probabilities. He notes that a certain information property is conserved, and that in a certain sense this is efficient. This implicitly raises the question of whether, or in what circumstances, it may be desirable to conserve such a property. This in turn leads to the question of whether time coherence should be a part of the Bayesian approach at all. In other words, for example, should the Bayesian theory allow one to reformulate models and priors distributions after seeing the data?

In my opinion, the answer is a clear "yes," and I regard this opinion as an essential ingredient in any sensible approach to inference and decision making. I call the overall process that of Bayesian data-analysis and postdata decision making. Some theory for this approach was presented in Hill (1985, 1987, 1988). For example, it is argued that the restricted likelihood principle of Hill (1987) still holds, even when the analysis includes the postdata selection of models, as in Hill (1985, p. 223). This does take us, to some extent, outside the classical version of Bayesian inference and decision-theory; however, the classical version often provides a first approximation to the data-analytic approach that I recommend. In my approach, which allows for data snooping of all sorts, it is still possible for one to persist with the original model (if any), provided that the data do not suggest the need for revision. Some related discussions concerning time coherency and/or generalizations of the Bayesian approach are in Diaconis and Zabell (1986), Goldstein (1983), and Lane and Sudderth (1985).

A minor technical comment is that an alternative derivation of the mathematical result of Zellner can be obtained from Jensen's inequality, as he suggests in connection with information theory.

Zellner is to be congratulated for clearly formulating the conservation property implicit in Bayes's theorem, and holding it up for our careful scrutiny. Although I have stated that I do not regard this property as essential to the Bayesian approach, there is another extremely important point implicit in Zellner's article, one with which I am in total agreement. If one does not wish to conserve this property, as is the case in all strictly non-Bayesian analyses of data, then it should be incumbent upon the statistician to state explicitly from whence the violation arises. Does it arise from exploratory data analysis, or is it built in in some other way. What does it represent?

An illustration occurs in the area of statistics known as "size-biased sampling," where it is routine for some statisticians to violate the likelihood principle, and thus also the conservation property, without ever changing the model. This is done with the purpose of obtaining unbiased estimates. A Bayesian analysis of the problem suggests, however, that to the extent such things are at all reasonable, it is because the model being used is not really thought to be appropriate. In this case presumably one should put forth a better model, one that builds in the effect of "size" explicitly, and then analyze the data using the more realistic model in a coherent way, rather than merely make a token adjustment for "size bias."

By careful consideration of the important questions raised by Professor Zellner, perhaps even the time-honored and powerful Bayesian approach, which is the only one we have that does not have built-in logical contradictions (due to the conditioning implicit in the use of model diagnostics), as discussed in Hill (1985, pp. 202–213; 1988), can be improved and made more realistic for applications.

ADDITIONAL REFERENCES

Diaconis, P., and Zabell, S. (1986), "Some Alternatives to Bayes's Rule," in *Proceedings of the Second University of California, Irvine, Conference on Political Economy*, eds. B. Grofman and G. Owen, Greenwich, CT: JAI Press, pp. 25–38.

Goldstein, M. (1983), "The Prevision of a Prevision," *Journal of the*

*Bruce M. Hill is Professor, Department of Statistics, University of Michigan, Ann Arbor, Michigan 48109.

American Statistical Association, 78, 817–819.
Hill, Bruce M. (1985), "Some Subjective Bayesian Considerations in the Selection of Models" (with discussion), *Econometric Reviews*, 4, 191–288.
——— (1987), "The Validity of the Likelihood Principle," *The American Statistician*, 41, 95–100.
——— (1988), "A Theory of Bayesian Data Analysis," unpublished manuscript, University of Michigan, Dept. of Statistics.
Lane, D., and Sudderth, W. (1985), "Coherent Predictions Are Strategic," *The Annals of Statistics*, 13, 1244–1248.

Comment

JOSÉ M. BERNARDO*

Professor Zellner provides an information-theoretical argument to justify updating beliefs using Bayes's theorem. His reasoning may be summarized as follows: Given the "input information" $l(\theta|y)$ and $\pi(\theta|I)$ and assuming the "output information" to consist of $\pi(\theta|D)$ and $p(y|I)$ where, by definition, $p(y|I)$ is the standard prior predictive, find the $\pi(\theta|D)$ that minimizes an information-based criterion. The force of the argument surely depends on the force of the assumptions—namely, on the support given for the choice of the criterion. I do not find convincing the proposed measure (2.2) of the information in $l(\theta|y)$, because axiomatics of information (e.g., see Good 1966) lead to log-probabilities, *not* log-likelihoods. Thus, although (2.3), (2.4), and (2.5) may be seen as (expected posterior) measures of information about θ in the prior, θ in the posterior, and y in the predictive, I fail to appreciate why (2.2) should be a good measure of information (information about what?).

The mathematics of the article are closely related to the following alternative formulation of the same problem: Treat inference about θ as a decision problem where the action space consists of the possible $\pi(\theta|D)$'s, assume a utility function $u[a, \theta] = u[\pi_\theta(\cdot|D), \theta]$ that describes the utility of the information processing rule (IPR) that leads to $\pi(\theta|D)$, and find the IPR that maximizes the expected utility.

There is a wide class of utility functions for which the answer is Bayes's theorem (see Good 1971); these functions are usually referred to as *proper* scoring rules. The best known of these is

$$u[\pi_\theta(\cdot|D), \theta] = A \log \pi_\theta(\cdot|D) + B(\theta).$$

The proof that the optimal IPR for this utility function is the Bayes theorem (Bernardo 1979; Savage 1971) is then a variation of the argument given in Section 3 of the article.

*José M. Bernardo is Professor of Statistics, Departamento de Estadística, Universidad de Valencia, 46071 Valencia, Spain.

ADDITIONAL REFERENCES

Bernardo, J. M. (1979), "Expected Information as Expected Utility," *The Annals of Statistics*, 7, 686–690.
Good, I. J. (1966), "A Derivation of the Probabilistic Explanation of Information," *Journal of the Royal Statistical Society*, Ser. B, 28, 578–581.
——— (1971), Discussion of "Measuring Information and Uncertainty," by R. J. Buehler, in *Foundations of Statistical Inference*, eds. V. P. Godambe and D. A. Sprott, Toronto: Holt, Rinehart & Winston, pp. 337–339.
Savage, L. J. (1971), "Elicitation of Personal Probabilities and Expectations," *Journal of the American Statistical Association*, 64, 783–801.

Comment

SOLOMON KULLBACK*

The notion of an information conservation principle has appeared in the statistical literature as a basis for the statistical concept of sufficiency.

Since one should expect that a measure of statistical information be positive, it would be appropriate that in (2.2), (2.3), (2.4), and (2.5) a minus sign be used before the integral sign. In this case, it is seen that the criterion functional (2.6) is the difference between input information and output information.

The introduction of Expression (2.7) and the discussion thereof is irrelevant. The statement that (2.7) can be interpreted as an information-theory divergence measure relating to the pdf's $\pi_p(\theta|y, I)p(y|I)$ and $\pi_a(\theta|I)f(y|\theta)$ is incorrect.

*Solomon Kullback is Professor Emeritus of Statistics, George Washington University, Washington, D.C. 20052.

Section 3, "Derivation of the Optimal Information Processing Rule," is not needed. If one writes the criterion function (2.6) as

$$\int_\theta \pi_p(\theta|D)\log(\pi_p(\theta|D)/\pi(\theta|D))d\theta,$$

where $\pi(\theta|D) = \pi_a(\theta|I)f(y|\theta)/p(y|I)$, then because of (2.1),

$$\int_\theta \pi(\theta|D)d\theta = 1, \qquad \int_\theta \pi_p(\theta|D) = 1,$$

and the criterion function as written here is the discrimination information between $\pi_p(\theta|D)$ and $\pi(\theta|D)$. It is a well-known fact in statistical information theory that the criterion function is thus nonnegative and equal to 0 iff $\pi_p(\theta|D) = \pi(\theta|D)$. Consequently, the result (2.8) is immediate without the argument in Section 3.

Example 4.6 on page 9 of Kullback (1959) may have some relevance to the article under discussion.

Reply

ARNOLD ZELLNER

I thank Professors Jaynes, Kullback, Hill, and Bernardo for their thoughtful comments on my article. Jaynes and Hill have pointed to the importance of having a framework, that of information processing, within the context of which it is possible to derive optimal information processing rules (IPR's) and to define the efficiency of various IPR's, a point also made by Seymour Geisser (personal communication, 1986). Further, Jaynes mentions that the present analysis provides "a close connection between entropy and Bayes's theorem" (p. 281), and Hill points out that "If one does not wish to conserve this property [the information-conservation principle], as is the case in all strictly non-Bayesian analyses of data, then it should be incumbent upon the statistician to state explicitly from whence the violation arises" (p. 281). He goes on to analyze the case of "size-biased sampling." I believe that these constructive comments are worthy of serious consideration by all.

Hill and Kullback point out that alternative proofs of the major mathematical result of the article are available, a point that was brought to my attention earlier by Udi Makov (personal communication, 1986), and by Hill and my colleague Robert McCulloch in personal conversations. I decided to publish the calculus-of-variations proof because it can be readily generalized to apply to other information processing problems, both static and dynamic. For example, if the prior and likelihood information are given different weights to reflect the possibly differing quality of these information inputs, then the information-criterion functional can be redefined accordingly and minimized to provide a solution that is different from Bayes's theorem. Also, the criterion functional in (2.6) can be minimized subject to additional side conditions, for example, that the information divergence between the output pdf, $\pi_p(\theta|D)$, and the likelihood function and/or prior pdf be less than or equal to given constants. The solution will then not be "too far" from the likelihood function and/or the prior pdf and will be in a form different from Bayes's theorem. Results such as these, as well as the possibility of analyzing "temporal problems" of the kind mentioned in Hill's comment, led me to present the calculus-of-variations proof. Also, although dynamic information processing problems are important and yet to be analyzed within the present approach, I believe it is useful to solve various static problems. In many disciplines (e.g., economics, engineering, and physics) both statics and dynamics play important roles. The same seems to be true in the information processing area.

Kullback points out that an information-conservation principle has been discussed in connection with the statistical concept of sufficiency. Non-Bayesian discussions of information and sufficient statistics abstract from information contained in prior distributions and thus differ from the discussion in my article. Bayesian discussions of sufficiency assume Bayes's theorem, whereas in my article Bayes's theorem is derived as an optimal IPR; then it is shown that information is conserved by using Bayes's rule. These are, in my opinion, very fundamental differences between former discussions of sufficiency and the analysis presented in my article. Further, I regard negative entropy as a measure of information in a distribution, since entropy is usually interpreted as a measure of disorder; that is, the higher the entropy is, the less informative a distribution is (e.g., see Jaynes 1983). For example, a uniform pdf for a scalar parameter θ has higher entropy than a highly peaked density for θ. Since the highly peaked density provides more information about the possible values of θ, it is more informative and thus the negative entropy is an appropriate measure of the information contained in a distribution. On the other hand, in information theory, entropy measures are employed to represent the expected information content of signals or messages prior to their arrival, a concept different from the information in a distribution. This distinction is important even though the solution to min $\Delta[\pi_p(\theta|D)]$ is the same as that for max$\{-\Delta[\pi_p(\theta|D)]\}$. Last, Kullback (1959, ex. 4.6) presented a measure of the information provided by an experiment that is somewhat different from the information-criterion functional used in my article.

Bernardo questions (2.2). Directly after (2.2)–(2.5) an example is provided showing that (2.2) provides a measure of the spread (or information) in a likelihood function. After

(2.8), an expression exactly in the form of (2.2) appears as a term in the information in $\pi_p^*(\theta|D)$ relative to $\pi_a(\theta|I)$. Further, (2.2) appears in the expression for the information in a posterior pdf relative to uniform measure. As regards Savage's and others' proofs that Bayes's theorem or the Bayesian IPR is an expected utility-maximizing solution, this is a fundamentally different result from that in my article, where no utility considerations enter and there is no assumption that the expected utility hypothesis is in some sense "valid" or "rational." My result deals with information processing, not utility-maximizing behavior.

Again, I thank the discussants for their thoughtful and stimulating comments and the editors for their assistance in publishing my article and the useful discussion of it.

[12]

BAYESIAN METHODS AND ENTROPY IN ECONOMICS AND ECONOMETRICS

Arnold Zellner
Graduate School of Business
University of Chicago
1101 East 58th Street
Chicago, Illinois 60627 USA

ABSTRACT. A discussion of some previous and current uses of Bayesian methods and entropy in economics and econometrics is presented.

1. Introduction

In this paper, an attempt will be made to introduce natural scientists to selected aspects of economic and econometric science. Then a brief overview of how Bayesian methods and entropy have been employed in economics and econometrics will be provided. While Bayesian methods have come to be used fairly broadly, entropy and entropy-based methods are just beginning to get more than isolated attention. Further, some of the author's work designed to produce a synthesis of Bayesian and entropy concepts in model-building and inference will be presented and discussed. That is, some explicit optimization procedures for producing (1) models for observations, (2) prior densities, (3) optimal information-processing rules, and (4) optimal inferences will be reviewed and brought together. In addition to axiomatic support for some of these procedures, which has appeared in the literature (see, *e.g.*, Shore and Johnson, 1980), it is emphasized that having inference components produced as solutions to explicit optimization problems means that they are explicit and reproducible. By altering conditions of the optimization problems, new solutions can be obtained, *e.g.*, new information-processing rules, *etc.*, and evaluated in practice.

The plan of the paper is as follows. In Section 2, an overview of economic and econometric science and the roles of Bayesian methods and entropy in it will be provided. Section 3 presents examples of how the principle of maximum entropy and related optimization procedures have been employed to produce models for observations, prior densities and information-processing rules. Also, optimization problems yielding point and interval estimates and predictions will be briefly reviewed. Finally, Section 4 contains a summary and some concluding remarks.

2. Economic and Econometric Sources

Pearson (1938) and Jeffreys (1967) stress the Unity of Science principle, namely that any area of study, say history, economics, physics, agriculture, *etc.*, can be a science if scientific methods are used in generating and interpreting data. In the last century or so, economists have come to use the methods of science much more generally with growing

W. T. Grandy, Jr. and L. H. Schick (eds.), Maximum Entropy and Bayesian Methods, 17–31.

success. Thus economic science has emerged and Nobel prizes are annually awarded to outstanding economists. Jeffreys (1967, p. 1) remarked that a main objective of science is learning from data and experience. This is achieved by (1) description and (2) induction, that is generalization to explain the variation of past data and to predict as yet unobserved data. There can be no question but that in the last century economists have become very serious about measurement problems and produced reams of data, relating to many aspects of economic behavior in all countries of the world. Much of these data come from censuses, scientifically designed surveys and administrative records. There is a growing field of experimental economics in which experiments are scientifically designed and carried out to produce data—see, *e.g.*, Aigner and Morris (1979).

In induction, that is explanation of the variation in past data and prediction of as yet unobserved data, economists have come to adopt the view that data are generated from stochastic, probabilistic models—see Haavelmo (1944). They have come to appreciate the important role of statistical inference in estimation, testing, prediction, and model-selection problems. Many more economists now understand the fundamental role of prediction in economic science—see, *e.g.*, Friedman (1953). The philosopher Feigl's (1953) definition of causality, "predictability according to a law or set of laws" and probabilistic confirmation procedures for establishing causality have come to be more widely appreciated—see *e.g.*, the papers in Aigner and Zellner (1989). New and old economic theories are now tested more strenuously with data than was the case in the past. Modern data generation and inference techniques, while far from perfect, are much better than they were fifty years ago. The laws of supply and demand have been quantitatively tested in many studies and found to be reliable. Also, economic science has been put to good use in solving a number of applied practical problems in industry and government. Thus economic science has progressed considerably in the last fifty years, particularly by bringing together data, mathematical economic theory, statistical methods and computers in a field of economics called econometrics. Indeed, almost all good modern economists use econometric techniques in their work to analyze data and draw conclusions from their data in a scientific manner.

From the early decades of this century up to the present, so-called "frequentist thinking", dominated econometric statistics—see *e.g.*, Haavelmo (1944) and Hood and Koopmans (1953). Laplace's, Edgeworth's, Jeffreys', Jaynes', Savage's, de Finetti's, and others' Bayesian work had little or no impact on economic and econometric research and practice until about the late 1950's and early 1960's. Since then there has been an upswing in Bayesian methodological and applied work in econometrics and statistics. According to Poirier's (1989) survey of the literature, Bayesian material has in recent years come to account for about 15% of the pages devoted to research in leading econometric and statistics journals. This upswing in Bayesian research and applications has taken place in part because Bayesian methods have produced many sensible non-Bayesian results and other results that are difficult to obtain by using non-Bayesian approaches—see Zellner (1984 and 1985) for further discussion of these issues. Also current, leading textbooks, *e.g.*, Judge *et al.* (1985), Green (1990), *etc.*, contain a good deal of Bayesian material. See Press (1989) for a listing of Bayesian computer programs which are important in facilitating applications of Bayesian methods.

On the philosophical side, progress has been slower with respect to Bayesian versus other approaches, mainly because the issues are deep and difficult and because not many economists, econometricians and statisticians have good backgrounds in the philosophy of science. For example, most econometricians are unaware of or have not read key works by Laplace, Jeffreys, Jaynes, de Finetti, Savage, *etc.* They are not generally familiar with different definitions of probability and their implications—see Jeffreys (1967, Ch. 7) for a perceptive discussion which most economists and statisticians have not read.

There has been some discussion of a distinction between uncertainty and probability or risk which is discussed in a paper by Grandy (1990) presented at this meeting. Friedman, Savage and others argue that there is no distinction between uncertainty and probability whereas Knight asserts that there is. Grandy (1990) in his work suggests that probabilities can be supplied to economic agents by having them maximize entropy subject to restrictions reflecting information which they have. This approach appears to be a useful one which will probably be employed by others to solve many economic problems. However, if there is no information available, Jeffreys' (1967) improper priors for representing "ignorance" may capture Knight's concept of uncertainty for with improper priors, say $p(\theta) \propto$ const., $-\infty < \theta < \infty$, $\Pr(a < \theta < b) = 0$ and $\Pr(a < \theta < b)/\Pr(c < \theta < d)$ is indeterminate. See also Cohen and Jaffray (1980) who find difficulty in defining "personal" probabilities under complete ignorance. This state of affairs may be close to Knight's state of uncertainty.

As mentioned above, Bayesian statistical methods have been developed which provide solutions to inference and decision problems in economics and econometrics which are as good or better than non-Bayesian solutions, when they exist—see, *e.g.*, Bawa, Brown, and Klein (1979), Fienberg and Zellner (1975), Zellner (1971, 1980, and 1985), Leamer (1978), Goel and Zellner (1986), Geisser *et al.* (1990), and the references in these works. For uses of Bayes' Theorem as a learning model in economic theory, see, *e.g.*, Boyer and Kihlstrom (1984) and Cyert and DeGroot (1987). Also, annual issues of the ASA/IMS Current Index to Statistics list current articles and books in statistics by author and by key words in titles. Hundreds of articles and books with the words Bayes or Bayesian in their titles are listed each year.

As regards entropy or maximum entropy concepts, they do not appear in many economics and econometrics textbooks. An early econometrics text which does utilize entropy concepts and methods is Davis (1941). In his book, *The Theory of Econometrics*, Davis gives a fascinating account of early econometric work and presents (p. 37ff) the number of ways P places in the income distribution can be assigned to N individuals, namely $Q = (N + P - 1)!/N!(P - 1)!$. He used Stirling's approximation to approximate $\log Q$, differentiated $\log Q$ with respect to P and with further approximations and an assumption obtained a general formula for the distribution of income in a society, studied general properties of the solution and explained how his solution can be used in practice. In addition, in Davis' "theory of budgets" (p. 170ff), he writes:

> "Now the concepts of potential energy and entropy are related in the field of physical phenomena by one common property; their integrals, taken over a path of variation in the variables of state, are exact in the sense in which this word was defined in Section 10 of Chapter 3. The recognition of the existence of such functional relationships between physical variables marked one of the major advances in science. It is conceivable that a similar recognition may be given to economic concepts other than that of utility, where it has already been applied with such conspicuous success. The place where such recognition may be expected is in the domain of budgets." (p. 171)

He goes on to develop a "theory of budgets" with an "isomorphism between the theory about to be set forth and the postulates of that profound subject [thermodynamics] in the physical sciences. In particular, the theory recognizes an isomorphism between money utility on the one hand and entropy on the other." (p. 171) See Davis (1941, 171–176) and Lisman (1949) for specific equations of the theory which parallel those of thermodynamics. Thus Davis (1941) sensed a need for entropy considerations in economics which was expressed later in an unpublished note by Jaynes (1982) as follows:

> "In a probabilistic model of the economy, we ought to include in the probability of any macroeconomic state an entropy factor exp(S) to take this multiplicity [number of different microeconomic ways in which it can be realized] into account. This is one of the factors—possibly the only variable factor—in the prior probability of that state. If we failed to do this in statistical mechanics we would get grossly, qualitatively wrong predictions, and the same may be true in macroeconomics." (p. 2)

Jaynes goes on to say that all of this "does not prove that entropy is the crucial missing factor; it may be some other recognized factor is even more important. But entropy is at least a promising candidate ..." (p. 2). He concludes that "To apply these ideas to real economic prediction would require the kind of judgment that comes from long familiarity with the subject matter. One needs to know which particular macroeconomic variables should be included to have a "full set"; and what is the underlying microeconomic "hypothesis space" that defines the multiplicity factors. On such matters the writer is willing to hazard some guesses, but feels the need of help. Therefore a realistic implementation of the thermodynamic analogy lies in the future." (p. 5)

Davis' (1941) work is an early attempt to exploit the "thermodynamic analogy" in the area of consumer behavior. While a bold and interesting effort, it did not lead to many verifiable predictions and currently is not central in current theoretical work to explain and predict consumers' behavior. Perhaps Davis' work needs to be reconsidered. In Maasoumi's (1990) Palgrave Dictionary paper entitled "Information Theory," interesting descriptive and statistical uses of entropy in economics and econometrics are described and referenced, see *e.g.*, Theil (1967), mainly motivated by the work of Shannon, Wiener, and Kullback in information theory. There is little, if anything, presented on the "thermodynamic analogy" which is central in Davis' (1941) and Jaynes' (1982) thinking.

In recent work by Golan (1988 and 1989), a descriptive, discrete stochastic model of an industry is developed and presented. His model "yields the most probable size distribution of firms given some boundary conditions such as limited resources." The model is used to analyze "the dynamics of firms in an industry" and "enables one to investigate the effects of various policy measures or technological changes in terms of the size distribution of the industry (economy)." Golan (1989, pp. 2–3). He also shows that his entropy-based approach, which leads him to a steady state distribution of firms by size specializes for a very large number of firms, all with the same constant returns to scale (CRS) technology, to the Bose-Einstein distribution (p. 17) and shows that many empirical firm size distributions, Yule, Pareto and Geometric, are limiting cases of the Bose-Einstein distribution function. Golan (1989) writes that his derived maximum entropy distributions "are general enough to include all the possible firm size distributions where the class of skew distributions (*e.g.*, economies with CRS) is only one of the possibilities." (p. 17). He also provides interesting empirical applications of his methods to the distribution of firms by size in the California wheat industry. Golan writes in his conclusions, "The ME approach enables one to explore economic variables and parameters that are otherwise left unexplored. These variables and parameters, in most cases reflect the physical constraints facing an economy." (p. 36).

In summary, entropy and entropy related concepts have been used in economics, mostly in descriptive and statistical information theory analyses—see Maasoumi (1990) for an overview of and references to this literature. Davis (1941) and Golan (1988 and 1989) are examples of work that attempt to develop and utilize the "thermodynamic analogy." However, from what is presented in Jaynes (1982), it may be that certain dynamic features of the analogy are unknown to economists at this time and that much more empirical and theoretical work needs to be done to get a "maximum entropy, thermodynamic model" that performs well in explaining and predicting the behavior of economic systems.

3. Maximum Entropy, Related Concepts and Econometric and Statistical Modeling and Inference

In this section, optimization problems are described which yield models for observations and prior probability density functions for their parameters. Examples are given to illustrate these techniques. Then an additional optizimation is introduced, the solution of which is an optimal information-processing rule (IPR) which links an output or post-data density to the input densities, the prior density for the parameters and the density function for the observations. As will be seen, entropy considerations play a role in this process.

To produce models for observations in an optimization framework, the well-known maximum entropy paradigm, discussed above, can be employed. For example, Cobb, Koppstein, and Chen (1983) point out that limiting stationary density functions for the diffusion equation, $dx_t = \mu(x_t)d_t + \sigma(x_t)dw_t$, where w_t = standard Wiener process, take the following form:

$$f_k(x|\beta) = \xi(\beta)\exp\left\{-\int[g_k(x)/v(x)]dx\right\}$$

where $2\mu(x) = g(x) - v'(x)$, $\sigma^2(x) = v(x)$ and

$$g_k(x) = \beta_0 + \beta_1 x + \ldots + \beta_k x^k.$$

Special cases of this density, useful in catastrophe theory, are given by Cobb, Koppstein, and Chen (1983) as: Type N_k: $v(x) = 1$, $-\infty < x < \infty$; Type G_k: $v(x) = x$, $0 < x < \infty$; Type I_k: $\text{var}(x) = x^2$, $0 < x < \infty$; and Type B_k: $\text{var}(x) = x(1-x)$, $0 < x < 1$. In Zellner and Highfield (1982), it is pointed out that these types of limiting densities are produced by maximizing entropy subject to particular side conditions as indicated below.

	Type	**Maximize Entropy Subject To**
1.	N_k	$\int x^r f(x)dx = \mu_r,\ r = 0,1,2,\ldots,k+1$, with $\mu_0 = 1$
2.	G_k	$\int x^r f(x)dx = \mu_r,\ r = 0,1,2,\ldots,k$, and $\int f(x)\log x\, dx = c$
3.	I_k	$\int x^r f(x)dx = \mu_r,\ 0,1,2,\ldots,k$, $\int f(x)\log x\, dx = c_1$ and $\int f(x)dx/x = c_2$

4. B_k $\int x^r(x)dx = \mu_r,\ 0,1,2,\ldots,k-1,\ \int f(x)\log x\, dx = c_1,$
and $\int f(x)\log(1-x)dx = c_2$

In addition to these maxent density functions and many others that can be derived, Ryu (1990) shows how maximizing entropy subject to side conditions can produce many kinds of regression models. Thus it appears that a wide range of models can be produced by maximizing entropy subject to appropriate side conditions as indicated above and in the previous section.

Having used maxent to produce a density function, say $f(y|\theta)$, where θ is a parameter or vector of parameters, there is a need now to define an optimization problem, the solution to which is a satisfactory prior density. Several approaches to this problem have been suggested in the literature. A brief review of the approach taken by Zellner (1971, 1977, and 1986a) will be given and related to others in the literature.

Let the information in a probability density function be defined as its negative entropy. Thus the information in the joint density for an observation y and a parameter $\theta, p(y,\theta)$, is

$$-H[p(y,\theta)] = \int\int p(y,\theta)\log p(y,\theta)dyd\theta = \int I(\theta)\pi(\theta)d\theta + \int \pi(\theta \log \pi(\theta)d\theta$$

where $p(y,\theta) = \pi(\theta)f(y|\theta)$ has been employed, where $\pi(\theta)$ is a prior density, $f(y|\theta)$ is the density for y given θ and

$$I(\theta) = \int f(y|\theta)\log f(y|\theta)dy$$

is the information in $f(y|\theta)$. Thus $-H[p(y,\theta)]$ breaks up into two parts, the first the prior average information in the data density, $\int I(\theta)\pi(\theta)d\theta$ plus the information in the prior density, $\int \pi(\theta)\log\pi(\theta)d\theta$.

Since in science there is a wish to feature the information in data, the following criterion functional $G[\pi(\theta)]$ is employed in Zellner (1971, 1977, and 1986a):

$$G[\pi(\theta)] = \int I(\theta)\pi(\theta)d\theta - \int \pi(\theta)\log\pi(\theta)d\theta$$

which is the difference between the prior average information in the data density minus the information in the prior density. On maximizing $G[\pi(\theta)]$ with respect to choice of $\pi(\theta)$ subject to $\int \pi(\theta)d\theta = 1$, the solution is the maximal data information prior (MDIP), $\pi_*(\theta)$ given by:

$$\pi_*(\theta) = c\exp\{I(\theta)\} \qquad a < \theta < b$$

where $c^{-1} = \int_a^b \exp\{I(\theta)\}d\theta$.

As pointed out in earlier work, if $I(\theta) = \text{const.}$, $\pi_*(\theta) = \text{const.}$, a uniform density. Further, $\pi_*(\theta)$ is invariant to linear transformations and the following general results, among others, have been obtained:

	Form of $f(y\|\theta)$	Form of $\pi_*(\cdot)$	
A.	$f(y-\theta)$	$\pi_*(\theta) \propto$ const.	$a < \theta < b$
B.	$\sigma^{-1}f(y/\sigma)$	$\pi_*(\sigma) \propto 1/\sigma$	$0 < a < \sigma < b$
C.	$\sigma^{-1}f[(y-\theta)/\sigma]$	$\pi_*(\theta,\sigma) \propto 1/\sigma$	$a_1 < \theta < b,\ 0 < a_2 < \sigma < b_2$

In addition MDIPs have been derived for autoregressive, binomial, multiple regression and other models' parameters—see Zellner (1977). Further, additional side conditions, reflecting available information, can be added to the above optimization problem, *e.g.*, given prior moments, *etc.*

Jaynes (1968) has suggested that a prior density be chosen so as to maximize entropy subject to side conditions. In particular, after seeing the criterion functional $G[\pi(\theta)]$ and results yielded by it, he suggested the possibility of maximizing $-\int \pi(\theta)\log\pi(\theta)d\theta$ subject to $\int \pi(\theta)d\theta = 1$ and $\int I(\theta)\pi(\theta)d\theta = c$, a given constant. While this gives solutions of the same form as above, there is a need to choose a value of c in order to evaluate Lagrange multipliers. Further, both approaches involve a dependence of the form of the prior on the form of the density which is true of many other approaches for generating priors. Many consider this dependence to be reasonable.

Lindley (1956) and Bernardo (1979) have utilized a different criterion functional in an effort to derive prior densities as solutions to an optimization scheme. Their criterion functional, denoted by $G_{LB}[\pi(\theta)]$ is, with $p(\theta,y) = g(\theta|y)h(y)$,

$$G_{LB}[\pi(\theta)] = \int \left[\int g(\theta|y)\log g(\theta|y)d\theta\right] h(y)dy - \int \pi(\theta)\log\pi(\theta)d\theta$$

Thus $G_{LB}[\pi(\theta)]$ is equal to the information in the posterior density, $g(\theta|y)$ averaged using $h(y)$ minus the information in $\pi(\theta)$. Here the objective is to have the average information in the posterior density be greatest relative to that in the prior, an objective that is different from that above in which the average data information is made as large as possible relative to that in the prior density. Also, with a little algebra, it is easy to show that

$$G[\pi(\theta)] = G_{LB}[\pi(\theta)] + \int h(y)\log h(y)dy - \int \pi(\theta)\log\pi(\theta)d\theta.$$

Thus $G[\pi(\theta)]$ is broader than $G_{LB}[\pi(\theta)]$ by the inclusion of additional terms shown above reflecting the information in $h(y)$ minus that in $\pi(\theta)$. Finally, while use of $G[\pi(\theta)]$ leads to easy solutions, it has been difficult to get exact, analytical solutions to the problem of maximizing $G_{LB}[\pi(\theta)]$ with respect to choice of $\pi(\theta)$.

Further, if we introduce a measure $m(y,\theta) = m(y|\theta)m(\theta)$ and write the "invariantized", negative entropy as

$$\int\int p(y,\theta)\log\frac{p(y,\theta)}{m(y,\theta)}dyd\theta = \int I_m(\theta)\pi(\theta)d\theta + \int \pi(\theta)\log\pi(\theta)/m(\theta)d\theta$$

where

$$I_m(\theta) = \int f(y|\theta)\log\frac{f(y|\theta)}{m(y|\theta)}dy,$$

then

$$G_m[\pi(\theta)] = \int I_m(\theta)\pi(\theta)d\theta - \int \pi(\theta)\log(\pi(\theta)/m(\theta))d\theta$$

can be maximized w.r.t. $\pi(\theta)$ subject to $\int \pi(\theta)d\theta = 1$ to obtain $\pi_*(\theta) = k\exp\{I_m(\theta)\}$, where k is a normalizing constant. Thus, with an appropriate choice of $m(y,\theta)$, an optimal choice of $\pi(\theta)$ can be derived subject to just the normalization condition or to it and other side conditions, say given moment conditions reflecting an investigator's given, initial information.

The solutions to the above optimization problems have limited invariance properties, *e.g.*, they are invariant to linear transformations of y and θ but, not, for example to all one-to-one continuous transformations. Jeffreys' rule for generating diffuse priors, $|F(\theta)|^{\frac{1}{2}}$, where $F(\theta)$ is the Fisher information matrix, is well known to be invariant to all continuous one-to-one transformations. However, in practice Jeffreys finds that for some problems, this broad invariant rule leads to "unsatisfactory" results, as for example in the case of a normal mean problem where it produces $\pi(\mu,\sigma) \propto 1/\sigma^2$. By restricting the class of transformations for which invariance is sought, Jeffreys (1967, pp. 182–183) obtains the diffuse prior $\pi(\mu,\sigma) \propto 1/\sigma$, which he deems "more satisfactory". Also, if one is interested in a parameter θ, $0 < \theta < \infty$, and $\eta = 0^5$ is either a meaningless parameter or a parameter of no interest, why should one restrict his choice of prior to obtain invariance for this transformation? (In a letter from George Barnard to the author, he pointed out that this point was made by W.S. Gosset in correspondence between him and R.A. Fisher many years ago.) More recently, Rao (1987) has written, "The choice of a metric naturally depends on a particular problem under investigation, and invariance may or may not be relevant." (p. 220) This discussion indicates that it is useful to consider the problem of deriving prior densities which are invariant to a specific, relevant transformation say a reciprocal transformation or to particular, relevant subsets of transformations, such as those considered by Jeffreys. For example consider the continuous, one-to-one transformation $\eta = h(\theta)$. Then we have the invariance condition, $\pi_1(\theta)d\theta = \pi_2(\eta)d\eta$ and since $d\eta = |g'(\theta)|d\theta$, $\pi_2(\eta) = \pi_1(\theta)/|g'(\theta)|$ or $\pi_1(\theta) = |g'(\theta)|\pi_2[g(\theta)]$ is the invariance condition or restriction on $\pi_1(\cdot)$ and $\pi_2(\cdot)$.

On defining the criterion functional G as follows,

$$G = \int \pi_1(\theta)I(\theta)d\theta - \int \pi_1(\theta)\log\pi_1(\theta)d\theta + \int \pi_2(\eta)I(\eta)d\eta - \int \pi_2(\eta)\log\pi_2(\eta)d\eta$$

where $I(\theta)$ and $I(\eta)$ are the negative entropies of $f(y|\theta)$ and $f(y|\eta)$, respectively, G can be maximized s.t. $\int_a^b \pi_1(\theta)d\theta = 1$ and $\int_{h(a)}^{h(b)} \pi_2(\eta)d\eta = 1$. The solution to this optimization problem is:

$$\pi_{1*}(\theta) = c\exp\{I(\theta) + 1/2\log|h'(\theta)|\}$$

and

$$\pi_{2*}(\eta) = \pi_{1*}[h^{-1}(\eta)]/|h'[h^{-1}(\eta)]|.$$

As an application of the above, consider $\eta = h(\theta) = 1/\theta$ with $0 < a < \eta < 1$ and $1 < \theta, 1/a$. For example, θ might be a Keynesian "multipler", and η the marginal propensity to save parameter. Further $y = \theta x + \epsilon$ is the simple regression model for income, y, given x, autonomous spending with $\epsilon a N(0,\sigma^2)$ error term. With this model for y, the negative entropies of $f(y|\theta, x)$ and of $f(y|\eta, x)$ can easily be evaluated and $|h'(\theta)| = 1/\theta^2$. Thus $\pi_{1*}(\theta) \propto 1/\theta$ and $\pi_{2*}(\eta) \propto 1/\eta$ and $\pi_{2*}(\eta) = \pi_{1*}(\theta)/|h'(\theta)| = (1/\theta)\theta^2 = \theta = 1/\eta$ and thus the invariance condition is satisfied. This solution can be compared to what results from taking $\pi_1(\theta) \propto$ const., namely $\pi_2(\eta) \propto 1/\eta^2$.

If more than one transformation is considered, say $\eta_1 = h_1(\theta)$, $\eta_2 = h_2(\theta), \ldots, \eta_m = h_m(\theta)$, then G can be defined as

$$G = \int \pi(\theta)I(\theta)d\theta - \int \pi(\theta)\log\pi(\theta)d\theta + \sum_{i=1}^{m}\left[\int \pi(\eta_i)I(\eta_i)d\eta_i - \int \pi(\eta_i)\log\pi(\eta_i)d\eta_i\right]$$
$$= (m+1)\left[\int \pi(\theta)I(\theta)d\theta - \int \pi(\theta)\log\pi(\theta)d\theta\right] + \sum_{i=1}^{m}\int \pi(\theta)\log|h_i'(\theta)|d\theta$$

on inserting the invariance conditions. Then maximizing G w.r.t. $\pi(\theta)$ s.t. $\int \pi(\theta)d\theta = 1$ yields:

$$\pi_*(\theta) \propto \exp\left\{I(\theta) + \sum_{i=1}^{m}\log|h_i'(\theta|/(m+1)\right\}.$$

As an example of the above approach, consider $f(y|\sigma) \propto \sigma^{-1}\exp\{-y^2/2\sigma^2\}$, $-\infty < y < \infty$, for which $I(\sigma) = \int f(y|\sigma)\log f(y|\sigma)dy =$ const. $- \log\sigma$. Now consider the transformations, $h_0(\sigma) = \log\sigma$ and $h_i(\sigma) = 1/\sigma^i$, $i = 1, \pm 2, \pm 3, \ldots$ Then

$$\pi_*(\sigma) \propto \exp\{I(\sigma) + \Sigma\log|g_i'(\sigma)|/(m+1)\} \propto \exp\{-\log\sigma + \text{const.} + 0\} \propto 1/\sigma$$

which is Jeffreys' rule that is invariant with respect to power transformations.

Having shown how to produce two components, a data density $f(y|\theta)$ and a prior density $\pi(\theta)$ as solutions to well-defined optimization problems, we now consider how to combine these "inputs" to obtain an output density which reflects the information in the inputs. Usually, this is done by use of Bayes' theorem which, of course depends on the product rule of probaility, $\Pr(A \cap B|I_0) = \Pr(A|I_0)\Pr(B|A, I_0)$, where I_0 denotes initial information. Jeffreys (1967, pp. 24–25) provides a proof of the product rule in connection with his Theorem 9 and then states, "The proof has assumed that the alternatives considered [elements of sets A, B, and $A \cap B$] are equally probably ... It has not been found possible to prove the theorem without using this condition ... Theorem 9 suggests the simplest general rule that they [probabilities conditional on *different* information sets] can follow if there is one at all." (p. 25) He then takes the more general proposition relating to probabilities conditional on different data as an *axiom*, his Axiom 7, which is in the form of the usual product rule.

Cox (1961), in contrast to many others, does not assume that the usual product rule holds but writes in his equation (3.1), p. 13, rewritten in the present notation, $\Pr(A \cap B|I_0) = F[\Pr(A|I_0), \Pr(B|A, I_0)]$ and notes that, "The form of the function F is in part arbitrary, but it can not be entirely so, because the equation must be consistent with Boolean algebra." (p. 13) After some ingenious analysis, he finds that the function F must satisfy a functional equation, which restricts the form of F and results in the usual product rule, with probabilities scaled zero to one under a further simplifying assumption—see Cox (1961, p. 16).

The upshot of this brief discussion is that the product rule of probability, as discussed by Cox and Jeffreys is difficult to prove under general conditions and thus is often introduced as an axiom. In what follows, an approach developed in Zellner (1988), will be described that can yield Bayes' theorem without use of the product rule. In this approach there are two inputs to an information-processing rule (IPR), $\pi(\theta|I_0)$ a prior density for the parameter θ and a data density $f(y|\theta, I_0)$, components which have been discussed above. The outputs of an IPR are $g(\theta|y, I_0)$, an output density or "post-data" density for θ and $h(y|I_0)$, a marginal density for y. $h(y|I_0)$ is called an "output" because it is the output of a definition, namely, $h(y|I_0) \equiv \int f(y|\theta, I_0)\pi(\theta|I_0)d\theta$.

On placing information measures on the outputs and inputs, the following criterion functional was utilized in Zellner (1988):

$$\Delta[g(\theta|y, I_0)] = \text{Output Information} - \text{Input Information}$$

It was suggested that $g(\theta|y, I_0)$ be determined as the solution to the following problem:

$$\min_g \Delta[g(\theta|y, I_0)] \quad \text{s.t.} \quad \int g(\theta|y, I_0)d\theta = 1$$

Minimization is suggested because there is a desire to keep the output information as close as possible to the input information so as to avoid a loss of information. For the information measures introduced in Zellner (1988), the solution to this problem, denoted by $g_*(\theta|y, I_0)$, was found to be

$$g_*(\theta|y, I_0) = \frac{\pi(\theta|I_0)f(y|\theta, I_0)}{h(y)}$$

which is Bayes' Rule. Further, it was shown that $\Delta[g_*(\theta|y, I_0)] = 0$. That is, when $g_*(\theta|y, I_0)$ is employed, Output Information = Input Information and thus Bayes' IPR is 100% efficient. This 100% efficiency property of Bayes' IPR stands in marked contrast to that of other IPR's which do not take account of all available information—see Hill (1988) and Jaynes (1988) for further discussion of this and other points. Finally, the functional $\Delta[g(\theta|y, I_0)]$ can be expressed as $\Delta[g(\theta|y, I_0)] = \int g(\theta|y, I_0) \log[g(\theta|y, I_0)/g_*(\theta|y, I_0)]d\theta$, that is the negative entropy of $g(\theta|y, I_0)$ relative to the measure $g_*(\theta|y, I_0)$. Thus the solution, obtained by taking $g(\theta|y, I_0) = g_*(\theta|y, I_0)$, is a maxent solution.

There are many possible variations of the above optimization problem that yield the optimal Bayesian IPR. One is to allow the qualities of the sample and of the prior information to vary from very poor to very good by introducing "quality scaling factors". That is, the explicit form of $\Delta[g(\theta|D)]$, where $D \equiv (y, I_0)$, is

$$\Delta[g(\theta|D)] = \int g(\theta|D)\log g(\theta|D)d\theta + \log h(y|I_0) - \int g(\theta|D)\log f(y|\theta, I_0)d\theta$$
$$- \int \pi(\theta|I_0)\log \pi(\theta|I_0)d\theta.$$

Now let $\pi(\theta|I_0) = q_1[\pi_a(\theta|I_0)]$ and $f(y|\theta, I_0) = q_2[f_a(y|I_0)]$ where $q_1[\cdot]$ and $q_2[\cdot]$ are functions which "quality adjust" the actually formulated densities, $\pi_a(\theta|I_0)$ and $f_a(y|I_0)$. Note the resemblance to robustness considerations. Then $\Delta[g(\theta|D)]$, shown above can be minimized subject to $\int g(\theta|D)d\theta = 1$ to yield an optimizing value $g_*(\theta|D)$, given by

$$g_*(\theta|D) \propto \exp\{\log q_1[\pi_a(\theta|I_0)] + \log q_2[f_a(y|\theta, I_0)]\}.$$

For example, if $q_1[\pi_a(\theta|I_0)] = \pi_a(\theta|I_0)]^{\lambda_1}$ and $q_2[f_a(y|\theta, I_0)] = [f_a(y|\theta, I_0)]^{\lambda_2}$, with $0 \leq \lambda_1, \lambda_2 \leq 1$,

$$g_*(\theta|D) \propto [\pi_a(\theta|I_0)]^{\lambda_1}[f_a(y|\theta, I_0)]^{\lambda_2}.$$

If the prior information is regarded as of extremely low quality ($\lambda_1 = 0$), $g_*(\theta|D) \propto [f_a(y|\theta, I_0)]^{\lambda_2}$ while if the sample information is of extremely low quality ($\lambda_2 = 0$), $g_*(\theta|D) \propto \pi_a(\theta|I_0)^{\lambda_1}$. Note further, that raising the prior density $\pi_a(\theta|I_0)$ and the data density, $f_a(y|\theta, I_0)$ to fractional powers, λ_1 and λ_2, respectively, will in many cases, *e.g.*, normal, beta, *etc.*, densities, increase the dispersion of these densities. Thus in this case, spreading out densities can be interpreted as a correction for the quality of information in prior and data densities.

Having the *exact, finite sample* posterior density, $g(\theta|D)$, for many different problems in econometrics has been widely recognized as a distinct advantage of the Bayesian approach. Further, the posterior density has been used to good advantage in econometrics and other fields to make probability statements about values of θ and/or of subsets of the elements of θ which are conditional on *given* sample and prior information. For example, if θ is a scalar parameter, $\Pr(A < \theta < B|D) = \int_A^B g(\theta|D)d\theta$, with A and B given values is a useful probability statement in many contexts. Further, for vector $\boldsymbol{\theta}' = (\boldsymbol{\theta}_1', \boldsymbol{\theta}_2')$ if $\boldsymbol{\theta}_1$ is of interest and $\boldsymbol{\theta}_2$ is a vector of nuisance parameters, the marginal posterior density of $\boldsymbol{\theta}_1$, $g(\boldsymbol{\theta}_1|D)$, is given by $g(\boldsymbol{\theta}_1|D) = \int g(\boldsymbol{\theta}_1, \boldsymbol{\theta}_2|D)d\boldsymbol{\theta}_2 = \int g(\boldsymbol{\theta}_1|\boldsymbol{\theta}_2, D)g(\boldsymbol{\theta}_2|D)d\boldsymbol{\theta}_2$, where $g(\boldsymbol{\theta}_1|\boldsymbol{\theta}_2, D)$ is the conditional posterior density for $\boldsymbol{\theta}_1$ given $\boldsymbol{\theta}_2$ and D. This important capability of eliminating nuisance parameters, here $\boldsymbol{\theta}_2$, has been extremely useful in economietrics.

In addition, conditional posterior densities, such as $q(\boldsymbol{\theta}_1|\boldsymbol{\theta}_2, D)$, have been used to explore how sensitive probability statements about possible values of $\boldsymbol{\theta}_1$, or of its elements, are to what is assumed about values of $\boldsymbol{\theta}_2$, a capability of great value.

As regards point estimates for θ, these have been generated in econometrics for all kinds of problems using a variety of loss functions. It is well known that for squared error loss, $L(\tilde{\theta}, \theta) = (\tilde{\theta} - \theta)^2$, the posterior mean of $g(\theta|D)$ is optimal in the sense of minimizing posterior expected loss. Further for absolute error loss $L(\tilde{\theta}, \theta) = |\tilde{\theta} - \theta|$, the posterior median is optimal while for "zero-one" loss, the posterior modal value is optimal. See Zellner (1986b) for estimation results for a class of asymmetric loss functions and Berger (1985) for further discussion of Bayesian decision theoretic principles of estimation.

Another most important use of a posterior density $g(\theta|D)$ is its role in deriving predictive densities, denoted by $p(y_f|D)$, where y_f is an as yet unobserved scalar or vector of observations assumed generated by $f(y_f|\theta, I_0)$. Since θ's value is usually unknown, $f(y_f|\theta, I_0)$ cannot be employed to make probability statements about y_f. However the joint density of y_f and θ is given by $f(y_f|\theta)g(\theta|D)$ and thus $p(y_f|D) = \int f(y_f|\theta)g(\theta|D)d\theta)$ is the integration that is performed to provide the predictive density which has been employed to make probability statements about values of y_f and to obtain point predictions.

For some recent applicaltions of these principles to forecasting countries' output growth rates and turning points in them—see Garcia-Ferrer *et al.* (1987), Hong and Gulati (1990), Zellner, Hong, and Min (1989).

Finally, Bayes' theorem can also be used to compute posterior odds relating to alternative hypotheses and models—see, *e.g.*, Jeffreys (1967), Zellner (1971 and 1984), Geisel (1975), Leamer (1978), Monahan (1983), and Min and Zellner (1990). In general the posterior odds for two mutually exclusive hypotheses or models is given by

$$\text{Posterior Odds} = \text{Prior Odds} \times \text{Bayes' Factor}.$$

This simple paradigm for converting prior odds into posterior odds has been applied by Jeffreys (1967) to many problems and by others in econometrics to variable and model selection problems and to problems involving the combining of models and their forecasts in econometrics—see Geisel (1975), Min and Zellner (1990) and Palm and Zellner (1990) for examples of these uses of posterior odds.

4. Summary and Concluding Remarks

From what has been presented above, it is apparent that Bayesian inference and decision techniques have assumed an important role in applied and theoretical econometric and statistical analyses. A new way of rationalizing Bayes' theorem in terms of information-processing has revealed a connection between Bayes' theorem and entropy-maximization. Prior densities have been derived by optimizing various entropy related criterion functionals. Other fruitful statistical uses of entropy in econometrics have been mentioned. However, a secure and useful entropy-related "thermodynamic analogy" has not as yet been dominant in economics in spite of the pioneering work of Davis (1941) and Golan (1988 and 1989).

REFERENCES

Aigner, D.J. and C.N. Morris (eds.): 1979, 'Experimental Design in Econometrics', *Annals J. of Econometrics* **11**, 205 pp.

Aigner, D.J. and A. Zellner (eds.): 1988, 'Causality', *Annals, J. of Econometrics* **39** 234 pp.

Bawa, V.S., S.J. Brown, and R.W. Klein: 1979, *Estimation Risk and Optimal Portfolio Choice*, North-Holland, Amsterdam.

Berger, J.O.: 1985, *Statistical Decision Theory* (2nd ed.), Springer-Verlag, New York.

Bernardo, J.: 1979, 'Reference Posterior Distributions for Bayesian Inference', *J. Royal Statistical Society* **B,41**, 113–147.

Boyer, M. and R.E. Kihlstrom (ed.): 1984, *Bayesian Models in Economic Theory*, North-Holland, Amsterdam.

Cobb, L., P. Koppstein, and N.H. Chen: 1983, 'Estimation and Moment Recursion Relations for Multimodal Distributions of the Exponential Family', *J. American Statistical Association* **78**, 124–130.

Cohen, M. and J.Y. Jaffray: 1980, 'Rational Behavior Under Complete Ignorance', *Econometrica* **48**, 1281–1300.

Cox, R.T.: 1961, *The Algebra of Probable Inference*, John Hopkins Press, Baltimore.

Cyert, R.M. and M.H. DeGroot: 1987, *Bayesian Analysis and Uncertainty in Economic Theory*, Chapman and Hall Ltd., London.

Davis, H.T.: 1941, *The Theory of Econometrics*, Principia Press, Bloomington, Indiana.

Fienberg, S.E. and A. Zellner: 1975, *Studies in Bayesian Econometrics and Statistics in Honor of Leonard J. Savage*, North-Holland, Amsterdam.

Feigl, H.: 1953, 'Notes on Causality', in *Readings in the Philosophy of Science*, H. Feigl and M. Brodbeck (eds.), Appleton-Century-Crofts, New York, 408–418.

Friedman, M.: 1953, *Essays in Positive Economics*, University of Chicago Press, Chicago.

Garcia-Ferrer, A.P., R.A. Highfield, F. Palm, and A. Zellner: 1987, 'Macroeconomic Forecasting Using Pooled International Data', *J. of Business and Economic Statistics* 5, 53–67.

Geisel, M.S.: 1975, 'Bayesian Comparisons of Simple Macroeconomic Models', *Studies in Bayesian Econometrics and Statistics in Honor of Leonard J. Savage*, S.E. Fienberg and A. Zellner (eds.), North-Holland, Amsterdam, 227–256.

Geisser, S., J. Hodges, S.J. Press, and A. Zellner (eds.): 1990, *Bayesian and Likelihood Methods in Statistics and Econometrics: Essays in Honor of George A. Barnard*, North-Holland, Amsterdam.

Goel, P.K. and A. Zellner (eds.): 1986, *Bayesian Inference and Decision Techniques: Essays in Honor of Bruno de Finetti*, North-Holland, Amsterdam.

Golan, A.: 1988, *A Discrete Stochastic Model of Economic Production and a Model of Fluctuations in Production-Theory and Empirical Evidence*, Ph.D. Thesis, Department of Agricultural Economics, University of Haifa, Haifa, Israel.

Grandy, C.: 1991, 'The Principle of Maximum Entropy and the Difference Between Risk and Uncertainty', *Proceedings, Tenth International MaxEnt Workshop*, Laramie, Wyoming, Kluwer, Dordrecht.

Green, W.H.: 1990, *Econometric Analysis*, Macmillan, New York.

Haavelmo, T.: 1944, 'The Probability Approach in Econometrics', Supplement to *Econometrica* **12**, 115 pp.

Hill, B.M.: 1988, 'Comment', *American Statistician* **42**, 281–282.

Hood, Wm.C. and T.C. Koopmans: 1953, *Studies in Econometric Method*, John Wiley & Sons, Inc., New York.

Jaynes, E.T.: 1968, 'Prior Probabilitites', *IEEE Transactions on Systems Science and Cybernetics*, **SSC–4**, 227–241.

Jaynes, E.T.: 1982a, 'How Should We Use Entropy in Economics? (Some Half-Baked Ideas in Need of Criticism)', manuscript.

Jaynes, E.T.: 1988, 'Comment', *American Statistician* **42**, 280–281.

Jeffreys, H.: 1967, *Theory of Probability* (3rd rev. ed.), Oxford University Press, London.

Judge, G.G., W.E. Griffiths, R.C. Hill, H. Lütkepohl, and T.C. Lee, *The Theory and Practice of Econometrics* (2nd ed.), John Wiley & Sons, Inc., New York.

Kloek, T. and Y. Haitovsky (eds.): 1988, *Competing Statistical Paradigms in Econometrics*, papers presented at the International Conference on the Foundations of Statistical Inference, Israel, 1985; *Annals, Journal of Econometrics* **37**, 192 pp.

Lindley, D.V.: 1956, 'On a Measure of Information Provided by an Experiment', *Annals of Mathematical Statistics* **27**, 986–1005.

Lisman, J.H.C.: 1949, 'Econometrics and Thermodynamics: A Remark on Davis' Theory of Budgets', *Econometrica* **17**, 59–62.

Maasoumi, E.: 1990, 'Information Theory', *The New Palgrave Econometrics*, J. Eatwell, M. Millgate, and P. Newman (eds.), W.W. Norton & Co., New York, 101–112.

Min, C.K. and A. Zellner: 1990, 'Bayesian and Non-Bayesian Methods for Combining Models and Forecasts with Applications to Forecasting International Growth Rates', manuscript, H.G.B. Alexander Research Foundation, University of Chicago.

Monahan, J.F.: 1983, 'Fully Bayesian Analysis of ARMA Time Series Models', *J. of Econometrics* **21**, 307–331.

Pearson, K.: 1938, *The Grammar of Science*, Everyman, London.

Poirier, D.J.: 1989, 'A Report from the Battlefront', *J. of Business and Economic Statistics* **7**, 137–140.

Press, S.J.: 1989, *Bayesian Statistics: Principles, Models and Applications*, John Wiley & Sons, Inc., New York.

Rao, C.R.: 1987, 'Differential Metrics in Probability Spaces', *Differential Geometry in Statistical Inference* **10**, Institute of Mathematical Statistics Lecture Notes-Monograph Series, Institute of Mathematical Statistics, Hayward, California, Chapter 5, 217–240.

Ryu, H.K.: 1990, *Orthogonal Basis and Maximum Entropy Estimation of Probability Density and Regression Functions*, doctoral dissertation, Department of Economics, University of Chicago.

Shore, J.E. and R.W. Johnson: 1980, 'Axiomatic Derivation of the Principle of maximum Entropy and the Principle of Cross-Entropy', *IEEE Transactions on Information Theory* **IT–26**, 26–37.

Theil, H.: 1967, *Economics and Information Theory*, North-Holland, Amsterdam.

Zellner, A.: 1971, *An Introduction to Bayesian Inference in Econometrics*, John Wiley & Sons, Inc., New York; reprinted in 1987 by Krieger Publishing Co., Malabar, Florida.

Zellner, A.: 1977, 'Maximal Data Information Prior Distributions', *New Developments in the Applications of Bayesian Methods*, A. Aykac and C. Brumat (eds.), North-Holland, Amsterdam, 211–232.

Zellner, A. (ed.): 1980, *Bayesian Analysis in Econometrics and Statistics: Essays in Honor of Harold Jeffreys*, North- Holland, Amsterdam.

Zellner, A.: 1984, *Basic Issues in Econometrics*, University of Chicago Press, Chicago.

Zellner, A.: 1985, 'Bayesian Econometrics', *Econometrica* **53**, 253–269.

Zellner, A.: 1986a: Presentation to Bayesian Study Year Meeting, Warwick University, Warwick, England.

Zellner, A.: 1986b: 'Bayesian Estimation and Prediction Using Asymmetric Loss Functions', *J. of the American Statistical Association* **81**, 446–451.

Zellner, A.: 1988, 'Optimal Information Processing and Bayes' Theorem', *American Statistician* **42**, 278–284.

Zellner, A. and R.A. Highfield: 1982, 'Calculation of Maximum Entropy Distributions and Approximation of Marginal Posterior Distributions', presented in NBER-NSF Seminar on Bayesian Inference, October, 1982 and published in *J. of Econometrics* **37** (1988), 195–210.

Zellner, A., C. Hong, and G.M. Gulati: 1990, 'Turning Points in Economic Time Series Loss Structures and Bayesian Forecasting', *Bayesian and Likelihood Methods in Statistics and Econometrics: Essays in Honor of George A. Barnard*, S. Geisser, J. Hodges, S.J. Press, and A. Zellner (eds.), North-Holland, Amsterdam, 371–396.

Zellner, A., C. Hong, and C. Min: 1989, 'Forecasting Turning Points in International Output Growth Rates Using Bayesian Exponentially Weighted Autoregresion, Time Varying

Parameter and Pooling Techniques', manuscript, Graduate School of Business, University of Chicago, to appear in *J. of Econometrics*.

PART IV

BAYESIAN METHODS

A

Estimation

INTERNATIONAL ECONOMIC REVIEW
Vol. 11, No. 3, October, 1970

ESTIMATION OF REGRESSION RELATIONSHIPS CONTAINING UNOBSERVABLE INDEPENDENT VARIABLES*

BY ARNOLD ZELLNER[1]

1. INTRODUCTION

IN A NUMBER OF INSTANCES, regression relationships are formulated with one or more unobservable variables in the set of independent variables. For example, in Crockett [1], consumption relationships are considered which contain Friedman's [2] permanent income variable. In the paper by Miller and Modigliani [7] concerning the measurement of the cost of capital, the key relationship which they consider contains "true unobservable expected earnings" as an independent variable. Last in Zellner [10], it is shown that equations of simultaneous equation models can be brought into a regression form involving some observable and some unobservable independent variables. Given that regression relations containing unobservable independent variables occur quite frequently, and in fact are a special case of "errors in the variables" models, it is important to have good methods for analyzing them. Previous analyses have almost always involved the use of an instrumental variable approach, an approach which leads to estimators with the desirable large sample property of consistency. However, it is not clear that the instrumental variable approach leads to asymptotically efficient estimators for all parameters of a model and the small sample properties of instrumental variable estimators are for the most part unknown.

In the present paper, we first consider the specification and interpretation of the models under consideration in Section 2. Then in Section 3 we apply a least squares approach to generate an estimator which, with a normality assumption, is a maximum likelihood estimator. The relationship of this estimator to certain instrumental variable estimators is set forth. Then in Section 4, a Bayesian analysis of the model is presented. Finally, in Section 5 some concluding remarks are presented.

2. SPECIFICATION AND EXPLANATION OF MODELS

The first model which we consider is the following one:

$$y_{1\alpha} = y_{2\alpha}^*\gamma + v_{1\alpha}\,, \tag{2.1}$$

$$y_{2\alpha} = y_{2\alpha}^* + v_{2\alpha}\,, \qquad \alpha = 1, 2, \cdots, n\,, \tag{2.2}$$

$$y_{2\alpha}^* = \beta_1 + \beta_2 x_{2\alpha} + \cdots + \beta_k x_{k\alpha}\,. \tag{2.3}$$

In (2.1) we have a regression relation with $y_{1\alpha}$ the α-th observation on the

* Manuscript received August 11, 1968.

[1] Research financed by the National Science Foundation under grant GS-1350. Some of the material contained herein was presented in lectures at the University of Hawaii during the summer of 1967. I would like to thank P. M. Laub, S. J. Press, and D. M. Wu for helpful comments.

dependent variable, $y_{2\alpha}^*$ the unobservable α-th value of an "independent" variable, γ a scalar coefficient, and $v_{1\alpha}$ a disturbance term. Equation (2.2) relates the observation $y_{2\alpha}$ to $y_{2\alpha}^*$ and a disturbance term $v_{2\alpha}$ and (2.3) gives an expression for $y_{2\alpha}^*$ in terms of observable independent variables $x_{2\alpha}, x_{3\alpha}, \cdots, x_{k\alpha}$. Our problem is to estimate γ and the β's from the observations. In Crockett's [1] analysis, $y_{1\alpha}$ is observed consumption, $y_{2\alpha}^*$, permanent income, and the x's are such variables as house value, educational attainment, age, etc., which are thought to be related to permanent income, and $y_{2\alpha}$ is measured income. In Miller's and Modigliani's work, $y_{2\alpha}^*$ is true anticipated earnings of the firm, $y_{2\alpha}$ is actual earnings, the x's are variables thought to be related to true earnings and $y_{1\alpha}$ is an observed variable related to the value of the firm.[2] With this said about the interpretation of (2.1)–(2.3), we can now express the system compactly as follows:

$$\underline{y}_1 = X\underline{\beta}\gamma + \underline{v}_1 \,, \tag{2.4}$$

$$\underline{y}_2 = X\underline{\beta} + {}_2\underline{v} \,, \tag{2.5}$$

where $\underline{y}_1$ and $\underline{y}_2$ are each $n \times 1$ vectors of observations, $\underline{v}_1$ and $\underline{v}_2$ are each $n \times 1$ vectors of disturbance terms, X is an $n \times k$ matrix, with rank k, with given elements, $\underline{\beta}$ is a $k \times 1$ vector of parameters and γ is a scalar parameter.

To show the relationship of (2.4) and (2.5) to "simultaneous equation" problems, let

$$\underline{y}_1 = \underline{y}_2\gamma + \underline{u}_1 \tag{2.6}$$

be a "structural" equation and (2.5) the reduced form equation for $\underline{y}_2$. Then on substituting for $\underline{y}_2$ in (2.6) from (2.5), we obtain (2.4) with $\underline{v}_1 = \underline{u}_1 + \gamma\underline{v}_2$.[3] For present purposes, we shall work with the system as set forth in (2.4) and (2.5), or with (2.4) elaborated as follows

$$\underline{y}_1 = X\underline{\beta}\gamma + W\underline{\delta} + v_1 \,, \tag{2.4a}$$

where W is an $n \times k_1$ matrix of observations on k_1 independent variables with rank k_1, and $\underline{\delta}$ is a $k_1 \times 1$ coefficient vector. W may be a submatrix of X.

We now turn to consider various methods of estimating the parameters of

[2] Actually, Miller and Modigliani have additional observable independent variables in (2.1).

[3] In general, if $\underline{y}_1 = Y_1\underline{\gamma}_1 + X_1\underline{\beta}_1 + \underline{u}_1$ denotes a structural equation and

$$(\underline{y}_1 \vdots\, Y_1) = (X_1 \vdots\, X_0)\begin{pmatrix} \Pi_{11} & \Pi_{10} \\ \Pi_{01} & \Pi_{00} \end{pmatrix} + (\underline{v}_1 \vdots\, V_1)$$

is the reduced form system for $\underline{y}_1$ and Y_1, then we have $\Pi_{11} - \Pi_{10}\underline{\gamma}_1 = \underline{\beta}_1$ and $\Pi_{01} - \Pi_{00}\underline{\gamma}_1 = 0$. Using these latter relations to substitute for Π_{11} and Π_{01}, the reduced form system can be written as

$$\underline{y}_1 = (X_1\Pi_{10} + X_0\Pi_{00})\underline{\gamma}_1 + X_1\underline{\beta}_1 + \underline{v}_1 \tag{a}$$

$$Y_1 = X_1\Pi_{10} + X_0\Pi_{00} + V_1 \,. \tag{b}$$

The unobservable independent variables in (a) are $X_1\Pi_{10} + X_0\Pi_{00}$.

these models with unobservable independent variables.

3. LEAST SQUARES ESTIMATION

3.1. *Estimation of the System (2.4)-(2.5).* For least squares analysis, we shall assume that the disturbance terms in (2.4) and (2.5) have the following properties:

$$E\underline{v}_1 = E\underline{v}_2 = \underline{0}\,, \tag{3.1}$$

and

$$E\begin{pmatrix} \underline{v}_1 \\ \cdots \\ \underline{v}_2 \end{pmatrix}(\underline{v}_1' : \underline{v}_2') = \begin{pmatrix} \sigma_{11}I_n & 0 \\ 0 & \sigma_{22}I_n \end{pmatrix}. \tag{3.2}$$

In (3.1) we are assuming that the disturbance terms have zero means while in (3.2) we are assuming that the elements of $\underline{v}_1$ and $\underline{v}_2$ have common variances, σ_{11} and σ_{22}, respectively, and that they are contemporaneously and serially uncorrelated. With this said about disturbance term properties, we now wish to minimize the following criterion function[4]

$$\begin{aligned} S &= \frac{1}{\sigma_{11}}(\underline{y}_1 - X\underline{\beta}\gamma)'(\underline{y}_1 - X\underline{\beta}\gamma) + \frac{1}{\sigma_{22}}(\underline{y}_2 - X\underline{\beta})'(\underline{y}_2 - X\underline{\beta}) \\ &= \frac{1}{\sigma_{11}}[(\underline{y}_1 - Z\underline{\beta})'(\underline{y}_1 - Z\underline{\beta}) + \lambda(\underline{y}_2 - X\underline{\beta})'(\underline{y}_2 - X\underline{\beta})] \end{aligned} \tag{3.3}$$

where $\lambda = \sigma_{11}/\sigma_{22}$ and $Z = X\gamma$. If we complete the square on $\underline{\beta}$ in (3.3), we are led to

$$\frac{1}{\sigma_{11}}[\underline{y}_1'\underline{y}_1 + \lambda\underline{y}_2'\underline{y}_2 - C'M^{-1}C + (\underline{\beta} - M^{-1}C)'M(\underline{\beta} - M^{-1}C)] \tag{3.4}$$

where $M = Z'Z + \lambda X'X$ and $C = Z'\underline{y}_1 + \lambda X'\underline{y}_2$. From the form of (3.4) and the fact that M is positive definite, the minimizing value for $\underline{\beta}$, say $\underline{\tilde{\beta}}$, is given by

$$\begin{aligned} \underline{\tilde{\beta}} &= (Z'Z + \lambda X'X)^{-1}(Z'\underline{y}_1 + \lambda X'\underline{y}_2) \\ &= \frac{1}{\lambda + \gamma^2}(\gamma\underline{\hat{\beta}}_1 + \lambda\underline{\hat{\beta}}_2) \end{aligned} \tag{3.5}$$

where $\underline{\hat{\beta}}_1 = (X'X)^{-1}X'\underline{y}$, which is an estimator for $\underline{\beta}\gamma$, and $\underline{\hat{\beta}}_2 = (X'X)^{-1}X'\underline{y}_2$, an estimator for $\underline{\beta}$.[5] On substituting from (3.5) in (3.4) and after a little algebra, we have

[4] Note that if the elements of $\underline{v}_1$ and $\underline{v}_2$ are normally distributed, the likelihood function is just

$$\frac{1}{\sigma_{11}^n}\lambda^{n/2}\exp\left\{-\frac{S}{2}\right\}$$

with S given in (3.3). Thus values of the parameters minimizing S will maximize the likelihood function.

[5] The variance-covariance matrix for $\underline{\tilde{\beta}}$, given γ and λ, is just $\text{Var}(\underline{\tilde{\beta}}) = (X'X)^{-1}\sigma_{22}/(1 + \gamma^2/\lambda)$. This should be compared with $\text{Var}(\underline{\hat{\beta}}_2) = (X'X)^{-1}\sigma_{22}$, where $\underline{\hat{\beta}}_2 = (X'X)^{-1}X'\underline{y}_2$.

$$\frac{1}{\sigma_{11}}\left[s_{11} + \lambda s_{22} + \frac{(\hat{\underline{y}}_1 - \gamma\hat{\underline{y}}_2)'(\hat{\underline{y}}_1 - \gamma\hat{\underline{y}}_2)}{1+\gamma^2/\lambda}\right], \tag{3.6}$$

where for $\alpha = 1, 2,$

$$\hat{\underline{y}}_\alpha = X\hat{\underline{\beta}}_\alpha\,, \quad \hat{\underline{\beta}}_\alpha = (X'X)^{-1}X'\underline{y}_\alpha \quad \text{and} \quad s_{\alpha\alpha} = (\underline{y}_\alpha - X\hat{\underline{\beta}})'(\underline{y}_\alpha - X\hat{\underline{\beta}}_\alpha)\,. \tag{3.7}$$

On differentiating (3.6) with respect to γ and setting the result equal to zero, we have

$$\gamma^2\hat{\underline{y}}_1'\hat{\underline{y}}_2 + \gamma(\lambda\hat{\underline{y}}_2'\hat{\underline{y}}_2 - \hat{\underline{y}}_1'\hat{\underline{y}}_1) - \lambda\hat{\underline{y}}_1'\hat{\underline{y}}_2 = 0 \tag{3.8}$$

as the necessary condition which γ must satisfy for (3.3) to be a minimum. It is to be noted that (3.8) is of exactly the same form as the necessary condition arising in the classical errors in the variables model (See Madansky [6] for example) except that here moments are in terms of "computed" values, $\hat{\underline{y}}_1$ and $\hat{\underline{y}}_2$, rather than observed values, $\underline{y}_1$ and $\underline{y}_2$. Solving the quadratic in (3.8), we are led to

$$\tilde{\gamma}_{(\lambda)} = \frac{\hat{\underline{y}}_1'\hat{\underline{y}}_1 - \lambda\hat{\underline{y}}_2'\hat{\underline{y}}_2 + \sqrt{(\lambda\hat{\underline{y}}_2'\hat{\underline{y}}_2 - \hat{\underline{y}}_1'\hat{\underline{y}}_1)^2 + 4\lambda(\hat{\underline{y}}_1'\hat{\underline{y}}_2)^2}}{2\hat{\underline{y}}_1'\hat{\underline{y}}_2} \tag{3.9}$$

as our estimator[6] for γ given λ. For a particular choice of λ, (3.9) can be computed to obtain an estimator for γ, say $\tilde{\gamma}_{(\lambda)}$, and this value, along with the selected value of λ, can be substituted in (3.5) to obtain an estimator for $\underline{\beta}$ say $\tilde{\underline{\beta}}_{(\lambda)}$.

The properties of estimators described in the preceding paragraph will depend on the value of λ/γ^2. In this regard, just as in the typical errors in the variables model, it is interesting to consider two extreme cases:

Case I. λ very large relative to γ^2:
For this case (3.5) reduces to

$$\begin{aligned}&\frac{s_{11}}{\sigma_{11}} + \frac{s_{22}}{\sigma_{22}} + \frac{1}{\sigma_{11}}\frac{(\hat{\underline{y}}_1 - \gamma\hat{\underline{y}}_2)'(\hat{\underline{y}}_1 - \gamma\hat{\underline{y}}_2)}{1+\gamma^2/\lambda}\\ &\quad \doteq \frac{s_{11}}{\sigma_{11}} + \frac{s_{22}}{\sigma_{22}} + \frac{1}{\sigma_{11}}(\hat{\underline{y}}_1 - \gamma\hat{\underline{y}}_2)'(\hat{\underline{y}}_1 - \gamma\hat{\underline{y}}_2)\end{aligned} \tag{3.10}$$

given that $\gamma^2/\lambda \doteq 0$. Then the estimator for γ is given by

[6] It can easily be established that $\tilde{\gamma}_{(\lambda)}$ is a consistent estimator for γ as follows: Since

$$\operatorname{plim}\frac{\hat{\underline{y}}_1'\hat{\underline{y}}_1}{n} = \gamma^2 Q\,, \quad \operatorname{plim}\frac{\hat{\underline{y}}_2'\hat{\underline{y}}_2}{n} = Q \quad \text{and} \quad \operatorname{plim}\frac{\hat{\underline{y}}_1'\hat{\underline{y}}_2}{n} = \gamma Q$$

where $Q = \lim_{n\to\infty} \underline{\beta}'X'X\underline{\beta}/n$, which is assumed to exist, we have, using Slutsky's Theoem,

$$\begin{aligned}\operatorname{plim}\tilde{\gamma}_{(\lambda)} &= \frac{\gamma^2 Q - \lambda Q + \sqrt{(\lambda Q - \gamma^2 Q)^2 + 4\lambda\gamma^2 Q^2}}{2\gamma Q}\\ &= \frac{\gamma}{2} - \frac{\lambda}{2\gamma} + \frac{1}{2\gamma}(\lambda+\gamma)^2 = \gamma \qquad \text{for } \gamma \neq 0\,.\end{aligned}$$

(3.11) $$\tilde{\gamma}_{(\infty)} = \frac{\hat{\underline{y}}_2'\hat{\underline{y}}_1}{\hat{\underline{y}}_2'\hat{\underline{y}}_2} = \frac{\hat{\underline{y}}_2'\underline{y}_1}{\hat{\underline{y}}_2'\hat{\underline{y}}_2} .$$

This is just the case in which $\hat{\underline{y}}_2 = X\hat{\underline{\beta}}_2$ is used to replace $X\underline{\beta}$ in $\underline{y}_1 = X\underline{\beta}\gamma + \underline{v}_1$ and $\underline{y}_1$ is regressed on $\hat{\underline{y}}_2$ to obtain an estimator for γ. It should be noted that in this case just $\underline{y}_2 = X\underline{\beta} + \underline{v}_2$ is used to estimate $\underline{\beta}$ rather than (3.5) which uses *both* $\underline{y}_1$ and $\underline{y}_2$. Further, noting that $Z = X\gamma$, (3.5) can be written as

(3.12) $$\begin{aligned}\tilde{\underline{\beta}}_{(\infty)} &= \lim_{\lambda\to\infty} \frac{1}{\gamma^2 + \lambda}(\gamma\hat{\underline{\beta}}_1 + \lambda\hat{\underline{\beta}}_2) \\ &= \hat{\underline{\beta}}_2\end{aligned}$$

since

$$\lim_{\lambda\to\infty} \frac{\lambda}{\gamma^2 + \lambda}\hat{\underline{\beta}}_2 = \hat{\underline{\beta}}_2 .$$

Case II. λ very small relative to γ^2:
Under this case, (3.6) can be expressed as

$$\begin{aligned}&\frac{s_{11}}{\sigma_{11}} + \frac{s_{22}}{\sigma_{22}} + \frac{1}{\sigma_{22}}\frac{(\hat{\underline{y}}_1 - \gamma\hat{\underline{y}}_2)'(\hat{\underline{y}}_1 - \gamma\hat{\underline{y}}_2)}{\gamma^2(1 + \lambda/\gamma^2)} \\ &\doteq \frac{s_{11}}{\sigma_{11}} + \frac{s_{22}}{\sigma_{22}} + \frac{1}{\sigma_{22}}\left(\hat{\underline{y}}_2 - \frac{1}{\gamma}\hat{\underline{y}}_1\right)'\left(\hat{\underline{y}}_2 - \frac{1}{\gamma}\hat{\underline{y}}_1\right)\end{aligned}$$

for $\lambda/\gamma^2 \doteq 0$. The value of γ which minimizes this last expression is given by[7]

(3.13) $$\begin{aligned}\tilde{\gamma}_{(0)} &= 1/(\hat{\underline{y}}_1'\hat{\underline{y}}_2/\hat{\underline{y}}_1'\hat{\underline{y}}_1) \\ &= \hat{\underline{y}}_1'\hat{\underline{y}}_1/\hat{\underline{y}}_2'\hat{\underline{y}}_1 .\end{aligned}$$

In (3.13), the reciprocal of the quantity $\hat{\underline{y}}_1'\hat{\underline{y}}_2/\hat{\underline{y}}_1'\hat{\underline{y}}_1$ is the estimator for γ when $\lambda/\gamma^2 \doteq 0$.

Thus (3.11) and (3.13) represent estimators for γ under extreme assumptions regarding λ/γ^2. In general the properties of these estimators will be different.[8] At present, it is important to emphasize that (3.11) and (3.13) involve extreme assumptions about the value of λ/γ^2 which if not in accord with the true value can affect the properties of estimators, particularly in "small sample" situations.[9]

In practical circumstances, we do not know the value of λ and thus (3.9) cannot be computed. However, we can obtain estimates of σ_{11} and σ_{22} quite readily from regressions of $\underline{y}_1$ on X and $\underline{y}_2$ on X, namely,

$$\hat{\sigma}_{11} = (\underline{y}_1 - X\hat{\underline{\beta}}_1)'(\underline{y}_1 - X\hat{\underline{\beta}}_1)/n$$

and

$$\hat{\sigma}_{22} = (\underline{y}_2 - X\hat{\underline{\beta}}_2)'(\underline{y}_2 - X\hat{\underline{\beta}}_2)/n$$

[7] The expression for $\tilde{\gamma}_{(0)}$ given in (3.13) can also, of course, be obtained by setting λ equal to zero in (3.9).

[8] Unless X has just one column in which case they are identical

[9] See Appendix A for some analysis bearing on this point.

where $\underline{\hat{\beta}}_i = (X'X)^{-1}X'\underline{y}_i$, $i = 1, 2$. Then an operational estimator for γ is:

$$\tilde{\gamma}_{(\hat{\lambda})} = \frac{\underline{\hat{y}}_1'\underline{\hat{y}}_1 - \hat{\lambda}\underline{\hat{y}}_2'\underline{\hat{y}}_2 + \sqrt{(\hat{\lambda}\underline{\hat{y}}_2'\underline{\hat{y}}_2 - \underline{\hat{y}}_1'\underline{\hat{y}}_1)^2 + 4\hat{\lambda}(\underline{\hat{y}}_1'\underline{\hat{y}}_2)^2}}{2\underline{\hat{y}}_1'\underline{\hat{y}}_2} \tag{3.14}$$

where $\hat{\lambda} = \hat{\sigma}_{11}/\hat{\sigma}_{22}$. The quantity $\tilde{\gamma}_{(\hat{\lambda})}$, given in (3.14), can readily be computed as a consistent estimate of γ and then substituted in (3.5), along with $\hat{\lambda}$, to obtain a consistent estimate of $\underline{\beta}$.

3.2. *Estimation of the System* (2.4a)–(2.5). We assume that the disturbance vectors $\underline{v}_1$ and $\underline{v}_2$ satisfy the assumptions shown in (3.1) and (3.2). Further we assume that $X = (X_1 \vdots W)$, that is that W in (2.4a) is a sub-matrix of X. Then we wish to minimize

$$\frac{1}{\sigma_{11}}[(\underline{y}_1 - Z\underline{\beta} - W\underline{\delta})'(\underline{y}_1 - Z\underline{\beta} - W\underline{\delta}) + \lambda(\underline{y}_2 - X\underline{\beta})'(\underline{y}_2 - X\underline{\beta})] \tag{3.15}$$

with respect to the elements of $\underline{\beta}, \underline{\delta}$ and γ where, as above, $Z = X\gamma$ and $\lambda = \sigma_{11}/\sigma_{22}$. Differentiating with respect to $\underline{\delta}$, we obtain for the minimizing value of $\underline{\delta}$,

$$\tilde{\underline{\delta}} = (W'W)^{-1}W'(\underline{y}_1 - Z\underline{\beta}) , \tag{3.16}$$

which, when substituted in (3.15), yields[10]

$$\frac{1}{\sigma_{11}}[(\underline{\tilde{y}}_1 - \tilde{Z}\underline{\beta})'(\underline{\tilde{y}}_1 - \tilde{Z}\underline{\beta}) + \lambda(\underline{y}_2 - X\underline{\beta})'(\underline{y}_2 - X\underline{\beta})] \tag{3.15a}$$

where

$$\underline{\tilde{y}}_1 = [I - W(W'W)^{-1}W']\underline{y}_1 \tag{3.17}$$

and

$$\begin{aligned} \tilde{Z} &= [I - W(W'W)^{-1}W']Z \\ &= [I - W(W'W)^{-1}W'][X_1\gamma \vdots W] \\ &= (\tilde{X}_1\gamma \vdots 0) \end{aligned} \tag{3.18}$$

with $\tilde{X}_1 = [I - W(W'W)^{-1}W']X_1$. Then, on partitioning the β vector, $\underline{\beta}' = (\underline{\beta}_1' \vdots \underline{\beta}_2')$, to conform with the partitioning of $\tilde{Z}$ in (3.18), the expression in (3.15a) becomes:

$$\frac{1}{\sigma_{11}}[(\underline{\tilde{y}}_1 - \tilde{Z}_1\underline{\beta}_1)'(\underline{\tilde{y}}_1 - \tilde{Z}_1\underline{\beta}_1) + \lambda(\underline{y}_2 - X_1\underline{\beta}_1 - W\underline{\beta}_2)'(\underline{y}_2 - X_1\underline{\beta}_1 - W\underline{\beta}_2)] . \tag{3.15b}$$

On minimizing (3.15b) with respect to $\underline{\beta}_2$, we have

$$\tilde{\underline{\beta}}_2 = (W'W)^{-1}W'(\underline{y}_2 - X_1\underline{\beta}_1) \tag{3.19}$$

which when substituted in (3.15b) yields

$$\frac{1}{\sigma_{11}}[(\underline{\tilde{y}}_1 - \tilde{Z}_1\underline{\beta}_1)'(\underline{\tilde{y}}_1 - \tilde{Z}_1\underline{\beta}_1) + \lambda(\underline{\tilde{y}}_2 - \tilde{X}_1\underline{\beta}_1)'(\underline{\tilde{y}}_2 - \tilde{X}_1\underline{\beta}_1)] \tag{3.15c}$$

where

[10] If W were not assumed to be a sub-matrix of X, the minimization of (3.15a) with respect to $\underline{\beta}$ and γ for given λ would proceed exactly as in the case of (3.3) above.

$$\tilde{\underline{y}}_2 = [I - W(W'W)^{-1}W']\underline{y}_2 \,. \tag{3.20}$$

It is seen that (3.15c) is in precisely the same form as (3.3) with $\tilde{\underline{y}}_1$ and $\tilde{\underline{y}}_2$ replacing $\underline{y}_1$ and $\underline{y}_2$, respectively, and $\tilde{Z}_1$ and $\tilde{X}_1$ replacing Z and X, respectively. Thus the minimizing value of $\underline{\beta}_1$ is

$$\tilde{\underline{\beta}}_1 = (\tilde{Z}_1'\tilde{Z}_1 + \lambda\tilde{X}_1'\tilde{X}_1)^{-1}(\tilde{Z}_1\tilde{\underline{y}}_1 + \lambda\tilde{X}_1\tilde{\underline{y}}_2) \tag{3.21}$$

while the minimizing value for γ is

$$\tilde{\gamma}_{(\lambda)} = \frac{m_{11} - \lambda m_{22} + \sqrt{(\lambda m_{22} - m_{11})^2 + 4\lambda m_{12}^2}}{2m_{12}} \tag{3.22}$$

where $m_{\alpha l} = \hat{\tilde{\underline{y}}}_\alpha'\hat{\tilde{\underline{y}}}_l$, α, $l = 1, 2$, with $\hat{\tilde{\underline{y}}}_\alpha = \tilde{X}_1(\tilde{X}_1'\tilde{X}_1)^{-1}\tilde{X}_1'\tilde{\underline{y}}_\alpha$ for $\alpha = 1, 2$.

As above, estimates of σ_{11} and σ_{22}, say $\hat{\sigma}_{11}$ and $\hat{\sigma}_{22}$, can be obtained from regressions of $\underline{y}_1$ on X and of $\underline{y}_2$ on X, respectively. These can be employed to obtain an estimate of $\lambda = \sigma_{11}/\sigma_{22}$, namely $\hat{\lambda} = \hat{\sigma}_{11}/\hat{\sigma}_{22}$. This value can be substituted in (3.23) to yield an estimate of γ, say $\tilde{\gamma}_{(\hat{\lambda})}$, and $\hat{\lambda}$ can be substituted in (3.21) to obtain an estimate of $\underline{\beta}_1$, $\tilde{\underline{\beta}}_1$, which can be substituted in (3.19) to provide an estimate of $\underline{\beta}_2$. Then having estimates of $\underline{\beta}_1$ and $\underline{\beta}_2$, (3.16) can be employed to obtain an estimate of $\underline{\delta}$. Thus all parameters can be estimated given the estimate $\hat{\lambda}$ of λ. As above, this procedure will produce usual instrumental variable estimates if it is assumed that $\lambda/\gamma^2 \to \infty$or $\lambda/\gamma^2 \to 0$.

4. BAYESIAN ESTIMATION

We now turn to the analysis of the system in (2.4)–(2.5) with the disturbance terms assumed to satisfy (3.1)–(3.2) and, in addition, to be normally distributed. Then the likelihood function is given by:

$$l(\underline{\beta}, \gamma, \sigma_1, \sigma_2 | \underline{y}) \propto \frac{1}{\sigma_1^n\sigma_2^n} \exp\left\{-\frac{1}{2\sigma_1^2}(\underline{y}_1 - X\underline{\beta}\gamma)'(\underline{y}_1 - X\underline{\beta}\gamma) - \frac{1}{2\sigma_2^2}(\underline{y}_2 - X\underline{\beta})'(\underline{y}_2 - X\underline{\beta})\right\} \tag{4.1}$$

where

$$\sigma_1 = \sqrt{\sigma_{11}}, \sigma_2 = \sqrt{\sigma_{22}} \quad \text{and} \quad \underline{y}' = (\underline{y}_1' \vdots \underline{y}_2') \,, \tag{4.2}$$

the sample information.

As regards prior *pdf* for the parameters, we shall first employ a "diffuse" prior *pdf* which has the property that it reproduces, in a sense to be explained below, the sampling theory results of the previous section. That is, we assume[11]

$$p(\underline{\beta}, \gamma, \sigma_1, \sigma_2) \propto 1/\sigma_1\sigma_2 \tag{4.3}$$

with $0<\sigma_1<\infty, 0<\sigma_2<\infty, -\infty < \beta_i < \infty, i = 1, 2, \cdots, k$, and $-\infty < \gamma < \infty$. Here we are assuming that $\log \sigma_1, \log \sigma_2, \gamma$ and the elements of $\underline{\beta}$ are uniformly and

[11] Use of the prior *pdf* in (4.3) here, of course, does not mean that it is appropriate for all situations. In some situations considerable prior information about the parameters may be available which can be represented by a prior *pdf* and incorporated in the analysis in place of (4.3).

independently distributed. With these assumptions and using Bayes' Theorem, the joint posterior *pdf* for the parameters is

$$p(\underline{\beta}, \gamma, \sigma_1, \sigma_2 \mid \underline{y}) \propto \frac{1}{\sigma_1^{n+1}\sigma_2^{n+1}} \exp\left\{-\frac{1}{2\sigma_1^2}[(\underline{y}_1 - Z\underline{\beta})'(\underline{y}_1 - Z\underline{\beta}) + \lambda(\underline{y}_2 - X\underline{\beta})'(\underline{y}_2 - X\underline{\beta})]\right\} \tag{4.4}$$

where, as above, $Z = X\gamma$ and $\lambda = \sigma_1^2/\sigma_2^2$. Thus, use of the prior assumptions in (4.3) leads to a joint posterior *pdf* with modal values for the parameters which are maximum likelihood estimates or very close to such estimates. Further, for given λ, say $\lambda = \hat{\lambda}$, the modal values of γ and $\underline{\beta}$ of the conditional posterior *pdf* will be exactly the estimates discussed in the previous section. While such close correspondences between sampling theory and Bayesian results are interesting, they do not of course necessarily recommend (4.2) as our prior assumptions in all circumstances and, further, modal values are not generally optimal point estimates relative to generally used loss functions.

On completing the square on $\underline{\beta}$ in the the exponent of (4.4), we have

$$p(\underline{\beta}, \gamma, \sigma_1, \sigma_2 \mid \underline{y}) \propto \frac{1}{\sigma_1^{n+1}} \frac{1}{\sigma_2^{n+1}} \exp\left\{-\frac{1}{2\sigma_1^2}[\underline{y}_1'\underline{y}_1 + \lambda \underline{y}_2'\underline{y}_2 - C'M^{-1}C + (\underline{\beta} - M^{-1}C)'M(\underline{\beta} - M^{-1}C)]\right\} \tag{4.5}$$

where $M = Z'Z + \lambda X'X$ and $C = Z'\underline{y}_1 + \lambda X'\underline{y}_2$.

We now integrate (4.5) with respect to $\underline{\beta}$ to obtain

$$\begin{aligned} p(\gamma, \sigma_1, \sigma_2 \mid \underline{y}) &\propto \frac{|M|^{-1/2}}{\sigma_1^{n-k+1}\sigma_2^{n+1}} \exp\left\{-\frac{1}{2\sigma_1^2}(\underline{y}_1'\underline{y}_1 + \lambda \underline{y}_2'\underline{y}_2 - C'M^{-1}C)\right\} \\ &\propto \frac{1}{(\gamma^2+\lambda)^{k/2}\sigma_1^{n-k+1}\sigma_2^{n+1}} \exp\left\{-\frac{1}{2\sigma_1^2}\left[s_{11} + \lambda s_{22} + \frac{\lambda(\hat{\underline{y}}_1 - \gamma\hat{\underline{y}}_2)'(\hat{\underline{y}}_1 - \gamma\hat{\underline{y}}_2)}{\lambda+\gamma^2}\right]\right\}, \end{aligned} \tag{4.6}$$

since $|M| = |(\gamma^2+\lambda)X'X| = (\gamma^2+\lambda)^k|X'X|$, and where $\hat{\underline{y}}_i = X\hat{\underline{\beta}}_i$, $\hat{\underline{\beta}}_i = (X'X)^{-1}X'\underline{y}_i$ and $s_{ii} = (\underline{y}_i - X\hat{\underline{\beta}}_i)'(\underline{y}_i - X\hat{\underline{\beta}}_i)$ for $i = 1, 2$. In (4.6) we now change variables from γ, σ_1 and σ_2 to γ, σ_1, and $\lambda = \sigma_1^2/\sigma_2^2$ to yield

$$p(\gamma, \sigma_1, \lambda \mid \underline{y}) \propto \frac{\lambda^{(n-2)/2}}{(\gamma^2+\lambda)^{k/2}\sigma_1^{2n-k+1}} \exp\left\{-\frac{1}{2\sigma_1^2}\left[s_{11} + \lambda s_{22} + \frac{\lambda(\hat{\underline{y}}_1 - \gamma\hat{\underline{y}}_2)'(\hat{\underline{y}}_1 - \gamma\hat{\underline{y}}_2)}{\lambda+\gamma^2}\right]\right\}. \tag{4.7}$$

Finally, we integrate (4.7) with respect to σ_1 to obtain

$$p(\gamma, \lambda \mid \underline{y}) \propto \left\{\frac{\lambda^{(n-2)/2}}{(\gamma^2+\lambda)^{k/2}}\right\} \frac{1}{\left\{s_{11} + \lambda s_{22} + \dfrac{\lambda(\hat{\underline{y}}_1 - \gamma\hat{\underline{y}}_2)'(\hat{\underline{y}}_1 - \gamma\hat{\underline{y}}_2)}{\lambda+\gamma^2}\right\}^{(2n-k)/2}} \tag{4.8}$$

which is the joint posterior *pdf* for γ and λ. For $k > 1$, numerical integration techniques can be utilized to evaluate the normalizing constant and to obtain marginal posterior *pdf*'s for γ and λ. It should be noted that in obtaining the marginal *pdf* for γ, we do not assign λ a value and proceed conditionally. Rather λ is integrated out of (4.8), an integration which can be given the well known interpretation of averaging a conditional posterior *pdf*, $p(\gamma \mid \lambda, \underline{y}_1, \underline{y}_2)$, with the marginal posterior *pdf* $p(\lambda \mid \underline{y}_1, \underline{y}_2)$ serving as the weighting function. In addition, it is interesting to observe that for given λ, the conditional posterior *pdf*, $p(\gamma \mid \lambda, \underline{y})$, will be dominated by the second factor for large n, a factor which has a modal value given by (3.9).

With respect to posterior inferences about $\underline{\beta}$, we note from the form of (4.5) that the conditional posterior *pdf* for $\underline{\beta}$ given γ, σ_1 and σ_2 is in the multivariate normal form with mean vector given by

$$E(\underline{\beta} \mid \sigma_1, \sigma_2, \gamma, \underline{y}) = M^{-1}C$$

$$= \frac{1}{\gamma^2 + \lambda}[\gamma(X'X)^{-1}X'\underline{y}_1 + \lambda(X'X)^{-1}X'\underline{y}_2] \tag{4.9}$$

and covariance matrix

$$\text{Var}\,(\underline{\beta} \mid \sigma_1, \sigma_2, \gamma, \underline{y}) = \frac{\sigma_2^2}{1 + \gamma^2/\lambda}(X'X)^{-1}\ . \tag{4.10}$$

The right hand side (*r.h.s.*) of (4.9) is in exactly the same form as the *r.h.s.* of (3.5) while the expression on the *r.h.s.* of (4.10) is precisely the variance-covariance matrix of the estimator in (3.5) for given γ and λ.

While the results in (4.9) and (4.10) are interesting, they cannot be used in practice since σ_1, σ_2 and γ are rarely known. To obtain operational procedures for making inferences about $\underline{\beta}$, at least two alternatives are possible. One is to go ahead conditionally, as in the sampling theory approach, by assuming $\sigma_1^2 = \hat{\sigma}_1^2$, $\sigma_2^2 = \hat{\sigma}_2^2$ and $\gamma = \hat{\gamma}$ where the "hatted" quantities are sample estimates, either derived from the posterior *pdf* for these parameters or else precisely the quantities used in the sampling approach described above. Since this is a conditional approach, it may leave something to be desired, particularly in small sample situations. A second alternative is to obtain the marginal *pdf* for a particular element of $\underline{\beta}$, say β_1 by integrating (4.5) with respect to all other elements of $\underline{\beta}$. Then, on introducing $\lambda = \sigma_1^2/\sigma_2^2$, σ_1 can be integrated out to yield $p(\beta_1, \lambda, \gamma \mid \underline{y})$. This trivariate *pdf* can be analyzed numerically[12] to provide the marginal posterior *pdf* for β_1. Similar operations can be performed to obtain the marginal posterior *pdf*'s for other elements of $\underline{\beta}$.[13]

While the prior *pdf* in (4.3) can be employed quite readily in the analysis of the model, we recognize that it does not have all the invariance properties of Jeffreys-Hartigan invariant prior *pdf*'s—see Jeffreys [4, (179)] and Hartigan [3]. For example, while the posterior *pdf* based on (4.3) is invariant with

[12] To avoid the trivariate integration and to use less conditioning information than in the first approach, $p(\beta_1, \lambda, \gamma \mid \underline{y})$ could be conditionalized by taking $\lambda = \hat{\lambda}$. Then the marginal *pdf* for β_1 can be obtained by integrating $p(\beta_1, \gamma \mid \lambda = \hat{\lambda}, \underline{y})$ with respect to γ.

[13] If (4.3) were replaced by $p(\underline{\beta}, \gamma, \sigma_1, \sigma_2) = p_1(\underline{\beta} \mid \sigma_1)p_2(\sigma_1)p_3(\sigma_2)p_4(\gamma)$ where $p_1(\underline{\beta} \mid \sigma_1)$ is multivariate normal with prior mean vector $\bar{\underline{\beta}}$ and prior covariance matrix $A^{-1}\bar{\sigma}_1^2$, with A given, and $p_2(\sigma_1)$ and $p_3(\sigma_2)$ are inverted gamma prior *pdf*'s, the marginal posterior *pdf* for λ and γ can be obtained using operations similar to those presented in the text

respect to changes in the units of measurement, it is not invariant with respect to any one to one differentiable transformation of the parameters.[14] To illustrate the analysis of the model using a particular invariant prior *pdf*, we shall employ Jeffreys' invariant prior *pdf* which is given by

$$p(\gamma, \underline{\beta}, \sigma_1, \sigma_2) \propto |\,\mathrm{Inf}\,|^{1/2} . \tag{4.11}$$

In (4.11) Inf denotes the well known information matrix which for the present model takes the following form:

$$\mathrm{Inf} = \left\{-E\frac{\partial^2 \log l}{\partial\theta_i\partial\theta_j}\right\}$$

$$= \begin{bmatrix} \frac{1}{\sigma_1^2}\underline{\beta}'X'X\underline{\beta} & \frac{\gamma}{\sigma_1^2}\underline{\beta}'X'X & 0 & 0 \\ \frac{\gamma}{\sigma_1^2}X'X\underline{\beta} & \left(\frac{\gamma^2}{\sigma_1^2}+\frac{1}{\sigma_2^2}\right)X'X & 0 & 0 \\ 0 & 0 & \frac{2n}{\sigma_1^2} & 0 \\ 0 & 0 & 0 & \frac{2n}{\sigma_2^2} \end{bmatrix} \tag{4.12}$$

where $\log l$ is the logarithm of the likelihood function, $\partial^2 \log l/\partial\theta_i\partial\theta_j$ is the (i, j)-th second partial derivative with respect to the parameters, $\underline{\theta}'=(\gamma, \underline{\beta}', \sigma_1, \sigma_2)$, and E denotes expectation with respect to the *pdf* for $\underline{y}' = (\underline{y}_1'\underline{y}_2')$ given $\underline{\theta}$. Then, for the present problem, Jeffreys' invariant prior *pdf* is

$$p(\gamma, \underline{\beta}, \sigma_1, \sigma_2) \propto \frac{(\underline{\beta}'X'X\underline{\beta})^{1/2}}{\sigma_1^2\sigma_2^2}\left(\frac{\gamma^2}{\sigma_1^2}+\frac{1}{\sigma_2^2}\right)^{(k-1)/2} \tag{4.13}$$

and the posterior *pdf* based on it is

$$\begin{aligned} p(\gamma, \underline{\beta}, \sigma_1, \sigma_2 \,|\, \underline{y}) &\propto |\,\mathrm{Inf}\,|^{1/2} l(\gamma, \underline{\beta}, \sigma_1, \sigma_2 \,|\, \underline{y}) \\ &\propto \frac{(\underline{\beta}'X'X\underline{\beta})^{1/2}}{\sigma_1^{n+2}\sigma_2^{n+2}}\left(\frac{\gamma^2}{\sigma_1^2}+\frac{1}{\sigma_2^2}\right)^{(k-1)/2} \exp\left\{-\frac{1}{2\sigma_1^2}[(\underline{y}_1 - X\underline{\beta}\gamma)'(\underline{y}_1 - X\underline{\beta}\gamma) \right. \\ &\left. + \lambda(\underline{y}_2 - X\underline{\beta})'(\underline{y}_2 - X\underline{\beta})]\right\} . \end{aligned} \tag{4.14}$$

We see from the first line of (4.14) that, in general, Jeffreys' prior *pdf* can be viewed as "standardizing" the likelihood function in large samples. Further, note that the factor $(\underline{\beta}'X'X\underline{\beta})^{1/2}$ in (4.14) which comes from the prior *pdf*, sets the posterior *pdf* identically equal to zero for $\underline{\beta} = \underline{0}$, a situation which we rule out a priori if we consider the model to be useful. Also, the factor $(\underline{\beta}'X'X\underline{\beta})^{1/2}/\sigma_1$ serves to standardize the conditional *pdf* for γ given $\underline{\beta}$, σ_1 and σ_2 while $(\gamma^2/\sigma_1^2 + 1/\sigma_2^2)^{k/2}$ is a factor from the prior which standardizes the conditional posterior *pdf* for $\underline{\beta}$ given γ, σ_1 and σ_2.

To derive the joint posterior *pdf* for γ and $\lambda = \sigma_1^2/\sigma_2^2$ from (4.14), we com-

[14] Of course, if γ, $\underline{\beta}$, σ_1 and σ_2 are the "natural" parameters for this model, then we would have little interest in reparameterizing the model.

plete the square on $\underline{\beta}$ in the exponent to obtain

$$(4.15)\quad p(\gamma, \underline{\beta}, \sigma_1, \sigma_2 \mid \underline{y}) \propto \left(\frac{\gamma^2}{\sigma_1^2}+\frac{1}{\sigma_2^2}\right)^{(k-1)/2} \frac{(\underline{\beta}'X'X\underline{\beta})^{1/2}}{\sigma_1^{n+2}\sigma_2^{n+2}} \exp\left\{-\frac{1}{2\sigma_1^2}[a+(\underline{\beta}-\bar{\underline{\beta}})'M(\underline{\beta}-\bar{\underline{\beta}})]\right\}$$

where

$$M = (\gamma^2 + \lambda)X'X;\ \bar{\underline{\beta}} = M^{-1}(\gamma X'\underline{y}_1 + \lambda X'\underline{y}_2) = (\gamma\hat{\underline{\beta}}_1 + \lambda\hat{\underline{\beta}}_2)/(\lambda+\gamma^2)\,,$$

and

$$(4.16)\quad \begin{aligned} a &= \underline{y}_1'\underline{y}_1 + \lambda \underline{y}_2'\underline{y}_2 - \bar{\underline{\beta}}'M\bar{\underline{\beta}} \\ &= s_{11} + \lambda s_{22} + \lambda(\hat{\underline{y}}_1 - \gamma\hat{\underline{y}}_2)'(\hat{\underline{y}}_1 - \gamma\hat{\underline{y}}_2)/(\lambda+\gamma^2)\,. \end{aligned}$$

The integration with respect to the elements of $\underline{\beta}$ can then be performed by noting that conditional on γ, σ_1 and σ_2, it is equivalent to finding the expectation of $(\underline{\beta}'X'X\underline{\beta})^{1/2}$ where $\underline{\beta}$ has a multivariate normal *pdf* with mean vector $\bar{\underline{\beta}}$ and covariance matrix $q(X'X)^{-1}$ with $q = \sigma_1^2/(\lambda+\gamma^2)$. If we let $\underline{\beta} = q^{1/2}T\underline{z}$ where T is a square non-singular matrix such that $T'X'XT = I_k$, we have[15]

$$(4.17)\quad \begin{aligned} &\int\cdots\int(\underline{\beta}'X'X\underline{\beta})^{1/2}\left|\frac{M}{\sigma_1^2}\right|^{1/2}\exp\left\{-\frac{1}{2\sigma_1^2}(\underline{\beta}-\bar{\underline{\beta}})'M(\underline{\beta}-\bar{\underline{\beta}})\right\}d\underline{\beta} \\ &= q^{1/2}\int\cdots\int(\underline{z}'\underline{z})^{1/2}\exp\left\{-\frac{1}{2}(z-\bar{z})'(z-\bar{z})\right\}d\underline{z} \end{aligned}$$

where $\bar{\underline{\beta}} = q^{1/2}T\bar{z}$. Note that $z'z$ will be distributed as non-central χ^2 with non-centrality parameter

$$(4.18)\quad \tau^2 = \bar{\underline{z}}'\bar{\underline{z}} = (\gamma\hat{\underline{y}}_1 + \lambda\hat{\underline{y}}_2)'(\gamma\hat{\underline{y}}_1 + \lambda\hat{\underline{y}}_2)/\sigma_1^2(\lambda+\gamma^2)\,.$$

Therefore (4.17) is proportional to

$$(4.19)\quad q^{1/2}e^{-\tau^2/2}\sum_{\alpha=0}^{\infty}\frac{\Gamma\left(\frac{k+1}{2}+\alpha\right)\Gamma\left(\alpha+\frac{1}{2}\right)}{2^\alpha(2\alpha)!\,\Gamma\left(\frac{k}{2}+\alpha\right)}(\tau^2)^\alpha$$

and the joint posterior *pdf* for γ, σ_1 and σ_2 is

$$(4.20)\quad p(\gamma, \sigma_1, \sigma_2|\underline{y}) \propto \frac{\exp\left\{-\frac{1}{2}\left(\frac{a}{\sigma_1^2}+\tau^2\right)\right\}}{\sigma_1^{n+2}\sigma_2^{n+2}\left(\frac{\gamma^2}{\sigma_1^2}+\frac{1}{\sigma_2^2}\right)}\cdot S$$

where a is given in (4.16), τ^2 in (4.18) and S denotes the series in (4.19). Noting that $a + \sigma_1^2\tau^2 = \underline{y}_1'\underline{y}_1 + \lambda\underline{y}_2'\underline{y}_2$ and changing variables to γ, σ_1 and λ, we have

$$(4.21)\quad p(\gamma, \sigma_1, \lambda|\underline{y}) \propto \frac{\lambda^{(n-1)/2}S}{\sigma_1^{2n+1}(\lambda+\gamma^2)}\exp\left\{-\frac{1}{2\sigma_1^2}(\underline{y}_1'\underline{y}_1 + \lambda\underline{y}_2'\underline{y}_2)\right\}.$$

We next integrate with respect to σ_1, taking account of the fact that terms of the series S have σ_1 appearing in them, to obtain our final result

[15] Note that $|M/\sigma_1^2|^{1/2} = \left(\frac{\gamma^2}{\sigma_1}+\frac{1}{\sigma_2^2}\right)^{k/2}|X'X|^{1/2}$.

$$p(\gamma, \lambda \mid \underline{y}) \propto \frac{\lambda^{(n-1)/2}}{(\lambda + \gamma^2)(\underline{y}_1'\underline{y}_1 + \lambda \underline{y}_2'\underline{y}_2)^n} \sum_{\alpha=0}^{\infty} b_\alpha \left[\frac{(\gamma\hat{\underline{y}}_1 + \lambda\hat{\underline{y}}_2)'(\gamma\hat{\underline{y}}_1 + \lambda\hat{\underline{y}}_2)}{(\underline{y}_1'\underline{y}_1 + \lambda\underline{y}_2'\underline{y}_2)(\lambda + \gamma^2)}\right]^\alpha \tag{4.22}$$

with

$$b_\alpha = \frac{\Gamma\left(\alpha + \frac{1}{2}\right)\Gamma\left(\frac{k+1}{2} + \alpha\right)}{2^n(2\alpha)!\,\Gamma\left(\frac{k}{2} + \alpha\right)} .$$

In (4.22), it is relevant to observe that for given λ, the quantity in square brackets under the summation sign achieves its maximum for a value of γ which is given precisely by (3.9). Numerical methods can be employed to evaluate the normalizing constant of (4.22) and to obtain marginal posterior *pdf*'s.

5. CONCLUDING REMARKS

In the present work, we have seen that there is a close link between the model considered in this paper and the classical errors in the variables model (*EVM*). In fact, certain widely employed instrumental variable estimators were shown to be limiting cases of an estimator which is precisely in the form of the slope coefficient estimator for the *EVM*. One difference in the present problem is that it is possible to use sample information to estimate the ratio of error variances. This estimate can be employed to construct an operational estimator.

Further, in the sampling theory estimation approach pursued above, an estimator for $\underline{\beta}$ was found which appears to be much better than the one that is currently often employed, namely $\hat{\underline{\beta}}_2 = (X'X)^{-1}X'\underline{y}_2$, since the former estimator is based on $2n$ observations rather than n and the sample information has been appropriately weighted.[16] In situations in which there is interest in estimating $\underline{\beta}$, as for example when it is desired to get good estimates of consuming units' permanent incomes, the estimator for $\underline{\beta}$ introduced above should yield more precise estimates than $\hat{\underline{\beta}}_2$.

In the Bayesian approach to the analysis of the model, it was found that posterior *pdf*'s can be readily computed. For two relatively diffuse prior *pdf*'s, properties of conditional posterior *pdf*'s for γ given λ and for $\underline{\beta}$ given γ and λ were explored. However, there is no need to use conditional results

[16] This point has relevance for distributed lag models of the following type:

$$\text{(i)} \quad C_t = kY_t^*(\underline{\theta}) + u_t ,$$
$$\text{(ii)} \quad Y_t = Y_t^*(\underline{\theta}) + v_t , \qquad t = 1, 2, \cdots, T$$

where C_t = measured consumption, Y_t = measured income, $Y_t^*(\underline{\theta})$ = expected income depending on parameters $\underline{\theta}$, and u_t and v_t are transitory consumption and income respectively. Given an expression for $Y_t^*(\underline{\theta})$ in terms of observable variables, it is clear that both the C_t and the Y_t series have information about $\underline{\theta}$. Yet usual analyses of the model, or others like it, utilize just the information in one of these series in the estimation of $\underline{\theta}$. I am indebted to H. Glejser for calling my attention to this application of the principles developed in this paper.

in the Bayesian approach. The marginal posterior *pdf* for parameters of the model, such as γ, can be computed and, given a loss function, it is well known that optimal point estimates can be obtained.

Last, while we did not treat the case in which the X matrix contains lagged values of the y's, that is the system is autoregressive, for given initial conditions, the analysis of the paper can be applied to yield estimates and posterior *pdf*'s. In this case, sampling theory estimators' finite sample properties appear difficult to establish. However, given a prior *pdf*, the Bayesian approach enables one to make finite sample posterior probability statements without much difficulty—see e.g., Zellner and Tiao [**11**], and Thornber [**9**] where this point is made in other contexts.

Postscript. Since this paper was written in 1967, the following interesting papers dealing with the finite sample properties of $\tilde{\gamma}_{(\infty)}$ have been published: T. Sawa, "The Exact Sampling Distribution of Ordinary Least Squares and Two-Stage Least Squares Estimators," *Journal of the American Statistical Association*, LXIV (September, 1969), 923–37, and K. Takeuchi, "Exact Sampling Moments of the Ordinary Least Squares, Instrumental Variable, and Two-Stage Least Squares Estimators," *International Economic Review*, XI (February, 1970), 1–12. In addition, the following unpublished paper dealing with the relation of simultaneous equation and errors in the variables models has recently come to the attention of the author: J. W. Hooper, "Comments on Some Limited Information Estimators," (undated).

The University of Chicago, U.S.A.

APPENDIX

In this appendix we present some additional points bearing on the approach we have taken in the analysis of our two equation model. First, we note that the equations

$$\underline{y}_1 = X\underline{\beta}\gamma + \underline{v}_1 \tag{A.1}$$

$$\underline{y}_2 = X\underline{\beta} + \underline{v}_2 \tag{A.2}$$

can be equivalently expressed, for $\gamma \neq 0$, as

$$\underline{y}_2 = X\underline{\xi}/\gamma + \underline{v}_2 \tag{A.3}$$

$$\underline{y}_1 = X\underline{\xi} + \underline{v}_1 \tag{A.4}$$

where $\underline{\xi} = \underline{\beta}\gamma$. The form of (A.1)-(A.2) suggest to some the use of the estimator $\gamma_{(\infty)}$ while the form of (A.3)-(A.4) might suggest the use of the estimator $\gamma_{(0)}$. The estimator $\gamma_{(\hat{\lambda})}$ is in a real sense a compromise between $\tilde{\gamma}_{(0)}$ and $\tilde{\gamma}_{(\infty)}$ since for any given sample of data $|\tilde{\gamma}_{(\infty)}| < |\tilde{\gamma}_{(\hat{\lambda})}| < |\tilde{\gamma}_{(0)}|$ is a set of inequalities satisfied by the estimates.

Richardson [8], employing Kabe's [5] approach based on the non-central Wishart distribution, has derived conditions for the existence of the moments of $\tilde{\gamma}_{(\infty)}$. For the present problem, it is easily established that the mean of $\tilde{\gamma}_{(\infty)}$ exists for $k > 1$. When the mean exists, Richardson's analysis shows that $\tilde{\gamma}_{(\infty)}$ is biased toward zero with the absolute size of the relative bias, $|(E\tilde{\gamma}_{(\infty)} - \gamma)/\gamma|$, being determined by $\underline{\beta}'X'X\underline{\beta}/\sigma_{22}$ and k, the number of elements in $\underline{\beta}$. For fixed k and $\underline{\beta}'X'X\underline{\beta}$, as σ_{22} increases in size, the bias increases. This result can be appreciated heuristically by noting that as σ_{22} increases (A.2) provides a poorer and poorer estimate, $\underline{\hat{y}}_2$ of $X\underline{\beta}$. Thus when $\underline{\hat{y}}_2 = X\underline{\hat{\beta}}_2$ is substituted for $X\underline{\beta}$ in (A.1), the measurement error variances for the elements of the independent variable vector, $\underline{\hat{y}}_2 = X\underline{\hat{\beta}}_2$ grow in size as σ_{22} increases and so does the bias of $\tilde{\gamma}_{(\infty)}$. Since $|\tilde{\gamma}_{(\infty)}| < |\tilde{\gamma}_{(\hat{\lambda})}|$ for any given sample, it may very well be that the estimator $\tilde{\gamma}_{(\hat{\lambda})}$ will have a distribution centered closer to the true value of the parameter than is that of $\tilde{\gamma}_{(\infty)}$ when sample information is limited.[17] Finally, while we do not wish to pursue the matter further here, it appears quite straightforward to derive the *pdf* for the sampling theory estimator $\tilde{\gamma}_{(0)}$. It should be noted that the mean of this estimator does not exist since $\underline{\hat{\beta}}_1'X'X\underline{\hat{\beta}}_1/\underline{\hat{\beta}}_1'X'X\underline{\hat{\beta}}_2$, for given $\underline{\hat{\beta}}_1$, is the reciprocal of a normally distributed random variable.

REFERENCES

[1] CROCKETT, J., "Technical Note," in I. Friend and R. Jones, *Proceedings of the Conference on Consumption and Saving*, University of Pennsylvania (1960), 213-22.

[2] FRIEDMAN, M., *A Theory of the Consumption Function* (Princeton: Princeton University Press, 1957).

[3] HARTIGAN, J., "Invariant Prior Distribution," *Annals of Mathematical Statistics*, XXXV (June, 1964), 836-45.

[4] JEFFREYS, H., *Theory of Probability*, 3rd ed. (Oxford: Clarendon Press, 1961).

[5] KABE, D. G., "A Note on the Exact Distributions of the GCL Estimators in Two Leading Overidentified Cases," *Journal of American Statistical Association*, LVIII (June, 1963), 533-7.

[6] MADANSKY, A., "The Fitting of Straight Lines when Both Variables are Subject to Error," *Journal of American Statistical Association*, LIV (March, 1959), 173-205.

[7] MILLER, M. H. and F. MODIGLIANI, "Some Estimates of the Cost of Capital to the Electric Utility Industry, 1954-57," *American Economic Review*, LVI (June, 1966), 333-91.

[8] RICHARDSON, D. H., "The Exact Distribution of a Structural Coefficient Estimator," Research Paper No. 16, University of Kansas (1967).

[9] THORNBER, H., "Finite Sample Monte Carlo Studies: An Autoregressive Illustration," *Journal of American Statistical Association*, LXII (September, 1967), 801-18.

[10] ZELLNER, A., "Bayesian Inference and Simultaneous Equation Econometric Models," Paper presented to the First World Congress of the Econometric Society, Rome (1965).

[11] ——— AND G. C. TIAO, "Bayesian Analysis of the Regression Model with Autocorrelated Errors," *Journal of American Statistical Association*, LIX (September, 1964), 763-78.

[17] However, the estimator $\tilde{\gamma}(\hat{\lambda})$ does not in general possess a finite mean.

[14]

Bayesian and Non-Bayesian Analysis of the Regression Model with Multivariate Student-t Error Terms

ARNOLD ZELLNER*

The linear multiple regression model is analyzed assuming the error vector has a multivariate Student-t distribution with zero location vector and scalar dispersion matrix; the multivariate Cauchy and normal distributions are special cases. It is found that the usual least squares coefficient estimate is the maximum likelihood estimate and the mean of the posterior distribution under a diffuse prior distribution. Inferences based on usual t- and F-statistics are shown valid for a range of error distributional assumptions including the multivariate-t assumption. Inferences about the scale parameter of the multivariate-t distribution can be made using an F-distribution rather than the usual χ^2 (or inverted χ^2) distribution.

1. INTRODUCTION

In many applications of the univariate linear regression model, error terms are assumed normally and independently distributed, each with zero mean and common variance. In this paper, a broader assumption is employed, namely, that the error terms have a joint multivariate Student-t distribution with zero location vector. With this assumption, the marginal distribution of each error is univariate Student-t, a distribution that includes the Cauchy and normal distributions as special cases. That error terms can have nonnormal distributions in practice has been appreciated by many, including Newcomb [8], Jeffreys [5, pp. 64–5; 6, pp. 214–6], Mandelbrot [7], Wise [12], Fama and Roll [4], Du Mouchel [3], Box and Tiao [2], Praetz [9] and Blattberg and Gonedes [1]. In the work of Newcomb and Jeffreys, discrete mixtures of distributions are employed. Mandelbrot, Wise, Fama, Roll and Du Mouchel consider the class of stable Paretian distributions, while Box and Tiao analyze a class of power distributions. Last, Praetz and Blattberg and Gonedes consider continuous mixtures of normal distributions that lead to errors that are independent, each with the same univariate Student-t distribution. Jeffreys [5, pp. 64–5] also reports results of analyses assuming error terms independent, each with the same univariate Student-t distribution.

By assuming that regression error terms have a joint multivariate Student-t distribution, it will be seen that inference procedures for the regression model are relatively simple. Also, certain effects of departures from normality can be readily analyzed.

The plan of this paper is as follows. In Section 2, the model is specified and several situations in which it is applicable are described. Then likelihood and sampling theory inference procedures are developed. A Bayesian analysis of the regression model with multivariate Student-t error terms under diffuse and natural conjugate priors is presented in Section 3 and the appendix. Last, Section 4 contains a summary of results and some concluding remarks.

2. LIKELIHOOD AND SAMPLING THEORY CONSIDERATIONS

Let the model for n observations, $\mathbf{y}' = (y_1, y_2, \ldots, y_n)$, be

$$\mathbf{y} = X\boldsymbol{\beta} + \mathbf{u} \ , \qquad (2.1)$$

where X is an $n \times k$ nonstochastic matrix with rank k, $\boldsymbol{\beta}$ is a $k \times 1$ vector of regression parameters with unknown values, and $\mathbf{u}$ is an $n \times 1$ random error vector. It is assumed that the joint probability density function (pdf) for the n elements of $\mathbf{u}$ is the multivariate Student-t pdf[1]

$$p(\mathbf{u}|\nu_o, \sigma) = \frac{g(\nu_o)}{(\sigma^2)^{n/2}} \frac{1}{\{\nu_o + \mathbf{u}'\mathbf{u}/\sigma^2\}^{(n+\nu_o)/2}} \qquad \begin{array}{l} \sigma, \nu_o > 0 \\ -\infty < u_i < \infty \\ i = 1, 2, \ldots, n \end{array} \qquad (2.2)$$

where $g(\nu_o) = \nu_o^{\nu_o/2}(\nu_o + n/2)/\pi^{n/2}\Gamma(\nu_o/2)$. As is well-known, the pdf in (2.2) is symmetric about the origin, its modal value. Its mean vector and covariance matrix are given by

$$E\mathbf{u} = 0 \quad \text{for } \nu_o > 1 \ , \quad \text{and}$$

$$E\mathbf{u}\mathbf{u}' = \nu_o\sigma^2/(\nu_o - 2)I_n \quad \text{for } \nu_o > 2 \ ; \qquad (2.3)$$

* Arnold Zellner is H.G.B. Alexander professor of economics and statistics, Graduate School of Business, University of Chicago, Chicago, IL 60637. Research was financed in part by National Science Foundation Grants GS-2347 and GS-40033, and by income from the H.G.B. Alexander Endowment Fund, Graduate School of Business, University of Chicago. The author is grateful to R.C. Blattberg, F. Palm, S. Stigler, W. Vandaele and the referees for helpful comments.

[1] See, e.g., [10, pp. 256–9 and 13, pp. 383–9] for a review of some properties of this pdf.

© Journal of the American Statistical Association
June 1976, Volume 71, Number 354
Theory and Methods Section

while the elements of $\mathbf{u}$ are uncorrelated, it is clear from the form of their joint pdf in (2.2) that they are not independent. When $\nu_o \to \infty$, the pdf in (2.3) approaches a multivariate normal form with zero mean vector and covariance matrix $\sigma^2 I_n$. For finite ν_o, u_i/σ has a univariate Student-*t* pdf with ν_o degrees of freedom. For small values of ν_o, the marginal distributions of elements of $\mathbf{u}$ will have heavy tails.[2]

It is interesting and useful to recognize the well-known fact that the multivariate Student-*t* pdf in (2.2) is a member of the class of distributions[3]

$$p(\mathbf{u}\,|\,\cdot) = \int_0^\infty p_N(\mathbf{u}\,|\,\tau)p(\tau\,|\,\cdot)d\tau\ , \qquad (2.4)$$

where

$$p_N(\mathbf{u}\,|\,\tau) = (2\pi\tau^2)^{-n/2}\exp\{-\mathbf{u}'\mathbf{u}/2\tau^2\} \qquad -\infty < u_i < \infty \qquad i = 1, 2, \ldots, n\ , \qquad (2.5)$$

and $p(\tau\,|\,\cdot)$ with $0 < \tau < \infty$, is a proper pdf for τ. When $p(\tau\,|\,\cdot)$ is chosen to be the inverted gamma pdf,

$$p_{IG}(\tau\,|\,\nu_o, \sigma) = [2/\Gamma(\nu_o/2)](\nu_o\sigma^2/2)^{\nu_o/2}\tau^{-(\nu_o+1)} \cdot \exp\{-\nu_o\sigma^2/2\tau^2\}\ , \qquad (2.6)$$

with $0 < \nu_o, \sigma, \tau < \infty$, the integration in (2.4) produces a marginal multivariate Student-*t* pdf for $\mathbf{u}$, $p(\mathbf{u}\,|\,\nu_o, \sigma)$ that is given in (2.2). Thus, it is possible to regard the error vector $\mathbf{u}$ in (2.1) as being randomly drawn from a multivariate normal distribution with a random standard deviation generated from (2.6).

Several situations in which the process in (2.1)–(2.2) may be useful in practice include the following. First, the observations $\mathbf{y}' = (y_1, y_2, \ldots, y_n)$ may be generated by a measuring instrument in a particular run. If the instrument's variability, represented by τ's value, has an unknown value within a run and is known to vary over runs, the model just presented in which τ is considered random can be useful. Second, if the elements of $\mathbf{y}$ are returns on particular stocks, in the "market-model" regression, $y_t = \beta_1 + \beta_2 x_t + u_t$, $t = 1, 2, \ldots, n$, where x_t is the market return, it may be reasonable to assume that the errors have a common random standard deviation τ with the inverted gamma pdf shown in (2.6). Third, in regressions for individuals or different groups of consumers or firms, common standard deviations of error terms may vary randomly across groups in accord with (2.6). If so, the regression model for a particular group would be in the form of (2.1)–(2.2).

Using (2.1) and (2.2), the likelihood function for the regression model is given by

$$p(\mathbf{y}\,|\,\boldsymbol{\beta}, \nu_o, \sigma) = [g(\nu_o)/(\sigma^2)^{n/2}]\{\nu_o + [\nu s^2 + (\boldsymbol{\beta} - \hat{\boldsymbol{\beta}})'X'X(\boldsymbol{\beta} - \hat{\boldsymbol{\beta}})]/\sigma^2\}^{-(n+\nu_o)/2}\ , \qquad (2.7)$$

where

$$\hat{\boldsymbol{\beta}} = (X'X)^{-1}X'\mathbf{y}\ , \quad \nu s^2 = (\mathbf{y} - X\hat{\boldsymbol{\beta}})'(\mathbf{y} - X\hat{\boldsymbol{\beta}})\ , \qquad (2.8)$$

and $\nu = n - k$. From (2.7) it is seen that $\hat{\boldsymbol{\beta}}$ and s^2 are sufficient statistics and that the least squares estimate, $\hat{\boldsymbol{\beta}}$, is the maximum likelihood (ML) estimate for $\boldsymbol{\beta}$. The ML estimate for σ^2 is

$$\hat{\sigma}^2 = (\mathbf{y} - X\hat{\boldsymbol{\beta}})'(\mathbf{y} - X\hat{\boldsymbol{\beta}})/n\ . \qquad (2.9)$$

Further, $\hat{\boldsymbol{\beta}}$ is the ML estimate for $\boldsymbol{\beta}$ for all likelihood functions that are monotonically decreasing functions of $\mathbf{u}'\mathbf{u} = (\mathbf{y} - X\boldsymbol{\beta})'(\mathbf{y} - X\boldsymbol{\beta})$.[4] In addition, when $\nu_o > 2$ and (2.3) is satisfied by the error vector $\mathbf{u}$, the conditions of the Gauss-Markov theorem are met. Thus, $\hat{\boldsymbol{\beta}}$ in (2.8) is a minimum variance linear unbiased estimator[5] and also a minimum variance unbiased estimator.

From (2.8) and (2.1), $\hat{\boldsymbol{\beta}} = \boldsymbol{\beta} + (X'X)^{-1}X'\mathbf{u}$, and for $\nu_o > 1$, $E\hat{\boldsymbol{\beta}} = \boldsymbol{\beta}$, since $E\mathbf{u} = \mathbf{0}$. Also, for $\nu_o > 2$, the covariance matrix for $\hat{\boldsymbol{\beta}}$, $V(\hat{\boldsymbol{\beta}})$, is

$$V(\hat{\boldsymbol{\beta}}) = E(\hat{\boldsymbol{\beta}} - \boldsymbol{\beta})(\hat{\boldsymbol{\beta}} - \boldsymbol{\beta})' = (X'X)^{-1}\nu_o\sigma^2/(\nu_o - 2)\ . \qquad (2.10)$$

Note that as $\nu_o \to \infty$, $\mathbf{u} \to N(\mathbf{0}, \sigma^2 I_n)$ and $V(\hat{\boldsymbol{\beta}}) \to (X'X)^{-1}\sigma^2$. Thus, for small and moderate values of ν_o, the variances of the elements of $\hat{\boldsymbol{\beta}}$ are inflated considerably, relative to those for large values of ν_o, in which case the multivariate Student-*t* distribution of $\hat{\boldsymbol{\beta}}$ is close to $N[\boldsymbol{\beta}, (X'X)^{-1}\sigma^2]$.

It is important to observe that even though the elements of $\mathbf{u}$ have the nonnormal pdf in (2.2) and are not independent, tests and intervals based on usual *t*- and *F*-statistics remain valid. That this is the case can be seen most easily by considering, for example, $t = (\hat{\beta}_i - \beta_i)/s(m^{ii})^{\frac{1}{2}}$, where m^{ii} is the (i, i)th element of $(X'X)^{-1}$ and $s^2 = \hat{\mathbf{u}}'\hat{\mathbf{u}}/(n - k)$. Given τ, the error vector $\mathbf{u}$ has the usual normal pdf, shown in (2.5), and, thus, t given τ has a univariate Student-*t* pdf with $n - k$ degrees of freedom. Since the joint pdf of t and τ, say, $p(t, \tau)$, can be written $p(t, \tau) = p(t\,|\,\tau)\,p_{IG}(\tau)$ and $p(t\,|\,\tau) = p(t)$, t and τ are independent. Thus, t is unconditionally univariate Student-*t* and usual probability statements based on this distribution will be appropriate. Similarly, any statistic, say, F, that has an F_{ν_1,ν_2} pdf given τ will also have a marginal F_{ν_1,ν_2} pdf, since the joint pdf for F and τ, $p(F, \tau)$, can be factored as $p(F, \tau) = p(F\,|\,\tau)p_{IG}(\tau)$. Since $p(F\,|\,\tau)$ does not depend on τ, F and τ are independent, and, thus, F has an un-

---

[2] When $\nu_o = 1$, (2.2) is the multivariate Cauchy pdf for which no moments exist.

[3] The class shown in (2.4), a "continuous mixture" of normal pdfs, can be extended to include discrete as well as continuous mixtures. Strawderman [11] has recently considered minimax estimation of the location parameters of such mixtures of normal distributions but does not treat the range of problems considered herein.

[4] That $\hat{\beta} = (X'X)^{-1}X'\mathbf{y}$ is the ML estimate when the $\mathbf{u}$ vector has a nonnormal distribution might appear to be in conflict with a result due to Gauss [6, p. 214]. This result, generalized to apply to the regression model, states that if the elements of $\mathbf{u}$ are *independently* distributed, each with zero mean and common finite variance, and if $\hat{\beta}$ is the ML estimate, then the elements of $\mathbf{u}$ are normally distributed. In this work, the independence assumption required to obtain Gauss' result is not satisfied.

[5] Note that if the elements of $\mathbf{u}$ are independent each with the same univariate Student-*t* pdf, as Jeffreys [5], Praetz [9], and Blattberg and Gonedes [1] assume, $\hat{\beta}$ is not the ML estimator. However, $\hat{\beta}$ is still a minimum variance linear unbiased estimator given that its second moments exist. Further, in this case, $E(\hat{\beta} - \beta)(\hat{\beta} - \beta)' = (X'X)^{-1}\sigma_u^2$, where $\sigma_u^2 = \nu_c\sigma_c^2/(\nu_c - 2)$, $\nu_c > 2$, is the common variance of the elements of $\mathbf{u}$ and ν_c and σ_c^2 are parameters of their common univariate Student-*t* pdf.

conditional F_{ν_1,ν_2} pdf. *This argument holds not only for* τ *having a marginal IG* pdf, $p_{IG}(\tau)$, *but also for any proper* pdf $p(\tau|\cdot)$ *for* τ *in* (2.4).

With respect to the ML estimator $\hat{\sigma}^2$ in (2.10), we have for $\nu_o > 2$, $E\hat{\sigma}^2 = E\mathbf{u}'[I_n - X(X'X)^{-1}X']\mathbf{u}/n = (n-k)\sigma_u^2/n$, where $\sigma_u^2 = \nu_o\sigma^2/(\nu_o - 2)$ is the common variance of the elements of $\mathbf{u}$—see (2.3). Thus, $\hat{\mathbf{u}}'\hat{\mathbf{u}}/(n-k)$ is an unbiased estimator for σ_u^2, while $(\nu_o - 2)\hat{\mathbf{u}}'\hat{\mathbf{u}}/\nu_o(n-k)$ is an unbiased estimator for σ^2. In the class of estimators $q\hat{\mathbf{u}}'\hat{\mathbf{u}}$, where q is a positive scalar, the minimal mean square error (MMSE) estimator for σ^2 is, with $\nu_o > 4$, $\tilde{\sigma}^2 = (\nu_o - 4)\hat{\mathbf{u}}'\hat{\mathbf{u}}/\nu_o(\nu + 2)$, while the MMSE estimator for σ_u^2 in this class is $\tilde{\sigma}_u^2 = (\nu_o - 4)\hat{\mathbf{u}}'\hat{\mathbf{u}}/(\nu_o - 2)(\nu + 2)$.

To test hypotheses and construct intervals for σ^2 or σ_u^2, note that given τ, $\mathbf{u}$ has the normal pdf shown in (2.7). Thus, given τ, $\hat{\mathbf{u}}'\hat{\mathbf{u}}/\tau^2 \equiv \omega/\nu\sigma^2$ has a χ^2 pdf with ν degrees of freedom. Let $x = \omega/\nu\sigma^2 = s^2/\sigma^2$, where $s^2 = \hat{\mathbf{u}}'\hat{\mathbf{u}}/\nu$. Since we know from (2.6) that $\sigma^2/\tau^2 \sim (\chi_{\nu_o}^2/\nu_o)^{-1}$, and since $x = (\omega/2\sigma^2)/(\sigma^2/\tau^2)$, it follows that $x = s^2/\sigma^2$ has an F-distribution with ν and ν_o degrees of freedom. This fact can be used to construct confidence intervals for and test hypotheses about σ^2. Since $\sigma_u^2 = \nu_o\sigma^2/(\nu_o - 2)$, where σ_u^2 is the common variance of the elements of $\mathbf{u}$, the distribution of $x = s^2/\sigma^2$ can also be employed to set up intervals and test hypotheses relating to σ_u^2 for any given $\nu_o > 2$.

Last, on examination of the likelihood function in (2.7), a maximum of the function with respect to $\boldsymbol{\beta}$, σ^2, *and* ν_o does not exist.[6] Thus, while ML estimates of $\boldsymbol{\beta}$ and σ^2 exist for any *given* ν_o, ML estimates of $\boldsymbol{\beta}$, σ^2 *and* ν_o do not exist. Thus, to use some of the results presented before, it is necessary to assign a value to ν_o that reflects an investigator's knowledge of the distributional properties of the regression error terms. Use of the usual normal assumption for the error terms is equivalent to assigning a large value to ν_o. When a small value for ν_o is thought to be more reasonable, it can be assigned, and the results shown before are available. In Section 3, it will be seen that it is possible to incorporate a prior pdf for ν_o in a Bayesian analysis.

3. BAYESIAN ANALYSIS

The model in (2.1), with error terms generated by (2.2) and likelihood function shown in (2.7), will be analyzed from the Bayesian point of view with a diffuse prior pdf for the parameters. Initially, it will be assumed that ν_o's value is given. Our diffuse prior pdf for the elements of $\boldsymbol{\beta}$ and σ^2 is

$$p(\boldsymbol{\beta}, \sigma^2) \propto 1/\sigma^2 \,, \quad \begin{array}{l} -\infty < \beta_i < \infty \,, \\ 0 < \sigma^2 < \infty \,, \end{array} \quad i = 1, 2, \ldots, k \,. \quad (3.1)$$

In (3.1), it is assumed that the elements of $\boldsymbol{\beta}$ and $\log \sigma^2$ are uniformly and independently distributed. Then the posterior pdf for the parameters is

$$\begin{aligned} p(\boldsymbol{\beta}, \sigma^2 | \mathbf{y}, \nu_o) &\propto (\sigma^2)^{-(n+2)}\{\nu_o + (\mathbf{y} - X\boldsymbol{\beta})'(\mathbf{y} - X\boldsymbol{\beta})/\sigma^2\}^{-(\nu_1+\nu_o)/2} \\ &\propto \frac{(\sigma^2)^{\nu_o/2-1}}{(\bar{\sigma}^2)^{(n+\nu_o)/2}}\{\nu_1 + (\boldsymbol{\beta} - \hat{\boldsymbol{\beta}})'X'X(\boldsymbol{\beta} - \hat{\boldsymbol{\beta}})/\bar{\sigma}^2\}^{-(\nu_1+k)/2} \\ &\propto \{A(\boldsymbol{\beta})\}^{-n/2}\left\{\frac{[\nu_o\sigma^2/A(\boldsymbol{\beta})]^{\nu_o/2-1}}{A(\boldsymbol{\beta})[1 + \nu_o\sigma^2/A(\boldsymbol{\beta})]^{(n+\nu_o)/2}}\right\} , \quad (3.2) \end{aligned}$$

where $A(\boldsymbol{\beta}) \equiv (\mathbf{y} - X\boldsymbol{\beta})'(\mathbf{y} - X\boldsymbol{\beta})$, $\bar{\sigma}^2 = (\nu_o\sigma^2 + \nu s^2)/\nu_1$, $\nu_1 = \nu + \nu_o$, $\nu = n - k$, $\hat{\boldsymbol{\beta}} = (X'X)^{-1}X'\mathbf{y}$ and $\nu s^2 = (\mathbf{y} - X\hat{\boldsymbol{\beta}})'(\mathbf{y} - X\hat{\boldsymbol{\beta}})$.

From the second line of (3.2), it is seen that the conditional posterior pdf for $\boldsymbol{\beta}$ given σ^2 and ν_o is in the form of a multivariate Student-t pdf with mean $\hat{\boldsymbol{\beta}}$, the least squares quantity and conditional posterior covariance matrix,

$$\begin{aligned} V(\boldsymbol{\beta}|\mathbf{y}, \sigma^2, \nu_o) &= (X'X)^{-1}\nu_1\bar{\sigma}^2/(\nu_1 - 2) \\ &= (X'X)^{-1}(\nu_o\sigma^2 + \nu s^2)/(\nu_o + \nu - 2) \,, \quad (3.3) \end{aligned}$$

provided that $\nu_1 > 2$. As $\nu_o \to \infty$, the conditional posterior pdf for $\boldsymbol{\beta}$ given σ^2 approaches a multivariate normal pdf with mean $\hat{\boldsymbol{\beta}}$ and covariance matrix $(X'X)^{-1}\sigma^2$, the usual result for the normal regression model with our diffuse prior pdf.

To obtain the marginal posterior pdf for $\boldsymbol{\beta}$, note from the third line of (3.2) that if we let $z = n\sigma^2/A(\boldsymbol{\beta})$, $0 < z < \infty$, then the joint posterior pdf for $\boldsymbol{\beta}$ and z is

$$p(\boldsymbol{\beta}, z|\mathbf{y}, \nu_o) \propto \{A(\boldsymbol{\beta})\}^{-n/2}\{z^{\nu_o/2-1}/(1 + \nu_o/n)^{(n+\nu_o)/2}\} \,. \quad (3.4)$$

From the form of (3.4), it is seen that $z = n\sigma^2/A(\boldsymbol{\beta})$ has an $F_{\nu_o,n}$ pdf and $\boldsymbol{\beta}$ and z are independent. Thus, the marginal posterior pdf for $\boldsymbol{\beta}$ is

$$\begin{aligned} p(\boldsymbol{\beta}|\mathbf{y}, \nu_o) &\propto \{A(\boldsymbol{\beta})\}^{-n/2} \\ &\propto \{\nu s^2 + (\boldsymbol{\beta} - \hat{\boldsymbol{\beta}})'X'X(\boldsymbol{\beta} - \hat{\boldsymbol{\beta}})\}^{-(\nu+k)/2} \,, \quad (3.5) \end{aligned}$$

which is in the form of a k-dimensional Student-t pdf and does not depend on the value given to ν_o. In fact, (3.5) is precisely the result that one obtains in the Bayesian analysis of the normal regression model with the diffuse prior for the parameters shown in (3.1). This result is the Bayesian analog of the sampling theory results reported in Section 2.

The marginal posterior pdf for σ^2 can be obtained by integrating the second line of (3.2) with respect to the elements of $\boldsymbol{\beta}$ using properties of the multivariate Student-t pdf. The result is[7]

$$p(\sigma^2|\mathbf{y}, \nu_o) \propto (\sigma^2/s^2)^{(\nu_o/2)-1}/(1 + (\nu_o\sigma^2/\nu s^2))^{(\nu_o+\nu)/2} \,. \quad (3.6)$$

From the form of (3.6), it is apparent that the marginal posterior pdf for σ^2/s^2 is an F pdf with ν_o and ν degrees

[6] That is, the necessary conditions on β, σ^2 and ν_o for a maximum of the likelihood function cannot be satisfied for $\nu_o \geq 1$. It may be possible to estimate ν_o from the data with repeated observations.

[7] If one has a prior pdf for ν_o, say, $p(\nu_o)$, with $0 < \nu_o < c$ (where c is a finite constant or ∞), it can be used to average (3.6) as $p(\sigma^2|\mathbf{y}) = \int_0^c p(\sigma^2|\mathbf{y}, \nu_o)p(\nu_o)d\nu_o$.

of freedom, a finding paralleling the sampling theory result of Section 2. From properties of the F pdf, the modal value of σ^2/s^2 is $(\nu/\nu_o)(\nu_o - 2)/(\nu + 2)$, when $\nu_o > 2$, and its mean is $\nu/(\nu - 2)$ when $\nu > 2$. Also, as $\nu_o \to \infty$, the posterior distribution of $\nu s^2/\sigma^2$ approaches a χ^2 pdf with ν degrees of freedom, a distributional result that holds for Bayesian analysis of the usual normal regression model with diffuse prior assumptions. This is reasonable, since as $\nu_o \to \infty$, the error terms' pdf approaches a normal form. Last, it was noted before that the posterior pdf for $z = n\sigma^2/(\mathbf{y} - X\boldsymbol{\beta})'(\mathbf{y} - X\boldsymbol{\beta})$ is $F_{\nu_o,n}$. The effect of integrating out the elements of $\boldsymbol{\beta}$ to obtain the marginal posterior pdf for σ^2/s^2 is to change the degrees of freedom in the F pdf from ν_o and n to ν_o and $\nu = n - k$, a reasonable change, since k β's have been integrated out.

In the preceding, we have employed the diffuse prior pdf given in (3.1). It is interesting to inquire about the form of the natural conjugate prior distribution for the present model. To answer this question, note that the likelihood function in (2.7) can be expressed as

$$p(\mathbf{y}|\boldsymbol{\beta}, \nu_o, \sigma^2) \propto \left\{\frac{(\nu_o\sigma^2/\nu s^2)^{\nu_o/2}}{(1 + \nu_o\sigma^2/\nu s^2)^{(\nu_o+\nu)/2}}\right\} \cdot \left\{\frac{1}{(\bar{\sigma}^2)^{k/2}[\nu_1 + (\boldsymbol{\beta} - \hat{\boldsymbol{\beta}})'X'X(\boldsymbol{\beta} - \hat{\boldsymbol{\beta}})/\bar{\sigma}^2]^{(\nu_1+k)/2}}\right\}, \quad (3.7)$$

where $\nu_1 = \nu_o + \nu$ and $\bar{\sigma}^2 = (\nu_o\sigma^2 + \nu s^2)/\nu_1$. From the form of (3.7), the natural conjugate prior distribution for σ^2 and $\boldsymbol{\beta}$ is the product of a marginal F pdf for σ^2 times a conditional k-dimensional multivariate Student-t pdf for $\boldsymbol{\beta}$ given σ^2, i.e.,

$$p(\boldsymbol{\beta}, \sigma^2|\cdot) = p_F(\sigma^2|\cdot)p_S(\boldsymbol{\beta}|\sigma^2, \cdot) ,$$

where $p_F(\sigma^2|\cdot)$ denotes an F pdf and $p_S(\boldsymbol{\beta}|\sigma^2, \cdot)$ a conditional multivariate Student-t pdf.[8] Given ν_o, the prior parameters of $p_F(\sigma^2|\cdot)$ are analogs of ν and s^2 appearing in the first factor of (3.7). Similarly, the prior parameters appearing in $p_S(\boldsymbol{\beta}|\sigma^2, \cdot)$ are analogs of $\hat{\boldsymbol{\beta}}$ and $X'X$. Thus, explicitly, the factors of the natural conjugate prior in (3.8) for the present model are given by

$$p_F(\sigma^2|s_a^2, \nu_a, \nu_o) \propto \frac{(\nu_o\sigma^2/\nu_a s_a^2)^{(\nu_o-2)/2}}{(1 + \nu_o\sigma^2/\nu_a s_a^2)^{(\nu_o+\nu_a)/2}} \qquad \begin{array}{l} 0 < \sigma < \infty \\ \nu_a, s_a > 0 \end{array} , \quad (3.8a)$$

and

$$p_S(\boldsymbol{\beta}|\sigma^2, \bar{\boldsymbol{\beta}}, A, \bar{\nu}_a) \propto (\bar{\sigma}_a^2)^{-k/2}\{\bar{\nu}_a + (\boldsymbol{\beta} - \bar{\boldsymbol{\beta}})'A(\boldsymbol{\beta} - \bar{\boldsymbol{\beta}})/\bar{\sigma}_a^2\}^{-(\bar{\nu}_a+k)/2} \qquad \begin{array}{c} A \text{ pds} \\ -\infty < \beta_i < \infty \\ i = 1, 2, \ldots, k \end{array} , \quad (3.8b)$$

where $\bar{\boldsymbol{\beta}}$ is the prior mean vector, $\bar{\nu}_a = \nu_o + \nu_a$ and $\bar{\sigma}_a^2 = (\nu_a s_a^2 + \nu_o\sigma^2)/\bar{\nu}_a$.

As with the natural conjugate for the usual normal regression model, it is seen that $\boldsymbol{\beta}$ and σ^2 are not independent in the natural conjugate prior distribution in (3.8). If the natural conjugate prior distribution is thought to represent the available prior information adequately, it can be used in obtaining posterior distributions (see the appendix). Of course, if (3.8) arises from the analysis of a previous sample, the likelihood function for the previous and current samples can be formulated and used in conjunction with (3.1) to obtain posterior distributions. Finally, when the number of parameters is not large, as, for example, in simple regression, use of numerical integration techniques to analyze posterior distributions permits a rather flexible choice of informative prior distributions—see, e.g., [14, 15, and 16], where this approach was utilized in analyzing small normal models with informative prior distributions.

4. SUMMARY AND CONCLUDING REMARKS

In this paper, the traditional multiple regression model has been analyzed under the assumption that error terms have a joint multivariate Student-t pdf with zero mean vector and covariance matrix in the form of a scalar times a unit matrix. Thus, the errors are uncorrelated but not independent. Under these assumptions, it was shown that the maximum likelihood (ML) estimator for the regression coefficient vector is the least squares estimator and that the ML estimator is a minimum variance linear unbiased estimator when relevant moments exist. Further, given existence of relevant moments, probability statements based on usual t- and F-statistics were found to hold even though the errors were assumed nonnormal. However, to make inferences about the scale parameter, it is necessary to use an F-distribution rather than the usual χ^2 distribution.

In the Bayesian analysis of the model with a diffuse prior pdf for the regression coefficients and multivariate Student-t error terms, it was found that the joint posterior distribution for the regression coefficients is in precisely the same multivariate Student-t form as arises from the usual normal model. However, the posterior distribution of the scale parameter σ^2 was found to be in the form of an F-distribution. When the degrees of freedom parameter in the error term distribution grows large, and thus the error terms' distribution approaches normality, the posterior distribution of $\nu s^2/\sigma^2$ approaches the usual χ^2 pdf with ν degrees of freedom. Last, a natural conjugate prior distribution for the "multivariate Student-t" regression model was presented.

With respect to inferences about the regression coefficients, it is relevant to appreciate that several important inference procedures for the normal regression model remain valid for the multivariate Student-t model. However, inferences about the scale parameter σ^2 will depend on the extent of the departure from normality, as measured by the value of the degrees of freedom parameter

[8] As $\nu_o \to \infty$, (3.8) converges to the usual normal-gamma conjugate prior distribution.

ν_o. Finally, it is conjectured that results similar to those presented before will be found for multivariate regression models with errors following a matrix Student-t distribution.

APPENDIX

In this appendix, the technical problems of analyzing the multivariate Student-t regression model with the natural conjugate prior distribution shown in (3.8) are considered. In particular, it will be shown how marginal posterior distributions for σ^2 and for individual regression coefficients can be computed. In this work, it is convenient to employ the joint pdf

$$p(\mathbf{y}, \tau \mid \boldsymbol{\beta}, \sigma^2) \propto \left[\frac{1}{\tau^n} \exp\left\{-\frac{1}{2\tau^2}(\mathbf{y} - X\boldsymbol{\beta})'(\mathbf{y} - X\boldsymbol{\beta})\right\}\right] \cdot \left[\frac{(\sigma^2)^{\nu_o/2}}{\tau^{\nu_o+1}} \exp\left\{-\frac{\nu_o\sigma^2}{2\tau^2}\right\}\right]. \quad \text{(A.1)}$$

When (A.1) is integrated with respect to τ, $0 < \tau < \infty$, the result is the likelihood function shown in (2.7).

As regards the natural conjugate prior distribution in (3.8), it is convenient for the present analysis to write it as

$$p(\boldsymbol{\beta}, \sigma^2, \Theta \mid \cdot) \propto p_F(\sigma^2 \mid \cdot)\left[\frac{1}{\Theta^k} \exp\left\{-\frac{1}{2\Theta^2}(\boldsymbol{\beta} - \bar{\boldsymbol{\beta}})'A(\boldsymbol{\beta} - \bar{\boldsymbol{\beta}})\right\}\right] \cdot \left[\frac{(\bar\sigma_a^2)^{\bar\nu_a/2}}{\Theta^{\bar\nu_a+1}} \exp\left\{-\frac{\bar\nu_a\bar\sigma_a^2}{2\Theta^2}\right\}\right], \quad \text{(A.2)}$$

where $p_F(\sigma^2 \mid \cdot)$ is given in (3.8a), $\bar\nu_a$ and $\bar\sigma_a^2$ have been defined in connection with (3.8b) and Θ is an auxiliary parameter satisfying $0 < \Theta < \infty$. When (A.2) is integrated with respect to Θ, the resulting marginal prior pdf for $\boldsymbol{\beta}$ and σ^2 is precisely that shown in (3.8).

On multiplying (A.1) and (A.2) and performing relevant integrations, marginal posterior distributions for σ^2 and for elements of $\boldsymbol{\beta}$ can be obtained. To obtain the marginal posterior pdf for σ^2, write the product of (A.1) and (A.2) as

$$p_F(\sigma^2 \mid \cdot)(\sigma^2)^{\nu_o/2}(\bar\sigma_a^2)^{\bar\nu_a/2}\left[\frac{1}{\tau^{n+\nu_o+1}} \exp\left\{-\frac{\nu s^2 + \nu_o\sigma^2}{2\tau^2}\right\}\right] \cdot \left[\frac{1}{\Theta^{k+\bar\nu_a+1}} \exp\left\{-\frac{\bar\nu_a\bar\sigma_a^2}{2\Theta^2}\right\}\right] \cdot \exp\left\{-\frac{1}{2\tau^2}[(\boldsymbol{\beta} - \hat{\boldsymbol{\beta}})'X'X(\boldsymbol{\beta} - \hat{\boldsymbol{\beta}}) + (\tau^2/\Theta^2)(\boldsymbol{\beta} - \bar{\boldsymbol{\beta}})'A(\boldsymbol{\beta} - \bar{\boldsymbol{\beta}})]\right\}. \quad \text{(A.3)}$$

Letting $\lambda = \tau^2/\Theta^2$, we can complete the square on $\boldsymbol{\beta}$ in the last exponential of (A.3) as

$$(\boldsymbol{\beta} - \hat{\boldsymbol{\beta}})'X'X(\boldsymbol{\beta} - \hat{\boldsymbol{\beta}}) + \lambda(\boldsymbol{\beta} - \bar{\boldsymbol{\beta}})'A(\boldsymbol{\beta} - \bar{\boldsymbol{\beta}}) = (\boldsymbol{\beta} - \bar{\boldsymbol{\beta}}_\lambda)'M(\boldsymbol{\beta} - \boldsymbol{\beta}_\lambda) + \hat{\boldsymbol{\beta}}'X'X\hat{\boldsymbol{\beta}} + \lambda\bar{\boldsymbol{\beta}}'A\bar{\boldsymbol{\beta}} - \bar{\boldsymbol{\beta}}_\lambda'M\bar{\boldsymbol{\beta}}_\lambda\,, \quad \text{(A.4)}$$

where $\bar{\boldsymbol{\beta}}_\lambda = M^{-1}(X'X\hat{\boldsymbol{\beta}} + \lambda A\bar{\boldsymbol{\beta}})$ with $M = X'X + \lambda A$. With (A.4) introduced in (A.3), (A.3) can be integrated with respect to the elements of $\boldsymbol{\beta}$ to yield

$$\frac{p_F(\sigma^2 \mid \cdot)(\sigma^2)^{\nu_o/2}(\bar\sigma_a^2)^{\bar\nu_a/2}}{\tau^{n+\nu_o+1}} \exp\left\{-\frac{\nu s^2 + \nu_o\sigma^2}{2\tau^2}\right\} \frac{1}{\Theta^{k+\bar\nu_a+1}} \exp\left\{-\frac{\bar\nu_a\bar\sigma_a^2}{2\Theta^2}\right\} \cdot |M/\tau^2|^{-\frac{1}{2}} \exp\left\{-\frac{1}{2\tau^2}[\hat{\boldsymbol{\beta}}'X'X\hat{\boldsymbol{\beta}} + \lambda\bar{\boldsymbol{\beta}}'A\bar{\boldsymbol{\beta}} - \bar{\boldsymbol{\beta}}_\lambda'M\bar{\boldsymbol{\beta}}_\lambda]\right\}. \quad \text{(A.5)}$$

We now change parameters in (A.5) from σ^2, τ and Θ to σ^2, τ and $\lambda = \tau^2/\Theta^2$. The Jacobian of this transformation is proportional to $\lambda^{\frac{1}{2}}/\tau$. Thus, (A.5) expressed in terms of σ^2, τ, and λ is

$$\frac{p_F(\sigma^2 \mid \cdot)(\sigma^2)^{\nu_o/2}(\bar\sigma_a^2)^{\bar\nu_a/2}}{\tau^{n+\nu_o+\bar\nu_a+3}} |M|^{-\frac{1}{2}}\lambda^{(k+\bar\nu_a+4)/2} \exp\left\{-\frac{1}{2\tau^2}[\nu s^2 + \nu_o\sigma^2 + \bar\nu_a\bar\sigma_a^2\lambda + \hat{\boldsymbol{\beta}}'X'X\hat{\boldsymbol{\beta}} + \lambda\bar{\boldsymbol{\beta}}'A\bar{\boldsymbol{\beta}} - \bar{\boldsymbol{\beta}}_\lambda'M\bar{\boldsymbol{\beta}}_\lambda]\right\}. \quad \text{(A.6)}$$

(A.6) can be integrated with respect to τ to obtain the marginal posterior pdf for σ^2 and λ, namely,

$$\frac{p_F(\sigma^2 \mid \cdot)(\sigma^2)^{\nu_o/2}(\bar\sigma_a^2)^{\bar\nu_a/2}|M|^{-\frac{1}{2}}\lambda^{(k+\bar\nu_a+4)/2}}{\{\nu s^2 + \nu_o\sigma^2 + \bar\nu_a\bar\sigma_a^2\lambda + \hat{\boldsymbol{\beta}}'X'X\hat{\boldsymbol{\beta}} + \lambda\bar{\boldsymbol{\beta}}'A\boldsymbol{\beta} - \bar{\boldsymbol{\beta}}_\lambda'M\bar{\boldsymbol{\beta}}_\lambda\}^{(n+\nu_o+\bar\nu_a+2)/2}}. \quad \text{(A.7)}$$

Using (3.8a), $\bar\sigma_a^2 = (\nu_a s_a^2 + \nu_o\sigma^2)/\bar\nu_a$ and $\nu_a = \nu_o + \nu_a$, (A.7) can be expressed as

$$p(\sigma^2, \lambda \mid \mathbf{y}, \cdot) = p_1(\sigma^2 \mid \lambda, \mathbf{y}, \cdot)p_2(\lambda \mid \mathbf{y}, \cdot)\,, \quad \text{(A.8a)}$$

with

$$p_1(\sigma^2 \mid \lambda, \mathbf{y}, \cdot) \propto \left[\frac{(1 + \lambda)\nu_o\sigma^2}{\nu s^2 + g(\lambda)}\right]^{(2\nu_o-2)/2} \Bigg/ \left[1 + \frac{(1 + \lambda)\nu_o\sigma^2}{\nu s^2 + g(\lambda)}\right]^{(2\nu_o+n+\nu_a+2)/2}, \quad \text{(A.8b)}$$

and

$$p_2(\lambda \mid \mathbf{y}, \cdot) \propto |M|^{-\frac{1}{2}}\lambda^{(k+\bar\nu_a+4)/2} \Big/ (1 + \lambda)^{(\nu_o-2)/2}[\nu s^2 + g(\lambda)]^{(n+\bar\nu_a+4)/2}\,, \quad \text{(A.8c)}$$

where $g(\lambda) = \lambda(\nu_a s_a^2 + \bar{\boldsymbol{\beta}}'A\bar{\boldsymbol{\beta}}) - \bar{\boldsymbol{\beta}}_\lambda'M\bar{\boldsymbol{\beta}}_\lambda + \hat{\boldsymbol{\beta}}'X'X\hat{\boldsymbol{\beta}}$. It is seen that the conditional posterior pdf for σ^2 given λ is in the form of an F pdf. Bivariate numerical integration techniques can be employed to obtain the marginal posterior pdf for σ^2 from (A.8a).

To obtain the posterior pdf for $\boldsymbol{\beta}$, note that (A.2) is proportional to

$$\frac{(\sigma^2)^{\nu_o-1}}{\Theta^{k+\bar\nu_a+1}\tau^{n+\nu_o+1}} \exp\left\{-\frac{1}{2\tau^2}[\nu s^2 + \nu_o\sigma^2 + (\tau^2/\Theta^2)(\nu_a s_a^2 + \nu_o\sigma^2) + c_\lambda + (\boldsymbol{\beta} - \bar{\boldsymbol{\beta}}_\lambda)'M(\boldsymbol{\beta} - \bar{\boldsymbol{\beta}}_\lambda)]\right\} \quad \text{(A.9)}$$

where $c_\lambda = \hat{\boldsymbol{\beta}}'X'X\hat{\boldsymbol{\beta}} + \bar{\boldsymbol{\beta}}'A\bar{\boldsymbol{\beta}} - \bar{\boldsymbol{\beta}}_\lambda'M\bar{\boldsymbol{\beta}}_\lambda$. In (A.9), change variables from σ^2, τ and Θ to σ^2, τ and $\lambda = \tau^2/\Theta^2$ and integrate out σ^2 and τ with the result,

$$p(\boldsymbol{\beta}, \lambda \mid \mathbf{y}, \cdot) \propto \frac{\lambda^{(k+\nu_o+4)/2}(1 + \lambda)^{\nu_o}}{\{\nu s^2 + \lambda\nu_a s_a^2 + c_\lambda + (\boldsymbol{\beta} - \bar{\boldsymbol{\beta}}_\lambda)'M(\boldsymbol{\beta} - \bar{\boldsymbol{\beta}}_\lambda)\}^{(n+k+\nu_o+1)/2}}. \quad \text{(A.10)}$$

From (A.10), it is seen that the conditional posterior pdf for $\boldsymbol{\beta}$ given λ is in the form of a multivariate Student-t pdf. Thus, (A.10) can be integrated analytically to obtain the joint posterior pdf for any element of $\boldsymbol{\beta}$, say, β_i, and λ, a bivariate pdf that can be analyzed numerically to obtain the marginal posterior pdf for β_i. For ease of numerical computation, it will be convenient to express $p(\beta_i, \lambda \mid \mathbf{y}, \cdot)$ in terms of β_i and $\phi = \lambda/(1 + \lambda)$, since the range of ϕ is given by $0 < \phi < 1$ given that $0 < \lambda < \infty$.

[Received May 1974. Revised November 1974.]

REFERENCES

[1] Blattberg, R.C. and Gonedes, N.J., "A Comparison of the Stable and Student Distributions as Statistical Models for Stock Prices," Unpublished manuscript, Graduate School of Business, University of Chicago, 1972.

[2] Box, G.E.P. and Tiao, G.C., "A Further Look at Robustness via Bayes Theorem," *Biometrika*, 49 (December 1962), 419–33.

[3] Du Mouchel, W., "Stable Distributions in Statistical Inference," Unpublished doctoral dissertation, Department of Statistics, Yale University, 1971.

[4] Fama, E.F. and Roll, R., "Parameter Estimates for Symmetric Stable Distributions," *Journal of the American Statistical Association*, 66 (June 1971), 331–8.

[5] Jeffreys, H., *Scientific Inference*, 2nd ed., Cambridge, England: Cambridge University Press, 1957.

[6] ———, *Theory of Probability*, 3rd ed., Oxford: Clarendon Press, 1967.

[7] Mandelbrot, B., "The Variation of Certain Speculative Prices," *Journal of Business*, 36 (October 1963), 394–419.

[8] Newcomb, S., "A Generalized Theory of the Combination of Observations So As to Obtain the Best Results," *American Journal of Mathematics*, 8 (June 1886), 343–66.

[9] Praetz, P.D., "The Distribution of Share Price Changes," *Journal of Business*, 45 (January 1972), 49–55.

[10] Raiffa, H. and Schlaifer, R., *Applied Statistical Decision Theory*, Boston: Graduate School of Business Administration, Harvard University, 1961.

[11] Strawderman, W.E., "Minimax Estimation of Location Parameters for Certain Spherically Symmetric Distributions," *Journal of Multivariate Analysis*, 4 (September 1974), 255–64.

[12] Wise, J., "Linear Estimators for Linear Regression Systems Having Infinite Residual Variances," Unpublished manuscript, Department of Economics, University of Hawaii, 1966.

[13] Zellner, A., "*An Introduction to Bayesian Inference in Econometrics*, New York: John Wiley & Sons, Inc., 1971.

[14] ——— and Geisel, M.S., "Analysis of Distributed Lag Models with Applications to Consumption Function Estimation," *Econometrica*, 38 (November 1970), 865–88.

[15] ——— and Williams, A.D., "Bayesian Analysis of the Federal Reserve-MIT-PENN Model's Almon Lag Consumption Function," *Journal of Econometrics* (October 1973), 267–300.

[16] ——— and Richard, J.F., "Use of Prior Information in the Analysis and Estimation of Cobb-Douglas Production Function Models," *International Economic Review*, 14 (February 1973), 107–19.

[15]

INTERNATIONAL ECONOMIC REVIEW
Vol. 20, No. 3, October, 1979

AN ERROR-COMPONENTS PROCEDURE (ECP) FOR INTRODUCING PRIOR INFORMATION ABOUT COVARIANCE MATRICES AND ANALYSIS OF MULTIVARIATE REGRESSION MODELS*

BY ARNOLD ZELLNER[1]

1. INTRODUCTION

The present paper takes up the problem of analyzing multivariate regression models when there is prior information about disturbance variances and correlations. For example, in a bivariate regression model it may be known that the variance of the first disturbance term is smaller than that of the second and that the two disturbance terms are positively correlated. Below a simple error-components procedure (ECP) for introducing such disturbance variance inequalities and positivity (or negativity) restrictions on disturbance correlations is described. It will be seen that introduction of such restrictions generally affects the precision with which regression coefficients can be estimated by reducing the number of free parameters in a disturbance covariance matrix. In general, an $m \times m$ *pds* disturbance covariance matrix has $m(m+1)/2$ distinct parameters. Use of the aforementioned variance and correlation restrictions can considerably reduce the number of free parameters in a covariance matrix.

In the literature, prior information about a disturbance covariance matrix has often been introduced by assuming that the matrix has a particular structure, e.g., the intra-class correlation structure in which all disturbances are assumed to have the same variance and to be equi-correlated in which case there are just two free parameters in the covariance matrix — see, e.g., Jeffreys [1967, p. 302]. The method for introducing prior information about a disturbance covariance matrix, described below, resembles the intra-class correlation approach but is less restrictive. For recent work on informative prior distributions for disturbance covariance matrices, see Lindley [1975] and Press [1975]. In these works, a hyper-distribution in the intra-class correlation form is assumed and a prior distribution for the parameters of the hyper-distribution is introduced. In contrast, in the present work, the form of the disturbance covariance matrix is first suitably simplified and then a prior distribution for the remaining free parameters is introduced.

The plan of the paper is as follows. In Section 2 a method of introducing variance inequalities and/or positivity (or negativity) restrictions on correlations is described. Sampling theory estimation procedures for regression systems

* Manuscript received August 5, 1977; revised June 12, 1978.

[1] Research financed by National Science Foundation Grant SOC 77-15662 and by income from the H.G.B. Alexander Endowment Fund, Graduate School of Business, U. of Chicago.

with restricted disturbance covariance matrices are presented and compared with procedures not incorporating such restrictions. From a Bayesian point of view, some of the results in this section can be viewed as establishing sampling properties of posterior measures that may be useful in approaching design problems. A Bayesian treatment of estimation problems is presented in Section 3. Last, in Section 4 some concluding remarks are presented.

2. COVARIANCE MATRIX RESTRICTIONS AND REGRESSION ANALYSIS: SAMPLING RESULTS

The approach to be taken in generating error covariance matrices with pre-assigned properties is one that involves close study of the error structure of a problem. It will be seen that certain assumptions about the error structure will imply variance inequalities and particular algebraic signs for error terms' correlations. To illustrate, we first consider a bivariate problem and then proceed to consider more general cases.

2.1. *Bivariate Seemingly Unrelated Regression* (*SUR*) *Case.* We consider a usual two-equation SUR system:

$$\begin{aligned} y_{1i} &= \underline{x}'_{1i}\,\underline{\beta}_1 + u_{1i} \\ y_{2i} &= \underline{x}'_{2i}\,\underline{\beta}_2 + u_{2i} \end{aligned} \qquad i = 1, 2,\ldots, n \tag{2.1}$$

where y_{1i} and y_{2i} are the i-th observations on two dependent variables, $\underline{x}'_{1i}$ and $\underline{x}'_{2i}$ are $1\times k_1$ and $1\times k_2$ non-stochastic vectors of independent variables, $\underline{\beta}_1$ and $\underline{\beta}_2$ are $k_1\times 1$ and $k_2\times 1$ vectors of regression coefficients with unknown values, and u_{1i} and u_{2i} are disturbance terms. We assume that the pairs (u_{1i}, u_{2i}) are independently drawn from a population with zero mean vector and 2×2 positive definite symmetric matrix, Σ with typical element σ_{ij}, $i, j=1, 2$.

In previous work it has been customary to consider the system (2.1) without restrictions on the values of elements of the disturbance covariance matrix, the σ_{ij}'s. In the present work, we introduce restrictions of the form $\sigma_{22}>\sigma_{11}$ and $0<\rho_{12}<1$, where $\rho_{12}=\sigma_{12}/\sqrt{\sigma_{11}\sigma_{22}}$ by introducing the following assumptions about the disturbance terms in (2.1):

$$u_{1i} = \eta_i \tag{2.2a}$$

$$u_{2i} = \eta_i + \varepsilon_{2i} \qquad i = 1, 2,\ldots, n. \tag{2.2b}$$

In (2.2), we assume that both disturbances contain a *common* random element, η_i, and that, in addition, u_{2i} contains a *specific* random component ε_{2i}. We assume that $E\eta_i=0$, $E\varepsilon_{2i}=0$, $\operatorname{Var}\eta_i=\sigma_{\eta\eta}$, $\operatorname{Var}\varepsilon_{2i}=\omega_{22}$ and $\operatorname{Cov}(\eta_i, \varepsilon_{2i})=0$ for $i=1, 2,\ldots, n$. Then the covariance matrix for the disturbance system in (2.2), denoted by Σ_r, is given by:

(2.3)
$$\Sigma_r = \begin{bmatrix} \sigma_{\eta\eta} & \sigma_{\eta\eta} \\ \sigma_{\eta\eta} & \sigma_{\eta\eta} + \omega_{22} \end{bmatrix}.$$

From (2.3), it is seen that the variance of u_{2i}, $\sigma_{\eta\eta}+\omega_{22}$, is indeed larger than the variance of u_{1i}, $\sigma_{\eta\eta}$, and that $\rho_{12}=\sqrt{\sigma_{\eta\eta}/(\sigma_{\eta\eta}+\omega_{22})}$ satisfies $0<\rho_{12}<1$. Thus use of the error structure (2.2) produces the restricted covariance matrix (2.3) that incorporates inequality restrictions on the variances and correlation coefficient. Further, Σ_r in (2.3) contains just *two* free parameters, $\sigma_{\eta\eta}$ and ω_{22}, whereas the unrestricted 2×2 covariance matrix $\Sigma=\{\sigma_{ij}\}$ contains *three* free parameters, σ_{11}, σ_{22} and σ_{12}. If it is desired to restrict ρ_{12} to the interval $-1<\rho_{12}<0$, this can be done by replacing (2.2b) with $u_{2i}=-\eta_i+\varepsilon_{2i}$. Also, if one wishes to drop the variance inequality restriction while maintaining the restriction $0<\rho_{12}<1$, this can be done by replacing (2.2a) with $u_{1i}=\eta_i+\varepsilon_{1i}$, where for all i, ε_{1i} has zero mean, constant variance and is uncorrelated with η_i and ε_{2i}. Below we take up the analysis of (2.1) subject to (2.2a–b).

We now wish to determine how use of the covariance matrix in (2.3) affects estimation of $\underline{\beta}_1$ and $\underline{\beta}_2$ in (2.1). We write (2.1) for all the observations as

(2.4a)
$$\begin{pmatrix} \underline{y}_1 \\ \underline{y}_2 \end{pmatrix} = \begin{pmatrix} X_1 & 0 \\ 0 & X_2 \end{pmatrix} \begin{pmatrix} \underline{\beta}_1 \\ \underline{\beta}_2 \end{pmatrix} + \begin{pmatrix} \underline{u}_1 \\ \underline{u}_2 \end{pmatrix}$$

or

(2.4b)
$$\underline{y} = Z\underline{\beta} + \underline{u}$$

where, for $\alpha=1, 2$, $\underline{y}_\alpha$ is an $n\times 1$ vector, X_α is a $n\times k_\alpha$ non-stochastic matrix of rank k_α and $\underline{u}_\alpha$ is an $n\times 1$ disturbance vector. In (2.4b) $\underline{y}'=(\underline{y}_1' \ \underline{y}_2')$, $\underline{\beta}'=(\underline{\beta}_1' \ \underline{\beta}_2')$, $\underline{u}'=(\underline{u}_1' \ \underline{u}_2')$ and Z represents the block diagonal matrix on the r.h.s. of (2.4a). The covariance matrix for $\underline{u}$ is assumed given by $\Omega=\Sigma_r\otimes I_n$. Then if Ω has a known value, as pointed out in Zellner [1962], the minimum variance linear unbiased (MVLU) estimator for $\underline{\beta}'=(\underline{\beta}_1' \ \underline{\beta}_2')$ is:

(2.5a)
$$\underline{\hat{\beta}}^* = (Z'\Omega^{-1}Z)^{-1}Z'\Omega^{-1}\underline{y}$$

with variance-covariance matrix

(2.5b)
$$V(\underline{\hat{\beta}}^*) = (Z'\Omega^{-1}Z)^{-1}.$$

Also $\underline{\hat{\beta}}^*$ in (2.5) is a maximum likelihood estimator for likelihood functions of the form $f(\underline{u}'\Omega^{-1}\underline{u})$ that are monotonically decreasing in $\underline{u}'\Omega^{-1}\underline{u}$. If $X_1=X_2=X$, $\underline{\hat{\beta}}^*$ in (2.5a) reduces algebraically to $\underline{\hat{\beta}}^*=(Z'Z)^{-1}Z'\underline{y}$, or $\underline{\hat{\beta}}_1^*=(X'X)^{-1}X'\underline{y}_1$ and $\underline{\hat{\beta}}_2^*=(X'X)^{-1}X'\underline{y}_2$, regardless of whether the covariance matrix Σ is restricted or not. However, whereas $\underline{\hat{\beta}}^*=(Z'Z)^{-1}Z'\underline{y}$ when Σ is diagonal, in the present problem the form of (2.3) is such that the restricted covariance matrix Σ_r is a priori non-diagonal, an implication of the error term specification in (2.2).

When Σ_r in (2.3) is inserted in (2.5a), after some simplification we obtain

$$(2.6)\qquad \underline{\hat{\beta}}^* = \begin{pmatrix} \underline{\hat{\beta}}_1^* \\ \underline{\hat{\beta}}_2^* \end{pmatrix} = \begin{pmatrix} \phi X_1'X_1 & -X_1'X_2 \\ -X_2'X_1 & X_2'X_2 \end{pmatrix}^{-1} \begin{pmatrix} \phi X_1'\underline{y}_1 - X_1'\underline{y}_2 \\ -X_2'\underline{y}_1 + X_2'\underline{y}_2 \end{pmatrix}$$

where $\phi=(\sigma_{\eta\eta}+\omega_{22})/\sigma_{\eta\eta}$. Since Σ_r's value is usually unknown in applications, following the approach set forth in Zellner [1962], a consistent estimate of Σ_r, say $\hat{\Sigma}_r$, can be inserted in (2.6) to yield an estimator $\underline{\tilde{\beta}}^*$. From the results in Zellner [1962], $\underline{\hat{\beta}}^*$ and $\underline{\tilde{\beta}}^*$ have the same asymptotic normal distribution and $\underline{\tilde{\beta}}^*$ is asymptotically efficient. Further, as pointed out Zellner [1962], it is possible to use $\underline{\tilde{\beta}}^*$ to obtain new residuals, another estimate of Σ_r, a new estimate of $\underline{\beta}$, etc. Ruble [1968] and Kmenta and Gilbert [1968] have pointed out that this iterative procedure produces the maximum likelihood estimate for $\underline{\beta}$, say $\underline{\hat{\beta}}_{ML}$. While $\underline{\tilde{\beta}}^*$ and $\underline{\hat{\beta}}_{ML}$ have the same asymptotic normal distribution, in finite samples their distributions are different with neither uniformly dominating the other in terms of mean-squared error; see, e.g., Kmenta and Gilbert [1968]. As is well known, with a diffuse prior *pdf*, the *ML* estimate is equal to or close to the modal value of a posterior *pdf*. On the other hand, $\underline{\hat{\beta}}^*$ will be shown to be equal to the mean of a conditional posterior *pdf* for $\underline{\beta}$, given Σ_r, when the prior *pdf* is diffuse. In addition, it is pertinent to point out that whether a consistent estimate of Σ_r or of Σ, the unrestricted covariance matrix is employed in (2.5), the resulting estimators for $\underline{\beta}$ will have the same asymptotic distribution given that the restrictions imposed on Σ, shown in (2.3), are true and provided that the independent variables do not include lagged values of the dependent variables, that is, that the model is a pure regression model with no autoregressive features.

To illustrate how use of (2.3) affects properties of regression coefficient estimators, we consider (2.6) in the special case in which $X_1'X_2=0$.[2] If $X_1'X_2=0$, then

$$(2.7a)\qquad \underline{\hat{\beta}}_1^* = \underline{\hat{\beta}}_1 - \phi^{-1}\underline{\hat{\beta}}_{12}$$

$$(2.7b)\qquad \underline{\hat{\beta}}_2^* = \underline{\hat{\beta}}_2 - \underline{\hat{\beta}}_{21}$$

where $\underline{\hat{\beta}}_1=(X_1'X_1)^{-1}X_1'\underline{y}_1$, $\underline{\hat{\beta}}_2=(X_2'X_2)^{-1}X_2'\underline{y}_2$, $\underline{\hat{\beta}}_{12}=(X_1'X_1)^{-1}X_1'\underline{y}_2$ and $\underline{\hat{\beta}}_{21}=(X_2'X_2)^{-1}X_2'\underline{y}_1$. The following properties of the estimators in (2.7) are noteworthy.

(i) The expression for $\underline{\hat{\beta}}_2^*$ does *not* involve elements of Σ_r.
(ii) $E\underline{\hat{\beta}}_1^*=\underline{\beta}_1$ and $E\underline{\hat{\beta}}_2^*=\underline{\beta}_2$ since, of course, $\underline{\hat{\beta}}_1^*$ and $\underline{\hat{\beta}}_2^*$ are MVLU estimators.
(iii) $V(\underline{\hat{\beta}}_1^*)=(X_1'X_1)^{-1}\omega_{22}/\phi=(X_1'X_1)^{-1}\omega_{22}\sigma_{\eta\eta}/(\sigma_{\eta\eta}+\omega_{22})$

and

(iv) $V(\underline{\hat{\beta}}_2^*)=(X_2'X_2)^{-1}\omega_{22}$.

On comparing $V(\underline{\hat{\beta}}_1^*)$ with $V(\underline{\hat{\beta}}_1)=(X_1'X_1)^{-1}\sigma_{\eta\eta}$ and $V(\underline{\hat{\beta}}_2^*)$ with $V(\underline{\hat{\beta}}_2)=(X_2'X_2)^{-1}\cdot(\omega_{22}+\sigma_{\eta\eta})$, we have

[2] Mehta and Swamy [1976], Revankar [1974], Srivastava [1970] and Zellner and Huang [1962] relax this assumption.

(2.8a) $$V(\underline{\hat{\beta}}_1^*) = \left(\frac{\sigma_{\eta\eta}}{\sigma_{\eta\eta} + \omega_{22}}\right) V(\underline{\hat{\beta}}_1)$$

and

(2.8b) $$V(\underline{\hat{\beta}}_2^*) = \left(\frac{\omega_{22}}{\sigma_{\eta\eta} + \omega_{22}}\right) V(\underline{\hat{\beta}}_2).$$

In (2.8), we have compared the covariance matrices of the MVLU estimators $\underline{\hat{\beta}}_1^*$ and $\underline{\hat{\beta}}_2^*$ with those of the equation-by-equation least squares estimators, $\underline{\hat{\beta}}_1$ and $\underline{\hat{\beta}}_2$. It is also of interest to compare the covariance matrices of $\underline{\hat{\beta}}_1^*$ and $\underline{\hat{\beta}}_2^*$ with those of the following estimators:

(2.9a) $$\underline{\tilde{\beta}}_1^u = \underline{\hat{\beta}}_1 - \frac{s_{12}}{s_{22}} \underline{\hat{\beta}}_{12}$$

(2.9b) $$\underline{\tilde{\beta}}_2^u = \underline{\hat{\beta}}_2 - \frac{s_{12}}{s_{11}} \underline{\hat{\beta}}_{21}$$

where $s_{\alpha l}$ is an estimator of $\sigma_{\alpha l}$, $\alpha, l=1, 2$. The estimators in (2.9) do not take account of the special structure of Σ_r in (2.3) and thus do not reflect the restrictions on the disturbance variance-covariance matrix. For example, $\underline{\tilde{\beta}}_2^u$ in (2.9b) depends on s_{12}/s_{11} whereas $\underline{\hat{\beta}}_2^*$ in (2.7b) does not.

To compare the estimators in (2.9) with those in (2.7), in Zellner [1963] it has been established that $\underline{\tilde{\beta}}_1^u$ and $\underline{\tilde{\beta}}_2^u$ are unbiased and their covariance matrices are given by:

(2.10a) $$V(\underline{\tilde{\beta}}_1^u) = (X_1'X_1)^{-1}\sigma_{11}(1 - \rho_{12}^2)\left(1 + \frac{1}{n' - 2}\right)$$

(2.10b) $$V(\underline{\tilde{\beta}}_2^u) = (X_2'X_2)^{-1}\sigma_{22}(1 - \rho_{12}^2)\left(1 + \frac{1}{n' - 2}\right)$$

where $n'=n-k_1-k_2>2$. The factor involving n' enters (2.10) reflecting the fact that the $s_{\alpha l}$ have been substituted for the $\sigma_{\alpha l}$ in the expression shown in (2.5). Now if σ_{11}, σ_{12} and ρ_{12} in (2.10) are evaluated using (2.3), we obtain with $n'>2$:

(2.11a) $$V(\underline{\tilde{\beta}}_1^u) = (X_1'X_1)^{-1}\sigma_{\eta\eta}[\omega_{22}/(\sigma_{\eta\eta} + \omega_{22})]\left(1 + \frac{1}{n' - 2}\right)$$

(2.11b) $$V(\underline{\tilde{\beta}}_2^u) = (X_2'X_2)^{-1}\omega_{22}\left(1 + \frac{1}{n' - 2}\right).$$

(2.11b) can be compared with $\mathrm{Var}(\underline{\hat{\beta}}_2^*)=(X_2'X_2)^{-1}\omega_{22}$. It is seen that the variances of the elements of $\underline{\tilde{\beta}}_2^u$ are inflated by the factor $1+1/(n'-2)$ that enters because the covariance matrix restrictions have not been incorporated in $\underline{\tilde{\beta}}_2^u$. In large samples, however, $\underline{\hat{\beta}}_2^*$ and $\underline{\tilde{\beta}}_2^u$ have the same variance-covariance matrix. Thus the results for this bivariate case with $X_1'X_2=0$ indicate very clearly the effects that covariance matrix restrictions, as implied by the form of (2.3), can have on the precision of estimating regression coefficients.

It should also be appreciated that the system in (2.1) subject to (2.2) can be expressed as $\underline{y}_1 = X_1\underline{\beta}_1 + \underline{\eta}$ and $\underline{y}_2 - \underline{y}_1 = X_2\underline{\beta}_2 - X_1\underline{\beta}_1 + \underline{\varepsilon}_2$ or

$$\text{(2.12a)} \qquad \begin{pmatrix} \underline{y}_1 \\ \underline{y}_2 - \underline{y}_1 \end{pmatrix} = \begin{pmatrix} X_1 & 0 \\ -X_1 & X_2 \end{pmatrix} \begin{pmatrix} \underline{\beta}_1 \\ \underline{\beta}_2 \end{pmatrix} + \begin{pmatrix} \underline{\eta} \\ \underline{\varepsilon}_2 \end{pmatrix}$$

or

$$\text{(2.12b)} \qquad \underline{z} = W\underline{\beta} + \underline{v}$$

where $\underline{z}' = (\underline{y}_1' \; \underline{y}_2' - \underline{y}_1')$, $\underline{\beta}' = (\underline{\beta}_1' \; \underline{\beta}_2')$, $\underline{v}' = (\underline{\eta}' \; \underline{\varepsilon}_2')$ and W represents the partitioned matrix on the r.h.s. of (2.12a). Then $E\underline{v} = \underline{o}$ and $E\underline{v}\,\underline{v}' = D_2 \otimes I_n$ where D_2 is a 2×2 diagonal matrix with elements $\sigma_{\eta\eta}$ and ω_{22}. Further,

$$\text{(2.13)} \qquad \underline{\hat{\beta}}^* = [W'(D_2^{-1} \otimes I_n)W]^{-1}W'(D_2^{-1} \otimes I_n)\underline{z}$$

is algebraically identical to the expression shown in (2.6). Writing the system as shown in (2.12) is convenient in analyzing systems with more than two equations and also in providing a Bayesian analysis of the system.

Also (2.12a) may be utilized to obtain the following estimators: $\hat{\sigma}_{\eta\eta} = (\underline{y}_1 - X_1\underline{\hat{\beta}}_1)'(\underline{y}_1 - X_1\underline{\hat{\beta}}_1)/n$ and $\hat{\omega}_{22} = (\underline{y}_2 - \underline{y}_1 - X_2\underline{\tilde{\beta}}_2 + X_1\underline{\tilde{\beta}}_1)'(\underline{y}_2 - \underline{y}_1 - X_2\underline{\tilde{\beta}}_2 + X_1\underline{\tilde{\beta}}_1)/n$ where $\underline{\hat{\beta}}_1 = (X_1'X_1)^{-1}X_1'\underline{y}_1$ and $\underline{\tilde{\beta}}_1$ and $\underline{\tilde{\beta}}_2$ are least squares quantities from a regression of $\underline{y}_2 - \underline{y}_1$ on X_2 and $-X_1$. Then $\hat{\varphi} = (\hat{\sigma}_{\eta\eta} + \hat{\omega}_{22})/\hat{\sigma}_{\eta\eta}$ can be substituted in (2.6) to obtain an operational estimator for $\underline{\beta}' = (\underline{\beta}_1' \; \underline{\beta}_2')$.

2.2. *Multivariate SUR Covariance Restricted (SUR–CR) Systems.* When we have more than two regression equations in a system, we can impose disturbance variance and correlation restrictions on the system by generalizing the error structure shown in (2.2). For example, if we have a standard SUR system, $\underline{y}_\alpha = X_\alpha\underline{\beta}_\alpha + \underline{u}_\alpha$, $\alpha = 1, 2, \ldots, m$ with $E\underline{u}_\alpha = \underline{o}$ and $E\underline{u}_\alpha\underline{u}_l' = \sigma_{\alpha l}I_n$, $\alpha, l = 1, 2, \ldots, m$ and wish to restrict elements of the $m \times m$ pds disturbance covariance matrix $\Sigma = \{\sigma_{\alpha l}\}$, we can, for example, proceed as follows. Write, as in (2.2),

$$\text{(2.14)} \qquad \begin{aligned} u_{1i} &= \eta_i \\ u_{2i} &= \eta_i + \varepsilon_{2i} \\ u_{3i} &= \eta_i + \varepsilon_{2i} + \varepsilon_{3i} \\ &\cdots\cdots\cdots\cdots \\ u_{mi} &= \eta_i + \varepsilon_{2i} + \varepsilon_{3i} + \cdots + \varepsilon_{mi} \end{aligned} \qquad i = 1, 2, \ldots, n.$$

With this specification, we have $\sigma_{11} < \sigma_{22} < \sigma_{33} < \cdots < \sigma_{mm}$ where $\sigma_{\alpha\alpha} = Eu_{\alpha i}^2$, for all i and α, and $0 < \rho_{\alpha l} < 1$, $\alpha, l = 1, 2, \ldots, m$, $\alpha \neq l$, where $\rho_{\alpha l}$ is the correlation coefficient for $u_{\alpha i}$ and u_{li}, for all i. Further, the form of (2.14) has the following implications for the magnitudes of the $\rho_{\alpha l}$ if it is assumed that the η_i's and the $\varepsilon_{\alpha i}$'s have zero means, constant variances, $\sigma_{\eta\eta}$ and $\omega_{\alpha\alpha}$, respectively, for all i and α, and are uncorrelated. Then the variance-covariance matrix for the disturbance terms in (2.14), denoted by Σ_r is given by

$$\Sigma_r = \begin{bmatrix} \sigma_{\eta\eta} & \sigma_{\eta\eta} & \sigma_{\eta\eta} \cdots \sigma_{\eta\eta} \\ \sigma_{\eta\eta} & c_2 & c_2 \cdots c_2 \\ \sigma_{\eta\eta} & c_2 & c_3 \cdots c_3 \\ \vdots & \vdots & \vdots \ddots \vdots \\ \sigma_{\eta\eta} & c_2 & c_3 \cdots c_m \end{bmatrix} \tag{2.15}$$

where $c_l = \sigma_{\eta\eta} + \sum_{\alpha=2}^{l} \omega_{\alpha\alpha}$, $l = 2, 3, \ldots, m$. From the form of (2.15), the following inequality restrictions on the disturbance correlation coefficients follow: $0 < \rho_{\alpha l} < 1$, for all values of α and l, $\alpha \neq l$, and

$$\rho_{\alpha l} > \rho_{\alpha l'} \quad \text{for} \quad l' > l > \alpha \tag{2.16a}$$

$$\rho_{\alpha l} > \rho_{\alpha' l} \quad \text{for} \quad l > \alpha > \alpha'. \tag{2.16b}$$

For example, in the case that $m = 4$, (2.16) provides the following inequalities for the $m(m-1)/2 = 6$ disturbance correlations:

$$\rho_{12} > \rho_{13} > \rho_{14} \quad \text{and} \quad \rho_{23} > \rho_{24}$$

$$\rho_{34} > \rho_{24} > \rho_{14} \quad \text{and} \quad \rho_{23} > \rho_{13}.$$

An interesting schematic to which these inequalities can be related is shown below:

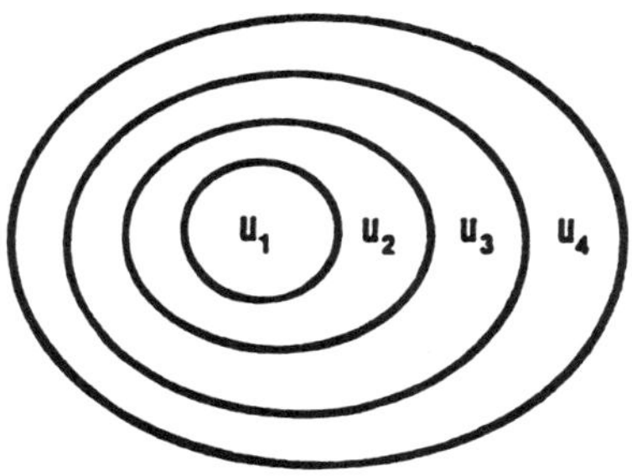

The inequalities in (2.16) are compatible with correlations for "nearest neighbors" being higher than those for "more distant neighbors," e.g., $\rho_{12} > \rho_{13} > \rho_{14}$, $\rho_{23} > \rho_{24}$ and so on.

A SUR system with an error structure in the form of (2.14) is given by

$$\begin{aligned} \underline{y}_1 &= X_1 \underline{\beta}_1 + \underline{\eta} \\ \underline{y}_2 &= X_2 \underline{\beta}_2 + \underline{\eta} + \underline{\varepsilon}_2 \\ \underline{y}_3 &= X_3 \underline{\beta}_3 + \underline{\eta} + \underline{\varepsilon}_2 + \underline{\varepsilon}_3 \\ &\cdots\cdots\cdots\cdots\cdots\cdots \\ \underline{y}_m &= X_m \underline{\beta}_m + \underline{\eta} + \underline{\varepsilon}_2 + \cdots + \underline{\varepsilon}_m \end{aligned} \tag{2.17}$$

with error term covariance matrix $\Sigma_r \otimes I_n$ where Σ_r is given in (2.15). While (2.17) could be analyzed as it stands, it is more convenient to make the following transformation of the system and then to analyze the transformed system:[3]

$$
\begin{aligned}
\underline{y}_1 &= X_1\underline{\beta}_1 + \underline{\eta} \\
\underline{y}_2 - \underline{y}_1 &= X_2\underline{\beta}_2 - X_1\underline{\beta}_1 + \underline{\varepsilon}_2 \\
\underline{y}_3 - \underline{y}_2 &= X_3\underline{\beta}_3 - X_2\underline{\beta}_2 + \underline{\varepsilon}_3 \\
&\cdots\cdots\cdots \\
\underline{y}_m - \underline{y}_{m-1} &= X_m\underline{\beta}_m - X_{m-1}\underline{\beta}_{m-1} + \underline{\varepsilon}_m
\end{aligned}
\tag{2.18a}
$$

or

$$
\begin{bmatrix} \underline{z}_1 \\ \underline{z}_2 \\ \underline{z}_3 \\ \vdots \\ \underline{z}_{m-1} \\ \underline{z}_m \end{bmatrix}
=
\begin{bmatrix}
X_1 & 0 & 0 & \cdots & 0 & 0 \\
-X_1 & X_2 & 0 & \cdots & 0 & 0 \\
0 & -X_2 & X_3 & \cdots & 0 & 0 \\
\vdots & \vdots & \ddots & \ddots & \vdots & \vdots \\
 & & & -X_{m-2} & X_{m-1} & 0 \\
0 & 0 & \cdots & 0 & -X_{m-1} & X_m
\end{bmatrix}
\begin{bmatrix} \underline{\beta}_1 \\ \underline{\beta}_2 \\ \underline{\beta}_3 \\ \vdots \\ \underline{\beta}_{m-1} \\ \underline{\beta}_m \end{bmatrix}
+
\begin{bmatrix} \underline{\eta} \\ \underline{\varepsilon}_2 \\ \underline{\varepsilon}_3 \\ \vdots \\ \underline{\varepsilon}_{m-1} \\ \underline{\varepsilon}_m \end{bmatrix}
\tag{2.18b}
$$

or

$$\underline{z} = W\underline{\beta} + \underline{v} \tag{2.18c}$$

where $\underline{z}' = (\underline{z}_1' \; \underline{z}_2' \cdots \underline{z}_m') = (\underline{y}_1' \; \underline{y}_2' - \underline{y}_1' \cdots \underline{y}_m' - \underline{y}_{m-1}')$, W represents the block bi-diagonal matrix on the r.h.s. of (2.18b), $\underline{\beta}' = (\underline{\beta}_1' \; \underline{\beta}_2' \cdots \underline{\beta}_m')$ and $\underline{v}' = (\underline{\eta}' \; \underline{\varepsilon}_1' \cdots \underline{\varepsilon}_m')$. Written in the form of (2.18), we have a set of regressions with uncorrelated error terms, that is

$$E\underline{v}\,\underline{v}' = D_m \otimes I_n \tag{2.19}$$

where D_m is an $m \times m$ diagonal matrix with elements $\sigma_{\eta\eta}, \omega_{22}, \omega_{33}, \ldots, \omega_{mm}$ on the main diagonal. Then for given D_m,

$$\underline{\hat{\beta}}^* = [W'(D_m^{-1} \otimes I_n)W]^{-1}W'(D_m^{-1} \otimes I_n)\underline{z} \tag{2.20}$$

is the MVLU estimator for $\underline{\beta}$ with variance covariance matrix $V(\underline{\hat{\beta}}^*) = [W'(D_m^{-1} \otimes I_n)W]^{-1}$. When the elements of D_m have unknown values,

$$\underline{\hat{\beta}}_a^* = [W'(\hat{D}_m^{-1} \otimes I_m)^{-1}W]^{-1}W'(\hat{D}_m^{-1} \otimes I_n)\underline{z} \tag{2.21}$$

where $\hat{D}_m$ is a consistent estimator of D_m is an asymptotically efficient estimator with large sample covariance matrix $[W'(\hat{D}_m^{-1} \otimes I_m)W]^{-1}$.

[3] Least squares residuals from (2.18a) can be employed to check that its error vectors are mutually uncorrelated.

3. BAYESIAN ANALYSIS

3.1. *Bayesian Analysis of Bivariate SUR Case.* The system that we shall analyze is the same one considered in Section 2, namely $\underline{y}_1 = X_1\underline{\beta}_1 + \underline{\eta}$ and $\underline{y}_2 = X_2\underline{\beta}_2 + \underline{\eta} + \underline{\varepsilon}_2$ or, in transformed form,

$$\begin{pmatrix} \underline{y}_1 \\ \underline{y}_2 - \underline{y}_1 \end{pmatrix} = \begin{pmatrix} X_1 & 0 \\ -X_1 & X_2 \end{pmatrix} \begin{pmatrix} \underline{\beta}_1 \\ \underline{\beta}_2 \end{pmatrix} + \begin{pmatrix} \underline{\eta} \\ \underline{\varepsilon}_2 \end{pmatrix} \tag{3.1a}$$

or

$$\underline{z} = W\underline{\beta} + \underline{v} \tag{3.1b}$$

where $\underline{z}' = (\underline{y}_1' \ \underline{y}_2' - \underline{y}_1')$, $\underline{\beta}' = (\underline{\beta}_1' \ \underline{\beta}_2')$, $\underline{v}' = (\underline{\eta}' \ \underline{\varepsilon}_2')$ and W represents the block triangular matrix on the r.h.s. of (3.1a). It is assumed that the elements of the vector $\underline{v}$ are normally and independently distributed with zero means and covariance matrix $D_2 \otimes I_n$ where D_2 is a 2×2 diagonal matrix with elements $\sigma_{\eta\eta} \equiv \sigma_\eta^2$ and $\omega_{22} \equiv \omega_2^2$, the common variances of the elements of $\underline{\eta}$ and $\underline{\varepsilon}_2$, respectively. Then the likelihood function for the system in (3.1) is

$$p(\underline{z}|\underline{\beta}, \sigma_\eta, \omega_2) \propto (\sigma_\eta\omega_2)^{-n} \exp\{-(\underline{z} - W\underline{\beta})'(D_2^{-1}\otimes I_n)(\underline{z} - W\underline{\beta})/2\} \tag{3.2}$$

If we employ the following diffuse prior *pdf*,

$$p(\underline{\beta}, \sigma_\eta, \omega_2) \propto 1/\sigma_\eta\omega_2, \qquad 0 < \sigma_\eta, \ \omega_2 < \infty \tag{3.3}$$
$$-\infty < \beta_{ij} < \infty.$$

Then the joint posterior *pdf* for $\underline{\beta}$, σ_η and ω_2 is:

$$p(\underline{\beta}, \sigma_\eta, \omega_2) \propto (\sigma_\eta\omega_2)^{-(n+1)} \exp\{-(\underline{z} - W\underline{\beta})'(D_2^{-1}\otimes I_n)(\underline{z} - W\underline{\beta})/2\}$$
$$\propto (\sigma_\eta\omega_2)^{-(n+1)} \exp\{-[\underline{\hat{u}}_*'\underline{\hat{u}}_* + (\underline{\beta} - \underline{\beta}^*)'W'(D_2^{-1}\otimes I_n)W(\underline{\beta} - \underline{\beta}^*)]/2\}$$

where

$$\underline{\beta}^* = [W'(D_2^{-1} \otimes I_n)W]^{-1}W'(D_2^{-1} \otimes I_n)\underline{z} \tag{3.4}$$

and

$$\underline{\hat{u}}_*'\underline{\hat{u}}_* = (\underline{z} - W\underline{\beta}^*)'(\underline{z} - W\underline{\beta}^*). \tag{3.5}$$

It is seen from the second line of (3.3) that the conditional posterior *pdf* for β given σ_η and ω_2 is multivariate normal with mean vector β^* and covariance matrix $[W'(D_2^{-1}\otimes I_n)W]^{-1}$. Note that the conditional mean, $\underline{\beta}^*$, is identical to the sampling theory estimate in (2.13).

To obtain the marginal posterior *pdf* for $\underline{\beta}$, it is convenient to write the first line of (3.3) as

$$(3.6)\quad p(\underline{\beta}, \sigma_\eta, \omega_2|\underline{z}) \propto (\sigma_\eta\omega_2)^{-(n+1)}\exp\{-[(\underline{z}_1 - X_1\underline{\beta}_1)'(\underline{z}_1 - X_1\underline{\beta}_1)/\sigma_\eta^2 + (\underline{z}_2 - X_2\underline{\beta}_2 + X_1\underline{\beta}_1)'(\underline{z}_2 - X_2\underline{\beta}_2 + X_1\underline{\beta}_1)/\omega_2^2]/2\}.$$

Then on integrating with respect to σ_η and ω_2, the following marginal posterior *pdf* for $\underline{\beta}' = (\underline{\beta}_1' \ \underline{\beta}_2')$ is obtained:

$$(3.7)\quad p(\underline{\beta}_1, \underline{\beta}_2|\underline{z}) \propto [(\underline{z}_1 - X_1\underline{\beta}_1)'(\underline{z}_1 - X_1\underline{\beta}_1)]^{-n/2} [\underline{z}_2 - X_2\underline{\beta}_2 + X_1\underline{\beta}_1)'(\underline{z}_2 - X_2\underline{\beta}_2 + X_1\underline{\beta}_1)]^{-n/2}.$$

Except for the appearance of $\underline{\beta}_2$ in the second factor on the r.h.s. of (3.7), this posterior *pdf* is in the form of a "double Student-*t*" *pdf*, Tiao and Zellner [1964], Zellner [1971]. The asymptotic expansion techniques employed in these references can be applied to (3.7) with the result that the leading normal term in the approximation has mean $\hat{\underline{\beta}} = [W'(\hat{D}_2^{-1}\otimes I_n)W]^{-1}W'(\hat{D}_2^{-1}\otimes I_n)\underline{z}$ and variance-covariance matrix $[W'(\hat{D}_2^{-1}\otimes I_n)W]^{-1}$ where $\hat{D}_2$ has diagonal elements given by $\hat{\sigma}_\eta^2 = (\underline{y}_1 - X_1\hat{\underline{\beta}}_1)'(\underline{y} - X_1\hat{\underline{\beta}}_1)/n$ and $\hat{\omega}_2^2 = (\underline{y}_2 - \underline{y}_1 - X_2\tilde{\underline{\beta}}_2 + X_1\tilde{\underline{\beta}}_1)'(\underline{y}_2 - \underline{y}_1 - X_2\tilde{\underline{\beta}}_2 + X_1\tilde{\underline{\beta}}_1)/n$ where $\hat{\underline{\beta}}_1 = (X_1'X_1)^{-1}X_1'\underline{y}_1$ and $\tilde{\underline{\beta}}_1$ and $\tilde{\underline{\beta}}_2$ are least squares quantities from a regression of $\underline{z}_2 = \underline{y}_2 - \underline{y}_1$ on X_2 and $-X_1$. Of course, the leading normal term of an asymptotic expansion is just an approximation to the posterior *pdf* in (3.7).

From the form of (3.7), the following exact statements can be made:

(i) The conditional posterior *pdf* for $\underline{\beta}_2$ given $\underline{\beta}_1$ is in the following multivariate Student-*t* form:

$$(3.8)\quad p(\underline{\beta}_2|\underline{\beta}_1, \underline{z}) \propto [\nu_2 s_*^2 + (\underline{\beta}_2 - \hat{\underline{\beta}}_2^c)'X_2'X_2(\underline{\beta}_2 - \hat{\underline{\beta}}_2^c)]^{-(\nu_2+k_2)/2}$$

where $\nu_2 = n - k_2$, $\nu_2 s_*^2 = (\underline{w} - X\hat{\underline{\beta}}_2^c)'(\underline{w} - X\hat{\underline{\beta}}_2^c)$, $\underline{w} = \underline{z}_2 + X_1\underline{\beta}_1 = \underline{y}_2 - \underline{y}_1 + X_1\underline{\beta}_1$, and

$$(3.9)\quad \hat{\underline{\beta}}_2^c = (X_2'X_2)^{-1}X_2'\underline{w} = \hat{\underline{\beta}}_2 - \hat{\underline{\beta}}_{21} + (X_2'X_2)^{-1}X_2'X_1\underline{\beta}_1$$

where $\hat{\underline{\beta}}_2 = (X_2'X_2)^{-1}X_2'\underline{y}_2$ and $\hat{\underline{\beta}}_{21} = (X_2'X_2)^{-1}X_2'\underline{y}_1$. $\hat{\underline{\beta}}_2^c$ in (3.9) is the conditional posterior mean of $\underline{\beta}_2$ given $\underline{\beta}_1$. If $X_2'X_1 = 0$, $\hat{\underline{\beta}}_2^c$ does not depend on $\underline{\beta}_1$ and is precisely equal to the sampling theory estimate shown in (2.7b). Further, the expression in the second line of (3.9) can be better understood if we note that $\underline{y}_2 = X_2\underline{\beta}_2 + \underline{\eta} + \underline{\varepsilon}_2 = X_2\underline{\beta}_2 + \underline{y}_1 - X_1\underline{\beta}_1 + \underline{\varepsilon}_2$ since $\underline{\eta} = \underline{y}_1 - X_1\underline{\beta}_1$. Thus given $\underline{\beta}_1$, we can analyze $\underline{w} = \underline{y}_2 - \underline{y}_1 + X_1\underline{\beta}_1 = X_2\underline{\beta}_2 + \underline{\varepsilon}_2$ as a usual multiple regression problem and (3.9) is the posterior mean given diffuse prior assumptions.

(ii) From (3.7), the conditional posterior *pdf* for $\underline{\beta}_1$ given $\underline{\beta}_2$ is in the "double-Student-*t*" form, that is

$$(3.10)\quad p(\underline{\beta}_1|\underline{\beta}_2, \underline{z}) \propto [\nu_1 s_1^2 + (\underline{\beta}_1 - \hat{\underline{\beta}}_1)'X_1'X_1(\underline{\beta}_1 - \hat{\underline{\beta}}_1)]^{-n/2}[\nu_1 s_0^2 + (\underline{\beta}_1 - \hat{\hat{\underline{\beta}}}_1)'X_1'X_1(\underline{\beta}_1 - \hat{\hat{\underline{\beta}}}_1)]^{-n/2}$$

where

$$\nu_1 = n - k_1, \quad \nu_1 s_1^2 = (\underline{z}_1 - X_1\hat{\underline{\beta}}_1)'(\underline{z}_1 - X_1\hat{\underline{\beta}}_1),$$

$$\nu_1 s_0^2 = (\underline{z}_2 - X_2\underline{\beta}_2 + X_1\hat{\hat{\underline{\beta}}}_1)'(\underline{z}_2 - X_2\underline{\beta}_2 + X_1\hat{\hat{\underline{\beta}}}_1)$$

and

$$\hat{\hat{\underline{\beta}}}_1 = -(X_1'X_1)^{-1}X_1'(\underline{z}_2 - X_2\underline{\beta}_2) \tag{3.11}$$
$$= \hat{\underline{\beta}}_1 - [\hat{\underline{\beta}}_{12} - (X_1'X_1)^{-1}X_1'X_2\underline{\beta}_2].$$

Thus, it is seen that the first factor on the r.h.s. of (3.10) is centered at $\hat{\underline{\beta}}_1 = (X_1'X_1)^{-1}X_1'\underline{y}_1$ while the second is centered at $\hat{\hat{\underline{\beta}}}_1$ as given in (3.11). Note that in large samples, $\hat{\underline{\beta}}_{12} = (X_1'X_1)^{1-}X_1'\underline{y}_2$ will be close (i.p.) to $(X_1'X_1)^{-1}X_1'X_2\underline{\beta}_2$.

(iii) From (3.7), the marginal posterior *pdf* for $\underline{\beta}_1$ is in the following form:

$$p(\underline{\beta}_1|\underline{z}) \propto [\nu_1 s_1^2 + (\underline{\beta}_1 - \hat{\underline{\beta}}_1)'X_1'X_1(\underline{\beta}_1 - \hat{\underline{\beta}}_1)]^{-(\nu_1+k_1)/2} \tag{3.12}$$
$$\cdot [\nu_a s_a^2 + (\underline{\beta}_1 - \tilde{\tilde{\underline{\beta}}}_1)'X_1'M_2X_1(\underline{\beta}_1 - \tilde{\tilde{\underline{\beta}}}_1)]^{-(\nu_a+k_1)/2}$$

where

$$\nu_a = n - k_1 - k_2, \quad \nu_a s_a^2 = \underline{z}_2'M_2\underline{z}_2 - \tilde{\tilde{\underline{\beta}}}_1'X_1'M_2X_1\tilde{\tilde{\underline{\beta}}}_1,$$

$$M_2 = I - X_2(X_2'X_2)^{-1}X_2',$$

and

$$\tilde{\tilde{\underline{\beta}}}_1 = -(X_1'M_2X_1)^{-1}X_1'M_2\underline{z}_2 \tag{3.13}$$
$$= -(X_1'M_2X_1)^{-1}X_1'M_2(\underline{y}_2 - \underline{y}_1).$$

Thus in (3.12), the first factor is centered at $\hat{\underline{\beta}}_1 = (X_1'X_1)^{-1}X_1'\underline{y}_1$ while the second is centered at $\tilde{\tilde{\underline{\beta}}}_1$ shown in (3.13). $\tilde{\tilde{\underline{\beta}}}_1$ is just the least squares quantity obtained from a regression of $\underline{z}_2 = \underline{y}_2 - \underline{y}_1$ on $-X_1$ after "taking out" the effects of X_2. The leading normal term of an asymptotic expansion of (3.12) is centered at

$$\tilde{\underline{\beta}}_1 = [X_1'X_1/s_1^2 + X_1'M_2X_1/s_a^2]^{-1}[X_1'X_1\hat{\underline{\beta}}_1/s_1^2 + X_1'M_2X_1\tilde{\tilde{\underline{\beta}}}_1/s_a^2]. \tag{3.14}$$

(iv) The joint posterior *pdf* for $\underline{\beta}' = (\underline{\beta}_1' \ \underline{\beta}_2')$, σ_η and ω_2, shown in (3.6) can be expressed in terms of $\underline{\beta}$, σ_η and θ where $\theta = \sigma_\eta^2/\omega_2^2$ as follows

$$p(\underline{\beta}, \sigma_\eta, \theta|\underline{z}) \propto \frac{\theta^{(n-2)/2}}{\sigma_\eta^{2n+3}} \exp\left\{-\frac{1}{2\sigma_\eta^2}[(\underline{z}_1 - X_1\underline{\beta}_1)'(\underline{z}_1 - X_1\underline{\beta}_1) \right. \tag{3.15}$$
$$\left. + \theta(\underline{z}_2 - X_2\underline{\beta}_2 + X_1\underline{\beta}_1)'(\underline{z}_2 - X_2\underline{\beta}_2 + X_1\underline{\beta}_1)]\right\}$$
$$\propto \frac{\theta^{(n-2)/2}}{\sigma_\eta^{2n+3}} \exp\left\{-\frac{1}{2\sigma_\eta^2}[\hat{\underline{u}}_\theta'\hat{\underline{u}}_\theta + (\underline{\beta} - \hat{\underline{\beta}}_\theta)'W_\theta'W_\theta(\underline{\beta} - \hat{\underline{\beta}}_\theta)]\right\}$$

where

$$\underline{\hat{\beta}}_\theta = (W_\theta' W_\theta)^{-1} W_\theta' \underline{z}_\theta, \quad \underline{\hat{u}}_\theta' \underline{\hat{u}}_\theta = (\underline{z}_\theta - W_\theta \underline{\hat{\beta}}_\theta)'(\underline{z}_\theta - W_\theta \underline{\hat{\beta}}_\theta),$$

$$\underline{z}_\theta' = (\underline{z}_1' \; \theta^{1/2} \underline{z}_2')$$

and

$$W_\theta = \begin{pmatrix} X_1 & 0 \\ -\theta^{1/2} X_1 & \theta^{1/2} X_2 \end{pmatrix}.$$

From the second line of (3.15), the conditional posterior *pdf* for $\underline{\beta}$ given σ_η and θ is multivariate normal with mean vector $\underline{\hat{\beta}}_\theta$ and covariance matrix $(W_\theta' W_\theta)^{-1}\sigma_\eta^2$. Further on integrating out σ_η, the result is

$$\text{(3.16)} \quad p(\underline{\beta}, \theta | \underline{z}) \propto \theta^{(n-2)/2}[\underline{\hat{u}}_\theta' \underline{\hat{u}}_\theta + (\underline{\beta} - \underline{\hat{\beta}}_\theta)' W_\theta' W_\theta (\underline{\beta} - \underline{\hat{\beta}}_\theta)]^{-(n+1)}.$$

From (3.16) it is apparent that the conditional posterior *pdf* for $\underline{\beta}$ given θ is in the multivariate Student-*t* form.

To obtain the marginal posterior *pdf* for a single element of $\underline{\beta}$, say β_{11} integrate (3.16) with respect to all elements of $\underline{\beta}$ other than β_{11}. The result yields $p(\beta_{11}, \theta | \underline{z})$, a bivariate posterior *pdf* that can be analyzed by use of numerical integration techniques.

(v) The marginal posterior *pdf* for θ can be obtained by integrating over $\underline{\beta}$ in (3.16) to yield

$$\text{(3.17)} \quad p(\theta | \underline{z}) \propto \theta^{(n-2)/2} |W_\theta' W_\theta|^{-1/2} / (\underline{\hat{u}}_\theta' \underline{\hat{u}}_\theta)^{n+1-(k_1+k_2)/2}$$

with $0 < \theta < \infty$.

(vi) By integrating $\underline{\beta}$ out of (3.3), the following posterior *pdf* for σ_η and ω_2 is obtained:

$$\text{(3.18)} \quad p(\sigma_\eta, \omega_2 | \underline{z}) \propto (\sigma_\eta \omega_2)^{-(n+1)} |W'(D_2^{-1} \otimes I_n) W|^{-1/2} \exp\{-\underline{\hat{u}}_*' \underline{\hat{u}}_* / 2\}$$

with $0 < \sigma_\eta, \omega_2 < \infty$. Bivariate numerical integration techniques can be employed to obtain the marginal posterior *pdfs* for σ_η and for ω_2 from (3.18).

3.2. *Bayesian Analysis of Multivariate SUR–CR Systems.* In this section we analyze the SUR–CR system shown in (2.18) that was obtained from (2.17), a system with error term variance and correlation restrictions already discussed in connection with (2.14). Given that the error vector $\underline{v}$ in (2.18c) is assumed drawn from a normal population with zero mean vector and covariance matrix given in (2.19), namely $D_m \otimes I_n$ where D_m is a diagonal matrix, the likelihood function is

$$\begin{aligned} \text{(3.19)} \quad p(\underline{z} | \underline{\beta}, D_m) &\propto |D_m^{-1} \otimes I_n|^{1/2} \exp\{-(\underline{z} - W\underline{\beta})'(D_m^{-1} \otimes I_n)(\underline{z} - W\underline{\beta})/2\} \\ &\propto |D_m^{-1} \otimes I_n|^{1/2} \exp\{-[(\underline{\beta} - \underline{\hat{\beta}}^*)' W'(D_m^{-1} \otimes I_n) W(\underline{\beta} - \underline{\hat{\beta}}^*) \\ &\quad + (\underline{z} - W\underline{\hat{\beta}}^*)'(D_m^{-1} \otimes I_n)(\underline{z} - W\underline{\hat{\beta}}^*)]/2\} \end{aligned}$$

where $\underline{\hat{\beta}}^* = [W'(D_m^{-1} \otimes I_n)W]^{-1} W'(D_m^{-1} \otimes I_n)\underline{z}$. If we employ the following diffuse prior *pdf*,

$$p(\underline{\beta}, D_m) \propto |D_m|^{-1/2}, \tag{3.20}$$

then the posterior *pdf* is

$$p(\underline{\beta}, D_m|\underline{z}) \propto |D_m|^{-(n+1)/2} \exp\{-[(\underline{\beta}-\underline{\hat{\beta}}^*)' W'(D_m^{-1} \otimes I_n) W (\underline{\beta}-\underline{\hat{\beta}}^*) + (\underline{z}-W\underline{\hat{\beta}}^*)'(D_m^{-1} \otimes I_n)(\underline{z}-W\underline{\hat{\beta}}^*)]/2\}. \tag{3.21}$$

From (3.21), the conditional posterior *pdf* for $\underline{\beta}$ given D_m is multivariate normal with mean $\underline{\hat{\beta}}^*$ and covariance matrix $[W'(D_m^{-1} \otimes I_n)W]^{-1}$. The mean of this conditional posterior *pdf* is identical to the MVLU estimate given D_m in (2.20). Also, $p(\underline{\beta}|D_m = \hat{D}_m, \underline{z})$, where $\hat{D}_m$ is an estimate of D_m, is normal with mean given by $\underline{\hat{\beta}}_a^* = [W'(\hat{D}_m^{-1} \otimes I_n)W]^{-1} W'(\hat{D}_m^{-1} \otimes I_n)\underline{z}$ and covariance matrix $[W'(\hat{D}_m^{-1} \otimes I_n)W]^{-1}$. The conditional mean $\underline{\hat{\beta}}_a^*$ is identical to the approximate MVLU estimate in (2.21).

4. CONCLUDING REMARKS

In this paper a simple, flexible error-components procedure (ECP) for introducing prior information about the values of error terms' correlations and variances was described. Use of such information usually results in a significant reduction in the number of free parameters in a covariance matrix. Further, it has been shown that such prior restrictions on covariance matrices affect the finite-sample precision with which regression coefficients in SUR systems can be estimated.

While the present paper dealt exclusively with pure regression systems, it should be appreciated that the ECP for introducing prior information about values of covariance matrices' elements can be applied not only in multivariate regression models but also in many other multivariate models, for example, multivariate regression models with serially correlated errors, structural econometric models, sets of transfer functions and sets of autoregressive-moving average processes. In models with serially correlated errors, the reduction in the number of free parameters in error covariance structures effected by the error-components procedure for introducing prior information can be even more dramatic than in the case of the multivariate regression model with serially uncorrelated errors. Further, while in the later model, the information matrix is block diagonal, indicating that asymptotically information about the regression coefficients is independent of information about the covariance matrix, this is not the case with time series and dynamic structural econometric models. Thus, it is to be anticipated that introducing prior information about values of covariance matrices' elements will have not only a finite sample but also a large sample impact on the precision with which structural and other coefficients can be estimated. Also, such information will affect properties of posterior odds ratios for hypotheses and of various sampling theory test procedures. Finally, procedures for appraising whether restrictions imposed on covariance matrices' elements by the ECP are in accord with information in the data are currently being considered.

H.G.B. Alexander Research Foundation, University of Chicago, U.S.A.

REFERENCES

JEFFREYS, H., *Theory of Probability*, 3rd Rev. ed. (London: Oxford University Press, 1967).

KMENTA, J. AND R. F. GILBERT, "Small Sample Properties of Alternative Estimators of Seemingly Unrelated Regressions," *Journal of the American Statistical Association*, 63 (December, 1968), 1180–1200.

LINDLEY, D. V., "Estimation of the Dispersion Matrix of a Multivariate Normal Distribution," presented to the 10th Meeting of the NBER–NSF Seminar on Bayesian Inference in Econometrics and Statistics, U. of Chicago (May, 1975).

MEHTA, J. S. AND P. A. V. B. SWAMY, "Further Evidence on the Relative Efficiencies of Zellner's Seemingly Unrelated Regression Estimator," *Journal of the American Statistical Association*, 71 (September, 1976), 634–639.

PRESS, S. J., "Simultaneous Bayesian Estimation of Multivariate Normal Parameters," manuscript, 1975, presented to the 10th Meeting of NBER–NSF Seminar on Bayesian Inference in Econometrics and Statistics, U. of Chicago (May, 1975).

REVANKAR, N. S., "Some Finite Sample Results in the Context of Two Seemingly Unrelated Regression Equations," *Journal of the American Statistical Association*, 69 (March, 1974), 187–190.

RUBLE, W. L., "Improving the Computation of Simultaneous Stochastic Linear Equations Estimates," Agricultural Economics Report No. 116 and Econometrics Special Report No. 1, Dept. of Agricultural Economics, Michigan State University, East Lansing, Michigan (October, 1968).

SRIVASTAVA, V. K., "The Efficiency of Estimating Seemingly Unrelated Regression Equations," *Annals of the Institute of Statistical Mathematics*, 22, No. 3 (1970), 483–493.

TIAO, G. C. AND A. ZELLNER, "Bayes' Theorem and the Use of Prior Knowledge in Regression Analysis," *Biometrika*, 51 (1964), 219–230.

ZELLNER, A., "An Efficient Method of Estimating Seemingly Unrelated Regressions and Tests for Aggregation Bias," *Journal of the American Statistical Association*, 57 (June, 1962), 348–368.

———, "Estimators for Seemingly Unrelated Regression Equations: Some Exact Finite Sample Results," *Journal of the American Statistical Association*, 58 (December, 1963), 977–992, "Corrigenda," 67 (March, 1972), 255.

———, *An Introduction to Bayesian Inference in Econometrica* (New York: John Wiley and Sons, Inc., 1971).

——— AND D. S. HUANG, "Further Properties of Efficient Estimators for Seemingly Unrelated Regression Equations," *International Economic Review*, 3 (September, 1962), 300–313.

[16]

Journal of Econometrics 25 (1984) 365–393. North-Holland

BAYESIAN ANALYSIS OF DICHOTOMOUS QUANTAL RESPONSE MODELS*

Arnold ZELLNER

University of Chicago, Chicago, IL 60637, USA

Peter E. ROSSI

Northwestern University, Evanston, IL 60201, USA

Received June 1982, final version received December 1983

Qualitative response models (QRM's) are analyzed from the Bayesian point of view, using diffuse and informative prior distributions. Exact finite-sample Bayesian and large-sample Bayesian and non-Bayesian estimation results are compared. In addition, the paper provides: (1) plots and discussion of the properties of likelihood functions for QRM's, (2) posterior distributions for logit models' derivatives and elasticities, (3) Bayesian prediction procedures for QRM's, (4) new estimates for the median and other fractiles of the logistic distribution, (5) posterior odds ratios for model selection problems, and (6) comparisons of two alternative Monte Carlo numerical integration procedures. It is concluded that asymptotic approximations are not accurate for small- to moderate-sized samples even when only a single input variable is used, and that operational Bayesian methods are available for providing both exact small-sample and large-sample approximate inferences for DRM's.

1. Introduction

Dichotomous qualitative response models (DQRM's) have been utilized in many econometric studies to explain the variation of dichotomous random variables – see, e.g., Laub (1971, 1972), Levedahl (1976), Izan (1978) and the extensive references in the review article by Amemiya (1981). In all these studies, the DQRM has been analyzed using non-Bayesian techniques, usually maximum-likelihood estimation and large-sample Neyman–Pearson testing procedures. Since it appears that no Bayesian estimation results for the DQRM have appeared in the econometric literature, in the present article exact and approximate Bayesian inference procedures are presented and applied in the analysis of DQRM's. Where appropriate, references to some Bayesian work in

*This is an extension of an original draft, July 1981, and of material presented at the Econometrics and Statistics Colloquium, University of Chicago, November 1981. Research financed by the H.G.B. Alexander Endowment Fund, Graduate School of Business, University of Chicago, and by the National Science Foundation. Part of this work was done while the first author was on leave at the Hoover Institution and the National Bureau of Economic Research, Stanford University.

the bioassay and related statistical literature – namely Freeman (1970), Ramsey (1972), Leonard (1972, 1979), Tsutakawa (1975), Stewart (1979) and Dempster Selwyn and Weeks (1981) – will be made.

In addition to comparing small-sample and large-sample estimation results for DQRM's, we (1) plot and discuss properties of likelihood functions, (2) derive and calculate posterior distributions for logit models' derivatives and elasticities, (3) discuss prediction procedures, (4) provide a new estimate for the median and other fractiles of the logistic distribution, (5) derive and interpret posterior odds ratios for model selection problems, (6) present new integral convergence results, (7) report exact analytical results for special, small-sample logit models, (8) introduce a multivariate Student-t importance function in Monte Carlo numerical integration, and (9) compare properties of two alternative Monte Carlo numerical integration procedures.

The plan of the paper is as follows. In section 2, DQRM's are described, and then Bayesian estimation and prediction procedures are presented including exact and approximate results utilizing diffuse and informative prior distributions. Bayesian procedures for analyzing hypotheses about parameters' values and for model selection problems are presented in section 3. Section 4 provides plots of likelihood functions and numerical analyses of posterior distributions. The latter shed some light on the quality of large-sample approximations to posterior distributions by comparing exact and approximate posterior distributions for various sample sizes. A summary of results and some concluding remarks are provided in section 5. Appendices include technical discussions of small-sample problems, integral convergence conditions and numerical integration techniques.

2. Bayesian estimation and prediction results for DQRM's

DQRM's are formulated to explain the variation of n independent dichotomous random variables, $\tilde{y}_i$, $i = 1, 2, \ldots, n$, where $\tilde{y}_i = 1$ with probability P_i and $\tilde{y}_i = 0$ with probability $1 - P_i$; thus $\mathrm{E}\tilde{y}_i = P_i$ and $\operatorname{var}\tilde{y}_i = P_i(1 - P_i)$. Given a sample of observations, $y' = (y_1, y_2, \ldots, y_n)$, the likelihood function is $\prod_{i=1}^{n} P_i^{y_i}(1 - P_i)^{1-y_i}$. This is the likelihood function for a so-called Poisson Scheme – see, e.g., Kenny and Keeping (1951, p. 47). In the DQRM, it is usually assumed that $P_i = F(x_i'\beta)$, where $F(\cdot)$ satisfies $0 < F(\cdot) < 1$, x_i' is a given $1 \times k$ non-stochastic vector, and β is a $k \times 1$ vector of parameters with unknown values. Since $\mathrm{E}\tilde{y}_i = P_i = F(x_i'\beta)$, $i = 1, 2, \ldots, n$, $F(\cdot)$ represents the locus of the P_i's. The likelihood function, denoted by $l(\beta \mid y)$, is

$$l(\beta \mid y) = \prod_{i=1}^{n} \left[F(x_i'\beta)\right]^{y_i} \left[1 - F(x_i'\beta)\right]^{1-y_i}. \tag{2.1}$$

Two widely utilized functional forms for $F(\cdot)$ in (2.1) are

$$P_i = F(x_i'\beta) = (2\pi)^{-\frac{1}{2}} \int_{-\infty}^{x_i'\beta} e^{-t^2/2}\,dt,$$

the probit model, and

$$P_i = F(x_i'\beta) = 1/\left(1 + e^{-x_i'\beta}\right),$$

the logit model. Of course, other functional forms for $F(\cdot)$, $0 < F(\cdot) < 1$ can be employed – see, e.g., Laub (1972), Zellner and Lee (1965) and Stewart (1979).

Given the likelihood function in (2.1) and a prior probability density function (pdf) for β, $\pi(\beta|I_0)$, where I_0 denotes the initial information, Bayes' Theorem yields the following posterior pdf for $\beta \subset R_\beta$, the parameter space

$$p(\beta|D) = c\pi(\beta|I_0)l(\beta|y). \tag{2.2}$$

In (2.2), $D = (I_0, y)$, the prior and sample information, and c, the normalizing constant, is given by

$$c^{-1} = \int_{R_\beta} \pi(\beta|I_0)l(\beta|y)\,d\beta.$$

The following procedures are available for analyzing the posterior pdf, $p(\beta|D)$ in (2.2), that is for evaluating the normalizing constant, computing moments, marginal posterior distributions, etc.: (1) analytical integration, (2) asymptotic expansions, (3) numerical integration (Simpson's Rule, etc.), and (4) Monte Carlo numerical integration. Numerical integration procedures will be discussed and applied in section 4. As regards analytical integration, this approach at present seems very difficult except in special small-sample cases (see appendix 1). In view of this, we shall first turn to some asymptotic expansion techniques to yield approximate posterior pdfs much in the fashion employed earlier in other contexts by Jeffreys (1967, app. A, pp. 428–431), Lindley (1961, 1981) and in the bioassay literature by Leonard (1972) and by Dempster, Selwyn and Weeks (1981) in a different context.

2.1. Case I: Diffuse prior, $\pi(\beta|I_0) \propto$ const.

In this case, the posterior pdf for β, $p(\beta|D)$, is

$$p(\beta|D) \propto l(\beta|y) = \exp\{\log l(\beta|y)\}. \tag{2.3}$$

With $L(\beta) \equiv \log l(\beta|y)$, expand $L(\beta)$ in a Taylor Series about the modal value of (2.3), $\hat{\boldsymbol{\beta}}$, which is the maximum-likelihood (ML) estimate [see Silvapulle (1981) for conditions insuring its existence], to obtain

$$\begin{aligned} p(\boldsymbol{\beta}|D) &\propto \exp\{L(\boldsymbol{\beta})\} \\ &\propto \exp\{-(\boldsymbol{\beta}-\hat{\boldsymbol{\beta}})'H(\boldsymbol{\beta}-\hat{\boldsymbol{\beta}})/2+R\} \\ &\propto \exp\{-(\boldsymbol{\beta}-\hat{\boldsymbol{\beta}})'H(\boldsymbol{\beta}-\hat{\boldsymbol{\beta}})/2\}\left[1+R+R^2/2!+\dots\right], \end{aligned} \tag{2.4}$$

where R is the remainder in the expansion,

$$H=-\left[\partial^2 L(\boldsymbol{\beta})/\partial\boldsymbol{\beta}\partial\boldsymbol{\beta}'\right]_{\boldsymbol{\beta}=\hat{\boldsymbol{\beta}}}=X'DX,$$

X is an $n\times k$ matrix of rank k with rows $\boldsymbol{x}_i'$, and D is a diagonal matrix with typical element

$$d_i=\left[y_i/F_i^2+(1-y_i)/(1-F_i)^2\right]f_i^2-(y_i-F_i)f_i'/F_i(1-F_i),$$

with

$$F_i\equiv F(\boldsymbol{x}_i'\hat{\boldsymbol{\beta}}),$$

$$f_i\equiv\left[\mathrm{d}F(z_i)/\mathrm{d}z_i\right]_{z_i=\boldsymbol{x}_i'\hat{\boldsymbol{\beta}}}\equiv f(\boldsymbol{x}_i'\hat{\boldsymbol{\beta}}),$$

$$f_i'\equiv\left[\mathrm{d}f(z_i)/\mathrm{d}z_i\right]_{z_i=\boldsymbol{x}_i'\hat{\boldsymbol{\beta}}}.$$

The leading term approximation in (2.4) gives

$$p(\boldsymbol{\beta}|D)\,\dot{\propto}\,\exp\{-(\boldsymbol{\beta}-\hat{\boldsymbol{\beta}})'H(\boldsymbol{\beta}-\hat{\boldsymbol{\beta}})/2\}, \tag{2.5}$$

that is $\boldsymbol{\beta}$ is approximately normally distributed with mean $\hat{\boldsymbol{\beta}}$ and covariance matrix H^{-1} for any differentiable $F(\cdot)$ such that $0<F(\cdot)<1$. This approximation is a special case of Jeffreys' (1967, p. 193) general result on the large-sample normality of posterior pdfs. Taking account of additional terms in (2.4) will probably improve the approximation. The convergence of the integral of (2.4) is shown in appendix 1.

Using the normal approximation in (2.5), approximate moments and marginal posterior pdfs are easily obtained from properties of the normal distribution. Further, if one is interested in the 'log-odds', which is given for the logit model by $\theta_i=\log[P_i/(1-P_i)]=\boldsymbol{x}_i'\boldsymbol{\beta}$, from the approximate posterior normality of $\boldsymbol{\beta}$,

$$p(\theta_i|D)\,\dot{=}\,\mathrm{N}(\boldsymbol{x}_i'\hat{\boldsymbol{\beta}},\boldsymbol{x}_i'H^{-1}\boldsymbol{x}_i), \tag{2.6}$$

that is θ_i is approximately normally distributed with mean $x_i'\hat{\beta}$ and variance $x_i'H^{-1}x_i$.[1] From (2.6), the approximate posterior pdf for P_i is

$$p(P_i|D) = \frac{(2\pi)^{-\frac{1}{2}}}{\hat{\phi}_i P_i(1-P_i)} \exp\left\{ -\left[\log\frac{P_i}{1-P_i} - x_i'\hat{\beta}\right]^2 \Big/ 2\hat{\phi}_i^2 \right\}, \tag{2.7}$$

where $\hat{\phi}_i^2 \equiv x_i'H^{-1}x_i$ and the median of $p(P_i|D)$ is approximately $\hat{P}_i = 1/(1+e^{-x_i'\hat{\beta}})$.

In addition, from (2.5) and for general $F(x_i'\beta)$, approximation results in Cramer (1946, p. 354) yield

$$\mathrm{E}(P_i|D) \doteq F(x_i'\hat{\beta}) \quad \text{and} \quad \mathrm{var}(P_i|D) \doteq a_i'H^{-1}a_i,$$

where $a_i = [\partial F(x_i'\beta)/\partial\beta]_{\beta=\hat{\beta}}$. With respect to the partial derivative,

$$\frac{\partial P_i}{\partial x_{ij}} = \frac{\partial}{\partial x_{ij}}\left(\frac{1}{1+e^{-x_i'\beta}}\right)$$

$$= \beta_j e^{-x_i'\beta}/\left(1+e^{-x_i'\beta}\right)^2 = \beta_j P_i(1-P_i),$$

its approximate posterior mean and variance are given by

$$\mathrm{E}\frac{\partial P_i}{\partial x_{ij}} \doteq \hat{\beta}_j F(x_i'\hat{\beta})[1 - F(x_i'\hat{\beta})]$$

$$\mathrm{var}\frac{\partial P_i}{\partial x_{ij}} \doteq \left[\frac{\partial}{\partial\beta}\left(\frac{\partial P_i}{\partial x_{ij}}\right)\right]'_{\beta=\hat{\beta}} H^{-1} \left[\frac{\partial}{\partial\beta}\left(\frac{\partial P_i}{\partial x_{ij}}\right)\right]_{\beta=\hat{\beta}}.$$

Last, the approximate posterior mean and variance of the elasticity,

$$\eta_{ij} \equiv (x_{ij}/P_i)(\partial P_i/\partial x_{ij}) = x_{ij}\beta_j(1-P_i),$$

are given by

$$\mathrm{E}\eta_{ij} \doteq x_{ij}\hat{\beta}_j[1 - F(x_i'\hat{\beta})],$$

$$\mathrm{var}\,\eta_{ij} \doteq \left(\frac{\partial\eta_{ij}}{\partial\beta}\right)'_{\beta=\hat{\beta}} H^{-1} \left(\frac{\partial\eta_{ij}}{\partial\beta}\right)_{\beta=\hat{\beta}}.$$

[1]See Leonard (1979) for an analysis that employs the θ_i's as basic parameters.

As regards prediction, if $\tilde{y}_*$ is a future, as yet unobserved random variable such that $\tilde{y}_* = 1$ with probability $P_* = F(\boldsymbol{x}_*'\boldsymbol{\beta})$ and $\tilde{y}_* = 0$ with probability $1 - P_* = 1 - F(\boldsymbol{x}_*'\boldsymbol{\beta})$, with $\boldsymbol{x}_*$ given, then $\mathrm{E}(y_*|\boldsymbol{\beta}) = F(\boldsymbol{x}_*'\boldsymbol{\beta})$ and given a posterior pdf for $\boldsymbol{\beta}$, $p(\boldsymbol{\beta}|D)$, the mean of the predictive pdf for $\tilde{y}_*$ is $\mathrm{E}(\tilde{y}_*|D) = \int F(\boldsymbol{x}_*'\boldsymbol{\beta})p(\boldsymbol{\beta}|D)\,\mathrm{d}\boldsymbol{\beta}$. This and other predictive moments of $\tilde{y}_*$ can be evaluated by use of numerical integration. An approximation to $\mathrm{E}(\tilde{y}_*|D)$ is $\mathrm{E}(y_*|D) \doteq F(\boldsymbol{x}_*'\hat{\boldsymbol{\beta}})$ and to $\mathrm{var}(y_*|D)$ is $\mathrm{var}(y_*|D) \doteq \boldsymbol{a}_*'H^{-1}\boldsymbol{a}_*$, where $\boldsymbol{a}_* = [\partial F(\boldsymbol{x}_*'\boldsymbol{\beta})/\partial\boldsymbol{\beta}]_{\boldsymbol{\beta}=\hat{\boldsymbol{\beta}}}$. These approximations may be improved by taking account of higher-order terms in the Taylor Series expansion of $F(\boldsymbol{x}_*'\boldsymbol{\beta})$ about $\hat{\boldsymbol{\beta}}$.

2.2. Case II: Informative priors, $\pi(\boldsymbol{\beta}|I_0)$

In this case the posterior pdf is given by

$$p(\boldsymbol{\beta}|D) \propto \pi(\boldsymbol{\beta}|I_0)l(\boldsymbol{\beta}|\boldsymbol{y}) \equiv q(\boldsymbol{\beta}), \tag{2.8}$$

where $l(\boldsymbol{\beta}|\boldsymbol{y})$ is given in (2.1). We first consider an expansion of (2.8) about the modal value, $\boldsymbol{\beta}_0$ of the posterior pdf; that is with $\log q(\boldsymbol{\beta}) \equiv Q(\boldsymbol{\beta})$,

$$p(\boldsymbol{\beta}|D) \propto \exp\{Q(\boldsymbol{\beta})\},$$

and

$$Q(\boldsymbol{\beta}) = Q(\boldsymbol{\beta}_0) - (\boldsymbol{\beta} - \boldsymbol{\beta}_0)'V_0(\boldsymbol{\beta} - \boldsymbol{\beta}_0)/2 + R_0,$$

where

$$V_0 \equiv -\left[\partial^2 Q(\boldsymbol{\beta})/\partial\boldsymbol{\beta}\partial\boldsymbol{\beta}'\right]_{\boldsymbol{\beta}=\boldsymbol{\beta}_0}$$

$$= -\left[\partial^2\left[\log\pi(\boldsymbol{\beta}|I_0) + \log l(\boldsymbol{\beta}|\boldsymbol{y})\right]/\partial\boldsymbol{\beta}\partial\boldsymbol{\beta}'\right]_{\boldsymbol{\beta}=\boldsymbol{\beta}_0}.$$

Then,

$$p(\boldsymbol{\beta}|D) \propto \exp\{-(\boldsymbol{\beta} - \boldsymbol{\beta}_0)'V_0(\boldsymbol{\beta} - \boldsymbol{\beta}_0)/2\}\left[1 + R_0 + R_0^2/2! + \dots\right]. \tag{2.9}$$

which gives an approximation for a wide range of prior pdfs.[2]

In the case that the prior pdf $\pi(\boldsymbol{\beta}|I_0)$ is normal with mean $\bar{\boldsymbol{\beta}}$ and covariance matrix A^{-1}, that is $\mathrm{N}(\bar{\boldsymbol{\beta}}, A^{-1})$, the general result in (2.9) applies with $V_0 = A + H_0$, where

$$H_0 \equiv -\left[\partial^2 \log l(\boldsymbol{\beta}|\boldsymbol{y})/\partial\boldsymbol{\beta}\partial\boldsymbol{\beta}'\right]_{\boldsymbol{\beta}=\boldsymbol{\beta}_0}.$$

[2] Ramsey (1972) analyzes the one independent variable case using a Dirchlet prior on the differences, $P(x_i) - P(x_{i-1}) > 0$, where $P(x_i) = \mathrm{E}\tilde{y}_i$, and $x_i - x_{i-1} > 0$ for all i. See also Kraft and van Eeden (1964).

As an alternative to the expansion about the modal value, in the case of a normal prior, $N(\bar{\beta}, A^{-1})$, the posterior pdf can be expressed as

$$p(\beta|D) \propto \exp\{-(\beta-\bar{\beta})'A(\beta-\bar{\beta})/2+\log l(\beta|y)\}$$

$$\propto \exp\{-[(\beta-\bar{\beta})'A(\beta-\bar{\beta})+(\beta-\hat{\beta})'H(\beta-\hat{\beta})]/2\} \times[1+R+R^2/2!+\dots] \tag{2.10}$$

$$\propto \exp\{-(\beta-\bar{\bar{\beta}})'G(\beta-\bar{\bar{\beta}})/2\}\times[1+R+R^2/2!+\dots],$$

where $\hat{\beta}$ is the ML estimate, H is defined after (2.4), R is the remainder in the Taylor Series expansion of $\log l(\beta|y)$, $G=A+H$, and

$$\bar{\bar{\beta}}=(A+H)^{-1}(A\bar{\beta}+H\hat{\beta}). \tag{2.11}$$

On comparing (2.9) and (2.10), it is seen that the leading normal terms are centered at β_0, the modal value of the posterior pdf in the case of (2.9) and at $\bar{\bar{\beta}}$, given in (2.11). The covariance matrices of the leading normal terms are $(A+H_0)^{-1}$ for (2.9) and $(A+H)^{-1}$ for (2.10), where H_0 and H are equal to $-[\partial^2\log l(\beta|y)/\partial\beta\partial\beta']$ evaluated at $\beta=\beta_0$ for H_0 and at $\beta=\hat{\beta}$ for H. Both procedures are computationally feasible since computer programs are available for obtaining β_0 and $\hat{\beta}$ as well as for other operations. For large sample conditions, $\beta_0 \doteq \bar{\bar{\beta}} \doteq \hat{\beta}$ and the forms of the leading normal terms in (2.9) and (2.10) will be about the same. Our preliminary view is that (2.9) will probably be a better approximation. Last, the leading normal terms of (2.9) and (2.10) can be employed to obtain posterior distributions and/or approximate moments of log-odds ratios, the probabilities $P_i=F(x_i'\beta)$ within and outside the sample, derivatives $\partial P_i/\partial x_{ij}$, and elasticities $\eta_{ij}=(x_{ij}/P_i)\partial P_i/\partial x_{ij}$ in much the same manner as shown above for the case of a diffuse prior.

Finally, in the logit model we have $\theta_i=\log[P_i/(1-P_i)]=\beta_0+\beta_1x_i$ or $x_i=(\theta_i-\beta_0)/\beta_1$. In some cases, it is relevant to fix the value of θ_i, say $\theta_i=0$, which implies $P_i=\frac{1}{2}$, and to estimate the associated x value, the median value or 50% lethal dosage (LED). More generally, let θ_0 be a given value for θ and assume that a posterior pdf for β_0 and β_1 is available. Then $x=(\theta_0-\beta_0)/\beta_1$ and the posterior pdf for x can be obtained from that of β_0 and β_1. If the latter is normal, then the posterior distribution of x is in the form of a ratio of correlated normal variables, can under certain conditions be markedly bimodal and fails to possess moments as noted in Tsutakawa (1975). For analysis of a bimodal posterior distribution, see, e.g., Zellner (1978, app.). When the distribution is not markedly bimodal, use of the loss function

$$L(x,\hat{x}) = [\beta_1(x-\hat{x})]^2 = (\beta_0+\beta_1\hat{x}-\theta_0)^2, \quad \theta_0 \text{ given},$$

leads to the point estimate $\hat{x}^*$ that minimizes posterior expected loss,

$$\begin{aligned}\hat{x}^* &= (\theta_0 \mathrm{E}\beta_1 - \mathrm{E}\beta_0\beta_1)/\mathrm{E}\beta_1^2 \\ &= \left[(\theta_0-\mathrm{E}\beta_0)/\mathrm{E}\beta_1 - \mathrm{cov}(\beta_0,\beta_1)/(\mathrm{E}\beta_1)^2\right]\left[1+\mathrm{var}\beta_1/(\mathrm{E}\beta_1)^2\right],\end{aligned}$$

where E is the posterior expectation, $\mathrm{var}\beta_1$ is the posterior variance of β_1, and $\mathrm{cov}(\beta_0,\beta_1)$ is the posterior covariance of β_0 and β_1. Of course, other loss functions can be employed if they are thought to be more appropriate and optimal point estimates can be obtained either analytically or numerically.[3]

3. Analysis of hypotheses and model selection procedures

In this section several frequently encountered testing and model selection problems[4] will be analyzed from the Bayesian point of view.

First, in applied analyses the algebraic sign of a coefficient, say β_j, an element of $\boldsymbol{\beta}$ is of great interest. That is, the hypotheses H_1: $\beta_j>0$ and H_2: $\beta_j<0$ are often important from a subject matter point of view. Given that we have the marginal posterior pdf for β_j, $p(\beta_j|D)$, it is direct to compute the posterior probability that $\beta_j>0$, $\Pr(\beta_j>0|D)=\int_0^\infty p(\beta_j|D)\,\mathrm{d}\beta_j$. Then $\Pr(\beta_j<0|D)=1-\Pr(\beta_j>0|D)$ and the posterior odds ratio for H_1 relative to H_2 is $K_{12}=\Pr(\beta_j>0|D)/\Pr(\beta_j<0|D)$. If $p(\beta_j|D)$ is $\mathrm{N}(\hat{\beta}_j,s_j^2)$, $z_j=(\beta_j-\hat{\beta}_j)/s_j$ is N(0, 1) and $\Pr(\beta_j>0|D)=\Pr(z_j>-\hat{\beta}_j/s_j|D)=1-\Phi(-\hat{\beta}_j/s_j)$, where $\Phi(\cdot)$ is the normal cumulative distribution function. This analysis can be extended to evaluate other posterior probabilities, e.g., $\Pr(\beta_j>0$ and $\beta_l>0|D)$, etc. by use of numerical integration techniques.

If $\boldsymbol{\beta}'=(\boldsymbol{\beta}_a',\boldsymbol{\beta}_b')$ and we consider the hypotheses H_1: $\boldsymbol{\beta}_b=\mathbf{0}$ and H_2: $\boldsymbol{\beta}_b\neq\mathbf{0}$, the posterior odds ratio for H_1 relative to H_2 is

$$K_{12}=\frac{\Pi_1}{\Pi_2}\cdot\frac{\int \pi_1(\boldsymbol{\beta}_a|H_1)\,l_1(\boldsymbol{\beta}_a|\boldsymbol{y})\,\mathrm{d}\boldsymbol{\beta}_a}{\int \pi_2(\boldsymbol{\beta}_a,\boldsymbol{\beta}_b|H_2)\,l_2(\boldsymbol{\beta}_a,\boldsymbol{\beta}_b|\boldsymbol{y})\,\mathrm{d}\boldsymbol{\beta}_a\,\mathrm{d}\boldsymbol{\beta}_b}, \tag{3.1}$$

[3] Freeman (1970) lets $\log P_i/(1-P_i)=\beta(x_i-\gamma)$ and considers sequential estimation of γ under an assumed known value for β and quadratic loss, $(\gamma-\hat{\gamma})^2$.

[4] See Cragg (1971) for an interesting use of posterior odds ratios in discriminating among variants of the Tobit model.

where Π_1/Π_2 is the prior odds ratio, $\pi_1(\cdot)$ and $\pi_2(\cdot)$ are the prior pdfs and $l_1(\cdot)$ and $l_2(\cdot)$ are the likelihood functions under H_1 and H_2, respectively. The integrals in (3.3) can be evaluated by numerical integration or approximated using the techniques discussed above.

A particular model selection problem is that of comparing and selecting forms for $F(\cdot)$ in $P_i = F(\boldsymbol{x}_i'\boldsymbol{\beta})$. For example, the issue might be a comparison of logit and probit models. Under the assumed form $F_1(\boldsymbol{x}_i'\boldsymbol{\beta}_1)$, denote the likelihood function by $l_1(\boldsymbol{\beta}_1|\boldsymbol{y})$ and the prior pdf for $\boldsymbol{\beta}_1$ by $\pi_1(\boldsymbol{\beta}_1)$ and similarly for $F_2(\boldsymbol{x}_i'\boldsymbol{\beta}_2)$, a second form for $F(\cdot)$. Then the posterior odds ratio for the model based on $F_1(\cdot)$ relative to that based on $F_2(\cdot)$ is

$$K_{12} = \frac{\Pi_1}{\Pi_2} \cdot \frac{\int \pi_1(\boldsymbol{\beta}_1) l_1(\boldsymbol{\beta}_1|\boldsymbol{y})\,\mathrm{d}\boldsymbol{\beta}_1}{\int \pi_2(\boldsymbol{\beta}_2) l_2(\boldsymbol{\beta}_2|\boldsymbol{y})\,\mathrm{d}\boldsymbol{\beta}_2}, \tag{3.2}$$

where Π_1/Π_2 is the prior odds ratio. The posterior odds ratio in (3.2) can be evaluated by numerical integration. In the special case in which $\pi_j(\boldsymbol{\beta}_j)$ is $\mathrm{N}(\bar{\boldsymbol{\beta}}_j, A_j^{-1})$ for $j = 1, 2$, and $l_j(\boldsymbol{\beta}_j|\boldsymbol{y}) \doteq l_j(\hat{\boldsymbol{\beta}}_j|\boldsymbol{y})\exp\{-Q(\boldsymbol{\beta}_j)/2\}$ for $j = 1, 2$, where $\hat{\boldsymbol{\beta}}_j$ is the ML estimate, $Q(\boldsymbol{\beta}_j) = (\boldsymbol{\beta}_j - \hat{\boldsymbol{\beta}}_j)'H_j(\boldsymbol{\beta}_j - \hat{\boldsymbol{\beta}}_j)$, and H_j is the negative of the Hessian matrix evaluated at the ML estimate, $\hat{\boldsymbol{\beta}}_j$ (3.2) is approximately given by

$$K_{12} \doteq \frac{\Pi_1}{\Pi_2} qLR \exp\{-(S_1 - S_2)/2\}. \tag{3.3}$$

In (3.3),

$$q = m_1/m_2 \quad \text{with} \quad m_j^2 = |A_j|/|A_j + H_j|, \quad j = 1, 2,$$

$$LR = l_1(\hat{\boldsymbol{\beta}}_1|\boldsymbol{y})/l_2(\hat{\boldsymbol{\beta}}_2|\boldsymbol{y}),$$

the likelihood ratio, and

$$S_j = \left(\hat{\boldsymbol{\beta}}_j - \bar{\bar{\boldsymbol{\beta}}}_j\right)' H_j \left(\hat{\boldsymbol{\beta}}_j - \bar{\bar{\boldsymbol{\beta}}}_j\right) + \left(\bar{\boldsymbol{\beta}}_j - \bar{\bar{\boldsymbol{\beta}}}_j\right)' A_j \left(\bar{\boldsymbol{\beta}}_j - \bar{\bar{\boldsymbol{\beta}}}_j\right), \tag{3.4}$$

with $\hat{\boldsymbol{\beta}}_j$ = ML estimate, $\bar{\boldsymbol{\beta}}_j$ = prior mean vector, and $\bar{\bar{\boldsymbol{\beta}}}_j = (A_j + H_j)^{-1}(A_j\bar{\boldsymbol{\beta}}_j + H_j\hat{\boldsymbol{\beta}}_j)$. The quantity S_j measures the extent to which the ML estimate and the prior mean vector are close to $\bar{\bar{\boldsymbol{\beta}}}_j$, the mean of a leading term approximation to the posterior pdf for $\boldsymbol{\beta}_j$. Also, by writing $H_j = K_j'K_j$, $A_j = C_j'C_j$, and

$$\boldsymbol{\varepsilon}_j = \begin{pmatrix} K_j\hat{\boldsymbol{\beta}}_j \\ C_j\bar{\boldsymbol{\beta}}_j \end{pmatrix} - \begin{pmatrix} K_j \\ C_j \end{pmatrix} \bar{\bar{\boldsymbol{\beta}}}_j, \tag{3.5}$$

it is seen that $S_j = \varepsilon_j'\varepsilon_j$, a 'residual sum of squares' interpretation that measures degree of compatibility between sample and prior information.

In summary, the posterior odds ratio in (3.2) reflects more than just the value of the LR. It incorporates other factors that are important in comparing alternative models. Given a loss structure it is possible to select a particular model by choosing so as to minimize expected loss if model choice is required. On the other hand, if the odds are not overwhelmingly in favor of one of the models, it will be wise to carry both of them along to be appraised further using additional data.

Last we note that if $\boldsymbol{\beta}_1$ and $\boldsymbol{\beta}_2$ have the same number of elements and $\pi_1(\boldsymbol{\beta}_1)$ and $\pi_2(\boldsymbol{\beta}_2)$ are in the Jeffreys form $\pi(\boldsymbol{\beta}_j) = c|I(\boldsymbol{\beta}_j)|^{\frac{1}{2}}$ for $\boldsymbol{\beta}_j \subset R_{\boldsymbol{\beta}_j}$, a large finite region, then K_{12} in (3.2) is given approximately by $K_{12} \doteq (\Pi_1/\Pi_2)LR$, where LR is the ratio of maximized likelihood functions. Numerical integration techniques and/or taking account of additional terms in the above asymptotic expansions may lead to better approximations to posterior odds ratios for the problems discussed above.

4. Numerical analysis of posterior distributions

In large samples, the posterior distribution of logit model parameters may be closely approximated by a leading normal term as in (2.4), (2.9) and (2.10). Using the asymptotic normal approximation, all relevant characteristics of the posterior distribution can be examined using well-known analytic results. It should be noted that the Newton–Raphson optimization routine which computes the posterior mode also provides, as a byproduct, the asymptotic covariance matrix.

For small samples, we must examine the posterior density to check the adequacy of the asymptotic normal approximation. If the asymptotic normal distribution poorly approximates the exact posterior density, numerical techniques must be used to analyze the density. That is, numerical integration techniques are employed to evaluate the normalizing constant and moments of a posterior density function as well as its marginal and conditional densities and probabilities associated with subregions of the parameter space.

Virtually any analysis of DQRM's posterior distributions requires an ability to evaluate multivariate integrals. In the absence of analytical formula for these integrals, we must resort to numerical methods. Methods of univariate numerical integration which can handle a wide variety of integrands have been developed and refined; Simpson's rule and quadrature methods are prominent among the univariate techniques. Extension of traditional approximation methods to multivariate integrals has been explored [see, e.g., Stroud (1971) and Dagenais and Liem (1981)]. However, the number of integrand evaluations required for a reasonable degree of accuracy may be very large and this number increases exponentially with the dimension of the integral.

The introduction of Monte Carlo integration techniques [see, e.g. Hammersley and Hanscomb (1979)] has made possible the evaluation of multivariate integrals at reasonable cost. Most Monte Carlo integration procedures have evolved from the simple uniform Monte Carlo procedure. In uniform Monte Carlo integration, the integral is viewed as finding the expectation of the integrand with respect to a uniform distribution of the arguments. To evaluate $I = \int_{R_\beta} f(\beta)\,\mathrm{d}\beta$ it seems reasonable to approximate I by drawing random β vectors from a uniform distribution over a bounded subset of R_β and then calculating the sample mean, $\hat{I} = \sum_{i=1}^{N} f(\beta^i)/N$.

If the β^i are drawn uniformly, many sample points will be associated with very small values of the integrand. If the integrand has large variation, the error of the integration procedure will be large. It is known that $\mathrm{var}(\hat{I}) \leq (\Delta f)^2/N$, where $\Delta f = f_{\max} - f_{\min}$. The major advantage of the uniform Monte Carlo procedure is that the number of sample points required to reach a given error bound does not depend on the dimension of the integral.

Monte Carlo integration procedures employing importance sampling have been introduced in an effort to reduce the number of sample points drawn that are associated with small values of the integrand. In our investigation we are interested in approximating the posterior expectation of a function, $h(\beta)$ [for example, $h(\beta) = \beta_i^r$], that is, with posterior kernel $k(\beta|D)$,

$$\mathrm{E}[h(\beta)|D] = \int_{R_\beta} h(\beta)k(\beta|D)\,\mathrm{d}\beta \Big/ \int_{R_\beta} k(\beta|D)\,\mathrm{d}\beta. \tag{4.1}$$

We approximate $\mathrm{E}[h(\beta)|D]$ by approximating the integrals in the numerator and denominator separately. Sample points are no longer drawn uniformly but are drawn from a density known as the importance function. To motivate this, write (4.1) as

$$\mathrm{E}[h(\beta)] = \int_{R_\beta} h(\beta)\frac{k(\beta|D)}{I(\beta)}I(\beta)\,\mathrm{d}\beta \Big/ \int_{R_\beta} \frac{k(\beta|D)}{I(\beta)}I(\beta)\,\mathrm{d}\beta.$$

We estimate $\mathrm{E}[h(\beta)|D]$ by drawing β^i from $I(\beta)$, the importance function, and computing the weighted average, $\sum_{i=1}^{N} h(\beta^i)w_i / \sum_{i=1}^{N} w_i$, where $w_i = k(\beta^i|D)/I(\beta^i)$.

The choice of an importance function that closely resembles the posterior kernel is critical for the success of the importance sampling approach. One must be particularly careful that the importance function is never substantially smaller than the kernel. In this study, we adopt a multivariate Student-t (MVS-t) importance function with mean set equal to the posterior mode and covariance matrix proportional to the asymptotic covariance matrix. Earlier, Kloek and van Dijk (1978) used a multivariate Cauchy importance function.

Choice of the degrees of freedom of the MVS-t distribution was made on the basis of the coefficient of variation in the weights. (See appendix 2 for details).

In a recent paper, Friedman and Wright (1980a) propose a new Monte Carlo technique which they have named 'pseudo-uniform' Monte Carlo integration. The Friedman–Wright algorithm partitions the region of integration into hyper-rectangles so as to minimize integrand variation in any hyper-rectangle. Once these hyper-rectangles are constructed, points are drawn uniformly within each rectangle. In our experience with pseudo-uniform Monte Carlo integration, a large number of integrand evaluations were required to construct the partition. Moreover, the partition must be changed for each moment integrand while all moments can be calculated simultaneously in an importance sampling approach (see appendix 2 for a comparison of numerical integration routines).

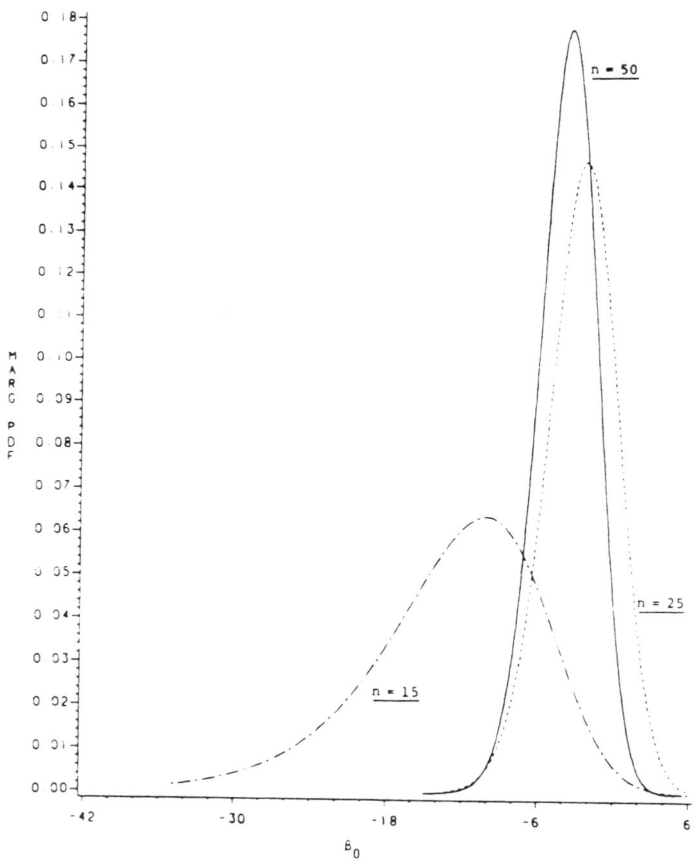

Fig. 1. Marginal posterior distribution of the intercept (β_0) under a diffuse prior; in the population logit model used to generate the data, $\beta_0 = -4.0$.

The essential computational problem in the analysis of the logit model is that the posterior distribution cannot be expressed in terms of a small set of sufficient statistics. The entire sample must be searched in order to evaluate the likelihood function at each point, $\boldsymbol{\beta}$. All numerical integration algorithms involve large numbers of integrand evaluations. Reduction in the cost of evaluating the posterior density function would substantially improve the performance of any numerical integration routine.

4.1. Numerical investigation of a two-parameter logit model

In order to assess the adequacy of the asymptotic approximation to the posterior density, we first examine data generated from a two-parameter logit model. We employ the logit population model, $P_i = F(\beta_0 + \beta_1 x_i)$ with $\beta_0 =$

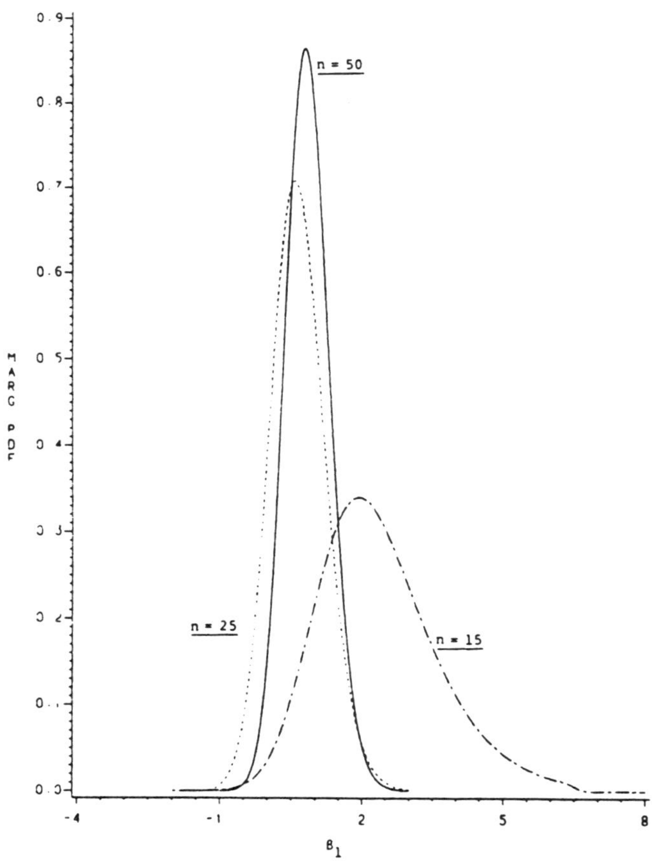

Fig. 2. Marginal posterior distribution of the slope (β_1) under a diffuse prior; in the population logit model used to generate the data, $\beta_1 = 0.8$.

Table 1

Central moments, skewness, and kurtosis of the posterior distributions of logit parameters under a diffuse prior.[a]

		μ_1	μ_2	μ_3	μ_4	S	K
		Marginal distribution of intercept (β_0)					
$n = 50$	Exact moments	−3.82	5.21	−2.56	85.7	−0.215	3.15
	Asymptotic moments	−3.50[b]	4.79	0.0	68.8	0.0	3.0
$n = 25$	Exact moments	−2.64	7.69	−4.40	184.0	−0.206	3.11
	Asymptotic moments	−2.23[b]	6.74	0.0	136.0	0.0	3.0
$n = 15$	Exact moments	−12.5	44.2	−164.0	6250.0	−0.557	3.19
	Asymptotic moments	−9.56[b]	32.2	0.0	3110.0	0.0	3.0
		Marginal distribution of slope (β_1)					
$n = 50$	Exact moments	0.845	0.221	0.0230	0.154	0.221	3.15
	Asymptotic moments	0.776[b]	0.203	0.0	0.124	0.0	3.0
$n = 25$	Exact moments	0.643	0.334	0.0503	0.354	0.261	3.17
	Asymptotic moments	0.549[b]	0.290	0.0	0.252	0.0	3.0
$n = 15$	Exact moments	2.35	1.56	1.06	7.71	0.546	3.19
	Asymptotic moments	1.80[b]	1.14	0.0	3.90	0.0	3.0

[a]The population model is $y_i = 1$ with prob. P_i, and 0 with prob. $1 - P_i$, where $P_i = F(\beta_0 + \beta_1 x_i)$. F is the logistic cumulative distribution function. Here $\beta_0 = -4.0$ and $\beta_1 = 0.8$. See footnote 5 for details of sample generation.

The moments of the exact posterior distribution were calculated using a bivariate version of Simpson's rule with 200 subintervals on each axis for a total of 40,401 integrand evaluations. Approximate moment calculations based on the leading normal approximation are given below the exact moment calculations.

S is a measure of skewness, $S = \mu_3/\sigma^3$, and K is a measure of kurtosis, $K = \mu_4/\sigma^4$.

[b]Maximum-likelihood estimates.

−4.0 and $\beta_1 = 0.8$. Diffuse priors are used in this analysis. The leading normal term of the asymptotic expansion of the posterior density, given by (2.4), was used to compute approximate posterior moments.

Data sets composed of 50, 25, and 15 observations were generated from the two-parameter model.[5] The independent variable values, x_i, were generated by taking independent draws from a normal distribution with mean 5.0 and standard deviation 0.75. By limiting the variation in the independent variable, we sought to subject the asymptotic normal approximation to a severe test.

In figs. 1 and 2, we present the exact marginal posterior distributions of β_0 and β_1. Marginal posterior density functions were computed using univariate Simpson's rule and spline interpolation between ordinate points. As the

[5]Values of the independent variable (or variables) were generated by draws from a normal random-number generator. Given the values of $\boldsymbol{\beta}$ and the independent variables, $P_i = [1 + \exp(-\boldsymbol{x}_i'\boldsymbol{\beta})]^{-1}$ was calculated, $i = 1, 2, \ldots, n$. To generate y_i with the appropriate binomial distribution, a uniform random number in (0, 1) is drawn. If P_i is less than or equal to the drawn number, y_i is set equal to 1; otherwise y_i is set equal to 0.

number of observations decreases, the marginal distributions become skewed and fat-tailed. For 50 observations, the marginal distributions resemble normal curves very closely. Examination of the central moments given in table 1 confirms the visual impression from figs. 1 and 2. For 50 observations, the moments based on the asymptotic normal approximation are close to moments calculated from the exact posterior distribution.

Fig. 3 presents the joint distribution of β_0 and β_1 for a sample of 50. It is immediately clear that β_0 and β_1 are inversely related; the asymptotic correlation coefficient is more negative than -0.9. By reparameterizing the model, we can substantially reduce this correlation. That is, we write

$$P_i = F(\mu + \beta_1(x_i - \bar{x})) \quad \text{where} \quad \mu = \beta_0 + \beta_1\bar{x},$$

with

$$\bar{x} \equiv \sum_{i=1}^{n} x_i/n.$$

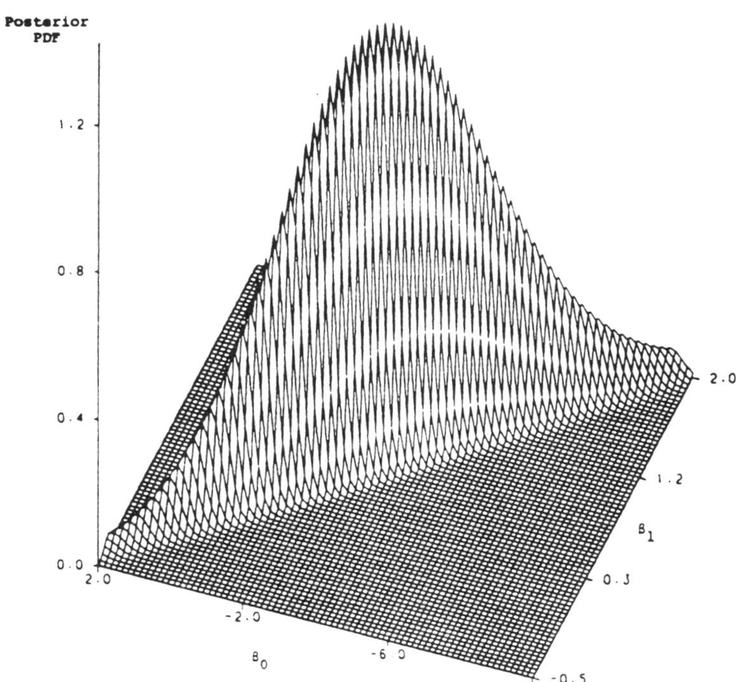

Fig. 3. Joint posterior distribution of the slope (β_1) and intercept (β_0) under a diffuse prior; 50 observations; in the population logit model, $\beta_0 = -4.0$, $\beta_1 = 0.8$.

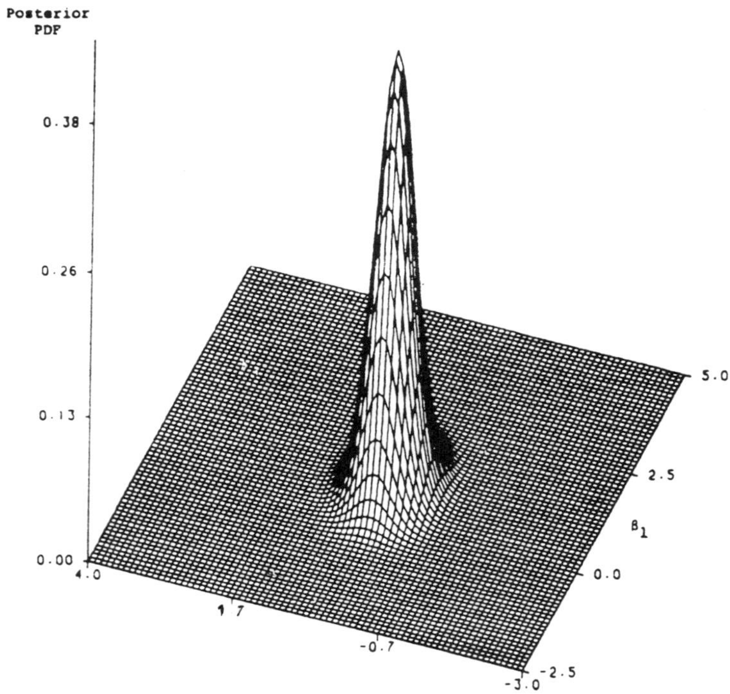

Fig. 4. Joint posterior distribution of the slope (β_1) and intercept (μ) with a diffuse prior and standardized independent variable; 50 observations; the population logit model: $P_i = F(\mu - \beta_1(x_{1i} - \bar{x}_1))$, $\mu = \beta_0 + \beta_1\bar{x}_1$, $\beta_0 = -4.0$, $\beta_1 = 0.8$.

In fig. 4, we graph the joint distribution of μ and β_1; μ and β_1 have a low posterior correlation and the posterior distribution takes on a spherical normal shape.

4.2. *Three-parameter logit model*

We now consider a logit model with two independent variables and three parameters,

$$P_i = F(\mu + \beta_1(x_{1i} - \bar{x}_1) + \beta_2(x_{2i} - \bar{x}_2)),$$

$$\mu = \beta_0 + \beta_1\bar{x}_1 + \beta_2\bar{x}_2,$$

$$\beta_0 = -3.5, \qquad \beta_1 = 0.4, \qquad \beta_2 = 0.15.$$

Three data sets with 100, 50 and 25 observations were generated by drawing x_{1i} and x_{2i} from a bivariate normal distribution with zero correlation[6] between x_1 and x_2 and identical variances, $\sigma_{x_1} = \sigma_{x_2} = 2.0$. (See footnote 5 for details on

[6] The results presented below changed little when x_1 and x_2 were drawn from a bivariate normal distribution with a correlation of 0.8.

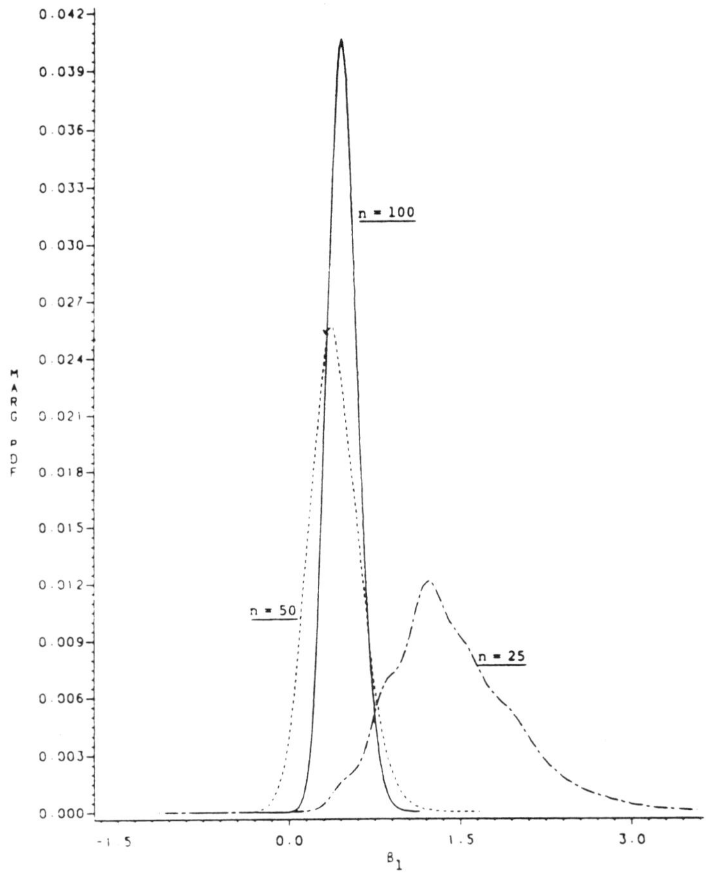

Fig. 5. Marginal posterior distribution of β_1 under a diffuse prior; the population logit model: $P_i = F(\mu + \beta_1(x_{1i} - \bar{x}_1) + \beta_2(x_{2i} - \bar{x}_2))$, $\beta_1 = 0.40$.

data generation.) A diffuse prior was employed and importance-sampling numerical integration techniques were used to analyze the posterior density. (See appendix 2 for details.)

Figs. 5 and 6 show the marginal densities for the slope coefficients obtained by bivariate Simpson's rule and spline interpolation.[7] As in the two-parameter case, the marginal densities exhibit considerable skewness and kurtosis for small n. For 100 and 50 observations, the marginal distributions of the slope coefficients closely resemble the normal curve. However, for 25 observations, the normal approximation does not represent the marginal densities very well. Examination of table 2 corroborates the visual evidence. The asymptotic approximate moments are very close to the exact moments for 100 and 50

[7]The unevenness in the marginal density for $n = 25$ is due to the small number of points at which the density is evaluated. Fifty bivariate integrations were performed, one for each ordinate.

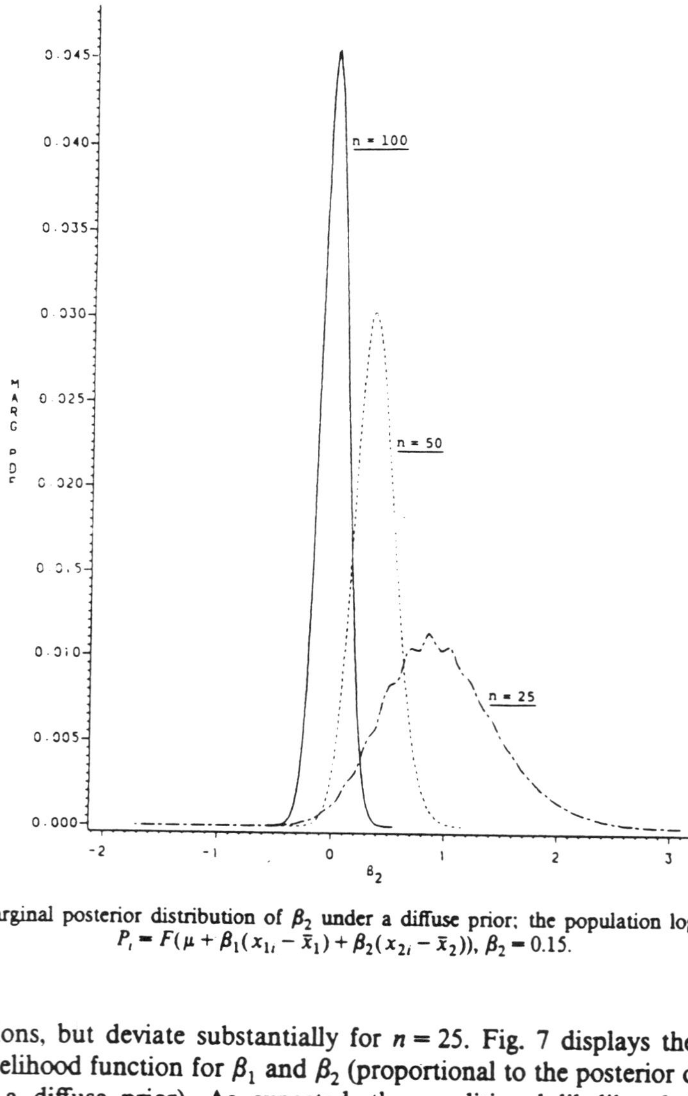

Fig. 6. Marginal posterior distribution of β_2 under a diffuse prior; the population logit model: $P_i = F(\mu + \beta_1(x_{1i} - \bar{x}_1) + \beta_2(x_{2i} - \bar{x}_2))$, $\beta_2 = 0.15$.

observations, but deviate substantially for $n = 25$. Fig. 7 displays the conditional likelihood function for β_1 and β_2 (proportional to the posterior distribution for a diffuse prior). As expected, the conditional likelihood function resembles a bivariate normal kernel.

The elasticity of P_i with respect to an independent variable x_{ij} in the logit model is given by $\eta_{ij} = x_{ij}\beta_j(1 - P_i)$. We examine the posterior distribution of η_{ij} evaluated at the mean of the independent variable for the one independent variable case, that is $\bar{\eta} = \bar{x}\beta_1(1 - \bar{P})$, where $\bar{P} = F(\beta_0 + \beta_1\bar{x})$. We transform from the distribution of (β_0, β_1) to the distribution of $(\beta_0, \bar{\eta})$ and integrate out β_0. Fig. 8 presents the posterior distribution of $\bar{\eta}$. Except for the minor side lobes, this distribution closely resembles a normal distribution.

Table 2

Central moments, skewness, and kurtosis of the posterior distributions of logit parameters under a diffuse prior.[a]

		μ_1	μ_2	μ_3	μ_4	S	K
		Marginal distribution of μ					
$n = 100$	Exact moments	−0.302	0.0504	−0.00200	0.00774	−0.177	3.05
	Asymptotic moments	−0.290[b]	0.0479	0.0	0.00688	0.0	3.0
$n = 50$	Exact moments	0.00214	0.104	−0.00332	0.0337	−0.0985	3.10
	Asymptotic moments	0.000871[b]	0.0999	0.0	0.0299	0.0	3.0
$n = 25$	Exact moments	−0.592	0.447	−0.0451	0.625	−0.151	3.13
	Asymptotic moments	−0.442[b]	0.367	0.0	0.404	0.0	3.0
		Marginal distribution of β_1					
$n = 100$	Exact moments	0.470	0.0172	0.000274	0.000827	0.121	2.79
	Asymptotic moments	0.444[b]	0.0168	0.0	0.000847	0.0	3.0
$n = 50$	Exact moments	0.400	0.0456	0.00275	0.00618	0.283	2.98
	Asymptotic moments	0.366[b]	0.0424	0.0	0.00539	0.0	3.0
$n = 25$	Exact moments	1.41	0.272	0.0889	0.250	0.626	3.37
	Asymptotic moments	1.12[b]	0.199	0.0	0.119	0.0	3.0
		Marginal distribution of β_2					
$n = 100$	Exact moments	−0.0102	0.0147	−0.000130	0.000663	−0.0731	3.09
	Asymptotic moments	−0.00898[b]	0.0132	0.0	0.000523	0.0	3.0
$n = 50$	Exact moments	0.367	0.0353	0.00197	0.00395	0.296	3.17
	Asymptotic moments	0.334[b]	0.0287	0.0	0.00247	0.0	3.0
$n = 25$	Exact moments	0.976	0.251	0.446	0.203	0.355	3.23
	Asymptotic moments	0.776[b]	0.200	0.0	0.120	0.0	3.0

[a] In the population logit model, $P_i = F(\mu + \beta_1(x_{1i} - \bar{x}_1) + \beta_2(x_{2i} - \bar{x}_2))$. Here $\mu = \beta_0 + \beta_1\bar{x}_1 + \beta_2\bar{x}_2$; $\beta_0 = -3.5$, $\beta_1 = 0.4$, $\beta_2 = 0.15$. See appendix 2 for details of sample generation. The moments of the exact posterior distribution were calculated using importance sampling Monte Carlo integration with a multivariate Student-t importance function. See appendix 2 for details. Moments based on the asymptotic normal approximation to the posterior density are given below the exact moments. See footnote a, table 1, for definition of skewness and kurtosis measures.

[b] Maximum-likelihood estimates.

5. Summary and concluding remarks

In this paper we have provided operational Bayesian techniques for analyzing a class of dichotomous quantal response models frequently employed in applied studies. Bayesian estimation and testing results were presented that are operational in small and large samples. This contrasts with sampling theory analyses that are usually based on an assumed large-sample normal distribution of maximum-likelihood estimators, a tenuous assumption in many circumstances. Given the availability of numerical integration techniques, the Bayesian approach, utilizing diffuse or informative prior distributions was shown to be operational and capable of providing analyses appropriate for

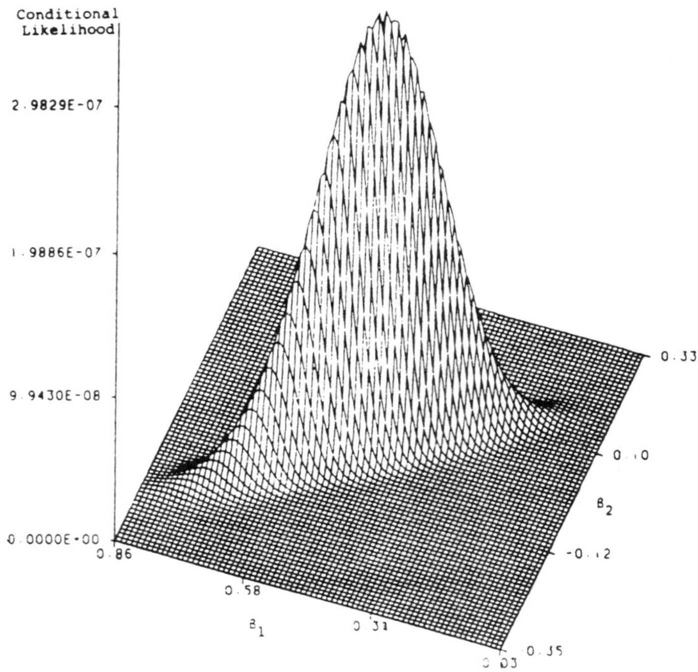

Fig. 7. Conditional likelihood function, $l(\beta_1, \beta_2 | D, \beta_0 = \hat{\beta}_0)$ where $\hat{\beta}_0$ is the maximum likelihood estimate; two independent variables and 100 observations.

small or large samples. Our numerical analyses of exact posterior density functions revealed that in small-sample situations normal asymptotic approximations do not work very well. However, for moderately large samples, the normal asymptotic approximation works well for the cases that we have examined. We have also demonstrated the feasibility of analysis of exact posterior distributions using numerical integration techniques. Such analysis yields finite-sample posterior distributions and odds ratios, optimal point estimates, predictions and intervals. Further work to extend these results to broader models involving both qualitative and quantitative economic variables is planned.

Appendix 1

Small-sample and integral-convergence results

In this appendix we present some exact results for a simple logit model, $P_i = 1/(1 + e^{-x_i\beta})$, where x_i and β are scalars. If our sample ($n = 2$) is $y_1 = 0$

for $x_1 = 1$ and $y_2 = 1$ for $x_2 = 2$, the likelihood function is

$$l(\beta \mid y) = (1 - P_1)P_2$$
$$= e^{-\beta}/(1 + e^{-\beta})(1 + e^{-2\beta}), \qquad -\infty < \beta < \infty, \tag{A.1}$$

which is plotted in fig. 9. It is seen that this likelihood function is unimodal and non-normally shaped. With a diffuse prior pdf $\pi(\beta) \propto$ const., $-\infty < \beta < \infty$, the posterior pdf is

$$p(\beta|D) = c\,e^{-\beta}/(1 + e^{-\beta})(1 + e^{-2\beta}), \qquad -\infty < \beta < \infty, \tag{A.2}$$

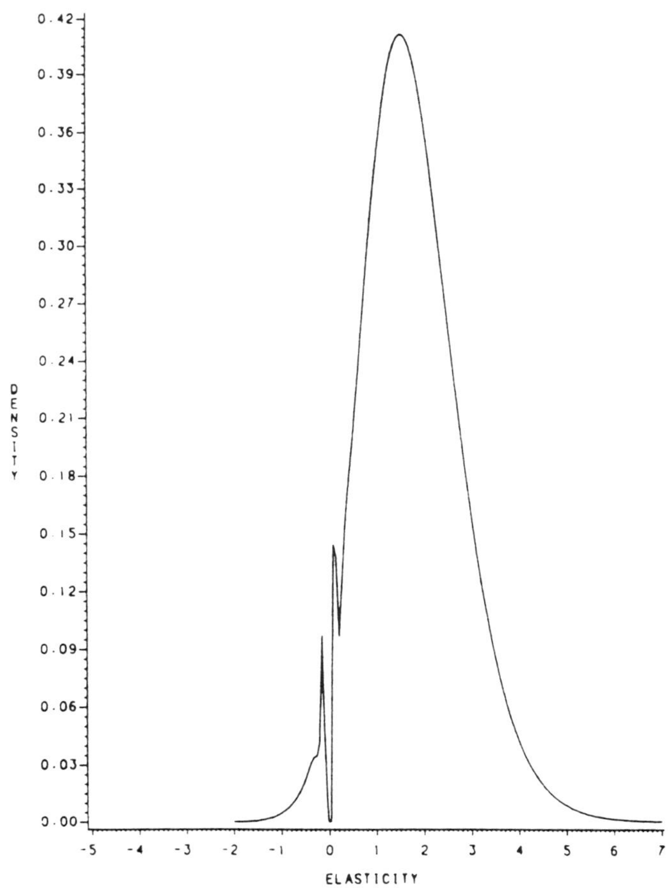

Fig. 8. Posterior distribution of elasticity of P_i with respect to independent variable evaluated at the mean of the independent variable; 50 observations; mean of this distribution is 1.78.

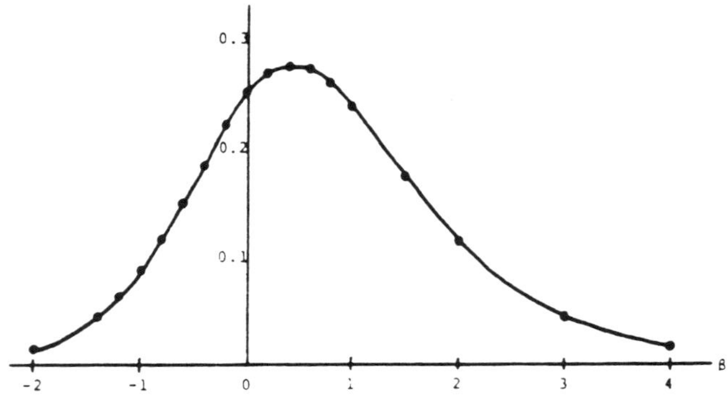

Fig. 9. Likelihood function ($n = 2$), $L(\beta) = e^{-\beta}/(1+e^{-\beta})(1+e^{-2\beta})$.

where c is the normalizing constant. To evaluate c, let $z = e^{-\beta}$, $0 < z < \infty$, with $dz/z = -d\beta$, and then

$$\begin{aligned} 1/c &= \int_{-\infty}^{\infty} \left[e^{-\beta}/(1+e^{-\beta})(1+e^{-2\beta})\right] d\beta \\ &= \int_0^{\infty} \frac{1}{(1+z)(1+z^2)} dz = \pi/4. \end{aligned} \tag{A.3}$$

Thus the normalized posterior pdf is

$$p(\beta|D) = \frac{4}{\pi} e^{-\beta}/(1+e^{-\beta})(1+e^{-2\beta}). \tag{A.4}$$

Further from $P_1 = 1/(1+e^{-\beta}) = 1/(1+z)$, the posterior mean of P_1 is

$$E(P_1|D) = (4/\pi)\int_0^{\infty} \frac{dz}{(1+z)^2(1+z^2)} = 2/\pi = 0.6366, \tag{A.5}$$

and from $P_2 = 1/(1+e^{-2\beta}) = 1/(1+z^2)$,

$$E(P_2|D) = (4/\pi)\int_0^{\infty} \frac{dz}{(1+z)(1+z^2)^2} = 1 - 1/\pi = 0.6817. \tag{A.6}$$

The integrals in (A.3), (A.5) and (A.6) were evaluated using results (nos. 83, 67, 69 and 89) of Pierce's (1929) table of integrals.

If instead of taking $-\infty < \beta < \infty$ in (A.2), we take $0 < \beta < \infty$, then with $z = e^{-\beta}$, $0 < z < 1$, and by similar integrations, the normalized posterior pdf is

$$p(\beta|D) = \left(\frac{4}{\ln 2 + \pi/2}\right) e^{-\beta}/(1 + e^{-\beta})(1 + e^{-2\beta}), \qquad 0 < \beta < \infty,$$

and

$$E(P_1|D) = 0.7479 \quad \text{and} \quad E(P_2|D) = 0.9147.$$

For comparative purposes, the MLE of β, determined numerically, is $\hat{\beta} = 0.42$, and the MLE's of P_1 and P_2 are $\hat{P}_1 = 0.60$ and $\hat{P}_2 = 0.70$.

If our sample were $y_1 = 0$ for $x_1 = -1$ and $y_2 = 1$ for $x_2 = 1$, then it is easy to show that the likelihood function is monotonically increasing in β, that is no finite maximum exists. Note that the function $P_i = 1/1 + e^{-x_i\beta}$ is constrained to equal $\frac{1}{2}$ at $x_i = 0$ and a large value for β produces a curve that best fits the data as shown in fig. 10. In this case with a prior $p(\beta) \propto$ const., $-\infty < \beta < \infty$, the posterior pdf is improper.

If single values of $\tilde{y}$ (= 1 or 0) are observed opposite $x = 1, 2, \ldots, n$, then the likelihood function is

$$l(\beta|y) = e^{-m\beta}/(1 + e^{-\beta})(1 + e^{-2\beta})\ldots(1 + e^{-n\beta}),$$

where m is the sum of the x values for which y values are equal to zero. Again letting $z = e^{-\beta}$ and using a uniform prior on β, $-\infty < \beta < \infty$, the posterior pdf for z is

$$p(z|D) \propto z^{m-1}/(1 + z)(1 + z^2)(1 + z^3)\ldots(1 + z^n), \qquad 0 < z < \infty. \tag{A.7}$$

Possibly the contour integration results in Whittaker and Watson (1963, p. 117 ff.) can be employed to integrate (A.7) in order to evaluate the normalizing constant and other features of the posterior pdf.

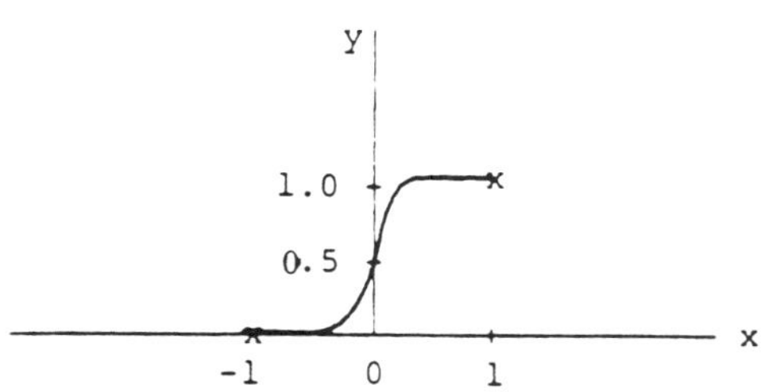

Fig. 10.

Integral-convergence conditions

We now seek to determine under what conditions integrals of the form $I = \int h(\boldsymbol{\beta})p(\boldsymbol{\beta})l(\boldsymbol{\beta}|D)\,\mathrm{d}\boldsymbol{\beta}$ converge. First, we consider cases in which the prior pdf, $p(\boldsymbol{\beta})$, is proper, that is $\int p(\boldsymbol{\beta})\,\mathrm{d}\boldsymbol{\beta} = 1$. Since $0 < l(\boldsymbol{\beta}|D) < 1$ for DQRM's, we have $\int h(\boldsymbol{\beta})p(\boldsymbol{\beta})l(\boldsymbol{\beta}|D)\,\mathrm{d}\boldsymbol{\beta} < \int h(\boldsymbol{\beta})p(\boldsymbol{\beta})\,\mathrm{d}\boldsymbol{\beta}$, and if this latter integral converges, the former converges. For example, if $p(\boldsymbol{\beta})$ is a multivariate normal pdf and $h(\boldsymbol{\beta}) = \beta_i^r$, $r = 0, 1, 2, \ldots$, $\int \beta_i^r p(\boldsymbol{\beta})\,\mathrm{d}\boldsymbol{\beta} < \infty$, the integral I converges. Also, if the elements of the prior have finite ranges, then the integral I converges for many choices of $h(\boldsymbol{\beta})$ including $h(\boldsymbol{\beta}) = \text{const}$. Last, if we regard a diffuse prior pdf as a 'spread out', *proper* normal pdf, then the integral I converges for many choices of $h(\boldsymbol{\beta})$ including $h(\boldsymbol{\beta}) = \text{const}$.

If we consider $\int p(\boldsymbol{\beta})l(\boldsymbol{\beta}|D)\,\mathrm{d}\boldsymbol{\beta}$, where $p(\boldsymbol{\beta})$ is an improper pdf, for example $p(\boldsymbol{\beta}) \propto \text{const.}$, $-\infty < \beta_i < \infty$, $i = 1, 2, \ldots, k$, then we have

$$\int l(\boldsymbol{\beta}|D)\,\mathrm{d}\boldsymbol{\beta} = \int \prod_{i=1}^{n} P_i^{y_i}(1 - P_i)^{1-y_i}\,\mathrm{d}\boldsymbol{\beta},$$

where for the logit model with $P_i = [1 + \exp\{-\boldsymbol{x}_i'\boldsymbol{\beta}\}]^{-1}$, $0 < P_i < 1$,

$$\int l(\boldsymbol{\beta}|D)\,\mathrm{d}\boldsymbol{\beta} = \int \exp\left\{-\sum_{i=1}^{m} \boldsymbol{x}_i'\boldsymbol{\beta}\right\} \Big/ \prod_{i=1}^{n}\left[1 + \exp\{-\boldsymbol{x}_i'\boldsymbol{\beta}\}\right]\mathrm{d}\boldsymbol{\beta}. \tag{A.8}$$

Here we have ordered the observations so that $y_i = 0$ for $i = 1, 2, \ldots, m$.

We first consider the two-parameter case, $\boldsymbol{x}_i'\boldsymbol{\beta} = \beta_0 + \beta_1 x_i$ with $0 < x_i < \infty$, $i = 1, 2, \ldots, n$. In this case the integrand of (A.8) is

$$\mathrm{e}^{-m\beta_0}\mathrm{e}^{-w\beta_1} \Big/ \prod_{i=1}^{n}\left[1 + \exp\{-(\beta_0 + \beta_1 x_i)\}\right], \tag{A.9}$$

where $w \equiv \sum_{i=1}^{m} x_i$, the sum of x_i's associated with zero values for the y_i's. We consider integration of (A.9) over the following four regions: (i) $0 < \beta_0, \beta_1 < \infty$, (ii) $-\infty < \beta_0, \beta_1 < 0$, (iii) $-\infty < \beta_0 < 0$ and $0 < \beta_1 < \infty$, and (iv) $0 < \beta_0 < \infty$ and $-\infty < \beta_1 < 0$.

For region (i), the integrand in (A.9) is less than $\exp\{-m\beta_0 - w\beta_1\}$ for $m > 0$. Since $\int_0^\infty\int_0^\infty \exp\{-m\beta_0 - w\beta_1\}\,\mathrm{d}\beta_0\mathrm{d}\beta_1 < \infty$, the integral of (A.9) over region (i) converges given $m > 0$.

For region (ii), the product in the denominator of (A.9) is greater than $\exp\{-(n\beta_0 + v\beta_1)\}$, where $v \equiv \sum_{i=1}^{n} x_i$. Thus (A.9) is less than $\exp\{-(m - n)\beta_0 - (w - v)\beta_1\}$. The integral of this last expression over region (ii) converges for $n > m$ and $v > w$, and thus the integral of (A.9) over region (ii) converges for $n > m$ and $v > w$.

For region (iii), $-\infty < \beta_0 < 0$ and $0 < \beta_1 < \infty$, write the product in the denominator of (A.9) as

$$\prod_{i=1}^{n} [1 + \exp\{-(\beta_0 + \beta_1 x_i)\}]$$

$$= 1 + \exp\{-\beta_0\} \sum_{j=1}^{n} \exp\{-\beta_1 s_{1_j}\}$$

$$+ \exp\{-2\beta_0\} \sum_{j=1}^{a_2} \exp\{-s_{2_j}\beta_1\}$$

$$+ \ldots + \exp\{-m\beta_0\} \sum_{j=1}^{a_m} \exp\{-s_{m_j}\beta_1\}$$

$$+ \ldots + \exp\{-n\beta_0\}\exp\{-s_n\beta_1\}, \tag{A.10}$$

where s_{m_j}, $j = 1, 2, \ldots, a_m$, are the distinct sums of m elements of $(x_1, x_2, \ldots, x_n)$ and $a_m = n(n-1)(n-2)\ldots(n-m+1)/m!$. Thus the product in (A.10) is greater than $\exp[-(m+1)\beta_0]\exp(-\beta_1 \sum_{i=1}^{m+1} x_{(i)})$, where the $x_{(i)}$'s are the ordered x_i's, that is, $0 < x_{(1)} < x_{(2)} < \ldots < x_{(n)}$. Then the integrand in (A.9) is less than $\exp(-m\beta_0 - w\beta_1)/\exp\{-(m+1)\beta_0 - \beta_1 L_{m+1}\} = \exp\{\beta_0 - (w - L_{m+1})\beta_1\}$, where $L_{m+1} \equiv \sum_{i=1}^{m+1} x_{(i)}$. The integral of $\exp\{\beta_0 - (w - L_{m+1})\beta_1\}$ over region (iii) converges if $w - L_{m+1} > 0$, that is, $\sum_{i=1}^{m} x_i > \sum_{i=1}^{m+1} x_{(i)}$.

For region (iv), $0 < \beta_0 < \infty$ and $-\infty < \beta_1 < 0$, note that the product in the denominator of (A.9) is greater than $\exp\{-(m-1)\beta_0 - \beta_1 U_{m-1}\}$, where $U_{m-1} = \sum_{i=0}^{m-2} x_{(n-i)}$. Thus the integrand of (A.9) is less than $\exp\{-m\beta_0 - w\beta_1\}/\exp\{-(m-1)\beta_0 - U_{m-1}\beta_1\} = \exp\{-\beta_0 - (w - U_{m-1})\beta_1\}$. The integral of this last expression over region (iv) converges if $w - U_{m-1} < 0$ or $\sum_{i=1}^{m} x_i < \sum_{i=0}^{m-2} x_{(n-i)}$ and thus, under this condition, the integral of (A.9) over region (iv) converges.

Since the integral of (A.9) over the region $-\infty < \beta_0, \beta_1 < \infty$, can be expressed as the sum of the four integrals defined over the regions (i)–(iv) and each of those converges, the integral of (A.9) over $-\infty < \beta_0, \beta_1 < \infty$ converges. The conditions sufficient for convergence, mentioned above, are: (a) $m > 0$, (b) $n > m$, (c) $\sum_{i=1}^{m} x_i > \sum_{i=1}^{m+1} x_{(i)} \equiv L_{m+1}$, and (d) $\sum_{i=1}^{m} x_i < \sum_{i=0}^{m-2} x_{(n-i)}$

$\equiv U_{m-1}$. These sufficient conditions can easily be verified in practice. Also, the same approach can be applied to establish convergence conditions in the k-dimensional case.

Appendix 2

Comparison of numerical integration techniques

Three multivariate numerical techniques were used in analyzing the posterior distributions of logit model parameters: (1) bivariate Simpson's rule, (2) Monte Carlo numerical integration with importance sampling, and (3) the 'uniform' Monte Carlo technique proposed by Friedman and Wright (1980a, b). In this appendix, we describe the implementation of these routines and compare their performance for a simple problem.

We compare the performance of the three integration techniques in the computation of posterior moments for parameters of a logit model with one independent variable, $P_i = F(\beta_0 + \beta_1 x_i)$. Analysis of the exact posterior distribution of (β_0, β_1) requires the use of a bivariate numerical integration routine. The first technique explored was bivariate Simpson's rule. A straightforward extension of univariate Simpson's rule techniques, the bivariate Simpson's routine, divides the (β_0, β_1) plane into an even grid. After the axes are subdivided, univariate Simpson's rule is performed in both directions to obtain the value of the integral. The region of integration was chosen to be $\{(\beta_0, \beta_1)$: $\hat{\beta}_0 - 6s_{\beta_0} \leq \beta_0 \leq \hat{\beta}_0 + 6s_{\beta_0}$; $\hat{\beta}_1 - 6s_{\beta_1} \leq \beta_1 \leq \hat{\beta}_1 + 6s_{\beta_1}\}$, where $\hat{\beta}_0, \hat{\beta}_1$ are the posterior modes and $s_{\hat{\beta}_0}, s_{\hat{\beta}_1}$ are asymptotic standard deviations. Our routine automatically computes the first four central moments, skewness, kurtosis, and writes out the marginal distributions for plotting. The routine is programmed in FORTRAN H Extended.

For higher-dimensional integrals, Simpson's rule proves impractical due to the geometric increase in the number of integrand evaluations required. Monte Carlo integration with importance sampling is a practical alternative. As pointed out above, the choice of the importance function is critical to the success of an importance sampling approach. A multivariate Student-t distribution was chosen for the importance function because of the ability of this family of distributions to exhibit thick and thin tails as well as the relative ease of drawing pseudo-random numbers from the MVS-t. The MVS-t density was centered at the posterior mode and given covariance matrix proportional to the covariance matrix of the asymptotic approximate distribution (the constant of proportionality is $\nu/(\nu-2)$ where ν is the degrees of freedom). MVS-t vectors were generated by drawing a vector of independent normal random variates, dividing each element of the normal random vector by the square root of an independent χ^2_ν random variate, and then transforming the resulting vector by premultiplication with the Cholesky root of the asymptotic covariance matrix.

The user of the importance sampling routine must specify the degrees of freedom parameter, ν, for the MVS-t importance function. Each posterior integral is of the form

$$I = \int_{R_\beta} h(\beta)\frac{k(\beta|D)}{I(\beta)}I(\beta)\ \mathrm{d}\beta \Big/ \int_{R_\beta} \frac{k(\beta|D)}{I(\beta)}I(\beta)\ \mathrm{d}\beta,$$

which is approximated by

$$\hat{I} = \sum h(\beta^i)w_i / \sum w_i,$$

where $k(\beta|D)$ is the posterior kernel, $w_i = k(\beta^i|D)/I(\beta^i)$ and β^i are the MVS-t random vectors. The coefficient of variation of the w_i measures the divergence between I and k. If the importance function has much thinner tails than the posterior kernel, the coefficient of variation of the weights would be large. Thus, the choice of ν is made by selecting various values of ν, computing the integral by importance sampling with a MVS-$t(\nu)$ importance function, and comparing the coefficient of variation of the weights.

The importance sampling routine is written in IBM 370/OS FORTRAN H Extended. The International Mathematical and Scientific Libraries (IMSL) subroutine GGNML (Version 8.0) was used to obtain standard normal random variates.

The third numerical integration technique explored was the uniform Monte Carlo method developed by Friedman and Wright (1980b). Friedman and Wright's routine, DIVONNE, partitions the region of integration into hyper-rectangles so as to minimize the integrand variation within each rectangle and draws points uniformly within each hyper-rectangle. Since the region must be repartitioned for each moment calculation, the uniform Monte Carlo technique requires a substantial number of functional evaluations to achieve any desired level of accuracy. In three or higher dimensions, substantial computing time is devoted to partitioning. The Friedman–Wright routine is designed for integrands with few known properties and possibly pathological, i.e., non-smooth behavior.

In table 3, we compare the performance of the three techniques in computation of the posterior moments of the slope and intercept parameters of the logit model with one independent variable. In order to facilitate comparison between techniques, each routine was set up so as to use approximately 30 secs of CPU time on the University of Chicago's Amdahl 470 V/7 computer. The Simpson's rule calculations are accurate to 5 significant digits. For a given amount of CPU time, the importance sampling routine far outperforms the uniform Monte Carlo program. In this application, several digits of accuracy in the calculation of the means and variance can be obtained with as few as 1000 points.

Table 3

Comparison of numerical integration routines for bivariate posterior integrals.[a]

	μ_1	μ_2	μ_3	μ_4
	Marginal distribution of intercept (β_0)			
Simpson's rule	−.3.82	5.21	−2.56	85.7
Importance sampling	−3.81	5.20	−2.16	83.4
Uniform Monte Carlo	−3.95	5.39	6.35	141.0
	Marginal distribution of slope (β_1)			
Simpson's rule	0.845	0.221	0.0230	0.154
Importance sampling	0.843	0.220	0.0188	0.149
Uniform Monte Carlo	0.861	0.193	0.0524	0.137

[a]All integrals are of the form $I = \int\int \beta_j^r p(\beta_j|D)\mathrm{d}\beta_0\mathrm{d}\beta_1$. A diffuse prior is used with a generated data set of 50 observations. The bivariate Simpson's rule routine used 200 subdivisions on each axis for a total of 40,401 integrand evaluations. The importance sampling routine used a multivariate Student-t distribution with 25 degrees of freedom as the importance function and 10,000 pts.

Importance sampling can be successfully applied in the analysis of the exact posterior density of the logit model's parameters with a reasonable degree of accuracy and at a low computational cost.

References

Amemiya, T., 1981, Qualitative response models: A survey, Journal of Economic Literature XIX, 1483–1536.

Cragg, J.G., 1971, Some statistical models for limited dependent variables with application to the demand for durable goods, Econometrica 39, 829–844.

Cramer, H., 1946, Mathematical methods of statistics (Princeton University Press, Princeton, NJ).

Dagenais, M.G. and T.C. Liem, 1981, Numerical approximations of marginals from 'well-behaved' joint posteriors, Journal of Statistical Computation and Simulation 12, 157–173.

Dempster, A.P., M.R. Selwyn and B.J. Weeks, 1981, Combining historical and randomized controls for assessing trends in proportions, Paper presented at American Statistical Association Meeting, Detroit, MI.

Freeman, P.R., 1970, Optimal Bayesian sequential estimation of the median effective dose, Biometrika 57, 79–89.

Friedman, J.H and M.H. Wright, 1980a, A nested partitioning procedure for numerical multiple integration, Working paper (Stanford Linear Accelerator Center, Stanford, CA).

Friedman, J.H. and M.H. Wright, 1980b, DIVONNE4: A program for multiple integration and adaptive importance sampling, Computation Research Group working paper (Stanford Linear Accelerator Center, Stanford, CA).

Hammersley, J.M. and D.C. Handscomb, 1979, Monte Carlo methods (Chapman and Hall, London).

Izan, H.Y., 1978, An empirical analysis of the economic effects of mandatory audit regulation, Doctoral dissertation (Graduate School of Business, University of Chicago, Chicago, IL).

Jeffreys, H., 1967, Theory of probability, 3rd rev. ed. (Oxford University Press, London).

Kenny, J.F. and E.S. Keeping, 1951, Mathematics of statistics: Part two, 2nd ed. (Van Nostrand, New York).

Kloek, T. and H. van Dijk, 1978, Bayesian estimates of equation system parameters: An application of integration by Monte Carlo, Econometrica 46, 1–19.

Kraft, C.H. and C. van Eeden, 1964, Bayesian bioassay, Annals of Mathematical Statistics 35, 886–890.

Laub, M.A., 1971, The dividend–earning relationship: A study of corporate quarterly panel data, 1947–65, Doctoral dissertation (Graduate School of Business, University of Chicago, Chicago, IL).

Laub, M.A., 1972, Some aspects of the aggregation problem in the dividend–earnings relationship, Journal of the American Statistical Association 67, 552–559.

Leonard, T., 1972, Bayesian methods for binomial data, Biometrika 59, 581–589.

Leonard, T., 1979, A general nonlinear filter with applications to time series and bioassay, Reported to the Econometrics and Statistics Colloquium, Nov. 21 (University of Chicago, Chicago, IL).

Levedahl, J.W., 1976, The effects of households' permanent and transitory income on both their timing of automobile purchases and their net outlay for automobile services, Doctoral dissertation (Department of Economics, University of Chicago, Chicago, IL).

Lindley, D.V., 1961, The use of prior probability distributions in statistical inference and decisions, in: J. Neyman, ed., Proceedings of the Fourth Berkeley Symposium on Mathematics, Statistics and Probabilities, Vol. I, 453–468.

Lindley, D.V., 1981, Approximate Bayesian methods, in: J.M. Bernardo et al., eds., Bayesian Statistics (University Press, Valencia, Spain) 223–237.

Pierce, B.O., 1929, A short table of integrals (Ginn and Co., Boston, MA).

Ramsey, F.L., 1972, A Bayesian approach to bioassay, Biometrics 28, 841–858.

Silvapulle, M.J., 1981, On the existence of maximum likelihood estimators for the binomial response models, Journal of the Royal Statistical Society B 43, 310–313.

Stewart, L., 1979, Multiparameter univariate Bayesian analysis, Journal of the American Statistical Association 74, 684–693.

Stroud, A.H. 1971, Approximate calculation of multiple integrals (Prentice-Hall, Englewood Cliffs, NJ).

Tsutakawa, R.K., 1975, Bayesian inference for bioassay, Technical report no. 52 (Mathematical Sciences) (University of Missouri, Columbia, MO).

Whittaker, E.T. and G.N. Watson, 1963, A course of modern analysis (Cambridge University Press, Cambridge).

Zellner, A., 1978, Estimation of functions of population means and regression coefficients including structural coefficients: A minimum expected loss (MELO) approach, Journal of Econometrics 8, 127–158.

Zellner, A. and T.H. Lee, 1965, Joint estimation of relationships involving discrete random variables, Econometrica 33, 382–394.

[17]

Bayesian Estimation and Prediction Using Asymmetric Loss Functions

ARNOLD ZELLNER*

Estimators and predictors that are optimal relative to Varian's asymmetric LINEX loss function are derived for a number of well-known models. Their risk functions and Bayes risks are derived and compared with those of usual estimators and predictors. It is shown that some usual estimators, for example, a scalar sample mean or a scalar least squares regression coefficient estimator, are inadmissible relative to asymmetric LINEX loss by providing alternative estimators that dominate them uniformly in terms of risk.

KEY WORDS: Asymmetric loss function; Inadmissibility; Estimation; Prediction; Risk function; Robustness.

1. INTRODUCTION

In some estimation and prediction problems, use of symmetric loss functions may be inappropriate, as has been recognized in the literature—see, for example, Ferguson (1967), Zellner and Geisel (1968), Aitchison and Dunsmore (1975), Varian (1975), and Berger (1980). That is, a given positive error may be more serious than a given negative error of the same magnitude or vice versa. The authors mentioned above, except for Varian, have considered asymmetric linear loss functions. Varian introduced a very useful asymmetric LINEX loss function that rises approximately exponentially on one side of zero and approximately linearly on the other side in his applied study of real estate assessment. Underassessment results in an approximately linear loss of revenue whereas overassessment often results in appeals with attendant, substantial litigation and other costs—see Varian (1975) for details. Also, in dam construction an underestimate of the peak water level is usually much more serious than an overestimate. Attention is directed herein at establishing properties of estimation and prediction procedures based on LINEX loss functions. For example, it is shown that the sample mean is an inadmissible estimator of the mean of a univariate normal distribution relative to a LINEX loss function. Also, results for estimating sets of normal means, multiple regression coefficients, and an exponential parameter are reported.

The plan of the article is as follows. In Section 2, LINEX loss functions are defined and described. Section 3 is devoted to estimation results for a scalar parameter and prediction of a scalar random variable. In Section 4 the analysis is extended to cover cases involving estimation of sets of parameters. A summary of results and some concluding remarks are presented in Section 5.

2. LINEX LOSS FUNCTIONS

Let $\Delta = \hat{\theta} - \theta$ denote the scalar estimation error in using $\hat{\theta}$ to estimate θ. Varian (1975) introduced the following convex loss function:

$$L(\Delta) = be^{a\Delta} - c\Delta - b, \qquad a, c \neq 0, b > 0. \quad (2.1a)$$

It is seen that $L(0) = 0$. Also, for a minimum to exist at $\Delta = 0$, we must have $ab = c$, and thus (2.1a) can be reexpressed as

$$L(\Delta) = b[e^{a\Delta} - a\Delta - 1], \qquad a \neq 0, b > 0. \quad (2.1b)$$

There are two parameters, a and b, involved in (2.1b) with b serving to scale the loss function and a serving to determine its shape. In Figure 1, values of $e^{a\Delta} - a\Delta - 1$ are plotted for selected values of a and Δ. It is seen that, for $a = 1$, the function is quite asymmetric with overestimation being more costly than underestimation. On the other hand, when $a < 0$, then (2.1b) rises almost exponentially when $\Delta = \hat{\theta} - \theta < 0$ and almost linearly when $\Delta = \hat{\theta} - \theta > 0$. For small values of $|a|$, the function is almost symmetric and not far from a squared error loss function. Indeed, on expanding $e^{a\Delta} \doteq 1 + a\Delta + a^2\Delta^2/2$, $L(\Delta) \doteq a^2\Delta^2/2$, a squared error loss function. Thus for small values of $|a|$, optimal estimates and predictions are not far different from those obtained with a squared error loss function. However, when $|a|$ assumes appreciable values, optimal point estimates and predictions will be quite different from those obtained with a symmetric squared error loss function—see Varian (1975) and below for examples.

For multiparameter estimation problems and multivariate prediction problems, there is a need to extend the form of (2.1). Let $\Delta_i = \hat{\theta}_i - \theta_i$ be the error in estimating θ_i by use of the estimate $\hat{\theta}_i$, $i = 1, 2, \ldots, k$. Then the separable extended LINEX loss function is given by

$$L(\mathbf{\Delta}) = \sum_{i=1}^{k} b_i(\exp(a_i\Delta_i) - a_i\Delta_i - 1),$$
$$a_i \neq 0, b_i > 0, i = 1, 2, \ldots, k, \quad (2.2)$$

where $\mathbf{\Delta}' = (\Delta_1, \Delta_2, \ldots, \Delta_k)$. This is a convex loss function such that $L(\mathbf{0}) = 0$, $L'(\mathbf{0}) = \mathbf{0}$, and $\mathbf{\Delta} = \mathbf{0}$ corresponds to a minimum. The function in (2.2) can be utilized in multiparameter estimation and multivariate prediction problems.

3. SCALAR PARAMETER AND UNIVARIATE PREDICTION PROBLEMS

We first consider estimation of a scalar parameter θ relative to the LINEX loss function (2.1). Assume that the posterior pdf for θ is $p(\theta \mid D)$, a proper pdf with $\theta \subset \Theta$, the parameter space, and where D denotes the sample and prior information. With E_θ denoting posterior expectation with respect to $p(\theta \mid D)$, the posterior expectation of the LINEX loss function in (2.1b) is

$$E_\theta L(\Delta) = b[e^{a\hat{\theta}}E_\theta e^{-a\theta} - a(\hat{\theta} - E_\theta\theta) - 1], \quad (3.1)$$

* Arnold Zellner is H.G.B. Alexander Distinguished Service Professor of Economics and Statistics, Graduate School of Business, University of Chicago, Chicago, IL 60637. This research was financed in part by the National Science Foundation and by income from the H.G.B. Alexander Endowment Fund, Graduate School of Business, University of Chicago.

Journal of the American Statistical Association
June 1986, Vol. 81, No. 394, Theory and Methods

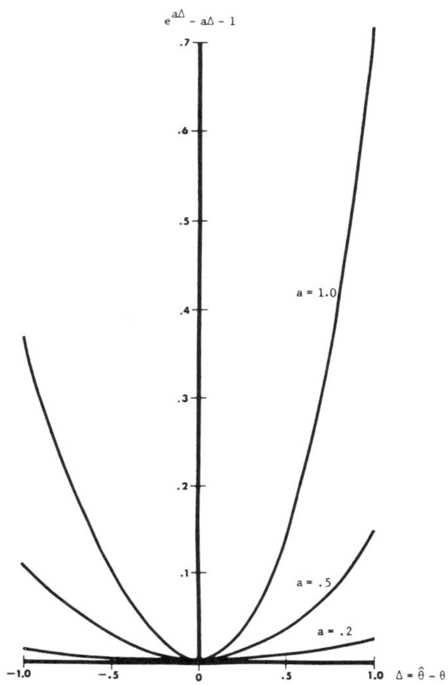

Figure 1. Plots of $e^{a\Delta} - a\Delta - 1$ for Selected Values of a.

and the value of $\hat{\theta}$ that minimizes (3.1), denoted by $\hat{\theta}_B$, is

$$\hat{\theta}_B = -(1/a)\log(E_\theta e^{-a\theta}), \tag{3.2}$$

provided, of course, that $E_\theta e^{-a\theta}$ exists and is finite. The general result in (3.2) can be employed in many problems to provide a point estimate that is optimal relative to a LINEX loss function. This involves evaluation of the moment-generating function for the posterior pdf, $E_\theta e^{-a\theta}$.

As mentioned earlier, when θ has a *normal* posterior pdf, say with mean m and variance v, (3.2) can be evaluated analytically. That is, $E_\theta e^{-a\theta} = \exp(-am + a^2v/2)$, and when this result is inserted in (3.2),

$$\hat{\theta}_B = m - av/2 \tag{3.3}$$

is the Bayesian estimate relative to the LINEX loss function in (2.1), a result paralleling that of Varian's (1975) result for prediction. Note that when $|a|v/2$ is small, $\hat{\theta}_B \doteq m$, the posterior mean that is optimal relative to squared error loss. However, when $|a|v/2$ is not small, $\hat{\theta}_B$ can depart markedly from the posterior mean.

An important case in which θ has a normal posterior distribution is the normal mean problem. For n observations, y_i, $i = 1, 2, \ldots, n$, independently drawn from a normal distribution with mean θ and known variance σ^2, the likelihood function is $l(\theta \mid \bar{y}, \sigma) \propto \exp\{-n(\theta - \bar{y})^2/2\sigma^2\}$, where $\bar{y}$ is the sample mean. If a diffuse prior, $p(\theta) \propto$ const., is employed, the posterior pdf, $p(\theta \mid D)$, is $N(\bar{y}, \sigma^2/n)$. Thus, using (3.3), the optimal estimate relative to the LINEX loss function (2.1) is

$$\hat{\theta}_B = \bar{y} - a\sigma^2/2n. \tag{3.4}$$

The risk functions for the estimator, $\hat{\theta}_B = \bar{y} - a\sigma^2/2$, and for $\bar{y}$, the sample mean, relative to the LINEX loss function in (2.1) are of interest. These risk functions, denoted by $R_L(\hat{\theta}_B)$ and $R_L(\bar{y})$, where the subscript L denotes risk relative to the LINEX loss function in (2.1b), are given by

$$R_L(\hat{\theta}_B) = bz, \qquad b > 0, \tag{3.5}$$

and

$$R_L(\bar{y}) = b(e^z - 1), \qquad b > 0, \tag{3.6}$$

where $z = a^2\sigma^2/2n$. On comparing (3.5) and (3.6), it is seen that risk is zero in both cases when $z = 0$. However, for $z > 0$, (3.6) is uniformly larger and thus the sample mean $\bar{y}$ is an inadmissible estimator relative to the LINEX loss function in (2.1). Tabulations of (3.5) and (3.6) are shown in Table 1. As can be seen from Table 1, if, for example, $z = a^2\sigma^2/2n = 1.0$, the risk associated with the sample mean, $\bar{y}$, is 71.8% larger than that associated with $\hat{\theta}_B$. Thus Table 1 shows substantial differences in the risks associated with $\bar{y}$ and $\hat{\theta}_B$ and, as pointed out previously, $\bar{y}$ is inadmissible.

It is also of interest to consider the risk functions of the estimators $\bar{y}$ and $\hat{\theta}_B = \bar{y} - av/2$ relative to squared error loss, $L(\theta, \hat{\theta}) = (\hat{\theta} - \theta)^2$. Given that the sampling distribution of $\bar{y}$ is given by $N(\theta, \sigma^2/n)$, these risk functions, denoted by $R_S(\bar{y})$ and $R_S(\hat{\theta}_B)$, with the subscript S denoting squared error loss, are

$$R_S(\bar{y}) = \sigma^2/n \tag{3.7}$$

and

$$R_S(\hat{\theta}_B) = \sigma^2/n + a^2v^2/4. \tag{3.8}$$

From (3.7)–(3.8), it is seen that relative to squared error loss, $\bar{y}$ uniformly dominates $\hat{\theta}_B$ and thus $\hat{\theta}_B = \bar{y} - av/2$ is inadmissible relative to squared error loss. Also note that if $v = \sigma^2/n$, as in (3.4), $R_S(\hat{\theta}_B) = \sigma^2/n[1 + a^2\sigma^2/4n]$ and for $a = 2$ and $\sigma^2/n = .2$, $R_S(\hat{\theta}_B) = 1.2\sigma^2/n$, a 20% inflation of risk relative to that for $\bar{y}$.

*Table 1. Tabulation of Risk Functions for $\hat{\theta}_B = \bar{y} - a\sigma^2/2n$ and $\bar{y}$ Relative to the LINEX Loss Function (2.1b)**

z	$R(\hat{\theta}_B)/b = z$	$R(\bar{y})/b = e^z - 1$
0	0	0
.2	.2	.221
.4	.4	.492
.6	.6	.822
.8	.8	1.223
1.0	1.0	1.718
1.2	1.2	2.320
1.4	1.4	3.055
1.6	1.6	3.953
1.8	1.8	5.050
2.0	2.0	6.389

* $z = a^2\sigma^2/2n$, where $\sigma^2/n = \text{var}(\bar{y})$, n = sample size, and a is a parameter in the LINEX loss function (2.1b). The risk functions, given in (3.5) and (3.6), have been scaled by division by the positive parameter b.

In summary, with respect to the normal mean problem using a diffuse prior, when our loss function is the LINEX function in (2.1), the estimator $\hat{\theta}_B$ in (3.4) dominates $\bar{y}$ in terms of risk. The differences in risk can be substantial when $|a|v/2$ is not small (see Table 2). On the other hand, when we use a squared error loss function, the sample mean's risk is uniformly lower than that of $\hat{\theta}_B$ in (3.4). Again the differences in risk can be substantial. These results indicate well the importance of employing an appropriate loss function and of making a good decision relative to it and not relative to some other loss function.

In (3.4), the Bayesian estimator for the mean relative to the LINEX loss function is seen to depend on σ^2. If σ^2 has an unknown value, σ^2 can be replaced by $s^2 = \sum_{i=1}^{n}(y_i - \bar{y})^2/v$ with $v = n - 1$ to obtain $\theta_B^* = \bar{y} - as^2/2n$. The risk of θ_B^* relative to the LINEX loss function in (2.1) is

$$R_L(\theta_B^*) = b[(1 + 2z/v)^{-v/2}e^z + z - 1],$$

where $z = a^2\sigma^2/2n$. When $v = n - 1 = 2$, $R_L(\theta_B^*)/b$ has been evaluated as follows: 0 for $z = 0$, 1.359 for $z = 1.0$, and 3.463 for $z = 2.0$, and when $v = n - 1 = 20$, 0 for $z = 0$, 1.048 for $z = 1.0$, and 2.193 for $z = 2.0$. On comparing these values with those in Table 2, it is seen that for $z \neq 0$, $R_L(\hat{\theta}_B) < R_L(\theta_B^*)$, indicating that estimation of σ^2 has increased risk. However, $R_L(\theta_B^*) < R_L(\bar{y})$ and thus $\hat{\theta}_B$ dominates $\bar{y}$. For $v = 2$, this last result can be shown by forming $\delta_2 = R_L(\bar{y}) - R_L(\theta_B^*)$ and expanding $e^Z = 1 + z + z^2/2! + z^3/3! + \cdots$. Since $\delta_v > \delta_2$ for $v > 2$, the general result follows.

Above, we have analyzed the scalar normal mean problem using a diffuse prior. Now assume that the prior for θ is $N(0, \tau^2)$, that is, a normal pdf with zero mean and variance τ^2. Then the posterior pdf for θ is normal with mean $m = \bar{y}/(1 + \lambda)$ and variance $v = \sigma^2/n(1 + \lambda)$ with $\lambda = \sigma^2/n\tau^2$. Then from (3.3), the Bayesian estimator relative to the LINEX loss function (2.1) is

$$\tilde{\theta}_B = \bar{y}/(1 + \lambda) - a\sigma^2/2n(1 + \lambda). \tag{3.9}$$

On comparing $\tilde{\theta}_B$ in (3.9) with $\hat{\theta}_B$ in (3.4), the estimator based on a diffuse prior, we have $\tilde{\theta}_B = \hat{\theta}_B/(1 + \lambda)$. Thus use of the $N(0, \tau^2)$ prior pdf has resulted in a shrinkage of $\hat{\theta}_B$ toward zero, the prior mean.

Shown in Table 2 are the risk functions and Bayes risks of the estimator $\tilde{\theta}_B$ in (3.9), the posterior mean, $\hat{\theta}_m = \bar{y}/(1 + \lambda)$, and the sample mean $\bar{y}$ relative to the LINEX loss function (2.1) and to a squared error loss function. As regards Bayes risks, that for $\tilde{\theta}_B$ is $bz/(1 + \lambda)$, which is smaller than that for the posterior mean, $b[e^{z/(1+\lambda)} - 1]$, which is smaller than that for the sample mean, $b[e^z - 1]$. On the other hand, for the squared error loss function the Bayes risk of the posterior mean, $\sigma^2/n(1 + \lambda)$, is smaller than that for the sample mean, σ^2/n, and also that for $\tilde{\theta}_B$, $[\sigma^2/n(1 + \lambda)] \times [1 + a^2/2(1 + \lambda)]$.

With respect to the risk functions shown in Table 2, under the LINEX loss function the sample mean has constant risk $b[e^z - 1]$, whereas $\tilde{\theta}_B$ and $\hat{\theta}_m$ have risk functions that depend on θ. It is the case that the risk functions cross, and thus none of the estimators uniformly dominates the others. The risk function $R_L(\tilde{\theta}_B)$ has a minimum at $\theta = (1/q)\ln c_1 = -a\sigma^2/2n(1 + \lambda)$ at which point risk equals $bz/(1 + \lambda)^2 = ba^2\sigma^2/2n(1 + \lambda)^2$. The risk function $R_L(\hat{\theta}_m)$ has a minimum at $\theta = (1/q)\ln c_2 = a\sigma^2/2n\lambda(1 + \lambda)$, where risk equals $b \ln c_2 = ba^2\sigma^2/2n(1 + \lambda)^2$. Thus the minimal values of $R_L(\tilde{\theta}_B)$ and $R_L(\hat{\theta}_m)$ are identical. Shown in Figure 2 are plots of $R_L(\tilde{\theta}_B)$ and $R_L(\hat{\theta}_m)$ and the constant risk of $\bar{y}$. In the vicinity of $\theta = 0$ and for $\theta < 0$, $R_L(\tilde{\theta}_B) < R_L(\hat{\theta}_m)$, whereas for large positive θ, $R_L(\tilde{\theta}_B) > R_L(\hat{\theta}_m)$. Also, for a wide range of values of θ about zero, $\tilde{\theta}_B$ and $\hat{\theta}_m$ have lower risk than the sample mean but neither uniformly dominates the sample mean.

In Table 2 the risk functions of $\tilde{\theta}_B$, $\hat{\theta}_m$, and $\bar{y}$ relative to a squared error loss function are presented. Risk function $R_S(\tilde{\theta}_3)$ has a minimum at $\theta = -a\sigma^2/n\lambda$, where its value is $\sigma^2/n(1 + \lambda)^2$. Risk function $R_S(\hat{\theta}_m)$ has a minimum at $\theta = 0$, the prior mean, where its value is $\sigma^2/n(1 + \lambda)^2$, the same as the minimal value of $R_S(\tilde{\theta}_B)$, and these minimal values are smaller than the constant risk associated with the sample mean, namely σ^2/n. The plots of these risk functions in Figure 3 show that they cross and thus none of these three estimators uniformly dominates the others. This is not surprising, since it is well known that $\bar{y}$ is an admissible estimator relative to quadratic loss.

The results presented above can be readily adapted to relate to estimation of a linear combination of regression coefficients. That is, let $\theta = \boldsymbol{l}'\boldsymbol{\beta}$ where $\boldsymbol{l}$ is a given $k \times 1$ vector of rank one and $\boldsymbol{\beta}$ is a $k \times 1$ vector of regression coefficients in a

Table 2. Risk Functions and Bayes Risks for Alternative Estimators for a Normal Mean With an $N(0, \tau^2)$ Prior pdf

Estimator	*Risk Function*	*Bayes Risk**
	A. Relative to the LINEX Loss Function (2.1)	
(3.9) $\tilde{\theta}_B = \bar{y}/(1 + \lambda) - a\sigma^2/2n(1 + \lambda)$	$R_L(\tilde{\theta}_B) = b[c_1e^{-q\theta} + q\theta + z/(1 + \lambda) - 1]$	$bz/(1 + \lambda)$
Posterior mean: $\hat{\theta}_m = \bar{y}/(1 + \lambda)$	$R_L(\hat{\theta}_m) = b[c_2e^{-q\theta} + q\theta - 1]$	$be^{z/(1+\lambda)} - 1$
Sample mean: $\bar{y}$	$R_L(\bar{y}) = b[e^z - 1]$	$b[e^z - 1]$
	B. Relative to Squared Error Loss, $(\hat{\theta} - \theta)^2$	
(3.9) $\tilde{\theta}_B = \bar{y}/(1 + \lambda) - a\sigma^2/2n(1 + \lambda)$	$R_S(\tilde{\theta}_B) = [\sigma^2/n + (\lambda\theta + a\sigma^2/n)^2]/(1 + \lambda)^2$	$\sigma^2/n(1 + \lambda)[1 + a^2/2(1 + \lambda)]$
Posterior mean: $\hat{\theta}_m = \bar{y}/(1 + \lambda)$	$R_S(\hat{\theta}_m) = [\sigma^2/n + \lambda^2\theta^2]/(1 + \lambda)^2$	$\sigma^2/n(1 + \lambda)$
Sample mean: $\bar{y}$	$R_S(\bar{y}) = \sigma^2/n$	σ^2/n

NOTE: The following quantities are employed: $q = a\lambda/(1 + \lambda)$, $\lambda = \sigma^2/n\tau^2$, $z = a^2\sigma^2/2n$, $c_1 = \exp(-\lambda z/(1 + \lambda)^2)$, and $c_2 = \exp(z/(1 + \lambda)^2)$.

* Bayes risk was computed using an $N(0, \tau^2)$ prior pdf.

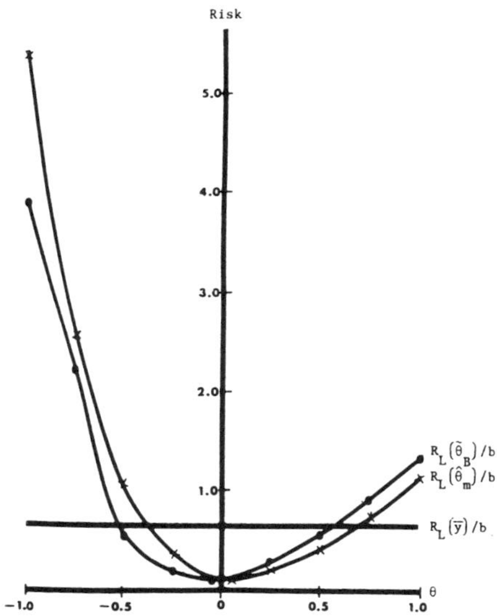

Figure 2. LINEX Risk Functions. The functions, given in Table 2, have been evaluated for a = 4.0, λ = σ²/nτ² = 1.0, and σ²/n = .0625.

normal multiple regression, $\mathbf{y} = X\boldsymbol{\beta} + \mathbf{u}$ with X an $n \times k$ nonstochastic matrix of rank k. If $\mathbf{u}$ has been drawn from a multivariate normal distribution with zero mean vector and covariance matrix $\sigma^2 I_n$, then the likelihood function is

$$p(\mathbf{y} \mid X, \beta, \sigma) \propto \sigma^{-n} \times \exp\{-[\nu s^2 + (\boldsymbol{\beta} - \hat{\boldsymbol{\beta}})'X'X(\boldsymbol{\beta} - \hat{\boldsymbol{\beta}})]/2\sigma^2\}$$

with $\hat{\boldsymbol{\beta}} = (X'X)^{-1}X'\mathbf{y}$, $\nu = n - k$, and $\nu s^2 = (\mathbf{y} - X\hat{\boldsymbol{\beta}})'(\mathbf{y} - X\hat{\boldsymbol{\beta}})$. With a diffuse prior distribution for $\boldsymbol{\beta}$, the conditional posterior distribution for $\boldsymbol{\beta}$ given σ^2 is a k-dimensional normal distribution with mean $\hat{\boldsymbol{\beta}}$ and covariance matrix $(X'X)^{-1}\sigma^2$. Thus $\theta = \boldsymbol{l}'\boldsymbol{\beta}$ given σ^2 has a univariate normal posterior pdf with mean $m = \boldsymbol{l}'\hat{\boldsymbol{\beta}}$ and variance $v = \boldsymbol{l}'(X'X)^{-1}\boldsymbol{l}\sigma^2$. Then the Bayesian estimate relative to the LINEX loss function (2.1) is

$$\hat{\theta}_B = \boldsymbol{l}'\hat{\boldsymbol{\beta}} - a\boldsymbol{l}'(X'X)^{-1}\boldsymbol{l}\sigma^2/2. \qquad (3.10)$$

The risk of $\hat{\theta}_B$ relative to LINEX loss is

$$R_L(\hat{\theta}_B) = bw, \qquad (3.11)$$

where $w = a^2\boldsymbol{l}'(X'X)^{-1}\boldsymbol{l}\sigma^2/2$. Also the risk of $\hat{\theta}_m = \boldsymbol{l}'\hat{\boldsymbol{\beta}}$, the posterior mean of θ, is

$$R_L(\hat{\theta}_m) = b(e^w - 1). \qquad (3.12)$$

Since $R_L(\hat{\theta}_m) > R_L(\hat{\theta}_B)$ for $w > 0$, the usual estimator $\hat{\theta}_m = \boldsymbol{l}'\hat{\boldsymbol{\beta}}$ is inadmissible relative to LINEX loss. Since further analysis of this problem with a normal prior for $\boldsymbol{\beta}$ is very similar to that presented before for the normal mean problem, it will not be presented.

We have considered several normal location problems. Now we use the LINEX loss function in connection with estimating a scalar parameter λ, $0 < \lambda < \infty$, of an exponential pdf, $p(x \mid \lambda) = \lambda e^{-\lambda x}$, $0 < x < \infty$. Given n independent observations $\mathbf{x}' = (x_1, x_2, \ldots, x_n)$ drawn from the exponential distribution, the likelihood function is $p(\mathbf{x} \mid \lambda) = \lambda^n e^{-n\bar{x}\lambda}$, where $\bar{x}$ is the sample mean. With a natural conjugate prior pdf, $p(\lambda \mid g_1, g_2) \propto \lambda^{g_1 - 1}e^{-g_2\lambda}$, $g_1, g_2 > 0$, the posterior pdf is $p(\lambda \mid D) \propto \lambda^{n+g_1-1}\exp(-(n\bar{x} + g_2)\lambda)$, a gamma pdf. Then from (3.2), the point estimate that is optimal relative to the LINEX loss function in (2.1) is, with $a > -(n\bar{x} + g_2)$,

$$\hat{\lambda}_B = (1/a)\log E_\lambda e^{-a\lambda}$$
$$= [(n + g_1)/a]\log[1 + a/(n\bar{x} + g_2)]. \qquad (3.13)$$

For a diffuse prior ($g_1 = g_2 = 0$), (3.13) specializes to

$$\bar{\hat{\lambda}}_B = (n/a)\log(1 + a/n\bar{x}), \qquad (3.14)$$

which can be compared with the posterior mean

$$\hat{\lambda}_m = \Gamma(n + g_1 + 1)/(n\bar{x} + g_2)\Gamma(n + g_1),$$

which specializes for $g_1 = g_2 = 0$ to $\bar{\lambda} = 1/\bar{x}$, the maximum likelihood estimate. When $a/n\bar{x}$ is small, $\log(1 + a/n\bar{x}) \doteq a/n\bar{x}$ and (3.14) reduces to $1/\bar{x}$. However, when $a/n\bar{x}$ is not small, (3.14) differs from the diffuse prior posterior mean $\bar{\lambda} = 1/\bar{x}$. In general, $0 < \bar{\hat{\lambda}}_B/\bar{\lambda} < 1$, with $\bar{\hat{\lambda}}_B$ given in (3.14).

Finally, we note that if we regard θ to be a future value of a variable with a *predictive* pdf $p(\theta \mid D)$ and $\hat{\theta}$ as a point prediction, the LINEX loss function in (2.1) is then a predictive loss function and (3.2) yields an optimal point prediction. If, for example, the predictive pdf is normal with mean m and variance v, Varian (1975) showed that the optimal point prediction is given by (3.3). For example, for a normal multiple regression model and a diffuse prior pdf, the predictive pdf for a future observation, say y_*, given σ is normal with mean m

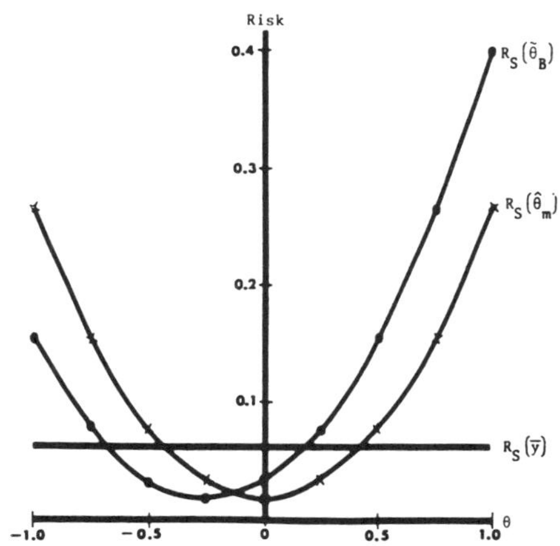

Figure 3. Squared Error Risk Functions. The functions, given in Table 2, have been evaluated for a = 4.0, λ = σ²/nτ² = 1.0, and σ²/n = .0625.

$= \mathbf{x}_*'\hat{\boldsymbol\beta}$ and variance $v = [1 + \mathbf{x}_*'(X'X)^{-1}\mathbf{x}_*]\sigma^2$, where $\hat{\boldsymbol\beta} = (X'X)^{-1}X\mathbf{y}$ and $\mathbf{x}_*$ is a given vector of independent variables in the relation $y_* = \mathbf{x}_*'\boldsymbol\beta + u_*$, with u_* an error term. Thus the optimal point prediction relative to LINEX loss is

$$\tilde{y}_* = \mathbf{x}_*'\hat{\boldsymbol\beta} - (a/2)[1 + \mathbf{x}_*'(X'X)^{-1}\mathbf{x}_*]\sigma^2. \quad (3.15)$$

Clearly the results for the normal mean problem can be used to compare the risks of $\tilde{y}_*$ and $\hat{y}_*$. One result that emerges from such a comparison is that the least squares predictor, $\hat{y}_* = \mathbf{x}_*'\hat{\boldsymbol\beta}$, is inadmissible relative to the LINEX loss function in (2.1). For problems in which a predictive pdf is nonnormal, the result in (3.2) can be employed with θ being the variable to be predicted and $\hat\theta_B$ the optimal point prediction.

4. MULTIPARAMETER PROBLEMS

Herein we consider the following model for the $n \times 1$ observation vector $\mathbf{y}$,

$$\mathbf{y} = \boldsymbol\theta + \mathbf{u} \quad (4.1)$$

with

$$\boldsymbol\theta = \mu\boldsymbol\iota + \mathbf{v}. \quad (4.2)$$

It is assumed that

$$p(\mathbf{y} \mid \boldsymbol\theta, \sigma) \sim N(\boldsymbol\theta, \sigma^2 I_n) \quad \text{and} \quad p(\boldsymbol\theta \mid \mu, \tau) \sim N(\mu\boldsymbol\iota, \tau^2 I_n),$$

where $\boldsymbol\theta' = (\theta_1, \theta_2, \ldots, \theta_n)$, $\boldsymbol\iota$ is an $n \times 1$ vector with all elements equal to 1, and μ is the mean of the hyperdistribution for $\boldsymbol\theta$. This model was introduced by Lindley (1962) in a discussion of Stein's (1962) results for the n-mean problem. The joint distribution of elements of $\boldsymbol\theta$ given $\mathbf{y}$, μ, σ^2, and τ^2 is

$$p(\boldsymbol\theta \mid \mathbf{y}, \mu, \tau^2, \sigma^2) \propto \exp\{-[(\mathbf{y} - \boldsymbol\theta)'(\mathbf{y} - \boldsymbol\theta) + \lambda(\boldsymbol\theta - \boldsymbol\iota\mu)'(\boldsymbol\theta - \boldsymbol\iota\mu)]/2\sigma^2\}, \quad (4.3)$$

where $\lambda = \sigma^2/\tau^2$. On completing the square on $\boldsymbol\theta$ and conditioning on $\mu = \bar{y}$, the sample mean, as Lindley (1962) does, the density of $\boldsymbol\theta$ is

$$p(\boldsymbol\theta \mid \lambda, \sigma, \mu = \bar{y}, \mathbf{y}) \propto \exp\{-(1 + \lambda) \times (\boldsymbol\theta - \hat{\boldsymbol\theta}_m)'(\boldsymbol\theta - \hat{\boldsymbol\theta}_m)/2\sigma^2\}, \quad (4.4)$$

with

$$\hat{\boldsymbol\theta}_m = (\mathbf{y} + \lambda\bar{y}\boldsymbol\iota)/(1 + \lambda) \quad (4.5)$$

the posterior mean and $\sigma^2/(1 + \lambda)I_n$ the posterior covariance matrix. Thus under these assumptions $\boldsymbol\theta$ has an $N(\hat{\boldsymbol\theta}_m, \sigma^2/(1 + \lambda)I_n)$ distribution.

Using these results and the extended loss function in (2.2), the optimal estimator of θ_i is

$$\hat\theta_i = (y_i + \lambda\bar{y})/(1 + \lambda) - a_i\sigma^2/2(1 + \lambda). \quad (4.6)$$

By direct evaluation, the risk function for $\hat\theta_i$ is

$$R_L(\hat\theta_i) = b_i[\exp\{-d_1(\theta_i - \bar\theta) + d_2 - q\} + d_1(\theta_i - \bar\theta) + q - 1], \quad (4.7)$$

where

$$d_1 = a_i\lambda/(1 + \lambda), \quad \bar\theta = \sum_{i=1}^{n} \theta_i/n, \quad q = a_i^2\sigma^2/2(1 + \lambda),$$

and

$$d_2 = (a_i^2/2)\text{var}(\hat\theta_{m_i})$$

with

$$\text{var}(\hat\theta_{m_i}) = \text{var}[(y_i + \lambda\bar{y})/(1 + \lambda)] = \sigma^2[1 + \lambda^2/n + 2\lambda/n]/(1 + \lambda)^2.$$

For comparison with (4.7), the risk of the posterior mean, $\hat\theta_{m_i}$, is

$$R_L(\hat\theta_{m_i}) = b_i[\exp\{-d_1(\theta_i - \bar\theta) + d_2\} + d_1(\theta_i - \bar\theta) - 1]. \quad (4.8)$$

Further, if a diffuse prior for $\boldsymbol\theta$ is employed, the posterior pdf for $\boldsymbol\theta$ is $N(\mathbf{y}, \sigma^2 I_n)$ and the diffuse prior LINEX estimator $\tilde\theta_i = y_i - a_i\sigma^2/2$ has risk function

$$R_L(\tilde\theta_i) = b_i(1 + \lambda)q. \quad (4.9)$$

Also, the diffuse prior posterior mean, y_i, has risk given by

$$R_L(y_i) = b_i[\exp\{(1 + \lambda)q\} - 1]. \quad (4.10)$$

It is seen that $R_L(\tilde\theta_i) < R_L(y_i)$ for $q > 0$ and that $R_L(\tilde\theta_i)$ and $R_L(y_i)$ do not depend on $\theta_i - \bar\theta$ whereas (4.7) and (4.8) do. The risk function in (4.7) has a minimum at $\theta_i - \bar\theta = d_2/d_1 - q$ with its value d_2. Equation (4.8) has a minimum at $\theta_i - \bar\theta = d_2/d_1$ with its value d_2. Thus minimal risks for the estimator $\hat\theta_i$ in (4.6) and the posterior mean $\hat\theta_{m_i}$ are the same and equal to $d_2 = b_i(a_i^2\sigma^2/2)(1 + \lambda^2/n + 2\lambda/n)/(1 + \lambda)^2$, which is smaller than (4.9) and (4.10) for $\lambda > 0$. When $\theta_i \gg \bar\theta$, the risk functions in (4.7) and (4.8) rise almost linearly, whereas when $\theta_i \ll \bar\theta$ these risk functions rise approximately in an exponential fashion. Since the risk functions for $\hat\theta_i$ and $\hat\theta_{m_i}$ cross, neither estimator dominates the other uniformly.

Employing the $N(\mu\boldsymbol\iota, \tau^2 I_n)$ prior pdf for $\boldsymbol\theta$, the Bayes risks of the estimators we consider have been computed and are shown below:

$$\text{BR}(\hat\theta_i) = b_i[\exp\{d_2 - q + d_1^2(1 - 1/n)\tau^2/2\} + q - 1], \quad (4.11)$$

$$\text{BR}(\hat\theta_{m_i}) = b_i[\exp\{d_2 + d_1^2(1 - 1/n)\tau^2/2\} - 1], \quad (4.12)$$

$$\text{BR}(\tilde\theta_i) = b_i(1 + \lambda)q, \quad (4.13)$$

$$\text{BR}(y_i) = b_i[\exp\{(1 + \lambda)q\} - 1]. \quad (4.14)$$

It is the case that $\text{BR}(\hat\theta_i) < \text{BR}(\hat\theta_{m_i})$ and $\text{BR}(\hat\theta_i) < \text{BR}(\tilde\theta_i) < \text{BR}(y_i)$. Since the Bayesian estimator $\hat\theta_i$ minimizes Bayes risk, it is not surprising that $\text{BR}(\hat\theta_i)$ is smallest.

5. SUMMARY AND CONCLUDING REMARKS

Asymmetric LINEX loss functions have been employed in the analysis of several central statistical estimation and prediction problems. Optimal estimators and predictors relative to LINEX loss and their associated risk functions have been derived. The analytical ease with which results can be obtained using asymmetric LINEX loss functions makes them attractive for use in applied problems and in assessing the effects of departures from assumed symmetric loss functions. For example, in the normal n-mean problem, it was straightforward to derive an estimator that is optimal relative to an extended LINEX loss function and to obtain its risk function and Bayes risk. Also, certain well-known estimators, for example, the sample mean, that are admissible relative to squared error loss were shown to be inadmissible relative to a LINEX loss function. The same result applies to the least squares estimator of a multiple regression coefficient or a linear combination of regression coefficients. As a referee has stated, ". . . the point that questions of admissibility may depend quite sensitively on features of the loss function, such as symmetry, is not generally appreciated . . . and implies that a lot more thought should be given to the choice of a loss function, rather than to blindly trust in squared error loss." See Zellner (1973) for an analysis of the effects of errors in specifying loss functions on solutions to control problems.

While the LINEX class of loss functions is convenient and useful, it is recognized that other asymmetric loss functions, for example, asymmetric linear and quadratic loss functions, are available and may be useful. Further study of the properties of alternative estimators relative to these and other types of asymmetric loss functions would be useful and is left for future research.

[*Received June 1985. Revised October 1985.*]

REFERENCES

Aitchison, J., and Dunsmore, I. R. (1975), *Statistical Prediction Analysis*, London: Cambridge University Press.

Berger, James O. (1980), *Statistical Decision Theory: Foundations, Concepts, and Methods*, New York: Springer-Verlag.

Ferguson, Thomas S. (1967), *Mathematical Statistics: A Decision Theoretic Approach*, New York: Academic Press.

Lindley, Dennis V. (1962), "Discussion on Professor Stein's Paper," *Journal of the Royal Statistical Society*, Ser. B, 24, 285–287.

Stein, Charles (1962), "Confidence Sets for the Mean of a Multivariate Normal Distribution," *Journal of the Royal Statistical Society*, Ser. B, 24, 265–296.

Varian, Hal R. (1975), "A Bayesian Approach to Real Estate Assessment," in *Studies in Bayesian Econometrics and Statistics in Honor of Leonard J. Savage*, eds. Stephen E. Fienberg and Arnold Zellner, Amsterdam: North-Holland, pp. 195–208.

Zellner, Arnold (1973), "The Quality of Quantitative Economic Policymaking When Targets and Costs of Change Are Misspecified," in *Selected Readings in Econometrics and Economic Theory: Essays in Honour of Jan Tinbergen*, ed. W. Sellekaerts, London: Macmillan, pp. 147–164 (reprinted in *Basic Issues in Econometrics*, by A. Zellner, Chicago: University of Chicago Press, 1984, pp. 169–183).

Zellner, Arnold, and Geisel, Martin S. (1968), "Sensitivity of Control to Uncertainty and Form of the Criterion Function," in *The Future of Statistics*, ed. Donald G. Watts, New York: Academic Press, pp. 269–289.

Journal of Econometrics 8 (1978) 127–158. © North-Holland Publishing Company

ESTIMATION OF FUNCTIONS OF POPULATION MEANS AND REGRESSION COEFFICIENTS INCLUDING STRUCTURAL COEFFICIENTS

A Minimum Expected Loss (MELO) Approach*

Arnold ZELLNER

Graduate School of Business, University of Chicago, Chicago, IL60637, USA

Received November 1975, final version received May 1977

Parameter estimates that minimize the posterior expectation of generalized quadratic loss functions are derived for a wide range of estimation problems encountered in econometrics and statistics including estimation of structural coefficients of linear structural econometric models and reciprocals and ratios of population means and regression or reduced form coefficients. These MELO estimates are simple in form and possess finite sampling moments and risk in contrast to other estimators, including maximum likelihood estimators, that possess no finite moments and have infinite risk relative to quadratic loss functions. Additional finite and large sample properties of MELO estimates are derived and discussed.

1. Introduction

In this paper, problems of estimating (a) reciprocals and ratios of population means and regression coefficients and (b) structural coefficients of linear structural econometric models are considered. For each problem in (a) and (b) formulae defining parameter estimates that Minimize Posterior Expected LOss, termed MELO estimates, are derived for specific loss functions and general forms for prior distributions and likelihood functions. Below these general results are specialized to provide explicit MELO estimates of parameters in normal likelihood functions when diffuse prior distributions are employed. When these MELO estimates are viewed as estimators, it is found that they have at least a finite second moment and thus finite risk relative to quadratic and quite a few other loss functions. In contrast, maximum likelihood (ML) estimators for the problems in (a) when likelihood functions are based on normal data distributions do not possess finite moments and have infinite risk relative to quadratic and other loss functions.

*Research financed by NSF Grants GS-40033 and SOC 7305547 and by income from the H.G.B. Alexander Endowment Fund, Graduate School of Business, University of Chicago. Many of the results in this paper were reported to the 9th NBER–NSF Seminar on Bayesian Inference in Econometrics, held at the University of Montreal, October 1974, and to the Indian Econometric Conference, New Delhi, January 1975.

With respect to problems in (b), it is well-known that many widely used estimators such as 2SLS, LIML, 3SLS, FIML, etc., can fail to possess finite moments[1] and thus can have infinite risk relative to quadratic and other loss functions. Further, it is the case that for problems in (a) and (b), means of posterior distributions, based on diffuse or natural conjugate prior distributions, often fail to exist. Thus, in these cases, the usual Bayesian point estimate, the posterior mean, is not available.

The plan of the paper is as follows. In section 2, problems of estimating the reciprocal of a population mean, the reciprocal of a regression coefficient and ratios of population means and regression coefficients are analyzed. The analysis of section 3 involves deriving MELO estimators for structural coefficients of linear structural econometric models. In section 4 further consideration is given to the properties of the MELO estimates and estimators derived in sections 2 and 3. Last, in section 5 some concluding remarks are presented.

2. MELO estimation of reciprocals and ratios of population means and regression coefficients

2.1. Reciprocals of population means and regression coefficients

Let our model for the observations be

$$y_i = \mu + u_i, \qquad i = 1, 2, \ldots, n, \tag{2.1}$$

where y_i is the ith observation, μ is the common mean of the observations, and u_i is the ith disturbance or error term. Our problem is to estimate $\theta = 1/\mu$.

The loss function that we shall employ is the following relative squared error loss function:[2]

$$L(\theta, \hat{\theta}) = (\theta - \hat{\theta})^2/\theta^2, \tag{2.2}$$

where $\hat{\theta} = \hat{\theta}(\underset{\sim}{y})$ is an estimate of θ. Note that for the relative squared error loss function, a *given* absolute error, $|\theta - \hat{\theta}|$, is considered to be more serious when the true value of θ is small than when it is large whereas with a squared error loss function, $(\theta - \hat{\theta})^2$, the same loss is experienced for a given absolute error whatever the magnitude of the true value of θ. Further, the

[1]See, e.g., Anderson–Sawa (1973), Bergstrom (1962), Hatanaka (1973), Sawa (1972).

[2]Relative squared error loss functions have been previously employed and discussed in the literature; see, e.g., De Groot (1970, p. 226) and Ferguson (1967, pp. 47, 26, 28). In Zellner–Vandaele (1975) it is shown that a Stein-like estimator for the mean of a multivariate normal pdf can be generated using a generalized relative squared error loss function and a diffuse, improper prior pdf.

loss function in (2.2) can be obtained as follows. From $\theta = 1/\mu$, we have $1-\theta\mu=0$. Letting $\varepsilon = 1-\hat{\theta}\mu$, $\varepsilon^2 = (1-\hat{\theta}\mu)^2 = (\theta-\hat{\theta})^2/\theta^2$. The quantity ε measures the extent to which the relation $1-\theta\mu = 0$ is in error when an estimate $\hat{\theta}$ is inserted for θ.

Given that we have a posterior pdf for μ, $p(\mu|\underline{y}, PI)$ that possesses finite first and second moments, we can compute posterior expected loss by expressing (2.2) as $L = (\theta-\hat{\theta})^2/\theta^2 = 1-2\mu\hat{\theta}+\mu^2\hat{\theta}^2$ and using the posterior pdf for μ to obtain

$$EL = 1-2\hat{\theta}E\mu+\hat{\theta}^2E\mu^2, \tag{2.3}$$

where $E \equiv$ posterior expectation. The value of $\hat{\theta}$ that minimizes (2.3), the MELO estimate, denoted by $\hat{\theta}^*$ is

$$\hat{\theta}^* = \frac{E\mu}{E\mu^2} = \frac{\bar{\mu}}{\bar{\mu}^2+\operatorname{var}\mu} = \frac{1}{\bar{\mu}}\ \frac{1}{1+\operatorname{var}\mu/\bar{\mu}^2}, \tag{2.4}$$

where $\bar{\mu} \equiv E\mu$, the posterior mean, and $\operatorname{var}\mu \equiv$ the posterior variance of μ.[3] Thus it is seen that the MELO estimate, $\hat{\theta}^*$, is equal to the reciprocal of the posterior mean, $1/\bar{\mu}$, times a factor $1/(1+\operatorname{var}\mu/\bar{\mu}^2)$ that has a value between zero and one. This factor depends on the squared coefficient of variation of the posterior pdf for μ. When the posterior pdf for μ is very sharp, as would be the case when the sample size n is large, $\hat{\theta}^*$ is close to $1/\bar{\mu}$. However, when $\operatorname{var}\mu/\bar{\mu}^2$ is large as would be the case when $\operatorname{var}\mu$ is large and/or $\bar{\mu}^2$ is small, $|\hat{\theta}^*| < 1/|\bar{\mu}|$. Thus the factor $1/(1+\operatorname{var}\mu/\bar{\mu}^2)$ can be viewed as acting to 'shrink' $1/\bar{\mu}$. Generally the estimate $\hat{\theta}^*$ in (2.4) will be employed when $|\bar{\mu}|/\sigma_\mu > 1$, a condition giving evidence that a value of μ in close proximity to zero is not very probable and also resulting in $\partial|\hat{\theta}^*|/\partial|\bar{\mu}| < 0$, a reasonable condition. When $|\bar{\mu}|/\sigma_\mu < 1$, the posterior pdf for θ will often show pronounced bimodality and the problem of obtaining a point estimate for θ is more complicated. In many cases prior information about the algebraic sign of μ, e.g. $\mu > 0$, is sufficient to provide a unimodal posterior pdf for θ – see the appendix for details.

We now turn to the evaluation of $\hat{\theta}^*$ in (2.4) in some special cases. Suppose that the y_i's in (2.1) are normally and independently distributed, each with mean μ and common known variance σ_0^2. Then the likelihood function is given by $l(\mu|\underline{y}, \sigma_0^2) \propto \exp\{-n(\mu-\bar{y})^2/2\sigma_0^2\}$, where '$\propto$' denotes proportionality and $\bar{y} = \sum_{i=1}^n y_i/n$, the sample mean. If we employ a diffuse prior pdf for $\mu, p(\mu) \propto$ const., $-\infty < \mu < \infty$, the posterior pdf for μ is in the following normal form, $p(\mu|\underline{y}, \sigma_0^2) \propto \exp\{-n(\mu-\bar{y})^2/2\sigma_0^2\}$, with posterior

[3] For the loss functions, $L(\theta, \hat{\theta}) = (\theta-\hat{\theta})^2/\theta^{2r}$, $r = 1, 2, \ldots$, where $\theta = 1/\mu$, the MELO estimate for θ is $\hat{\theta}^* = E\mu^{2r-1}/E\mu^r$, $r = 1, 2, \ldots$, where $E \equiv$ posterior expectation operator.

mean, $\bar{\mu} = \bar{y}$, and posterior variance, $\text{var}\,\mu = \sigma_0^2/n$.[4] Thus, for this set of assumptions, the MELO estimate $\hat{\theta}^*$ in (2.4) is given by

$$\hat{\theta}^* = \frac{1}{\bar{y}}\left(\frac{1}{1+\sigma_0^2/n\bar{y}^2}\right), \tag{2.5}$$

which viewed as an estimator has finite moments (see section 4) and hence bounded risk relative to the loss function in (2.2). Further, $\tilde{\theta} = 1/\bar{y}$ is the ML estimator for $\theta = 1/\mu$. Since $\tilde{\theta}$ is the reciprocal of a normal variable, $\bar{y}$, its moments do not exist and it has infinite risk relative to a quadratic loss function, the loss function in (2.2), and many other loss functions.

When the observations in (2.1) are normally and independently distributed, each with mean μ and variance σ^2 and both μ and σ^2 have unknown values, we shall analyze the problem of estimating $\theta = 1/\mu$ under the assumption that our prior information about parameters' values is diffuse.[5] That is, we shall employ a standard diffuse prior pdf, namely, $p(\mu,\sigma) \propto 1/\sigma$, $-\infty < \mu < \infty$ and $0 < \sigma < \infty$. When this prior pdf is combined with the normal likelihood function to obtain the joint posterior pdf for μ and σ and then σ is integrated out, the marginal posterior pdf for μ is well-known to be in the following univariate Student-t form,

$$p(\mu|\underline{y}, PI) \propto \{vs^2 + n(\mu-\bar{y})^2\}^{-(v+1)/2},$$

where $v = n-1$ and $vs^2 = \sum_{i=1}^{n}(y_i-\bar{y})^2$. For this posterior pdf, we have $E\mu = \bar{y}$ and $\text{var}\,\mu = vs^2/n(v-2)$ for $v > 2$. Thus from (2.4), the MELO estimate for $\theta = 1/\mu$, is given by

$$\hat{\theta}^* = \frac{1}{\bar{y}}\left(\frac{1}{1+vs^2/n(v-2)\bar{y}^2}\right), \qquad v = n-1 > 2. \tag{2.6}$$

In (2.6), the MELO estimate $\hat{\theta}^*$ is in the form of a product of the reciprocal of the sample mean, $1/\bar{y}$, times a shrinking factor that has a value between zero and one.[6] Further (2.6) viewed as an estimator has finite moments and hence bounded risk. Additional properties of the estimators in (2.5) and (2.6) will be established in section 4. Among these is the property that these

[4]Since μ has a normal posterior pdf, the posterior mean of $\theta = 1/\mu$ does not exist. Further, if a proper natural conjugate prior pdf for μ were employed, the posterior pdf would be normal and the posterior mean of $\theta = 1/\mu$ would still fail to exist. However, (2.4) can be readily evaluated with either a diffuse or a natural conjugate prior distribution.

[5]The analysis can easily be extended to the case of a proper natural conjugate prior distribution for μ and σ or to certain priors that impose restrictions on the range of μ, e.g. $0 < \mu < \infty$.

[6]The condition $|\bar{\mu}|/\sigma_\mu > 1$, mentioned above, here specializes to $|\bar{y}|/(s/n^{1/2}) > (v/(v-2))^{1/2}$, obviously related to a sampling theory pre-test of the hypothesis $\mu = 0$.

estimators have the same *large sample* normal distribution as the ML estimator and hence are consistent and asymptotically efficient.

The problem of estimating the reciprocal of a regression coefficient often arises in practice. For example, in the simple Haavelmo consumption model, the reciprocal of the 'multiplier' is equal to the marginal propensity to save. Similarly in a simple quantity theory model the reciprocal of the velocity coefficient is a parameter of the money demand function. If our model for the observations is a simple regression model, $y_i = \pi x_i + u_i$, $i = 1, 2, \ldots, n$, where x_i is an independent variable and we wish to estimate $\theta = 1/\pi$ using the loss function in (2.2), we first derive the posterior pdf for π using whatever prior distribution and likelihood function that are considered appropriate. Let this posterior pdf be denoted by $p(\pi|\underline{y}, PI)$ and assume that it possesses finite first and second moments. Then using $\theta = 1/\pi$, the posterior expectation of the loss function in (2.2) can be expressed as $EL = 1 - 2\hat{\theta}E\pi + \hat{\theta}^2 E\pi^2$ and the MELO estimate for $\theta = 1/\pi$ is given by

$$\hat{\theta}^* = \frac{1}{\bar{\pi}}\left(\frac{1}{1 + \operatorname{var}\pi/\bar{\pi}^2}\right), \tag{2.7}$$

where $\bar{\pi} = E\pi$ and $\operatorname{var}\pi$ denote the posterior mean and variance of π, respectively. The MELO estimate in (2.7) is seen to be equal to the product of the reciprocal of the posterior mean of π, $1/\bar{\pi}$, and a 'shrinking' factor that has a value between zero and one.

To evaluate (2.7) for a particular case, assume that the data process, $y_i = \pi x_i + u_i$ is a simple normal regression, that is the u_i's are normally and independently distributed, each with zero mean and variance σ^2. If our prior information regarding the values of π and σ is vague, we can employ the following diffuse prior pdf, $p(\pi, \sigma) \propto 1/\sigma$, $-\infty < \pi < \infty$ and $0 < \sigma < \infty$.[7] The joint posterior pdf for π and σ is then given by

$$p(\pi, \sigma|\underline{y}, PI) \propto \sigma^{-(n+1)} \exp\{-[\nu s^2 + (\pi - \hat{\pi})^2 m_{xx}]/2\sigma^2\},$$

where $\nu = n - 1$, $\hat{\pi} = \sum_{i=1}^n x_i y_i / m_{xx}$, $m_{xx} = \sum_{i=1}^n x_i^2$, and $\nu s^2 = \sum_{i=1}^n (y_i - \hat{\pi}x_i)^2$. The marginal posterior pdf for π is then

$$p(\pi|\underline{y}, PI) \propto \{\nu s^2 + (\pi - \hat{\pi})^2 m_{xx}\}^{-(\nu+1)/2},$$

with posterior mean $\bar{\pi} = \hat{\pi}$ and $\operatorname{var}\pi = \nu s^2/(\nu - 2)m_{xx}$ for $\nu > 2$. Thus for

[7]The analysis can easily be extended to the case of a proper natural conjugate prior pdf for π and σ or to certain priors that limit the range of π, e.g., $0 < \pi < \infty$.

these assumptions, (2.7) becomes

$$\hat{\theta}^* = \frac{1}{\hat{\pi}}\left(\frac{1}{1+\nu s^2/(\nu-2)m_{xx}\hat{\pi}^2}\right), \qquad \nu = n-1 > 2. \tag{2.8}$$

It is seen that the MELO estimate in (2.8) is the product of $1/\hat{\pi}$, the ML estimator, and a 'shrinking' factor that has a value between zero and one.[8] It is well-known that the ML estimator $\tilde{\theta} = 1/\hat{\pi}$ does not possess finite moments, see e.g. Bergstrom (1962), and has infinite risk relative to quadratic and other loss functions. On the other hand, as shown in section 4, the MELO estimator in (2.8) has finite moments and bounded risk relative to quadratic and other loss functions. Also, as the sample size gets large, the MELO estimator in (2.8) and the ML estimator have the same large sample normal distribution.

2.2. *Ratios of regression coefficients and population means*

If our problem is to estimate the ratio of two parameters, say the ratio of two multiple regression coefficients, $\theta = \beta_1/\beta_2$, where β_1 and β_2 are multiple regression coefficients, we have $\beta_1 - \beta_2\theta = 0$. If an estimate of θ is inserted in this last expression, $\varepsilon = \beta_1 - \beta_2\hat{\theta}$, where ε measures the extent to which the 'restriction' $\beta_1 - \beta_2\theta = 0$ fails to be satisfied. Letting our loss L be $L = \varepsilon^2 = (\beta_1 - \beta_2\hat{\theta})^2$, we have

$$L = (\beta_1 - \beta_2\hat{\theta})^2 = \beta_2^2(\theta - \hat{\theta})^2. \tag{2.9}$$

In (2.9) loss is given by the product of β_2^2 and $(\theta-\hat{\theta})^2$, a generalized quadratic loss function. Note that for a *given* value of $|\theta-\hat{\theta}|$, L is larger when β_2^2 is large than when it is small. The larger β_2^2, ceteris paribus, the smaller $|\theta|$ and (2.9) provides a greater loss for a given absolute error $|\theta-\hat{\theta}|$. This contrasts with the implications of a squared error loss function, $(\theta-\hat{\theta})^2$ that provides the same loss for a given absolute error no matter what the value of θ. Thus to a large extent (2.9) behaves in a fashion similar to the relative squared error loss function introduced above.

Given that we have a posterior pdf for β_1 and β_2 that possesses finite first and second moments, we can compute posterior expected loss and obtain a MELO estimate for $\theta = \beta_1/\beta_2$ as follows:

$$EL = E(\beta_1 - \beta_2\hat{\theta})^2 = E\beta_1^2 - 2\hat{\theta}E\beta_1\beta_2 + \hat{\theta}^2 E\beta_2^2, \tag{2.10}$$

[8]Here the condition for $\partial|\hat{\theta}^*|/\partial|\hat{\pi}| < 0$ is $|\hat{\pi}|/sm_{xx}^{-1/2} > ((\nu-2)/\nu)^{1/2}$ obviously related to a sampling theory pre-test of the hypothesis $\pi = 0$. See Zellner (1976b, p. 622) for a discussion of pre-testing.

where E denotes the expectation with respect to the posterior pdf for β_1 and β_2. From (2.10), the $\hat{\theta}$ that minimizes EL, denoted by $\hat{\theta}^*$, is given by

$$\begin{aligned}\hat{\theta}^* &= E\beta_1\beta_2/E\beta_2^2 \\ &= [\bar{\beta}_1\bar{\beta}_2+\operatorname{cov}(\beta_1,\beta_2)]/[\bar{\beta}_2^2+\operatorname{var}\beta_2] \\ &= (\bar{\beta}_1/\bar{\beta}_2)[1+\operatorname{cov}(\beta_1,\beta_2)/\bar{\beta}_1\bar{\beta}_2]/(1+\operatorname{var}\beta_2/\bar{\beta}_2^2), \qquad (2.11)\end{aligned}$$

where $\bar{\beta}_1$ and $\bar{\beta}_2$ are posterior means, $\operatorname{var}\beta_2$ is the posterior variance of β_2 and $\operatorname{cov}(\beta_1,\beta_2)$ is the posterior covariance of β_1 and β_2. Given a particular posterior pdf for β_1 and β_2, the MELO estimate in (2.11) can be evaluated as shown below.[9]

If instead of the ratio of multiple regression coefficients, we consider the ratio of two population means, $\theta = \mu_1/\mu_2$, we have $\mu_1-\mu_2\theta = 0$ and $\varepsilon = \mu_1 - \mu_2\hat{\theta}$, where $\hat{\theta}$ is an estimate of θ. Then just as above we take our loss to be

$$L = \varepsilon^2 = (\mu_1-\mu_2\hat{\theta})^2 = \mu_2^2(\theta-\hat{\theta})^2.$$

Given a posterior pdf for μ_1 and μ_2 that has finite first and second moments, we have $EL = E\mu_1^2-2\hat{\theta}E\mu_1\mu_2+\hat{\theta}^2E\mu_2^2$ and thus the MELO estimate, $\hat{\theta}^*$, for $\theta = \mu_1/\mu_2$ is

$$\begin{aligned}\hat{\theta}^* &= E\mu_1\mu_2/E\mu_2^2 \\ &= (\bar{\mu}_1/\bar{\mu}_2)[1+\operatorname{cov}(\mu_1,\mu_2)/\bar{\mu}_1\bar{\mu}_2]/(1+\operatorname{var}\mu_2/\bar{\mu}_2^2), \qquad (2.12)\end{aligned}$$

where $\bar{\mu}_1$ and $\bar{\mu}_2$ are posterior means for μ_1 and μ_2, respectively, and $\operatorname{cov}(\mu_1,\mu_2)$ and $\operatorname{var}\mu_2$ are the posterior covariance of μ_1 and μ_2 and the posterior variance of μ_2, respectively.

Ratios of parameters are often also encountered in analyses of simple simultaneous equation models. For example, let $y_{1t} = \gamma y_{2t}+u_{1t}$ and $y_{2t} = \beta x_t + u_{2t}$ be two structural equations. The reduced form equations for this system are $y_{1t} = \pi_1 x_t + v_{1t}$ and $y_{2t} = \pi_2 x_t + v_{2t}$ with $\pi_1 = \beta\gamma$ and $\pi_2 = \beta$. Thus the structural coefficient γ is given by $\gamma = \pi_1/\pi_2$, the ratio of two regression coefficients of a bivariate regression system, the reduced form equations. From $\gamma = \pi_1/\pi_2$, we have $\pi_1-\pi_2\gamma = 0$, the restriction on π_1 and π_2. Let $\hat{\gamma}$ be an estimate of γ and let our loss be $L = \varepsilon^2 = (\pi_1-\pi_2\hat{\gamma})^2 = \pi_2^2(\gamma-\hat{\gamma})^2$. Then

[9]For $\bar{\beta}_1,\bar{\beta}_2 > 0$, $\partial\hat{\theta}^*/\partial\bar{\beta}_1 > 0$ and $\partial\hat{\theta}^*/\partial\bar{\beta}_2 < 0$ if $(\bar{\beta}_1/\bar{\beta}_2)/2 < \hat{\theta}^*$, with $\hat{\theta}^*$ given in (2.11). This last inequality is equivalent to $\frac{1}{2} < [1+\operatorname{cov}(\beta_1,\beta_2)/\bar{\beta}_1\bar{\beta}_2]/[1+\operatorname{var}\beta_2/\bar{\beta}_2^2]$, a condition on the factor that multiplies $\bar{\beta}_1\bar{\beta}_2$ in the second line of (2.11) and on the posterior pdf for β_1 and β_2. If $\operatorname{cov}(\beta_1,\beta_2)=0$, the condition reduces to $\operatorname{var}\beta_2/\bar{\beta}_2^2<1$ or $\|\bar{\beta}^2\|(\operatorname{var}\beta_2)^{1/2}>1$, a condition implying that β_2's value is probably not close to zero.

$EL = E\pi_1^2 - 2\hat{\gamma}E\pi_1\pi_2 + \hat{\gamma}^2\pi_2^2$ and the MELO estimate for $\gamma = \pi_1/\pi_2$ is

$$\begin{aligned}\hat{\gamma}^* &= E\pi_1\pi_2/E\pi_2^2 \\ &= (\bar{\pi}_1/\bar{\pi}_2)[1+\text{cov}(\pi_1,\pi_2)/\bar{\pi}_1\bar{\pi}_2]/(1+\text{var}\,\pi_2/\bar{\pi}_2^2),\end{aligned} \tag{2.13}$$

where $\bar{\pi}_1$ and $\bar{\pi}_2$ are posterior means for π_1 and π_2, respectively, and $\text{cov}(\pi_1,\pi_2)$ and $\text{var}\,\pi_2$ are the posterior covariance of π_1 and π_2 and the posterior variance of π_2, respectively.

To illustrate applications of the above analysis, we first consider a normal multiple regression problem, $\underline{y} = X\underline{\beta}+\underline{u}$, where $\underline{y}$ is an $n\times 1$ vector, X is a given $n\times k$ matrix with rank k, $\underline{\beta}$ is a $k\times 1$ vector of regression coefficients with unknown values, and $\underline{u}$ is an $n\times 1$ disturbance vector. It is assumed that the elements of $\underline{u}$ have been independently drawn from a normal distribution with zero mean and variance σ^2 whose value is not known. Under these assumptions, the likelihood function is

$$l(\underline{\beta},\sigma|\underline{y},X) \propto \sigma^{-n}\exp\{-[vs^2+(\underline{\beta}-\hat{\underline{\beta}})'X'X(\underline{\beta}-\hat{\underline{\beta}})]/2\sigma^2\},$$

where $v = n-k$, $\hat{\underline{\beta}} = (X'X)^{-1}X'\underline{y}$ and $vs^2 = (\underline{y}-X\hat{\underline{\beta}})'(\underline{y}-X\hat{\underline{\beta}})$. If our prior information about the parameters' values is vague[10] and we represent it by the usual following diffuse prior, $p(\underline{\beta},\sigma)\propto 1/\sigma$, $0<\sigma<\infty$ and $-\infty<\beta_i<\infty$, $i=1,2,\ldots,k$, the joint posterior pdf for $\underline{\beta}$ and σ is

$$p(\underline{\beta},\sigma|\underline{y},X,PI) \propto \sigma^{-(n+1)}\exp\{-[vs^2+\underline{\beta}-\hat{\underline{\beta}})'X'X(\underline{\beta}-\hat{\underline{\beta}})]/2\sigma^2\}.$$

On integrating this last expression with respect to σ, $0<\sigma<\infty$, the marginal posterior pdf for $\underline{\beta}$ is in the following well-known multivariate Student-t form,

$$p(\underline{\beta}|\underline{y},X,PI) \propto \{vs^2+(\underline{\beta}-\hat{\underline{\beta}})'X'X(\underline{\beta}-\hat{\underline{\beta}})\}^{-(v+k)/2}.$$

Then the posterior mean and covariance matrix for $\underline{\beta}$ are $E\underline{\beta}=\hat{\underline{\beta}}$ and $V(\underline{\beta}) = (X'X)^{-1}vs^2/(v-2)$, for $v>2$. From these results we know all that is needed to evaluate the MELO estimate in (2.11) for $\theta=\beta_1/\beta_2$, where β_1 and β_2 are the first and second elements of the regression coefficient vector $\underline{\beta}$. Thus, for this problem, (2.11) specializes as follows:

$$\hat{\theta}^* = (\hat{\beta}_1/\hat{\beta}_2)[1+m^{12}\bar{s}^2/\hat{\beta}_1\hat{\beta}_2]/(1+m^{22}\bar{s}^2/\hat{\beta}_2^2), \qquad v=n-k>2, \tag{2.14}$$

[10] The analysis can easily be extended to the case in which a proper natural conjugate prior distribution is employed.

where $\hat{\beta}_1$ and $\hat{\beta}_2$ are the first two elements of $\hat{\underline{\beta}} = (X'X)^{-1}X'\underline{y}$, m^{ij} is the (i,j)th element of $(X'X)^{-1}$ and

$$\bar{s}^2 = \nu s^2/(\nu-2) = (\underline{y}-X\hat{\underline{\beta}})'(\underline{y}-X\hat{\underline{\beta}})/(\nu-2).$$

It is seen that the MELO estimator $\hat{\theta}^*$ for $\theta = \beta_1/\beta_2$ in (2.14) is equal to $\hat{\beta}_1/\hat{\beta}_2$, the ML estimator, times a factor that depends on the relative second-order posterior moments of β_1 and β_2. It is well-known that the moments of the ML estimator $\tilde{\theta} = \hat{\beta}_1/\hat{\beta}_2$ do not exist whereas the moments of the MELO estimator in (2.14) do exist. Thus the MELO and ML estimators have very different finite sample properties; however as the sample size gets large, their large sample distributions become identical – see section 4.

The problem of estimating the ratio of two population means, $\theta = \mu_1/\mu_2$, can be solved using analysis similar to that employed above. As a specific example, let $y_{1i} = \mu_1 + u_{1i}$ and $y_{2i} = \mu_2 + u_{2i}$, $i = 1,2,\ldots,n$, where μ_1 and μ_2 are population means, y_{1i} and y_{2i} are observations and u_{1i} and u_{2i} are disturbance terms. Here we shall assume that the u_{1i}'s and u_{2i}'s are normally and independently distributed with zero means, that the u_{1i}'s have common variance σ_1^2 and u_{2i}'s have common variance σ_2^2. Under these assumptions, the likelihood function is

$$l(\mu_1,\mu_2,\sigma_1,\sigma_2|\underline{y}_1,\underline{y}_2) \propto \sigma_1^{-n}\sigma_2^{-n}\exp\{-[\nu s_1^2+n(\mu_1-\bar{y}_1)^2]/2\sigma_1^2\}$$
$$\times \exp\{-[\nu s_2^2+n(\mu_2-\bar{y}_2)^2]/2\sigma_2^2\},$$

where $\nu = n-1$, $\bar{y}_1 = \sum_{i=1}^n y_{1i}/n$, $\bar{y}_2 = \sum_{i=1}^n y_{2i}/n$, $\nu s_1^2 = \sum_{i=1}^n (y_{1i}-\bar{y}_1)^2$ and $\nu s_2^2 = \sum_{i=1}^n (y_{2i}-\bar{y}_2)^2$.

If we employ the following diffuse prior pdf, $p(\mu_1,\mu_2,\sigma_1,\sigma_2) \propto 1/\sigma_1\sigma_2$, $0<\sigma_i<\infty$ and $-\infty<\mu_i<\infty$, $i=1,2$, the marginal posterior pdf for μ_1 and μ_2 is given by

$$p(\mu_1,\mu_2|\underline{y}_1,\underline{y}_2,PI) \propto \{\nu s_1^2+n(\mu_1-\bar{y}_1)^2\}^{-(\nu+1)/2}$$
$$\times \{\nu s_2^2+n(\mu_2-\bar{y}_2)^2\}^{-(\nu+1)/2},$$

with posterior means, $E\mu_1 = \bar{y}_1$ and $E\mu_2 = \bar{y}_2$ and posterior variances, $\operatorname{var}\mu_1 = \nu s_1^2/n(\nu-2)$ and $\operatorname{var}\mu_2 = \nu s_2^2/n(\nu-2)$ for $\nu > 2$. For this problem the posterior covariance of μ_1 and μ_2 is zero. Thus the MELO estimator for $\theta = \mu_1/\mu_2$ is given from (2.12) as follows:

$$\hat{\theta}^* = (\bar{y}_1/\bar{y}_2)\{1/(1+\nu s_2^2/n(\nu-2)\bar{y}_2^2)\}, \qquad \nu = n-1 > 2. \tag{2.15}$$

It is seen that (2.15) is in the form of a product of the ML estimator, $\tilde{\theta} = \bar{y}_1/\bar{y}_2$, times a factor that depends on the coefficient of variation of the posterior pdf for μ_2. While the ML estimator $\bar{\theta} = \bar{y}_1/\bar{y}_2$ does not possess finite moments, the MELO estimator in (2.15) has finite moments and hence bounded risk relative to quadratic and other loss functions.

As a last case, consider the simple bivariate normal regression system, $y_{1t} = \pi_1 x_t + v_{1t}$ and $y_{2t} = \pi_1 x_t + v_{1t}$, $t = 1,2,\ldots,T$, where the pairs of error terms (v_{1t}, v_{2t}) are normally and independently distributed, each with zero mean vector and common 2×2 pds covariance matrix Ω. Given that we employ a diffuse prior pdf for π_1, π_2 and the distinct elements of Ω, $p(\pi_1, \pi_2, \Omega) \propto |\Omega|^{-1}$, it is known that the marginal posterior pdf for π_1 and π_2 is in the bivariate Student-t form with posterior mean $\underline{\hat{\pi}}' = (\hat{\pi}_1, \hat{\pi}_2)$ where

$$\hat{\pi}_i = \sum_{t=1}^{T} x_t y_{it} / \sum x_t^2, \qquad i = 1,2,$$

and posterior covariance matrix $\bar{S}/m_{xx}$, where $m_{xx} = \sum x_t^2$ and $\bar{S}$ is a symmetric 2×2 matrix with typical element,

$$\bar{s}_{ij} = \sum (y_{it} - x_t \hat{\pi}_i)(y_{jt} - x_t \hat{\pi}_j)/(v-2), \qquad i,j = 1,2,$$

with $v = n - 1 > 2$. On inserting these posterior moments in (2.13), the MELO estimate is

$$\hat{\gamma}^* = (\hat{\pi}_1/\hat{\pi}_2)(1 + \bar{s}_{12}/m_{xx}\hat{\pi}_1\hat{\pi}_2)/(1 + \bar{s}_{22}/m_{xx}\hat{\pi}_2^2). \tag{2.16}$$

Again it is the case that the MELO estimate $\hat{\gamma}^*$ is the product of the ML estimate, $\hat{\pi}_1/\hat{\pi}_2$, times a 'correction' factor.

For the reader's convenience, some of the results in this section are presented in tabular form in table 1.

3. MELO estimates for structural coefficients

In this section we derive MELO estimates for structural coefficients of linear structural econometric models. Initially we consider a single structural equation and then go on to consider joint estimation of sets of structural equations' parameters.

3.1. 'Single equation analysis

Let a structural equation of a model, say the first equation that is assumed

Table 1

Tabular summary of MELO estimates.

Problem	Loss function	MELO estimate[a]	Eq. in text
1. Reciprocal mean, $\theta = 1/\mu$	$L = ((\theta - \hat{\theta})/\theta)^2$	A. $\hat{\theta}^* = (1/\bar{\mu})(1/(1 + \text{var}\,\mu/\bar{\mu}^2))$	(2.4)
		B. $\hat{\theta}^* = (1/\bar{y})(1/[1 + vs^2/n(v-2)])$	(2.6)
2. Reciprocal of regression coefficient, $\theta = 1/\pi$	$L = ((\theta - \hat{\theta})/\theta)^2$	A. $\hat{\theta}^* = (1/\bar{\pi})(1/(1 + \text{var}\,\pi/\bar{\pi}^2))$	(2.7)
		B. $\hat{\theta}^* = (1/\hat{\pi})(1/[1 + vs^2/(v-2)m_{xx}\hat{\pi}^2])$	(2.8)
2. Ratio of multiple regression coefficients, $\theta = \beta_1/\beta_2$	$L = \beta_2^2(\theta - \hat{\theta})^2$	A. $\hat{\theta}^* = (\bar{\beta}_1/\bar{\beta}_2)[1 + \text{cov}(\beta_1, \beta_2)/\bar{\beta}_1\bar{\beta}_2]/(1 + \text{var}\,\beta_1/\bar{\beta}_2^2)$	(2.11)
		B. $\hat{\theta}^* = (\hat{\beta}_1/\hat{\beta}_2)(1 + m^{12}\bar{s}^2/\hat{\beta}_1\hat{\beta}_2)/(1 + m^{22}\bar{s}^2/\hat{\beta}_2^2)$	(2.14)
4. Ratio of population means, $\theta = \mu_1/\mu_2$	$L = \mu_2^2(\theta - \hat{\theta})^2$	A. $\hat{\theta}^* = (\bar{\mu}_1/\bar{\mu}_2)[1 + \text{cov}(\mu_1, \mu_2)/\bar{\mu}_1\bar{\mu}_2]/(1 + \text{var}\,\mu_2/\bar{\mu}_2^2)$	(2.12)
		B. $\hat{\theta}^* = (\bar{y}_1/\bar{y}_2)/[1 + vs_2^2/n(v-2)\bar{y}_2^2]$	(2.15)
5. Ratio of bivariate regression coefficients, $\gamma = \pi_1/\pi_2$	$L = \pi_2^2(\gamma - \hat{\gamma})^2$	A. $\hat{\gamma}^* = (\bar{\pi}_1/\bar{\pi}_2)[1 + \text{cov}(\pi_1, \pi_2)/\bar{\pi}_1\bar{\pi}_2]/(1 + \text{var}\,\pi_2/\bar{\pi}_2^2)$	(2.13)
		B. $\hat{\gamma}^* = (\hat{\pi}_1/\hat{\pi}_2)(1 + \bar{s}_{12}/m_{xx}\hat{\pi}_1\hat{\pi}_2)/(1 + \bar{s}_{22}/m_{xx}\hat{\pi}_2^2)$	(2.16)

[a] Case A is the general MELO estimate expressed in terms of posterior moments. Case B is a special case of Case A in which normal data processes and diffuse prior distributions are employed.

identified, be given by

$$\underset{n\times 1}{\underline{y}_1} = \underset{n\times m_1}{Y_1}\ \underset{m_1\times 1}{\underline{\gamma}_1} + \underset{n\times k_1}{X_1}\ \underset{k_1\times 1}{\underline{\beta}_1} + \underset{n\times 1}{\underline{u}_1}, \tag{3.1}$$

where $\underline{y}_1$ and Y_1 are an $n\times 1$ vector and an $n\times m_1$ matrix, respectively, of observations on $m = m_1+1$ endogenous variables, X_1 is an $n\times k_1$ matrix of rank k_1 of observations on k_1 predetermined variables, $\underline{\gamma}_1$ and $\underline{\beta}_1$ are $m_1\times 1$ and $k_1\times 1$ vectors of structural coefficients and $\underline{u}_1$ is an $n\times 1$ vector of structural disturbances. X_1 is a sub-matrix of the $n\times k$ matrix X of observations on all predetermined variables in the model, that is $X = (X_1 \vdots X_0)$ where X_0 is $n\times k_0$, $k = k_1+k_0$, and it is assumed that X has rank k.

The reduced form equations for $\underline{y}_1$ and Y_1 are given by

$$(\underset{n\times 1}{\underline{y}_1} \vdots \underset{n\times m_1}{Y_1}) = \underset{n\times k}{X}\ (\underset{k\times 1}{\underline{\pi}_1} \vdots \underset{k\times m_1}{\Pi_1}) + (\underset{n\times 1}{\underline{v}_1} \vdots \underset{n\times m_1}{V_1}), \tag{3.2}$$

where $\underline{\pi}_1$ is a $k\times 1$ vector and Π_1 is a $k\times m_1$ matrix of reduced form coefficients and $\underline{v}_1$ is an $n\times 1$ vector and V_1 is an $n\times m_1$ matrix of reduced form disturbances.

On multiplying both sides of (3.2) on the right by $(1 \vdots -\underline{\gamma}_1')'$ and comparing the result with (3.1), we obtain the well-known results

$$\underline{\pi}_1 = \Pi_1\underline{\gamma}_1 + \begin{pmatrix}\underline{\beta}_1\\ \underline{0}\end{pmatrix}, \tag{3.3a}$$

and

$$\underline{v}_1 - V_1\underline{\gamma}_1 = \underline{u}_1, \tag{3.3b}$$

where the zero vector in (3.3a) is $(k-k_1)\times 1$ with $k-k_1 \geqq m_1$ the necessary order condition for identification implied by the usual rank condition for identification of $\underline{\gamma}_1$ and $\underline{\beta}_1$. In (3.3a) we have the restrictions on the reduced form coefficients that involve the parameter vectors $\underline{\gamma}_1$ and $\underline{\beta}_1$ that we wish to estimate.

To formulate a loss function for the problem of estimating $\underline{\gamma}_1$ and $\underline{\beta}_1$ multiply both sides of (3.3a) on the left by $X = (X_1 \vdots X_0)$ to obtain $X\underline{\pi}_1 = X\Pi_1\underline{\gamma}_1 + X_1\underline{\beta}_1$, or $X\underline{\pi}_1 = \bar{Z}_1\underline{\delta}_1$, where $\bar{Z}_1 = (X\Pi_1 \vdots X_1)$ and $\underline{\delta}_1' = (\underline{\gamma}_1' \vdots \underline{\beta}_1')$. If $\hat{\underline{\delta}}_1$ is an estimate of $\underline{\delta}_1$, let $\underline{\varepsilon} = X\underline{\pi}_1 - \bar{Z}_1\hat{\underline{\delta}}_1$ and define the loss function as follows:

$$\begin{aligned} L = \underline{\varepsilon}'\underline{\varepsilon} &= (X\underline{\pi}_1 - \bar{Z}_1\hat{\underline{\delta}}_1)'(X\underline{\pi}_1 - \bar{Z}_1\hat{\underline{\delta}}_1) \\ &= (\underline{\delta}_1 - \hat{\underline{\delta}}_1)'\bar{Z}_1'\bar{Z}_1(\underline{\delta}_1 - \hat{\underline{\delta}}_1), \end{aligned} \tag{3.4}$$

where in going from the first to the second line of (3.4) $X\underline{\pi}_1 = \bar{Z}_1\underline{\delta}_1$ has been employed. Thus the loss function in (3.4) is quadratic in $\underline{\delta}_1 - \hat{\underline{\delta}}_1$ with a pds matrix, $\bar{Z}_1'\bar{Z}_1$, where $\bar{Z}_1 = (X\Pi_1 \vdots X_1)$.

Given a posterior pdf for $\Pi = (\underline{\pi}_1 \vdots \Pi_1)$ that possesses finite first and second moments, posterior expected loss, evaluated from the first line of (3.4) is

$$EL = E\underline{\pi}_1'X'X\underline{\pi}_1 - 2\hat{\underline{\delta}}_1'E\bar{Z}_1'X\underline{\pi}_1 + \hat{\underline{\delta}}_1'E\bar{Z}_1'\bar{Z}_1\hat{\underline{\delta}}_1, \tag{3.5}$$

and the value of $\hat{\underline{\delta}}_1, \hat{\underline{\delta}}_1^*$ that minimizes expected loss is

$$\begin{aligned}\hat{\underline{\delta}}_1^* &= (E\bar{Z}_1'\bar{Z}_1)^{-1}E\bar{Z}_1'X\underline{\pi}_1 \\ &= \begin{bmatrix} E\Pi_1'X'X\Pi_1 & E\Pi_1'X'X_1 \\ EX_1'X\Pi_1 & X_1'X_1 \end{bmatrix}^{-1} \begin{bmatrix} E\Pi_1'X'X\underline{\pi}_1 \\ EX_1'X\underline{\pi}_1 \end{bmatrix}.\end{aligned} \tag{3.6}$$

$\hat{\underline{\delta}}_1^*$ in (3.6) is the MELO estimate relative to the loss function in (3.4) and whatever posterior pdf is employed in evaluating the posterior moments in (3.6).[11]

As an explicit practical example illustrating application of (3.6), we write the reduced form system in (3.2) as

$$Y = X\Pi + V, \tag{3.7}$$

where $Y = (\underline{y}_1 \vdots Y_1)$, $\Pi = (\underline{\pi}_1 \vdots \Pi_1)$ and $V = (\underline{v}_1 \vdots V_1)$, and assume that the rows of V are normally and independently distributed, each with zero mean vector and common pds $m \times m$ covariance matrix Ω, with $m = m_1 + 1$. If we employ a diffuse prior pdf for Π and Ω, $p(\Pi, \Omega) \propto |\Omega|^{-(m+1)/2}$, it is well-known [see e.g. Zellner (1971a, p. 229)] that the marginal posterior pdf for Π is in the following matrix Student-t form:

$$p(\Pi|Y, PI) \propto |S + (\Pi - \hat{\Pi})'X'X(\Pi - \hat{\Pi})|^{-n/2}, \tag{3.8}$$

where $\hat{\Pi} = (X'X)^{-1}X'Y$, the posterior mean of Π, and

$$S = (Y - X\hat{\Pi})'(Y - X\hat{\Pi}) = \hat{V}'\hat{V}.$$

Letting

$$\Pi - \hat{\Pi} = (\underline{\pi}_1 - \hat{\underline{\pi}}_1, \underline{\pi}_2 - \hat{\underline{\pi}}_2, \ldots, \underline{\pi}_m - \hat{\underline{\pi}}_m),$$

[11] $\hat{\underline{\delta}}_1^*$ is an estimate of $\underline{\delta}_1$ that appears in $X\underline{\pi}_1 = \bar{Z}_1\underline{\delta}_1$, a set of n exact equations. In general if we are interested in estimating a vector $\underline{\theta}$ appearing in $\underline{\eta} = A\underline{\theta}$, where $\underline{\eta}$ is an $n \times 1$ vector and A is an $n \times k$ matrix of rank k, let $\hat{\underline{\theta}}$ be any estimate of $\underline{\theta}$. Then if our loss function is $L = (\underline{\eta} - A\hat{\underline{\theta}})'Q(\underline{\eta} - A\hat{\underline{\theta}}) = (\underline{\theta} - \hat{\underline{\theta}})'A'QA(\underline{\theta} - \hat{\underline{\theta}})$, where Q is an $n \times n$ pds matrix, the MELO estimate for θ is $\hat{\underline{\theta}}^* = (EA'QA)^{-1}EA'Q\underline{\eta}$ where $E \equiv$ posterior expectation operator.

and

$$(\underline{\pi}-\hat{\underline{\pi}})' = (\underline{\pi}_1' - \hat{\underline{\pi}}_1', \underline{\pi}_2' - \hat{\underline{\pi}}_2', \dots, \underline{\pi}_m' - \hat{\underline{\pi}}_m'),$$

the posterior covariance matrix for $\underline{\pi}$ is [see e.g. Box–Tiao (1973, p. 477)]

$$E(\underline{\pi}-\hat{\underline{\pi}})(\underline{\pi}-\hat{\underline{\pi}})' = \bar{S} \otimes (X'X)^{-1}, \tag{3.9}$$

where $\bar{S} = (Y - X\hat{\Pi})'(Y - X\hat{\Pi})/(\nu - 2)$, with $\nu = n - k - (m-1) > 2$.

Using the above results to evaluate the posterior expectations in (3.6), we have

$$\begin{aligned} E\Pi_1'X'X\Pi_1 &= E(\Pi_1 - \hat{\Pi}_1 + \hat{\Pi}_1)'X'X(\Pi_1 - \hat{\Pi}_1 + \hat{\Pi}_1) \\ &= \hat{\Pi}_1'X'X\hat{\Pi}_1 + E(\Pi_1 - \hat{\Pi}_1)'X'X(\Pi_1 - \hat{\Pi}_1) \\ &= \hat{\Pi}_1'X'X\hat{\Pi}_1 + k\bar{S}_{22}, \end{aligned} \tag{3.10}$$

and

$$\begin{aligned} E\Pi_1'X'X\underline{\pi}_1 &= E(\Pi_1 - \hat{\Pi}_1 + \hat{\Pi}_1)'X'X(\underline{\pi}_1 - \hat{\underline{\pi}}_1 + \hat{\underline{\pi}}_1) \\ &= \hat{\Pi}_1'X'X\hat{\underline{\pi}}_1 + E(\Pi_1 - \hat{\Pi}_1)'X'X(\underline{\pi}_1 - \hat{\underline{\pi}}_1) \\ &= \hat{\Pi}_1'X'X\hat{\underline{\pi}}_1 + k\bar{\underline{s}}_{12}, \end{aligned} \tag{3.11}$$

where the partitionings, $\Pi = (\underline{\pi}_1 \vdots \Pi_1)$, $\Pi = (\hat{\underline{\pi}}_1 \vdots \hat{\Pi}_1)$ and

$$\bar{S} = (Y - X\hat{\Pi})'(Y - X\hat{\Pi})/(\nu - 2) = \begin{pmatrix} \bar{s}_{11} & \bar{\underline{s}}_{12}' \\ \bar{\underline{s}}_{12} & S_{22} \end{pmatrix}, \tag{3.12}$$

have been employed.[12] Substituting the results (3.10) and (3.11) into (3.6), we have

$$\hat{\underline{\delta}}_1^* = \begin{bmatrix} \hat{\Pi}_1'X'X\hat{\Pi}_1 + k\bar{S}_{22} & \hat{\Pi}_1'X'X_1 \\ X_1'X\hat{\Pi}_1 & X_1'X_1 \end{bmatrix}^{-1} \begin{bmatrix} \hat{\Pi}_1'X'X\hat{\underline{\pi}}_1 + k\bar{\underline{s}}_{12} \\ X_1'X\hat{\underline{\pi}}_1 \end{bmatrix}. \tag{3.13}$$

[12] In the derivation reported in (3.10), note that $E(\Pi_1 - \hat{\Pi}_1)'X'X(\Pi_1 - \hat{\Pi}_1) = \{E(\underline{\pi}_i - \hat{\underline{\pi}}_i)'X'X(\underline{\pi}_j - \hat{\underline{\pi}}_j)\} = \{\operatorname{tr} X'X \operatorname{cov}(\underline{\pi}_i, \underline{\pi}_j)\} = \{k\bar{s}_{ij}\} = k\bar{S}_{22}$. Similar operations yield $E(\Pi_1 - \hat{\Pi}_1)'X'X(\underline{\pi}_1 - \hat{\underline{\pi}}_1) = k\bar{\underline{s}}_{12}$, with $\bar{S}_{22}$ and $\bar{\underline{s}}_{12}$ defined in (3.12).

With $\hat{\underline{\delta}}_1^* = (\hat{\underline{\gamma}}_1'^*, \hat{\underline{\beta}}_1'^*)'$, we obtain the following expressions for $\hat{\underline{\gamma}}_1^*$ and $\hat{\underline{\beta}}_1^*$ from (3.13):

$$\hat{\underline{\gamma}}_1^* = (M_1 + k\bar{S}_{22})^{-1}(M_1\hat{\underline{\gamma}}_1 + k\bar{S}_{22}\hat{\hat{\underline{\gamma}}}_1), \tag{3.14}$$

where $M_1 = \hat{\Pi}_1' X[I - X_1(X_1'X_1)^{-1}X_1']X\hat{\Pi}_1$, $\hat{\underline{\gamma}}_1$ is the 2SLS estimator for $\underline{\gamma}_1$. $\hat{\hat{\underline{\gamma}}} = \bar{S}_{22}^{-1}\bar{\underline{s}}_{12} = (\hat{V}_1'\hat{V}_1)^{-1}\hat{V}_1\hat{\underline{v}}_1$, and

$$\hat{\underline{\beta}}_1^* = (X_1'X_1)^{-1}X_1'(X\hat{\underline{\pi}}_1 - X\hat{\Pi}_1\hat{\underline{\gamma}}_1^*). \tag{3.15}$$

From (3.14), it is seen that the MELO estimate for $\underline{\gamma}_1$ is a matrix weighted average of the 2SLS estimate, $\hat{\underline{\gamma}}_1$, and the estimate $\hat{\hat{\underline{\gamma}}}_1$ formed from reduced form residuals. Also (3.14) can be expressed as a matrix weighted average cf the DLS and 2SLS estimates of $\underline{\gamma}_1$ – see Zellner (1976a).

Surprisingly, the estimate $\hat{\underline{\delta}}_1^*$ for $\underline{\delta}_1$ in (3.13) is in the form of a '$\mathscr{R}$-class' estimate with $\mathscr{R} = \mathscr{R}^*$ where $\mathscr{R}^* = 1 - k/(v-2)$, with $v = n - k - (m-1) > 2$. Hence, the present analysis yields an optimal value for $\mathscr{R}$ that is less than one, the value that yields the 2SLS estimate.[13] Hatanaka (1973, pp. 12–14) has shown that in general the first two finite-sample moments of $\mathscr{R}$-class estimators exist when $\mathscr{R} < 1$ and other conditions, satisfied above, are met.[14] Thus $\hat{\underline{\delta}}_1^*$ in (3.13) has finite first and second moments and bounded risk relative to quadratic loss functions. Further it is to be noted that $\mathscr{R}^* \to 1$ and $n^{1/2}(\mathscr{R}^* - 1) \to 0$ as $n \to \infty$. Thus $\hat{\underline{\delta}}_1^*$ is consistent and asymptotically equivalent to other consistent and asymptotically normal $\mathscr{R}$-class estimators. For the reader's benefit, the following table provides values of $\mathscr{R}^*$ for various values of k, the number of variables in X and $v - 2 = n - k - (m-1) - 2$.

From the table it is interesting to observe that $\mathscr{R}^*$ is substantially below one in a variety of circumstances. Also when $k = v - 2$, $\mathscr{R}^* = 0$, which yields the 'direct' or 'ordinary' least squares estimator for $\underline{\delta}_1$. When $k > v - 2 = n - k - (m-1) - 2$ or $2k + m - 1 > n$, a condition under which the number of parameters is large relative to the sample size n, $\mathscr{R}^* < 0$.

We further note that the derivation of $\hat{\underline{\delta}}_1^*$ in (3.13) is appropriate when X contains lagged endogenous variables, given that initial values of such variables are taken as given in formulating the normal likelihood function for

[13]When $\mathscr{R} > 1$, Sawa (1972) showed that $\mathscr{R}$-class estimators for structural coefficients in an equation containing two endogenous variables do not possess finite moments of any order. See also Mariano (1973). In Nagar (1959) and Sawa (1972), 'approximately optimal' values of $\mathscr{R}$ that are functions of parameters with unknown values are presented.

[14]These other conditions, in terms of our notation, are: (i) $k - k_1 - m_1 \geqq 0$ (order condition for identification); (ii) $n - k - m_1 \geqq 0$; and (iii) $n - k_1 - m_1 \geqq 2$. With the assumption $v = n - k - (m-1) > 2$ satisfied, Hatanaka's conditions are also satisfied. Hatanaka further assumes that X does not contain lagged endogenous variables.

Table 2

Values of $\mathcal{R}^*$ for selected values of k and $\nu-2$.

Values of k	Values of $\nu-2$										
	1	2	3	4	5	6	10	20	50	...	∞
1	0	1/2	2/3	3/4	4/5	5/6	9/10	19/20	49/50	...	1
2	−1	0	1/3	1/2	3/5	2/3	4/5	9/10	24/25	...	1
3	−2	−1/2	0	1/4	2/5	1/2	7/10	17/20	47/50	...	1
4	−3	−1	−1/3	0	1/5	1/3	3/5	4/5	23/25	...	1
5	−4	−3/2	−2/3	−1/5	0	1/6	1/2	3/4	9/10	...	1

the reduced form system, $Y = X\Pi + V$. In addition, it is possible to employ a broadened loss function $L = (\underline{\delta}_1 - \hat{\underline{\delta}}_1)'\bar{Z}_1' Q_1 \bar{Z}_1(\underline{\delta} - \hat{\underline{\delta}}_1)$, where Q_1 is a given pds matrix in place of (3.4).[15]

3.2. 'Full-system' analysis

We now turn to derive MELO estimates of structural coefficients, assumed identified, appearing in all equations of a linear simultaneous equation model. Let the αth structural equation be given by

$$\underline{y}_\alpha = Y T_\alpha \underline{\gamma}_\alpha + X_\alpha \underline{\beta}_\alpha + \underline{u}_\alpha, \qquad \alpha = 1, 2, \ldots, g, \tag{3.16}$$

where $\underline{y}_\alpha$ is an $n \times 1$ vector of observations on an endogenous variable whose coefficient is equal to one, Y_α is an $n \times m_\alpha$ matrix of observations on m_α other endogenous variables appearing in the αth equation with $m_\alpha \times 1$ coefficient vector $\underline{\gamma}_\alpha$, X_α is an $n \times k_\alpha$ matrix of observations on k_α predetermined variables with $k_\alpha \times 1$ coefficient vector $\underline{\beta}_\alpha$, and $\underline{u}_\alpha$ is an $n \times 1$ disturbance vector. As usual, it is assumed that X_α is a submatrix of X, the $n \times k$ matrix of observations on all k predetermined variables that has rank k. Using the reduced form equations for $\underline{y}_\alpha$ and Y_α, $\underline{y}_\alpha = X\underline{\pi}_\alpha + \underline{v}_\alpha$ and $Y_\alpha = X\Pi_\alpha + V_\alpha$, it is possible to express (3.16) as

$$\underline{y}_\alpha = X\Pi_\alpha \underline{\gamma}_\alpha + X_\alpha \underline{\beta}_\alpha + \underline{v}_\alpha, \tag{3.17a}$$

or $\qquad \alpha = 1, 2, \ldots, g.$

$$X\underline{\pi}_\alpha = X\Pi_\alpha \underline{\gamma}_\alpha + X_\alpha \underline{\beta}_\alpha, \tag{3.17b}$$

From (3.17), we can write

$$\begin{pmatrix} X\underline{\pi}_1 \\ X\underline{\pi}_2 \\ \vdots \\ X\underline{\pi}_g \end{pmatrix} = \begin{pmatrix} X\Pi_1 & 0 & \cdots & 0 \\ 0 & X\Pi_2 & \cdots & 0 \\ \vdots & \vdots & \ddots & \vdots \\ 0 & 0 & \cdots & X\Pi_g \end{pmatrix} \begin{pmatrix} \underline{\gamma}_1 \\ \underline{\gamma}_2 \\ \vdots \\ \underline{\gamma}_g \end{pmatrix} + \begin{pmatrix} X_1 & 0 & \cdots & 0 \\ 0 & X_2 & \cdots & 0 \\ \vdots & & \ddots & \vdots \\ 0 & 0 & \cdots & X_g \end{pmatrix} \begin{pmatrix} \underline{\beta}_1 \\ \underline{\beta}_2 \\ \vdots \\ \underline{\beta}_g \end{pmatrix}, \tag{3.18a}$$

[15] In Zellner–Park (1975), the MELO estimate $\hat{\underline{\delta}}_1^*$ in (3.13) is computed for equations of Klein's Model I and of the Girshick–Haavelmo demand and supply model for food and compared with DLS and 2SLS estimates. The MELO estimates tend to lie between the DLS and 2SLS estimates.

or

$$\underline{\bar{w}} = \bar{W}\underline{\gamma} + \bar{X}\underline{\beta}, \tag{3.18b}$$

or

$$\underline{\bar{w}} = \bar{Z}\underline{\delta}. \tag{3.18c}$$

where $\underline{\bar{w}}$ denotes the partitioned vector on the l.h.s. of (3.18a), $\bar{W}$ and $\bar{X}$ the first and second block diagonal matrices on the r.h.s. of (3.18a) $\underline{\gamma}' = (\underline{\gamma}'_1, \underline{\gamma}'_2, \ldots, \underline{\gamma}'_g)$, $\underline{\beta}' = (\underline{\beta}'_1, \underline{\beta}'_2, \ldots, \underline{\beta}'_g)$, $\bar{Z} = (\bar{W} \vdots \bar{X})$ and $\underline{\delta}' = (\underline{\gamma}' \vdots \underline{\beta}')$.

As our loss function for estimating the parameter vector $\underline{\delta}$ in (3.18), we let $\underline{\varepsilon} = \underline{\bar{w}} - \bar{Z}\underline{\hat{\delta}}$ where $\underline{\hat{\delta}}$ is an estimate of $\underline{\delta}$ and form

$$L = (\underline{\bar{w}} - \bar{Z}\underline{\hat{\delta}})'Q(\underline{\bar{w}} - \bar{Z}\underline{\hat{\delta}})$$

$$= (\underline{\delta} - \underline{\hat{\delta}})'\bar{Z}'Q\bar{Z}(\underline{\delta} - \underline{\hat{\delta}}), \tag{3.19}$$

where Q is a pds matrix that may have elements that are parameters with unknown values. Given a posterior pdf for the reduced form coefficients and elements of Q, we compute the posterior expectation of (3.19) and find the minimizing value of $\underline{\delta}$ as follows where E is the posterior expectation operator:

$$EL = E\underline{\bar{w}}'Q\underline{\bar{w}} - 2\underline{\hat{\delta}}'E\bar{Z}'Q\underline{\bar{w}} + \underline{\hat{\delta}}'\bar{Z}Q\bar{Z}\underline{\hat{\delta}}. \tag{3.20}$$

By differentiating (3.20) with respect to $\underline{\hat{\delta}}$, we find the minimum expected loss estimate to be

$$\underline{\hat{\delta}}^* = (E\bar{Z}'Q\bar{Z})^{-1}E\bar{Z}'Q\underline{\bar{w}}. \tag{3.21}$$

The MELO estimate $\underline{\hat{\delta}}^*$ is a general solution for our problem of estimating the structural coefficient vector $\underline{\delta}$ given the loss function in (3.19) and a posterior pdf for the reduced form coefficients and the elements Q. Below we shall consider the case $Q = \Omega_g^{-1} \otimes I_n$ where Ω_g is the $g \times g$ pds reduced form disturbance covariance matrix.

To evaluate (3.21) with $Q = \Omega_g^{-1} \otimes I_n$, consider the reduced form system, $Y = X\Pi + V$, where $Y = (\underline{y}_1, \underline{y}_2, \ldots, \underline{y}_g)$ is an $n \times g$ matrix of observations on all endogenous variables in the system, X is an $n \times k$ matrix of rank k of observations on all predetermined variables, Π is a $k \times g$ matrix of reduced form coefficients and V is an $n \times g$ matrix of disturbances. It is assumed that the n rows of V have been independently drawn from a g-dimensional normal distribution with zero mean vector and $g \times g$ pds covariance matrix Ω_g. If we employ the following diffuse prior pdf for Π and Ω_g,

$p(\Pi, \Omega_g) \propto |\Omega_g|^{-(g+1)/2}$, the joint posterior pdf for Π and Ω_g is given by [see e.g. Zellner (1971a, p. 227)]

$$p(\Pi, \Omega_g | Y, PI) \propto |\Omega_g|^{-(n+g+1)/2} \times \exp\{-\tfrac{1}{2}\operatorname{tr}[\hat{\Omega}_g + (\Pi - \hat{\Pi})' X' X (\Pi - \hat{\Pi})]\Omega_g^{-1}\}, \tag{3.22}$$

where $\hat{\Pi} = (X'X)^{-1}X'Y$ and $\hat{\Omega}_g = (Y - X\hat{\Pi})'(Y - X\hat{\Pi})$. From (3.22), it is seen that the conditional posterior pdf for Π given Ω_g is in the multivariate normal form with mean $\hat{\Pi}$ and covariance matrix $\Omega_g \otimes (X'X)^{-1}$. This fact will be employed to evaluate the posterior expectations in (3.16) conditional on Ω_g. That is, with respect to $E\bar{Z}'Q\bar{Z}$ on the r.h.s. of (3.16), the (α, l)th submatrix of $\bar{Z}'Q\bar{Z} = \bar{Z}'(\Omega_g^{-1} \otimes I_n)\bar{Z}$ is $\bar{Z}_\alpha' \bar{Z}_l \omega^{\alpha l}$, where $\omega^{\alpha l}$ is the (α, l)th element of Ω_g^{-1}, $\bar{Z}_\alpha = (X\Pi_\alpha \vdots X_\alpha)$ and $\bar{Z}_l = (X\Pi_l \vdots X_l)$. Then the conditional posterior mean of the (α, l)th submatrix is given by

$$E\bar{Z}_\alpha' \bar{Z}_l \omega^{\alpha l} = [\hat{Z}_\alpha' \hat{Z}_l + E(\bar{Z}_\alpha - \hat{Z}_\alpha)' (\bar{Z}_l - \hat{Z}_l)\omega^{\alpha l}], \tag{3.23}$$

where E is the conditional (given Ω_g) posterior expectation operator, $\bar{Z}_\alpha = (X\hat{\Pi}_\alpha \vdots X_\alpha)$ and $\hat{Z}_l = (X\hat{\Pi}_l \vdots X_l)$ with $\hat{\Pi}_\alpha = (X'X)^{-1}X'Y_\alpha$ and $\hat{\Pi}_l = (X'X)^{-1}X'Y_l$ the posterior means of Π_α and Π_l, respectively. Further, the vector $E\bar{Z}'Q\underline{\bar{w}}$ on the r.h.s. of (3.21) is

$$E\bar{Z}'(\Omega_g^{-1} \otimes I_n)\underline{\bar{w}} = E\left\{\sum_{l=1}^{g} \bar{Z}_\alpha' \underline{\bar{w}}_l \omega^{\alpha l}\right\}, \qquad \alpha = 1, 2, \ldots, g,$$

$$= \left\{\sum_{l=1}^{g} [\hat{Z}_\alpha' \hat{w}_l + E(\bar{Z}_\alpha - \hat{Z}_\alpha)'(\underline{\bar{w}}_l - \underline{\hat{w}}_l)]\omega^{\alpha l}\right\}, \tag{3.24}$$

where $\underline{\hat{w}}_l = X\underline{\hat{\pi}}_l$, with $\hat{\pi}_l = (X'X)^{-1}X'\underline{y}_l$. Thus the optimal estimate $\underline{\hat{\delta}}^*$ given Ω_g is

$$\underline{\hat{\delta}}^* = \{[\hat{Z}_\alpha' \hat{Z}_l + E(\bar{Z}_\alpha - \hat{Z}_\alpha)'(\bar{Z}_l - \hat{Z}_l)]\omega^{\alpha l}\}^{-1} \times \left\{\sum_{l=1}^{g} [\hat{Z}_\alpha' \underline{\hat{w}}_l + E(\bar{Z}_\alpha - \hat{Z}_\alpha)'(\underline{\bar{w}}_l - \underline{\hat{w}}_l)]\omega^{\alpha l}\right\}, \tag{3.25}$$

where the entry in the first pair of curly braces is a typical submatrix, $\alpha, l = 1, 2, \ldots, g$, and in the second pair, a typical element of a vector, $\alpha = 1, 2, \ldots g$. Given that the conditional posterior pdf for the reduced form coefficients, Π, given Ω_g is multivariate normal with mean $\hat{\Pi} = (X'X)^{-1}X'Y$ and covariance matrix $\Omega_g \otimes (X'X)^{-1}$, the conditional posterior expectations in (3.25) can be evaluated to give

$$\underline{\hat{\delta}}^* = \left\{ \left[\left(\begin{array}{c|c} \hat{Y}_\alpha \hat{Y}_l & Y'_\alpha X_l \\ \hline X'_\alpha Y_l & X'_\alpha X_l \end{array} \right) + k \begin{pmatrix} \Omega_{\alpha l} & 0 \\ 0 & 0 \end{pmatrix} \right] \omega^{\alpha l} \right\}^{-1}$$

$$\times \left\{ \sum_{l=1}^{g} \left[\begin{pmatrix} \hat{Y}'_\alpha \\ X'_\alpha \end{pmatrix} \underline{y}_l + k \begin{pmatrix} \underline{\omega}_{\alpha l} \\ \underline{0} \end{pmatrix} \omega^{\alpha l} \right] \right\}, \qquad (3.26)$$

where $\hat{Y}_\alpha = X\hat{\Pi}_\alpha$, $\hat{Y}_l = X\hat{\Pi}_l$, $\omega^{\alpha l}$ is the (α, l)th element of Ω_g^{-1}, $\Omega_{\alpha l}$ is a submatrix of Ω_g that is equal to the sampling covariance between corresponding rows of $V_\alpha = Y_\alpha - X\Pi_\alpha$ and $V_l = Y_l - X\Pi_l$, and $\underline{\omega}_{\alpha l}$ is a vector of sampling covariances between elements of $\underline{v}_\alpha = \underline{y}_\alpha - X\underline{\pi}_\alpha$ and corresponding rows of $V_l = Y - X\Pi_l$.

From (3.22), the marginal posterior pdf for the distinct elements of Ω_g is well-known to be in the form of an inverted Wishart pdf – see e.g. Zellner (1971a, p. 227), and this result will be employed in future work to evaluate the following posterior expectations $E\omega^{\alpha l}$, $E\Omega_{\alpha l}$ and $E\underline{\omega}_{\alpha l}\omega^{\alpha l}$ that are needed to provide a MELO estimate of $\underline{\delta}$ that is not conditional upon Ω_g being given.[16]

4. Properties of MELO estimates and estimators

In this section, selected properties of MELO estimates and estimators presented above will be established. We first consider properties of MELO estimates and then go on to consider large and finite sample properties of the MELO estimators for the estimation problems considered in previous sections.[17]

4.1. Properties of MELO estimates

For the MELO estimates presented in section 2, it is direct to establish that as the sample size grows large, each MELO estimate approaches its corresponding maximum likelihood (ML) estimate. Also, this property holds for the MELO structural coefficient estimates in section 3. Thus, when non-dogmatic prior distributions are employed, as the sample size grows, the MELO estimates approach corresponding ML estimates. In this connection, it is well-known that when the sample size grows, in general posterior distributions derived from non-dogmatic prior distributions and well-behaved

[16] Of course sample estimates of $\Omega_{\alpha l}, \omega^{\alpha l}$, and $\underline{\omega}_{\alpha l}$ can be inserted in (3.26) to give an approximation to the unconditional MELO estimate of $\underline{\delta}$.

[17] See Hill (1975) for a thoughtful discussion of the relevance of sampling properties of Bayesian estimators. In Zellner–Park (1975) further analysis of the finite sample properties of MELO estimators is presented.

likelihood functions, approach a normal distribution's form centered at the ML estimate.[18]

In very small sample situations, it is the case that posterior distributions for the problems considered in previous sections can have more than one pronounced mode.[19] Examination of posterior distributions' shapes will indicate to an investigator whether more than one important mode is present. In such cases, it is necessary to be extremely careful and thoughtful in choosing a point estimate. In the appendix to this paper, a problem of point estimation when a posterior distribution has more than one pronounced model is analyzed.

In connection with estimation of $\theta = 1/\mu$, the reciprocal of a population mean, it is apparent from (2.5) and (2.6) that the MELO estimate, $\hat{\theta}^*$, satisfies $|\hat{\theta}^*| < |\tilde{\theta}|$ where $\tilde{\theta} = 1/\bar{y}$, the ML estimate. For (2.5) and (2.6), $\hat{\theta}^*/\tilde{\theta} = 1/(1+Z^2)$ where Z^2 is the square of the coefficient of variation of the posterior pdf for μ. Thus the extent to which $\tilde{\theta}^*$ differs from $\tilde{\theta}$ depends just on the relative precision of the posterior pdf for μ. Similar observations apply to (2.8), the MELO estimate for a reciprocal of a regression coefficient.

With respect to estimation of ratios of parameters, the relation of the MELO estimates in (2.12) and (2.13) to ML estimates is readily apparent. E.g. for (2.12), we have $\hat{\theta}^* = \tilde{\theta}_{\mathrm{ML}}[1+\mathrm{cov}(\beta_1,\beta_2)/\hat{\beta}_1\hat{\beta}_2]/[1+\mathrm{var}\,\beta_2/\hat{\beta}_2^2]$, where $\tilde{\theta}_{\mathrm{ML}} = \hat{\beta}_1/\hat{\beta}_2$, the ML estimate, $\mathrm{cov}(\beta_1,\beta_2) = m^{12}\bar{s}^2$ is the posterior variance of β_1 and β_2, and $\mathrm{var}\,\beta_2 = m^{22}\bar{s}^2$ is the posterior variance of β_2. Then, with $\rho_{12} = \mathrm{cov}(\beta_1,\beta_2)/(\mathrm{var}\,\beta_1\ \mathrm{var}\,\beta_2)^{1/2}$, $\hat{\theta}^*/\tilde{\theta}_{\mathrm{ML}} = (1+\rho_{12}\phi_1\phi_2)/(1-\phi_2^2)$, where ϕ_1 and ϕ_2 are the posterior coefficients of variation for β_1 and β_2, respectively. Thus, depending on the values of ρ_{12}, ϕ_1 and ϕ_2, the ratio $\hat{\theta}^*/\tilde{\theta}_{\mathrm{ML}}$ can assume the values shown below:

	Condition on ρ_{12}, ϕ_1 and ϕ_2	Range of ratio $\hat{\theta}^*/\tilde{\theta}_{\mathrm{ML}}$
(a)	$\rho_{12} = 0, 0 < \phi_2^2, \phi_1^2 < \infty$	$0 < \hat{\theta}^*/\tilde{\theta}_{\mathrm{ML}} < 1$
(b)	$\phi_2^2 \leqq \rho_{12}\phi_1\phi_2 < \infty$	$1 \leqq \hat{\theta}^*/\tilde{\theta}_{\mathrm{ML}} < \infty$
(c)	$-1 < \rho_{12}\phi_1\phi_2 < \phi_2^2$	$0 < \hat{\theta}^*/\tilde{\theta}_{\mathrm{ML}} < 1$
(d)	$-\infty < \rho_{12}\phi_1\phi_2 \leqq -1$	$-\infty < \hat{\theta}^*/\tilde{\theta}_{\mathrm{ML}} \leqq 0$

When $\rho_{12} = 0$, line (a) of the table indicates $0 < \hat{\theta}^*/\tilde{\theta}_{\mathrm{ML}} < 1$. From line (b) of the table, it is seen that $\hat{\theta}^*$ is 'expanded' relative to $\tilde{\theta}_{\mathrm{ML}}$ when $\phi_2^2 < \rho_{12}\phi_1\phi_2 < \infty$. This latter condition can be satisfied when $\hat{\beta}_1$ and $\hat{\beta}_2$ have the same algebraic sign and $\rho_{12} > 0$ or $\hat{\beta}_1$ and $\hat{\beta}_2$ have opposite algebraic signs and $\rho_{12} < 0$. In either case, we have $|\phi_2| < |\rho_{12}||\phi_1|$, that is the relative variability

[18]See e.g. Jeffreys (1967), Le Cam (1958), Lindley (1961), Zellner (1971a).

[19]This is also true for the sampling distributions of ML estimators, e.g. $\hat{\theta}_{\mathrm{ML}} = 1/\bar{y}$, where $\bar{y}$ is a sample mean, or $\hat{\theta}_{\mathrm{ML}} = \hat{\beta}_1/\hat{\beta}_2$, where $\hat{\beta}_1$ and $\hat{\beta}_2$ are ML regression coefficient estimators.

of the posterior pdf for β_2 is less than that for β_1. Thus, heuristically, adjusting $\hat{\beta}_1$ *upward* relative to $\hat{\beta}_2$ guards against an underestimate of the ratio β_1/β_2. Under the condition of line (c), $0 < \hat{\theta}^*/\tilde{\theta}_{ML} < 1$, there is a 'shrinking' of the ML estimate. Last, in line (d), with $-\infty < \rho_{12}\phi_1\phi_2 < -1$, $\tilde{\theta}^*$ and $\tilde{\theta}_{ML}$ can have different algebraic signs. In this case, if ϕ_1 and ϕ_2 have the same algebraic sign, then $\rho_{12} < 0$, and the coefficients of variation ϕ_1 and ϕ_2 are rather large. Thus given the negative posterior correlation of β_1 and β_2, there is a high probability of obtaining the 'wrong' algebraic sign in estimating β_1/β_2 by use of the ML estimate and θ^* has an algebraic sign different from $\hat{\theta}_{ML}$. However, this will just occur for large relative variability of one or both of the marginal posterior pdfs for β_1 and β_2.

For the MELO structural coefficient estimates in (3.6), (3.13), (3.21) and (3.26), as the sample size grows these estimates approach ML estimates since they are functions of posterior means that approach ML estimates under general conditions.

In small samples, (3.14) indicates that a particular MELO estimate of the coefficient vector $\underline{\gamma}_1$ can be expressed as a matrix weighted average of the 2SLS estimate and $\hat{\underline{\gamma}}_1 = (\hat{V}_1'\hat{V}_1)^{-1}\hat{V}_1'\underline{\hat{v}}_1$, or of the 2SLS and DLS estimates [Zellner (1976a)]. Properties of matrix weighted averages of two vectors can be studied conveniently using the curve décolletage [Dickey (1975) and Leamer (1973)]. The small sample problem of multi-modal posterior distributions and point estimation for structural coefficients will not be treated herein but deserves attention.

4.2. *Large sample properties of MELO estimators*

When the MELO estimates presented above are viewed as estimators, it is the case that for each problem considered the MELO estimator has the same asymptotic distribution as the ML estimator.

The MELO estimator for the reciprocal of a population mean, $\theta = 1/\mu$, is given in (2.4). From (2.4), we can write

$$n^{1/2}(\hat{\theta}^* - \theta) = n^{1/2}(1/\bar{\mu} - \theta) - (n^{1/2}/\bar{\mu})(\text{var}\,\mu/\bar{\mu}^2)(1 + \text{var}\,\mu/\bar{\mu}^2), \quad (4.1)$$

where $\hat{\theta}^*$ is the MELO estimator (2.4), n is the sample size, $\bar{\mu}$ is the mean and var μ is the variance of the posterior pdf for μ. The probability limit (plim) of the second term on the r.h.s. of (4.1) is zero, given that plim $\bar{\mu} \neq 0$. Thus $n^{1/2}(\hat{\theta}^* - \theta)$ and $n^{1/2}(1/\bar{\mu} - \theta)$ have the same asymptotic distribution by use of the convergence result in Cramer (1946, p. 254). Further, with $\bar{\mu} = \bar{y} + \Delta$, where $\bar{y}$ is the sample mean and

$$\Delta = \bar{\mu} - \bar{y}, \qquad 1/\bar{\mu} = 1/\bar{y} - \Delta/\bar{y}(1 + \Delta/\bar{y}),$$

and

$$n^{1/2}(1/\bar{\mu}-\theta) = n^{1/2}(1/\bar{y}-\theta)-n^{1/2}\Delta/\bar{y}(1+\Delta/\bar{y}).$$

Given that $\operatorname{plim}\bar{y} \neq 0$ and $\operatorname{plim} n^{1/2}\Delta = 0$, $n^{1/2}(1/\bar{\mu}-\theta)$, $n^{1/2}(1/\bar{y}-\theta)$, and $n^{1/2}(\hat{\theta}^*-\theta)$ all have the same asymptotic normal distribution under usual central limit theorem conditions – see e.g. Dhrymes (1974) and Theil (1971).

For the MELO reciprocal mean estimator in (2.6), we have

$$n^{1/2}(\hat{\theta}^*-\theta) = n^{1/2}(1/\bar{y}-\theta)-(n^{1/2}/\bar{y})\,(\bar{s}^2/n\bar{y}^2)/(1+\bar{s}^2/n\bar{y}^2),$$

$$\bar{s}^2 = vs^2/(v-2), \qquad vs^2 = \sum_{i=1}^{n}(y_i-\bar{y})^2,$$

and

$$v = n-1 > 2.$$

Then given $\operatorname{plim}\bar{y} \neq 0$ and applying the convergence result in Cramer (1946, p. 254), $n^{1/2}(\hat{\theta}^*-\theta)$ and $n^{1/2}(1/\bar{y}-\theta)$ have the same asymptotic distribution that will be normal under usual conditions of central limit theorems. Since proofs that the asymptotic distributions of the MELO estimators in (2.7) and (2.8) and of the ML estimator $1/\hat{\pi}$ are the same follow along similar lines as those presented above, they will not be presented.

With respect to the MELO estimator for the ratio of two parameters, e.g. (2.11) that is the estimator for the ratio of two regression coefficients, we have

$$n^{1/2}(\hat{\theta}^*-\theta) = n^{1/2}(\bar{\beta}_1/\bar{\beta}_2-\theta)+(n^{1/2}\bar{\beta}_1/\bar{\beta}_2) \times \left(\frac{\operatorname{cov}(\beta_1,\beta_2)/\bar{\beta}_1\bar{\beta}_2-\operatorname{var}\beta_2/\bar{\beta}_2^2}{1+\operatorname{var}\beta_2/\bar{\beta}_2^2}\right), \tag{4.2}$$

where $\hat{\theta}^*$ is shown in (2.11), $\theta = \beta_1/\beta_2$, $\bar{\beta}_1$ and $\bar{\beta}_2$ are posterior means, and $\operatorname{cov}(\beta_1,\beta_2)$ and $\operatorname{var}\beta_2$ are the posterior covariance of β_1 and β_2 and the posterior variance of β_2, respectively. Given that $\operatorname{plim}\bar{\beta}_2 \neq 0$ and that $\operatorname{plim} n^{1/2}\operatorname{cov}(\beta_1,\beta_2) = \operatorname{plim} n^{1/2}\operatorname{var}\beta_2 = 0$, $n^{1/2}(\hat{\theta}^*-\theta)$ and $n^{1/2}(\bar{\beta}_1/\bar{\beta}_2-\theta)$ have the same asymptotic distribution that is the same as the asymptotic normal distribution for $n^{1/2}(\hat{\beta}_1/\hat{\beta}_2-\theta)$, where $\hat{\beta}_1/\hat{\beta}_2$ is the ML estimator for θ.[20]

[20]The proof that $n^{1/2}(\hat{\theta}^*-\theta)$, $n^{1/2}(\bar{\beta}_1/\bar{\beta}_2-\theta)$ and $n^{1/2}(\hat{\beta}_1/\hat{\beta}_2-\theta)$ all have the same asymptotic normal distribution, given $\operatorname{plim}\bar{\beta}_2 = \operatorname{plim}\hat{\beta}_2 \neq 0$, is direct and thus is not presented.

As regards the structural coefficient estimator $\hat{\underline{\delta}}_1^*$ in (3.6), we have

$$\begin{aligned}\hat{\underline{\delta}}_1^* &= (\hat{Z}_1'\hat{Z}_1+\Delta_1)^{-1}(\hat{Z}_1'X\underline{\hat{\pi}}_1+\underline{\Delta}_2)\\ &= [I+(\hat{Z}_1'\hat{Z}_1)^{-1}\Delta_1]^{-1}[(\hat{Z}_1'\hat{Z}_1)^{-1}\hat{Z}_1'X\underline{\hat{\pi}}_1+(\hat{Z}_1'\hat{Z}_1)^{-1}\underline{\Delta}_2]\\ &= \hat{\underline{\delta}}_1-[I+(\hat{Z}_1'\hat{Z}_1)^{-1}\Delta_1]^{-1}(\hat{Z}_1'\hat{Z}_1)^{-1}[\Delta_1\hat{\underline{\delta}}_1-\underline{\Delta}_2],\end{aligned} \tag{4.3}$$

where $\hat{\underline{\delta}}_1 = (\hat{Z}_1'\hat{Z}_1)^{-1}\hat{Z}_1'X\underline{\hat{\pi}}_1$ is the 2SLS estimator, $\hat{Z}_1 = (\hat{Y}_1 \vdots X_1)$ with $\hat{Y}_1 = X\hat{\Pi}_1$ where $\hat{\Pi}_1 = (X'X)^{-1}X'Y_1$, $\Delta_1 = E\bar{Z}_1'\bar{Z}_1-\hat{Z}_1'\hat{Z}_1$, and $\underline{\Delta}_2 = E\bar{Z}_1'X\underline{\pi}_1-\hat{Z}_1'X\underline{\hat{\pi}}_1$, with E the posterior expectation operator and $\bar{Z}_1 = (X\Pi_1 \vdots X_1)$. Then $n^{1/2}(\hat{\underline{\delta}}_1^*-\underline{\delta}_1) = n^{1/2}(\hat{\underline{\delta}}_1-\underline{\delta}_1)+n^{1/2}\Delta_3$, where Δ_3 represents the second term on the r.h.s. of the third line of (4.3). Since plim $n^{1/2}\Delta_3 = 0$,[21] $n^{1/2}(\underline{\delta}_1^*-\underline{\delta}_1)$ and $n^{1/2}(\hat{\underline{\delta}}_1-\underline{\delta}_1)$ have the same asymptotic normal distribution.[22] Further since the 2SLS and limited-information ML estimators have the same asymptotic normal distribution, this is also the asymptotic distribution of $n^{1/2}(\hat{\underline{\delta}}_1^*-\underline{\delta}_1)$. With respect to the particular structural coefficient estimator in (3.13), note that it is an $\mathscr{R}$-class estimator with $\mathscr{R} = \mathscr{R}^* = 1-k/(n-k-m+1)$. Thus plim $n^{1/2}(\mathscr{R}^*-1) = 0$ and under this condition it is known that $\mathscr{R}$-class estimators, including the limited information ML estimator, all have the same asymptotic normal distribution.[23]

4.3. Some finite sample properties of MELO estimators

In this section some finite sample properties of the MELO estimators, presented above, are established.[24] In particular, the existence of the moments of certain MELO estimators will be demonstrated. Also, for the reciprocal mean problem, an exact expression for the MELO estimator's average risk is derived. Finally, considerations bearing on the admissibility of certain MELO estimators are presented.

The MELO estimator (2.5) for $\theta = 1/\mu$, with the value of σ^2 assumed known, is $\hat{\theta}^* = \bar{y}/(\bar{y}^2+\sigma_0^2/n) = (n^{1/2}/\sigma_0)Z/(1+Z^2)$, where $Z = n^{1/2}\bar{y}/\sigma_0$. Then it is easily established that $|\hat{\theta}^*| \leqq n^{1/2}/2\sigma_0$. Thus the rth absolute moment of $\hat{\theta}^*$ is

$$E|\hat{\theta}^*|^r = \int_{-\infty}^{\infty} |\hat{\theta}^+|^r p(\bar{y}|\mu,\sigma_0/n^{1/2})\,\mathrm{d}\bar{y} < (n^{1/2}(2\sigma_0)^r < \infty,$$

[21]Note that plim $n^{1/2}(\hat{Z}_1'\hat{Z}_1)^{-1}\Delta_1\hat{\underline{\delta}}_1 =$ plim $n^{1/2}(\hat{Z}_1'\hat{Z}_1)^{-1}\underline{\Delta}_2 = 0$ assuming that plim $\hat{Z}_1'\hat{Z}_1/n = M$, a pds matrix and since plim $\Delta_1/n^{1/2} = 0$ and plim $\underline{\Delta}_2/n^{1/2} = \underline{0}$.

[22]For a proof that $n^{1/2}(\hat{\underline{\delta}}_1-\underline{\delta}_1)$ has an asymptotic normal distribution, see e.g. Dhrymes (1974) and Theil (1971).

[23]See, e.g., Theil (1971, pp. 505–506).

[24]Further properties including exact finite sample distributions, exact and approximate risk functions, etc., are presented in Zellner–Park (1975).

for $0 < \sigma_0/n^{1/2} < \infty$, where $p(\bar{y}|\mu, \sigma_0/n^{1/2})$ is the normal pdf for $\bar{y}$ with mean μ and standard deviation $\sigma_0/n^{1/2}$. Since the absolute moments of $\hat{\theta}^*$ exist and are finite, $E(\hat{\theta}^*)^r < \infty$, $r = 1, 2, \ldots$. Further, the risk of $\hat{\theta}^*$ relative to the loss function, $L = ((\theta - \hat{\theta}^*)/\theta)^2 = (1 - \mu\hat{\theta}^*)^2$ is finite for $0 < \sigma_0/n^{1/2} < \infty$ and $-\infty < \mu < \infty$.[25]

In estimating $\theta = 1/\mu$ when σ^2 has an unknown value, the MELO estimator, presented in (2.6), is $\hat{\theta}^* = \bar{y}/[\bar{y}^2 + vs^2/n(v-2)] = (1/\bar{c})\bar{z}/(1+\bar{z}^2)$, where $\bar{c}^2 = vs^2/n(v-2)$, $v = n-1 > 2$, and $\bar{z} = \bar{y}/\bar{c}$. Then $|\hat{\theta}^*| \leqq 1/2\bar{c}$ and

$$E|\hat{\theta}^*|^r = \int_0^{\infty} \int^{\infty} |\hat{\theta}^*|^r p(\bar{y}|\mu, \sigma/n^{1/2}) g(x|v)\,\mathrm{d}\bar{y}\,\mathrm{d}x$$

$$< \left(\frac{v-2}{2\sigma^2/n}\right)^{r/2} \int_0^{\infty} x^{-r/2} g(x|v)\,\mathrm{d}x < \infty,$$

for $0 < \sigma/n^{1/2} < \infty$, where $x = vs^2/\sigma^2$ has a χ^2 pdf with $v = n-1$ degrees of freedom, denoted by $g(x|v)$. Thus $E|\hat{\theta}^*|^r$ and $E(\hat{\theta}^*)^r$, $r = 1, 2, \ldots$, exist and are finite. This implies that the risk of $\hat{\theta}^* = \bar{y}/[\bar{y}^2 + vs^2/n(v-2)]$, $v > 2$, relative to $L = (\theta - \hat{\theta}^*)^2/\theta^2 = (1 - \mu\hat{\theta}^*)^2$ exists and is finite for $0 < \sigma/n^{1/2} < \infty$ and $-\infty < \mu < \infty$.

To obtain the average risk (AR) of the estimator $\hat{\theta}^* = \bar{y}/(\bar{y}^2 + \sigma_0^2/n)$, the following integral was evaluated:

$$AR = \int_{-\infty}^{\infty} \int_{-\infty}^{\infty} (1 - \mu\hat{\theta}^*)^2 p(\bar{y}|\mu, \sigma_0/n^{1/2})\,\mathrm{d}\bar{y}\,\mathrm{d}\mu = \pi\sigma_0/n^{1/2}, \qquad (4.4)$$

where the improper prior pdf for μ, $p(\mu)\,\mathrm{d}\mu \propto \mathrm{d}\mu$, $-\infty < \mu < \infty$ has been employed, $p(\bar{y}|\mu, \sigma_0/n^{1/2})$ is the normal pdf for $\bar{y}$ with mean μ and variance σ_0^2/n, and $\pi = 3.14159$.[26]

[25]The risk of $\hat{\theta}^*$ relative to the quadratic loss function, $L = (\hat{\theta}^* - \theta)^2$ is finite for $0 < |\mu| < \infty$ and $0 < \sigma_0/n^{1/2} < \infty$.

[26]To evaluate (4.4), write $\hat{\theta}^* = \bar{y}/(\bar{y}^2 + \sigma_0^2/n) = (n^{1/2}/\sigma_0)(w + \bar{\mu})/[1 + (w\bar{\mu})^2]$, where $w = n^{1/2}(\bar{y} - \mu)/\sigma_0$ is $N(0, 1)$ and $\bar{\mu} = n^{1/2}\mu/\sigma_0$. Then $L = (1 - \mu\hat{\theta}^*)^2 = \{[1 + w(w + \bar{\mu})]/[1 + (w + \bar{u})^2]\}^2$ and (4.4) becomes

$$AR = \sigma_0/(2\pi n)^{1/2} \int_{-\infty}^{\infty} \int_{-\infty}^{\infty} \{[1 + w(w + \bar{\mu})]/[1 + (w + \bar{\mu})^2\}^2\,\mathrm{d}^{-w^2/2}\,\mathrm{d}w\,\mathrm{d}\bar{\mu},$$

where $\mathrm{d}\mu = (\sigma_0/n^{1/2})\,\mathrm{d}\bar{\mu}$ has been employed. Now make the following change of variables, $x = w + \mu$ and $w = w$, that has Jacobian equal to one, to obtain

$$AR = \sigma_0/(2\pi n)^{1/2} \int_{-\infty}^{\infty} \int_{-\infty}^{\infty} [(1 + wx)/(1 + x^2)]\,\mathrm{e}^{-w^2/2}\,\mathrm{d}w\,\mathrm{d}x$$

$$= \sigma_0/n^{1/2} \int_{-\infty}^{\infty} [1/(1 + x^2)]\,\mathrm{d}x.$$

In the class of estimators for $\theta = 1/\mu$, $\hat{\theta} = \hat{\theta}(\bar{y}, \sigma/n^{1/2})$, with the value of $\sigma/n^{1/2}$ known, such that AR is finite relative to the loss function $L = (\theta - \hat{\theta})^2/\theta^2 = (1 - \mu\hat{\theta})^2$ and improper prior pdf $\rho(\mu)\,d\mu \propto d\mu$, $-\infty < \mu < \infty$, $\hat{\theta}^* = \bar{y}/(\bar{y}^2 + \sigma_0^2/n)$ is the estimator in this class that minimizes AR and is thus admissible. That is, there cannot be another estimator in this class that has risk below that for $\hat{\theta}^*$ over a portion (or all) of the parameter space and risk not above that for $\hat{\theta}^*$ over the remaining portion of the parameter space. If there were such an estimator, it would have smaller AR than that for $\hat{\theta}^*$ which is impossible since $\hat{\theta}^*$ is the estimator in the class that minimizes AR.

The MELO estimator for $\theta = 1/\mu$ when σ^2 has an unknown value is from (2.6), $\hat{\theta}^* = \bar{y}/(\bar{y}^2 + \bar{s}^2/n)$, where $\bar{s}^2 = vs^2/(v-2)$, with $v = n-1 > 2$ and $vs^2 = \sum_{i=1}^n (y_i - \bar{y})^2$. By a direct calculation, the average risk (AR) of $\hat{\theta}^*$ relative to the loss function in (2.2) is found to be[27]

$$AR = (\pi/2n^{1/2})\left[\left(\frac{2}{v-2}\right)^{1/2} \Gamma\left(\frac{v+1}{2}\right) + \left(\frac{v-2}{2}\right)^{1/2} \Gamma\left(\frac{v-2}{2}\right)\right] \Big/ \Gamma(v/2), \qquad v = n-1 > 2, \tag{4.5}$$

which is finite. Thus the MELO estimator $\hat{\theta}^* = \bar{y}/(\bar{y}^2 + \bar{s}^2/n)$ is admissible.

In this last expression, note that $1 + x^2 = (x-i)(x+i)$, where $i = (-1)^{1/2}$, and thus there is a pole of order one in the upper half plane at $x = i$. Using a contour integration around a half circle in the upper half plane of the complex domain with center at the origin, the value of the integral

$$\int_{-\infty}^{\infty} [1/(1+x^2)]\,dx$$

is equal to $2\pi i$ times the residue at i, namely $\frac{1}{2}i$, that is $2\pi i/2i = \pi$ by the Cauchy Residue Theorem. Or more simply,

$$\int_{-\infty}^{\infty} (1+x^2)^{-1}\,dx = \pi,$$

using properties of the univariate Cauchy pdf. Thus $AR = \pi\sigma_0/n^{1/2}$ as stated in (4.4).

[27]To derive (4.5), write

$$\hat{\theta}^* = \bar{y}(\bar{y}^2 + \bar{s}^2)/n) = (n^{1/2}/\sigma)\,(w+\bar{\mu})^2 + z/(v-2)],$$

where $\bar{\mu} = n^{1/2}\mu/\sigma$, $w = (\bar{y} - \bar{\mu})/(\sigma/n^{1/2})$, and $z = (v-2)\bar{s}^2/\sigma^2 = vs^2/\sigma^2$. Note that w and z are independently distributed with w having a $N(0,1)$ pdf and z a χ^2 pdf with $v = n-1$ degrees of freedom. The loss function can be expressed as follows:

$$L = \left(\frac{\hat{\theta}^* - \theta}{\theta}\right)^2 = (\hat{\theta}^*\mu - 1)^2 = \left[\frac{z/(v-2) + w(w+\bar{\mu})}{z/(v-2) + (2+\bar{\mu})^2}\right]^2.$$

Some finite sample properties of MELO estimators for ratios of parameters are presented in Zellner–Park (1975. In the same paper attention is given to the properties of the MELO structural coefficient estimator $\hat{\underline{\delta}}_1^*$, shown in (3.13). Using Kadane's (1971) asymptotic small-σ approximations to the moments of $\mathscr{R}$-class estimators, it is found that the approximate risk of the MELO estimator $\hat{\underline{\delta}}_1^*$ relative to the loss function in (3.4) is smaller than that of the 2SLS estimator when the latter's moments exist under a range of conditions frequently encountered in practice.

5. Summary and conclusions

In this paper a minimum expected loss (MELO) approach was utilized to generate point estimates for a number of related estimation problems including reciprocals and ratios of parameters and structural coefficients of linear structural econometric models. For these problems, well-defined and reasonable loss functions were formulated and estimates that minimize posterior expected loss were derived. These MELO estimates have relatively simple forms and thus are quite operational. In the case of 'single-equation' estimation of structural coefficients, it was found that with a normal reduced form system and a diffuse prior pdf for its parameters, the MELO estimate for the structural coefficient vector is an $\mathscr{R}$-class estimate with a value of $\mathscr{R}$ less than one in finite samples. Further, a 'systems' MELO estimate for structural coefficients was derived. These MELO point estimates are useful in

---

From $p(\mu,\sigma)\propto 1/\sigma$, $p(\bar{\mu},\sigma)\propto 1/n^{1/2}$ and thus average risk is given by

$$AR = (1/(2\pi n)^{1/2})\int_0^{\infty}\int_{-\infty}^{\infty}\int_{-\infty}^{\infty} L\,\mathrm{e}^{-w^2/2}p(z)\,\mathrm{d}w\,\mathrm{d}\bar{\mu}\,\mathrm{d}z,$$

with $p(z)$ the χ_ν^2 pdf. In this last expression, let $x = w+\bar{\mu}$, $w = w$ and $z = z$, a change of variables with Jacobian equal to one, to obtain

$$AR = (1/(2\pi n)^{1/2})\int_0^{\infty}\int_{-\infty}^{\infty}\int_{-\infty}^{\infty}\frac{[z/(\nu-2)+wx]^2}{[x/(\nu-2)+x^2]^2}\mathrm{e}^{-w^2/2}p(z)\,\mathrm{d}w\,\mathrm{d}x\,\mathrm{d}z$$

$$= (1/n^{1/2})\int_0^{\infty}\left[\int_{-\infty}^{\infty}\frac{[z/(\nu-2)]^2+x^2}{[z/(\nu-2)+x^2]^2}\mathrm{d}x\right]p(z)\,\mathrm{d}z.$$

The integral in square brackets can be evaluated using properties of the univariate Student-t pdf with three degrees of freedom to yield

$$AR = (\pi/2)\int_0^{\infty}(z/(\nu-2))^{1/2}+((\nu-2)/z)^{1/2}p(z)\,\mathrm{d}z.$$

This last integral can be evaluated given that $p(z)$ is a χ^2 pdf with ν d.f. to yield the expression shown in (4.5).

providing information about parameters' values and can supplement other measures describing properties of posterior distributions.

When the MELO point estimates are regarded as point estimators, it was pointed out that these estimators' moments and risk exist and are finite. These properties contrast markedly with those of some other estimators for these problems that do not possess finite moments and have infinite risk relative to quadratic and other loss functions. For example, maximum likelihood estimators for reciprocals and ratios of means and of regression coefficients do not possess finite moments under a wide range of distributional assumptions relating to data processes. In the case of estimating structural coefficients, it is well-known that many commonly used estimators fail to possess moments under frequently encountered conditions. That ML and some other estimators for the problems considered in this paper can or do have infinite risk relative to quadratic and other loss functions is of interest particularly in view of the widespread use of quadratic loss functions and risk considerations in choosing point estimators, for example in the Gauss–Markov Theorem, in Charles Stein's work, in Monte Carlo experiments, etc. In situations in which ML and other estimators have infinite risk whereas MELO estimators have finite risk, the former estimators are clearly inadmissible. MELO estimators based on proper prior distributions are admissible relative to the loss functions for which they have been derived. Some MELO estimators based on improper priors are admissible while others may be only under special conditions [Hill (1975)]. In large samples, the sampling properties of the ML, MELO and other consistent estimators for particular problems considered in this paper are identical.

Further, it is recognized that (1) not all agree that admissibility is an overriding or even important criterion in choosing point estimates; (2) ML and perhaps other estimators might perform better relative to performance criteria that are not sensitive to the existence or non-existence of sampling moments; and (3) while point estimates are valuable, it must be recognized that they alone are not adequate solutions to most estimating problems. Thus it is fortunate that posterior distributions are available for the problems considered in this paper that reflect the prior and sample information much better than do point estimates alone.

Appendix

A.1. Bimodality of the posterior pdf for $\theta = 1/\mu$[28]

To illustrate a case in which a posterior pdf is bimodal, consider $y_i = \mu + \varepsilon_i$ where the ε_i's are NID$(0, \sigma^2)$ with the value of σ^2 assumed known. With a

[28]Analysis of the bimodality of the sampling distribution of the maximum likelihood estimator for θ, $\hat{\theta} = 1/\bar{y}$, with $\bar{y} \sim N(\mu, \sigma^2/n)$ is similar to that presented below.

diffuse prior pdf for μ, $p(\mu) \propto$ const., $-\infty < \mu < \infty$, the posterior pdf for μ is $p(\mu|\sigma, \bar{y}) \propto \exp\{-n(\mu-\bar{y})^2/2\sigma^2\}$, a normal pdf. Making a change of variable $\theta = 1/\mu$, the posterior pdf for θ is

$$p(\theta|\sigma, \bar{y}) \propto \theta^{-2} \exp\left\{-\frac{z^2}{2}\left(\frac{\theta-\hat{\theta}}{\theta}\right)^2\right\}, \tag{A.1}$$

where $\hat{\theta} = 1/\bar{y}$ and $z = n^{1/2}\bar{y}/\sigma$. Note that z is the ratio of the posterior mean for μ, $\bar{y}$, to its posterior standard deviation, $\sigma/n^{1/2}$. Also z is a sampling theory test statistic for testing the hypothesis $\mu = 0$.

For convenience in analyzing the modes of (A.1), let $\eta = (\hat{\theta}-\theta)/\theta$. Then

$$\theta^{-2} \exp\left\{-\frac{z^2}{2}\left(\frac{\theta-\hat{\theta}}{\theta}\right)^2\right\} \propto (1+\eta)^2 \exp\{-z^2\eta^2/2\}. \tag{A.2}$$

From the form of (A.2), when z^2 has a small value, it is clear from a plot of the two factors on the r.h.s. of (A.2) that there will be pronounced bimodality. On taking the logarithm of the expression on the r.h.s. of (A.2) and differentiating it, the two modal values of η, η_1 and η_2, can be determined. They are $\eta_1, \eta_2 = (1/2)(-1 \pm (1+8/z^2)^{1/2})$. From $\eta = (\hat{\theta}-\theta)/\theta$, $\theta = \hat{\theta}/(1+\eta)$ and thus the modal values for θ are

$$\theta_1 = \hat{\theta}\left(\frac{2}{1+(1+8/z^2)^{1/2}}\right) \quad \text{and} \quad \theta_2 = \hat{\theta}\left(\frac{2}{1-(1+8/z^2)^{1/2}}\right). \tag{A.3}$$

It is clear that $\theta_1 > 0$ and $\theta_2 < 0$. Also, the modal values of the posterior pdf can be quite different from the ML estimate $\hat{\theta}$, particularly when $z = n^{1/2}\bar{y}/\sigma$ is small in absolute value. For example, when $z = 1$, $\theta_1 = \hat{\theta}/2$ and $\theta_2 = -\hat{\theta}$. The MELO estimate for θ is $\hat{\theta}^* = (1/\bar{y})(1+\sigma^2/n\bar{y}^2)^{-1} = \hat{\theta}z^2/(1+z^2)$. Thus when $z^2 = 1$, $\hat{\theta}^* = \hat{\theta}/2 = \theta_1$, a modal value. As z^2 gets large, θ_1 and $\hat{\theta}^*$ both approach $\hat{\theta}$ while θ_2 approaches $-\infty$.

On inserting the modal values for η in the expression on the r.h.s. of (A.2), denoted by $f(\eta)$, we have

$$f(\eta_1)/f(\eta_2) = \left(\frac{1+(1+8/z^2)^{1/2}}{1-(1+8/z^2)^{1/2}}\right)^2 \exp\left\{\frac{z^2}{2}(1+8/z^2)^{1/2}\right\}. \tag{A.4}$$

When $z^2 = 1$, $f(\eta_1)/f(\eta_2) = 4e^{3/2} = 17.93$, the ratio of the ordinates of the posterior pdf at the modal values, η_1 and η_2 or θ_1 and θ_2. Thus even for $z^2 = 1$, one mode is much higher than the other. When z^2 grows in value, the relative heights of the modes grows. As $z^2 \to 0$, $f(\eta_1)/f(\eta_2) \to 1$.

A.2. Point estimation of θ with bimodal distribution

As shown above, (A.1) is bimodal with one modal value positive and the other negative. Further, as $|\theta| \to 0$, $p(\theta|\sigma, \bar{y}) \to 0$ and the posterior probability that $\theta = 1/\mu > 0$ is

$$\int_0^\infty p(\mu|\sigma, \bar{y})\,\mathrm{d}\mu.$$

Conditional upon $\mu > 0$ or $\theta > 0$, the MELO estimate of θ relative to $(\theta - \hat{\theta})^2/\theta^2$ is given by $\hat{\theta}^*_+ = E(\mu|\sigma, \mu > 0)/E(\mu^2|\sigma, \mu > 0)$ where E is the posterior expectation operator. An explicit expression for $\hat{\theta}^*_+$ is[29]

$$\hat{\theta}^*_+ = \frac{\bar{y}[1 - F(-z)] + (\sigma/n^{1/2})P(z)}{\bar{y}^2[1 - F(-z)] + 2\bar{y}(\sigma/n^{1/2})P(z) + \sigma^2/n[1 - F(-z) - zP(z)}$$

$$= \hat{\theta}^* \frac{(1+z^2)(1+Rz)}{(1+z^2)(1+Rz) - 1} = \hat{\theta}\frac{z(1+Rz)}{z(1+Rz)+R}, \tag{A.5}$$

where $z = n^{1/2}\bar{y}/\sigma$, $P(z) = (1/(2\pi)^{1/2})\exp\{-z^2/2\}$, $F(-z) = \int_{-\infty}^{-z} P(z)\,\mathrm{d}z$, $\hat{\theta}^* = (1/\bar{y})/(1 + \sigma^2/n\bar{y}^2)$, $\hat{\theta} = 1/\bar{y}$, and $R = [1 - F(-z)]/P(z)$ Mills' ratio. From (A.5), as $z \to \infty$, $\hat{\theta}^*_+/\hat{\theta}^* \to \hat{\theta}^*_+/\hat{\theta} \to 1$. This is reasonable since as $z \to \infty$, the mode situated over positive values of θ becomes very sharp and centered at the ML estimate $\hat{\theta} = 1/\bar{y}$. On the other hand, for $z = n^{1/2}\bar{y}/\sigma = 1$, $\hat{\theta}^*_+ = 1.4326\hat{\theta}^* = 0.7163\hat{\theta}$ from (A.5) and $\hat{\theta}^* = 0.5\hat{\theta}$. Thus with $z = 1$, introducing the conditional information that $\mu > 0$ or $\theta > 0$, results in the MELO estimate $\hat{\theta}^*_+$ being expanded relative to $\hat{\theta}^*$, the MELO estimate not incorporating the information $\mu > 0$. However, with $z = 1$, $\hat{\theta}^*_+$ is still smaller in value than the ML estimate $\hat{\theta}$. Similar calculations can be performed in the case that the conditioning information is $\mu < 0$. In this way MELO point estimates, $\hat{\theta}^*_+$ and $\hat{\theta}^*_-$ can be obtained. With a 2×2 loss structure and posterior probabilities for the hypotheses, $\mu > 0$ and $\mu < 0$, one can choose between $\hat{\theta}^*_+$ and $\hat{\theta}^*_-$ on the basis of which act has lower expected loss. In case

[29]The expression for $\hat{\theta}^*_+$ in (A.5) was obtained by straightforward evaluation of the following integrals:

$$\hat{\theta}^*_+ = \int_0^\infty \mu p(\mu|\sigma, \bar{y})\,\mathrm{d}\mu \bigg/ \int_0^\infty \mu^2 p(\mu|\sigma, \bar{y})\,\mathrm{d}\mu,$$

with

$$p(\mu|\sigma, \bar{y}) = \frac{1}{(2\pi)^{1/2}(\sigma/n^{1/2})}\exp\left\{-\frac{n}{2\sigma^2}(\mu - \bar{y})^2\right\}.$$

no loss structure is available, one has posterior probabilities that $\theta > 0$ and $\theta < 0$ as well as the point estimates $\hat{\theta}^*_+$ and $\hat{\theta}^*_-$.

References

Anderson, T.W. and T. Sawa, 1973, Distributions of estimates of coefficients of a single equation in a simultaneous system and their asymptotic expansions, Econometrica 41, 683–714.

Bergstrom, A.R., 1962, The exact sampling distributions of least squares and maximum likelihood estimators of the marginal propensity to consume, Econometrica 30, 480–490.

Box, G.E.P. and G.C. Tiao, 1973, Bayesian inference in statistical analysis (Addison-Wesley, Reading, MA).

Cramer, H., 1946, Mathematical methods of statistics (Princeton University Press, Princeton, NJ).

DeGroot, M.H., 1970, Optimal statistical decisions (McGraw-Hill, New York).

Dhrymes, P.J., 1974, Econometrics: Statistical foundations and applications (Springer-Verlag, New York).

Dickey, J.M., 1975, Bayesian alternatives to the F test and least-squares estimate in the normal linear model, in: S.E. Fienberg and A. Zellner, eds., Studies in Bayesian econometrics and statistics in honor of Leonard J. Savage (North-Holland, Amsterdam), 515–554.

Ferguson, T.S., 1967, Mathematical statistics: A decision theoretical approach (Academic Press, New York).

Goldberger, A.S., 1964, Econometric theory (Wiley, New York).

Hatanaka, M., 1973, On the existence and the approximation formulae for the moments of the $\mathscr{k}$-class estimators, Economic Studies Quarterly 24, 1–15.

Hill, B.M., 1975, On coherence, inadmissibility and inference about many parameters in the theory of least squares, in: S.E. Fienberg and A. Zellner, eds., Studies in Bayesian econometrics and statistics in honor of Leonard J. Savage (North-Holland, Amsterdam), 555–584.

Jeffreys, H., 1967, Theory of probability, 3rd rev. ed. (Clarendon Press, Oxford).

Kadane, J.B., 1971, Comparison of $\mathscr{k}$-class estimators when the disturbances are small, Econometrica 39, 723–738.

Kappenman, R.F. and S. Geisser, 1970, Bayesian and fiducial solutions for the Fieller–Creasy problem, Sankhya 32, 331–340.

Leamer, E.E., 1973, Multicollinearity: A Bayesian interpretation, Review of Economics and Statistics 55, 371–380.

Le Cam, L., 1958, Les propriétés asymptotiques des solutions de Bayes, Publ. Inst. Statist. (University of Paris) 7, 17–35.

Lindley, D.V., 1961, The use of prior probability distributions in statistical inference and decisions, in: J. Neyman, ed., Proc. Fourth Berkeley Symp. Math. Statist. and Probab., Vol. I, 453–468.

Madansky, A., 1964, On the efficiency of three-stage least-squares estimation, Econometrica 32, 51–56.

Mariano, R.S., 1973, Approximations to the distribution functions of Theil's $\mathscr{k}$-class estimators, Econometrica 41, 715–722.

Nagar, A.L., 1959, The bias and moment matrix of the general $\mathscr{k}$-class estimators of the parameters in simultaneous equations, Econometrica 29, 575–595.

Press, S.J., 1969, The t-ratio distribution, Journal of the American Statistical Association 64, 242–252.

Sargan, J.D., 1964, Three-stage least-squares and full maximum likelihood estimates, Econometrica 32, 77–81.

Sawa, T., 1972, Finite-sample properties of the $\mathscr{k}$-class estimators, Econometrica 40, 653–680.

Theil, H., 1971, Principles of econometrics (Wiley, New York).

Zellner, A., 1971a, An introduction to Bayesian inference in econometrics (Wiley, New York).

Zellner, A., 1971b, Bayesian and non-Bayesian analysis of log-normal distribution and log-normal regression, Journal of the American Statistical Association 66, 327–330.

Zellner, A., 1976a, A note on the relationship of minimum expected loss (MELO) and other structural coefficient estimates, manuscript (H.G.B. Alexander Research Foundation, University of Chicago, Chicago, IL).

Zellner, A., 1976b, Reply, Econometrica 44, 619–624.

Zellner. A. and S.B. Park, 1975, Further properties of minimum expected loss (MELO) estimators for nonlinear functions of parameters including structural coefficients, Manuscript (H.G.B. Alexander Research Foundation, University of Chicago, Chicago, IL).

Zellner, A. and W. Vandaele, 1975, Bayes–Stein estimators for *k*-means, regression and simultaneous equation models, in: S.E. Fienberg and A. Zellner, eds., Studies in Bayesian econometrics and statistics in honor of Leonard J. Savage (North-Holland, Amsterdam) 627–654.

[19]

Minimum Expected Loss (MELO) Estimators for Functions of Parameters and Structural Coefficients of Econometric Models

ARNOLD ZELLNER and SOO-BIN PARK*

Further sampling properties of previously derived (Zellner 1975) MELO estimates for reciprocals and ratios of parameters and for structural coefficients of econometric models are presented. Approximate small-sample moments and risk functions of MELO estimators for these problems are derived and compared with those of other well-known estimators. The exact finite sample distribution of the MELO reciprocal mean estimator is derived and compared with that of the maximum likelihood estimator. Properties of alternative structural coefficient estimators are compared with those of the MELO estimator both theoretically and in terms of two applications.

KEY WORDS: Bayes estimators; Ratio estimation; Simultaneous equations estimation; Maximum likelihood estimators.

1. INTRODUCTION

In this article we present sampling properties of new estimates for the reciprocals and ratios of means and regression coefficients and for coefficients of linear structural econometric models. See Chow (1960) and Kappenman, Geisser, and Antle (1970) for examples of reciprocal and ratio estimation. The estimates considered herein were derived in Zellner (1978) by minimizing the posterior expectation of specific generalized quadratic loss functions, and hence are termed minimum expected loss (MELO) estimates. Sampling properties of these and other estimates are of interest in connection with design problems and in characterizing properties of estimation procedures before data are taken, a consideration of importance to many, including subjective Bayesians—see, e.g., Hill (1975). In Zellner (1978) it was noted that maximum likelihood (ML) estimators for the problems mentioned above do not possess finite moments and have infinite risk relative to quadratic and many other loss functions, whereas MELO estimators for these problems have finite moments and finite risk. The present article reports additional sampling properties of MELO estimates and applications of MELO estimation to two well-known structural econometric models, Klein's Model I and the Girshick-Haavelmo food model.

* Arnold Zellner is H.G.B. Alexander Professor of Economics and Statistics, Graduate School of Business, University of Chicago, Chicago, IL 60637. Soo-Bin Park is Professor, Department of Economics, Carleton University, Ottawa, Ontario, Canada. Research was financed in part by National Science Foundation Grant GS 40033, and by income from the H.G.B. Alexander Endowment Fund, Graduate School of Business, University of Chicago. The second author received financial support from Carleton University and the Canada Council while visiting the University of Chicago on leave from Carleton University. This work was reported to the Tenth Meeting of the NBER-NSF Seminar on Bayesian Inference in Econometrics, held at the University of Chicago, May 9–10, 1975.

2. FINITE SAMPLE PROPERTIES OF MELO ESTIMATORS FOR RECIPROCALS

We assume that the observations y_i, $i = 1, 2, \ldots, n$, are independently drawn from a normal population with mean μ ($\mu \neq 0$) and variance σ^2. Our problem is to estimate $\theta = 1/\mu$. As is well-known, the ML estimator for θ,

$$\tilde{\theta} = 1/\bar{y} \,, \tag{2.1}$$

where $\bar{y} = \sum_{i=1}^{n} y_i/n$, does not possess finite moments and has infinite risk relative to quadratic and many other loss functions. As pointed out in Zellner (1978), a MELO estimate of θ relative to $L_r(\theta, \hat{\theta}) = [(\theta - \hat{\theta})/\theta]^2 = (1 - \mu\hat{\theta})^2$ is

$$\hat{\theta} = E_\mu \mu / E_\mu \mu^2 = \frac{1}{\bar{\mu}}\left(\frac{1}{1 + \operatorname{var}\mu/\bar{\mu}^2}\right), \tag{2.2}$$

where $E_\mu \equiv$ posterior expectation operator and $\bar{\mu}$ and $\operatorname{var}\mu$ are the posterior mean and variance of μ, respectively. If $\sigma^2 = \sigma_o^2$, a known value, and a diffuse prior pdf for μ is employed in the normal mean problem, then $\bar{\mu} = \bar{y}$, $\operatorname{var}\mu = \sigma_o^2/n$, and from (2.2),

$$\hat{\theta} = \frac{1}{\bar{y}}\,\frac{1}{1 + \sigma_o^2/(n\bar{y}^2)} = \frac{z}{c(1 + z^2)}\,, \tag{2.3}$$

where $c^2 = \sigma_o^2/n$ and $z = \bar{y}/c$ has a $N(\mu/c, 1)$ sampling pdf. Thus, the MELO estimator in (2.3) is equal to the ML estimator, $1/\bar{y}$, multiplied by a "shrinking factor" $z^2/(1 + z^2)$. Note that $z = n^{\frac{1}{2}}\bar{y}/\sigma_o$ is a sampling theory test statistic for testing $\mu = 0$; and also the reciprocal of the coefficient of variation of the posterior pdf for μ. As pointed out in Zellner (1978), $|\hat{\theta}| \leq \frac{1}{2}c$ and thus $\hat{\theta}$ possesses finite moments of all orders and finite risk relative to $L_r(\theta, \hat{\theta})$ and many other loss functions. Also, when $|z| < 1$, the posterior pdf for θ shows pronounced bimodality and special treatment of this problem within the MELO estimation framework is required—see the Appendix of Zellner (1978).

Reprinted from: © Journal of the American Statistical Association
March 1979, Volume 74, Number 365
Theory and Methods Section
Pages 185–193

By use of the δ-method (see, e.g., Jeffreys 1967, pp. 428–431), the approximate first and second moments of the MELO estimator $\hat{\theta}$ in (2.3) are

$$E\hat{\theta} \doteq \theta/(1 + 2a_o) \quad \text{and} \quad E\hat{\theta}^2 \doteq \theta^2(1 + a_o)/(1 + 2a_o)^2 , \qquad (2.4)$$

where $a_o = c^2\theta^2$, the squared coefficient of variation of $\bar{y}$. From (2.4), the approximate relative bias of $\hat{\theta}$ is $E\hat{\theta}/\theta \doteq 1/(1 + 2a_o)$, which approaches one as n grows. Further, in Zellner (1978) it is shown that $\sqrt{n}(\hat{\theta} - \theta)$ and $\sqrt{n}(\tilde{\theta} - \theta)$, where $\tilde{\theta} = 1/\bar{y}$, have the same asymptotic normal distribution. Also from (2.4), $\text{var}\hat{\theta} \doteq \theta^2 a_o/(1 + 2a_o)^2 \doteq \theta^2 a_o$.

Using the approximate moments in (2.4), the approximate risk function for $\hat{\theta}$ in (2.3) relative to $L_r(\theta, \hat{\theta})$ is

$$E(1 - \hat{\theta}/\theta)^2 \doteq 1 - (1 + 3a_o)/(1 + 2a_o)^2 . \qquad (2.5)$$

Shown in Table 1 are the approximate risk function (2.5)

1. Approximate and Exact Risk Functions for MELO Estimator $\hat{\theta}$ in (2.3) Relative to the Loss Function, $L = [(\theta - \hat{\theta})/\theta]^2$

	Expected loss		
Value of $\sqrt{n}\lvert\mu\rvert/\sigma_o$	(A) Exact[a]	(B) Approximate[b]	(C) Difference (A) − (B)
0	1.0000	1.0000	.0000
.2	.9789	.9708	.0081
.4	.9183	.8916	.0267
.6	.8256	.7828	.0428
.8	.7115	.6657	.0458
1.0	.5882	.5556	.0326
1.5	.3085	.3460	−.0375
2.0	.1408	.2222	−.0814
2.5	.0772	.1506	−.0734
3.0	.0641	.1074	−.0433
4.0	.0577	.0617	−.0040
5.0	.0435	.0398	.0037
6.0	.0291	.0277	.0014
7.0	.0217	.0204	.0013

[a] Evaluated by numerical integration.
[b] Evaluated by use of (2.5).

and the exact risk function, evaluated by application of Simpson's rule for numerical integration carried out in double precision (15+ significant digits on the Xerox Sigma-9 computer) with a tolerance of 0.0001 that provides results accurate to at least three significant figures. It is seen that the exact and approximate risk functions are not far different, are symmetric with a maximum value of one, and fall sharply toward zero as $\sqrt{n}|\mu|/\sigma_o$ grows in magnitude. Note that the risk of $\hat{\theta}$ in (2.2) relative to $L(\hat{\theta}, \theta) = (\theta - \hat{\theta})^2$ can be obtained from $E(\theta - \hat{\theta})^2 = \theta^2 EL_r(\theta, \hat{\theta})$ and is finite for $0 < \sqrt{n}|\mu|/\sigma_o < \infty$.

The sampling distribution of $\hat{\theta}$ in (2.3) is obtained from the known normal pdf for z by a change of variable. We have:

$$p_1(\hat{\theta}|\mu, c) = (1/(2\pi)^{\frac{1}{2}})\left(\frac{1 - g}{2c\hat{\theta}^2 g}\right)\exp\left\{-\frac{1}{2}\left(\frac{1 - g}{2c\hat{\theta}} - \frac{\mu}{c}\right)^2\right\}$$
$$\text{for} \quad |z| < 1 , \quad |\hat{\theta}| < \tfrac{1}{2}c ; \qquad (2.6a)$$

$$p_2(\hat{\theta}|\mu, c) = (1/(2\pi)^{\frac{1}{2}})\left(\frac{1 + g}{2c\hat{\theta}^2 g}\right)\exp\left\{-\frac{1}{2}\left(\frac{1 + g}{2c\hat{\theta}} - \frac{\mu}{c}\right)^2\right\}$$
$$\text{for} \quad |z| > 1 , \quad |\hat{\theta}| < \tfrac{1}{2}c ; \qquad (2.6b)$$

where $g \equiv (1 - 4c^2\hat{\theta}^2)^{\frac{1}{2}}$. Then the unconditional pdf for $\hat{\theta}$ is given by:

$$p(\hat{\theta}|\mu, c) = \Pr(|z| < 1)p_1(\hat{\theta}|\mu, c) + \Pr(|z| > 1)p_2(\hat{\theta}|\mu, c) . \qquad (2.7)$$

Plots of (2.7) for selected values of the parameters $\theta = 1/\mu$ and $c = \sigma_o/\sqrt{n}$ are shown in the figure along

Sampling Pdf's for MELO Estimator $\hat{\theta}$ in (2.3) and ML Estimator $\tilde{\theta}$ in (2.1) for $\theta = 1/\mu$

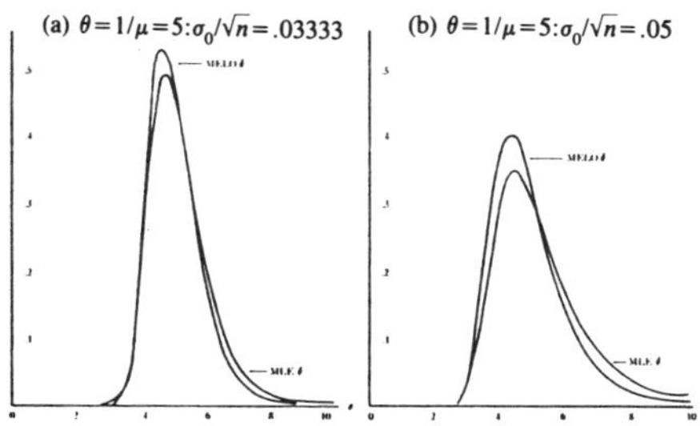

with plots of the pdf of the ML estimator $\tilde{\theta} = 1/\bar{y}$. The MELO estimator's pdf's have higher modes and thinner right-hand tails than the corresponding pdf's for the ML estimator. As noted above, these differences will disappear as the sample size grows. However, for the conditions portrayed in the figure, the MELO estimator's pdf's are more closely concentrated about the true value, $\theta = 5$, than are the corresponding pdf's for the ML estimator. In very small-sample situations, the ML estimator has a markedly bimodal pdf—see the Appendix of Zellner (1978) for some analysis of this problem, while the MELO estimator's pdf is U-shaped.

When the value of σ^2 is not known, the MELO estimate for θ, $\hat{\theta}^*$, relative to $L_r(\theta, \hat{\theta}^*) = [(\theta - \hat{\theta}^*)/\theta]^2$, can be obtained by using the marginal posterior pdf for μ to evaluate the moments in (2.2). When a diffuse prior pdf for μ and σ^2 is employed, the marginal posterior pdf for μ is in the univariate Student-t form with mean $\bar{\mu} = \bar{y}$ and $\text{var}\mu = \nu s^2/[n(\nu - 2)]$, with $\nu = n - 1 > 2$ and $\nu s^2 = \sum_{i=1}^{n} (y_i - \bar{y})^2$. Then, for these assumptions, the MELO estimator is:

$$\hat{\theta}^* = \frac{1}{\bar{y}}\frac{1}{1 + \nu s^2/[n(\nu - 2)\bar{y}^2]} , \quad \nu = n - 1 > 2 . \qquad (2.8)$$

In Zellner (1978), it is shown that the moments of $\hat{\theta}^*$ exist and are finite. The approximate first and second moments of $\hat{\theta}^*$ are

$$E\hat{\theta}^* \doteq \theta/(1 + qa) \quad \text{and} \quad E\hat{\theta}^{*2} \doteq \theta^2(1 + a)/(1 + qa)^2 , \qquad (2.9)$$

where $q = 1 + (n-1)/(n-3)$ and $a = \sigma^2/(n\mu^2)$. The approximate relative bias of $\hat{\theta}^*$ is $E\hat{\theta}^*/\theta \doteq 1/(1+qa)$, while the approximate variance is $\text{var}\hat{\theta}^* \doteq \theta^2 a/(1+qa)^2$. Using the results in (2.9), the approximate risk function for $\hat{\theta}^*$ relative to $L_r(\theta, \hat{\theta}^*)$ is

$$E[(\theta - \hat{\theta}^*)/\theta]^2 \doteq 1 - [1 + (2q-1)a]/(1+qa)^2 , \quad (2.10)$$

which is similar in form to (2.5). In Table 2, (2.10) has been evaluated for $n = 20$ and $n = 30$. On comparison with the approximate risk values of Table 1 for the MELO estimator $\hat{\theta}$, based on a known value of σ^2, it is seen that entries in Table 2 are slightly larger than the corresponding entries in Table 1.

2. Approximate Risk Functions for MELO Estimator $\hat{\theta}^*$ in (2.8) Relative to the Loss Function $L = [(\theta - \hat{\theta}^*)/\theta]^2$

Value of $\sqrt{n}\lvert\mu\rvert/\sigma$	Approximate expected loss[a]	
	$n = 20$	$n = 30$
0	1.0000	1.0000
.2	.9719	.9715
.4	.8953	.8939
.6	.7892	.7868
.8	.6739	.6709
1.0	.5643	.5610
1.5	.3530	.3504
2.0	.2267	.2250
2.5	.1533	.1523
3.0	.1091	.1085
4.0	.0624	.0622
5.0	.0401	.0400
6.0	.0279	.0278
7.0	.0205	.0204

[a] Calculated from (2.10) in text.

As regards finite sample pdf's, it is interesting to observe from (2.8) that

$$w \equiv \bar{\theta}/\hat{\theta}^* - 1 = (n-1)/[(n-3)x] ,$$

where $x = n\bar{y}^2/s^2$ has a noncentral F distribution, $F'(1, n-1, n\mu^2/\sigma^2)$—see Kendall and Stuart (1973). Using this fact, the pdf for w is given by:

$$p(w \mid n, \mu/\sigma) = \frac{\exp\{-(n/2)(\mu/\sigma)^2\}}{(n-3)^{\frac{1}{2}}} \sum_{r=0}^{\infty} \cdot \frac{(\mu/\sigma)^{2r}\left[\dfrac{n}{2(n-3)}\right]^r}{r!B\left(r+\dfrac{1}{2}, \dfrac{n-1}{2}\right)} \frac{(1/w)^{r+\frac{1}{2}}}{[1 + (1/w)/(n-3)]^{r+n/2}} \quad (2.11)$$

for $0 < w < \infty$ and $n > 3$. The pdf for $\bar{\theta}/\hat{\theta}^*$, $1 < \bar{\theta}/\hat{\theta}^* < \infty$, the ratio of the ML estimator $\bar{\theta} = 1/\bar{y}$ to the MELO estimator $\hat{\theta}^*$ shown in (2.8), can be obtained directly from (2.11).

With respect to the finite sample pdf for $\hat{\theta}^*$ in (2.8), the conditional pdf for $\hat{\theta}^*$ given $\bar{c}^2 = \nu s^2/[n(\nu-2)]$, $\nu > 2$, is in precisely the same form as (2.7) with $\bar{c}^2$ replacing c^2. To obtain the marginal pdf for $\hat{\theta}^*$, $p(\hat{\theta}^* \mid \mu, c)$, the following integration must be performed numerically:

$$p(\hat{\theta}^* \mid \mu, c) = \int_0^{\infty} p(\hat{\theta}^* \mid \bar{c}, \mu, c) f(\bar{c} \mid \sigma, n) d\bar{c} ,$$

where $f(\bar{c} \mid \sigma, n)$ is the pdf for $\bar{c} = \nu s^2/(\nu - 2)$ which is readily obtained from the χ^2 pdf with $n-1$ degrees of freedom for $\nu s^2/\sigma^2$.

We now consider MELO and ML estimators for the reciprocal of a regression coefficient, a problem very similar to estimating the reciprocal of a population mean Let our simple regression model be $y_i = \pi x_i + v_i$ $i = 1, 2, \ldots, n$, where the v_i's are independently drawn from a normal population with zero mean and variance σ_v^2. The ML estimator of $\gamma = 1/\pi$ is $\tilde{\gamma} = 1/\hat{\pi}$, where $\hat{\pi} = \sum_{i=1}^n x_i y_i/m_{xx}$, with $m_{xx} = \sum_{i=1}^n x_i^2$. As is well-known, $\tilde{\gamma} = 1/\hat{\pi}$ does not possess finite moments; see, e.g., Bergstrom (1962). On the other hand, a MELO estimator, $\hat{\gamma}$, for $\gamma = 1/\pi$, derived in Zellner (1978), relative to $L_r(\gamma, \hat{\gamma}) = [(\gamma - \hat{\gamma})/\gamma]^2$ and a diffuse prior for π and σ_v^2, is

$$\hat{\gamma} = \frac{1}{\hat{\pi}} \frac{1}{1 + \nu s^2/[m_{xx}(\nu-2)\hat{\pi}^2]}, \quad \nu = n - 1 > 2 , \quad (2.12)$$

where $\nu s^2 = \sum_{i=1}^n (y_i - x_i\hat{\pi})^2$ and $\nu = n - 1$. It is seen that (2.12) is in the form of (2.8). Approximate moments and risk for $\hat{\gamma}$ can be evaluated in precisely the same manner as for (2.8).

Similarly, if we have a usual normal linear multiple regression model, with $k \times 1$ ML coefficient estimator $\hat{\boldsymbol{\beta}} = (X'X)^{-1}X'\mathbf{y}$ for $\boldsymbol{\beta}$, and wish to estimate, say, $\gamma_1 = 1/\beta_1$, where β_1 is the first element of $\boldsymbol{\beta}$, the ML estimator is $\tilde{\gamma}_1 = 1/\hat{\beta}_1$, which does not possess moments, whereas a MELO estimator for γ_1 based on a relative squared error loss function and a diffuse prior pdf for the parameters is

$$\hat{\gamma}_1 = \frac{1}{\hat{\beta}_1} \frac{1}{1 + \nu s^2 m^{11}/[(\nu-2)\hat{\beta}_1^2]}, \quad \nu = n - k > 2 , \quad (2.13)$$

where $\nu s^2 = (\mathbf{y} - X\hat{\boldsymbol{\beta}})'(\mathbf{y} - X\hat{\boldsymbol{\beta}})$, $\nu = n - k$ and m^{11} is the (1, 1) element of $(X'X)^{-1}$. Expression (2.13) is in precisely the form of (2.8), and its approximate moments and risk can be evaluated in the manner shown previously. Also, the distributional properties of (2.12) and (2.13) are very similar to those of (2.8).

While we have employed diffuse priors above to generate MELO estimates of $\theta = 1/\mu$, it should be appreciated that the posterior moments of μ derived from a marginal posterior pdf for μ, based on an informative prior pdf, can be substituted in (2.2) to yield a MELO estimate for θ.

3. PROPERTIES OF MELO ESTIMATORS FOR RATIOS OF PARAMETERS

If the parameter ϕ is equal to the ratio of two other parameters, $\phi = \mu_1/\mu_2$, then $\mu_2\phi - \mu_1 = 0$. Let $\epsilon = \mu_2\hat{\phi}$

$- \mu_1$, where $\hat{\phi}$ is an estimate of ϕ, and assume that the loss function is $L = \epsilon^2 = (\mu_2\hat{\phi} - \mu_1)^2 = \mu_2^2(\phi - \hat{\phi})^2$, a generalized quadratic loss function. Given a posterior pdf for $\boldsymbol{\mu}' = (\mu_1, \mu_2)$ that possesses finite second moments, the posterior expectation of L is $E_{\boldsymbol{\mu}}L = \hat{\phi}^2 E_{\boldsymbol{\mu}}\mu_2^2 - 2\hat{\phi}E_{\boldsymbol{\mu}}\mu_1\mu_2 + E_{\boldsymbol{\mu}}\mu_1^2$. Then the MELO estimate of ϕ, given in Zellner (1978), is

$$\hat{\phi} = E_{\boldsymbol{\mu}}\mu_1\mu_2/E_{\boldsymbol{\mu}}\mu_2^2 = \frac{\bar{\mu}_1}{\bar{\mu}_2}\,\frac{1 + \operatorname{cov}(\mu_1, \mu_2)/(\bar{\mu}_1\bar{\mu}_2)}{1 + \operatorname{var}\mu_2/\bar{\mu}_2^2}, \quad (3.1)$$

where $\bar{\mu}_1$ and $\bar{\mu}_2$ are posterior means of μ_1 and μ_2, respectively, $\operatorname{cov}(\mu_1, \mu_2)$ is the posterior covariance of μ_1 and μ_2, and $\operatorname{var}\mu_2$ is the posterior variance of μ_2. The MELO estimate $\hat{\phi}$ in (3.1) is applicable to a broad range of ratio estimation problems. Below, (3.1) is evaluated for several problems, and selected sampling properties of particular MELO ratio estimates are derived. Zellner (1978) shows that these MELO estimators have finite moments and asymptotic normal distributions identical to those of ML estimators for these problems.

Assume that the pairs of observations (y_{1i}, y_{2i}), $i = 1, 2, \ldots, n$, are independently drawn from a bivariate normal distribution with mean vector $\boldsymbol{\mu}' = (\mu_1, \mu_2)$ and diagonal covariance matrix with elements σ_1^2 and σ_2^2 and that we wish to estimate $\phi = \mu_1/\mu_2$. When σ_2^2 has a known value and the prior pdf for $\boldsymbol{\mu}$ and σ_1 is diffuse; that is, $p(\boldsymbol{\mu}, \sigma_1) \propto 1/\sigma_1$, the posterior moments appearing in (3.1) are readily available and yield the following MELO estimator:

$$\hat{\phi}_1 = \frac{\bar{y}_1}{\bar{y}_2}\,\frac{1}{1 + \sigma_2^2/(n\bar{y}_2^2)}. \quad (3.2)$$

Note that for this problem, $\operatorname{cov}(\mu_1, \mu_2) = 0$. In (3.2), $\hat{\phi}_1$ is equal to the ML estimator $\bar{y}_1/\bar{y}_2$, which possesses no finite moments, times a "shrinking factor" that assumes values between zero and one and is a function of $\sqrt{n}\bar{y}_2/\sigma_2$, the reciprocal of the coefficient of variation of the marginal posterior pdf for μ_2, and a sampling theory test statistic for testing $\mu_2 = 0$.

The first two approximate moments of $\hat{\phi}$ in (3.2) are $E\hat{\phi}_1 \doteq \phi/(1 + 2a_2)$ and

$$E\hat{\phi}_1^2 \doteq \phi(1 + a_1)(1 + a_2)/(1 + 2a_2)^2, \quad (3.3)$$

where $a_i = \sigma_i^2/(n\mu_i^2)$, $i = 1, 2$. Also,

$$\operatorname{var}\hat{\phi}_1 \doteq \phi^2(a_1 + a_2 + a_1a_2)/(1 + 2a_2)^2$$

and the approximate relative bias of $\hat{\phi}_1$ is $E\hat{\phi}_1/\phi = 1/(1 + 2a_2)$. Using the results in (3.3), the approximate risk of $\hat{\phi}_1$ relative to the loss function $L = \mu_2^2(\phi - \hat{\phi}_1)^2$ is

$$EL \doteq \mu_2^2\phi^2\left[1 - \frac{2(1 + 2a_2) - (1 + a_1)(1 + a_2)}{(1 + 2a_2)^2}\right]. \quad (3.4)$$

In Table 3 the approximate risk computed from (3.4) is compared with the exact risk calculated by numerical integration as explained above. Maximum risk is experienced as $|\mu_2| \to 0$, and risk falls rapidly as $\sqrt{n}|\mu_2|/\sigma_2$ grows in value. Also, calculated approximate and exact risk values are not very different.

3. Exact and Approximate Risk Functions for MELO Estimator $\hat{\theta}_1$ in (3.2) Relative to Loss Function, $L = \mu_2^2(\phi - \hat{\phi}_1)^2$

Values of $\sqrt{n}\lvert\mu_2\rvert/\sigma_2$	Expected loss (A) Exact[a]	(B) Approximate[b]	(C) Difference (A) − (B)
0	1.0000	1.0000	0.0000
.2	.9792	.9712	0.0080
.4	.9193	.8932	0.0261
6	.8279	.7863	0.0416
.8	.7157	.6718	0.0439
1.0	.5948	.5644	0.0304
1.5	.3232	.3622	−0.0390
2.0	.1649	.2444	−0.0795
2.5	.1092	.1772	−0.0680
3.0	.1014	.1371	−0.0357
4.0	.0987	.0953	0.0034
5.0	.0848	.0755	0.0093
6.0	.0701	.0646	0.0055
7.0	.0626	.0581	0.0045

[a] Computed by numerical integration.
[b] Computed from (3.4) in text.
Note: $\mu_1 = 1.0$ and $\sigma_1/\sqrt{n} = .2$ are employed in these calculations.

The sampling pdf for $\hat{\phi}_1$ in (3.2) can be obtained using results for the pdf of the MELO reciprocal estimator (2.3) since $\hat{\phi}_1 = \bar{y}_1[\bar{y}_2/(\bar{y}_2^2 + \sigma_2^2/n)]$. Note that $\bar{y}_1$ is independent of the factor in square brackets that is in precisely the form of (2.3) and has a pdf in the form of (2.7). Also, from (3.2), the pdf of $\hat{\phi}_1/\tilde{\phi}$, where $\tilde{\phi} = \bar{y}_1/\bar{y}_2$, the ML estimator, can be obtained from that of the noncentral χ^2 pdf for $n\bar{y}_2^2/\sigma_2^2$.

When σ_2^2 has an unknown value, the MELO estimator for $\phi = \mu_1/\mu_2$ relative to the loss function $L = \mu_2^2(\phi - \hat{\phi}_2)^2$ and the diffuse prior pdf $p(\boldsymbol{\mu}, \sigma_1, \sigma_2) \propto 1/\sigma_1\sigma_2$ is obtained from (3.1) by inserting moments of the marginal posterior pdf for $\boldsymbol{\mu}' = (\mu_1, \mu_2)$ that is in the form of a product of univariate Student-t pdf's for μ_1 and μ_2. This results in the following MELO estimator for ϕ:

$$\hat{\phi}_2 = \frac{\bar{y}_1}{\bar{y}_2}\,\frac{1}{1 + \nu s_2^2/[(\nu - 2)n\bar{y}_2^2]}, \quad \nu = n - 1 > 2, \quad (3.5)$$

where $\nu s_2^2 = \sum_{i=1}^{n}(y_{2i} - \bar{y}_2)^2$. The approximate first and second moments of $\hat{\phi}_2$ are:

$$E\hat{\phi}_2 \doteq \phi/(1 + qa_2) \quad \text{and}$$

$$E\hat{\phi}_2^2 \doteq \phi^2(1 + a_1)(1 + a_2)/(1 + qa_2)^2, \quad (3.6)$$

where a_1 and a_2 have been defined above and $q = 1 + (n - 1)/(n - 3)$. Further,

$$\operatorname{var}\hat{\phi}_2 \doteq \phi^2(a_1 + a_2 + a_1a_2)/(1 + qa_2)^2$$

and the approximate risk of $\hat{\phi}_2$ in (3.5) relative to the loss function $L = \mu_2^2(\phi - \hat{\phi}_2)^2$ is:

$$EL \doteq \mu_2^2\phi^2\left[1 - \frac{2(1 + qa_2) - (1 + a_1)(1 + a_2)}{(1 + qa_2)^2}\right]. \quad (3.7)$$

In Table 4, (3.7) is evaluated for several sets of parameter

4. Approximate Risk Functions for Estimator $\hat{\phi}_2$ in (3.5) Relative to Loss Function, $L = \mu_2^2(\phi - \hat{\phi}_2)^2$

Values of $\sqrt{n}\lvert\mu_2\rvert/\sigma_2$	Approximate risk[a]	
	$\sqrt{n}\mu_1/\sigma_1 = 4$	$\sqrt{n}\mu_1/\sigma_1 = 5$
0	1.0000	1.0000
.2	9721	.9718
.4	8963	.8954
.6	7920	.7901
.8	6798	.6766
1.0	5743	.5695
1.5	3749	.3661
2.0	2589	.2467
2.5	1932	.1784
3.0	1543	.1378
4.0	1142	.0955
5.0	0954	.0754
6.0	0852	.0646
7.0	.0791	.0580

[a] Computed from (3.7) in text with $n = 30$.

values. Maximum risk is encountered as $|\mu_2| \to 0$. As $\sqrt{n}|\mu_2|/\sigma_2$ increases in value, approximate risk declines rather sharply. On comparing values of the approximate risk in Table 4 with comparable values in Table 3, it is seen that those in Table 4 are slightly larger.

If the pairs of observations (y_{1i}, y_{2i}), $i = 1, 2, \ldots, n$, are normally and independently distributed with mean vector $\boldsymbol{\mu}' = (\mu_1, \mu_2)$ and common positive definite symmetric (pds) covariance matrix having known elements σ_1^2, σ_2^2, and σ_{12}, the MELO estimator for $\phi = \mu_1/\mu_2$ relative to the loss function $L = \mu_2^2(\phi - \hat{\phi}_3)^2$ and a diffuse prior pdf for $\boldsymbol{\mu}$, obtained from (3.1) by inserting moments of the bivariate normal posterior pdf for $\boldsymbol{\mu}$, is:

$$\hat{\phi}_3 = \frac{\bar{y}_1}{\bar{y}_2}\,\frac{1 + \sigma_{12}/(n\bar{y}_1\bar{y}_2)}{1 + \sigma_2^2/(n\bar{y}_2^2)} = \frac{\bar{y}_1\bar{y}_2 + \sigma_{12}/n}{\bar{y}_2^2 + \sigma_2^2/n}\,. \qquad (3.8)$$

The estimator $\hat{\phi}_3$ differs from $\hat{\phi}_1$ in (3.2) in that $\sigma_{12}/(n\bar{y}_1\bar{y}_2)$ appears in (3.8) but not in (3.2). Also, $\hat{\phi}_3$ is the product of the ML estimator, $\bar{y}_1/\bar{y}_2$, times a factor that can act to shrink, expand, or even change the algebraic sign of $\bar{y}_1/\bar{y}_2$. The approximate first two moments of $\hat{\phi}_3$ are

$$E\hat{\phi}_3 \doteq \phi(1 + 2a_{12})/(1 + 2a_2)$$

and

$$E\hat{\phi}_3^2 \doteq \phi^2[(1 + a_1)(1 + a_2) + a_{12}(6 + 5a_{12})]/(1 + 2a_2)^2\,, \qquad (3.9)$$

where $a_i = \sigma_i^2/(n\mu_i)^2$, $i = 1, 2$, and $a_{12} = \sigma_{12}/(n\mu_1\mu_2)$. Except for the appearance of terms involving a_{12}, (3.9) is exactly the same as (3.3). Also, $\mathrm{var}\hat{\phi}_3 \doteq \phi^2[a_1 + a_2 + a_1a_2 + a_{12}(2 + a_{12})]/(1 + 2a_2)^2$ and the approximate risk of $\hat{\phi}_3$ relative to $L = \mu_2^2(\phi - \hat{\phi}_3)^2$ is

$$EL \doteq \mu_2^2\phi^2 \cdot\left[1 - \frac{2(1 + 2a_2)(1 + 2a_{12}) - (1 + a_1)(1 + a_2) - a_{12}(6 + 5a_{12})}{(1 + 2a_2)^2}\right], \qquad (3.10)$$

which is similar to (3.4) except that (3.10) involves a_{12}.

In (3.8), $\hat{\phi}_3$ is based on assumed known values for σ_1^2, σ_2^2, and σ_{12}. If these parameters have unknown values and if a diffuse prior pdf for them and $\boldsymbol{\mu}$ is employed, then it is known that the marginal posterior pdf for $\boldsymbol{\mu}$ is in the bivariate Student-t form with known posterior moments that can be substituted in (3.1) to yield the following MELO estimate for $\phi = \mu_1/\mu_2$:

$$\hat{\phi}_4 = \frac{\bar{y}_1}{\bar{y}_2}\,\frac{1 + \bar{s}_{12}/(n\bar{y}_1\bar{y}_2)}{1 + \bar{s}_2^2/(n\bar{y}_2^2)} = \frac{\bar{y}_1\bar{y}_2 + \bar{s}_{12}/n}{\bar{y}_2^2 + \bar{s}_2^2/n}\,, \quad n > 4\,, \qquad (3.11)$$

where

$$\bar{s}_2^2 = \sum_{i=1}^{n} (y_{2i} - \bar{y}_2)^2/(n - 4)$$

and

$$\bar{s}_{12} = \sum_{i=1}^{n} (y_{1i} - \bar{y}_1)(y_{2i} - \bar{y}_2)/(n - 4)\,.$$

The approximate first two moments of $\hat{\phi}_4$ are

$$E\hat{\phi}_4 \doteq \phi(1 + ba_{12})/(1 + ba_2) \qquad (3.12a)$$

and

$$E\hat{\phi}_4^2 \doteq \phi^2\{(1 + a_1)(1 + a_2) + a_{12}[2(1 + b) + a_{12}(1 + b^2)] + n(b - 1)^2(a_1a_2 + a_{12}^2)/(n - 1)^2\}/(1 + ba_2)^2\,, \qquad (3.12b)$$

where $b = 1 + (n - 1)/(n - 4)$. The approximate risk of $\hat{\phi}_4$ relative to the loss function $L = \mu_2^2(\phi - \hat{\phi}_4)^2$, is:

$$EL \doteq \mu_2^2\phi^2\left[1 - \frac{1}{(1 + ba_2)^2}\{2(1 + ba_2)(1 + ba_{12}) - (1 + a_1)(1 + a_2) - a_{12}[2(1 + b) + a_{12}(1 + b^2)] - n(b - 1)^2(a_1a_2 + a_{12}^2)/(n - 1)^2\}\right]\cdot \qquad (3.13)$$

A slight variant of the problem of estimating the ratio of two population means is the problem of estimating the ratio of two multiple regression coefficients, $\alpha = \beta_1/\beta_2$. It is assumed that β_1 and β_2 are elements of a $k \times 1$ regression coefficient vector $\boldsymbol{\beta}$ in $\mathbf{y} = X\boldsymbol{\beta} + \mathbf{u}$, where $\mathbf{y}$ is an $n \times 1$ vector, X is a given $n \times k$ matrix of rank k, and $\mathbf{u}$ is an $n \times 1$ error vector assumed drawn from a multivariate normal distribution with zero mean vector and covariance matrix $\sigma^2 I_n$. With a diffuse prior pdf for $\boldsymbol{\beta}$ and σ, $p(\boldsymbol{\beta}, \sigma) \propto 1/\sigma$, it is well-known that the marginal posterior pdf for $\boldsymbol{\beta}$ is in the multivariate Student-t form with posterior mean $\hat{\boldsymbol{\beta}} = (X'X)^{-1}X'\mathbf{y}$ and posterior covariance matrix $(X'X)^{-1}\nu s^2/(\nu - 2)$, where

$$\nu s^2 = (\mathbf{y} - X\hat{\boldsymbol{\beta}})'(\mathbf{y} - X\hat{\boldsymbol{\beta}})$$

for $\nu = n - k > 2$. Then the MELO estimate of $\alpha = \beta_1/\beta_2$ relative to the loss function $L = \beta_2^2(\alpha - \hat{\alpha})^2$ is:

$$\hat{\alpha} = \frac{\hat{\beta}_1}{\hat{\beta}_2}\,\frac{1 + m^{12}\nu s^2/[(\nu - 2)\hat{\beta}_1\hat{\beta}_2]}{1 + m^{22}\nu s^2/[(\nu - 2)\hat{\beta}_2^2]}\,, \quad \nu = n - k > 2\,, \qquad (3.14)$$

where $\hat{\beta}_1$ and $\hat{\beta}_2$ are the posterior means of β_1 and β_2, respectively, and m^{ij} is the (i, j)th element of $(X'X)^{-1}$. $\hat{\beta}_1/\hat{\beta}_2$ is the ML estimator for α that possesses no finite moments, whereas it is shown in Zellner (1978) that $\hat{\alpha}$

possesses finite moments. Furthermore, $\hat{\alpha}$ and the ML estimator have the same asymptotic normal distribution. Approximate moments and the approximate risk of $\hat{\alpha}$ can be evaluated and will be similar in form to those associated with the estimator $\hat{\phi}_4$ in (3.11).

Finally, as noted at the end of Section 2, MELO estimates for ratios of parameters can obviously be generated not only when diffuse prior pdf's are employed, but also when informative prior pdf's are utilized.

4. MELO ESTIMATION OF A STRUCTURAL EQUATION'S COEFFICIENTS

Consider an equation, say the first, of a linear structural econometric model:

$$\underset{n\times 1}{\mathbf{y}_1} = \underset{n\times m_1}{Y_1}\ \underset{m_1\times 1}{\boldsymbol{\gamma}_1} + \underset{n\times k_1}{X_1}\ \underset{k_1\times 1}{\boldsymbol{\beta}_1} + \underset{n\times 1}{\mathbf{u}_1}, \tag{4.1}$$

where $(\mathbf{y}_1\ \ Y_1)$ is an $n \times m$ matrix of observations on $m = m_1 + 1$ endogenous variables, X_1 is an $n \times k_1$ matrix of observations on k_1 predetermined variables, $\boldsymbol{\gamma}_1$ and $\boldsymbol{\beta}_1$ are $m_1 \times 1$ and $k_1 \times 1$ vectors of structural coefficients, respectively, and $\mathbf{u}_1$ is an $n \times 1$ disturbance vector. It is assumed that the parameters of (4.1) are identified by usual zero restrictions that imply that certain variables appearing in the system of which (4.1) is the first equation do not appear in (4.1). Specifically, the variables appearing in the entire system are $(\mathbf{y}_1\ \ Y_1\ \ Y_o)$ and $X = (X_1\ \ X_o)$, and Y_o and X_o are assumed not to appear in (4.1), where X is of dimensionality $n \times k$ and rank k, X_o is $n \times k_o$, $k = k_1 + k_o$, and Y_o is $n \times m_o$.

The reduced form equations for the endogenous variables $(\mathbf{y}_1\ \ Y_1)$ appearing in (4.1) are:

$$(\mathbf{y}_1\ \ Y_1) = X(\boldsymbol{\pi}_1\ \ \Pi_1) + (\mathbf{v}_1\ \ V_1) \tag{4.2a}$$

or

$$\underset{n\times m}{Y} = \underset{n\times k}{X}\ \underset{k\times m}{\Pi} + \underset{n\times m}{V}, \tag{4.2b}$$

where $Y \equiv (\mathbf{y}_1\ \ Y_1)$, $\Pi \equiv (\boldsymbol{\pi}_1\ \ \Pi_1)$, and $V \equiv (\mathbf{v}_1\ \ V_1)$, with V a matrix of reduced form disturbance terms. Multiplying (4.2a) on the right by $(1\ -\boldsymbol{\gamma}_1')'$ and comparing the result with (4.1) leads to the following well-known restrictions:

$$\boldsymbol{\pi}_1 = \Pi_1\boldsymbol{\gamma}_1 + \begin{pmatrix}\boldsymbol{\beta}_1\\ \mathbf{0}\end{pmatrix}, \tag{4.3}$$

where the zero vector on the right side of (4.3) is a $(k - k_1) \times 1$ or $k_o \times 1$ vector with $k_o \geq m_1$, the necessary order condition for identification of the structural coefficients. Multiply both sides of (4.3) on the left by $X = (X_1\ \ X_o)$ to obtain:

$$\begin{aligned} X\boldsymbol{\pi}_1 &= X\Pi_1\boldsymbol{\gamma}_1 + X_1\boldsymbol{\beta}_1 \\ &= \bar{Z}_1\boldsymbol{\delta}_1 , \end{aligned} \tag{4.4}$$

where $\bar{Z}_1 \equiv (X\Pi_1\ \ X_1)$ and $\boldsymbol{\delta}_1' = (\boldsymbol{\gamma}_1'\ \ \boldsymbol{\beta}_1')$. Note that the $n \times (m_1 + k_1)$ matrix $\bar{Z}_1$ has full column rank given that Π_1, a $k \times m_1$ matrix, has rank m_1, the rank condition for identification of the structural coefficients.

The loss function that was used in Zellner (1978) to obtain a MELO estimate of $\boldsymbol{\delta}_1$ is obtained as follows. Let $\mathbf{d}_1$ be any estimate of $\boldsymbol{\delta}_1$ and write $\boldsymbol{\varepsilon} = X\boldsymbol{\pi}_1 - \bar{Z}_1\mathbf{d}_1$, where the $n \times 1$ vector $\boldsymbol{\varepsilon}$ measures the extent to which the restrictions in (4.4) fail to be satisfied. Then the loss function, L, is taken to be $L = \boldsymbol{\varepsilon}'\boldsymbol{\varepsilon}$, or

$$\begin{aligned} L &= (X\boldsymbol{\pi}_1 - \bar{Z}_1\mathbf{d}_1)'(X\boldsymbol{\pi}_1 - \bar{Z}_1\mathbf{d}_1) \\ &= (\boldsymbol{\delta}_1 - \mathbf{d}_1)'\bar{Z}_1'\bar{Z}_1(\boldsymbol{\delta}_1 - \mathbf{d}_1) , \end{aligned} \tag{4.5}$$

where in going from the first to the second line of (4.5), $X\boldsymbol{\pi}_1 = \bar{Z}_1\boldsymbol{\delta}_1$ from (4.4) has been employed. We see that (4.5) is a quadratic loss function in $\boldsymbol{\delta}_1 - \mathbf{d}_1$ with the matrix of the quadratic form being $\bar{Z}_1'\bar{Z}_1$.

If we have a posterior pdf for $\Pi = (\boldsymbol{\pi}_1\ \ \Pi_1)$, the reduced form coefficient matrix shown in (4.2) that possesses finite first and second moments, we can use this posterior pdf to evaluate the posterior expectation of the loss function in (4.5) as follows:

$$EL = E\boldsymbol{\pi}_1'X'X\boldsymbol{\pi}_1 - 2\mathbf{d}_1'E\bar{Z}_1'X\boldsymbol{\pi}_1 + \mathbf{d}_1'E\bar{Z}_1'\bar{Z}_1\mathbf{d}_1 , \tag{4.6}$$

where E denotes the posterior expectation operator. On minimizing (4.6) with respect to the vector $\mathbf{d}_1$, it is found that

$$\mathbf{d}_1^* = (E\bar{Z}_1'\bar{Z}_1)^{-1}E\bar{Z}_1'X\boldsymbol{\pi}_1 \tag{4.7}$$

is the MELO estimate.

To evaluate (4.7), we must have a specific posterior pdf for Π. If we assume that (4.2b) is a usual normal reduced form system, with the rows of V normally and independently distributed, each with zero mean vector and $m \times m$ pds covariance matrix Ω, and if we employ a diffuse prior pdf, $p(\Pi, \Omega) \propto |\Omega|^{-(m+1)/2}$, it is well-known that the marginal posterior pdf is in the following matrix Student-t form—see, e.g., Zellner (1971, p. 229):

$$p(\Pi | Y) \propto |S + (\Pi - \hat{\Pi})'X'X(\Pi - \hat{\Pi})|^{-n/2} , \tag{4.8}$$

where $\hat{\Pi} = (X'X)^{-1}X'Y$ and $S = (Y - X\hat{\Pi})'(Y - X\hat{\Pi})$. Zellner (1978) shows that when the posterior pdf in (4.8) is used to evaluate the posterior expectations in (4.7), the result is

$$\begin{aligned} \mathbf{d}_1^* &= \begin{pmatrix}\tilde{\boldsymbol{\gamma}}_1\\ \tilde{\boldsymbol{\beta}}_1\end{pmatrix} \\ &= \begin{pmatrix}\hat{Y}_1'\hat{Y}_1 + k\bar{S}_{22} & \hat{Y}_1'X_1\\ X_1'\hat{Y}_1 & X_1'X_1\end{pmatrix}^{-1}\begin{pmatrix}\hat{Y}_1'\hat{\mathbf{y}}_1 + k\bar{\mathbf{s}}_{12}\\ X_1'\hat{\mathbf{y}}_1\end{pmatrix} \end{aligned} \tag{4.9}$$

where $\hat{Y}_1 = X(X'X)^{-1}X'Y_1$, $\hat{\mathbf{y}}_1 = X(X'X)^{-1}X'\mathbf{y}_1$, and

$$\begin{aligned} \bar{S} &= (Y - X\hat{\Pi})'(Y - X\hat{\Pi})/(\nu - 2) \\ &= \begin{pmatrix}\bar{s}_{11} & \bar{\mathbf{s}}_{12}'\\ \bar{\mathbf{s}}_{21} & \bar{S}_{22}\end{pmatrix}, \end{aligned} \tag{4.10}$$

with $\nu = n - k - m_1 > 2$, a condition required for the existence of the second moments of the posterior pdf in (4.8).

As noted in Zellner (1978), the MELO estimator in (4.9) is in the form of a k-class estimator with a value for k, k*, given by:

$$k^* = 1 - \frac{k}{\nu - 2} = 1 - \frac{k}{n - k - m_1 - 2}. \qquad (4.11)$$

Since $\nu > 2$, for finite n, $k^* < 1$.[1] As $n \to \infty$, $k^* \to 1$ and thus $\mathbf{d}_1{}^*$ in (4.9) is a consistent estimator. Furthermore, $\sqrt{n}(k^* - 1) \to 0$ as $n \to \infty$ and thus from known properties of k-class estimators, $\sqrt{n}(\mathbf{d}_1{}^* - \boldsymbol{\delta}_1)$ approaches a limiting normal distribution and is asymptotically efficient.

Hatanaka's (1973) results on the existence of the moments of k-class estimators show that the second sampling moments of $\mathbf{d}_1{}^*$ in (4.9) exist and are finite. Thus the risk of $\mathbf{d}_1{}^*$ relative to the loss function in (4.5) exists and is finite in contrast to the infinite risk of the ML estimator. Further, Kadane's (1971) approximate moment expressions for k-class estimators yield:

$$E\mathbf{d}_1{}^* = \boldsymbol{\delta}_1 + \sigma^2 r_{k^*} Q\mathbf{q} + O(\sigma^3) \qquad (4.12)$$

and

$$\begin{aligned} E(\mathbf{d}_1{}^* - \boldsymbol{\delta}_1)(\mathbf{d}_1{}^* - \boldsymbol{\delta}_1)' &= \sigma^2 Q + \sigma^4[\{(1 - 2r_{k^*})\,\mathrm{tr}(C_1Q) \\ &\quad + \mathrm{tr}(C_2Q)\}Q + \{(r_{k^*} - 1)^2 + 2s_{k^*} + 2\}QC_1Q \\ &\quad + (s_{k^*} - r_{k^*} + 1)QC_2Q] + O(\sigma^6), \qquad (4.13) \end{aligned}$$

where k* is given in (4.11), $r_{k^*} \equiv (1 - k^*)n + k^*k - m_1 - k_1 - 1$, $s_{k^*} \equiv k^*(k^* - 1)(n - k)$, $Q \equiv (\bar{Z}_1{}'\bar{Z}_1)^{-1}$, σ^2 is the common (small) variance of the elements of $\mathbf{u}_1$ in (4.1), and $\mathbf{q}$, $C_1 = \mathbf{q}\mathbf{q}'$ and the positive semidefinite matrix C_2 are defined in Kadane (1971, p. 726). Using (4.13), the risk function of $\mathbf{d}_1{}^*$, $E(\mathbf{d}_1{}^* - \boldsymbol{\delta}_1)'Q^{-1}(\mathbf{d}_1{}^* - \boldsymbol{\delta}_1) = \mathrm{tr}E(\mathbf{d}_1{}^* - \boldsymbol{\delta}_1)(\mathbf{d}_1{}^* - \boldsymbol{\delta}_1)'Q^{-1}$, is given by:

$$\begin{aligned} E(\mathbf{d}_1{}^* - \boldsymbol{\delta}_1)'Q^{-1}(\mathbf{d}_1{}^* - \boldsymbol{\delta}_1) &= \mathrm{tr}E(\mathbf{d}_1{}^* - \boldsymbol{\delta}_1)(\mathbf{d}_1{}^* - \boldsymbol{\delta}_1)'Q^{-1} \\ &= \sigma^2(m_1 + k_1) + \sigma^4[\{(1 - 2r_{k^*})(m_1 + k_1) \\ &\quad + (r_{k^*} - 1)^2 + 2s_{k^*} + 2\}\,\mathrm{tr}QC_1 + \{(m_1 + k_1) \\ &\quad + s_{k^*} - r_{k^*} + 1\}\,\mathrm{tr}QC_2] + O(\sigma^6), \qquad (4.14) \end{aligned}$$

where k* is given in (4.11) and the other quantities in (4.14) have been defined above. The leading term in (4.14), $\sigma^2(m_1 + k_1)$, involves $m_1 + k_1$, the number of estimated structural coefficients.

Using (4.14) and a similar expression for the risk of the two-stage least squares estimator (2SLS), when its moments exist, we have:

$$\begin{aligned} EL_{2SLS} - EL_{MELO} &\doteq \sigma^4(1 - k^*)(n - k)[\{4(m_1 + k_1) \\ &\quad + 6 - k(2 + a)\}\,\mathrm{tr}QC_1 + (1 + k^*)\,\mathrm{tr}QC_2], \qquad (4.15) \end{aligned}$$

where $a \equiv (n - k + 2)/(n - k - m_1 - 2) > 1$ for finite n. Sufficient conditions for (4.15) to be nonnegative for finite n are $-1 < k^* < 1$ and $m_1 + k_1 > [(2 + a)k - 6]/4$. Note that even when these conditions are not satisfied, (4.15) can be larger than zero.

To illustrate application of the MELO estimator in (4.9), we present some numerical estimation results for two small well-known econometric models, Klein's Model I (1950, pp. 58–66) in Table 5 and the Girshick–Haavelmo supply and demand model for food (Girshick and Haavelmo 1953) in Table 6. In Table 5, the values of k*, computed from (4.11), are $\frac{1}{9}$ for the consumption equation and $\frac{1}{5}$ for the other two structural equations. These values are markedly different from 1, that associated with the 2SLS estimate. The MELO estimates are generally between the direct least squares (DLS) and 2SLS estimates and closer to the former than the latter in this example. For example, in the consumption equation, the consistent MELO estimate of the coefficient of P_{-1} is 0.09812, which is between the DLS estimate, 0.08988, and the 2SLS estimate, 0.2162, and much closer to the former. This is one of several cases reported in the table in which there is a substantial numerical difference between the MELO and 2SLS coefficient estimates. (See Zellner (1976) for analysis of the relationship of MELO, 2SLS, and DLS estimates.) Also, MELO estimates' large-sample standard errors are smaller, and in a few

5. DLS, 2SLS, and MELO Estimates for Klein's Model I Based on U.S. Annual Data, 1921–1941

Equation	*Variable*	*DLS*[b]	*2SLS*[b]	*MELO*[b]
Consumption	$W_1 + W_2$	.7962	.8102	.7975
		(.03594)	(.04025)	(.03599)
	P	.1929	.0173	.1810
		(.08207)	(.1180)	(.08397)
	P_{-1}	.08988	.2162	.09812
		(.08156)	(.1073)	(.08271)
	1	16.237	16.555	16.25[illegible]
		(1.1721)	(1.3208)	(1.1743)
$\hat{\sigma}^2$[a]		.8514	1.0441	.8523
k		$k = 0$	$k = 1$	$k^* = 1/9$
Investment	P	.4796	.1502	.4479
		(.08738)	(.1732)	(.09316)
	P_{-1}	.3330	.6159	.3603
		(.09075)	(.1628)	(.09498)
	K_{-1}	−.1118	−.1578	−.1162
		(.02405)	(.03613)	(.02452)
	1	10.126	20.278	11.104
		(4.9175)	(7.5427)	(5.0279)
$\hat{\sigma}^2$[a]		.8249	1.3832	.8301
k		$k = 0$	$k = 1$	$k^* = 1/5$
Demand for Labor	$Y + T - W_2$	.4395	.4389	.4394
		(.02916)	(.03563)	(.03017)
	$(Y + T - W_2)_{-1}$	.1461	.1467	.1462
		(.03367)	(.03884)	(.03446)
	t	.1302	.1304	.1303
		(.02798)	(.02914)	(.02877)
	1	1.4970	1.5003	1.4975
		(1.1427)	(1.1478)	(1.1434)
$\hat{\sigma}^2$[a]		.4764	.4764	.4764
k		$k = 0$	$k = 1$	$k^* = 1/5$

[a] $\hat{\sigma}^2$ is the sum of squared residuals divided by sample size.

[b] Figures in parentheses below the coefficient estimates are large sample standard errors. In the case of DLS, both the coefficient estimates and their associated standard errors are inconsistent estimates.

[1] Nagar (1959) and Sawa (1972) have results indicating that use of a value of k less than one is beneficial in terms of mean squared error.

6. DLS, 2SLS, and MELO Estimates for the Girshick-Haavelmo Model Based on U.S. Annual Data, 1922–1941

Equation[a]	*Variable*	*DLS*[b]	*2SLS*[b]	*MELO*[b]
Retail demand for food	y_2	−.3361	−.4863	−.3569
		(.0570)	(.1698)	(.0776)
	y_3	.2771	.2902	.2722
		(.0308)	(.0515)	(.0342)
	z_9	−.2341	−.3166	−.2438
		(.0491)	(.1081)	(.0582)
	z_8	.06379	.09657	.07375
		(.0285)	(.0401)	(.0301)
	1	103.97	115.564	105.692
		(4.576)	(12.884)	(6.045)
$\hat{\sigma}^2$[a]		.7208	1.0290	.7325
k		$k = 0$	$k = 1$	$k^* = 6/11$
Retail supply of food	y_2	.1388	.1633	.1562
		(.0454)	(.0894)	(.0564)
	y_4	.5589	.6366	.5891
		(.0536)	(.1047)	(.0664)
	z_8	.3076	.3372	.3222
		(.0440)	(.0489)	(.0454)
	1	25.292	14.419	20.235
		(7.083)	(8.761)	(7.721)
$\hat{\sigma}^2$[a]		.9867	1.1138	1.0121
k		$k = 0$	$k = 1$	$k^* = 6/11$
Farm food supply	y_5	.03670	.4351	.1628
		(.0992)	(.2317)	(.1321)
	z_6	.06550	−.2146	−.02319
		(.1028)	(.1920)	(.1216)
	z_8	−.2019	−.1930	−.1990
		(.1624)	(.2183)	(.1689)
	1	96.464	84.772	92.763
		(8.531)	(12.745)	(9.193)
$\hat{\sigma}^2$[a]		16.057	29.010	17.356
k		$k = 0$	$k = 1$	$k^* = 7/12$
Wholesale demand	y_2	2.3014	2.7878	2.5019
		(.2189)	(.3454)	(.2654)
	z_8	.4306	.6191	.5083
		(.2050)	(.2477)	(.2165)
	1	−139.146	−190.097	−160.150
		(23.045)	(36.268)	(27.899)
$\hat{\sigma}^2$[a]		23.150	28.887	24.141
k		$k = 0$	$k = 1$	$k^* = 7/12$

[a] $\hat{\sigma}^2$ is the sum of squared residuals divided by sample size.
[b] Figures in parentheses below the coefficient estimates are large-sample standard errors. In the case of DLS, both the coefficient estimates and their associated standard errors are inconsistent estimates.

cases substantially smaller, than corresponding 2SLS estimates' standard errors.

Shown in Table 6 are DLS, 2SLS, and MELO estimates of four equations of the Girshick–Haavelmo model estimated with 20 annual observations for the U.S., 1922–1941, in Girshick and Haavelmo (1953). The k^* values for the MELO estimates are, from (4.11), equal to 6/11 for two equations and equal to 7/12 for the remaining two equations. As with Klein's Model I, these values are quite different from 1, the value associated with the 2SLS estimate. In Table 6 the MELO estimates, almost without exception, lie between the DLS and 2SLS estimates. In some cases, there are large differences in the estimates. For example, in the retail demand for food equation, the 2SLS estimate of the coefficient of y_2, the price variable, is -0.4863, while the MELO and DLS estimates are -0.3569 and -0.3361, respectively. In addition, just as with Klein's Model I, the large-sample consistent standard errors associated with the MELO estimates are all smaller than the corresponding large-sample 2SLS standard errors.

5. CONCLUDING REMARKS

We have considered finite sampling properties of some MELO estimates for reciprocals and ratios of population means and regression coefficients and for structural coefficients of an equation of a linear simultaneous equation model. It was pointed out that maximum likelihood estimators for many of these problems do not possess finite sample moments and have infinite risk relative to quadratic and other loss functions. In contrast, the MELO estimators considered in this article possess at least finite first and second sampling moments and thus have finite risk relative to quadratic and other loss functions. Some other exact and approximate sampling results characterizing MELO estimators' properties were presented.

In estimation of structural coefficients of an equation of a linear simultaneous equation model, it is known that the ML estimator does not possess finite moments and that under certain conditions the 2SLS estimator does not possess finite sample moments. In these cases, the MELO estimator's finite risk stands in marked contrast to the infinite risk of the ML and 2SLS estimators. Approximate moment and risk expressions for the MELO estimator were presented. Calculation of two numerical examples indicated that MELO structural coefficient estimates tended to fall between DLS and 2SLS estimates. Further, the consistent large-sample standard errors for the MELO estimates were without exception smaller, and often much smaller, than the corresponding consistent large-sample standard errors associated with the 2SLS estimates.

Finally, it is important to emphasize that point estimation is just part of the problem of inference. In future work, it would be desirable to study more fully the properties of posterior distributions for the estimation problems considered in this article.

[*Received December 1975. Revised August 1978.*]

REFERENCES

Bergstrom, A.R. (1962), "The Exact Sampling Distributions of Least Squares and Maximum Likelihood Estimators of the Marginal Propensity to Consume," *Econometrica*, 30, 480–490.

Chow, Gregory C. (1960), "Statistical Demand Functions for Automobiles and Their Use for Forecasting," in *The Demand for Durable Goods*, ed. Arnold C. Harberger, Chicago: University of Chicago Press, Ch. IV, 149–178.

Girshick, M.A., and Haavelmo, Trygve (1953), "Statistical Analysis of the Demand for Food: Examples of Simultaneous Estimation of Structural Equations," in *Studies in Econometric Method*, eds. William C. Hood and Tjalling C. Koopmans, New York: John Wiley & Sons, 92–111.

Hatanaka, M. (1973), "On the Existence and the Approximation Formulae for the Moments of the k-Class Estimators," *Economic Studies Quarterly*, 24, 1–15.

Hill, Bruce M. (1975), "On Coherence, Inadmissibility and Inference About Many Parameters in the Theory of Least Squares," in *Studies in Bayesian Econometrics and Statistics in Honor of Leonard J. Savage*, eds. Stephen E. Fienberg and Arnold Zellner, Amsterdam: North-Holland Publishing Co., 555–584.

Jeffreys, Harold (1967), *Theory of Probability*, Oxford: The Clarendon Press.

Kadane, Joseph B. (1971), "Comparison of k-Class Estimators When the Disturbances Are Small," *Econometrica*, 39, 723–737.

Kappenman, Russell F., Geisser, Seymour, and Antle, Charles E. (1970), "Bayesian and Fiducial Solutions for the Fieller-Creasy Problem," *Sankhyā*, 32B, 331–340.

Kendall, Maurice G., and Stuart, Alan (1958), *The Advanced Theory of Statistics, Volume 1: Distribution Theory*, London: Charles W. Griffin & Co.

———, and Stuart, Alan (1973), *The Advanced Theory of Statistics, Volume 2: Inference and Relationship*, 3rd ed., London: Charles W. Griffin & Co.

Klein, Lawrence R. (1950), *Economic Fluctuations in the United States, 1921–1941*, New York: John Wiley & Sons.

Nagar, A.L. (1959), "The Bias and Moment Matrix of the General *k*-Class Estimators of the Parameters in Simultaneous Equations," *Econometrica*, 27, 575–595.

Sawa, T. (1972), "Finite Sample Properties of the k-Class Estimators," *Econometrica*, 40, 653–680.

Zellner, Arnold (1971), *An Introduction to Bayesian Inference in Econometrics*, New York: John Wiley & Sons.

——— (1976), "A Note on the Relationship of Minimum Expected Loss (MELO) and Other Structural Coefficient Estimates," unpublished manuscript, H.G.B. Alexander Research Foundation, Graduate School of Business, University of Chicago.

——— (1978), "Estimation of Functions of Population Means and Regression Coefficients Including Structural Coefficients: A Minimum Expected Loss (MELO) Approach," *Journal of Econometrics*, 8, 127–158.

[20]

Bayesian method of moments (BMOM) analysis of mean and regression models*

1. INTRODUCTION

In the traditional likelihood and Bayesian approaches, it is usually assumed that enough information is available to formulate a likelihood function and, in the Bayesian approach, a prior density for the parameters of the selected likelihood function – see e.g. Jeffreys (1988), Box and Tiao (1973), Berger (1985), Geisser (1993), Press (1989), and Zellner (1971). However, if not enough information is available to specify a form for the likelihood function, then clearly there will be problems in both the traditional likelihood and Bayesian approaches. In situations like this, some resort to non-likelihood-based methods, say least squares regression, method of moments or boot-strap approaches that are 'data-based'. That is, least squares is usually justified by its producing the best fit to a given sample of data with no appeal to sampling properties. However, if appropriate sampling assumptions are made, unbiasedness and a variance–covariance matrix of the least squares estimator can be produced without specifying a complete likelihood function. Similarly, the well known Gauss–Markov theorem's assumptions provide certain optimality properties for the least square estimator without introducing a likelihood function. These general results, obtained under certain sampling assumptions, have been widely utilized. However, they do not yield conditional probability statements about possible values of parameters based on a *single set of given observations.* In the present paper, it will be shown how such conditional results can be obtained without introducing sampling assumptions, likelihood functions and prior densities. Moments of parameters and future values of variables will be derived and shown to have certain optimal properties based on just a single set of observed data.

Further, to obtain a posterior distribution for parameters given just their moments, a maximum entropy (maxent) approach, see e.g. Jaynes (1982a, b) and Cover and Thomas (1991) will be utilized since this provides a most conservative choice of density that incorporates information in moment side conditions. However, if enough information is available to formulate a tentative likelihood function and a prior density for its parameters, the results of a traditional Bayesian analysis can be compared and/or

* Research for this paper was financed in part by the National Science Foundation and by income from the H.G.B. Alexander Endowment Fund, Graduate School of Business, University of Chicago. Some of this work was reported to the University of Chicago Econometrics and Statistics Colloquium in November 1994, the Conference on Informational Aspects of Bayesian Statistics in Honor of H. Akaike, Japan, December 1993, the Department of Statistics Seminar, Iowa State University, April 1994, and the 1994 Conference in Honor of Seymour Geisser, Taiwan, December 1994. Thanks to participants and to a referee for helpful comments. Published in J. C. Lee, W. D. Johnson and A. Zellner (eds), *Prediction and Modelling Honouring Seymour Geisser*, New York: Springer-Verlag, 61–74.

combined with those yielded by the BMOM approach. See Green and Strawderman (1994) for an application of the BMOM approach in the analysis of a natural resource model with an unknown likelihood function.

The plan of the paper is as follows. In Section 2, an analysis of a simple scalar mean process is presented since this is a central, important case and analysis of it reveals well the essential features of the BMOM approach. Then in Section 3, results for multiple regression and autoregressive models are given. Section 4 includes a summary of results and indications of future research.

2. Analysis of a scalar mean problem

In this section, we assume that n given observations $\mathbf{y}' = (y_1, y_2, \ldots, y_n)$ have been obtained that relate to a scalar mean, θ, as follows:

$$y_i = \theta + u_i \qquad i = 1, 2, \ldots, n \tag{2.1}$$

where the u_i's are unobserved, *realized* error terms; see Chaloner and Brant (1988), Zellner (1975) and Zellner and Moulton (1985) for traditional Bayesian analyses of realized error terms. Note that we have made the important assumption that the errors, say measurement errors, are additive. Since θ and the u_i's are unobserved quantities with unknown values, we shall assume that we can view their possible values probabilistically. That is, given the data $\mathbf{y}$, we assume that the means of θ and the u_i's as well as other features of their distributions exist but have unknown values. Note that a *realized* error term, u_i, usually does not have a zero mean. Operations and assumptions introduced below will enable us to express these posterior moments and other quantities as functions of the given data.

2.1 *First-order posterior and predictive moments*

Given the observations' values in (2.1) for both θ and the realized error terms, the u_i's have fixed unknown values that we shall regard as subjectively random just as was done in Chaloner and Brant (1988), Zellner (1975), and Zellner and Moulton (1985). Here, in contrast to the work in these papers, we shall assume that not enough information is available to formulate a likelihood function and a prior density and thus it is not possible to use Bayes's theorem. From (2.1) we have

$$\bar{y} = \theta + \bar{u} \tag{2.2}$$

where $\bar{y} = \Sigma^n_{i=1} y_i/n$, a given sample mean and $\bar{u} = \Sigma^n_{i=1} u_i/n$, the mean of the realized error terms. The symbol E denotes a posterior expectation operator, so-called because it is utilized after the data have been observed. From (2.2), we have

$$\bar{y} = E\theta \mid D + E\bar{u} \mid D \tag{2.3}$$

where D denotes the given data, here $(y_1, y_2, \ldots y_n)$

We now introduce the following assumption.

Assumption 1: $$E\bar{u}|D = \Sigma^n_{i=1} Eu_i|D/n = 0 \tag{2.4}$$

This assumption indicates that we believe that there is nothing systematic in the realized error terms and thus the expectation of their mean is equal to zero. If, for example, we believe that the i'th observation is an additive outlier, we would have $u_i = \eta + \varepsilon_i$ with the mean of the parameter η, $E\eta \mid D \neq 0$ and then $E\bar{u} \mid D \neq 0$.

Given Assumption I in (2.4), we have from (2.3),

$$E\theta \mid D = \bar{y} \tag{2.5}$$

that is the posterior mean of θ is the sample mean $\bar{y}$. Note that this result has been obtained without selecting a likelihood function and a prior density for its parameters as is done in many analyses.

Given the result in (2.5), we have, on taking the expectation of both sides of (2.1), the posterior expectation of u_i, that the i'th realized error term is:

$$Eu_i \mid D = y_i - E\theta \mid D = y_i - \bar{y} \tag{2.6}$$

which is the deviation of y_i from the mean $\bar{y}$. On summing both sides of (2.6) and dividing by n, we have $\Sigma^n_{i=1} Eu_i \mid D/n = \Sigma^n_{i=1}(y_i - \bar{y})/n = 0$, the 'sample analogue' of Assumption I.

Further, note that if we seek an optimal point estimate, $\tilde{\theta}$, relative to a quadratic loss function $L(\theta, \tilde{\theta}) = (\theta - \tilde{\theta})^2$, we have $EL(\theta, \tilde{\theta}) = E(\theta - E\theta)^2 + (\tilde{\theta} - E\theta)^2$ and as is well known, taking $\tilde{\theta} = E\theta$ is optimal. Thus the estimate, $\bar{y}$ in (2.5) is optimal relative to squared error loss.

If y_{n+1} is a future, as yet unobserved value satisfying $y_{n+1} = \theta + u_{n+1}$, where u_{n+1} is a future, as yet unrealized error term, we can write $Ey_{n+1} \mid D = E\theta \mid D + Eu_{n+1} \mid D$ and assume that $Eu_{n+1} \mid D = 0$. Then

$$Ey_{n+1} \mid \mathrm{D} = E\theta \mid D = \bar{y} \tag{2.7}$$

which is the mean of the future observation y_{n+1} given the data $\mathbf{y}$ and the information in Assumption I. $\bar{y}$ is an optimal point prediction for y_{n+1} relative to a squared error predictive loss function.

The first-order moments in (2.5), (2.6) and (2.7) are identical to those obtained in a traditional normal likelihood function, diffuse prior analysis. However (2.5), (2.6) and (2.7) have been obtained without an i.i.d. normality assumption and without an improper diffuse prior.

2.2 *Second-order posterior and predictive moments*

Having derived the posterior and predictive means in (2.5) and (2.7), we now turn to consider derivation of second-order posterior and predictive moments.

From (2.1) in vector notation $\mathbf{y} = \iota\theta + \mathbf{u}$ where $\iota' = (1, 1, \ldots, 1)$ a $1 \times n$ vector of ones, we have $\hat{\mathbf{u}} = \mathbf{y} - \iota\bar{y} = [I - \iota(\iota'\iota)^{-1}\iota']\mathbf{y} = [I - \iota'=(\iota'\iota)^{-1}\iota']\mathbf{u}$ and thus

$$\mathbf{u} - \hat{\mathbf{u}} = \iota(\iota'\iota)^{-1}\iota'\mathbf{u} = \iota(\iota'\iota)^{-1}\iota'(\mathbf{u} - \hat{\mathbf{u}}) \tag{2.8}$$

where $\iota'\hat{\mathbf{u}} = 0$ has been employed in (2.8). Then,

$$V(\mathbf{u} \mid D) = E(\mathbf{u} - \hat{\mathbf{u}})(\mathbf{u} - \hat{\mathbf{u}})' \mid D = \iota(\iota'\iota)^{-1}\iota' E(\mathbf{u} - \hat{\mathbf{u}})\,(\mathbf{u} - \hat{\mathbf{u}})' \mid D\iota(\iota'\iota)^{-1}\iota' \tag{2.9}$$

is a functional equation that $V(\mathbf{u} \mid D)$ must satisfy, where D denotes given sample and prior information. Thus we introduce the following assumption.

Assumption II: $V(\mathbf{u} \mid \sigma^2, D) = E(\mathbf{u} - \hat{\mathbf{u}})(\mathbf{u} - \hat{\mathbf{u}})' \mid \sigma^2, D = \iota(\iota'\iota)^{-1}\iota'\sigma^2$ (2.10)

with σ^2 a positive scalar parameter which satisfies (2.9). Note from $y_i = \theta + u_i$ and $y_i = \bar{y} + \hat{u}_i$ that $u_i - \hat{u}_i = \bar{y} - \theta$. Also, $u_j - \hat{u}_j = \bar{y} - \theta$ and thus $u_i - \hat{u}_i = u_j - \hat{u}_j$ for all i, j. Thus it is not surprising that in (2.10) all variances are the same and all correlations equal 1. Then from $\bar{y} - \theta = (\iota'\iota)^{-1}\iota'(\mathbf{y} - \iota\theta) = (\iota'\iota)^{-1}\iota'(\mathbf{u} - \hat{\mathbf{u}})$, we have

$$\begin{aligned} V(\theta \mid \sigma^2, D) &= E(\theta - \bar{y})^2 \mid \sigma^2, D \\ &= (\iota'\iota)^{-1}\iota' E(\mathbf{u} - \hat{\mathbf{u}})\,(\mathbf{u} - \hat{\mathbf{u}})\iota(\iota'\iota)^{-1} \mid \sigma^2, D \\ &= (\iota'\iota)^{-1}\sigma^2 = \sigma^2/n \end{aligned} \tag{2.11}$$

where Assumption II has been used for $E(\mathbf{u} - \hat{\mathbf{u}})\,(\mathbf{u} - \hat{\mathbf{u}})' \mid \sigma^2, D$ in (2.11). Thus σ^2/n is the posterior variance for θ given the parameter σ^2 and the data.

In addition, $E(\mathbf{u} - \hat{\mathbf{u}})'\,(\mathbf{u} - \hat{\mathbf{u}}) \mid D, \sigma^2 = \text{tr}\,\iota(\iota'\iota)^{-1}\iota'\sigma^2 = \sigma^2$ from (2.10). Then $E(\mathbf{u} - \hat{\mathbf{u}})'\,(\mathbf{u} - \hat{\mathbf{u}}) \mid D/n = E\mathbf{u}'\mathbf{u}/n \mid D - \hat{\mathbf{u}}'\hat{\mathbf{u}}/n, = E\sigma^2 \mid D/n$ where $E\mathbf{u} \mid D = \hat{\mathbf{u}}$ has been used. If we define $E\sigma^2 \mid D = E\Sigma_{i=1}^{n}(u_i - E\bar{u})^2/n \mid D = E\mathbf{u}'\mathbf{u}/n \mid D$, since $E\bar{u} \mid D = 0$ by Assumption I, we have $E\sigma^2 \mid D - \hat{\mathbf{u}}'\hat{\mathbf{u}}/n = E\sigma^2/n \mid D$ or

$$E\sigma^2 \mid D = \hat{\mathbf{u}}'\hat{\mathbf{u}}/(n-1) \equiv s^2 \tag{2.12}$$

which is the posterior mean of σ^2 and thus $V(\theta \mid D) = s^2/n$.

For the future observation $Y_{n+1} = \theta + u_{n+1}$, $Ey_{n+1} = \bar{y}$, given $E\theta = \bar{y}$ and $Eu_{n+1} \mid D, \sigma^2 = 0$. Also, $y_{n+1} - \bar{y} = \theta - \bar{y} + u_{n+1}$ and

$$E[(y_{n+1} - \bar{y})^2 \mid \sigma^2, D] = E(\theta - \bar{y})^2 \mid \sigma^2, D + Eu_{n+1}^2 \mid \sigma^2, D = \sigma^2(1 + 1/n) \tag{2.13}$$

given that u_{n+1} and $\theta - \bar{y}$ are uncorrelated, (2.11) and $Eu_{n+1}^2 \mid D, \sigma^2 = \sigma^2$.

A traditional diffuse prior, normal likelihood approach produces results identical to those in (2.11) and (2.13). However, the traditional approach yields $E\sigma^2 \mid D = \nu s^2/(\nu - 2)$ for $\nu > 2$ where $\nu = n - 1$ rather than (2.12), $E(\sigma^2 \mid D) = s^2$. Also $V(\theta \mid D) = s^2/n$ in the BMOM approach, whereas in the traditional Bayesian approach $V(\theta \mid D) = \nu s^2/(\nu - 2)n$. Further, (2.1), (2.12) and (2.13) are obtained without an i.i.d. normality assumption and an improper prior density.

2.3 Derivation of posterior and predictive densities

Above we have derived the first two posterior moments of θ given σ^2, namely $E\theta \mid D = \bar{y}$ and $\text{Var}(\theta \mid \sigma^2, D) = \sigma^2/n$. Also, $E\sigma^2 \mid D = s^2 = \hat{\mathbf{u}}'\hat{\mathbf{u}}/(n-1)$, $V(\theta \mid D) = s^2/n$, $Ey_{n+1} \mid D = \bar{y}$ and $\text{Var}(y_{n+1} \mid \sigma^2, D) = (1 + 1/n)\sigma^2$. It is well known that maxent densities can be derived that incorporate the information in moment conditions; see e.g. Jaynes (1982a, b), Cover and Thomas (1991) an) Zellner (1991, 1993). That is, we choose a density, say $f(x)$, to maximize $H(f) = -\int f(x)\ell n f(x)dx$ subject to $\int x^i f(x)dx = \mu_i$, $i = 0, 1, 2, \ldots$. with $\mu_0 = 1$, where we have utilized uniform measure in defining entropy, $H(f)$. See Shore and Johnson (1980) for consistency, invariance and other desirable properties of this entropy-based procedure. Here we have the following results.

(a) The proper maxent posterior density for θ given σ^2 is a normal density with mean $\bar{y}$ and variance σ^2/n, i.e. $g_N(\theta \mid \sigma^2, D) \sim N(\bar{y}, \sigma^2/n)$.

(b) The proper maxent predictive density for y_{n+1} given σ^2 and D with mean $\bar{y}$ and variance $\sigma^2(1 + 1/n)$ is a normal density $h_N(y_{n+1} \mid \sigma^2, D)$ with these moments, $N[\bar{y}, \sigma^2(1 + 1/n)]$.

(c) The proper maxent posterior density for σ^2 with $E\sigma^2 = s^2$ is the exponential density $g_e(\sigma^2 \mid D) = (1/s^2)\exp\{-\sigma^2/s^2\}$, $0 < \sigma^2 < \infty$, with s^2 given in (2.12).

(d) From (a) and (c), the joint posterior density for θ and σ^2 is

$$f(\theta, \sigma^2 \mid D) = g_N(\theta \mid \sigma^2, D)g_e(\sigma^2 \mid D) \tag{2.14}$$

which is a maxent density given that θ/σ and σ are assumed independent.[1] On integrating over σ^2, $0 < \sigma^2 < \infty$, the marginal posterior density for θ is – see Appendix for the derivation –

$$g(\theta \mid D) = \sqrt{n/2s^2}\,\exp\left\{-\sqrt{2n}\,|\theta - \bar{y}|/s\right\} \qquad -\infty < \theta < \infty \tag{2.15}$$

This double-exponential or Laplace[2] marginal posterior density for θ is symmetric about $\bar{y}$, the mean, median and modal value, with variance equal to s^2/n and has thick tails relative to normal or Student-t densities. By a change of variable, $z = \sqrt{n}(\theta - \bar{y})/s$, a standardized form of (2.15) is: $g(z \mid D) = (1/\sqrt{2})\exp\{-\sqrt{2}\,|z|\}$, $-\infty < z < \infty$. See Appendix for further properties of this density.

(e) From (b) and (c), the joint density for y_{n+1} and σ^2 is

$$h_N(y_{n+1} \mid \sigma^2, D)g_e(\sigma^2 \mid D) \tag{2.16}$$

where $h_N \sim N[\bar{y}, \sigma^2(1 + 1/n)]$ and g_e is the exponential density shown in (c). On integrating (2.16) with respect to σ^2, $0 < \sigma^2 < \infty$, the marginal predictive density of y_{n+1} is in the following double-exponential form,

$$h_e(y_{n+1} \mid D) = \left(1/s_e\sqrt{2}\right)\exp\left\{\sqrt{2}\,|y_{n+1} - \bar{y}|/s_e\right\} \qquad -\infty < y_{n+1} < \infty \tag{2.17}$$

where $s_e^2 = (1 + 1/n)s^2$. The mean, median and modal value of (2.17) are all equal to $\bar{y}$ and its variance is s_e^2. As with the density in (2.15), the tails of the predictive density can be rather thick.

(f) The proper maxent posterior density for $u_i = y_i - \theta$, the i'th realized error term with mean $\hat{u}_i$ and variance σ^2/n is given by

$$g_N(u_i \mid \sigma, D) = \sqrt{n/2\pi\sigma^2}\exp\left\{-\left(u_i - \hat{u}_i\right)^2 n/2\sigma^2\right\} \quad -\infty < u_i < \infty \qquad (2.18)$$

(g) The marginal posterior density for u_i, obtained by integration from the joint density, $g_N(u_i \mid \sigma, D)g_e(\sigma^2 \mid D)$ is

$$g_N(u_i \mid D) = \sqrt{n/2s^2}\exp\left\{-\sqrt{2n} \mid u_i - \hat{u}_i \mid / s\right\} \quad -\infty < u_i < \infty \qquad (2.19)$$

a double-exponential density centered at $\hat{u}_i$, the i'th residual.

(h) If it is known that $y_i > 0$ for all i, as with 'time to failure' or 'duration' data, $0 < \theta < \infty$ and the maxent proper density for θ subject just to $E\theta = \bar{y} > 0$ is the following exponential density:

$$g_1(\theta \mid D) = (1/\bar{y})\exp\{-\theta/\bar{y}\} \qquad 0 < \theta < \infty \qquad (2.20)$$

The posterior and predictive densities above can readily be implemented in practice. As indicated below, they can be compared and/or combined with posterior and predictive densities derived from assumed likelihood functions and prior densities using Bayes's theorem.

3 BMOM analysis of multiple regression and autoregression models

The BMOM analysis of multiple regression and autoregression models is quite similar to that utilized in analysing the scalar mean model in the previous section. The given $n \times 1$ observation vector $\mathbf{y}$ is assumed to satisfy the following n well known equations:

$$\mathbf{y} = X\boldsymbol{\beta} + \mathbf{u} \qquad (3.1)$$

where X is a given $n \times k$ matrix of rank k, $\boldsymbol{\beta}$ is a $k \times 1$ vector of regression coefficients with unknown values and $\mathbf{u}$ is an $n \times 1$ vector of realized error terms. If (3.1) relates to an autoregression, it is assumed that the initial values of the output variable, say $(y_0, y_{-1}, \ldots, y_{-(q-1)})$, for a q'th order autoregression are known and are elements of the X matrix. Given that it is assumed that we do not have enough information to formulate a likelihood function with much confidence, the BMOM approach can be employed to derive posterior and predictive moments and densities.

3.1 Derivation of first-order posterior and predictive moments

To derive first-order moments, we multiply both sides of (3.1) by $(X'X)^{-1}X'$ and take the posterior expectation of both sides to obtain, as in connection with (2.2),

$$(X'X)^{-1}X'\mathbf{y} = E\boldsymbol{\beta} \mid D + (X'X)^{-1}X'E\mathbf{u} \mid D \tag{3.2}$$

Now paralleling Assumption I above in (2.4), the following Assumption I' is introduced.

Assumption I':
$$(X'X)^{-1}X'Eu \mid D = \mathbf{0} \tag{3.3}$$

If one of the columns of X is a column of ones, that is there is an intercept in (3.10), (3.3) includes Assumption I, namely $E\bar{u} \mid D = \Sigma^n{}_1 Eu_i \mid D/n = 0$. Further, (3.3) implies that there is nothing 'systematic' in $E\mathbf{u} \mid D$ that is correlated with the columns of X. For example, this would not be the case if it were believed that some important variable or variables had been omitted from the relation in (3.1) or if it were believed that the values of the independent variables in X were measured with error. In these two cases, the u_i's would contain components that would be expected to be correlated with columns of X and (3.3) would not be expected to hold. However, if it is believed that (3.3) is valid, then clearly from (3.2) the posterior mean, $E\boldsymbol{\beta}$ is given by

$$E\boldsymbol{\beta} \mid D = (X'X)^{-1}X'\mathbf{y} = \hat{\boldsymbol{\beta}} \tag{3.4}$$

which is the least squares quantity, a rather simple result.[3] Further, given a quadratic loss function, $L(\boldsymbol{\beta}, \tilde{\boldsymbol{\beta}}) = (\boldsymbol{\beta} - \tilde{\boldsymbol{\beta}})'Q(\boldsymbol{\beta} - \tilde{\boldsymbol{\beta}})$, where Q is a given positive definite symmetric matrix, the value of $\tilde{\boldsymbol{\beta}}$ that minimizes posterior expected loss is the posterior mean in general and given in (3.4) for this specific problem.

From (3.4), the mean of the realized error vector is given by

$$E\mathbf{u} \mid D = \mathbf{y} - XE\boldsymbol{\beta} \mid D = \mathbf{y} - X\hat{\boldsymbol{\beta}} = \hat{\mathbf{u}} \tag{3.5}$$

where $\hat{\mathbf{u}}$ is the least squares residual vector and we have $X'\hat{\mathbf{u}} = 0$, the sample analogue of Assumption I' in (3.3).

As regards the predictive mean, we have for a future observation, y_f, assumed to satisfy $y_f = \mathbf{x}'_f\boldsymbol{\beta} + u_f$, with the $1 \times k$ vector $\mathbf{x}'_f$ given and u_f a random, as yet unrealized error term assumed drawn from a distribution with a zero mean (equal to $E\bar{u} \mid D = 0$, as assumed above in Assumption I'). Then the predictive mean is

$$Ey_f \mid D = x'_f E\boldsymbol{\beta} \mid D = \mathbf{x}'_f\hat{\boldsymbol{\beta}} = \hat{y}_f \tag{3.6}$$

which is just the least squares point prediction. Given a squared error predictive loss function, the predictive mean in (3.6) minimizes the expectation of such a predictive loss function.

Having obtained the first-order moments above, we now turn to derive second–order moments.

3.2 Derivation of second-order posterior and predictive moments

From (3.1) and (3.4), we have $\boldsymbol{\beta} - E\boldsymbol{\beta} \mid D = (X'X)^{-1}X'(\mathbf{u} - \hat{\mathbf{u}})$, where $\hat{\mathbf{u}} = \mathbf{y} - X\hat{\boldsymbol{\beta}}$ with $X'\hat{\mathbf{u}} = 0$. Thus the posterior covariance matrix, denoted by $V(\boldsymbol{\beta} \mid D)$, is

$$V(\boldsymbol{\beta} \mid D) = E(\boldsymbol{\beta} - \hat{\boldsymbol{\beta}})(\boldsymbol{\beta} - \hat{\boldsymbol{\beta}})' \mid D = (X'X)^{-1}X'E(\mathbf{u} - \hat{\mathbf{u}})(\mathbf{u} - \hat{\mathbf{u}})' \mid DX(X'X)^{-1} \quad (3.7)$$

To evaluate (3.7) another assumption is needed, paralleling that in (2.10). Given that $\hat{\mathbf{u}} = (I - X(X'X)^{-1}X')\mathbf{y} = (I - X(X'X)^{-1}X')\mathbf{u}$, we have that

$$\mathbf{u} - \hat{\mathbf{u}} = X(X'X)^{-1}X'(\mathbf{u} - \hat{\mathbf{u}}) \quad (3.8)$$

and

$$E(\mathbf{u} - \hat{\mathbf{u}})(\mathbf{u} - \hat{\mathbf{u}})' \mid D = X(X'X)^{-1}X'E(\mathbf{u} - \hat{\mathbf{u}})(\mathbf{u} - \hat{\mathbf{u}})' \mid DX(X'X)^{-1}X' \quad (3.9)$$

is a functional equation that must be satisfied. Note that only k elements of $\mathbf{u}$ are free. Thus, we introduce the following assumption, the analogue of (2.10):

Assumption II': $V(\mathbf{u} \mid \sigma^2, D) = E(\mathbf{u} - \hat{\mathbf{u}})(\mathbf{u} - \hat{\mathbf{u}})' \mid \sigma^2, D = X(X'X)^{-1}X'\sigma^2$ (3.10)

where σ^2 is a positive constant.[4] On inserting the expression in (3.10) in (3.9), it is seen that the functional equation is satisfied for any given value of σ^2. Substituting from (3.10) in (3.7), we have

$$V(\boldsymbol{\beta} \mid \sigma^2, D) = (X'X)^{-1}\sigma^2 \quad (3.11)$$

which is the posterior covariance matrix for $\boldsymbol{\beta}$ given σ^2 and D, the data.

To obtain a value for the posterior expectation of σ^2, from (3.10), $E(\mathbf{u} - \hat{\mathbf{u}})'(\mathbf{u} - \hat{\mathbf{u}}) \mid \sigma^2, D = \text{tr}\, X(X'X)^{-1}X'\sigma^2 = k\sigma^2$. Then $E\mathbf{u}'\mathbf{u}/n \mid D - \hat{\mathbf{u}}'\hat{\mathbf{u}}/n = kE\sigma^2 \mid D/n$ and if we define $E\sigma^2 \mid D = E\Sigma_{i=1}^{n}(u_i - E\bar{u})^2/n \mid D = E\mathbf{u}'\mathbf{u}/n \mid D$, since $E\bar{u} \mid D = 0$ from Assumption I', $E\sigma^2 \mid D - \hat{\mathbf{u}}'\hat{\mathbf{u}},/n = kE\sigma^2 \mid D/n$ or

$$E\sigma^2 \mid D = \hat{\mathbf{u}}'\hat{\mathbf{u}}/(n-k) \equiv s^2 \quad (3.12)$$

which is the posterior mean of σ^2.

For the future observation, $y_f = \mathbf{x}_f'\boldsymbol{\beta} + u_f$, $Ey_f \mid D = x_f'\hat{\boldsymbol{\beta}} \equiv \hat{y}_f$, as shown above. Then $y_f - \hat{y}_f = \mathbf{x}'_f(\boldsymbol{\beta} - \hat{\boldsymbol{\beta}}) + u_f$ and

$$E(y_f - \hat{y}_f)^2 \mid \sigma^2, D = (1 + \mathbf{x}'_f(X'X)^{-1}\mathbf{x}_f)\sigma^2 \quad (3.13)$$

given that u_f and the elements of $\boldsymbol{\beta} - \hat{\boldsymbol{\beta}}$ have zero covariances and $Eu_f \mid D = 0$.

Given the above posterior and predictive moments, the following are maxent posterior and predictive densities that incorporate the information in these moments.

(a') The proper maxent posterior density for $\boldsymbol{\beta}$ given σ^2 and the data with mean $\hat{\boldsymbol{\beta}} = (X'X)^{-1}X'\mathbf{y}$ and covariance matrix $(X'X)^{-1}\sigma^2$, is a multivariate normal density, $g_N(\boldsymbol{\beta} \mid \sigma^2, D) \sim MVN(\boldsymbol{\beta}, (X'X)^{-1}\sigma^2)$.

(b') The proper maxent predictive density for y_f given σ^2, $\mathbf{x}_f$ and the data is a normal density, $h_N(y_f \mid \sigma^2, D) \sim N[(\hat{y}_f, (1 + \mathbf{x}'_f(X'X)^{-1}\mathbf{x}_f)\sigma^2)]$.

(c') The proper maxent density for σ^2 with $E\sigma^2 = s^2 = \hat{\mathbf{u}}'\hat{\mathbf{u}}/(n-k)$ is the exponential density $g_e(\sigma^2 \mid D) = (1/s^2)\exp(-\sigma^2/s^2)$, $0 < \sigma^2 < \infty$.

(d') From (a') and (c'), the joint posterior density for $\boldsymbol{\beta}$ and σ^2 is[5]

$$f(\boldsymbol{\beta}, \sigma^2 \mid D) = g_N(\boldsymbol{\beta} \mid \sigma^2, D)g_e(\sigma^2 \mid D) \tag{3.14}$$

which is a maxent density given that $\boldsymbol{\beta}/\sigma$ and σ are independent. The marginal posterior density of a single element of $\boldsymbol{\beta}$, say β_i, can be obtained by integrating (3.14) with respect to the remaining elements of $\boldsymbol{\beta}$ and then with respect to σ^2. The result is the following double exponential density for β_i:

$$g(\beta_i \mid D) = \left(1/s_i\sqrt{2}\right)\exp\left\{-\sqrt{2}\mid\beta_i - \hat{\beta}_i\mid/s_i\right\} \quad -\infty < \beta_i < \infty \tag{3.15}$$

where $\hat{\beta}_i$ is the i'th element of $\hat{\boldsymbol{\beta}}$ and s_i^2 is the (i, i)'th element of $(X'X)^{-1}s^2$. Also, from (3.14), the marginal distribution of $\eta = \boldsymbol{\ell}'\boldsymbol{\beta}$, where $\boldsymbol{\ell}$ is a given vector of rank one, is

$$g(\eta \mid D) = \left(1/s_i\sqrt{2}\right)\exp\left\{-\sqrt{2}\mid\eta - \hat{\eta}\mid/s_\eta\right\} \quad -\infty < \eta < \infty \tag{3.16}$$

where $\hat{\boldsymbol{\eta}} = \boldsymbol{\ell}'\hat{\boldsymbol{\beta}}$ and $s_\eta^2 = \boldsymbol{\ell}'(X'X)^{-1}\boldsymbol{\ell}s^2$

(e') From (b') and (c'), the joint density for y_f and σ is

$$h_N(y_f \mid \sigma^2, D)g_e(\sigma^2 \mid D) \tag{3.17}$$

where $h_N \sim N[\hat{y}_f, (1 + x'_f(X'X)^{-1}x_f)\sigma^2]$ and g_e is the exponential density shown in (c'). On integrating (3.17) with respect to σ^2, the marginal predictive density for y_f is the following double-exponential density:

$$h_f(y_f \mid D) = \left(1/s_e\sqrt{2}\right)\exp\left\{-\sqrt{2}\mid y_f - \hat{y}_f\mid/s_e\right\} \quad -\infty < y_f < \infty \tag{3.18}$$

The mean of this density is $\hat{y}_f$ and its variance is $s_e^2 = [1 + \mathbf{x}'_f(X'X)^{-1}\mathbf{x}_f]s^2$.

(f') The proper maxent posterior density for u_i with given mean $\hat{u}_i$ and given variance $v_i = \mathbf{x}'_i(X'X)^{-1}\mathbf{x}_i\sigma^2$ is

$$g_N\left(u_i \mid \sigma^2, D\right) = 1/\sqrt{2\pi v_i}\,\exp\left\{-\left(u_i - \hat{u}_i\right)^2/2v_i\right\} \tag{3.19}$$

(g') The marginal posterior density for u_i, obtained by integrating the joint density $g_N(u_i \mid \sigma, D)g_e(\sigma^2 \mid \mathrm{D})$ with respect to σ is:

$$g_e(u_i \mid D) = \left[1/\sqrt{2v_i}\right]\exp\left\{-\sqrt{2} \mid u_i - \hat{u}_i \mid / \sqrt{v_i}\right\} \qquad -\infty < u_i < \infty \tag{3.20}$$

a double-exponential density.

(h') If it is known that $\theta = \ell'\boldsymbol{\beta}$ is strictly positive, where ℓ' is a given $1 \times k$ vector of rank one, the maxent density for θ subject just to $E\theta \mid D = \ell'\hat{\boldsymbol{\beta}} > 0$, is the following exponential density:

$$g(\theta \mid D) = (1/\ell'\hat{\boldsymbol{\beta}})\exp\{-\theta/\ell'\hat{\boldsymbol{\beta}}\} \qquad 0 < \theta < \infty \tag{3.21}$$

3.3 Use of additional prior information

If in addition to the sample information $\mathbf{y} = X\boldsymbol{\beta} + \mathbf{u}$, we represent additional prior information by use of a conceptual sample, $\mathbf{y}_c = X_c\boldsymbol{\beta} + \mathbf{u}_c$, where $\mathbf{y}_c$ is an $n_c \times 1$ vector of conceptual data points, X_c is an $n_c \times k$ given matrix, $\boldsymbol{\beta}$ is a $k \times 1$ vector of regression parameters and $\mathbf{u}_c$ is an $n_c \times 1$ vector of realized conceptual error terms, we can write

$$\begin{pmatrix} y_c \\ y \end{pmatrix} = \begin{pmatrix} X_c \\ X \end{pmatrix}\beta + \begin{pmatrix} u_c \\ u \end{pmatrix} \tag{3.22a}$$

or

$$\mathbf{w} = W\boldsymbol{\beta} + \boldsymbol{\varepsilon} \tag{3.22b}$$

Then making Assumption I' and II' relative to the system in (3.22b), the posterior mean of $\boldsymbol{\beta}$ is

$$\begin{aligned} \hat{\boldsymbol{\beta}} \equiv E\boldsymbol{\beta} \mid D' &= (W'W)^{-1}W'w \\ &= (X'_cX_c + X'X)^{-1}(X'_cy_c + X'y) \\ &= (X'_cX_c + X'X)^{-1}(X'_cX_c\hat{\beta}_c + X'X\hat{\beta}) \end{aligned} \tag{3.23}$$

where, when X_c is of full column rank, $\hat{\boldsymbol{\beta}}_c = (X'_cX_c)^{-1}X'_c\mathbf{y}_c$ is a prior mean vector, $\hat{\boldsymbol{\beta}} = (X'X)^{-1}X'\mathbf{y}$, the least squares estimate and X'_cX_c is assigned a value by the investigator. Then, with Assumption II' relating to (3.22b), we have

$$E(\boldsymbol{\beta} - E\boldsymbol{\beta})(\boldsymbol{\beta} - E\boldsymbol{\beta})' \mid \mathrm{D}, \sigma^2) = (W'W)^{-1}\sigma^2 \tag{3.24}$$

and

$$E\sigma^2 \mid \mathrm{D} = (w - W\tilde{\boldsymbol{\beta}})'(\mathbf{w} - W\tilde{\boldsymbol{\beta}})/(n + n_c - k). \tag{3.25}$$

Also maxent distributions are available for this system that incorporates a conceptual sample.

Above are some results of applying the BMOM approach to analyse data assumed generated by a multiple regression or an autoregression model. In addition, posterior and predictive intervals can easily be computed from the above posterior and predictive densities by the procedures described in the Appendix. Generally these intervals are broader than corresponding intervals based on the conditional normal posterior and predictive densities, derived above, with $\sigma^2 = s^2$. Also they are broader than conventional intervals based on marginal Student-t densities based on posterior and predictive densities based on normal likelihood functions and diffuse prior densities for their parameters. See Appendix A for one such comparison.[6]

4. Summary and concluding remarks

In the preceding sections, posterior and predictive moments for parameters and future, as yet unobserved observations have been derived using one given set of data and a few simple assumptions. There was no need to formulate a density function for as yet unobserved data as a basis for a likelihood function nor to introduce a prior density for its parameters. Also, no use was made of Bayes's theorem in deriving posterior and predictive moments.

Then, proper posterior and predictive densities were derived by maximizing entropy subject to the derived moments. These densities for location or regression coefficients with doubly infinite ranges are in the double exponential or Laplacian form while the maxent posterior density for variance parameters are in the exponential form. For location parameters with strictly positive values, the maxent posterior densities subject just to a first moment constraint are in the exponential form. Similar results were obtained for predictive densities for as yet unobserved observations.

It should be appreciated that these maxent predictive densities, in the double exponential or exponential form can serve as models for as yet unobserved data and employed in calculation of posterior odds in model comparison and selection problems using new data. In particular such models can be compared to those derived using a particular likelihood function, a prior density for its parameters and Bayes's theorem. The two different posterior densities, based on analyses of a given sample of data, can be employed as prior densities in the calculation of posterior odds using a second sample of data. The posterior odds can then be used to choose between the two models or to combine them using the approach described in Min and Zellner (1993) and Palm and Zellner (1992). In future work, such calculations will be reported. See Zellner (1995) for some BMOM results relating to multivariate regression and simultaneous equations models.

Notes

1. A maxent density without this independence assumption has also been derived.
2. See Stigler (1986, p. 111) for a fascinating discussion of Laplace's derivation of this distribution using the 'principle of insufficient reason'.
3. If, for example, rather than (3.1), we have $\mathbf{y} = X\boldsymbol{\beta} + Z\boldsymbol{\gamma} + \boldsymbol{\varepsilon}$, that is $\mathbf{u} = Z\boldsymbol{\gamma} + \boldsymbol{\varepsilon}$, $\hat{\boldsymbol{\beta}} = (X'X)^{-1}X'\mathbf{y} = \boldsymbol{\beta} + (X'X)^{-1}X'Z\boldsymbol{\gamma} + (X'X)^{-1}X'\boldsymbol{\varepsilon}$ and $\hat{\boldsymbol{\beta}} = E\boldsymbol{\beta} \mid D + (X'X)^{-1}X'ZE\boldsymbol{\gamma} \mid D$ given $(X'X)^{-1}X'E\boldsymbol{\varepsilon} \mid \mathrm{D} = \mathbf{0}$. Thus $E\boldsymbol{\beta} \mid \mathrm{D} \neq \boldsymbol{\beta}$ for $X'Z \neq 0$ and $E\boldsymbol{\gamma} \mid D \neq 0$.
4. If we write $\mathbf{y} = XHH^{-1}\boldsymbol{\beta} + \mathbf{u} = Z\boldsymbol{\theta} + \mathbf{u}$, where H is a square $k \times k$ non-singular matrix such that $H'X'XH = I_k$ and $\boldsymbol{\theta} = H^{-1}\boldsymbol{\beta}$, a $k \times 1$ vector, then $Z'\mathbf{y} = \boldsymbol{\theta} + Z'\mathbf{u}$ and $E\boldsymbol{\theta} \mid D = Z'\mathbf{y}$ given $Z'E\mathbf{u} \mid D = 0$. Also, if we assume that the k error terms in $Z'\mathbf{u}$ have equal variances and are mutually uncorrelated, this implies (3.10) and (3.11).
5. To compute the joint density of the elements of $\boldsymbol{\beta}$, draw σ^2 from $g_e(\sigma^2 \mid \mathrm{D})$ and insert the drawn value in $g_N(\boldsymbol{\beta} \mid \sigma^2, D)$ and draw $\boldsymbol{\beta}$ from this multivariate normal density. Thus draws from the joint density $f(\boldsymbol{\beta}, \sigma^2 \mid \mathrm{D})$ are obtained by repeated use of this procedure.
6. In recent work with Benjamin Sacks and Julian Tobias, information in higher order moments of σ^2 has been utilized in computing maxent densities for σ^2 and elements of β. [Footnote added 1996]

References

Berger, J.O. (1985), *Statistical Decision Theory*, 2nd edn, New York: Springer-Verlag.

Box, G.E.P. and G.C. Tiao (1973), *Bayesian Inference in Statistical Analysis*, Reading, MA: Addison-Wesley.

Chaloner, K. and R. Brant (1988), 'A Bayesian Approach to Outlier Detection and Residual Analysis', *Biometrika*, **75**, 651–9.

Cover, T.M. and J.A. Thomas (1991), *Elements of Information Theory*, New York: John Wiley.

Geisser, S. (1993), *Predictive Inference: An Introduction*, New York: Chapman & Hall.

Gradshteyn, I.S. and J.M. Ryzhik (1980), *Table of Integrals, Series and Products*, (corrected and enlarged edition prepared by A. Jeffrey), San Diego, CA: Academic Press.

Green, E.J. and W.E. Strawderman (1994), 'A Bayesian Growth and Yield Model for Slash Pine Plantations', Department of Natural Resources and Department of Statistics, Rutgers University, New Brunswick, NJ, to appear in *Applied Statistics*, 1996.

Jaynes, E.T. (1982a), *Papers on Probability, Statistics and Statistical Physics*, Dordrecht, Netherlands: Reidel.

Jaynes, E.T. (1982b), 'On the Rationale of Maximum-Entropy Methods', *Proc. of the IEEE*, **70**, 939–52.

Jeffreys, H. (1988), *Theory of Probability* (3rd reprinted edn, 1st edition, 1939), Oxford: Oxford University Press.

Min. C-k. and A. Zellner (1993), 'Bayesian and Non-Bayesian Methods for Combining Models and Forecasts with Applications to Forecasting International Growth Rates', *Journal of Econometrics*, **56**, 89–118.

Palm, F.C. and A. Zellner (1992), 'To Combine or Not to Combine? Issues of Combining Forecasts', *Journal of Forecasting*, **11**, 687–701.

Press, S.J. (1989), *Bayesian Statistics*, New York: John Wiley & Sons.

Shore, J.E. and R.W. Johnson (1980), 'Axiomatic Derivation of the Principle of Maximum Entropy and the Principle of Minimum Cross-Entropy', *IEEE Transactions*, vol. IT-26, no. 1, 26–37.

Stigler, S.M. (1986), *The History of Statistics*, Cambridge, MA: Harvard University Press.

Zellner, A. (1971), *An Introduction to Bayesian Inference in Econometrics*, New York: John Wiley, reprinted in 1987 by Krieger Publishing, Malabar, Florida, and by Wiley, 1996.

Zellner, A. (1975), 'Bayesian Analysis of Regression Error Terms', *Journal of the American Statistical Association*, **70**, 138–44.

Zellner, A. (1991), 'Bayesian Methods and Entropy in Economics and Econometrics', in W.T. Grandy and L.H. Schick (eds), *Maximum Entropy and Bayesian Methods*, Dordrecht, Netherlands: Kluwer, pp. 17–32.

Zellner, A. (1993), 'Prior Information, Model Formulation and Bayesian Analysis', paper presented at the Conference on Informational Aspects of Bayesian Statistics in Honor of H. Akaike, Fuji Conference Center, Japan, December 1993. A short version of the paper is in the ASA's 1993 *Proceedings of the Bayesian Statistics Science Section*, pp. 202–7 and the full paper is in *Journal of Econometrics*, **75** (1), 1996, 51–68.

Zellner, A. (1995), 'The Finite Sample Properties of Simultaneous Equations' Estimates and Estimators: Bayesian and Non-Bayesian Approaches', paper presented at the Conference in Honor of Professor Carl F. Christ, 21–22 April 1995 at Johns Hopkins University and to appear in the conference volume; Chapter 22 in this volume edited by L.B. Klein.

Zellner, A. and B.R. Moulton (1985), 'Bayesian Regression Diagnostics with Applications to International Consumption and Income Data', *Journal of Econometrics*, **29**, 187–211.

Appendix Derivation of double-exponential density in (2.15)

From (2.14), $f(\theta, \sigma^2 \mid D) = g_N(\theta \mid \sigma^2, D) g_e(\sigma^2 \mid D)$ with $g_N(\theta \mid \sigma^2, D) = \sqrt{(n/2\pi\sigma^2)} \times \exp\{-n(\theta - \bar{y})^2/2\sigma^2\}$ and $g_e(\sigma^2 \mid D) = (1/s^2)\exp\{-\sigma^2/s^2\}$. Then, with $a \equiv 1/s^2$ and $b \equiv n(\theta - \bar{y})^2/2$

$$
\begin{aligned}
\int_0^\infty f(\theta, \sigma^2 \mid D) d\sigma^2 &= \sqrt{n/2\pi}a \int_0^\infty \frac{1}{\sigma} \exp\{-[b/\sigma^2 + a\sigma^2]\} d\sigma^2 \\
&= \sqrt{2n/\pi}a \int_0^\infty \exp\{-[b/\sigma^2 + a\sigma^2]\} d\sigma^2 \\
&= \sqrt{n/2\pi}a \left[\frac{1}{2}\sqrt{\pi}/a \exp\{-2\sqrt{ab}\}\right] \qquad (A.1)
\end{aligned}
$$

where the integral has been evaluated using a result given in Gradshteyn and Ryzhik (1980, p. 307, entry 3.325). On inserting the above values for a and b in (A.1), we have

$$g(\theta \mid D) = \sqrt{n/2s^2}\,\exp\left\{-2\sqrt{n/2}\,|\theta - \bar{y}|/s\right\} \quad -\infty < \theta < \infty \tag{A.2}$$

Further, letting $w = \sqrt{(n/2)}\,(\theta - \bar{y})/s$, a 'standardized' form of the density is

$$g(w \mid D) = \exp\{-2\,|\,w\,|\} \qquad -\infty < w < \infty \tag{A.3}$$

Note that $\int_0^\infty \exp\{-2w\}dw = [-1/2\exp(-2w)]_0^\infty = 1/2$ and, given symmetry of $g(w \mid D)$, $\int_{-\infty}^\infty g(w \mid D)dw = 1$. Also, the symmetry of $g(w \mid D)$ implies that all odd order moments abut zero are zero; that is, $\int_{-\infty}^\infty w^{2r+1} g(w \mid D)dw = 0$ for $r = 0, 1, 2, \ldots$. The even order moments are given by

$$\begin{aligned}
\mu_{2r} &= 2\int_0^\infty w^{2r} e^{-2w}\,dw \\
&= \int_0^\infty (x/2)^{2r} e^{-x}\,dx = \Gamma(2r+1)2^{2r} \qquad r = 1, 2, \ldots \\
&= (2r)!/2^{2r}
\end{aligned} \tag{A.4}$$

where $x = 2w$ and $\Gamma(\cdot)$ is the, gamma function. For example $\mu_2 = 1/2$, $\mu_4 = 3/2$, etc.

Note that $w = \sqrt{n}\,(\theta - \bar{y})/s\sqrt{2} = z/\sqrt{2}$, where $z = \sqrt{n}(\theta - \bar{y})/s$ is a standardized normal variable for the normal *conditional* posterior density for θ given $\sigma^2 = s^2$, $g_N(\theta \mid \sigma^2 = s^2, \bar{y}) \sim N(\bar{y}, s^2/n)$. For the conditional density, $E(z^2 \mid \sigma^2 = s^2, D) = 1$, while for the unconditional density, $E(z^2 \mid D) = 2E(w^2 \mid D) = 1$ which, surprisingly, is equal to the conditional variance. However, $E(z^4 \mid \sigma^2 = s^2, D) = 3$ in the conditional normal density for z while using the marginal density $Ez^4 \mid D = 4Ew^4 \mid D = 4\ 3/2 = 6$, two times the value of that for the conditional density. The fourth moment of z divided by the squared second moment, denoted by $\mu_4/\mu_2^2 = 6$ and the 'excess' over that for the conditional normal density for z is $6 - 3 = 3$. Thus the marginal double-exponential density for z is quite leptokurtic.

Note that $z = \sqrt{n}(\theta - \bar{y})/s$ has a univariate Student-t posterior density with $\nu = n - 1$ degrees of freedom if a standard normal likelihood function for θ and σ and a diffuse prior $\pi(\theta, \sigma) \propto 1/\sigma$ were combined using Bayes's theorem.* For this Student-t posterior density, we have: $Ez \mid D = 0$ for $\nu > 0$, $Ez^2 \mid D = \nu/\nu - 2)$ for $\nu > 2$, and $Ez^4 \mid D = 3\nu^2/(\nu - 2)(\nu - 4)$ for $\nu > 4$. Thus the excess for the Student-t-based posterior density for z is: $\mu_4/\mu_2^2 - 3 = 6/(\nu - 4)$, for $\nu > 4$, which for $\nu > 6$ is considerably less than 3, the excess of the double-exponential density for z. Thus, for $n - 1 = \nu > 6$ or

* This well known result, probably first derived by Jeffreys, is presented in many works including Jeffreys (1988), Berger (1985), Box and Tiao (1973), Press (1989), and Zellner (1971).

$n > 7$, the double-exponential density is more leptokurtic. Also, as ν grows in value, the excess for the Student-t posterior density goes to zero whereas that for the double-exponential density has a constant value equal to 3.

Since $z = \sqrt{n}(\theta - \bar{y})/s = \sqrt{2}w$, the double-exponential posterior density for z is from (A.3):

$$p(z \mid D) = \left(1/\sqrt{2}\right)\exp\left\{-\sqrt{2}\mid z \mid\right\} \quad -\infty < z < \infty \tag{A.5}$$

Then for $c > 0$, $P_c \equiv Pr(c < z < \infty \mid D) = \int_c^\infty f(z \mid D)dz = (1/2)\exp\{-\sqrt{2}c\}$ and thus $\ell n 2P_c = -\sqrt{2}c$. For example for $P_c = 0.025$, $\ell n\ 0.05 = -2.9957 = -1.4142c$ and $c = 2.118$. Thus $0.025 = Pr\{2.118 < z < \infty\}$ and from the symmetry of the density in (A.5), we have

$$Pr\{-2.118 < z < 2.118 \mid D\} = 0.95 \tag{A.6}$$

This 95 per cent interval for $z = \sqrt{n}(\theta - \bar{y})/s$, -2.118 to 2.118 implies that a 95 per cent interval for θ is $\bar{y} \pm 2.118\ s/\sqrt{n}$. This interval is somewhat broader than a 95 per cent interval based on the conditional normal posterior density for θ with $\sigma^2 = s^2$, namely $\bar{y} \pm 1.96\ s/\sqrt{n}$, the widths being $4.24\ s/\sqrt{n}$ in the former case and $3.92\ s/\sqrt{n}$ in the latter, about an 8 per cent difference.

[21]

The finite sample properties of simultaneous equations estimates and estimators: Bayesian and non-Bayesian approaches*

I. INTRODUCTION

For many years, Carl Christ (1960, 1966, 1979, 1994) has been concerned about properties of procedures for estimating parameters of econometric models, especially simultaneous equations models (SEMs). He has appreciated the importance of finite sample properties of estimation procedures and been aware of alternative criteria for evaluating estimation performance. However, his previous work and writings have been mainly concerned with properties of sampling theory approaches with little attention given to Bayesian estimation procedures. In this paper I shall mention some general Bayesian estimation results and then go on to consider some specific old and new Bayesian estimation procedures for SEMs.

With respect to sampling theory, likelihood and traditional Bayesian approaches to estimation, strong assumptions are often made about how the observations have been generated, often with little or no support from economic theory or any other source, except 'tradition.' For example, as a student asked some years ago, 'Is it necessary to assume that God runs the US economy through a given historical period over and over again in order to justify sampling properties of estimators?' Also, in formulating a likelihood function, normality of the observations is often assumed without much justification. Then too, in analyses of SEMs, asymptotic theory is usually employed to justify procedures applied to finite samples of data. Many are aware of the dangers involved and no one has expressed them more eloquently than Phillips (1983) who remarks that, 'For the process by which asymptotic machinery works inevitably washes out sensitivities that are present and important in finite samples' (p. 508). For example, in Zellner (1978) it is shown analytically that certain simple econometric estimators have a bimodal density with no finite moments in finite samples but a unimodal normal density with finite moments in large samples. The current practical relevance of these issues has been emphasized by Kim and Pagan (1994) in their concluding section:

> The estimation of the parameters of such specifications [derived from 'inter-temporal optimization, stochastic driving forces, and full information upon the part of agents'] has also been done in standard ways that resemble much econometric work with macroeconomic models of the past, although in section 3 we have raised some questions about the ability to

* Research for this paper was financed in part by the National Science Foundation and by income from the H.G.B. Alexander Endowment Fund, Graduate School of Business, University of Chicago. Thanks to the editor, Patrick K. Asea and the referees for very helpful comments. To be published in a conference volume edited by L. R. Klein

> effectively estimate the parameters of many small stochastic general equilibrium models, given the sample sizes generally available for this purpose. (p. 40)

Further, with respect to what happens as the sample size actually increases, rather than what is assumed to happen as it increases, Phillips (1983) writes, 'When a model grows in complexity with increases in the availability of data, traditional asymptotic methods of statistical analysis inevitably become self-defeating. To extend Fisher's own metaphor, by the time the cannon has rumbled into place, and been loaded and sighted to fire, the sparrow will have flown away' (p. 509, fn. omitted).

These considerations regarding sampling assumptions and asymptotic approximations raised by Phillips (1983), Kim and Pagan (1994) and many others pin-point a need for an estimation approach that is based on given, available data and which does not require the often insecure assumptions underlying sampling, likelihood and asymptotic approaches. In this connection, the finite sample Bayesian method of moments (BMOM) approach, which is based on a given set of data and does not use sampling assumptions, has been developed and applied to location and multiple regression models in Zellner (1994b) and a natural resource model in Green and Strawderman (1994). Herein this approach will be extended to apply to unrestricted reduced form equations and to produce optimal, finite sample estimates of structural coefficients relative to several reasonable loss functions. Results so obtained will be compared with those derived earlier using traditional Bayesian methods and sampling theory estimation procedures. Further, some Monte Carlo experimental results relating to Bayesian and non-Bayesian estimation procedures will be reviewed.

The plan of the paper is as follows. In Section II, some Bayesian and non-Bayesian inference results for SEMs are reviewed and then the BMOM approach for analysing reduced form systems is presented. Optimal BMOM point estimates of structural coefficients are derived and discussed in Section III. These results are compared with those provided by the traditional Bayesian approach and sampling theory approaches. A review of several Monte Carlo experiments, Diebold and Lamb (1993), Park (1982), Tsurumi (1990) and Zellner (1971), in which properties of Bayesian and non-Bayesian estimators are compared is presented in Section IV. Finally, in Section V, a summary is presented and a new answer to Christ's (1960) question, 'Any Verdict Yet?' is provided.

II. Bayesian and non-Bayesian inference procedures for simultaneous equations models (SEMS)

In this section, we shall consider some Bayesian and non-Bayesian procedures for estimating parameters of SEMs. First, we shall briefly review the status of finite sample analysis of SEMs in Section 2.1. Then we shall apply the Bayesian method of moments (BMOM) approach to analyse unrestricted reduced form systems in Section 2.2 first without and then with the use of added prior information. It will be seen that this approach provides clear-cut analytical finite sample results that are quite operational.

2.1 Brief review of finite sample analysis of SEMs

To estimate the parameters of a single equation of an SEM, there are many sampling theory estimators available, namely OLS, LIML, 2SLS, other K-Class estimators, Double

K-Class estimators, Fuller's Modified LIML, etc. While these estimators, with the exception of OLS, are, under certain conditions, consistent, asymptotically normally distributed and efficient, they have no known exact finite sample justification. Surprisingly, some leading authorities and many practitioners recommend and/or use the inconsistent OLS method in finite samples since it may have better sampling properties and be more robust in small samples than available consistent estimators; see e.g. Christ (1960, 1966), and Greene (1993). For example, in small samples some consistent estimators' sampling densities have heavy tails and thus their use leads to many outlying estimates – see below for some examples. Further, LIML estimators are known not to have finite moments in general and other estimators do not have finite moments in certain cases, which means that these estimators are inadmissible relative to quadratic and many other loss functions. Even with bounded loss functions suggested by Barnard (1975) and Christ (1979), estimators whose densities have heavy tails will not be entirely satisfactory. Thus, after many years of research and application, there is still a great need for finite sample optimal estimation procedures for SEMs.

Nagar (1959) and Sawa (1972,1973), recognizing this need for optimal finite sample estimators, worked to determine optimal values of K within the K-class of estimators. While Nagar derived optimal values of K, it turned out that his 'optimal' values depend on parameters with unknown values that have to be estimated and when estimated, there is no guarantee that the estimator is indeed optimal. Sawa (1973) considered a linear combination of the consistent 2SLS and inconsistent OLS estimators in an attempt to get an approximately optimal estimator. However, the rationale for entertaining a class of estimators that are linear combinations of an inconsistent and a consistent estimator is unclear. Years ago, Tinbergen used OLS in fitting his models and, in more recent years, many have used OLS in estimating structural econometric models. See below for optimal Bayesian estimates that can be exactly equal to the OLS estimate in small samples.

On the work that has been done on derivation of the finite sample distributions of estimators, e.g., Bergstrom (1962), Basmann (1961, 1963), Phillips (1983), Zellner (1978), etc., several have noted that estimators' first and higher moments are often not finite and that for the Anderson–Rubin LIML estimator they are not finite in general. As a 65th birthday present for Ted Anderson, in Zellner (1983), written for his *Festschrift* volume, I showed that the LIML estimator considered conditional on a favorable outcome of a rank test for over-identifying restrictions does have finite moments. Since use of this pre-test has been recommended on other grounds by many, including Koopmans (1953), it is indeed interesting that its use has an important impact on the LIML estimator's properties. In the simplest case in which a structural coefficient is equal to the reciprocal of a reduced form coefficient, $\gamma = 1/\pi$, the estimator $\hat{\gamma} = 1/\hat{\pi}$, where $\hat{\pi}$ is a normally distributed least squares estimator the moments of $\hat{\gamma}$ are not finite. However, if we perform a test of $\pi = 0$ and reject and consider the distribution of $\hat{\gamma}$ subject to this outcome of the pre-test, this distribution has finite moments. In contrast, Bayesian MELO estimators for this problem and others, Zellner (1978, 1986a), and Judge et al. (1985, 635–7) have finite first and second moments in general.

Further, in Zellner (1978, appendix) and Nelson and Startz (1990), it has been pointed out that finite sample distributions of some usual estimators, LIML, 2SLS and other instrumental variable estimators, can be bimodal. This result holds for the simple case considered above, $\hat{\gamma} = 1/\hat{\pi}$ when $\hat{\pi}$ has a high probability of assuming positive and negative values. Use of information about the algebraic sign of $\hat{\pi}$, $0 < \hat{\pi} < \infty$ produces a unimodal density in this and other cases.

With respect to exact finite sample estimation results so far obtained, Robinson (1979) has pointed out that the exact results have in the main, if not entirely, been obtained under a finite sample normality assumption, which is rather restrictive although generalization to elliptically symmetric densities such as multivariate Student-*t* and matrix Student-*t* seems to be possible. However, even in the normality case, the resulting estimators' densities are usually difficult to derive, very complicated and depend on parameters with unknown values[1] which may explain why many use approximate asymptotic analysis. Finally, and most disturbing, is the fact that the estimators considered in this finite sample work are not known to be optimal in any sense in finite samples.

In the traditional Bayesian approach, finite sample results have been obtained by use of various numerical integration procedures in cases in which integrations cannot be performed analytically, some of which are reviewed in Drèze and Richard (1983). See also Geweke (1989, 1994a, b), Kloek and van Dijk (1978), Zellner, Bauwens and van Dijk (1988), and Zellner et al. (1994). With respect to Bayesian estimation, it is well known that Bayesian estimates that are optimal relative to specific loss functions can be readily computed, e.g. posterior modal values for 0–1 loss functions, medians for absolute error loss functions, means for squared error loss functions, corrected means for asymmetric loss functions (see, e.g., Varian, 1974 and Zellner, 1986b), etc. Also complete posterior densities are available which can be compared and/or combined with the BMOM maxent densities in Zellner (1994b) and below. In addition, in both Bayesian and non-Bayesian analyses, it would be worthwhile to have more analysis of diagnostic checking of relations and procedures for combining the results of diagnostic checks and estimation procedures – see e.g. Reynolds (1980), Griffiths and Dao (1980) and Judge et al. (1985, 117ff.) for examples of such Bayesian procedures.

Last, the sampling assumptions underlying assumed likelihood functions, estimators' distributions and properties and asymptotic analyses are very strong, as pointed out above, and deserve much more attention. Such sampling assumptions are avoided in the BMOM approach described in Green and Strawderman (1994), Zellner (1994b), and below.

2.2 *Unrestricted reduced form estimation by the Bayesian method of moments*

Herein it will be assumed that an *m*-equation reduced form system has been formulated and a given set of satisfactory data is available to estimate reduced form parameters. Further, it is assumed that not enough information is available to specify sampling properties of the data and, in particular, a likelihood function. Without a likelihood function, it is not possible to pursue the traditional Bayesian analysis or maximum likelihood analysis since both of these approaches require use of a likelihood function. The new Bayesian method of moments (BMOM) approach in Zellner (1994b), utilized in Green and Strawderman (1994), permits an analysis of the given data to be performed without specifying a likelihood function or introducing any sampling assumptions. It thus fills a need that researchers who don't have enough information to specify a likelihood function face; see, e.g. Green and Strawderman (1994) for a specific example.

The BMOM approach is called 'Bayesian' since it permits inverse probability statements to be made from a given sample of data, a major objective of Bayes's original paper. It utilizes background information, embodied in two assumptions[2] presented below, that permits 'post-data' or posterior moments of parameters' subjective densities to be evaluated quite easily without use of Bayes's theorem. Given the moments of parameters' subjective densities, specific proper densities with the given moments can be obtained by, for example, the principle of maximum entropy. See, e.g., Jaynes (1982a, b), Shore and Johnson (1980), Cover and Thomas (1991), Soofi (1994), Zellner (1993), and Zellner and Highfield (1988) for discussion of this principle, its properties, applications, and references to the literature. Briefly, application of the principle of maximum entropy provides a proper density for parameters with the given moments that is most spread out relative to an underlying measure, say a uniform measure which will be used below in the analysis of an unrestricted reduced form system.

We express a usual m-equation reduced form system as follows:

$$\underset{n\times 1}{y_i} = \underset{n\times k}{X} \; \underset{k\times 1}{\pi_i} + \underset{n\times 1}{\text{v}_i} \qquad i = 1, 2, \ldots, m \tag{2.1a}$$

or

$$\begin{pmatrix} y_1 \\ y_2 \\ \vdots \\ y_m \end{pmatrix} = \begin{pmatrix} X & 0 & \cdots & 0 \\ 0 & X & \cdots & 0 \\ \vdots & \vdots & \ddots & \vdots \\ 0 & 0 & \cdots & X \end{pmatrix} \begin{pmatrix} \pi_1 \\ \pi_2 \\ \vdots \\ \pi_m \end{pmatrix} + \begin{pmatrix} \text{v}_1 \\ \text{v}_2 \\ \vdots \\ \text{v}_m \end{pmatrix} \tag{2.1b}$$

or

$$y = Z\pi + \text{v} \tag{2.1c}$$

where Z denotes the block diagonal matrix on the right side of (2.1b), $y' = (y'_1, y'_2, \ldots, y'_m)$, $\pi' = (\pi'_1, \pi'_2, \ldots, \pi'_m)$ and $\text{v}' = (\text{v}'_1\ \text{v}'_2, \ldots, \text{v}'_m)$. We assume that y and Z, the data, are *fixed* and *known* and that the rank of X is k, that is $\rho(X) = k$. The realized error vector v as well as the parameter vector have *fixed, unknown* values that we regard as subjectively random, as is usually done in Bayesian analyses; see, e.g. Chaloner and Brant (1988), Zellner and Moulton (1985), Zellner (1975), and Zellner (1994b), for traditional Bayesian and BMOM analyses of realized regression error terms. With the data given and π and v having fixed unknown values, considered subjectively random, we have the following theorem.

Theorem I: With y, Z and v given quantities, and $\rho(X) = k$, given the assumption $Z'E\mathrm{v} = 0$, or $X'E\mathrm{v}_i = 0$, $i = 1, 2, \ldots, m$, where *E denotes the posterior expectation given the observed data*; it follows that $E\pi = (Z'Z)^{-1}Z'y$, or

$$E\pi_i = (X'X)^{-1} X'y_i \qquad i = 1, 2, \ldots, m \tag{2.2}$$

Proof: From (2.1c), $(Z'Z)^{-1}Z'y = \pi + (Z'Z)^{-1}Z'\mathrm{v}$ and $(Z'Z)^{-1}Z'y = E\pi + (Z'Z)^{-1}Z'E\mathrm{v}$. Given that $Z'E\mathrm{v} = 0$ or, equivalently, $X'E\mathrm{v}_i = 0$, $i = 1, 2, \ldots, m$, then $E\pi = (Z'Z)^{-1}Z'y$ and $E\pi_i = (X'X)^{-1}X'y_i$, $i = 1, 2, \ldots, m$.

Note that the assumption $X'E\mathrm{v}_i = 0$ implies that the columns of X are orthogonal to $E\mathrm{v}_i$, the expected value of the *realized* error vector v_i given the data. That is, it is assumed that there is nothing systematic in $E\mathrm{v}_i$ that is correlated with the columns of X.[3] If a variable or variables correlated with those in X had been left out or if there were measurement errors in the variables in X, then it would not be appropriate to assume $X'E\mathrm{v}_i = 0$ and the analysis would have to be broadened to accommodate these possibilities.

Next we indicate that $E\pi = (Z'Z)^{-1}Z'y$ is an optimal point estimate relative to quadratic loss in the following corollary.

Corollary 1: $E\pi = (Z'Z)^{-1}Z'y$ is an optimal point estimate relative to a quadratic loss function $L(\tilde{\pi}, \pi) = (\tilde{\pi} - \pi)'Q(\tilde{\pi} - \pi)$, where Q is a given positive definite symmetric matrix and $\tilde{\pi}$ is some estimate.

Proof: $EL(\tilde{\pi}, \pi) = E(\tilde{\pi} - \pi)'Q(\tilde{\pi} - \pi) = E[\tilde{\pi} - E\pi - (\pi - E\pi)]'Q[\tilde{\pi} - E\pi - (\pi - E\pi)] = E(\tilde{\pi} - E\pi)'Q(\tilde{\pi} - E\pi) + E(\pi - E\pi)'Q(\pi - E\pi)$. Thus given that Q is pds, taking $\tilde{\pi} = E\pi$, the posterior mean, is optimal in the sense of minimizing posterior expected loss.

Further we have the following result for the mean of the realized error vector, $E\mathrm{v}$.

Corollary 2: The posterior mean of v, $E\mathrm{v}$, is from (2.1c),

$$E\mathrm{v} = y - ZE\pi = y - Z\hat{\pi} = \hat{\mathrm{v}} \tag{2.3a}$$

or

$$E\mathrm{v}_i = y_i - XE\pi_i = y_i - X\hat{\pi}_i = \hat{\mathrm{v}}_i \qquad i = 1, 2, \ldots, m \tag{2.3b}$$

where $\hat{\mathrm{v}}' = (\hat{\mathrm{v}}'_1, \hat{\mathrm{v}}'_2, \ldots, \hat{\mathrm{v}}'_m)$ is a vector of least squares residuals. Using the argument in Corollary 1, $E\mathrm{v} = \hat{\mathrm{v}}$ is an optimal point estimate of the realized error vector v relative to quadratic loss functions, $(\tilde{\mathrm{v}} - \mathrm{v})'Q(\tilde{\mathrm{v}} - \mathrm{v})$, with Q pds.

We now turn to the evaluation of second moments of π for which the following theorem is relevant.

Theorem II: Given that $E(\mathrm{v} - \hat{\mathrm{v}})(\mathrm{v} - \hat{\mathrm{v}})' = E\Omega \otimes X(X'X)^{-1}X'$ or $E(\mathrm{v}_i - \hat{\mathrm{v}}_i)(\mathrm{v}_j - \hat{\mathrm{v}}_j)' = E\omega_{ij}X(X'X)^{-1}X'$, where $E\Omega$ is an $m \times m$ pds matrix with elements $E\omega_{ij} = E\mathrm{v}_i'\mathrm{v}_j$, then the posterior covariance matrix for π is

$$V(\pi) = E\Omega \otimes (X'X)^{-1} \\ = \hat{\Omega} \otimes (X'X)^{-1} \tag{2.4}$$

where $\hat{\Omega} = E\Omega$ with $\hat{\omega}_{ij} = E\omega_{ij} = \hat{v}_i'\hat{v}_j/(n-k)$, $i, j = 1, 2, \ldots, m$.

Proof: From $y = Z\pi + v$ and $y = ZE\pi + Ev = Z\hat{\pi} + \hat{v}$, $\pi - \hat{\pi} = -(Z'Z)^{-1}Z'(v - \hat{v})$, since $Z'\hat{v} = 0$. Thus,

$$\begin{aligned} E(\pi - \hat{\pi})(\pi - \hat{\pi})' &= (Z'Z)^{-1}Z'E(v - \hat{v})(v - \hat{v})'Z(Z'Z)^{-1} \\ &= (Z'Z)^{-1}Z'[E\Omega \otimes X(X'X)^{-1}X']Z(Z'Z)^{-1} \\ &= E\Omega \otimes (X'X)^{-1} \\ &= \hat{\Omega} \otimes (X'X)^{-1} \end{aligned}$$

With respect to the assumption in Theorem II, $V(v) = E(v - \hat{v})(v - \hat{v})' = E\Omega \otimes X(X'X)^{-1}X'$, note that $v - \hat{v} = -Z(\pi - \hat{\pi}) = -Z(Z'Z)^{-1}Z'(v - \hat{v})$ and thus $E(v - \hat{v})(v - \hat{v})' = Z(Z'Z)^{-1}Z'E(v - \hat{v})(v - \hat{v})'Z(Z'Z)^{-1}Z'$ is a functional equation that $V(v)$ must satisfy. The above form for $V(v)$ satisfies this functional equation. Further let $E\omega_{ij} \equiv Ev'_iv_j/n = (E(v_i - \hat{v}_i)'(v_j - \hat{v}_j) + \hat{v}'_i\hat{v}_j)/n = (k\, E\omega_{ij} + \hat{v}'_i\hat{v}_j)/n$ and thus $E\omega_{ij} \equiv \hat{v}'_i\hat{v}_j/(n-k) = \hat{\omega}_{ij}$.

In summary BMOM results for URRF estimation are:

(a) $$E\pi = (Z'Z)^{-1}Z'y = \hat{\pi} \text{ or } E\pi_i = (X'X)^{-1}X'y_i = \hat{\pi}_i \tag{2.5}$$

(b) $$V(\pi) = \hat{\Omega} \otimes (X'X)^{-1} \text{ or } \mathrm{Cov}(\pi_i, \pi_j) = (X'X)^{-1}\hat{\omega}_{ij} \tag{2.6}$$

with $\hat{\omega}_{ij} = \hat{v}'_i\hat{v}_j/(n-k)$ where

(c) $$Ev_i = y_i - XE\pi_i = y_i - X\hat{\pi}_i = \hat{v}_i \tag{2.7}$$

(d) $$E\Omega = \hat{\Omega} = \{\hat{\omega}_{ij}\} \text{ with } \hat{\omega}_{ij} = \hat{v}_i'\hat{v}_j/(n-k) \tag{2.8}$$

Various proper maximum entropy (maxent) densities for π and Ω with the above given moments are available, see, e.g., Jaynes (1982a, b), Shore and Johnson (1980), Cover and Thomas (1991), Soofi (1994), Zellner (1993), and Zellner and Highfield (1988). For example, the proper density that maximizes entropy subject to (a) and (b) is a multivariate normal density with mean $E\pi$ in (a) and variance–covariance matrix $V(\pi)$ in (b). Further additional information can be introduced by use of a 'conceptual sample', in the form of (2.1c), namely $y_i^o = X^o\pi_i + vi^o$, $i = 1, 2, \ldots, m$, much the same as in 'mixed estimation', Theil (1971, p. 347ff.) except that here y_i^o, X^o and vi^o are *given fixed* quantities.

On combining this conceptual sample with the actual sample,

$$\begin{pmatrix} y_i^o \\ y_i \end{pmatrix} = \begin{pmatrix} X^o \\ X \end{pmatrix}\pi_i + \begin{pmatrix} v_i^o \\ v_i \end{pmatrix} \tag{2.9a}$$

or

$$w_i = W\pi_i + \varepsilon_i \tag{2.9b}$$

with $w'_i = (y_i^{o\prime}\ y'_i)$, $W' = (X^{o\prime}\ X')$ and $\varepsilon'_i = (v_i^{o\prime}\ v'_i)$. Then with assumptions similar to those introduced above, we have

$$\hat{\pi}_i \equiv E\pi_i = (W'W)^{-1}W'w_i = (X^{o\prime}X^o + X'X)^{-1}(X^{o\prime}y_i^o + X'y_i) \qquad (2.10)$$
$$= (X^{o\prime}X^o + X'X)^{-1}(X^{o\prime}X^o\hat{\pi}_{oi} + X'X\hat{\pi}_i)$$

where $\hat{\pi}_{oi} = (X^{o\prime}X^o)^{-1}X^{o\prime}y_i^o$ is a given prior value for π_i and $\hat{\pi}_i = (X'X)^{-1}X'y_i$ is a vector of least squares quantities. Thus the posterior mean in (2.10) is a matrix-weighted average of the prior mean $\hat{\pi}_{oi}$ and $\hat{\pi}_i$ the least squares estimate. Further, with an assumption analogous to that in Theorem 2 we have $\mathrm{Cov}(\pi_i, \pi_j) = (W'W)^{-1}E\omega_{ij} = (X^{o\prime}X^o + X'X)^{-1}\hat{\varepsilon}'_i\hat{\varepsilon}_j/(n' - k)$ where $\hat{\varepsilon}_i = w_i - W\tilde{\pi}_i$ and $\tilde{\varepsilon}_j = w_j - W\tilde{\pi}_j$, $i, j = 1, 2, \ldots, m$ and $n' = n + n_o$, where n_o is the number of conceptual observations.

III. Optimal finite sample estimates of structural parameters

In this section we consider the estimation of the parameters in a single, identified equation of a SEM, a central problem in econometrics for many years. After giving various interpretations of the model, finite sample, optimal BMOM estimates of the structural coefficients are obtained relative to three different loss functions without using sampling assumptions, likelihood functions or prior densities. The optimal estimates so obtained are then compared with those obtained using a normal likelihood function, diffuse prior density and Bayes's theorem. It is then shown how the analysis can be extended to provide optimal estimates of parameters of several or all equations of a SEM.

3.1 Estimation parameters of an identified structural equation using various loss functions

We consider an identified equation, say the first of an SEM, namely

$$\underset{n\times 1}{y_1} = \underset{n\times m_1}{Y_1}\ \underset{m_1\times 1}{\gamma_1} + \underset{n\times k_1}{X_1}\ \underset{k_1\times 1}{\beta_1} + \underset{n\times 1}{u_1} \qquad (3.1)$$

where γ_1 and β_1 are structural coefficient vectors, u_1 is a given vector of realized error terms, y_1 and Y_1 are a vector and matrix of given observations on the included endogenous variables and X_1 is a given matrix of predetermined variables. $Y = (y_1\ Y_1\ Y_o)$ and $X = (X_1\ X_o)$, with X of full column rank, are matrices of observations on all endogenous and predetermined variables, respectively, appearing in the complete SEM. The reduced form equations for y_1 and Y_1 are

$$\underset{n\times 1}{y_1} = \underset{n\times k}{X}\ \underset{k\times 1}{\pi_1} + \underset{n\times 1}{v_1} \qquad (3.2)$$

$$\underset{n\times m_1}{Y_1} = \underset{n\times k}{X}\ \underset{k\times m_1}{\Pi_1} + \underset{n\times m_1}{V_1} \qquad (3.3)$$

On substituting from (3.3) in (3.1), we have

$$y_1 = X\Pi_1\gamma_1 + X_1\beta_1 + v_1 \tag{3.4a}$$
$$= \bar{Z}_1\delta_1 + v_1 \tag{3.4b}$$

with $\delta_1' = (\gamma_1'\ \beta_1')$ and $\bar{Z}_1 = (X\Pi_1\ X_1)$ which is the restricted reduced form equation for y_1. Note that $X\pi_1 = \bar{Z}_1\delta_1$ are the restrictions imposed on (3.2) and that $\rho(\bar{Z}_1) = m_1 + k_1$ is a rank condition for identification of δ_1.

Various-interpretations of the model are available, as has been pointed out in the literature. From (3.4a), it can be viewed as an unobservable regressor model, $X\Pi_1$ being the unobserved regressors. From (3.3) and (3.4a) it can be viewed as a non-linear in the parameters model. Also, as pointed out in Zellner et al. (1994), (3.3) and (3.4a) are in the form of a multivariate errors-in-the-variables model, a rational expectations model and a consumption based asset-pricing model.

With this said about interpretations of this model, we shall now derive optimal finite sample estimates of $\delta_1' = (\gamma_1'\ \beta_1')$ relative to three loss functions, a goodness-of-fit loss function, L_g, a precision of estimation loss function, L_p and a 'balanced' or 'extended' loss function, $L_b == \omega L_g + (1-\omega)L_p$ with $0 \le \omega \le 1$. The goodness-of-fit loss function, L_g, is similar to the sum of squared error terms used to produce least squares estimates. The precision of estimation loss function, L_p, is a generalized quadratic loss function relating to parameter estimation. As explained below, other loss functions can also be employed.

3.2 *Estimation relative to 'goodness-of-fit' loss function,* $\mathrm{L_g}$

The convex loss function reflecting goodness of fit of the relation in (3.4) is:

$$L_g = (y_1 - \bar{Z}_1\hat{\delta}_1)'(y_1 - \bar{Z}_1\hat{\delta}_1) \tag{3.5}$$

where $\hat{\delta}_1$ is some estimate, y_1 is a given vector of observations and $\bar{Z}_1 = (X\Pi_1\ X_1)$ is assumed of full column rank. Taking the posterior expectation of L_g, we have:

$$EL_g = \hat{\delta}_1' E\bar{Z}_1'\bar{Z}_1\hat{\delta}_1 - 2\hat{\delta}'_1 E\bar{Z}_1' y_1 + y_1'y_1$$

On completing the square on $\hat{\delta}_1$, the result is:

$$EL_g = (\hat{\delta}_1 - \hat{\delta}_1^*)'E\bar{Z}_1'\bar{Z}_1(\hat{\delta}_1 - \hat{\delta}_1^*) + y_1'y_1 - \hat{\delta}_1^{*\prime}E\bar{Z}_1'\bar{Z}_1\hat{\delta}_1^* \tag{3.6}$$

with

$$\hat{\delta}_1^* = \left(E\bar{Z}'_1\bar{Z}_1\right)^{-1}E\bar{Z}'_1 y_1$$
$$= \begin{pmatrix} E\Pi'_1 X' X\Pi_1 & E\Pi'_1 X' X_1 \\ X'_1 XE\Pi_1 & X'_1 X_1 \end{pmatrix}^{-1} \begin{pmatrix} E\Pi'_1 X' y_1 \\ X'_1 y_1 \end{pmatrix} \tag{3.7}$$

Theorem III: From (3.6), taking $\hat{\delta}_1 = \hat{\delta}_1^*$, with $\hat{\delta}_1^*$ given in (3.7) provides minimal expected loss given that $E\bar{Z}_1'\bar{Z}_1$ is pds.

On inserting $\hat{\delta}_1 = \hat{\delta}_1^*$ in (3.6), minimal expected loss is:

$$\begin{aligned} E(L_g \mid \hat{\delta}_1 = \hat{\delta}_1^*) &= y'_1 y_1 - \hat{\delta}_1^{*\prime} E\bar{Z}_1'\bar{Z}_1 \hat{\boldsymbol{\delta}}_1^* \\ &= y_1'y_1 - y_1'E\bar{Z}_1(E\bar{Z}_1'\bar{Z}_1)^{-1}E\bar{Z}_1'y_1 \\ &= \mathbf{y}_1'[I - E\bar{Z}_1(E\bar{Z}_1'\bar{Z}_1)^{-1}\, E\bar{Z}_1']\mathbf{y}_1 \end{aligned} \qquad (3.8)$$

If some other value of $\hat{\delta}_1$, say $\hat{\delta}^0$, is considered, from (3.6),

$$E(L_g \mid \hat{\delta}_1 = \hat{\delta}_1^0) - E(L_g \mid \hat{\delta}_1 = \hat{\delta}_1^*) = (\hat{\delta}_1^0 - \hat{\delta}_1^*)' E\bar{Z}_1'\bar{Z}_1(\hat{\delta}_1^0 - \hat{\delta}_1^*) > 0$$

given that $E\bar{Z}_1'\bar{Z}_1$ is pds.

3.3 *Estimation relative to a precision of estimation loss function, L_p*

The precision of estimation loss function that we consider is the following generalized quadratic loss function:

$$L_p = (\delta_1 - \hat{\delta}_1)'\bar{Z}_1'\bar{Z}_1(\delta_1 - \hat{\delta}_1) \qquad (3.9a)$$

$$= (X\pi_1 - \bar{Z}_1\hat{\delta}_1)'(X\pi_1 - \bar{Z}_1\hat{\delta}_1) \qquad (3.9b)$$

since $X\pi_1 = \bar{Z}_1\delta_1$. Then posterior expected loss is given by:

$$EL_p = \hat{\delta}_1' E\bar{Z}_1'\bar{Z}_1\hat{\delta}_1 - 2\hat{\delta}'_1 E\bar{Z}_1'X\pi_1 + E\pi_1'X'X\pi_1$$

On completing the square on $\hat{\delta}_1$, we have:

$$EL_p = (\hat{\delta}_1 - \hat{\delta}_1^+)'E\bar{Z}_1'\bar{Z}_1(\hat{\delta}_1 - \hat{\delta}_1^+) + E\pi_1'X'X\pi_1 - \hat{\delta}_1^{+\prime}E\bar{Z}_1'\bar{Z}_1\hat{\delta}_1^+ \qquad (3.10)$$

with

$$\begin{aligned} \hat{\delta}_1^+ &= \left(E\bar{Z}'_1\,\bar{Z}_1\right)^{-1} E\bar{Z}'_1\, X\pi_1 \\ &= \begin{pmatrix} E\Pi'_1 X' X\Pi_1 & E\Pi'_1 X' X_1 \\ X'_1 X E\Pi_1 & X'_1 X_1 \end{pmatrix}^{-1} \begin{pmatrix} E\Pi'_1 X' \pi_1 \\ X'_1 \pi_1 \end{pmatrix} \end{aligned} \qquad (3.11)$$

Theorem IV: From (3.10), taking $\hat{\delta}_1 = \hat{\delta}_1^+$ given in (3.11) provides minimal expected loss.

On inserting $\hat{\delta}_1 = \hat{\delta}_1^+$ in (3.10), minimal expected loss is:

$$\begin{aligned} EL_p \mid \hat{\delta}_1 = \hat{\delta}_1^+ &= E\pi'_1 X'X\pi_1 - \hat{\delta}_1^{+\prime}E\bar{Z}_1'\bar{Z}_1\hat{\delta}_1^+ \\ &= E\pi_1'X'X\pi_1 - E\pi_1'X'\bar{Z}_1(E\bar{Z}_1'\bar{Z}_1)^{-1}E\bar{Z}_1'X\pi_1 \end{aligned} \qquad (3.12)$$

For any other estimate, $\hat{\delta}_1^0$, from (3.10),

$$E(L_p \mid \hat{\delta}_1 = \hat{\delta}_1^0) - E(L_p \mid \delta_1 = \hat{\delta}_1^+) = (\hat{\delta}_1^0 - \hat{\delta}_1^+)' E\bar{Z}_1'\bar{Z}_1(\hat{\delta}_1^0 - \hat{\delta}_1^+) > 0 \qquad (3.13)$$

3.4 Estimation relative to a balanced-loss function:[4] $\mathrm{L_b} = \omega \mathrm{L_g} + (1-\omega)\mathrm{L_{p'}}, 0 \leq \omega \leq 1$
The balanced loss function that we employ is:

$$L_b = \omega(y_1 - \bar{Z}_1\hat{\delta}_1)'(y_1 - \bar{Z}_1\hat{\delta}_1) + (1-\omega)(X\pi_1 - \bar{Z}_1\hat{\delta}_1)'(X\pi_1 - \bar{Z}_1\hat{\delta}_1) \qquad (3.14)$$

Posterior expected loss is given by:

$$EL_b = \hat{\delta}_1' E\bar{Z}_1'\bar{Z}_1\hat{\delta}_1 - 2\hat{\delta}_1'[\omega E\bar{Z}_1' y_1 + (1-\omega)E\bar{Z}_1' X\pi_1] + \omega y_1' y_1 + (1-\omega))E\pi_1' X' X\pi_1$$

On completing the square on $\hat{\delta}_1$, the result is:

$$EL_b = (\hat{\delta}_1 - \hat{\delta}_1^x)' E\bar{Z}_1'\bar{Z}_1(\hat{\delta}_1 - \hat{\delta}_1^x) + \omega y_1' y_1 + (1-\omega)E\pi_1' X' X\pi_1 - \hat{\delta}_1^{x\prime} E\bar{Z}_1'\bar{Z}_1\hat{\delta}_1^x \qquad (3.15)$$

where

$$\hat{\delta}_1^x = (E\bar{Z}_1'\bar{Z}_1)^{-1}[\omega E\bar{Z}_1' y_1 + (1-\omega)E\bar{Z}_1' X\pi_1] \qquad (3.16a)$$

or

$$\hat{\delta}_1^x = \omega\hat{\delta}_1^* + (1-\omega)\hat{\delta}_1^x \qquad (3.16b)$$

with $\hat{\delta}_1^*$ given in (7) and $\hat{\delta}_1^+$ in (11).

Theorem V: From (3.15), taking $\hat{\delta}_1 = \hat{\delta}_1^x$ given in (3.16) provides minimal expected loss.

For any choice of $\hat{\delta}_1$, say $\hat{\delta}_1^0$, we have from (3.15):

$$E(L_b \mid \hat{\delta}_1 = \hat{\delta}_1^0) - E(L_b \mid \hat{\delta}_1 = \hat{\delta}_1^x) = (\hat{\delta}_1^0 - \hat{\delta}_1^x)' E\bar{Z}'_1\bar{Z}_1(\hat{\delta}_1^0 - \hat{\delta}_1^x) > 0$$

Above, optimal estimates for three specific loss functions have been derived. To derive estimates for many other loss functions, note that one complete posterior density for δ_1 in (3.4b) can be obtained as follows. Using results in Zellner (1994b), analyse (3.4b) conditional on Π_1 as a regression model to obtain a normal maxent density for δ_1 given Π_1, σ_{11} and the data, D: $N\{\bar{\delta}_1, (\bar{Z}_1'\bar{Z}_1)^{-1}\omega_{11} \mid \Pi_1, \omega_{11}, D)$, where $\bar{\delta}_1 = (\bar{Z}_1'\bar{Z}_1)^{-1}\bar{Z}_1' y_1$ and ω_{11} is a variance parameter that has an exponential maxent density given Π_1 and the data, $f(\omega_{11} \mid \Pi_1, D) \propto \exp\{-\omega_{11}/\bar{\omega}_{11}\}$ with $\bar{\omega}_{11} = (y_1 - \bar{Z}_1\bar{\delta}_1)'(y_1 - \bar{Z}_1\bar{\delta}_1)/(n - m_1 - k_1)$. Further, from (2.5) and (2.6), the maxent posterior density for the elements of Π_1 is multivariate normal given the data, $f_N(\Pi_1 \mid D)$. Then drawing a Π_1 from $f_N(\Pi_1 \mid D)$, inserting it in the exponential density for ω_{11}, drawing a value of ω_{11} from its exponential density, substituting the drawn Π_1 and ω_{11} in the normal conditional density for δ_1 and drawing a value of δ_1 from it provides a draw from the marginal density of δ_1. Repeating this process several thousand times will provide a good approximation to the densities, moments, fractiles, etc. of the elements of δ_1. This output can be employed to compute optimal estimates for a wide range of statistical and subject matter loss functions.

3.5 *Explicit forms for optimal estimates*

Using the moments of π given in Theorems I and II above, $\bar{\delta}_1^*$, $\bar{\delta}_1^+$ and $\bar{\delta}_1^x$ in (3.7), (3.11) and (3.16), respectively, were evaluated with results given in Table 1 in the column labelled BMOM approach. It is seen that for the goodness-of-fit and balanced-loss functions, the optimal point estimates are in the form of 'double K class' estimates while for the case of the precision of estimation case, the optimal estimate is in the form of a 'single K class' estimate with a value of K less than one for finite sample size.[5] For this last case, the estimate reduces to OLS for $K = 0$, that is $1 - k/(n - k) = 0$ or $n = 2k$. For example, if we have X with $k = 10$ columns or predetermined variables, with a sample size of $n = 20$, OLS is optimal with respect to the precision of estimation loss function. Shown below are optimal values of K for various values of k/n for the BMOM estimate in line 2 of Table 1:

Value of k/n:	0.50	0.45	0.40	0.35	0.30	0.25	0.20	0.15	0.10
Optimal value of K:	0	0.18	0.33	0.46	0.57	0.67	0.75	0.82	0.89

From the above, we see that there is a broad range of values for k/n for which the optimal value of K is much below one, the value producing the 2SLS estimate. However, as $k/n \to 0$, $K \to 1$. Similarly the optimal estimates for the goodness-of-fit and balanced-loss functions are somewhat different from 2SLS and other commonly used estimates in finite samples.

Also shown in Table 1 are the optimal point estimates for the three loss functions produced using a traditional normal reduced form likelihood function, a diffuse prior for its parameters and Bayes's theorem. It is seen that the point estimates are in the same forms as the corresponding BMOM estimates but the values of K_1 and K_2 differ. In all cases, the value of K_1 is somewhat smaller than in the BMOM case and for a broad range of finite sample cases will be much below one in value.

We now turn to indicate how optimal BMOM point estimates can be obtained for parameters of several equations.

3.6 *Estimation of parameters of several equations*

As in connection with (3.1)–(3.4), we denote the i'th structural equation as

$$y_i = \bar{Z}_i\delta_i + v_i \qquad i = 1, 2, \ldots, m \tag{3.17}$$

where $\bar{Z}_i = (X\Pi_i\ X_i)$ and $\delta_i' = (\gamma_i'\ \beta_i')$. We can represent the set of equations in (3.17) as follows:

Table 1 *Bayesian minimum expected loss (MELO) structural coefficient estimates*[a]

Loss function	BMOM approach	Diffuse prior, normal LF traditional Bayesian approach[b]
1. Goodness of fit: $L_g = (y_1 - \bar{Z}_1\hat{\delta}_1)'(y_1 - \bar{Z}_1\hat{\delta}_1)$	$K_1 = 1 - k/(n-k)$ $K_2 = 1$	$K_1 = 1 - k/(n-k-m_1-2)$ $K_2 = 1$
2. Precision of estimation: $L_p = (\hat{\delta}_1 - \delta_1)'\bar{Z}'_1\bar{Z}_1(\hat{\delta}_1 - \delta_1)$	$K_1 = K_2 = 1 - k/(n-k)$	$K_1 = K_2 = 1 - k/(n-k-m_1-2)$
3. Balanced or extended loss function: $L_b = \omega L_g + (1-\omega)L_p$	$K_1 = 1 - k/(n-k)$ $K_2 = 1 - (1-\omega)k/(n-k)$	$K_1 = 1 - k/(n-k-m_1-2)$ $K_2 = 1 - (1-\omega)k/(n-k-m_1-2)$

[a] The MELO estimates are in the form of K class or double K class estimates, i.e.

$$\delta_1(K_1, K_2) = \begin{bmatrix} Y'_1 Y_1 - K_1 \hat{V}'_1 \hat{V}_1 & Y'_1 X_1 \\ X'_1 Y_1 & X'_1 X_1 \end{bmatrix}^{-1} \begin{bmatrix} (Y_1 - K_2 \hat{V}_1)' y_1 \\ X'_1 y_1 \end{bmatrix}$$

The parameters, $\delta'_1 = (\gamma'_1 \beta'_1)$ in the following structural equation are estimated by $\bar{\delta}_1 (K_1, K_2)$: $\underset{n\times 1}{y_1} = \underset{n\times m_1}{Y_1}\ \underset{m_1\times 1}{\gamma_1} + \underset{n\times k_1}{X_1}\ \underset{k_1\times 1}{\beta_1} + \underset{n\times 1}{u_1}$,

with $Y_1 = X\Pi_1 + V_1$, the RF equations for Y_1 and $\hat{V}_1 = Y_1 - X\hat{\Pi}_1$ with $\underset{n\times m_1}{\hat{V}} = \underset{n\times m_1}{Y_1} - \underset{n\times k}{X}\ \underset{k\times m_1}{\hat{\Pi}_1}$

and $\hat{\Pi}_1 = (X'X)^{-1}X'Y_1$.

[b] See Zellner (1978, 1986a) for derivation of these results.

$$\begin{pmatrix} y_1 \\ y_2 \\ \vdots \\ y_m \end{pmatrix} = \begin{pmatrix} \bar{Z}_1 & 0 & \cdots & 0 \\ 0 & \bar{Z}_2 & \cdots & 0 \\ \vdots & \vdots & \ddots & \vdots \\ 0 & 0 & \cdots & \bar{Z}_m \end{pmatrix} \begin{pmatrix} \delta_1 \\ \delta_2 \\ \vdots \\ \delta_m \end{pmatrix} + \begin{pmatrix} v_1 \\ v_2 \\ \vdots \\ v_m \end{pmatrix} \tag{3.18}$$

or

$$w = \bar{W}\delta + v \tag{3.19}$$

Then if we use a multi-equation goodness-of-fit loss function,

$$L = (w - \bar{W}\hat{\delta})'Q(w - \bar{W}\hat{\delta}) \tag{3.20}$$

posterior expected loss is given by:

$$EL = \hat{\delta}'E\bar{W}'Q\bar{W}\hat{\delta} - 2\hat{\delta}'E\bar{W}'Qw + w'EQw$$

and the optimal point estimate $\hat{\delta}^{opt}$ is

$$\delta^{opt} = (E\bar{W}'Q\bar{W})^{-1}E\bar{W}'Qw \tag{3.21}$$

δ^{opt} can be evaluated using BMOM or traditional Bayesian posterior moments. Similar optimal 'systems' estimates can also be derived using joint precision of estimation and balanced loss functions.

We now turn to a review of the results of several Monte Carlo experiments designed to compare the relative performance of Bayesian and non-Bayesian point estimation procedures.

IV. Performance of Bayesian and non-Bayesian estimation procedures

In this section results of several Monte Carlo experiments designed to compare the performance of Bayesian and non-Bayesian procedures are reviewed. These experiments test performance of various procedures in repeated experimental trials. To the extent that these trials resemble what is actually done in using these estimation procedures, the results are relevant in establishing their operating characteristics and in helping to answer Carl Christ's (1960) query, 'Any Verdict Yet?'

4.1 The Tsurumi (1990) Experiments

In these experiments Tsurumi employed a version of the model used in Cragg's (1967) experiments, a three-equation linear SEM. The estimators considered included OLS, 2SLS, LIML, Fuller (1977) modified LIML (FLIML), Zellner's minimum expected loss (ZM) and extended MELO (ZEM), conditional LI Bayesian (CLIB), unconditional LIB (ULIB) and mode of posterior density (PMOD). The ZM and ZEM estimators are the K and double K class estimates described in Table 1, lines 2 and 3 in the third column with $\omega = 0.75$. Each of these estimation procedures was employed in estimating five coefficients in the first equation of the three-equation system, namely

$$y_{1t} = \gamma_{12}y_{2t} + \beta_{11}x_{1t} + \beta_{13}x_{3t} + \beta_{15}x_{5t} + \beta_{17}x_{7t} + u_{1t} \qquad t = 1, 2, \ldots, T \quad (4.1)$$

Tables 2, 3, 4, and 5, taken from Tsurumi (1990), display the main results of his study. To evaluate alternative estimation procedures, he used a sum of relative mean absolute deviations (SRMAD) criterion, defined in footnote 1 of Table 2 as well as a sum of relative mean squared errors that produced similar results. The results indicate that the Bayesian estimators outperform the non-Bayesian estimators according to the criterion SRMAD.

Table 2 Sums of relative mean absolute deviations: no multicollinearity; $R^2 = 0.9$

	$\rho_{12}^2 = 0$ SRM AD	$\rho_{12}^2 = 0.3$ SRM AD	$\rho_{12}^2 = 0.5$ SRM AD	$\rho_{12}^2 = 0.7$ SRM AD	$\rho_{12}^2 = 0.9$ SRM AD
			$n = 100$		
OLS	5.000(1)	7.317(9)	8.573(9)	9.582(9)	10.49(9)
2SLS	5.071(5)	5.041(4)	5.063(4)	5.088(4)	5.115(4)
LIML	5.074(6)	5.039(2)	5.061(2)	5.086(2)	5.112(2)
FLIM	5.066(3)	5.073(6)	5.073(6)	5.100(6)	5.127(6)
ZM	5.056(2)	5.073(7)	5.115(8)	5.100(6)	5.199(8)
ZEM	5.067(4)	5.044(5)	5.069(5)	5.096(5)	5.125(5)
CLIB	5.074(6)	5.038(1)	5.060(1)	5.085(1)	5.112(2)
ULIB	5.074(6)	5.039(2)	5.061(2)	5.086(2)	5.111(1)
PMOD	5.122(9)	5.082(8)	5.082(7)	5.104(7)	5.139(7)
			$n = 20$		
OLS	5.001(1)	5.288(9)	5.507(9)	5.718(9)	5.929(9)
2SLS	5.131(5)	5.147(3)	5.180(3)	5.213(2)	5.251(2)
LIML	5.179(9)	5.200(7)	5.243(7)	5.285(6)	5.330(6)
FLIM	5.107(4)	5.144(2)	5.183(4)	5.220(3)	5.261(5)
ZM	5.022(3)	5.184(6)	5.312(8)	5.439(8)	5.569(8)
ZEM	5.017(2)	5.054(1)	5.096(1)	5.137(1)	5.183(1)
CLIB	5.154(7)	5.163(5)	5.193(5)	5.224(5)	5.258(3)
ULIB	5.179(9)	5.200(7)	5.242(6)	5.285(6)	5.330(6)
PMOD	5.135(6)	5.154(4)	5.178(2)	5.221(4)	5.260(4)

Notes

1. SRM AD = the sum of relative MADs:

$$\text{SRMAD} = \Sigma_{i=1}^{5} \text{MAD}(\hat{\theta}_i^{(j)}) / \min_p \text{MAD}(\hat{\theta}_i^{(p)})$$

   where MAD $(\hat{\theta}_i^{(j)})$ is the mean absolute deviation of the estimate of θ_i by the jth estimation method.
2. The numbers in parentheses are the ranks of the SRMADs.
3. $\rho_{12} = \text{Corr}(y_{2t}, u_{1t})$, the correlation coefficient between the right hand side endogenous variable and the error term.
4. R^2 = the coefficient of determination of the reduced form equation for the right hand side endogenous variable, y_{2t}.
5. For each case, results are based on 400 replications.

Source: Tsurumi (1990).

Table 3 Sums of relative mean absolute deviations: multicollinearity; $R^2 = 0.9$

	$\rho_{12}^2 = 0$ SRM AD	$\rho_{12}^2 = 0.3$ SRM AD	$\rho_{12}^2 = 0.5$ SRM AD	$\rho_{12}^2 = 0.7$ SRM AD	$\rho_{12}^2 = 0.9$ SRM AD
			$n = 100$		
OLS	5.000(1)	9.083(8)	11.193(8)	12.682(8)	14.150(8)
2SLS	12.291(7)	7.604(6)	7.295(5)	7.159(5)	7.257(5)
LIML	41.487(9)	68.342(9)	22.575(9)	19.495(9)	31.733(9)
FLIM	8.331(4)	6.402(2)	7.022(4)	7.556(6)	8.155(6)
ZM	6.601(2)	7.109(5)	8.440(7)	9.429(7)	10.389(7)
ZEM	6.888(3)	5.474(1)	5.777(1)	6.044(1)	6.362(3)
CLIB	14.146(8)	8.040(7)	7.323(6)	6.900(4)	6.809(4)
ULIB	10.192(6)	6.629(4)	6.378(2)	6.198(2)	6.238(2)
PMOD	10.192(5)	6.549(3)	6.451(3)	6.362(3)	6.112(1)
			$n = 20$		
OLS	5.469(2)	6.901(5)	7.836(6)	8.554(8)	9.335(8)
2SLS	9.302(7)	8.094(7)	8.301(7)	8.307(5)	8.496(5)
LIML	54.670(9)	32.307(9)	21.398(9)	20.936(9)	35.815(9)
FLIM	5.938(4)	6.782(3)	7.552(3)	8.145(4)	8.727(6)
ZM	5.470(3)	6.847(4)	7.751(5)	8.448(6)	9.203(7)
ZEM	5.000(1)	5.396(1)	5.616(1)	5.792(1)	5.987(1)
CLIB	10.977(8)	9.017(8)	8.806(8)	8.508(7)	8.432(4)
ULIB	6.858(5)	6.273(2)	6.573(2)	6.899(2)	7.104(2)
PMOD	7.228(6)	7.225(6)	7.641(4)	7.947(3)	8.171(3)

Notes: As for Table 2.

Source: Tsurumi (1990).

Tsurumi concludes:

1. The sampling results show that while no single estimator dominates all the other estimators at all times, the conditional and unconditional Bayesian estimators and Zellner's extended minimum expected loss estimator tend to outperform the other estimators when the degree of simultaneity increases. (p.179)
2. Except for one case in which alternative estimators' performance was about the same, except for OLS which exhibited poorer performance, '... ZEM is always in the top three rankings [see Table 5] ... In many cases, ZEM has the least SRMAD's [sum of relative mean absolute deviations].' (p.190)[6]

Thus Tsurumi's experiments provide evidence that Bayesian estimators' performance relative to the loss functions that he used, SRMAD, SRMSE, etc. is rather good relative to non-Bayesian estimators' performance.

Table 4 Sums of relative mean absolute deviations: multicollinearity; R^2 = 0.35

	$\rho_{12}^2 = 0$ SRM AD	$\rho_{12}^2 = 0.3$ SRM AD	$\rho_{12}^2 = 0.5$ SRM AD	$\rho_{12}^2 = 0.7$ SRM AD	$\rho_{12}^2 = 0.9$ SRM AD
			$n = 100$		
OLS	5.009(1)	6.868(4)	8.024(5)	8.868(6)	9.644(6)
2SLS	13.456(7)	9.228(7)	9.110(7)	9.190(7)	9.652(7)
LIML	108.13(9)	62.284(9)	35.206(9)	31.591(9)	38.200(9)
FLIM	6.869(4)	6.910(5)	7.882(3)	8.635(4)	9.395(3)
ZM	5.795(2)	6.833(3)	7.916(4)	8.730(5)	9.487(5)
ZEM	6.241(3)	5.407(1)	5.656(1)	5.917(1)	6.447(1)
CLIB	16.806(8)	10.053(8)	9.638(8)	9.330(8)	9.721(8)
ULIB	8.566(5)	6.476(2)	6.991(2)	7.624(2)	8.445(2)
PMOD	9.484(6)	7.377(6)	8.040(6)	8.597(3)	9.398(4)
			$n = 20$		
OLS	5.382(2)	6.096(4)	6.633(4)	7.110(4)	7.760(5)
2SLS	8.713(7)	7.877(7)	8.007(7)	7.934(7)	8.301(7)
LIML	88.585(9)	36.099(9)	20.193(9)	21.552(9)	16.646(9)
FLIM	5.736(4)	6.187(5)	6.678(5)	7.123(5)	7.723(3)
ZM	5.388(3)	6.095(3)	6.632(3)	7.102(3)	7.749(4)
ZEM	5.000(1)	5.252(1)	5.423(1)	5.662(1)	6.267(1)
CLIB	10.005(8)	8.620(8)	8.603(8)	8.288(7)	8.542(8)
ULIB	6.419(5)	5.940(2)	6.235(2)	6.675(2)	7.399(2)
PMOD	6.807(6)	6.675(6)	6.930(6)	7.241(6)	7.796(6)

Notes: As for Table 2.

Source: Tsurumi (1990).

Table 5 Top three ranked estimators for each case of R^2, and n *with or without multicollinearity*[a]

	$\rho_{12}^2 = 0$	$\rho_{12}^2 = 0.3$	$\rho_{12}^2 = 0.5$	$\rho_{12}^2 = 0.7$	$\rho_{12}^2 = 0.9$
			Without multicollinearity		
$R_2 = 0.9$	(1)OLS	(1)CLIB	(1)CLIB	(1)CLIB	(1)ULIB
$n = 100$	(2)ZM	(2)LIML	(2)LIML	(2)LIML	(2)LIML
	(3)FLIM	(3)ULIB	(3)ULIB	(3)ULIB	(3)CLIB
$R^2 = 0.9$	(1)OLS	(1)ZEM	(1)ZEM	(1)ZEM	(1)ZEM
$n = 20$	(2)ZEM	(2)FLIM	(2)PMOD	(2)2SLS	(2)2SLS
	(3)ZM	(3)2SLS	(3)2SLS	(3)FLIM	(3)CLIB
$R^2 = 0.35$)	(1)OLS	(1)ZEM	(1)ZEM	(1)ZEM	(1)ZEM
$n = 100$	(2)ZM	(2)FLIM	(2)PMOD	(2)PMOD	(2)PMOD
	(3)ZEM	(3)PMOD	(3)FLIM	(3)CLIB	(3)FLIM
$R_2 = 0.35$	(1)ZEM	(1)ZEM	(1)ZEM	(1)ZEM	(1)ZEM
$n = 20$	(2)0LS	(2)FLIM	(2)FLIM	(2)PMOD	(2)PMOD
	(3)ZEM	(3)PMOD	3)PMOD	(3)FLIM	(3)FLIM

Table 5 continued

	$\rho^2_{12} = 0$	$\rho^2_{12} = 0.3$	$\rho^2_{12} = 0.5$	$\rho^2_{12} = 0.7$	$\rho^2_{12} = 0.9$
		With multicollincarity			
$R^2 = 0.9$	(1)OLS	(1)ZEM	(1)ZEM	(1)ZEM	(1)PMOD
$n = 100$	(2)ZM	(2)FLIM	(2)ULIB	(2)ULIB	(2)ULIB
	(3)ZEM	(3)PMOD	(3)PMOD	(3)PMOD	(3)ZEM
$R^2 = 0.9$	(1)ZEM	(1)ZEM	(1)ZEM	(1)ZEM	(1)ZEM
$n = 20$	(2)OLS	(2)ULIB	(2)ULIB	(2)ULIB	(2)ULIB
	(3)ZM	(3)FLIM	(3)FLIM	(3)PMOD	(3)PMOD
$R^2 = 0.35$	(1)OLS	(1)ZEM	(1)ZEM	(1)ZEM	(1)ZEM
$n = 100$	(2)ZM	(2)ULIB	(2)ULIB	(2)ULIB	(2)ULIB
	(3)ZEM	(3)ZM	(3)FLIM	(3)PMOD	(3)FLIM
$R^2 = 0.35$	(1)ZEM	(1)ZEM	(1)ZEM	(1)ZEM	(1)ZEM
$n = 20$	(2)OLS	(2)ULIB	(2)ULIB	(2)ULIB	(2)ULIB
	(3)ZM	(3)ZM	(3)ZM	(3)ZM	(3)FLIM

Note: Based on results in Tables 2–4.

Source: Tsurumi (1990).

4.2 *The Diebold–Lamb (1993) experiments*

The Diebold–Lamb (1993) Monte Carlo experiments employed a version of the widely used Nerlove agricultural supply response model from which data were generated and analysed. Their key supply response parameter, α, is related to reduced form coefficients as follows, $\alpha = \pi_2/(1 - \pi_3 - \pi_4)$. They compared the performance of $\hat{\alpha} = \hat{\pi}_2/(1 - \hat{\pi}_3 - \hat{\pi}_4)$, with the $\hat{\pi}$'s least squares estimates, and the Bayesian minimum expected loss (MELO) estimate, a special case of the estimate that is optimal relative to a precision of estimation loss function, L_p in Table 1. Specifically, the loss function employed is $L_p(\alpha, \tilde{\alpha})^2 = \delta_2^2(\alpha - \tilde{\alpha}^2) = (\delta_1 - \delta_2\tilde{\alpha})^2$ where $\delta_1 \equiv \pi_2$ and $\delta_2 \equiv 1 - \pi_3 - \pi_4$ and $\tilde{\alpha}$ is some estimate of α. The value of $\tilde{\alpha}$ that minimizes posterior expected loss is $\tilde{\alpha}_{MELO} = E\delta_1\delta_2/E\delta_2^2$ where $E \equiv$ posterior expectation. Diebold and Lamb used a normal reduced form system and a diffuse prior to obtain the posterior moments needed to compute the MELO estimate, $\tilde{\alpha}^*$ and gave it an interesting 'shrinkage' interpretation.

Table 6 Minimum expected loss estimation, $T = 25$, $\gamma = 1$, mean-squared error reduction relative to least squares

	$\rho = 0.50$	$\rho = 0.625$	$\rho = 0.75$	$\rho = 0.90$
$\sigma_\varepsilon = 1, \sigma_u = 5$	52%	45%	39%	35%
$\sigma_\varepsilon = 3, \sigma_u = 3$	37%	31%	21%	12%
$\sigma_\varepsilon = 2, \sigma_u = 2$	32%	24%	17%	6%
$\sigma_\varepsilon = 3, \sigma_u = 5$	32%	12%	17%	6%
$\sigma_\varepsilon = 5, \sigma_u = 5$	32%	29%	17%	6%

Source: Diebold and Lamb (1993).

Shown in Tables 6 and 7, taken from Diebold and Lamb (1993), are some of their experimental results that strongly favor the Bayesian MELO estimation approach.

Table 7 Empirical distributions of estimated supply response $\sigma_\varepsilon = 1$, $\sigma_u = 5$, $\rho = 0.625$

	$\gamma = 0.5$		$\gamma = 0.75$		$\gamma = 1.0$	
	$\hat{\alpha}_{LS}$	$\hat{\alpha}_{MELO}$	$\hat{\alpha}_{LS}$	$\hat{\alpha}_{MELO}$	$\hat{\alpha}_{LS}$	$\hat{\alpha}_{MELO}$
			$T = 25$			
MEAN	1.89	1.49	1.95	1.60	1.89	1.60
VAR	34.12	14.63	16.19	6.26	5.23	2.74
MSE	34.13	14.89	16.19	6.42	5.24	2.90
MIN	–25.32	–13.60	–44.51	–7.37	–6.68	–3.24
25%	–0.99	–0.84	0.06	–0.02	0.61	0.52
50%	1.60	1.32	1.73	1.50	1.73	1.54
75%	3.93	3.58	3.39	3.01	2.88	2.53
MAX	54.37	23.43	43.76	15.66	23.19	10.54
			$T = 50$			
MEAN	2.00	1.80	2.07	1.90	1.98	1.87
VAR	12.53	8.54	4.79	3.50	1.83	1.53
MSE	12.57	8.58	4.80	3.51	1.83	1.55
MIN	–22.73	–15.19	–10.58	–6.46	–4.33	–2.77
25%	–0.03	– 0.06	0.74	0.68	1.13	1.08
50%	1.65	1.56	1.85	1.72	1.88	1.80
75%	3.84	3.63	3.27	3.05	2.77	2.64
MAX	49.71	16.35	27.12	13.06	12.27	8.87
			$T = 100$			
MEAN	1.85	1.79	1.97	1.91	1.93	1.89
VAR	4.47	4.12	1.69	1.57	0.70	0.66
MSE	4.49	4.16	1.69	1.57	0.70	0.67
MIN	–4.81	–4.50	–2.31	–2.19	–0.85	–0.81
25%	0.43	0.43	1.05	1.03	1.34	1.31
50%	1.78	1.70	1.93	1.89	1.91	1.88
75%	3.24	3.11	2.78	2.68	2.45	2.38
MAX	9.52	9.35	6.51	6.35	5.13	5.01

Source: Diebold and Lamb (1993).

Diebold's and Lamb's conclusions follow:

1. 'MELO is always best [in terms of MSE], often by a large margin' (p. 10). In Table 6, results for sample size $T = 25$ indicate reductions in MSE of from 6 per cent to 52 per cent associated with the use of the MELO estimator relative to it, the usual sampling theory estimator.

2. '... the MELO estimator displays some bias, but the variability of the MELO estimator is substantially less than that of the LS estimator. This variance reduction results [in] overall superiority of the MELO estimator, whether assessed by MSE or the various order statistics ... For example, for $T = 25$, MELO estimation reduces MSE from 5.93 to 3.10' (p. 11).

If other estimation criteria were employed, say 0–1 loss or degree of concentration, then of course Bayesian estimates that are optimal relative to them, e.g. the posterior modal value for 0–1 loss or the posterior median for absolute error loss, etc. should be employed in comparing Bayesian and non-Bayesian estimation procedures.

4.3 The Park (1982) experiments

These experiments involved appraising the finite sample performance of OLS, 2SLS and MELO estimators with data generated from a three-equation linear model similar to that employed by Cragg (1967). The MELO estimator employed is a single K class estimator with $K = 1 - k/(n - k - m_1 - 2)$ as shown in Table 1.

Park's major conclusions are:

1. The MELO estimates appear to have more pronounced biases in estimating structural coefficients than the 2SLS estimates. However, MELO is found to outperform 2SLS according to several criteria including MSE and MAE in a wide range of situations' (p. 295).
2. (i) Ranking the estimators by the estimated absolute bias (table 2) yields an ordering of 2SLS, MELO and OLS. Pairwise comparisons between MELO and 2SLS show that MELO tends to have a significantly larger bias than 2SLS.
(ii) Ranking by the variance reverses the ordering: OLS, MELO, and 2SLS. Pairwise comparisons show that MELO often has a significantly smaller variance than 2SLS.
(iii) Judging by the MSE criterion (table 3), MELO performs better than 2SLS. Further, a small variance of OLS helps to offset a substantial part of its bias.
(iv)Ranking by MAE suggests an ordering of MELO, OLS, and 2SLS in the majority of experiments. Pairwise t tests show that MELO clearly surpasses 2SLS in terms of MAE.
(v) Judging by the absolute sampling errors (table 4), MELO performs better than OLS or 2SLS. The frequency with which the MELO estimate came closer to the true values of the structural parameters than other estimates, however, varied over the different coefficients in the experiments.
(vi) All three estimation methods produce biased estimates of the variances of structural disturbances (table 5). Both MELO and OLS estimates appear to be negatively biased while the 2SLS estimates tend to be positively biased.
(vii) The asymptotic standard errors of all three estimators tend to underestimate the variations in the coefficients estimates. The degree of underestimation appears to be more pronounced in MELO and OLS than in 2SLS.

Although sampling experiments provide insights into the behavior of estimators under a variety of circumstances, the experimental results may be specific to the model employed. Furthermore, the parameter space for a given model is so large that the results may provide only a limited picture. With all these limits of experimental approach, the findings emerged from the experiments generate the hypothesis that for a small model with a sample of moderate size, MELO performs as well as, if not better than, 2SLS in terms of the criteria considered. (p. 310)

4.4 The Zellner (1971, 276–87) experiments

The simple model used in these experiments is: $y_{1t} = \gamma y_{2t} + u_{1t}$ and $y_{2t} = \beta x_t + u_{2t}$, $t = 1, 2, \ldots, T$ with x_t exogenous and (u_{1t}, u_{2t}) i.i.d. bivariate normal with zero means and non-diagonal 2×2 pds covariance matrix. The reduced form equations are: $y_{1t} = \pi_1 x_t + v_{1t}$ and $y_{2t} = \pi_2 x_t + v_{2t}$ and the estimation of $\gamma = \pi_1/\pi_2$ was considered using two estimators. The first is the sampling theory FIML, 3SLS, 2SLS, etc. estimator $\hat{\gamma} = \hat{\pi}_1/\hat{\pi}_2$ with $\hat{\pi}_i = \sum_1^T x_t y_{it} / \sum_1^T x_t^2$, $i = 1, 2$. The second is $\tilde{\gamma}_B$ = modal value of posterior density of γ based on a normal likelihood function and a diffuse prior density for the reduced form parameters.

Some typical small sample results of these experiments are shown in Figures 1 and 2. Note that both the $\hat{\gamma}$ and $\tilde{\gamma}_B$ estimates for sample size $T = 20$ are well-centered around the true value of γ, namely 2. However, the dispersion of the Bayesian modal values, the $\tilde{\gamma}_B$'s, is much less than that of the sampling theory estimates, the $\hat{\gamma}$'s. The conclusions reached from this and other experiments are:

> The experiments reported above indicate that the differences in the sampling properties of Bayesian and sampling theory estimators in 'small-sample' situations are quite striking. Under the conditions we have examined Bayesian procedures produced better results than sampling-theory estimation procedures. With respect to point estimation, Bayesian estimates tended to be more highly concentrated about the true value of the parameter being estimated than sampling-theory estimates, particularly in small-sample situations. In this connection it appears relevant to note that the sampling-theory estimators considered are usually given a large-sample justification. Thus there is no assurance that these estimators will perform well in small-sample situations. With respect to interval estimation intervals computed from posterior

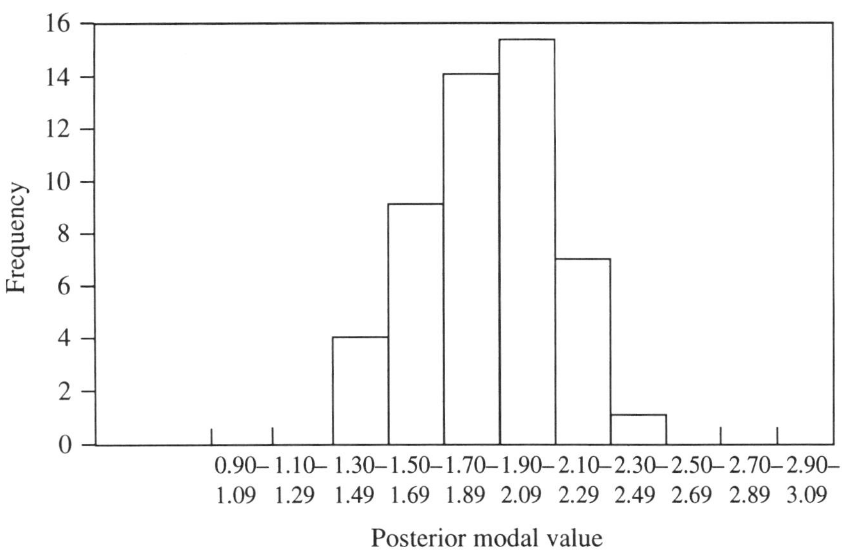

Source: Zellner (1971, 280–81)

Figure 1 Frequency distribution of Bayesian modal values of posterior distributions for γ (Run I: T = 20)

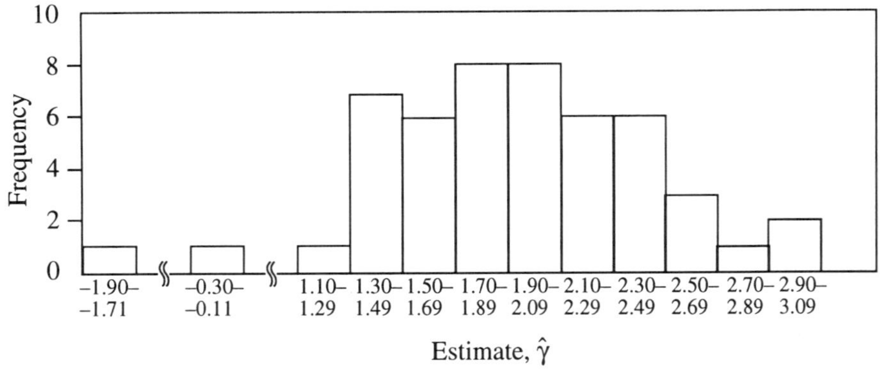

Source: As for Figure 1

Figure 2 Frequency distribution of sampling estimates of γ (Run I:T = 20)

> pdf's were found to have rather good sampling properties. On the other hand, *approximate* sampling-theory confidence intervals, often used in practice, were found to be deficient in small-sample situations. In large-sample situations, as theory predicts, Bayesian and sampling-theory procedures performed about equally well in terms of the criterion of performance in repeated samples. (pp. 286–7)

These experimental results provide evidence that Bayesian estimates of SEMs' parameters have rather good sampling properties. With more reasonable prior information incorporated in the BMOM approach or in the traditional Bayesian approach further improvement in estimation precision can be realized. Further, it must be emphasized that given a specific loss function, the Bayesian estimate that is optimal for it is the relevant estimate to study in Monte Carlo experiments, a point sometimes overlooked in evaluating Bayesian and other estimators. Then too, given that Bayes's risk is finite for a given loss function, it is known that the Bayesian estimator that minimizes Bayes's risk is admissible. Finally, see Zellner and Park (1979) for Bayesian MELO, OLS and 2SLS estimates of the parameters of Klein's Model I and of the Girschick–Haavelmo model of the supply and demand for food. The MELO estimates tend to be between the OLS and 2SLS estimates. In Zellner (1980) it is shown that the MELO estimate for the precision of estimation loss function, L_p, is a matrix weighted average of the OLS and 2SLS estimates.

V. Summary and conclusions

In this paper a new approach, the Bayesian method of moments (BMOM) approach has been utilized to produce optimal structural coefficient estimates relative to three different loss functions that are easily computed. Indeed they take the form of *K* class or double *K* class estimates with given values of the *K* quantities. Also, these estimates have been related to Bayesian minimum expected loss (MELO) and extended MELO estimates derived in a traditional Bayesian fashion using normal likelihood functions and diffuse priors. Then some Monte Carlo evidence was reviewed showing that Bayesian estimators performed better than non-Bayesian estimators.

Further, as briefly explained above, to obtain complete finite sample posterior densities for parameters, the BMOM approach provides results that can be used in a maximum entropy framework to derive maxent posterior densities of structural parameters – see Zellner (1994b) and Green and Strawderman (1994) for examples of this approach applied to regression models. Also in the traditional Bayesian approach various numerical procedures are available to compute finite sample posterior densities, moments, medians, modal values, intervals, etc. – see e.g., Drèze and Richard (1983), Kloek and van Dijk (1978), Geweke (1989, 1994a, b), van Dijk (1985), Poirier (1991), and Zellner et al. (1994).

In connection with Christ's (1960) query 'Any Verdict Yet?' he wrote, 'In summary, it is not yet clear that the least squares method for structural estimation is dead and should be discarded. It is now clear, however, that even for small samples least squares sometimes will not do as well as simultaneous equations methods' (p. 845). In terms of the BMOM and traditional Bayesian MELO estimates relative to a precision of estimation, generalized quadratic loss function, it is relevant to determine under what conditions these MELO estimates reduce to OLS and when they will depart from OLS. Also, as shown above, the Bayesian approach has provided a clear-cut justification for some *ad hoc* *K* class and double *K* class estimates. Given the other positive features of the BMOM and traditional Bayesian approaches stated in many works and above, the answer to Christ's query has to be, 'Yes. The verdict is that Bayesian methods produce very good finite sample and large sample estimation and other inference results.'

Notes

1. In a Bayesian context, such 'nuisance' parameters are integrated out, thereby giving a useful marginal density for the parameter or parameters of interest. See e.g., Johnston (1984, 510ff.), Judge et al. (1985, ch. 4) and Zellner (1971) for material on the Bayesian approach.
2. If additional information is available, it can be formalized, introduced as an additional assumption(s) and will lead to different results that can formally be compared with those based on just two assumptions.
3. Note that if one column of X is an $n \times 1$ column of 1's, denoted by ι, from $X'Ev_i = 0$, we have $E\iota'v_i/n = E\bar{v}_i = 0$, where $\bar{v}_i$ is the arithmetic mean of the elements of the vector v_i.
4. See Zellner (1986a, 1994a) for previous work using such 'two part', 'extended' or 'balanced' loss functions.
5. See, e.g., Christ (1966), Theil (1971) and Judge et al. (1985) for discussions of *K* class estimators from the sampling theory point of view and Swamy and Mehta (1983) for consideration of their properties in 'undersized' samples.
6. Tsurumi (1990) also reports that, 'The results when MAD's are used are quite similar to those when MSE's are used ...' (p. 189).

References

Barnard, G. (1975), 'Comment', in *Report of the Tenth NBER-NSF Seminar on Bayesian Inference in Econometrics*, University of Chicago, 9–10 May.

Basmann, R.L. (1961), 'A Note on the Exact Finite Sample Frequency Functions of Generalized Classical Linear Estimators in Two Leading Overidentified Cases', *Journal of the American Statistical Association*, **56**, 619–36.

Basmann, R.L. (1963), 'A Note on the Exact Finite Sample Frequency Functions of Generalized Classical Linear Estimators in a Leading Three Equation Case', *Journal of the American Statistical Association*, **58**, 161–71.

Bergstrom, A.R. (1962), 'The Exact Sampling Distributions of Least Squares and Maximum Likelihood Estimators of the Marginal Propensity to Consume', *Econometrica*, **30**, 480–90.

Chaloner, K. and R. Brant (1988), 'A Bayesian Approach to Outlier Detection and Residual Analysis', *Biometrika*, **75**, 651–9.

Christ, C.F. (1960), 'Simultaneous Equations Estimation: Any Verdict Yet?' *Econometrica*, **28**, 835–45.
Christ, C.F. (1966), *Econometric Models and Methods*, New York: John Wiley.
Christ, C.F. (1979), 'Comment', *Journal of the American Statistical Association*, **74**, 645–6.
Christ, C.F. (1994), 'The Cowles Commission's Contributions to Econometrics at Chicago, 1939–1955', *Journal of Economic Literature*, **XXXII**, 30–50.
Cover, T.M. and J.A. Thomas (1991), *Elements of Information Theory*, New York: John Wiley.
Cragg, J.G. (1967), 'On the Relative Small-Sample Properties of Several Structural-Equation Estimators', *Econometrica*, **35**, 89–110.
Diebold, F.X. and R.L. Lamb (1993), 'Why are Estimates of Agricultural Supply Response So Variable?' Economics and Econometrics Working Paper 93–151, University of Chicago.
Drèze, J.H. and J.-F. Richard (1983), 'Bayesian Analysis of Simultaneous Equation Systems', in Z. Griliches and M.D. Intriligator (eds), *Handbook of Econometrics:* Vol. 1, Amsterdam: North-Holland, Ch. 9, pp. 517–98.
Fuller, W.A. (1977), 'Some Properties of a Modification of the Limited Information Estimator', *Econometrica*, **45**, 939–53.
Geweke, J. (1989), 'Bayesian Inference in Econometric Models Using Monte Carlo Integration', *Econometrica*, **57**, 1317–39.
Geweke, J. (1994a), 'Monte Carlo Simulation and Numerical Integration', Working Paper 526, Research Department, Federal Reserve Bank of Minneapolis.
Geweke, J. (1994b), 'Bayesian Comparison of Econometric Models', Working Paper 532, Research Department, Federal Reserve Bank of Minneapolis.
Green, E.J. and W.E. Strawderman (1994), 'A Bayesian Growth and Yield Model for Slash Pine Plantations', Department of Natural Resources and Department of Statistics, Rutgers University, New Brunswick, NJ, to appear in *Applied Statistics*, 1996.
Greene, W. H. (1993), *Econometric Analysis* (2nd edn), New York: Macmillan.
Griffiths, W. E. and D. Dao (1980), 'A Note on a Bayesian Estimator in an Autocorrelated Error Model', *Journal of Econometrics*, **12**, 389–92.
Jaynes, E.T. (1982a), *Papers on Probability, Statistics and Statistical Physics*, Dordrecht, Netherlands: Reidel.
Jaynes, E.T. (1982b), 'On the Rationale of Maximum-Entropy Methods', *Proc. of the IEEE*, **70**, 939–52.
Johnston, J. (1984), *Econometric Methods* (3rd edn), New York: McGraw-Hill.
Judge, G.G., W.E. Griffiths, R. Carter Hill, H. Lütkepohl and T.-C. Lee (1985), *The Theory and Practice of Econometrics* (2nd edn), New York: John Wiley.
Kim, K. and A.R. Pagan (1994), 'The Econometric Analysis of Calibrated Macroeconomic Models', ms, Victoria University of Wellington, Australian National University and University of Rochester.
Kloek, T. and H.K. van Dijk (1978), 'Bayesian Estimates of Equation System Parameters: An Application of Integration by Monte Carlo', *Econometrica*, **46**, 1–20.
Koopmans, T. C. (1953), 'Identification Problems in Economic Model Construction', in Wm. C. Hood and T.C. Koopmans (eds), *Studies in Econometric Method, New* York: John Wiley, ch. II, pp. 27–48.
Nagar, A.L. (1959), 'The Bias and Moment Matrix of the General k-Class Estimators of the Parameters in Simultaneous Equations', *Econometrica*, **27**, 575–95.
Nelson, C.R. and R. Startz (1990), 'Some Further Results on the Exact Small Sample Properties of the Instrumental Variables Estimator', *Econometrica*, **58**, 967–76.
Park, S.-B. (1982), 'Some Sampling Properties of Minimum Expected Loss (MELO) Estimators of Structural Coefficients', *Journal of Econometrics*, **18**, 295–311.
Phillips, P.C.B. (1983), 'Exact Small Sample Theory in the Simultaneous Equations Model', Z. Griliches and M. D. Intriligator (eds), *Handbook of Econometrics*, Vol. 1, Amsterdam: North-Holland, ch. 8, pp. 449–516.
Poirier, D.J. (ed.) (1991), *Bayesian Empirical Studies in Economics and Finance*, Annals Issue of the *Journal of Econometrics*, **49**, July/August.
Poirier, D.J. (1992), 'A Return to the Battlefront', *Journal of Business and Economic Statistics*, **10**, 473–4.
Reynolds, R. (1980), 'A Study of the Relationship Between Health and Income', doctoral dissertation, Dept of Economics, University of Chicago.
Robinson, P. M. (1979), 'Comment', *Journal of the American Statistical Association*, **74**, 646–7.
Sawa, T. (1972), 'Finite-Sample Properties of the k-Class Estimators', *Econometrica*, **40**, 653–80.
Sawa, T. (1973), 'The Mean Square Error of a Combined Estimator and Numerical Comparison with the TSLS Estimator', *Journal of Econometrics*, **1**, 115–32.
Shore, J.E. and R.W. Johnson (1980), 'Axiomatic Derivation of the Principle of Maximum Entropy and the Principle of Minimum Cross-Entropy', *IEEE Transactions*, Vol. IT-26, no. 1, 26–37.
Soofi, E.S. (1994), 'Capturing the Intangible Concept of Information', *Journal of the American Statistical Association*, **89** (428), 1243–54.

Swamy, P.A.V.B. and J.S. Mehta (1983), 'Further Results on Zellner's Minimum Expected Loss and Full Information Maximum Likelihood Estimators for Undersized Samples', *Journal of Business and Economic Statistics*, **1**, 154–62.

Theil, H. (1971), *Principles of Econometrics*, New York: John Wiley.

Tsurumi, H. (1990), 'Comparing Bayesian and Non-Bayesian Limited Information Estimators', in S. Geisser, J.S. Hodges, S.J. Press, and A. Zellner (eds), *Bayesian and Likelihood Methods in Statistics and Econometrics*, Amsterdam: North-Holland, pp. 179–207.

Van Dijk, H.K. (ed.) (1985), *Bayesian Analysis of Some Econometric and Statistical Models*, Annals Issue of the *Journal of Econometrics*, **29**, July/August.

Varian, H. R. (1974), 'A Bayesian Approach to Real Estate Assessment', in S.E. Fienberg and A. Zellner (eds), *Studies in Bayesian Econometrics and Statistics*,, Amsterdam: North-Holland, pp. 195–208.

Zellner, A. (1971), *An Introduction to Bayesian Inference in Econometrics*, New York: John Wiley, reprinted by Krieger Publishing Co., 1987 and by Wiley, 1996.

Zellner, A. (1975), 'Bayesian Analysis of Regression Error Terms', *Journal of the American Statistical Association*, **70**, 138–44.

Zellner, A. (1978), 'Estimation of Functions of Population Means and Regression Coefficients Including Structural Coefficients: A Minimum Expected Loss (MELO) Approach', *Journal of Econometrics*, **8**, 127–58.

Zellner, A. (1980), 'A Note on the Relationship of Minimum Expected Loss (MELO) and Other Structural Coefficient Estimates', *Review of Economics and Statistics*, **LXII**, 481–4.

Zellner, A. (1983), 'Canonical Representation of Linear Structural Econometric Models, Rank Tests for Identification and Existence of Estimators' Moments', invited paper in S. Karlin, T. Amemiya and L.A. Goodman (eds), *Studies in Econometrics, Time Series and Multivariate Statistics*, New York: Academic Press, pp. 227–40.

Zellner, A. (1986a), 'Further Results on Bayesian Minimum Expected Loss (MELO) Estimates and Posterior Distributions for Structural Coefficients', in D. Slottje (ed.), *Advances in Econometrics*, **5**, 171–82.

Zellner, A. (1986b), 'Bayesian Estimation and Prediction Using Asymmetric Loss Functions', *Journal of the American Statistical Association*, **81**, 446–51.

Zellner, A. (1993), 'Models, Prior Information and Bayesian Analysis', presented at Conference on Informational Aspects of Bayesian Statistics in Honor of H. Akaike, December 1993 and in L. Bauwens, W. Polasek, H. K. van Dijk (eds), special issue of *Journal of Econometrics*; November 1996, 51–68; and Chapter 8 in this volume.

Zellner, A. (1994a), 'Bayesian and Non-Bayesian Estimation Using Balanced Loss Functions', in S.S. Gupta and J.O. Berger (eds), *Statistical Decision Theory and Related Topics V*, New York: Springer-Verlag, ch. 28, pp. 377–90.

Zellner, A. (1994b), 'Bayesian Method of Moments/Instrumental Variable (BMOM/IV) Analysis of Mean and Regression Models', H.G.B. Alexander Research Foundation ms, Graduate School of Business, University of Chicago, in J.C. Lee, A. Zellner and W.O. Johnson (eds), *Modelling and Prediction: Honoring Seymour Geisser*, New York: Springer-Verlag, 61–74; Chapter 21 in this volume.

Zellner, A., L. Bauwens and H.K. van Dijk (1988), 'Bayesian Specification Analysis and Estimation of Simultaneous Equation Models Using Monte Carlo Methods', *Journal of Econometrics*, **38**, 39–72.

Zellner, A. and R.A. Highfield (1988), 'Calculation of Maximum Entropy Distributions and Approximation of Marginal Posterior Distributions', *Journal of Econometrics*, **37**, 195–210.

Zellner, A., C. Min. D. Dallaire, and J. Currie (1994), 'Bayesian Analysis of Simultaneous Equation, Asset-Pricing and Related Models Using Markov Chain Monte Carlo Techniques and Convergence Checks', H.G.B. Alexander Research Foundation ms, Graduate School of Business, University of Chicago.

Zellner, A. and B.R. Moulton (1985), 'Bayesian Regression Diagnostics with Applications to International Consumption and Income Data', *Journal of Econometrics*, **29**, 187–211.

Zellner, A. and S.B. Park (1979), 'Minimum Expected Loss (MELO) Estimators for Functions of Parameters and Structural Coefficients of Econometric Models', *Journal of the American Statistical Association*, **74**, 185–93.

DESCRIPTION

This is an ex-library book and may have the usual library/used-book markings inside.This book has hardback covers. In good all round condition. No dust jacket.

ORDER NUMBER

PO NYAATP08

BOOK NUMBER

2297904

BAY NUMBER

2676-13

AUTHOR	Zellner, Arnold
TITLE	Bayesian Analysis in Econometrics and

B

Testing and Model Selection

[22]

3.7 Posterior Odds Ratios for Regression Hypotheses: General Considerations and Some Specific Results

1 Introduction

Much work reported in the econometric literature has been directed at comparing Bayesian and non-Bayesian *estimation* results while much less work has been devoted to comparing Bayesian and non-Bayesian *analyses of alternative hypotheses*. This is not to say that the latter topic has been entirely neglected in the literature; see, for example, the references at the end of this paper, which include important works by Jeffreys, DeGroot, Dickey, Geisel, Good, Jaynes, Leamer, Lempers, Lindley, Raiffa, Schlaifer, Schwarz, Whittle, and others.[1] However, I believe it fruitful to open up discussion of some issues which appear relevant and important for practical analyses of hypotheses because they are not treated well and/or systematically in most statistics and econometric textbooks and in other portions of the literature.

What are some of the issues regarding analyses of hypotheses? First, there is the basic issue of whether to associate probabilities with hypotheses, that is, whether it is meaningful in a theory for analyzing hypotheses to quantify a statement like, "The parameter's value is probably zero." Most sampling theory analyses of hypotheses do not or cannot quantify such a statement. Second, the issue of what concept of probability is to be employed in quantifying such a statement arises. Most Bayesians regard probability as a measure of a degree of reasonable belief and not a frequency concept utilized widely in non-Bayesian analyses of hypotheses (see Jeffreys 1967, chap. 7 for a cogent discussion of this issue). Third, there is the issue of whether a separate theory of testing is needed; perhaps it is possible "to roll together significance testing and estimation in a single process," as sug-

Some minor changes have been made in the article, "Posterior Odds Ratios for Regression Hypotheses: General Considerations and Some Specific Results," which was an invited paper presented at the Econometric Society meeting, December 28–30, 1979, Atlanta, Georgia.

1. Jeffreys's name is placed first in this list because of his pioneering work on the philosophy, development, and applications of posterior odds ratios; see Jeffreys (1957, 1967).

gested by Good (1980). This suggestion seems questionable and fails to take account of a fundamental difference between estimation and significance testing problems. In the former, no particular value or values of the parameters are singled out for special attention, whereas in the latter, particular values, often zero, are given special attention. This is not to deny that estimation problems are often erroneously treated as testing problems.[2] However, as Jeffreys (1963) states, "Every quantitative law in physics implies a series of significance tests that have rejected numerous possible modifications of the law" (p. 403). In econometrics, too, most economic theories such as Friedman's theory of the consumption function imply hypotheses which can and have been tested. For example, is the elasticity of permanent consumption with respect to permanent income equal to one? Fourth, there is the issue of sharp nulls versus nonsharp nulls, that is, with respect to a regression coefficient $\beta = 0$, a sharp null hypothesis or $-\epsilon < \beta < \epsilon$, with $\epsilon > 0$ and small, a nonsharp null hypothesis. Good theories of testing should be able to provide results for both types of null hypotheses. Fifth, there is the issue of the use of prior information in testing. Use of one-tailed rather than two-tailed tests involves use of some prior information about the parameter space. Good testing procedures should be able to accommodate this kind and other kinds of prior information about parameters' possible values in analyzing hypotheses. Sixth, the use of power functions for tests requires prior information regarding parameters' values under the alternative hypothesis which is usually summarized poorly, if at all in practice. Seventh, choices of significance levels when the sample size is large and the use of "*P*-values" or "tail-areas" have to be more clearly explained.[3] Eighth, in some problems, asymptotically equivalent sampling theory tests yield conflicting results in finite samples (see Berndt and Savin

2. Some years ago, a student who estimated a demand function for tin was shocked by the finding that his estimate of the price elasticity was "not significantly different from zero at the 5% significance level." Since there was no special reason for concentrating attention on the zero value, his problem was an estimation and not a testing problem. His negative point estimate accompanied by a broad confidence interval, showed that his estimate was not very precise, not that the parameter's value is necessarily equal to zero.

3. In conversation with the author, Dennis Lindley has pointed out that very few, if any, statistics textbooks explain how the significance level for a given test is to be changed as the sample size changes. The same can be said about econometrics textbooks. For large sample sizes, say 5,000, as encountered in survey data, empirical workers usually lament that use of usual t-statistics and the 5% significance level shows that parameters' values are usually significantly different from zero. Many of them sense that they should not be using the 5% level but do not know how to obtain a more appropriate significance level. On this point of large t-values when the sample size is large, see Jeffreys's (1967, pp. 435–36) cogent discussion which suggests that they may be caused by departures from assumptions, particularly independence assumptions, associated with hypotheses when the sample size is large.

1977 and Savin 1976), an issue which deserves attention. Ninth, the rationale for the 5% "accept-reject syndrome" which afflicts econometrics and other areas requires immediate attention. If a test statistics' value is just a shade larger than a 5% critical value, does it make sense to reject the null hypothesis or is it better practice to be able to calculate a result which says that the data information does not convincingly support either of the hypotheses? Tenth, does it make sense to test a null hypothesis with no alternative hypothesis present? Jeffreys (1967) and others point out that this is not good practice since there will always be some functions of the observations, that is, test statistics, which will assume unusual values under the null hypothesis.[4] Eleventh, there is the issue of how to combine analyses of hypotheses with estimation and prediction procedures.

The above are just some of the issues and problems which arise in analyzing hypotheses in econometrics and other areas. While not all of these issues can be thoroughly treated in one paper, what will be done is first, in section 2, to provide some information on what kinds of hypotheses are considered by economists and econometricians and how they are analyzed. In section 3, Bayesian methods for analyzing hypotheses will be described along with some applied simple examples. Several regression hypotheses are analyzed in section 4, while in section 5 some concluding remarks are provided.

2 Hypotheses and Procedures for Analyzing Them in Economics and Econometrics

Since there is much controversy about the kinds of hypotheses and procedures for analyzing them actually used by economists and econometricians, it was decided to attempt to get some facts relating to these issues by conducting a small survey of a portion of the 1978 economic and econometric literature. For each of the following five journals, *American Economic Review*, *Journal of Political Economy*, *International Economic Review*, *Journal of Econometrics*, and *Econometrica*, a 1978 issue was randomly selected. In the issues so selected, 22 full-length articles which

4. Jeffreys remarks, "The fundamental idea, and one that I should naturally accept, is that a law [or model] should not be accepted on data that themselves show large departures from its predictions. But this requires a quantitative criterion of what is to be considered a large departure. The probability of getting the whole of an actual set of observations, given the law [or model], is ridiculously small. . . . If mere improbability of the observations, given the hypothesis, was the criterion, any hypothesis would be rejected. Everybody rejects the conclusion, but this can mean only that improbability of the observations, given the hypothesis, is not the criterion, and some other must be provided. The principle of inverse probability [Bayes's theorem] does this at once" (1967, pp. 384–85).

contained analyses of quantitative data were studied with results summarized in table 1.

As can be seen in table 1, of the 22 quantitative articles appearing in 1978 issues of the journals listed above, 18 or 82 percent included analyses of hypotheses. Thus, analysis of hypotheses is a central activity in current applied economic and econometric research, hardly a surprising result. Apparently, applied workers do not regard (rightly or wrongly) their problems as amenable to analysis in just an estimation framework.

Of the 18 quantitative articles involving analyses of hypotheses, standard sampling theory hypothesis testing and/or *P*-value techniques were employed. In 7 of the 18 articles, or 39 percent, the 5% level of significance was employed throughout. In 2 articles both the 5% and 1% levels of signifi-

Table 1 Survey of 22 Quantitative Articles in Leading Economic Journals in 1978 with Respect to Hypothesis Testing

Item	Number of Articles	Proportion
1. Includes hypothesis testing	18	.82[a]
2. Significance level employed		
5%	7	.39
5% and 1%	2	.11
1%	1	.06
Various levels[b]	6	.33
Not applicable	2	.11
3. Includes discussion of power	1	.06
4. Types of hypotheses[c]		
Sharp	7	.39
Nonsharp	9	.50[d]
Both	2	.11

NOTE: Only full-length articles which contained analyses of quantitative data were included. From each of the five journals, *American Economic Review*, *Journal of Political Economy*, *International Economic Review*, *Journal of Econometrics*, and *Econometrica*, we randomly selected an issue in 1978 and categorized the quantitative articles in those issues.

[a]Proportion calculated with respect to 22 articles. The other proportions are calculated with respect to 18 articles.

[b]Either report of *P*-values or various significance levels were used within the same article.

[c]A sharp hypothesis involves assigning a specific value to a parameter, e.g., H_0: $\mu = 0$ A nonsharp hypothesis does not involve assigning a specific value to a parameter, e.g., H_0: $\mu \geq 0$.

[d]The high proportion of nonsharp null hypotheses is due to (i) theories that only specify the algebraic sign of a coefficient and (ii) testing of nonnested models. An example of (ii) is testing which of two different nonnested theoretical distributions better fits the data.

cance were employed. In 6 articles various significance levels and/or *P*-values were used in the same article. An explicitly stated significance level was not mentioned in 2 of the 18 articles. In none of the papers was there a discussion of the relation between choice of significance levels and sample sizes which varied considerably both within and across studies; nor was there any discussion of the meaning of *P*-values.

Of the 18 articles in which hypotheses were analyzed and tested, there was just *one* which included a discussion of power. This is surprising in view of the fact that statistics and econometrics textbooks emphasize the importance of errors of the first and second kind and power considerations in general and in choice of a level of significance. Also, in almost all 18 papers, the usual accept-reject procedure was followed without attention given to the actual values of test statistics. For example, in one leading study, a *t*-statistic's value just barely exceeded the 5% critical value and the null hypothesis was rejected with no discussion of the "strength of the evidence."

Finally, the results in table 1 indicate that both sharp and nonsharp null hypotheses are considered in current applied economic and econometric research. The high proportion of nonsharp null hypotheses encountered is due to use of economic theories which specify only the algebraic signs of coefficients and not their exact values and to analyses involving comparisons of nonnested alternative models. An example of the latter analyses is an effort to determine which of two different distributions is better supported by the data.

To summarize, this small survey has uncovered the following points:

1. There is widespread use of the 5% and 1% levels of significance in testing with no systematic relation between choice of significance level and sample size.

2. Where *P*-values were employed, their meaning and rationale were not explained.

3. Power considerations are not generally discussed in empirical studies involving tests of hypotheses.

4. Both sharp and nonsharp hypotheses are utilized in empirical work, with the former generally being analyzed by use of Neyman-Pearson testing theory. Nonsharp null hypotheses were not analyzed very well in general.

5. There was very little formal or informal use of prior information regarding the values of parameters involved in the testing problems considered in these 18 articles.

6. Practically no attention was given in these 18 articles to the effects of tests or pretests on the properties of subsequent tests and/or computed

coefficient estimates. That is, the problems of pretest biases and lack of independence of various test statistics were not mentioned in most of the studies under review.

The results of this small survey, probably not surprising to many, suggest that there is room for improvement in analyses of hypotheses in economics and econometrics. In the following sections Bayesian results, which may be helpful in this regard, will be presented.[5]

3 Bayesian Analyses of Hypotheses

In the Bayesian approach to comparing and testing hypotheses, it is considered meaningful and useful to employ probabilities to represent degrees of reasonable belief associated with alternative hypotheses. The inductive calculus, primarily Bayes's theorem, is utilized to calculate the effects of the information in new data on initial or prior probabilities associated with alternative hypotheses in a logically consistent and operational manner. It should be appreciated that such calculations can in certain circumstances yield surprising and dramatic changes in initial probabilities. Also, while not discussed herein, Bayesian principles can be employed to design experiments to provide maximal discrimination between or among alternative hypotheses (see, e.g., Box and Hill 1967).

In the case in which a parameter θ, $-\infty < \theta < \infty$, is being considered and hypotheses do not specify a particular value for θ but rather just a range of possible values, for example, $\theta > 0$ versus $\theta < 0$ or $\theta > 1.0$ versus $\theta < 1.0$, examples of nonsharp hypotheses, Bayesians employ the following "Laplacian" approach, to use Jaynes's (1980) terminology in using data to modify initial probabilities associated with alternative hypotheses. Let $p(\theta \mid I_0)$ be the prior probability density function (pdf), where I_0 denotes the prior information. Then *before* observing the data, denoted by y, the initial probabilities associated with, for example, $H_1: \theta > 0$ and $H_2: \theta < 0$ are obviously given by $p_1^{(0)} = \int_0^\infty p(\theta \mid I_0)d\theta$ and $p_2^{(0)} = 1 - p_1^{(0)} = \int_{-\infty}^0 p(\theta \mid I_0)d\theta$ for H_1 and H_2, respectively. The prior odds ratio for H_1 and H_2 is, by definition, $K_{12}^{(0)} = p_1^{(0)}/p_2^{(0)}$. Above, we have assumed that $p(\theta \mid I_0)$ is a proper prior pdf. If we have little prior information regarding θ's possible values and employ Jeffreys's prior for representing knowing little, $p_J(\theta \mid I_0)d\theta \propto d\theta$, $-\infty < \theta < \infty$, an improper prior,[6] then, as Jeffreys (1967,

5. In view of the controversies about alternative testing procedures in the statistical literature, it is surprising that there were no discussions of alternative testing methodologies in the papers surveyed.

6. See A. Rényi (1970) for an axiom system for probability theory, a generalization of Kolmogorov's, which accommodates unbounded measures and improper prior pdf's used in quantum theory, statistical mechanics, and mathematical statistics. In particular, Rényi's theory, as well as other results, shows that Bayes's theorem is valid when improper prior pdf's are employed.

pp. 117 ff.) points out, $K_{12}^{(0)}$ is indeterminate in value, his representation of the assumed state of ignorance. Whether the prior is proper or improper, Bayes's theorem can be employed to obtain the posterior pdf for θ, $p(\theta \mid D)$ where $D \equiv (I_0, y)$ as follows

$$(3.1) \qquad p(\theta \mid D) = cp(\theta \mid I_0)\ell(\theta \mid y) \qquad -\infty < \theta < \infty$$

where the normalization constant c is given by $1/c = \int_{-\infty}^{\infty} p(\theta \mid I_0)\ell(\theta \mid y)d\theta$, with $\ell(\theta \mid y)$ being the likelihood function. As is well known, the posterior pdf $p(\theta \mid D)$ incorporates both the information in the prior pdf and in the data y. Given that we have the posterior pdf in (3.1), it is possible to calculate the following posterior probabilities, p_1 associated with $H_1 : \theta > 0$ and p_2 associated with $H_2 : \theta < 0$:

$$(3.2a) \qquad p_1 = Pr(\theta > 0 \mid D) = \int_0^{\infty} p(\theta \mid D)d\theta$$

and

$$(3.2b) \qquad p_2 = Pr(\theta < 0 \mid D) = \int_{-\infty}^{0} p(\theta \mid D)d\theta$$

and the posterior odds ratio, K_{12} for H_1 relative to H_2 is just $K_{12} = p_1/p_2$, or

$$(3.3) \qquad K_{12} = Pr(\theta > 0 \mid D)/Pr(\theta < 0 \mid D) = \int_0^{\infty} p(\theta \mid D)d\theta \Big/ \int_{-\infty}^{0} p(\theta \mid D)d\theta .$$

Some simple but important examples illustrating uses of (3.3) and related odds ratios will now be presented. First, consider the usual linear normal regression model $y = X\beta + u$, where y is $n \times 1$, X is a given $n \times k$ matrix of rank k, β is a $k \times 1$ vector of regression coefficients, and u is an $n \times 1$ disturbance vector. It is assumed that the elements of u have been independently drawn from a normal distribution with zero mean and variance σ^2. The likelihood function is

$$(3.4) \qquad \ell(\beta, \sigma \mid D) \propto \sigma^{-n}\exp\{-(y - X\beta)'(y - X\beta)/2\sigma^2\} \propto \sigma^{-n}\exp\{-[\nu s^2 + (\beta - \hat{\beta})'X'X(\beta - \hat{\beta})]/2\sigma^2\},$$

where $\hat{\beta} = (X'X)^{-1}X'y$, $\nu s^2 = (y - X\hat{\beta})'(y - X\hat{\beta})$, and $\nu = n - k$. If we employ a standard diffuse prior pdf for β and σ, $p(\beta, \sigma) \propto 1/\sigma$, $0 < \sigma < \infty$ and $-\infty < \beta_i < \infty$, $i = 1, 2, \ldots, k$, then it is well known that the marginal posterior pdf for β_i is in the univariate Student-t form, that is,

$$(3.5) \qquad (\beta_i - \hat{\beta}_i)/s_{\hat{\beta}_i} = t,$$

where β_i and $\hat{\beta}_i$ are the i'th elements of β and $\hat{\beta}$, respectively, $s_{\hat{\beta}_i}^2 = m^{ii}s^2$, with m^{ii} the i–ith element of $(X'X)^{-1}$, and t is a random variable with a univariate Student-t pdf with $\nu = n - k$ degrees of freedom. Suppose that we are interested in the following hypotheses, $H_1 : \beta_i > 0$ and $H_2 : \beta_i < 0$. From (3.5), $\beta_i = \hat{\beta}_i + s_{\hat{\beta}_i} t$ and thus $\beta_i > 0$ is equivalent to $\hat{\beta}_i + s_{\hat{\beta}_i} t > 0$ or $t > -\hat{\beta}_i/s_{\hat{\beta}_i}$. Thus the posterior probability that $\beta_i > 0$ is $Pr(\beta_i > 0 | D) = Pr\{t > -\hat{\beta}_i/s_{\hat{\beta}_i} | D\}$, a probability which can be evaluated using tables of the Student-t distribution with ν degrees of freedom. Similarly, the hypothesis $\beta_i < 0$ is equivalent to $\beta_i = \hat{\beta}_i + s_{\hat{\beta}_i} t < 0$ or $t < -\hat{\beta}_i/s_{\hat{\beta}_i}$ and its posterior probability is $Pr(\beta_i < 0 | D) = Pr(t < -\hat{\beta}_i/s_{\hat{\beta}_i} | D)$. Then the posterior odds ratio for $H_1 : \beta_i > 0$ and $H_2 : \beta_i < 0$ is

$$K_{12} = \frac{Pr(\beta_i > 0 | D)}{Pr(\beta_i < 0 | D)} = \frac{Pr(t > -\hat{\beta}_i/s_{\hat{\beta}_i} | D)}{Pr(t < -\hat{\beta}_i/s_{\hat{\beta}_i} | D)}, \tag{3.6}$$

where, it should be noted $\hat{\beta}_i/s_{\hat{\beta}_i}$ is the usual sample t-statistic. Shown in table 2 are values of K_{12} and $Pr(\beta_i > 0 | D)$ for selected values of $\hat{\beta}_i/s_{\hat{\beta}_i}$ using a value of $\nu = n - k = 20$.

From table 2, it is seen that when $\hat{\beta}_i/s_{\hat{\beta}_i} = 0$, $Pr(\beta_i > 0 | D) = .5000$ and $K_{12} = 1.00$ while when $\hat{\beta}_i/s_{\hat{\beta}_i} = 2.00$, $Pr(\beta_i > 0 | D) = 0.9704$ and $K_{12} = 32.8$. These results are useful in appraising the implications of sample evidence for the algebraic sign of a regression coefficient. It should be appreciated that the above analysis can also be carried through using informative prior pdf's for $\boldsymbol{\beta}$ and σ, for the regression model with multivariate Student-t disturbance terms (see Zellner 1976), and other models. In addition, it is possible to compute $Pr\{\beta_i > 0$ and $\beta_j > 0 | D\}$ from the bivariate posterior distribution for β_i and β_j; see, for example, Davis (1978) for this and more general results. Last, below it is shown how to analyze three hypotheses, namely, $\beta_i > 0$, $\beta_i < 0$ and the sharp null hypothesis $\beta_i = 0$.

A second example of nonsharp null hypotheses involves study of the properties of solutions to a second-order autoregressive process, $y_t = \alpha_1 y_{t-1} + \alpha_2 y_{t-2} + \epsilon_t$, $t = 1, 2, \ldots, T$ where the ϵ_t's are assumed independently drawn from a normal distribution with zero mean and variance σ^2. Given two initial values of y_t and a diffuse prior for α_1, α_2, and σ, it is shown in Zellner (1971, pp. 194 ff.) that the marginal posterior pdf for α_1 and α_2 is

Table 2

	Values of $\hat{\beta}_i/s_{\hat{\beta}_i}$, $\nu = 20$					
	0.00	0.50	1.00	1.50	2.00	2.50
$Pr(\beta_i > 0 \mid D)$	0.5000	0.6887	0.8354	0.9254	0.9704	0.9894
K_{12}[a]	1.00	2.20	5.06	12.4	32.8	93.3

[a] $K_{12} = Pr(\beta_i > 0 | D)/Pr(\beta_i < 0 | D)$.

in the bivariate Student-t form and how numerical integration techniques can be employed to compute posterior probabilities that the solution to the model is (i) explosive and nonoscillatory, or (ii) nonexplosive and nonoscillatory, or (iii) nonexplosive and oscillatory or (iv) explosive and oscillatory. These posterior probabilities are just volumes under the posterior pdf, $p(\alpha_1, \alpha_2 | D)$ over regions in the $\alpha_1 - \alpha_2$ plane associated with solution properties (i)–(iv). This analysis can be performed using not only a diffuse prior pdf but also informative prior pdf's.

The next example, presented and analyzed by Jaynes (1976, pp. 181 ff.), involves study of two independent samples of data to determine whether the population mean for the first is larger than that of the second. That is, we have n_1 independent observations drawn from $N(\mu_1, \sigma^2)$ and n_2 independent observations drawn from $N(\mu_2, \sigma^2)$, and the nonsharp hypotheses are $H_1: \mu_1 > \mu_2$ and $H_2: \mu_1 < \mu_2$. Using a diffuse prior pdf for μ_1, μ_2 and σ, $p(\mu_1, \mu_2, \sigma) \propto 1/\sigma$, Jaynes derived the marginal posterior pdf for μ_1 and μ_2, $p(\mu_1, \mu_2 | D)$ which is in the bivariate Student-t form. From this posterior pdf, he computed the posterior probability that $\mu_1 > \mu_2$, that is, $Pr(\mu_1 > \mu_2 | D) = \int_{-\infty}^{\infty} \int_{\mu_2}^{\infty} p(\mu_1, \mu_2 | D) d\mu_1, d\mu_2$, and contrasted this solution to the problem with a sampling theory analysis involving a test of $\mu_1 = \mu_2$ versus $\mu_1 \neq \mu_2$, which appeared in a leading engineering statistics textbook. This example illustrates the importance of obtaining the solution to the question asked, $\mu_1 > \mu_2$ versus $\mu_1 < \mu_2$ rather than the solution to some other question, for example, $\mu_1 = \mu_2$ versus $\mu_1 \neq \mu_2$.[7]

Another example of nonsharp hypotheses involves the comparison of two hypothesized distributions for a set of independent observations, x_1, $x_2, \ldots, x_n$. Under H_1, the pdf for the observations is $f_1[(x - \mu)/\sigma]$, where μ is a location parameter and σ is a scale parameter, while under H_2 it is $f_2[(x - \mu)/\sigma]$. For example, $f_1(\cdot)$ might be a normal pdf while $f_2(\cdot)$ might be a double-exponential distribution. Using Bayes's theorem, it is well known that the posterior odds ratio K_{12} for this problem is given by:

$$K_{12} = \frac{\Pi_1 \iint \ell_1(\mu, \sigma | x) p_1(\mu, \sigma) d\mu d\sigma}{\Pi_2 \iint \ell_2(\mu, \sigma | x) p_2(\mu, \sigma) d\mu d\sigma}, \tag{3.7}$$

7. Jaynes (1976) analyzes several other specific, important problems from Bayesian and non-Bayesian points of view and presents an interesting review of historical developments in statistics and physics. He comments, "In most real applications, it is just the specific case at hand that is of concern to us; and it is hard to see how frequency statements about a mythical population or an imaginary experiment can be considered any more 'objective' than the Bayesian statements. Finally, no statistical method which fails to provide any way of taking prior information into account can be considered a full treatment of the problem [of comparing and testing alternative hypotheses]" (pp. 193–94).

where Π_1/Π_2 is the prior odds ratio, $x' = (x_1, x_2, \ldots, x_n)$, $p_1(\mu, \sigma)$ and $p_2(\mu, \sigma)$ are the prior pdf's, and $\ell_1(\mu, \sigma | x) = \Pi_{i=1}^n f_1[(x_i - \mu)/\sigma]$ and $\ell_2(\mu, \sigma | x) = \Pi_{i=1}^n f_2[(x_i - \mu)/\sigma]$ are the likelihood functions under H_1 and H_2, respectively. Among others, Dyer (1972, 1973) has studied the use of (3.7) with $\Pi_1/\Pi_2 = 1$ and, for $i = 1, 2, p_i(\mu, \sigma) = c/\sigma$, with c a positive constant, $-\infty < \mu < \infty$ and $0 < \sigma < \infty$ for discriminating between alternative distributions. Under these assumptions, he pointed out that a theorem from Zidek (1969) shows that K_{12} is a uniformly most powerful invariant test statistic for discriminating between $f_1(\cdot)$ and $f_2(\cdot)$. Also, he and others have reported Monte Carlo results showing that K_{12} performs well or better than other test statistics for discriminating between various $f_1(\cdot)$ and $f_2(\cdot)$. See also Blattberg and Gonedes (1974), who have applied (3.7) to discriminate between various hypothesized distributions for stock returns.

In Thornber (1966), Geisel (1970, 1973), Gaver and Geisel (1974), Lempers (1971), Zellner (1971), and Leamer (1978), odds ratios for nonnested normal linear regression models have been studied and applied. For example, Geisel computed posterior odds ratios to evaluate variants of money-multiplier and Keynesian-multiplier models studied earlier by Friedman and Meiselman. Zellner (1971, pp. 306 ff.) derived and studied the posterior odds ratio for two nonnested linear normal multiple regression models, $y = X_1\beta_1 + u_1$ and $y = X_2\beta_2 + u_2$ employing natural conjugate prior distributions for regression coefficients and disturbance terms' variances. It was found that the posterior odds ratio's value reflects (*a*) the prior odds ratio's value, (*b*) relative precision of prior and posterior distributions for parameters, (*c*) relative goodness of fit of the two models, and (*d*) extent to which prior and sample information regarding parameters' values are in agreement. It is thus seen that the value of the posterior odds ratio for the two models is conditioned by a range of relevant considerations and not just by relative goodness of fit.[8]

In summary, posterior probabilities can be computed for a number of nonsharp hypotheses frequently encountered in applied work. The posterior probabilities so computed reflect sample and prior information, as little or as much of the latter as an investigator decides to use. Further, posterior probabilities for investigators with different prior pdf's can be computed and compared; see, for example, Leamer (1978), Zellner and Geisel (1970), Zellner and Richard (1973), and Zellner and Williams (1973) for illustrations of this point. In addition, if an investigator has a loss structure giving the consequences of errors in choosing between or among hypotheses, it is well known that posterior probabilities can be employed to

8. See also Zellner (1978), where these points are made in another context.

evaluate expected losses associated with different choices of hypotheses and to choose among them in such a way as to minimize expected loss; see, for example, Blackwell and Girshick (1954), Chernoff and Moses (1959), DeGroot (1970), Zellner (1971) and Leamer (1978).

Having discussed various examples of nonsharp hypotheses, we now turn our attention to a consideration of sharp hypotheses.

As indicated by the results of the survey discussed in section 2, economists often consider sharp null hypotheses, for example, $\beta_i = 0$ versus $\beta_i \neq 0$, where β_i is a regression coefficient or $\eta = 1$ versus $\eta \neq 1$, where η is a returns-to-scale parameter, or $\boldsymbol{\beta} = \mathbf{0}$ versus $\boldsymbol{\beta} \neq \mathbf{0}$ where $\boldsymbol{\beta}$ is a vector of slope coefficients in a regression. Much work has been done to derive posterior odds ratios for sharp null hypotheses and alternative hypotheses, either sharp or nonsharp; see, for example, Jeffreys (1957, 1967, 1980), who is a pioneer in this area, Dickey (1971, 1975, 1977, 1980), Geisel (1970), Good (1950, 1965), Griffiths and Dao (1980), Leamer (1978), Lempers (1971), Lindley (1957), Mayer (1979), Reynolds (1982), Smith and Spiegelhalter (1980), Thornber (1966), Whittle (1951), Zellner (1971, 1978), and Zellner and Siow (1979, 1980).

First, it has been generally noted that in comparing two sharp, *simple* hypotheses, say $H_1 : \boldsymbol{\theta} = \boldsymbol{a}$ and $H_2 : \boldsymbol{\theta} = \boldsymbol{b}$, where $\boldsymbol{a}$ and $\boldsymbol{b}$ are given vectors, the posterior odds ratio is given by:

$$K_{12} = (\Pi_1/\Pi_2)f(\boldsymbol{y} \mid \boldsymbol{\theta} = \boldsymbol{a})/g(\boldsymbol{y} \mid \boldsymbol{\theta} = \boldsymbol{b}), \tag{3.8}$$

where Π_1/Π_2 is the prior odds ratio, $f(\boldsymbol{y} \mid \boldsymbol{\theta} = \boldsymbol{a})$ is the pdf for $\boldsymbol{y}$, the data under H_1 and $g(\boldsymbol{y} \mid \boldsymbol{\theta} = \boldsymbol{b})$ is the pdf for $\boldsymbol{y}$ under H_2. In many problems, $f(\cdot)$ and $g(\cdot)$ have the same form. From (3.8) it is seen that the so-called Bayes factor is just the likelihood ratio, $f(\boldsymbol{y} \mid \boldsymbol{\theta} = \boldsymbol{a})/g(\boldsymbol{y} \mid \boldsymbol{\theta} = \boldsymbol{b})$, and thus well-known optimal properties of likelihood ratio tests for simple hypotheses (see, e.g., Silvey 1970, pp. 97 ff.) are enjoyed by posterior odds ratios.

In many situations, the hypotheses do not specify all values of the parameters. For example, if $\boldsymbol{\theta}' = (\boldsymbol{\theta}_1', \boldsymbol{\theta}_2')$ and the hypotheses are $H_1 : \boldsymbol{\theta}_1 = \boldsymbol{a}_1$ and $\boldsymbol{\theta}_2$ unrestricted, where $\boldsymbol{a}_1$ is a given vector and $H_2 : \boldsymbol{\theta}_1$ and $\boldsymbol{\theta}_2$ unrestricted, the posterior odds ratio, based on data $\boldsymbol{y}$, is

$$K_{12} = (\Pi_1/\Pi_2) \int p_1(\boldsymbol{y} \mid \boldsymbol{\theta}_1 = \boldsymbol{a}_1, \boldsymbol{\theta}_2)p(\boldsymbol{\theta}_2 \mid \boldsymbol{\theta}_1 = \boldsymbol{a}_1)d\boldsymbol{\theta}_2 / \int p_2(\boldsymbol{y} \mid \boldsymbol{\theta}_1, \boldsymbol{\theta}_2)p(\boldsymbol{\theta}_1, \boldsymbol{\theta}_2)d\boldsymbol{\theta}_1 d\boldsymbol{\theta}_2. \tag{3.9}$$

In (3.9) for $i = 1, 2$, $p_i(\boldsymbol{y} \mid \cdot)$ is the likelihood function, $p(\boldsymbol{\theta}_2 \mid \boldsymbol{\theta}_1 = \boldsymbol{a}_1)$ is the conditional prior pdf for $\boldsymbol{\theta}_2$ given $\boldsymbol{\theta}_1 = \boldsymbol{a}_1$, $p(\boldsymbol{\theta}_1, \boldsymbol{\theta}_2)$ is the joint prior pdf for $\boldsymbol{\theta}_1$ and $\boldsymbol{\theta}_2$ under H_2 and the integrations are performed over the region of $\boldsymbol{\theta}_2$ in the numerator and over the region of $\boldsymbol{\theta}_1$ and $\boldsymbol{\theta}_2$ in the denominator. K_{12} in (3.9) is seen to be a product of the prior odds ratio, Π_1/Π_2 and the Bayes factor, which here is the ratio of averaged likelihood functions with prior pdf's serving as the weight functions. To appreciate how (3.9) relates to problems encountered in practice, some specific examples will be provided.

Among many analyses of sharp null hypotheses frequently encountered in practice, Jeffreys (1967, pp. 268 ff.) considers hypotheses relating to a mean θ, $H_1: \theta = 0$ and $H_2: \theta \neq 0$ under the assumption that normal independent observations are available, each with a common standard deviation with unknown value. That is, $y_i = \theta + \epsilon_i$, $i = 1, 2, \ldots, n$ with the ϵ_i's assumed independently drawn from a normal distribution with zero mean and variance σ^2 with unknown value. For this and similar problems, Jeffreys noted that old-line physicists and astronomers usually computed something like $t = \bar{y}/s_{\bar{y}}$, where $\bar{y}$ is the sample mean, $\Sigma_{i=1}^n y_i/n$ and $s_{\bar{y}}$ is the standard error of the mean, where $s_{\bar{y}}^2 = \Sigma_{i=1}^n (y_i - \bar{y})^2/n(n-1)$. If $|t| < 2$, usually the measurements were taken as supporting $H_1: \theta = 0$; if $2 < |t| < 3$, some limited support for $H_2: \theta \neq 0$ is provided by the data; if $|t| > 3$, then strong support for $H_2: \theta \neq 0$ is provided by the data. All of this was done without use of formal statistical theory and thus Jeffreys (1979) mentioned that years ago he saw a need to produce a connected theory which would rationalize these rough rules used in applied work. In approaching this problem, Jeffreys (1967, p. 252) recognized that there might be little or much prior information available. He decided to develop his theory of significance testing for cases in which there is little prior information, as in "early stages of a subject." (p. 252). Herein, this course will also be followed because it leads to results which can be used in practice and which can be readily compared with non-Bayesian testing procedures. However, as Jeffreys recognized, if more prior information is available, it can be used in formulating and computing posterior odds ratios, as shown in work by Dickey (1971, 1975, 1977, 1980), Zellner (1971, pp. 307 ff.) and others.

For hypotheses $H_1: \theta = 0$ and $0 < \sigma < \infty$ and $H_2: \theta \neq 0$ and $0 < \sigma < \infty$ about a normal mean and standard deviation, Jeffreys takes the prior odds ratio Π_1/Π_2 equal to one and considers the following posterior odds ratio

$$K_{12} = \int_0^\infty p(y \mid \theta = 0, \sigma) g(\sigma) d\sigma \Big/ \int_{-\infty}^\infty \int_0^\infty p(y \mid \theta, \sigma) f(\theta \mid \sigma) g(\sigma) d\sigma d\theta. \tag{3.10}$$

In (3.10), the likelihood functions are $p(y \mid \theta = 0, \sigma) = (2\pi\sigma^2)^{-n/2} \exp\{-n(\bar{y}^2 + \hat{\sigma}^2)/2\sigma^2\}$ and $p(y \mid \theta, \sigma) = (2\pi\sigma^2)^{-n/2} \exp\{-n[(\theta - \bar{y})^2 + \hat{\sigma}^2]/2\sigma^2\}$, where $\bar{y}$ is the sample mean and $n\hat{\sigma}^2 = \Sigma_{i=1}^n (y_i - \bar{y})^2$, $g(\sigma) \propto 1/\sigma$, the diffuse prior on σ, and $f(\theta/\sigma)$ is the prior on θ/σ. Jeffreys (1967, p. 268)

Table 3 Values of t^2 Associated with Corresponding Values of K_{12} and $\nu = n-1$ from (3.11)

	K_{12}				
ν	1	$10^{-1/2}$	10^{-1}	$10^{-3/2}$	10^{-2}
9[a]	3.5	7.7	13.3	...	...
15	3.8	7.1	11.1	15.9	21.5
20	4.0	7.0	10.6	14.5	18.9
50	4.6	7.4	10.0	12.8	16.0
100	5.2	7.7	10.3	12.8	15.5
200	5.7	8.2	10.7	13.1	15.6
500	6.8	9.1	11.4	13.8	16.2
1,000	7.4	9.7	12.0	14.3	16.6
2,000	8.1	10.4	12.7	15.0	17.3
5,000	9.0	11.3	13.6	15.9	18.2
10,000	9.7	12.0	14.3	16.6	18.9
50,000	11.3	13.6	15.9	18.2	20.5
100,000	12.0	14.3	16.6	18.9	21.2

[a]For $\nu = 9$, Jeffreys has used his exact result for K_{12} to compute the following t^2 values: 3.8 for $K_{12} = 1$, 7.7 for $K_{12} = 10^{-1/2}$, and 13.1 for $K_{12} = 10^{-1}$. It is seen that the exact results are in good agreement with the approximate results even though $\nu = 9$ is small. Jeffreys (1967, p. 439) tabulates exact values for $\nu = 1,2,3, \ldots, 9$.

presents several ingenious arguments for using $f(\theta \mid \sigma) = 1/\pi\sigma(1 + \theta^2/\sigma^2)$, a Cauchy prior for θ given σ which is centered at zero, the value suggested by H_1 when there is little prior information, other than that θ's value may be zero.[9] If these likelihood functions and prior pdf's are substituted in (3.10) and the integrations are performed, exactly in terms of the numerator and approximately in terms of the denominator, the result is:

$$K_{12} \doteq (\pi\nu/2)^{1/2}/(1 + t^2/\nu)^{(\nu-1)/2}, \tag{3.11}$$

where $\nu = n - 1$, $t = \sqrt{n}\,\bar{y}/s$, the usual t-statistic with $\bar{y} = \Sigma_{i=1}^{n} y_i/n$ and $s^2 = \Sigma_{i=1}^{n} (y_i - \bar{y})^2/\nu$ and the error of the approximation is of order $1/n$ of the whole expression.

In table 3 values of K_{12} from (3.11) for selected values of ν and t^2 from Jeffreys's (1967, p. 439) table are presented. From table 3, it is seen that

9. Use of the Cauchy prior results in $K_{12} \doteq 1$ when $n = 1$ and $K_{12} \doteq 0$ when $\bar{y} \neq 0$ and $\hat{\sigma} \doteq 0$ when $n \geq 2$. Also, use of the Cauchy prior avoids some difficulties associated with the use of a normal prior when the sample size is small, see Jeffreys (1967), p. 273. Jeffreys (1967, p. 251) points out that use of $p(\theta) \propto$ const, $-\infty < \theta < \infty$ is in conflict with the prior information that θ's value is probably small, as suggested by the null hypothesis and leads to $K_{12} = \infty$, an unsatisfactory result. On the other hand, when the Cauchy prior is employed and $\bar{y} \gg 0$, the posterior pdf under $\theta \neq 0$ is very similar to what is obtained in estimation using a diffuse prior pdf for θ.

with $\nu = 20$, $K_{12} \doteq 1.0$ when $t^2 = 4.0$ or $t = \pm 2$. Thus a value of $t = \pm 2$ when $\nu = 20$ does not provide convincing support for either $\theta = 0$ or $\theta \neq 0$. On the other hand when $\nu = 20$ and $t^2 = 10.6$ or $t = \pm 3.26$, $K_{12} \doteq 10^{-1}$; that is, the posterior odds ratio is 10:1 against $\theta = 0$. Further as ν grows in value, the value of t^2 associated with $K_{12} = 1$ grows in value. For example, when $\nu = 5{,}000$, $K_{12} \doteq 1$ when $t^2 = 9.0$ or $t = \pm 3.0$, which corresponds to many sample theorists' usual informal lowering of the significance level as the sample size grows in order to maintain a balance between probabilities of errors of the first and second kind. Also, this result is related to Lindley's (1957) paradox.

Jeffreys points out that for moderately large ν, K_{12} in (3.11) can be further approximated by $K_{12} \doteq (\pi\nu/2)^{1/2}\exp\{-t^2/2\}$ from which the dependence of K_{12} on ν and t is clearly seen. Further, as is evident from (3.11), for given ν, K_{12} is a monotonically decreasing function of t^2 and hence a monotonically increasing function of the P-value associated with $t = \sqrt{n}\,\bar{y}/s$, $t \geq 0$. From a numerical analysis of (3.11) for $\nu = 20$, the empirical relation connecting $\ell n K_{12}$ and $\ell n P$, where $P \equiv P$-value is:

$$\text{(3.12)} \quad \ell n K_{12} = \underset{(.0162)}{2.52} + \underset{(.00284)}{0.877}\ell n P + \hat{u}, \qquad r^2 = 0.9998,$$

based on 21 pairs of values of P and K_{12}, the latter ranging from 0.01 to 1.0. From (3.12), it is seen that the elasticity of the odds ratio K_{12} with respect to the P-value is 0.877 and thus high P-values are associated with high values for K_{12}. This may account for the widespread use of P-values in practice. That is, they can often be justified in terms of their relation to posterior odds ratios based on particular priors. However, since posterior odds ratios have a direct interpretation, their use is favored.

Jeffreys (1967, pp. 384 ff.) provides strong criticism[10] of the usual sampling interpretation of P-values, denoted by P in the following words,

> If P is small, that means that there have been unexpectedly large departures from prediction [under the null hypothesis]. But why should these be stated in terms of P? The latter gives the probability of departures, measured in a particular way, equal to *or greater than* the observed set, and the contribution from the actual value [of the test statistic] is nearly always negligible. *What the use of* P *implies, therefore, is that a hypothesis that may be true may be rejected because it has not predicted observable results that have not occurred.* This seems a remarkable procedure. On the face of it the fact that such results have not occurred might more reasonably be taken as evidence for the law [or null hypothesis], not against it. The same applies to all the current significance tests based on P integrals. [P. 385]

10. See also Jaynes (1976) and note 7, above.

289 Posterior Odds Ratios for Regression Hypotheses: General Considerations and Some Specific Results

Jeffreys (1967, pp. 395 ff.) is also critical of Neyman-Pearson testing procedures. While he lauds Neyman and Pearson for their giving attention to explicitly formulated alternative hypotheses and their detailed analysis of errors of the first and second kind, he remarks,

> Now Pearson and Neyman proceed by working out the above risks [of errors] for different values of the new parameter, and call the result the power function of the test, the test itself being in terms of the P integral. But if the actual value [of the parameter or parameters] is unknown the value of the power function is also unknown; the total risk of errors of the second kind must be compounded of the power functions over the possible values, with regard to their risk of occurrence. [Pp. 395–96]

Thus problems associated with use of P-values and in assessing "total risk of errors of the second kind" without a prior distribution are considered by Jeffreys to be critical defects of Neyman-Pearson testing theory. In addition, Jeffreys (1967, pp. 396–97) shows that for two hypotheses, regarding a scalar parameter α, $H_1 : \alpha = 0$ and $H_2 : \alpha \neq 0$, if it is known that both hypotheses have probability 1/2 and if under H_2, α is known to be uniformly distributed over its range, then the expectation of the total fraction of mistakes of the first and second kind will be made a minimum by use of the critical value $K_{12} = 1$; that is, if $K_{12} > 1$, choose H_1 and if $K_{12} < 1$ choose H_2. This is an early statement of a Bayesian decision theoretic result relevant for choosing hypotheses.

Above, two hypotheses were considered, namely, $\theta = 0$ and $\theta \neq 0$. In Zellner and Siow (1979) attention is given to the following three hypotheses: $H_1 : \theta = 0$, $H_2 : \theta > 0$, and $H_3 : \theta < 0$, with prior probabilities Π_1 Π_2, and Π_3, respectively, and $\Pi_1 + \Pi_2 + \Pi_3 = 1$. The odds ratios for these three hypotheses are:[11]

$$K_{12} = (\Pi_1/\Pi_2)p(y\,|\,\theta = 0)/\int_0^\infty p(y\,|\,\theta)p_2(\theta)d\theta, \tag{3.13}$$

$$K_{13} = (\Pi_1/\Pi_3)p(y\,|\,\theta = 0)/\int_{-\infty}^0 p(y\,|\,\theta)p_3(\theta)d\theta, \tag{3.14}$$

and

$$K_{23} = (\Pi_2/\Pi_3)\int_0^\infty p(y\,|\,\theta)p_2(\theta)d\theta/\int_{-\infty}^0 p(y\,|\,\theta)p_3(\theta)d\theta, \tag{3.15}$$

11. Similar odds ratios can be obtained for three hypotheses relating to a regression coefficient's value, namely, $\beta_i = 0$, $\beta_i > 0$, and $\beta_i < 0$.

where y is a vector of observations, $p(y|\theta)$ is the likelihood function and $p_2(\theta)$ and $p_3(\theta)$ are the prior pdf's for θ under H_2 and H_3, respectively. Given values of the Π's, the prior pdf's and likelihood functions (3.13)–(3.15) can readily be computed as shown in the Zellner and Siow (1979) analysis of the normal mean problem with σ's value known and unknown. In this paper, Jeffreys's Cauchy priors for θ/σ, defined over $\theta > 0$ and $\theta < 0$, were employed along with a diffuse prior for σ and the following approximate odds ratios were derived.

$$K_{12} \doteq (\Pi_1/\Pi_2)(1/2)(\pi\nu/2)^{1/2}/(1 + t^2/\nu)^{(\nu-1)/2}F(t), \tag{3.16}$$

$$K_{13} \doteq (\Pi_1/\Pi_3)(1/2)(\pi\nu/2)^{1/2}/(1 + t^2/\nu)^{(\nu-1)/2}F(-t), \tag{3.17}$$

and

$$K_{23} \doteq (\Pi_2/\Pi_3)F(t)/F(-t), \tag{3.18}$$

where $t = \sqrt{n}\,\bar{y}/s$, the usual t-statistic, $\nu = n - 1$, and $F(\cdot)$ is the cumulative normal distribution. Tabulation of the odds ratio in (3.16) in table 4 reveals that by taking $\Pi_1 = .50$, $\Pi_2 = .40$, and $\Pi_3 = .10$, $K_{12} \doteq 1$ when $\nu = 20$ and $t = \sqrt{n}\,\bar{y}/s$ is slightly larger than 1.70, not far from the one-tailed 5% critical value, 1.725 used in non-Bayesian one-tailed testing of $H_1: \theta = 0$ versus $H_2: \theta > 0$ when $\nu = 20$. Of course other values of the prior probabilities Π_1, Π_2, and Π_3 and different prior pdf's can be employed if they are thought to be more appropriate.

In table 5, taken from Zellner and Siow (1979), values of $K_{23} \div \Pi_2/\Pi_3$ for various values of $t = n^{1/2}\bar{y}/s$ are shown. For example, it is seen that when $t = 1.00$, Bayes's factor in favor of H_2: $\theta > 0$ versus H_3: $\theta < 0$ is 5.30. Appropriate choice of the value of Π_2/Π_3 can be made to reflect initial information about H_2 and H_3.

In addition to the hypotheses considered above, Jeffreys (1967) has developed and applied Bayesian posterior odds ratios for hypotheses relating to a binomial parameter, contingency tables, Poisson parameters, equality of normal means with standard deviations assumed equal or unequal, standard deviations and correlations of normal populations, and other problems. In addition, he has provided a large sample approximation to the odds ratio for two hypotheses, namely, $-2\ell n K_{12} \doteq \chi^2_q - q\ell n\, n$ where q is the number of restrictions implied by the null hypotheses H_1 relative to the alternative hypothesis, H_2, n = sample size, and $\chi^2_q = -2\ell n$LR, where LR denotes the usual likelihood ratio test statistic. Similar large-sample approximations were developed later and independently by Lindley (1961) and Schwarz (1978). Thus, it may be concluded that not only has the theory of Bayesian analyses of hypotheses been developed, but also applications of it

291 Posterior Odds Ratios for Regression Hypotheses: General Considerations and Some Specific Results

Table 4 Approximate Values of K_{12} Computed from Equation (3.16) with Prior Odds Ratio $\pi_1/\pi_2 = 1.25$

Degrees of freedom,	Values of $z = n^{1/2}\bar{y}/s$											
$\nu = n-1$	1.20	1.40	1.50	1.60	1.70	1.80	1.90	2.00	2.50	3.00	3.50	4.00
15	1.80	1.40	1.22	1.07	.925	.801	.690	.593	.267	.113	.0465	.0188
20	2.05	1.57	1.36	1.18	1.02	.873	.746	.634	.266	.103	.0374	.0132
50	3.12	2.35	2.02	1.72	1.46	1.23	1.03	.860	.311	.096	.0258	.0062
100	4.36	3.26	2.79	2.37	2.00	1.68	1.39	1.15	.392	.110	.0257	.0050
200	6.13	4.57	3.90	3.31	2.78	2.32	1.92	1.58	.522	.139	.0299	.0052
500	9.66	7.18	6.12	5.18	4.35	3.63	3.00	2.46	.795	.205	.0418	.0068
1000	13.6	10.1	8.64	7.31	6.14	5.11	4.22	3.45	1.11	.282	.0566	.0089
2000	19.3	14.3	12.2	10.3	8.66	7.21	5.95	4.87	1.56	.395	.0784	.0122
5000	30.6	22.6	19.3	16.3	13.7	11.4	9.39	7.70	2.46	.620	.122	.0189

NOTE: K_{12} is the posterior odds ratio relating to hypotheses H_1: $\theta = 0$ and H_2: $\theta > 0$, where θ is the mean of normal distribution with variance σ^2. Under H_1 the prior pdf employed is $p(\sigma) \propto 1/\sigma$ while under H_2 the prior pdf $p(\theta|\sigma) = (2/\pi)/(1+\theta^2/\sigma^2)$, $0 < \theta < \infty$, multiplied by $1/\sigma$ was used.

Table 5 Approximate Values of $K_{23} \doteq \Pi_2/\Pi_3 \doteq F(t)/F(-t)$

Values of	Values of $t = n^{1/2}\bar{y}/s$								
	0	.25	.50	.75	1.00	1.25	1.50	1.75	2.00
$F(t)/F(-t)$	1.00	1.49	2.24	3.41	5.30	8.47	14.0	24.0	43.0

NOTE: K_{23} is the posterior odds ratio in favor of H_2: $\theta > 0$ relative to H_3: $\theta < 0$, Π_2/Π_3 is the prior odds ratio, and $F(\cdot)$ denotes the cumulative normal distribution function.

to a number of important small-sample and large-sample problems have been developed.

In the next section some recent extensions of Jeffreys's analyses to selected regression hypotheses will be described.

4 Analyses of Selected Regression Hypotheses

Let the regression model for the $n \times 1$ observation vector y be:

$$y = \iota\alpha + X\beta + u, \tag{4.1}$$

where ι is an $n \times 1$ vector with all elements equal to one, α and β are a scalar parameter and a $k \times 1$ vector of regression coefficients, respectively, $(\iota \vdots X)$ is an $n \times (k + 1)$ given matrix of rank $k + 1$, and u is an n $\times$ 1 vector of disturbance terms. It is assumed that the variables in X are measured in terms of deviations from their respective sample means and thus $\iota'X = \mathbf{0}'$. Further, the elements of u are assumed independently drawn from a normal population with zero mean and finite variance σ^2 with an unknown value.

In Zellner and Siow (1980), the following two hypotheses are considered:

$$H_1: \beta = \mathbf{0}, \quad -\infty < \alpha < \infty \text{ and } 0 < \sigma < \infty \tag{4.2}$$

$$H_2: \beta \neq 0, \quad -\infty < \alpha < \infty \text{ and } 0 < \sigma < \infty. \tag{4.3}$$

The likelihood functions under these two hypotheses are:

$$\begin{aligned} p(y \mid \alpha, \sigma, H_1) &\propto \sigma^{-n}\exp\{-(y - \alpha\iota)'(y - \alpha\iota)/2\sigma^2\} \\ &\propto \sigma^{-n}\exp\{-[\nu_1 s_1^2 + n(\alpha - \bar{y})^2]/2\sigma^2\} \end{aligned} \tag{4.4}$$

and

$$\begin{aligned} p(y \mid \alpha, \beta, \sigma, H_2) &\propto \sigma^{-n}\exp\{-(y - \alpha\iota - X\beta)' \\ &\qquad (y - \alpha\iota - X\beta)/2\sigma^2\} \\ &\propto \sigma^{-n}\exp\{-[\nu_2 s_2^2 + n(\alpha - \bar{y})^2 \\ &\qquad + (\beta - \hat{\beta})'X'X(\beta - \hat{\beta})]/2\sigma^2\} \end{aligned} \tag{4.5}$$

where the proportionality constant is $(2\pi)^{-n/2}$ in each case,

293 Posterior Odds Ratios for Regression Hypotheses: General Considerations and Some Specific Results

$$\bar{y} = \sum_{i=1}^{n} y_i/n,\ \nu_1 s_1^2 = \sum_{i=1}^{n} (y_i - \bar{y})^2,\ \nu_1 = n - 1,$$
$$\hat{\beta} = (X'X)^{-1}X'y,\ \nu_2 s_2^2 = (y - \bar{y}\iota - X\hat{\beta})'(y - \bar{y}\iota - X\hat{\beta}),$$
$$\text{and } \nu_2 = n - k - 1.$$

With prior odds ratio equal to one, the posterior odds ratio for H_1 and H_2, denoted by K_{12} is

$$K_{12} = \frac{\int\int p(y|\alpha, \sigma, H_1)p(\alpha, \sigma|H_1)d\alpha\, d\sigma}{\int\cdots\int p(y|\alpha, \beta, \sigma, H_2)p(\alpha, \beta, \sigma|H_2)d\alpha\, d\beta\, d\sigma} \tag{4.6}$$

where the integrations for α and the elements of β are to be performed over the range $-\infty$ to $+\infty$ while those for σ are over the range $0 < \sigma < \infty$. As regards the prior pdf's in (4.6), Zellner and Siow (1980) employed the following Jeffreys-like priors:

$$p(\alpha, \sigma|H_1) \propto 1/\sigma \qquad -\infty < \alpha < \infty \text{ and } 0 < \sigma < \infty \tag{4.7}$$

and

$$p(\alpha, \beta, \sigma) \propto f(\beta|\sigma)/\sigma \quad -\infty < \alpha < \infty \text{ and } 0 < \sigma < \infty \tag{4.8a}$$

with

$$f(\beta|\sigma) = c|X'X/n\sigma^2|^{1/2}/(1 + \beta'X'X\beta/n\sigma^2)^{(k+1)/2} \qquad -\infty < \beta_i < \infty \qquad i = 1, 2, \ldots, k \tag{4.8b}$$

where the proportionality constants in (4.7) and (4.8a) are the same, $c = \Gamma[(k + 1)/2]/\pi^{(k+1)/2}$ is the normalizing constant for the proper k-dimensional Cauchy pdf for β given σ which is centered at $\mathbf{0}$, the value suggested by the null hypothesis H_1. The matrix of the pdf in (4.8b), $X'X/n\sigma^2$ is suggested by the form of the information matrix and/or by consideration of standardized coefficients for the regression model.

When (4.7) and (4.8) are inserted in (4.6) and the integrations are performed, exactly in terms of the numerator and approximately in terms of the denominator (see Zellner and Siow 1980 for details), the result is:

$$\begin{aligned} K_{12} &\doteq a(\nu_2/2)^{k/2}(\nu_2 s_2^2/\nu_1 s_1^2)^{(\nu_2-1)/2} \\ &= a(\nu_2/2)^{k/2}/[1 + (k/\nu_2)F_{k,\nu_2}]^{(\nu_2-1)/2} \\ &= a(\nu_2/2)^{k/2}(1 - R^2)^{(\nu_2-1)/2}, \end{aligned} \tag{4.9}$$

where $a = \pi^{1/2}/\Gamma[(k + 1)/2]$, $F_{k,\nu_2} = \hat{\beta}'X'X\hat{\beta}/ks_2^2$, $R^2 = \hat{\beta}'X'X\hat{\beta}/(\nu_2 s_2^2 + \hat{\beta}'X'X\hat{\beta})$, and $\nu_2 s_2^2 + \hat{\beta}'X'X\hat{\beta} = \nu_1 s_1^2$ have been employed in going from the first to the second line of (4.9). As shown in Zellner and Siow (1980), the error of the approximation in (4.9) is of order $1/n$ of the whole expression. In the first line of (4.9) K_{12} has been expressed in terms of $\nu_2 s_2^2/\nu_1 s_1^2$, the ratio of the sum of squared least squares residuals under H_1 and H_2 while in the second and third lines, K_{12} has been expressed in terms of the usual F-statistic, F_{k,ν_2}, and the sample squared multiple correlation coefficient, respectively. Also, from (4.9), the following large-sample Jeffreys-like approximation can be obtained:

$$-2\ell n K_{12} \doteq \chi_k^2 - k\ell n \nu_2, \tag{4.10}$$

where $\chi_k^2 = \hat{\beta}'X'X\hat{\beta}/s_2^2$.

Shown in table 6, taken from Zellner and Siow (1980), is a tabulation of K_{12} values from (4.9) for various values of k and R^2 and the associated values of F_{k,ν_2} for $\nu_2 = 20$ and $\nu_2 = 100$. Also shown in the table are .05 and .01 critical values of F_{k,ν_2} and associated values of R^2. For example, when $\nu_2 = 20$ and $k = 1$, $K_{12} \doteq 1$ when $F_{1,20} = t_{20}^2 = 4.0$ or $R^2 = .16$. Under these conditions as R^2 or $F_{1,20}$ grows in value, K_{12} declines in value. For $R^2 = .49$ or $F_{1,20} = 18.9$, the associated value of K_{12} is 10^{-2}, strong evidence against $H_1 : \beta = 0$. With $\nu_2 = 20$, as k, the number of regression coefficients, grows, a higher value of R^2 is needed to have $K_{12} = 1$. When $k = 6$, an $R^2 = .48$ or $F_{6,20} = 3.1$ provides $K_{12} \doteq 1$. Generally, the R^2 values associated with $K_{12} = 1$ for $\nu_2 = 20$ are not far different from the .05 critical values of R^2 provided in the last column of the table. However, when $\nu_2 = 100$, part B of the table shows that the values of R^2 associated with $K_{12} = 1$ when $k > 1$ are larger than the R^2 values corresponding to the .01 critical values of the F-statistic. Thus for large ν_2 values, the .05 critical values of F and their associated R^2 values are poor approximations to indifference values ($K_{12} = 1$) of these posterior odds ratios. Further, for $\nu_2 = 100$, (4.10) provides results similar to those in table 6 in terms of indifference values for R^2. For example, with $\nu_2 = 100$, $k = 6$, and $\chi_6^2 = 27.6$ or $R^2 = .22$, (4.10) yields $K_{12} \doteq 1$, which is precisely the value of R^2 yielding $K_{12} \doteq 1$ in table 6. However, for $\nu_2 = 20$, the approximation in (4.10), while not far off, is not as accurate; see table 7 for more details.

Now let (4.1) be rewritten as

$$\underset{n\times 1}{y} = \underset{n\times 1}{\iota}\underset{1\times 1}{\alpha} + \underset{n\times k_1}{X_1}\underset{k_1\times 1}{\beta_1} + \underset{n\times k_2}{X_2}\underset{k_2\times 1}{\beta_2} + \underset{n\times 1}{u} \tag{4.11}$$

with $k = k_1 + k_2$, $X = (X_1 \vdots X_2)$, and $\beta' = (\beta_1', \beta_2')$. Consider the hypotheses

$$H_A : \beta_2 = \mathbf{0} \tag{4.12}$$

$$H_B : \beta_2 \neq \mathbf{0} \tag{4.13}$$

with α, the elements of $\boldsymbol{\beta}_1$, and σ unrestricted under H_A and H_B. Using prior assumptions similar to those presented above, Zellner and Siow (1980) derived the following posterior odds ratio, K_{AB} for H_A and H_B with their prior odds ratio assumed equal to one:

$$(4.14)\qquad \begin{aligned} K_{AB} &\doteq b(\nu_B/2)^{k_2/2}(\nu_B s_B^2/\nu_A s_A^2)^{(\nu_B-1)/2} \\ &= b(\nu_B/2)^{k_2/2}[(1-R_B^2)/(1-R_A^2)]^{(\nu_B-1)/2} \\ &= b(\nu_B/2)^{k_2/2}/[1+(k_2/\nu_B)F_{k_2,\nu_B}]^{(\nu_B-1)/2} \end{aligned}$$

Table 6 Values of R^2 and F_{k,ν_2} Associated with Particular Values of K_{12} and k in (4.9) for $\nu_2 = 20$ and $\nu_2 = 100$

		K_{12}					.01 and .05 Critical Values of F and Associated R^2s	
k	Value of:	1	$10^{-1/2}$	10^{-1}	$10^{-3/2}$	10^{-2}	.01	.05
				A. $\nu_2 = 20$				
1	R^2	.16	.26	.35	.42	.49	.29	.18
	$F_{1,20}$	4.0	7.0	10.6	14.5	18.9	8.10	4.35
2	R^2	.27	.35	.43	.49	.55	.37	.26
	$F_{2,20}$	3.7	5.5	7.5	9.7	12.3	5.85	3.49
3	R^2	.35	.42	.48	.54	.60	.43	.32
	$F_{3,20}$	3.5	4.8	6.3	8.0	9.9	4.94	3.10
4	R^2	.40	.47	.53	.58	.63	.47	.36
	$F_{4,20}$	3.4	4.4	5.7	7.0	8.6	4.43	2.87
5	R^2	.45	.51	.57	.61	.66	.51	.40
	$F_{5,20}$	3.2	4.2	5.2	6.4	7.7	4.10	2.71
6	R^2	.48	.54	.59	.64	.68	.54	.44
	$F_{6,20}$	3.1	3.9	4.9	5.9	7.1	3.87	2.60
				B. $\nu_2 = 100$				
1	R^2	.050	.072	.093	.11	.13	.065	.380
	$F_{1,100}$	5.2	7.7	10.3	12.8	15.5	6.90	3.94
2	R^2	.089	.11	.13	.15	.17	.088	.058
	$F_{2,100}$	4.9	6.2	7.5	8.8	10.3	4.82	3.09
3	R^2	.12	.14	.16	.18	.20	.11	.075
	$F_{3,100}$	4.6	5.5	6.4	7.4	8.3	3.98	2.70
4	R^2	.15	.17	.19	.21	.23	.12	.090
	$F_{4,100}$	4.4	5.1	5.9	6.6	7.3	3.51	2.46
5	R^2	.18	.20	.21	.23	.25	.14	.10
	$F_{5,100}$	4.3	4.9	5.5	6.1	6.7	3.20	2.30
6	R^2	.20	.22	.24	.25	.27	.15	.12
	$F_{6,100}$	4.2	4.7	5.2	5.7	6.2	2.99	2.19

NOTE: $F_{k,\nu_2} = (\nu_2/k)R^2/(1-R^2)$, with $\nu_2 = n-k-1$.

where $b = \pi^{1/2}/\Gamma[(k_2 + 1)/2]$, R_A^2 and R_B^2 are the squared sample multiple correlation coefficients under H_A and H_B, respectively, $F_{k_2, \nu_B} = \hat{\beta}_2' V' V \hat{\beta}_2 / k_2 s_B^2$, the usual F-statistic with k_2 the number of elements in β_2, $V = [I - X_1(X_1'X_1)^{-1}X_1']X_2$, $\nu_A = n - k_1 - 1$, $\nu_B = n - k_1 - k_2 - 1 = n - k - 1$, and $\nu_A s_A^2$ and $\nu_B s_B^2$ the sum of squared least squares residuals under H_A and H_B, respectively. It is seen that the third line of (4.14) is in a form exactly similar to that of the second line of (4.9). Thus table 6 can be employed to obtain values of K_{AB} associated with values of F_{k_2, ν_B} if in using the table, ν_B is identified with ν_2 and k_2 is identified with k.

To illustrate application of the above analysis, the Hald data, 13 observations on a dependent variable and four independent variables given in Draper and Smith (1966, App. B) and Geisser and Eddy (1979, p. 158) have been analyzed. The problem is to compare all possible regression models, that is, the model with a constant term and all four independent variables and models including any three variables, any two variables, any one variable, and none of the four variables. In toto, there are 16 possible models, each assumed to be linear in the independent variables. By use of the posterior odds ratio in (4.14), the results in table 6, from Zellner and Siow (1980), were obtained.

In the second column of table 7, values of K_{AB}, the posterior odds ratio for a particular subset regression (H_A), relative to the regression including all four independent variables and a constant (H_B) are presented. The largest value of K_{AB}, 3.9 is encountered for model (1,2,c) while the next largest values are 3.3 for (1,2,3,c) and (1,2,4,c), 2.8 for (1,3,4,c), 1.8 for (1,4,c) and .90 for (2,3,4,c). That overwhelming evidence in favor of a particular subset regression is not obtained is not surprising with just 13 observations. On the other hand, a number of subset regressions have very small values for their posterior odds ratios.

Also shown in table 7 are the results of applying several other criteria to the Hald data. As is well known, use of R^2 results in choice of the model with the largest number of free parameters here (1,2,3,4,c), for which $R^2 = .9824$, considered by most to be a defect of R^2. Interestingly, the six models other than (1,2,3,4,c) with the highest posterior odds ratios are the ones with the highest R^2 values, undoubtedly a reflection of the fact that K_{AB} can be expressed as a function of R^2; see (4.14). However, note that model (3,4,c) has a posterior odds ratio of only .14 while its $R^2 = .9353$. Use of the standard error of regression and the Geisser-Eddy predictive sample reuse (PSR) criterion leads to about the same set of preferred models as does use of the posterior odds ratios. Similarly Aitken's R^2-adequate criterion at the 5% significance level also selects about the same set of preferred subset regressions. In this last approach, a significance level must be chosen; see

297 Posterior Odds Ratios for Regression Hypotheses: General Considerations and Some Specific Results

Table 7 Posterior Odds Ratios, K_{AB}, and Other Model Selection Criteria Computed Using the Hald Data (n = 13)

Independent Variables, H_A	Posterior Odds Ratio, K_{AB}[b]	R^2	Standard Error of Regression	Geisser-Eddy PSR Criterion[c]	Aitken's R^2 Adequate Criterion (.05)[d]
1,2,3,c[a]	3.3	.9823	2.312	−33.50	adequate
1,2,4,c	3.3	.9823	2.309	−33.22	adequate
1,3,4,c	2.8	.9813	2.377	−33.64	adequate
2,3,4,c	0.90	.9728	2.864	−36.26	adequate
1,2,c	3.9	.9787	2.406	−33.18	adequate
1,3,c	4.1×10^{-4}	.5482	11.077	−53.23	—
1,4,c	1.8	.9725	2.734	−35.08	adequate
2,3,c	1.1×10^{-2}	.8470	6.445	−46.86	—
2,4,c	1.2×10^{-3}	.6800	9.321	−51.62	—
3,4,c	1.4×10^{-1}	.9353	4.192	−40.95	—
1,c	3.2×10^{-4}	.5339	10.727	−51.93	—
2,c	8.5×10^{-4}	.6663	9.077	−50.58	—
3,c	8.7×10^{-5}	.2859	13.278	−54.83	—
4,c	9.2×10^{-4}	.6745	8.964	−49.84	—
c	8.9×10^{-5}	0	15.044	—	—
1,2,3,4,c	1.0	.9824	2.446	−34.84	adequate

[a]1,2,3,c indicates that H_A in this case is a linear regression involving the independent variables x_1, x_2, x_3 and a constant term.

[b]In each case, K_{AB} is the posterior odds ratio for H_A, as shown in the first column, relative to H_B: 1,2,3,4,c, the model with all four independent variables and a constant term, and the prior odds ratio's value has been taken equal to one.

[c]W. F. Eddy provided these to the author; reported values in Geisser and Eddy (1979), table 4, are erroneous.

[d]Reported in Aitken (1974). Models indicated to be "adequate" passed the 5% R^2-adequate criterion while the others did not.

Aitken (1974, p. 226) for discussion of the effects of varying the significance level.

While the results of applying various criteria to the Hald data are not qualitatively very different, except for use of R^2, it appears that the posterior odds ratio metric seems preferable for the problem of reflecting the effects of both sample and prior information as it impinges on the evaluation of alternative subset regressions. Also, posterior odds ratios provide posterior probabilities associated with alternative subset regressions which can be employed to average parameter estimates and forecasts from alternative models as pointed out in the literature. Finally, it is not suggested that posterior odds ratios be used mechanically in variable selection problems.

The present approach can be elaborated to take account of more subject matter information through use of more informative priors; see Dickey (1975, 1980), of possible autocorrelation, and Gaver and Geisel (1974), Griffiths and Dao (1980), Fomby and Guilkey (1978), and Mayer (1979) and of other possible departures from standard assumptions regarding error terms' properties. In addition, posterior odds ratios can be formulated and computed for a wide range of problems, as mentioned earlier and not just for linear regression problems.

Some results presented by Zellner and Siow (1978) for a two-equation regression system will now be reviewed. The model for the observations is:

$$y_1 = X_1\beta_1 + u_1 \tag{4.15a}$$

$$y_2 = X_2\beta_2 + u_2, \tag{4.15b}$$

where for $i = 1, 2$, y_i is an $n_i \times 1$ vector, X_i is an $n_i \times k$ given matrix of rank k, β_i is a $k \times 1$ vector of regression parameters, u_i is an $n_i \times 1$ vector of disturbance terms. It is assumed that the elements of u_1 and u_2 have been independently drawn from normal populations with zero means and variances σ_1^2 and σ_2^2, respectively. The likelihood function is then,

$$\begin{aligned} p(y_1, y_2 | \beta_1, \beta_2, \sigma_1 \sigma_2) \propto \sigma_1^{-n_1}\sigma_2^{-n_2}\exp\{-[(y_1 - X_1\beta_1)'(y_1 - X_1\beta_1)/2\sigma_1^2 \\ + (y_2 - X_2\beta_2)'(y_2 - X_2\beta_2)/2\sigma_2^2]\}. \end{aligned} \tag{4.16}$$

Given a prior pdf for the parameters of (4.16), the associated posterior distribution can be obtained and used to investigate nonsharp null hypotheses about parameters' values, for example, $H: c_1 < \phi < c_2$ and its negation $\bar{H}: 0 < \phi < c_1$ or $c_2 < \phi < \infty$, where c_1 and c_2 are given positive numbers and $\phi = \sigma_2^2/\sigma_1^2$.

With a diffuse prior pdf for the parameters of (4.16),

$$p(\beta_1, \beta_2, \sigma_1, \sigma_2) \propto 1/\sigma_1\sigma_2, \tag{4.17}$$

where, for $i = 1, 2$, $0 < \sigma_i < \infty$ and $-\infty < \beta_{ij} < \infty$, $j = 1, 2, \ldots, k$,

$$\begin{aligned} p(\beta_1, \beta_2, \sigma_1, \sigma_2 | D) \propto \prod_{i=1}^{2} \sigma_i^{-(n_i+1)} \\ \times \exp\{-[\nu_i s_i^2 + (\beta_i - \hat{\beta}_i)'X_i'X_i(\beta_i - \hat{\beta}_i)]/2\sigma_i^2\}, \end{aligned} \tag{4.18}$$

where $D = (y_1, y_2, I)$, the data and prior information, $\hat{\beta}_i = (X_i'X_i)^{-1}X_i'y_i$, $\nu_i s_i^2 = (y_i - X_i\hat{\beta}_i)'(y_i - X_i\hat{\beta}_i)$, and $\nu_i = n_i - k$ for $i = 1, 2$. On integrating (4.18) with respect to β_1 and β_2, the result is:

$$p(\sigma_1, \sigma_2 | D) \propto \prod_{i=1}^{2} \sigma_i^{-(\nu_i+1)}\exp\{-\nu_i s_i^2/2\sigma_i^2\}, \tag{4.19}$$

the product of two inverted gamma pdf's for σ_1 and σ_2. As shown in Zellner (1971, p. 378), (4.19) implies that

$$F = (\sigma_2^2/s_2^2)/(\sigma_1^2/s_1^2) \tag{4.20}$$

has an F_{ν_1,ν_2} posterior pdf.[12] Then

$$Pr\{c_1 < \sigma_2^2/\sigma_1^2 < c_2 \mid D\} = Pr\{(s_1^2/s_2^2)c_1 < F < (s_1^2/s_2^2)c_2 \mid D\} \tag{4.21}$$

which can be evaluated using tables of the F-distribution or by univariate numerical integration. Also, $F' = (\sigma_1^2/s_1^2)/(\sigma_2^2/s_2^2)$ has an F_{ν_2,ν_1} posterior pdf and thus inferences about $\lambda = \sigma_1^2/\sigma_2^2$ can also be made easily.

If (β_1, σ_1) and (β_2, σ_2) are independently distributed, each with a natural conjugate prior pdf, then it is straightforward to show that $\lambda = \sigma_1^2/\sigma_2^2$ has a posterior pdf in the F-form and thus posterior probabilities such as $Pr(a_1 < \lambda < a_2 \mid D)$ can be easily evaluated. If in (4.15), $\beta_1 = \beta_2 = \beta$ and a diffuse prior pdf $p(\beta, \sigma_1, \sigma_2) \propto 1/\sigma_1\sigma_2$ is employed, Zellner (1971, p. 107) has derived the posterior pdf for $\lambda = \sigma_1^2/\sigma_2^2$. Also, Zellner and Siow (1978) show how to obtain the posterior pdf for λ when the model in (4.15) is parameterized in terms of β_1, $\delta = \beta_2 - \beta_1$, σ_1, and σ_2 and an informative prior pdf for δ is employed.

If interest centers on $\delta = \beta_1 - \beta_2$, its marginal posterior pdf can be derived using several different prior pdf's. For example, with respect to the system in (4.15), if it is assumed that $\sigma_1 = \sigma_2 = \sigma$ and the diffuse prior pdf $p(\beta_1, \beta_2, \sigma) \propto 1/\sigma$ is employed, it is shown in Zellner and Siow (1978) that F, given by

$$F = (\delta - \hat{\delta})'M(\delta - \hat{\delta})/ks^2, \tag{4.22}$$

where $\hat{\delta} = \hat{\beta}_1 - \hat{\beta}_2$, $M^{-1} = (X_1'X_1)^{-1} + (X_2'X_2)^{-1}$, and $\nu s^2 = (y_1 - X_1\hat{\beta}_1)'(y_1 - X_1\hat{\beta}_1)$ with $\nu = n - 2k$ has an $F_{k,\nu}$ posterior pdf, the Bayesian version of what has come to be known as the Chow (1960) test statistic. Using (4.22), various posterior probabilities relating to δ's possible values can be evaluated. Also, this analysis is easily generalizable to the case in which the prior pdf for β_1, β_2, and σ is in the natural conjugate form. Further, in the case $\delta = \beta_2 - \beta_1$ and $\sigma_1 \neq \sigma_2$, the posterior pdf for δ can also be obtained.

In the case of sharp hypotheses regarding parameters of the system in (4.15), Zellner and Siow (1978) considered the hypotheses shown in the following table:

12. This result is Bayesian analogue of the Goldfeld and Quandt (1965) sampling theory test for the equality of disturbance variances.

	$\beta_1 = \beta_2$	$\beta_1 \neq \beta_2$
$\sigma_1 = \sigma_2$	H_A	H_B
$\sigma_1 \neq \sigma_2$	H_C	H_D

For example, employing Jeffreys-like prior pdf's for the parameters under H_A and H_B and a prior odds ratio of one, the following approximate posterior odds ratio was obtained:

$$K_{AB} \doteq a(\nu_B/2)^{k/2}/[1 + (k/\nu_B)F_{k,\nu_B}]^{(\nu_B - 1)/2}, \tag{4.23}$$

where $a = \pi^{1/2}/\Gamma[(k+1)/2]$, $\nu_B = n - 2k$, $F_{k,\nu_B} = \hat{\delta}'A\hat{\delta}/ks_B^2$, the usual F-statistic with $\hat{\delta} = \hat{\beta}_1 - \hat{\beta}_2$, $A = V'V$, with $V = Z - XP$, where $Z' = (0' \vdots X_2')$, $X' = (X_1' \vdots X_2')$, and $P = (X'X)^{-1}X'Z$, and $\nu_B s_B^2 = (y_1 - X_1\hat{\beta}_1)'(y_1 - X_1\hat{\beta}_1) + (y_2 - X_2\hat{\beta}_2)'(y_2 - X_2\hat{\beta}_2)$, the sum of squared least squares residuals under H_B.

It is seen that (4.23) is in the form of the second line of (4.9) and thus table 5 can be utilized to obtain values of K_{AB} associated with particular values of F_{k,ν_B}, k, and ν_B given that one takes $\nu_B = \nu_2$ in using the table. Additional odds ratios pertaining to other pairs of hypotheses, for example, H_C versus H_D have been derived.

In connection with (4.23), and other posterior odds ratios in this form, for given k and ν_B, K_{AB} is a monotonic function of F_{k,ν_B}. This fact makes it easy to obtain the cumulative sampling distribution of K_{AB} under the null hypothesis $H_A : \beta_1 = \beta_2$ and $\sigma_1 = \sigma_2$. That is, under the null hypothesis, F_{k,ν_B} has a well-known sampling distribution which permits evaluation of $Pr(c_1 < F_{k,\nu_B} < c_2)$ with c_1 and c_2 given positive constants. Let b_1 and b_2 be the values of K_{AB} associated with $F_{k,\nu_B} = c_1$ and $F_{k,\nu_B} = c_2$, respectively. Then $Pr(b_2 < K_{AB} < b_1) = Pr(c_1 < F_{k,\nu_B} < c_2)$, and probability statements can be made regarding the sampling properties of K_{AB} by use of the cumulative F-distribution with k and ν_B degrees of freedom. Further, any test procedure in the form reject H_A if $F_{k,\nu_B} > F_c$, where F_c is a critical value, is equivalent to reject H_A if $K_{AB} < K_c$, where K_c is the value of K_{AB} associated with $F_{k,\nu_B} = F_c$. Thus any "optimal" sampling properties of such F-tests will be enjoyed by use of the posterior odds ratio K_{AB} in the way described above. However, as pointed out above, the sampling theory rationalization of testing procedures has some controversial aspects and is not needed to justify the computation of posterior odds ratios for a given sample of data. For a given sample of data, the inductive probability calculus, in particular Bayes's theorem, provides a justification for computing logically consistent posterior probabilities and odds ratios.

5 Concluding Remarks

In what has been presented, some Bayesian results for comparing and testing both sharp and nonsharp hypotheses have been reviewed. Bayesian posterior odds ratios have been derived for a number of hypotheses encountered in econometrics and other sciences. Their use permits investigators to incorporate prior information formally and flexibly in appraising alternative hypotheses. Moreover, if an explicit loss structure is available, it is well known that it is possible to use posterior probabilities to choose among alternative hypotheses in such a way as to minimize posterior expected loss and average risk. Further, as has been pointed out in the literature, posterior probabilities can be employed to obtain Bayesian pretest estimates and combined predictions from alternative models.[13] In addition, several optimal sampling properties associated with use of posterior odds ratios in testing have been mentioned.

As regards sampling theory test procedures, some controversial points regarding their use and justification have been reviewed. In particular, Jeffreys's critiques of the use of *P*-values and sampling theory power functions were presented. It is thought that these issues merit further consideration by sampling theorists and others.

The results of the survey of testing procedures in use, presented in section 2, revealed that there is much room for improvement in analyzing alternative hypotheses in applied studies. Further development and use of posterior odds ratios will be helpful in producing improved analyses of alternative hypotheses and models in econometrics and other sciences.

References

Aitken, M. A. (1974). "Simultaneous Inference and the Choice of Variable Subsets in Multiple Regression." *Technometrics*, 16:221–27.

Berndt, E. R., and N. E. Savin (1977). "Conflict among Criteria for Testing Hypotheses in the Multivariate Linear Regression Model." *Econometrica*, 45:1263–77.

Blackwell, D., and M. A. Girshick (1954). *Theory of Games and Statistical Decisions*. New York: John Wiley & Sons.

Blattberg, R. C., and N. J. Gonedes (1974). "A Comparison of the Stable and Student Distributions as Statistical Models for Stock Prices." *Journal of Business*, 47:244–80.

13. See, e.g., Geisel (1970, 1973), Griffiths and Dao (1980), Zellner (1979), and Zellner and Vandaele (1975).

Box, G. E. P., and W. J. Hill (1967). "Discrimination among Mechanistic Models." *Technometrics*, 9:57–71.

Chernoff, H., and L. E. Moses (1959). *Elementary Decision Theory*. New York: John Wiley & Sons.

Chow, G. C. (1960). "Tests of Equality between Sets of Coefficients in Two Linear Regressions." *Econometrica*, 28:591–605.

Davis, W. M. (1978). "Bayesian Analysis of the Linear Model Subject to Linear Inequality Constraints." *Journal of the American Statistical Association*, 73:573–79.

DeGroot, M. H. (1970). *Optimal Statistical Decisions*. New York: McGraw-Hill Book Co.

Dickey, J. M. (1971). "The Weighted Likelihood Ratio, Linear Hypotheses on Normal Location Parameters." *Annals of Mathematical Statistics*, 42:204–23.

———. (1975). "Bayesian Alternatives to the *F*-test and Least-Squares Estimates in the Normal Linear Model." In S. E. Fienberg and A. Zellner, eds., *Studies in Bayesian Econometrics and Statistics in Honor of Leonard J. Savage*, pp. 515–54. Amsterdam: North-Holland Publishing Co.

———. (1977). "Is the Tail Area Useful as an Approximate Bayes Factor?" *Journal of the American Statistical Association*, 72:138–42.

———. (1980). "Approximate Coherence for Regression Model Inference with a New Analysis of Fisher's Broadbalk Wheatfield Example." In A. Zellner, ed., *Bayesian Analysis in Econometrics and Statistics: Essays in Honor of Harold Jeffreys*, pp. 333–54. Amsterdam: North-Holland Publishing Co.

Draper, N. R., and H. Smith (1966). *Applied Regression Analysis*. New York: John Wiley & Sons.

Dyer, A. R. (1972). "A Comparison of Classification and Hypothesis Testing Procedures for Separate Families of Hypotheses." Ph.D. diss., Department of Statistics, University of Chicago.

———. (1973). "Discrimination Procedures for Separate Families of Hypotheses." *Journal of the American Statistical Association*, 68:970–74.

Fomby, T. B., and D. K. Guilkey (1978). "On Choosing the Optimal Level of Significance for the Durbin-Watson Test and the Bayesian Alternative." *Journal of Econometrics*, 8:203–13.

Gaver, K. M., and M. S. Geisel (1974). "Discriminating among Alternative Models: Bayesian and Non-Bayesian Methods." In P. Zarembka, ed., *Frontiers of Econometrics*, pp. 49–77. New York: Academic Press.

Geisel, M. S. (1970). "Comparing and Choosing among Parametric Statistical Models: A Bayesian Analysis with Macroeconomic Applications." Ph.D. diss., Graduate School of Business, University of Chicago.

———. (1973). "Bayesian Comparisons of Simple Macroeconomic Models." *Journal of Money, Credit and Banking*, 5:751–72. Reprinted in S. E. Fienberg and A. Zellner, eds., *Studies in Bayesian Econometrics and Statistics in Honor of Leonard J. Savage*, pp. 227–56. Amsterdam: North-Holland Publishing Co., 1975.

Geisser, S., and W. F. Eddy (1979). "A Predictive Approach to Model Selection." *Journal of the American Statistical Association*, 74:153–60.

Goldfeld, S. M., and R. E. Quandt (1965). "Some Tests for Homoscedasticity." *Journal of the American Statistical Association*, 60:539–47.

Good, I. J. (1950). *Probability and the Weighing of Evidence*. New York: Hafner Publishing Co.

———. (1965). *The Estimation of Probabilities*. Research Monograph no. 30. Cambridge, Mass.: MIT Press.

———. (1980). Comment presented at the International Meeting on Bayesian Statistics, Valencia, Spain, May 28–June 2, 1979. In J. M. Bernardo, M. H. DeGroot, D. V. Lindley, and A. F. M. Smith, eds., *Bayesian Statistics: Proceedings of the First International Meeting Held in Valencia (Spain), May 28 to June 2, 1979*, pp. 636–37. Valencia, Spain: University Press.

Griffiths, W., and D. Dao (1980). "A Note on a Bayesian Estimator in an Autocorrelated Error Model." *Journal of Econometrics*, 12: 389–92.

Jaynes, E. T. (1976). "Confidence Intervals Vs. Bayesian Intervals." In W. L. Harper and C. A. Hooker, eds., *Foundations of Probability Theory, Statistical Inference, and Statistical Theories*, pp. 175–213, followed by discussion. Dordrecht, Holland: D. Reidel Publishing Co.

———. (1980). Comment presented at the International Meeting on Bayesian Statistics, Valencia, Spain, May 28–June 2, 1979. In J. M. Bernardo, M. H. DeGroot, D. V. Lindley, and A. F. M. Smith, eds., *Bayesian Statistics: Proceedings of the First International Meeting Held in Valencia (Spain), May 28 to June 2, 1979*, pp. 618–29. Valencia, Spain: University Press.

Jeffreys, H. (1957). *Scientific Inference*. 2d ed.; 1st ed. 1931, Cambridge, England: University Press.

———. (1963). "Review of L. J. Savage et al., *The Foundations of Statistical Inference (1962)*." *Technometrics*, 5:407–10.

———. (1967). *Theory of Probability*. 3d rev. ed.; 1st ed. 1939. London: Oxford University Press.

———. (1979). Personal communication to author.

———. (1980). "Some General Points in Probability Theory." In A. Zellner, ed., *Bayesian Analysis in Econometrics and Statistics: Essays in Honor of Harold Jeffreys*, pp. 451–53. Amsterdam: North-Holland Publishing Co.

Leamer, E. E. (1978). *Specification Searches: Ad Hoc Inference with Nonexperimental Data*. New York: John Wiley & Sons.

Lempers, F. B. (1971). *Posterior Probabilities of Alternative Linear Models*. Rotterdam: University of Rotterdam Press.

Lindley, D. V. (1957). "A Statistical Paradox." *Biometrika*, 44:187–92. See also M. S. Bartlett's comment on Lindley's paper in *Biometrika*, 44:533–34.

———. (1961). "The Use of Prior Probability Distributions in Statistical Inference and Decision." In J. Neyman, ed., *Proceedings of the Fourth Berkeley Symposium on Mathematical Statistics and Probability*, 1:453–68. Berkeley: University of California Press.

Mayer, F. (1979). "Sur l'Utilisation du critere de decision MELO dans le choix d'un seuil de signification optimal." Department of Economics, University of Montreal.

Raiffa, H., and R. Schlaifer (1961). *Applied Statistical Decision Theory*. Boston: Graduate School of Business Administration, Harvard University.

Rényi, A. (1970). *Foundations of Probability*. San Francisco: Holden-Day

Reynolds, R. A. (1982). "Posterior Odds for the Hypothesis of Independence between Stochastic Regressors and Disturbances." *International Economic Review*, 23:479–90.

Savin, N. E. (1976). "Conflict among Testing Procedures in a Linear Regression Model with Autoregressive Disturbances." *Econometrica*, 40:1303–15.

Schwarz, G. (1978). "Estimating the Dimension of a Model." *Annals of Statistics*, 6: 461–64.

Silvey, S. D. (1970). *Statistical Inference*. Baltimore: Penguin Books.

Smith, A. F. M., and D. J. Spiegelhalter (1980). "Bayes Factors and Choice Criteria for Linear Models." *Journal of the Royal Statistical Society*, Ser. B, 42: 213–20.

Thornber, E. H. (1966). "Applications of Decision Theory to Econometrics." Ph.D. diss., Department of Economics, University of Chicago.

Whittle, P. (1951). *Hypothesis Testing in Time Series Analysis*. Uppsala: Almqvist and Wiksells Boktryckeri AB.

Zellner, A. (1971). *An Introduction to Bayesian Inference in Econometrics*. New York: John Wiley & Sons.

———. (1976). "Bayesian and Non-Bayesian Analysis of the Regression Model with Multivariate Student-*t* Errors." *Journal of the American Statistical Association*, 71:400–405.

———. (1978). "Jeffreys-Bayes Posterior Odds Ratio and the Akaike Information Criterion for Discriminating between Models." *Economics Letters*, 1:337–42.

———. (1979). "Statistical Analysis of Econometric Models." *Journal of the American Statistical Association*, 74:628–43, followed by discussion.

Zellner, A., and M. S. Geisel (1970). "Analysis of Distributed Lag Models with Applications to Consumption Function Estimation." *Econometrica*, 38:865–88.

Zellner, A., and J. F. Richard (1973). "Use of Prior Information in the Analysis and Estimation of Cobb-Douglas Production Function Models." *International Economic Review*, 14:107–19.

Zellner, A., and A. Siow (1978). "Uses of Posterior Odds Ratios in Econometrics with Applications to Hypotheses in Regression Analysis." Paper presented at the 16th meeting of the NBER-NSF Seminar on Bayesian Inference in Econometrics, University of Chicago, June 2–3, 1978; see A. Zellner, ed., *Reports of the NBER-NSF Seminar on Bayesian Inference in Econometrics*, pp. 227–28. H. G. B. Alexander Research Foundation, Graduate School of Business, University of Chicago, 1979.

———. (1979). "On Posterior Odds Ratios for Sharp Null Hypotheses and One-Sided Alternatives." H. G. B. Alexander Research Foundation, Graduate School of Business, University of Chicago.

———. (1980). "Posterior Odds Ratios for Selected Regression Hypotheses." Paper presented at the International Meeting on Bayesian Statistics, Valencia, Spain, May 28–June 2, 1979. In J. M. Bernardo, M. H. DeGroot, D. V. Lindley, and A. F. M. Smith, eds. *Bayesian Statistics: Proceedings of the First International Meeting Held in Valencia (Spain), May 28 to June 2, 1979*, pp. 585–603. Valencia, Spain: University Press.

Zellner, A., and W. H. Vandaele (1975). "Bayes-Stein Estimators for *k*-Means, Regression, and Simultaneous Equation Models." In S. E. Fienberg and A. Zellner, eds., *Studies in Bayesian Econometrics and Statistics in Honor of Leonard J. Savage*, pp. 627–53. Amsterdam: North-Holland Publishing Co.

Zellner, A., and A. Williams (1973). "Bayesian Analysis of the Federal Reserve–MIT–Penn Model's Almon Lag Consumption Function." *Journal of Econometrics*, 1:267–99.

Zidek, J. V. (1969). "A Representation of Bayes Invariant Procedures in Terms of Haar Measure." *Annals of the Institute of Statistical Mathematics*, 21:291–308.

[23]

Posterior Odds Ratios for Selected Regression Hypotheses

A. ZELLNER and A. SIOW
University of Chicago

SUMMARY

Bayesian posterior odds ratios for frequently encountered hypotheses about parameters of the normal linear multiple regression model are derived and discussed. For the particular prior distributions utilized, it is found that the posterior odds ratios can be well approximated by functions that are monotonic in usual sampling theory F statistics. Some implications of this finding and the relation of our work to the pioneering work of Jeffreys and others are considered. Tabulations of odds ratios are provided and discussed.

Keywords: BAYESIAN ODDS RATIOS; HYPOTHESIS TESTING; REGRESSION HYPOTHESES; REGRESSION MODEL.

1. INTRODUCTION

In this paper we derive posterior odds ratios for selected sharp hypotheses which are frequently encountered in regression analysis[1]. Our approach involves use of generalized forms of Jeffreys's prior distributions that he regards as appropriate when there is little previous information, that is "...in the early stages of a subject...," Jeffreys (1967, p. 252). Of course if more information is available, more informative prior distributions can of course be employed as has been done by Dickey (1971, 1975, 1977), Leamer (1978), Zellner (1971, p. 307 ff.) and others. Herein, we shall emphasize the situation in which little is known and, as will be seen resulting posterior odds can be expressed in terms of usual t or F statistics and degrees of freedom. Thus the

1. See Jaynes (1976) for valuable analyses of a number of important practical examples illustrating the need for care in formulating relevant hypotheses and using appropriate techniques in order to obtain sensible results.

results to be presented provide a direct small-sample link between Bayesian posterior odds ratios and non-Bayesian test statistics as in the previous work of Jeffreys (1957, 1967, 1978), Lindley (1957) and others. Also, some large sample connections between Bayesian posterior odds ratios and non-Bayesian large sample test statistics are developed which are special cases of the general results of Lindley (1961) and Schwarz (1978).

Several, including Thornber (1966), Geisel (1970), Geisel and Gaver (1974), Leamer (1978), and Lempers (1971) have considered posterior odds ratios for regression hypotheses when little information is available. Our approach differs from those utilized in these works in that we employ prior distributions different from those employed in these works.

Since our approach is an extension of that originally presented by Jeffreys (1967, Ch.V.), we present a brief review of Jeffreys's related results in Section 2. In Section 3 posterior odds ratios for several important regression hypotheses are derived. Section 4 presents some numerical evaluations of the posterior odds ratios derived in Section 3 while a summary of results and some concluding remarks are given in Section 5.

2. REVIEW OF JEFFREYS'S RESULTS

Jeffreys (1967, Ch.V) has derived posterior odds ratios for a number of important testing problems in which little prior information is available and the issue is whether a parameter's value is equal to zero, a sharp null hypothesis. A sharp null hypothesis of "no effect" is frequently encountered and thus it is important to have an analysis of it. Jeffreys refers to such an analysis as "significance testing" and contrasts it with an estimation approach in which no special value of the parameter, for example zero is singled out for special attention. Also, he (1967, p. 251) points out that his estimation prior probability density function (pdf) for representing "knowing little", for example a uniform prior *pdf* is inappropriate for a significance testing situation in which little is known about a parameter's value[2].

To be specific, consider Jeffreys's (1967, p. 268 ff.) analysis of the normal mean problem,

$$y_i = \lambda + u_i \qquad i = 1,2,\ldots,n \tag{2.1}$$

where the y_i's are observations and the u_i's are unobserved errors assumed independently drawn from a normal population with zero mean and standard

2. In regression analysis when we delimit the number of regressors to be finite, we are obviously using sharp null hypotheses about the values of the coefficients of omitted variables.

deviation σ, $0 < \sigma < \infty$ which has an unknown value. The two hypotheses which Jeffreys considers are:

$$H_i : \lambda = 0 \text{ and } H_2 : \lambda \neq 0. \tag{2.2}$$

As regards prior *pdf*'s, under H_1 Jeffreys utilizes

$$p(\sigma | H_1) \propto 1/\sigma \qquad 0 < \sigma < \infty \tag{2.3}$$

Under H_2, Jeffreys (1967, p. 268) remarks that, "From consideration of similarity it [the prior *pdf* for λ under H_2] must depend on σ, since there, is nothing in the problem except σ to give a scale for λ". His prior under H_2 is

$$p(\lambda,\sigma)d\lambda d\sigma \propto f\left(\frac{\lambda}{\sigma}\right) \frac{d\lambda}{\sigma} \frac{d\sigma}{\sigma} \tag{2.4}$$

where $\int_{-\infty}^{\infty} f(\lambda/\sigma)\, d\lambda/\sigma = 1$. Then with prior odds 1:1, the posterior odds ratio, K_{12} is:

$$K_{12} = \frac{\int_0^{\infty}\sigma^{-n-1} \exp\{-n(\bar{y}^2 + \hat{\sigma}^2)/2\sigma^2\}d\sigma}{\int_{-\infty}^{\infty} \int_0^{\infty} F(\lambda/\sigma)\, \sigma^{-n-2} \exp\{-n[(\lambda-\bar{y})^2 + \hat{\sigma}^2]/2\sigma^2\}d\sigma d\lambda} \tag{2.5}$$

where $y = \Sigma_{i=1}^{n} y_i/n$ and $n\hat{\sigma}^2 = \Sigma_{i=1}^{n} (y_i-\bar{y})^2$.

From detailed consideration of (2.5) in the case $n = 1$ in which no decision regarding H_1 and H_2 can be made ($K_{12} = 1$), Jeffreys finds "that the consideration that one observation shall give an indecisive result is satisfied if $f(v)$ [with $v = \lambda/\sigma$] is any even function with integral 1." (p. 269). Further, the condition that $K_{12} = 0$ for $n \geq 2$ when $\hat{\sigma} = 0$ and $y \neq 0$ requires that the denominator of (2.5) diverge. This will occur if and only if $\int_0^{\infty} f(v)v^{n-1}dv$ diverges (p.269). As Jeffreys notes, "the simplest function satisfying this condition for $n > 1$ and also satisfying (3) [$\int_{-\infty}^{\infty} f(v)dv = 1$] is $f(v) = 1/\pi(1 + v^2)$." Thus his form for $f(\lambda/\sigma)$ is

$$f\left(\frac{\lambda}{\sigma}\right) \frac{d\lambda}{\sigma} = \frac{1}{\pi} \frac{1}{1 + \lambda^2/\sigma^2} \frac{d\lambda}{\sigma} \qquad -\infty < \lambda < \infty \tag{2.6}$$

a *pdf* in the univariate Cauchy form centered at zero. With respect to this point, Jeffreys (1967, p.251) states, "We must... say that the mere fact that it has been suggested that λ is zero corresponds to some presumption that it is fairly small". After pointing to unsatisfactory features of a normal prior *pdf* for λ,[3] he writes, "The chief advantage of the form [(2.6)] what we have chosen is that in any significance test it leads to the conclusion that if the null hypothesis [$\lambda = 0$] has a small posterior probability, the posterior probability of the parameters is nearly the same as in the estimation problem. Some difference remains but it is only a trace". (p. 273).

When (2.6) is substituted in (2.5) and the integrations are performed, approximately in terms of the denominator, Jeffreys obtains (1967, p. 272):

$$K_{12} \doteq (\pi\nu/2)^{1/2} \,/\, (1 + t^2/\nu)^{(\nu-1)/2} \qquad (2.7)$$

where $\nu = n\text{-}1$ and $t = \sqrt{n}\, y/s$, with $s^2 = \Sigma_{i=1}^{n} (y_i\text{-}y)^2/\nu$ and the error of the approximation "is of the order of $1/n$ of the whole expression". Also Jeffreys (p.274) provides an exact expression for K_{12}. Shown below are values of K_{12} for selected values of ν and t^2 taken from Jeffreys's table (p.439):

From Table 2.1, it is seen that when $\nu = 20$, $K_{12} = 1$ when $t^2 = 4.0$ while for $\nu = 5{,}000$, $K_{12} = 1$ when $t^2 = 9$. It is thus seen that as ν increases in value, a larger value of t^2 is required for indifference ($K_{12} = 1$) between H_1 and H_2. This corresponds to a sampling theorist's usual lowering of the significance level as ν grows in value and also bears a direct relationship to Lindley's Paradox (1957). Also note that in contrast to DeGroot's (1973) result, the tail area or "*p*-value" associated with the *t*-value is *not* equal to the posterior probability on the null hypothesis[4]. For example, with $\nu = 20$ and $t = 2.0$, the "*p*-value" is approximately .025 and yet $K_{12} = 1$ or the posterior probability on $H_1 : \lambda = 0$ is ½. Finally, as Jeffreys (1967, p. 272) remarks, the variation of K_{12} with t is much more important than its variation with ν. For moderately large ν, $K_{12} \doteq (\pi\nu/2)^{1/2}\exp(-t^2/2)$, from which the dependence of K_{12} on ν and t is clearly seen.

In a brief treatment of regression, Jeffreys (1967, pp. 324-326) remarks that "...The whole of the tests related to the normal law of error can be adapted immediately to tests concerning the introduction of a new function to

3. Jeffreys (1967, p.273) points out that if the prior *pdf* for $v = \lambda/\sigma$ were $p(v) \propto \exp\{-cv^2\}$, where c is some given positive constant, the posterior odds ratio for $\lambda = 0$ and $\lambda \neq 0$ "... would never be less than some positive function of n [the sample size] however closely the observations agreed among themselves". Also, on this same page he points out a second defect of this normal form for the prior *pdf*.

4. It appears that DeGroot (1973) obtains his result that the tail area associated with a sampling theory test statistics's value is equal to the posterior probability on the null hypothesis by use of a very special prior *pdf* on his parameter θ. His prior probabilities on θ's possible values are fixed even though a given departure of θ from its null value of zero implies differing departures of the underlying location parameter's value from zero as n, the sample size changes.

TABLE 2.1

Values of t^2 Associated with Corresponding Values of K_{12} and $\nu = n-1$ from (2.7)

	K_{12}				
ν	1	$10^{-1/2}$	10^{-1}	$10^{-3/2}$	10^{-2}
9*	3.5	7.7	13.3	...	...
15	3.8	7.1	11.1	15.9	21.5
20	4.0	7.0	10.6	14.5	18.9
50	4.6	7.4	10.0	12.8	16.0
100	5.2	7.7	10.3	12.8	15.5
200	5.7	8.2	10.7	13.1	15.6
500	6.8	9.1	11.4	13.8	16.2
1,000	7.4	9.7	12.0	14.3	16.6
2,000	8.1	10.4	12.7	15.0	17.3
5,000	9.0	11.3	13.6	15.9	18.2
10,000	9.7	12.0	14.3	16.6	18.9
50,000	11.3	13.6	15.9	18.2	20.5
100,000	12.0	14.3	16.6	18.9	21.2

represent a series of measures''. (p. 325). He considers the important special case for which the hypothesis is that an added term's coefficient is equal to zero and points out that (2.7) is the approximate posterior odds ratio for this problem where t is the usual t-statistic relating to the added term's coefficient and ν is the degrees of freedom associated with the t-statistic. Below it will be seen that Jeffreys's result is included in our general results as a special case.

3. POSTERIOR ODDS RATIOS FOR SELECTED REGRESSION HYPOTHESES

Let our regression model for the nx1 observation vector $\mathbf{y}$ be:

$$\mathbf{y} = \alpha\iota + X\beta + \mathbf{u} \tag{3.1}$$

* For $\nu = 9$, Jeffreys has used his exact result for K_{12} to compute the following t^2 values: 3.8 for $K_{12} = 1$, 7.7 for $K_{12} = 10^{-1/2}$, and 13.1 for $K_{12} = 10^{-1}$. It is seen that the exact results are in good agreement with the approximate results even though $\nu = 9$ is small. Jeffreys (1967, p. 439) tabulates exact values for $\nu = 1,2,3,\ldots,9$.

where ι is an nx1 vector with all elements equal to one, α and β are a scalar parameter and a kx1 vector of parameters with unknown values, $(\iota:X)$ is an nx$(k+1)$ given matrix of rank $k+1$ and $\mathbf{u}$ is an $n+1$ vector of error terms. It is assumed that the variables in X are measured in terms of deviations from their respective sample means and thus $\iota'X = \mathbf{0}'$. Further, the elements of $\mathbf{u}$ are assumed independently drawn from a normal population with zero mean and finite variance σ^2 with unknown value.

We initially consider the following two hypotheses:

$$H_1: \beta = 0, -\infty < \alpha < \infty \text{ and } 0 < \sigma < \infty \tag{3.2}$$

$$H_2: \beta \neq \mathbf{0}, -\infty < \alpha < \infty \text{ and } 0 < \sigma < \infty. \tag{3.3}$$

The likelihood functions under these two hypotheses are given by:

$$p(\mathbf{y}|\alpha,\sigma,H_1) \propto \sigma^{-n}\exp\{-(\mathbf{y}-\alpha\iota)'(\mathbf{y}-\alpha\iota)/2\sigma^2\} \tag{3.4}$$
$$\propto \sigma^{-n}\exp\{-[\nu_1 s_1^2 + n(\alpha y)^2]/2\sigma^2\}$$

and

$$p(\mathbf{y}|\alpha,\beta,\sigma,H_2) \propto \sigma^{-n}\exp\{-(\mathbf{y}-\alpha\iota-X\beta)'(\mathbf{y}-\alpha\iota-X\beta)/2\sigma^2\} \tag{3.5}$$
$$\propto \sigma^{-n}\exp\{-[\nu_2 s_2^2 + n(\alpha-\bar{y})^2 + (\beta-\hat{\beta})'X'X(\beta-\hat{\beta})]/2\sigma^2\}$$

where the proportionality constant is $(2\pi)^{-n/2}$ in each case,

$$y = \Sigma_{i=1}^n y_i/n,\ \nu_1 s_1^2 = \Sigma_{i=1}^n (y_i y)^2,\ \nu_1 = n-1$$
$$\hat{\beta} = (X'X)^{-1}X'\mathbf{y},\ \nu_2 s_2^2 = (\mathbf{y}y\iota-X\hat{\beta})'(\mathbf{y}y\iota-X\hat{\beta}), \text{ and } \nu_2 = n-k-1.$$

The following prior assumptions will be utilized in obtaining a posterior odds ratio. First we place equal prior probabilities of ½ on both hypotheses and thus the prior odds ratio is 1:1. Second, under H_1 we employ a diffuse prior distribution for α and σ, that is,

$$p(\alpha,\sigma|H_1) \propto 1/\sigma \quad -\infty < \alpha < \infty \text{ and } 0 < \sigma < \infty. \tag{3.6}$$

Under H_2 we utilize the following prior *pdf*

$$p(\alpha,\beta,\sigma|H_2) \propto f(\beta|\sigma)/\sigma \quad -\infty < \alpha < \infty \text{ and } 0 < \sigma < \infty \tag{3.7a}$$

with

$$f(\beta|\sigma) = c|X'X/n\sigma^2|^{1/2}/(1 + \beta'X'X\beta/n\sigma^2)^{(k+1)/2} \quad -\infty < \beta_i < \infty \quad (3.7b)$$
$$i = 1,2,...k$$

where $c = \Gamma\,[(k+1)/2]/\pi^{(k+1)/2}$.

In (3.6) and (3.7a) the factors of proportionality are assumed the same. Further, in (3.7b) it has been assumed that the prior *pdf* for β given σ is in the form of a k-dimensional multivariate Cauchy probability density function with zero location vector and matrix $X'X/n$, a matrix suggested by the form of the information matrix.

The posterior odds ratio, K_{12} for H_1 and H_2 with the prior odds ratio 1:1, is:

$$K_{12} = \frac{\int p(\mathbf{y}|\alpha,\sigma,H_1)p(\alpha,\sigma|H_1)d\alpha d\sigma}{\int p(\mathbf{y}|\alpha,\beta,\sigma,H_2)p(\alpha,\beta,\sigma|H_2)d\alpha d\beta d\sigma} \quad (3.8)$$

Explicitly, the integration in the numerator of (3.8) is performed as follows. The integral to be evaluated is:

$$I_N = \int_0^\infty \int_{-\infty}^{\infty} \sigma^{-(n+1)}\exp\{-[\nu_1 s_1^2 + n(\alpha-\bar{y})^2]/2\sigma^2\}d\alpha d\sigma$$

Using properties of the univariate normal *pdf*, integrate with respect to α to obtain:

$$I_N = (2\pi/n)^{1/2}\int_0^\infty \sigma^{-(\nu_1+1)}\exp\{-\nu_1 s_1^2/2\sigma^2\}d\sigma \quad (3.9a)$$
$$= (2\pi/n)^{1/2}\Gamma(\nu_1/2)\,(2/\nu_1 s_1^2)^{\nu_1/2}/2$$

where the integration over σ was performed by utilizing well-known properties of the inverted gamma *pdf*—see, e.g. Zellner (1971, p. 371)—.

The integral in the denominator of (3.8), denoted by I_D will be evaluated as follows:

$$I_D = \int f(\beta|\sigma,H_2)\sigma^{-(n+1)}\exp\{-[\nu_2 s_2^2 + n(\alpha-\bar{y})^2$$
$$+ (\beta-\hat{\beta})'X'X(\beta-\hat{\beta})]/2\sigma^2\}d\alpha d\beta d\sigma$$

with $f(\beta|\sigma)$ given in (3.7b). First integrate over α using properties of the univariate normal *pdf* to obtain:

$$I_D = (2\pi/n)^{1/2}\int f(\beta|\sigma,H_2)\sigma^{-n}\exp\{-[\nu_2 s_2^2 + (\beta-\hat{\beta})'X'X(\beta-\hat{\beta})]/2\sigma^2\}d\beta d\sigma.$$

On inserting $f(\beta|\sigma,H_2)$ from (3.7a) and performing the integration with respect to the elements of β approximately,[5]

$$I_D \doteq (2\pi/n)^{1/2}c(2\pi/\nu_2)^{k/2} \int_0^\infty \frac{1}{\sigma^n} \frac{1}{\{1+\hat{\beta}'X'X\hat{\beta}/n\sigma^2\}^{(k+1)/2}} \exp\{-\nu_2 s_2^2/2\sigma^2\}d\sigma$$

Then,

$$I_D \doteq (2\pi/n)^{1/2}c(2\pi/\nu_2)^{k/2}\Gamma(\nu_1/2)1/2(2/\nu_2 s_2^2)^{\nu_1/2}/(1+\hat{\beta}'X'X\hat{\beta}/\nu_2 s_2^2)^{(k+1)/2} \quad (3.9b)$$

where the integration over σ has beeen performed approximately.

Using the results in (3.9a) and (3.9b), the approximate posterior odds ratio for H_1 vs. H_2 is given by:

$$K_{12} \doteq (1/c)(\nu_2/2\pi)^{k/2}(\nu_2 s_2^2/\nu_1 s_1^2)^{\nu_1/2}(1+\hat{\beta}'X'X\hat{\beta}/\nu_2 s_2^2)^{(k+1)/2}$$

$$= a(\nu_2/2)^{k/2}(\nu_2 s_2^2/\nu_1 s_1^2)^{(\nu_2-1)/2} \quad (3.10)$$

with $a = \pi^{1/2}/\Gamma[(k+1)/2]$, since $\nu_2 s_2^2 + \hat{\beta}'X'X\hat{\beta} = \nu_1 s_1^2$. Alternatively, K_{12} in (3.10) can be expressed as:

$$K_{12} \doteq a(\nu_2/2)^{k/2}/[1+(k/\nu_2)F_{k,\nu_2}]^{(\nu_2-1)/2} \quad (3.11)$$

or

$$K_{12} \doteq a(\nu_2/2)^{k/2}(1-R^2)^{(\nu_2-1)/2} \quad (3.12)$$

where $F_{k,\nu_2} = \hat{\beta}'X'X\hat{\beta}/ks_2^2$ and $R^2 = \hat{\beta}'X'X\hat{\beta}/(\nu_2 s_2^2 + \hat{\beta}'X'X\hat{\beta})$, the usual "F-statistics" and the squared sample multiple correlation coefficient, respectively. Further, a large sample approximation to $-2\ell n K_{12}$ is given by:

$$-2\ell n K_{12} \doteq \chi_k^2 - k\ell n \nu_2 \quad (3.13)$$

5. This approximate integration can be viewed as finding the mean of $f(\beta|\sigma, H_2)$ a bounded function of β. Cramér (1946, p. 353 ff.) indicates that the error of the approximation is $0(n^{-1})$ in line with Jeffreys's remark cited in Section II. Thus if the posterior odds ratio $K_{12} = I_N/I_D$ and if the integral I_N is evaluated exactly and $I_D = \hat{I}_D + 0(n^{-1})$, where $\hat{I}_D$ is the approximate value of I_D, $K_{12} = I_N/[\hat{I}_D + 0(n^{-1})]$ or $I_N/\hat{I}_D = K_{12}[1+0(n^{-1})]$ and thus the error in using I_N/I_D is $K_{12}.0(n^{-1})$, as pointed out above by Jeffreys.

where $\chi_k^2 = \hat{\beta}' X' \hat{\beta} / s_2^2$.[6]

We now consider a hypothesis relating to a subvector of β in (3.1). Rewrite (3.1) as

$$\mathbf{y} = \alpha\iota + X_1\beta_1 + X_2\beta_2 + \mathbf{u} \tag{3.14}$$

where $X = (X_1 \vdots X_2)$ with X_1 and X_2 k_1x1 and k_2x1 vectors, respectively and $k_1 + k_2 = k$. All other assumptions made in connection with (3.1) apply to (3.14). For convenience, we shall reparametrize (3.14) as follows:

$$\mathbf{y} = \alpha\iota + X_1\eta + V\beta_2 + \mathbf{u} \tag{3.15}$$

where $V = [I - X_1(X_1'X_1)^{-1}X_1']X_2$ and $\eta = \beta_1 + (X_1'X_1)^{-1}X_1'X_2\beta_2$. Note that $X_1'V = 0$

A posterior odds ratio relating to the following two hypotheses will be derived:

$$H_A : \beta_2 = \mathbf{0} \tag{3.16}$$

and

$$H_B : \beta_2 \neq \mathbf{0} \tag{3.17}$$

with α, the elements of η and σ unrestricted under both hypotheses.

The likelihood functions under these two hypotheses are given by:

$$p(\mathbf{y} | \alpha, \sigma, \eta, \mathrm{H}_A) \propto \sigma^{-n} \exp\{\nu_A s_A^2 + n(\alpha - \bar{y})^2 + (\eta - \hat{\eta})' X_1' X_1 (\eta - \hat{\eta})] / 2\sigma^2\} \tag{3.18}$$

and

$$p(\mathbf{y} | \alpha, \sigma, \eta, \beta_2, H_B) \propto \sigma^{-n} \exp\{-[\nu_B s_B^2 + n(\alpha - \bar{y})^2 + \\ + (\eta - \hat{\eta})' X_1' X_1 (\eta - \hat{\eta}) + (\beta_2 - \hat{\beta}_2)' V' V (\beta_2 - \hat{\beta}_2)] / 2\sigma^2\} \tag{3.19}$$

where

6. To obtain (3.13), write (3.11) as $K_{12} \doteq a(\nu_2/2)^{k/2} \exp\{\nu^2 - 1)/2 \, \ell n \, [1 + (k/\nu^2) F_{k,\nu_2}]\}$ and expand the logarithmic factor in the exponential as $\ell n(1+x) \doteq x$. The result is $K_{12} \doteq a(\nu_2/2)^{k/2} \exp\{-kF_{k,\nu_2}/2\}$. Then $-2\ell n K_{12} \doteq \chi_k^2 - k\ell n \nu_2$, where $\chi_k^2 = kF_{k,\nu_2}$ and terms not depending on ν_2 have been dropped in this large-ν_2 approximation. Further, under $\beta = 0$ the approximate sampling *pdf* for $-2\ell n K_{12}$ can be obtained from that of χ_k^2. Also, again under $\beta = \mathbf{0}$ the approximate cumulative sampling *pdf* for K_{12} in (3.11) can be obtained from that of F_{k,ν_2}. That is, since K_{12} is a one-to-one monotonic function of F_{k,ν_2} for fixed k and ν_2, $\Pr(F_{k,\nu_2} > x) = \Pr(K_{12} < x')$, where x' is the value of K_{12} associated with $F_{k,\nu_2} = x$.

$$\bar{y} = \Sigma_{i=1}^{n} y_i/n,\ \hat{\eta} = (X_1'X_1)^{-1}X_1'\mathbf{y}, \qquad \nu_A s_A^2 = (\mathbf{y}-\bar{y}\iota - X_1\hat{\eta})'(\mathbf{y}-\bar{y}\iota - X_1\hat{\eta}),$$

$$\nu_A = n-k_1-1,\ \hat{\beta}_2 = (V'V)^{-1}V'\mathbf{y}, \quad \nu_B s_B^2 = (\mathbf{y}-\bar{y}\iota - X_1\hat{\eta} - V\hat{\beta}_2)'(\mathbf{y}-\bar{y}\iota - X_1\hat{\eta} - V\hat{\beta}_2)$$

and $\nu_B = n-k_1-k_2-1$.

Under H_A, we employ the following diffuse prior *pdf* for the parameters:

$$p_A(\alpha,\sigma,\eta \mid H_A) \propto |X_1'X_1|^{1/2}/\sigma \quad -\infty < \alpha,\eta_i < \infty \quad i=1,2,\ldots,k_1 \tag{3.20}$$

$$0<\sigma<\infty$$

while under H_B the prior *pdf* is:

$$p_B(\alpha,\sigma,\eta,\beta_2 \mid H_B) \propto |X_1'X_1|^{1/2} h(\beta_2 \mid \sigma)/\sigma \tag{3.21a}$$

$$-\infty < \alpha,\ \eta_i < \infty, \qquad i=1,2,\ldots,k_1$$
$$0<\sigma<\infty$$

with

$$h(\beta_2 \mid \sigma) = c_B |V'V/n\sigma^2|^{1/2}/(1+\beta_2'V'V\beta_2/n\sigma^2)^{(k_2+1)/2} \tag{3.21b}$$

$$-\infty < \beta_{2i} < \infty \qquad i=1,2,\ldots,k_2.$$

In (3.20) and (3.21a) the factor of proportionality is taken to be the same. In (3.21b), the prior *pdf* for β_2 given σ is the form of a k_2-dimensional multivariate Cauchy *pdf* with zero location vector and matrix $V'V/n$, a matrix suggested by the form of the information matrix.

The posterior odds ratio, K_{AB} for H_A and H_B with the prior odds ratio 1:1 is:

$$K_{AB} = \frac{\int p(\mathbf{y} \mid \alpha,\sigma,\eta,H_A) p_A(\alpha,\sigma,\eta \mid H_A) d\alpha d\sigma d\eta}{\int p(\mathbf{y} \mid \alpha,\sigma,\eta,\beta_2,H_B) p_B(\alpha,\sigma,\eta,\beta_2 \mid H_B) d\alpha d\sigma d\eta d\beta_2} \tag{3.22}$$

On applying integration techniques similar to those employed above (see Appendix), the following approximate expression for K_{AB} is obtained:

$$\begin{aligned} K_{AB} &\doteq b(\nu_B/2)^{k_2/2} (\nu_B s_B^2/\nu_A s_A^2)^{(\nu_B-1)/2} \\ &= b(\nu_B/2)^{k_2/2} [(1-R_B^2)/(1-R_A^2)]^{(\nu_B-1)/2} \\ &= b(\nu_B/2)^{k_2/2}/[1+(k_2/\nu_B)F_{k_2,\nu_B}]^{(\nu_B-1)/2} \end{aligned} \tag{3.23}$$

where $b = \pi^{1/2}/\Gamma[(k_2+1)/2]$, R^2_A and R^2_B are the squared sample multiple correlation coefficientes under H_A and H_B and $F_{k_2\nu_B} = \hat{\beta}_2' V' V\hat{\beta}_2/k_2 s^2_B$, the usual "$F$-statistic". Also, if ν_B is large, the following approximate result is available:

$$-2\ell n K_{AB} \doteq \chi^2_{k_2} - k_2 \ell n \nu_B = \nu_B(R^2_B - R^2_A)/(1-R^2_B) - k_2 \ell n \nu_B \tag{3.24}$$

with $\chi^2_{k_2} = \hat{\beta}_2' V' V\hat{\beta}_2/s^2_B$.

We now consider the following four hypotheses relating to β_1 and β_2 in (3.14), each assumed to have the same prior probability:

$$H_1 : \beta_1 = \mathbf{0} \text{ and } \beta_2 = \mathbf{0}, \tag{3.25a}$$

$$H_2 : \beta_1 \neq \mathbf{0} \text{ and } \beta_2 \neq \mathbf{0}, \tag{3.25b}$$

$$H_3 : \beta_1 \neq \mathbf{0} \text{ and } \beta_2 = \mathbf{0}, \tag{3.25c}$$

and

$$H_4 : \beta_1 = \mathbf{0} \text{ and } \beta_2 \neq \mathbf{0}, \tag{3.25d}$$

The posterior odds ratio for H_1 and H_2, K_{12}, given in (3.11) is:

$$K_{12} \doteq a(\nu_2/2)^{k/2}/[1+(k/\nu_2)F_{k\nu_2}]^{(\nu_2-1)/2}, \tag{3.26}$$

where $a = \pi^{1/2}/\Gamma[(k+1)/2]$ and $\nu_2 = n\text{-}k\text{-}1$. This odds ratio has been derived employing the prior assumptions in (3.6) and (3.7), the latter involving a multivariate Cauchy prior *pdf* for β_1 and β_2 given σ. The posterior odds ratio for H_3 and H_2, K_{32} is identical to K_{AB} in (3.23), namely

$$K_{32} \doteq b(\nu_2/2)^{k_2/2}/[1+(k^2/\nu_2)F_{k_2\nu_2}]^{(\nu_2-1)/2} \tag{3.27}$$

where $b = \pi^{1/2}/\Gamma[(k_2+1)/2]$ and $\nu_2 = \nu_B = n\text{-}k\text{-}1$. K_{32} also can be obtained by using the conditional prior *pdf* for β_1 given $\beta_2 = \mathbf{0}$ and σ associated with the multivariate Cauchy *pdf* in (3.7b) under H_2 along with uniform independent priors for α and log σ. Similarly, the posterior odds ratio for H_4 and H_2,K_{42} can be obtained and is:

$$K_{42} \doteq q(\nu_2/2)^{k_1/2}/[1 + (k_1/\nu_2)F_{k_1\nu_2}]^{(\nu_2-1)/2} \tag{3.28}$$

where $q = \pi^{1/2}/\Gamma[(k_1+1)/2]$. Last, from (3.27) and (3.28), the posterior odds ratio for H_3 and H_4,K_{34} is:

$$K_{34} = K_{32}/K_{42} \doteq g(\nu_2/2)^{(k_2-k_1)/2} \left(\frac{1+(k_1/\nu_2)F_{k_1\nu_2}}{1+(k_2/\nu_2)F_{k_2\nu_2}} \right)^{(\nu_2-1)/2} \quad (3.29)$$

where $g = \Gamma[(k_1+1)/2]/\Gamma[(k_2+1)/2]$.

The posterior odds ratios in (3.26)-(3.29) can be helpful in screening sets of variables, X_1 and X_2 for inclusion a in regression in situations in which there is little prior information and the initial presumption is that neither set of variables probably belongs in the regression. A special case of the above analysis is one in which X_1 and X_2 are vectors and thus β_1 and β_2 are scalars. In this case, we are screening individual variables por possible inclussion in the regression. Further, elaboration of the hypotheses in (3.25) to relate individual coefficients is possible and would lead to posterior odds ratios useful in determining which individual variables to include in a regression.

To gain greater familiarity with the odds ratios derived above, we now turn to consider some numerical evaluations of them.

4. NUMERICAL EVALUATION OF SELECTED ODDS RATIOS

In this Section, we provide some numerical evaluations of the odds ratios derived in Section III. First, note that when $k = 1$, the posterior odds ratio K_{12} in (3.11) for the hypotheses $\beta = \mathbf{0}$ and $\beta \neq \mathbf{0}$ reduces to $K_{12} \doteq (\pi\nu_2/2^{1/2}/(1 + t^2/\nu^2)^{(\nu_2-1)/2}$, with $\nu^2 = n - 2$ which is exactly in the form of Jeffreys's odds ratio in (2.7). Thus the numerical results in Table 2.1 apply directly to the case of simple regression. From Table 2.1, it is seen that for $\nu_2 = 20$, $K_{12} \doteq 1$ when $t^2 = 4.0$ where $t^2 = \hat{\beta}^2\Sigma(x_i - \bar{x})^2/s^2$ is the square of the usual t-statistic. Since $r^2 = t^2/(\nu^2 + t^2)$, a value of $r^2 = 1/6$ corresponds to $t^2 = 4.0$ and $K_{12} \doteq 1$ for $\nu_2 = 20$. For $\nu_2 = 5{,}000$ and $t^2 = 9.0$ (or $r^2 = .0018$), $K_{12} \doteq 1$. Thus indifference, ($K_{12} \doteq 1$) is achieved for a larger value of t^2 (or a lower value of r^2) with $\nu_2 = 5{,}000$ as compared with $\nu_2 = 20$. For $\nu_2 = 20$, $K_{12} \doteq 1/100$, that is the odds are 100:1 against $\beta = 0$ when $t^2 = 18.9$ or $r^2 = .486$. For $\nu_2 = 5{,}000$, $K_{12} \doteq 1/100$ when $t^2 = 18.2$ or $r^2 = .00377$. Thus with $\nu_2 = 5{,}000$, a value of $t^2 = 18.2$ (or equivalently, $r^2 = .00377$) strongly favors the hypotheses $\beta \neq 0$. Since values of ν_2 in the vicinity of several thousand are frequently encountered in analyses of cross-section or survey data, these results are relevant for applied work. In particular, they point (a) the need for absolutely larger t-values for indifference ($K_{12} = 1$) as ν_2 increases and (b) recognition that for large values of ν_2, small values of r^2 can be consistent with strong evidence *against* $\beta = 0$. These results, it must be emphasized, apply in situations in which we have little prior information about β's value under the hypotheses $\beta \neq 0$. If more information is available, suitable prior *pdf*'s reflecting it would have to be introduced, as pointed out by Jeffreys (1967, p. 252).

TABLE 4.1
Values of R^2 and F_{k,ν_2} Associated with Particular Values of K_{12} and k in (3.12) for $\nu_2 = 20$ and $\nu_2 = 100$*

A. $\nu_2 = 20$

k	Value of:	K_{12}: 1	$10^{-1/2}$	10^{-1}	$10^{-3/2}$	10^{-2}	.01 and .05 Critical Values of F and Associated R^2's: .01	.05
1	R^2	.16	.26	.35	.42	.49	.29	.18
	$F_{1,20}$	4.0	7.0	10.6	14.5	18.9	8.10	4.35
2	R^2	.27	.35	.43	.49	.55	.37	.26
	$F_{2,20}$	3.7	5.5	7.5	9.7	12.3	5.85	3.49
3	R^2	.35	.42	.48	.54	.60	.43	.32
	$F_{3,20}$	3.5	4.8	6.3	8.0	9.9	4.94	3.10
4	R^2	.40	.47	.53	.58	.63	.47	.36
	$F_{4,20}$	3.4	4.4	5.7	7.0	8.6	4.43	2.87
5	R^2	.45	.51	.57	.61	.66	.51	.40
	$F_{5,20}$	3.2	4.2	5.2	6.4	7.7	4.10	2.71
6	R^2	.48	.54	.59	.64	.68	.54	.44
	$F_{6,20}$	3.1	3.9	4.9	5.9	7.1	3.87	2.60
				B. $\nu_2 = 100$				
1	R^2	.050	.072	.093	.11	.13	.065	.038
	$F_{1,100}$	5.2	7.7	10.3	12.8	15.5	6.90	3.94
2	R^2	.089	.11	.13	.15	.17	.088	.058
	$F_{2,100}$	4.9	6.2	7.5	8.8	10.3	4.82	3.09
3	R^2	.12	.14	.16	.18	.20	.11	.075
	$F_{3,100}$	4.6	5.5	6.4	7.4	8.3	3.98	2.70
4	R^2	.15	.17	.19	.21	.23	.12	.090
	$F_{4,100}$	4.4	5.1	5.9	6.6	7.3	3.51	2.46
5	R^2	.18	.20	.21	.23	.25	.14	.10
	$F_{5,100}$	4.3	4.9	5.5	6.1	6.7	3.20	2.30
6	R^2	.20	.22	.24	.25	.27	.15	.12
	$F_{6,100}$	4.2	4.7	5.2	5.7	6.2	2.99	2.19

*Note that $F_{k,\nu_2} = (\nu_2/k)R^2/(1-R^2)$, with $\nu_2 = n-k-1$.

In Table 4.1, we have evaluated the posterior odds ratio K_{12} for $H_1 : \beta = \mathbf{0}$ vs. $H_2 : \beta \neq \mathbf{0}$ for $\nu_2 = 20$ and $\nu_2 = 100$, where $\nu_2 = n\text{-}k\text{-}1$ for selected values of k, the number of elements in β and selected values of R^2, the sample squared multiple correlation coefficient. Also shown in the table are values of associated F-statistics, $F_{k,\nu_2} = (\nu_2/k)R^2/(1\text{-}R^2)$, and .01 and .05 critical values of the F statistic as well as the R^2 values associated with these critical values. From the first line of Table 4.1, we see that for $k = 1$ and $\nu_2 = n\text{-}k\text{-}1 = 20$, $K_{12} \doteq 1$ when $R^2 = .16$ and $F_{1,20} = t^2_{20} = 4.00$. Note that for these conditions the sampling theorists's .05 critical value of F is $F_{1,20}(.05) = (2.086)^2 = 4.35$ with an associated $R^2 = .18$. Thus the 5% F value is somewhat larger than the Bayesian indifference ($K_{12} = 1$) value of 4.0. Alternatively, an $R^2 = .16$ leads to $K_{12} \doteq 1$ while an $R^2 = .18$ is associated with the sampling theorists's .05 critical value of F. On the other hand, a .01 critical value of F is 8.10, with an associated $R^2 = .29$ which is far from the F value 4.0, or $R^2 = .16$ which yields $K_{12} \doteq 1$.

TABLE 4.2

Values of χ^2_k and Associated R^2's for $K_{12} \doteq 1$ Using Approximate Formula (3.13) for $\nu_2 = 20$ and $\nu_2 = 100$ and Selected Values for k

	$\nu_2 = 20$		$\nu_2 = 100$		.05 and .01 Critical Values of χ^2_k	
k	χ^2_k	R^{2*}	χ^2_k	R^{2*}	$\chi^2_k(.05)$	$\chi^2_k(.01)$
1	3.00	.13	4.6	.044	3.84	6.63
2	5.99	.23	9.2	.084	5.99	9.21
3	8.99	.31	13.8	.12	7.81	11.30
4	12.0	.37	18.4	.16	9.49	13.30
5	15.0	.43	23.0	.19	11.10	15.10
6	18.0	.47	27.6	.22	12.60	16.80

* Note that $\chi^2_k = \hat{\beta}'X'X\hat{\beta}/s^2$ and $R^2 = \chi^2_k/(\nu_2 + \chi^2_k)$, where $\nu_2 = n\text{-}k\text{-}1$.

In Table 4.2 values of χ_k^2 and associated values of R^2 which correspond to $K_{12} \doteq 1$ have been tabulated for $\nu_2 = 20$ and $\nu_2 = 100$ and $k = 1,2,\ldots,6$, to gain some information on the quality of the approximation in (3.13). The entries in Table 4.2 have been computed from the large sample approximate formula (3.13), that is $-2\ell n K_{12} \doteq \chi_k^2 - k\ell n \nu_2$. Also shown in Table 4.2 are the .05 and .01 critical values of χ_k^2. For $k = 1$ and $\nu_2 = 20$, the indifference values of χ_1^2 and R^2 are 3.00 and .13, respectively. The latter value can be compared with the more accurate indifference value of $R^2 = .16$ given in Table 4.1. The difference in these values arises because the results in Table 4.2 are based on a cruder approximation than those in Table 4.1. For $\nu_2 = 100$, the corresponding indifference ($K_{12} = 1$) values of R^2 in Tables 4.1 and 4.2 are fairly similar in value. Also, from Table 4.2, the relation of the crude indifference values of χ_k^2 can be compared with the .05 and .01 sampling theory critical values of χ_k^2. For $\nu_2 = 100$, it is seen that for $k = 1$, the indifference value of χ_1^2, namely 4.6 falls between the .05 critical value, 3.84 and the .01 critical value, 6.63. For $k > 2$ and $\nu_2 = 100$, the Bayesian indifference values of χ_k^2 are all larger than the .01 critical values of χ_k^2.

As regards other posterior odds ratios derived in Section III, it is the case that the numerical results in Tables 2.1, 4.1 and 4.2, can be utilized to evaluate them provided that the degrees of freedom and k parameters are suitably reinterpreted. For example, the expressions for K_{AB} in (3.23) and (3.24) can be evaluated if in using the tables, k is replaced by k_2, $F_{k,\nu}{}^2$ is replaced by F_{k_2,ν_2} (note $\nu_2 = \nu_B = n$-k-1) and χ_k^2 is replaced by χ_{k2}^2. Similarly, the odds ratios K_{32} and K_{42} in (3.27) and (3.28), respectively can be implemented using the results in the tables by similar redefinitions. Last, the odds ratio, K_{32} in (3.29) can be evaluated from results given in Table 4.1 by use of $K_{32} = K_{32}/K_{42}$, where values of K_{32} and K_{42} can be obtained from entries in Table 4.1. Finally, it is to be noted that in the expression for K_{32} in (3.29), there is an interesting allowance for the possibly differing numbers of parameters in β_1 and β_2.

5. SUMMARY AND CONCLUDING REMARKS

In this paper we have derived approximate posterior odds ratios for sharp null hypotheses which are frequently encountered in regression analyses. These posterior odds ratios are appropriate when little is known regarding parameter values and special attention is given to specific values, e.g. zero values of the regression coefficients. With slight modifications, other special values can be incorporated in the analysis by reparametrizing to convert the null hypotheses to involve zero values. In our work we have employed asymptotic expansions of certain integrals which are very convenient, yield results which can be compared directly with sampling theory analyses, and are quite accurate, as shown in Jeffreys's work. With some extra computational cost, a numerical integration approach, suggested by Dickey (1971) could be applied to obtain slightly more accurate results.

In line with Jeffreys's, Lindley's and some others's previous results, we have found that sampling theorists's usual .05 critical values of test statistics can be far from a Bayesian posterior odds indifference value of one under a variety of circumstances. Whether this finding is interpreted as a systematic flaw in sampling theory practice is of course critically dependent on the nature of the usually implicit loss structure used in sampling theory testing. Cases in which sampling theorists mechanically employ a 5% significance level no matter what the sample size and/or the number of parameters are interpreted as flawed analyses. If sampling theorists and Bayesians carefully consider the underlying loss structure in choosing between or among hypotheses, the above analysis indicates that there can be a compatibility between Bayesian and sampling theory results in testing but, of course their interpretations will differ radically.

While, as pointed out above there can be some degree of compatibility between Bayesian and sampling theory testing results, the direct interpretation of sample evidence, as reflected in F statistics or R^2 values in terms of posterior odds ratios stands in marked contrast to sampling theorists's and others's unclear interpretations of sample evidence in terms of "p-values", and/or values of R^2 or of $\bar{R}^2$, the adjusted coefficient of determination. As mentioned above, a p-value associated with the value of a test statistic is not at all an accurate measure of the posterior probability associated with a null hypothesis. However, it should be noted that most of the posterior odds ratios derived above are monotonically increasing functions of the p-values associated with t or F statistics involved in the posterior odds ratios. Thus there is some rationale for considering p-values; however, since posterior odds ratios have a direct interpretation and explicitly reflect the prior information employed, their use is preferable to the use of p-values. Also, posterior odds ratios on alternative hypotheses can be employed, as described below to average estimates (and predictions) over alternative hypotheses when posterior odds ratios do not yield a clear-cut choice of a particular hypothesis.

In terms of the hypotheses considered above, it is possible to use their associated posterior odds ratios to obtain optimal (relative to quadratic loss) Bayesian "pre-test" point estimates --see Zellner and Vandaele (1974, pp. 640-641). For example with respect to the hypotheses $H_1 : \beta = \mathbf{0}$ and $H_2 : \beta \neq \mathbf{0}$, the point estimate that is optimal relative to quadratic loss is $\bar{\beta} = P_1\mathbf{0} + (1\text{-}P_1)\widetilde{\beta} = (1\text{-}P_1)\widetilde{\beta} = (1+K_{12})^{-1}\tilde{\beta}$, where P_1 is the posterior probability for H_1, $K_{12} = P_1/(1\text{-}P_1)$ is the posterior odds ratio for H_1 and H_2, and $\widetilde{\beta}$ is the posterior mean for β under H_2. With the prior *pdf* (3.7) which we have employed under H_2, $\widetilde{\beta}$ will be close to $\tilde{\beta}$, the least-squares estimate. Thus $\bar{\beta} \doteq$

$(1+K_{12})^{-1}\hat{\beta}$ where K_{12} given in (3.11) is a function of the usual F statistic. Also note that the "shrinkage factor", $(1+K_{12})^{-1}$ is between zero and one with a value near zero when K_{12} is large and a value near one when K_{12} is small. This shrinkage factor can be compared with others which have appeared in the sampling theory literature —see Zellner and Vandaele (1975, p. 639).

Finally, it would be interesting to compare the posterior odds ratios derived above with others based on more informative prior distributions.

APPENDIX

Herein we evaluate the integrals appearing in equation (3.22) of the text. The integral in the numerator, denoted by I_N is:

$$I_N \propto \int |X_1'X_1|^{1/2}\sigma^{-(n+1)}\exp\{-[\nu_A s_A^2 + n(\alpha\bar{y})^2 + (\eta-\hat{\eta})'X_1'X_1(\eta-\hat{\eta})]/2\sigma^2\}d\alpha d\eta d\sigma. \tag{A.1}$$

where ν_A, s_A^2, $\bar{y}$ and $\hat{\eta}$ have been defined in the text in connection with (3.18). We can integrate over α and the k_1 elements of η using properties of univariate and multivariate normal *pdf*'s, respectively to obtain:

$$I_N \propto (2\pi)^{(k_1+1)/2}n^{-1/2}\int_0^\infty \sigma^{-\nu_A}\exp\{-\nu_A s_A^2/2\sigma^2\}d\sigma. \tag{A.2}$$

Using properties of the inverted gamma *pdf*, the integral in (A.2) can be evaluated to yield:

$$I_N \propto (2\pi)^{(k_1+1)/2}n^{-1/2}\Gamma(\nu_A/2)(1/2)(2/\nu_A s_A^2)^{\nu_A/2}. \tag{A.3}$$

The integral in the denominator of (3.22) in the text, denoted by I_D is:

$$I_D \propto \int h(\beta_2|\sigma)|X_1'X_1|^{1/2}\sigma^{-(n+1)}\exp\{-[\nu_B s_B^2 + n(\alpha-\bar{y})^2 + (\eta-\hat{\eta})'X_1'X_1(\eta-\hat{\eta}) + (\beta_2-\hat{\beta}_2)'V'V(\beta_2-\hat{\beta}_2)]/2\sigma^2\}d\alpha d\eta d\sigma \tag{A.4}$$

where $h(\beta_2|\sigma)$ is given in (3.21b) of the text and ν_B, s_B^2, $\hat{\eta}$, $\hat{\beta}_2$ and V have been defined in connection with (3.19). The integration over α and η can be performed exactly using properties of normal distributions to yield:

$$I_D \propto (2\pi)^{(k_1+1)/2}n^{-1/2}\int h(\beta_2|\sigma)\sigma^{-(n-k_1)}\exp\{-[\nu_B s_B^2 + (\beta_2-\hat{\beta}_2)'V'V(\beta_2-\hat{\beta}_2)]/2\sigma^2\}d\beta_2 d\sigma \tag{A.5}$$

The integration over β_2 can be done approximately by noting that is it equivalent to obtaining the expectation of the bounded function $h(\beta_2|\sigma)$. Following Jeffreys's approach and also Cramér's (1946, p. 353) approximation results, we have on integrating approximately with respect to the k_2 elements of β_2

$$I_D \propto (2\pi)^{(k+1)/2} n^{-(k_2+1)/2} c_B \int_0^\infty (1 + \hat{\beta}_2' V' V \hat{\beta}_2 / n\sigma^2)^{-(k_2+1)/2} \cdot \sigma^{-(n-k_1)} \exp\{-\nu_B s_B^2 / 2\sigma^2\} d\sigma$$

Large values of the second two factors in the integrand of this last expression are near $\nu_B s_B^2/n$. If, as Jeffreys (1967, p. 272) does, we substitute $\sigma^2 = \nu_B s_B^2/n$ in the first, slowly varying factor of the integrand, and then integrate with respect to σ, the result is:

$$I_D \propto c_B 2\pi^{(k+1)/2} n^{-(k_2+1)/2} (1 + \hat{\beta}_2' V' V \hat{\beta}_2 / \nu_B s_B^2)^{-(k_2+1)/2} \times \Gamma(\nu_A/2)(1/2)(2/\nu_B s_B^2)^{\nu_A/2} \tag{A.6}$$

Then, using (A.3) and (A.6)

$$K_{AB} \doteq \frac{I_N}{I_D} = \frac{1}{c_B} \left(\frac{n}{2\pi}\right)^{k_2/2} \left(\frac{\nu_B s_B^2}{\nu_A s_A^2}\right)^{\nu_A/2} (1 + \hat{\beta}_2' V' V \hat{\beta}_2 / \nu_B s_B^2)^{(k_2+1)/2} \tag{A.7}$$

Now c_B, the normalizing constant in (3.21b) is

$c_B = \Gamma[(k_2+1)/2]/\pi^{(k_2+1)/2}$ and $\nu_A s_A^2 = \nu_B s_B^2 + \hat{\beta}_2' V' V \hat{\beta}_2$. Thus,

$$K_{AB} \doteq b(\nu_B/2)^{k_2/2} (\nu_B s_B^2 / \nu_A s_A^2)^{(\nu_B-1)/2} \tag{A.8}$$

with $b = \pi^{1/2}/\Gamma[(k_2+1)/2]$ and where in going from (A.7) to (A.8) $(n/2)^{k_2/2}$ has been replaced by $(\nu_B/2)^{k_2/2}$ which to the order of the approximation is equivalent. (A.8) is exactly the expression in (3.23) in the text. Further, on substituting $\nu_B s_B^2/\nu_A s_A^2 = (1-R_B^2)/(1-R_A^2)$ in (A.8), the second line of (3.23) is obtained. Finally, from $\nu_A s_A^2 = \nu_B s_B^2 + \hat{\beta}_2' V' V \hat{\beta}_2$, $\nu_A s_A^2/\nu_B s_B^2 = 1 + \hat{\beta}_2' V' V \hat{\beta}_2/\nu_B s_B^2 = 1 + (k_2/\nu_B) F_{k_2,\nu_B}$, which when utilized in (A.8) gives the third line of (3.23).

ACKNOWLEDGEMENTS

Research financed by income from the H.G.B. Alexander Endowment Fund, Graduate School of Business, U. of Chicago and by National Science Foundation Grant, SOC 77-15662.

REFERENCES

CRAMER, H. (1946) *Mathematical Methods of Statistics*. Princeton: Princeton University Press.

DEGROOT, M.H. (1973) Doing What Comes Naturally: Interpreting a Tail Area as a Posterior Probability or as a Likelihood Ratio, *J. Amer. Statist. Assoc.* **68**, 966-969.

DICKEY, J.M. (1971) The Weighted Likelihood Ratio, Linear Hypotheses on Normal Location Parameters *Ann. Math. Statist.* **42**, 204-223.

— (1975) Bayesian Alternatives to the F-test and Least-Squares Estimates in the Normal Linear Model. In *Studies in Bayesian Econometrics and Statistics in Honor of Leonard J. Savage*, (S.E. Fienberg and A. Zellner eds.), 515-554 Amsterdam: North-Holland.

— (1977) Is the Tail Area Useful as an Approximate Bayes factor? *J. Amer. Statist. Assoc.* **72**, 138-142.

GAVER, K.M. and M.S. GEISEL (1974) Discriminating Among Alternative Models: Bayesian and Non-Bayesian Methods, in *Frontiers of Econometrics,* (P. Zarembka ed.) New York: Academic Press.

GEISEL, M.S. (1970) Comparing and Choosing Among Parametric Statistical Models: A Bayesian Analysis with Macroeconomic Application. Ph. D. Thesis. University of Chicago.

JAYNES, E.T. (1976) Confidence Intervals Vs. Bayesian Intervals, in *Foundations of Probability. Theory, Statistical Inference, and Statistical Theories of Science*, (W.L. Harper and C.A. Hooker eds.) 175-213. Dordrecht-Holland: D. Reidel.

JEFFREYS, H. (1957), *Scientific Inference* (2nd ed.) Cambridge: University Press.

— (1967) *Theory of Probability* (3rd rev. ed.), Oxford: University Press.

(1980) "Some General Points in Probability Theory", in *Bayesian Analysis in Econometrics and Statistics: Essays in Honor of Harold Jeffreys* (A. Zellner ed.) 451-453. Amsterdam: North-Holland.

LEAMER, E.E. (1978) *Specification Searches*, New York: Wiley

LEMPERS, F.B. (1971), *Posterior Probabilities of Alternative Linear Models,* Rotterdam: University Press.

LINDLEY, D.V. (1957) A Statistical Paradox, *Biometrika* **44**, 187-192.

— (1961) The Use of Prior Probability Distributions in Statistical Inference and Decisions, in *Proc. Fourth Berkeley Symp.* (J. Neyman, ed.) 453-468. Berkeley: University of California Press.

SCHWARZ, G. (1978) Estimating the Dimension of a Model, *Ann. Statist.* **6**, 461-464.

THORNBER, E.H. (1966) Applications of Decision Theory to Econometrics, Ph. D. thesis, University of Chicago.

ZELLNER, A. (1971), *An Introduction to Bayesian Inference in Econometrics,* New York: Wiley

— and W. VANDAELE (1975) Bayes-Stein Estimators for k-means, Regression and Simultaneous Equation Models, in *Studies in Bayesian Econometrics and Statistics in Honor of Leonard J. Savage* (S.E. Fienberg and A. Zellner eds.) 627-653 Amsterdam: North-Holland.

Economics Letters 1 (1978) 337–342

JEFFREYS–BAYES POSTERIOR ODDS RATIO AND THE AKAIKE INFORMATION CRITERION FOR DISCRIMINATING BETWEEN MODELS *

Arnold ZELLNER
University of Chicago, Chicago, IL 60637, USA

Received January 1979

The Jeffreys–Bayes posterior odds criterion (POC) and an Akaike information criterion (AIC) for discriminating between two regression models are compared. It is found that the AIC is a truncated version of the POC that omits important factors.

Jeffreys has done pioneering work showing how posterior odds ratios can be employed to discriminate between or among alternative hypotheses or models. See in particular Jeffreys (1967, ch. V). Applications and extensions of Jeffreys' analysis to the problem of discriminating among alternative regression models have been made by Thornber (1966), Geisel (1970), Lempers (1971), Zellner (1971) and others. In the present note, a comparison of a Jeffreys–Bayes posterior odds ratio and the Akaike information criterion [Akaike (1972, 1974), Sawa (1976)], based on the Kullback–Leibler information measure, is made for the problem of discriminating between two nested regression models. It will be seen that the Akaike information criterion is a truncated version of a posterior odds ratio that omits factors that are important in discriminating between models.

Assume that we are interested in the following multiple regression equation:

$$\underset{n\times 1}{\boldsymbol{y}} = \underset{n\times k_1}{X_1}\ \underset{k_1\times 1}{\boldsymbol{\beta}_1} + \underset{n\times k_2}{X_2}\ \underset{k_2\times 1}{\boldsymbol{\beta}_2} + \underset{n\times 1}{\boldsymbol{u}}\ , \tag{1}$$

where $\boldsymbol{y}$ is a vector of observations on the dependent variable, X_1 and X_2 are matrices of observations on independent variables of ranks k_1 and k_2, respectively, $\boldsymbol{\beta}_1$ and $\boldsymbol{\beta}_2$ are regression coefficient vectors whose elements have unknown values and $\boldsymbol{u}$ is a vector of errors. Assume that the elements of $\boldsymbol{u}$ have been independently drawn from a normal distribution with zero mean and variance σ^2. For simplicity,

* This paper arose from discussions at the 14th meeting of the NBER-NSF Seminar on Bayesian Inference held at the Bell Laboratories, Holmdel, NJ, June 3–4, 1977. Research was financed by income from the H.G.B. Alexander Endowment Fund, Graduate School of Business, University of Chicago.

we shall assume that σ^2 has a known value. [1]

We are interested in the following hypotheses relating to $\boldsymbol{\beta}_2$:

$$H_0:\ \boldsymbol{\beta}_2 = \mathbf{0} \quad \text{and} \quad H_A:\ \boldsymbol{\beta}_2 \neq \mathbf{0}\,. \tag{2}$$

Since, in general, the posterior odds ratio relating to two mutually exclusive hypotheses is equal to the prior odds ratio times the ratio of averaged likelihood functions, where the averaging is done using the prior distributions for the parameters – see e.g. Zellner (1971, ch. X), we have in the present instance:

$$K_{0A} = \frac{\Pi_0}{\Pi_A}\ \frac{\int p(\boldsymbol{\beta}_1)\, l(\boldsymbol{\beta}_1 | \boldsymbol{y}, \sigma)\, \mathrm{d}\boldsymbol{\beta}_1}{\int p(\boldsymbol{\beta}_1, \boldsymbol{\beta}_2)\, l(\boldsymbol{\beta}_1, \boldsymbol{\beta}_2 | \boldsymbol{y}, \sigma)\, \mathrm{d}\boldsymbol{\beta}_1 \mathrm{d}\boldsymbol{\beta}_2}\,, \tag{3}$$

where Π_0/Π_A is the prior odds ratio, $p(\boldsymbol{\beta}_1)$ and $p(\boldsymbol{\beta}_1, \boldsymbol{\beta}_2)$ are prior distributions under H_0 and H_A, respectively, and

$$l(\boldsymbol{\beta}_1 | \boldsymbol{y}, \sigma) = (\sqrt{2\pi})^{-n} \sigma^{-n} \exp - \{[\nu_1 s_1^2 + (\boldsymbol{\beta}_1 - \hat{\boldsymbol{\beta}}_1)' X_1' X_1 (\boldsymbol{\beta}_1 - \hat{\boldsymbol{\beta}}_1)]/2\sigma^2\}\,, \tag{4a}$$

$$l(\boldsymbol{\beta} | \boldsymbol{y}, \sigma) = (\sqrt{2\pi})^{-n} \sigma^{-n} \exp\{-[\nu_2 s_2^2 + (\boldsymbol{\beta} - \tilde{\boldsymbol{\beta}})' Z' Z (\boldsymbol{\beta} - \tilde{\boldsymbol{\beta}})]/2\sigma^2\}\,, \tag{4b}$$

with $\hat{\boldsymbol{\beta}}_1 = (X_1' X_1)^{-1} X_1' \boldsymbol{y}$, $\nu_1 s_1^2 = (\boldsymbol{y} - X_1 \hat{\boldsymbol{\beta}}_1)'(\boldsymbol{y} - X_1 \hat{\boldsymbol{\beta}}_1)$, $\nu_1 = n - k_1$, $\boldsymbol{\beta}' = (\boldsymbol{\beta}_1', \boldsymbol{\beta}_2')$, $Z = (X_1 \vdots X_2)$, $\tilde{\boldsymbol{\beta}} = (Z'Z)^{-1} Z' \boldsymbol{y}$, $\nu_2 s_2^2 = (\boldsymbol{y} - Z\tilde{\boldsymbol{\beta}})'(\boldsymbol{y} - Z\tilde{\boldsymbol{\beta}})$, and $\nu_2 = n - k_1 - k_2$.

On substituting from (4) in (3),

$$K_{0A} = (\Pi_0/\Pi_A) \exp\{-(\nu_1 s_1^2 - \nu_2 s_2^2)/2\sigma^2\} \times \frac{\int p(\boldsymbol{\beta}_1) \exp\{-(\boldsymbol{\beta}_1 - \hat{\boldsymbol{\beta}}_1)' X_1' X_1 (\boldsymbol{\beta}_1 - \hat{\boldsymbol{\beta}}_1)/2\sigma^2\}\, \mathrm{d}\boldsymbol{\beta}_1}{\int p(\boldsymbol{\beta}) \exp\{-(\boldsymbol{\beta} - \tilde{\boldsymbol{\beta}})' Z' Z (\boldsymbol{\beta} - \tilde{\boldsymbol{\beta}})/2\sigma^2\}\, \mathrm{d}\boldsymbol{\beta}}\,. \tag{5}$$

Given the prior distributions $p(\boldsymbol{\beta}_1)$ and $p(\boldsymbol{\beta})$, (5) can be evaluated to yield a value for K_{0A} – for one example, see Zellner (1971, p. 309). If our prior distributions are relatively flat over the regions in which the exponential factors in the integrands assume appreciable values, we can approximate the integrals' values by expanding $p(\boldsymbol{\beta}_1)$ around $\hat{\boldsymbol{\beta}}_1$ and $p(\boldsymbol{\beta})$ around $\tilde{\boldsymbol{\beta}}$ to linear terms, that is $p(\boldsymbol{\beta}_1) \doteq p(\hat{\boldsymbol{\beta}}_1) + (\boldsymbol{\beta}_1 - \hat{\boldsymbol{\beta}}_1)' f(\hat{\boldsymbol{\beta}}_1)$ and $p(\boldsymbol{\beta}) \doteq p(\tilde{\boldsymbol{\beta}}) + (\boldsymbol{\beta} - \tilde{\boldsymbol{\beta}})' g(\tilde{\boldsymbol{\beta}})$, where $f(\hat{\boldsymbol{\beta}}_1)$ and $g(\tilde{\boldsymbol{\beta}})$ are vectors of first derivatives of $p(\boldsymbol{\beta}_1)$ and $p(\boldsymbol{\beta})$, evaluated at $\hat{\boldsymbol{\beta}}_1$ and $\tilde{\boldsymbol{\beta}}$, respectively. On inserting these approximations in the integrals in (5) and performing the integrations over the elements of $\boldsymbol{\beta}_1$ and $\boldsymbol{\beta}$ from $-\infty$ to ∞, the result is

$$K_{0A} \doteq \frac{\Pi_0}{\Pi_A} \exp\{-(\nu_1 s_1^2 - \nu_2 s_2^2)/2\sigma^2\}\, \frac{p(\hat{\boldsymbol{\beta}}_1)}{p(\tilde{\boldsymbol{\beta}})} \cdot \frac{|X_1' X_1/\sigma^2|^{-1/2}}{|Z'Z/\sigma^2|^{-1/2}}\, \frac{(\sqrt{2\pi})^{k_1}}{(\sqrt{2\pi})^{k}}$$

$$\doteq \frac{\Pi_0}{\Pi_A}\, \frac{p(\hat{\boldsymbol{\beta}}_1)}{p(\tilde{\boldsymbol{\beta}})} \left| \frac{\hat{V}_{12}' \hat{V}_{12}}{n\sigma^2} \right|^{1/2} \cdot \left(\frac{n}{2\pi}\right)^{k_2/2} \exp\{-(\nu_1 s_1^2 - \nu_2 s_2^2)/2\sigma^2\}\,, \tag{6a}$$

[1] In the appendix we relax this condition.

where $\hat{V}'_{12}\hat{V}_{12} = X'_2X_2 - X'_2X_1(X'_1X_1)^{-1}X'_1X_2$ and $|Z'Z| = |X'_1X_1||\hat{V}'_{12}\hat{V}_{12}|$ have been employed, or

$$K_{0A} \doteq \frac{\Pi_0 p(\hat{\boldsymbol{\beta}}_1)}{\Pi_A p(\tilde{\boldsymbol{\beta}})}\left(\frac{|X'_1X_1|}{|Z'Z|}\right)^{-1/2}(1/2\pi\sigma^2)^{k_2/2}\exp\{-(\nu_1 s_1^2 - \nu_2 s_2^2)/2\sigma^2\}\ . \tag{6b}$$

It is seen that K_{0A} in (6b) depends on the following factors:

(i) Π_0/Π_A, the prior odds ratio,
(ii) $p(\hat{\boldsymbol{\beta}}_1)/p(\tilde{\boldsymbol{\beta}})$, the prior pdfs evaluated at the maximum likelihood estimates, $\hat{\boldsymbol{\beta}}_1$ and $\tilde{\boldsymbol{\beta}}$,
(iii) $(|X'_1X_1|/|Z'Z|)^{-1/2}$, a factor involving the determinants of the design matrices,
(iv) $(1/2\pi\sigma^2)^{k_2/2}$, and
(v) $\exp\{-(\nu_1 s_1^2 - \nu_2 s_2^2)/\sigma^2\}$, the likelihood ratio.

While much could be said about each of the factors (i)–(v), just brief comments will be presented. Factor (ii) reflects the extent to which prior information is in agreement with sample information in the sense that $p(\hat{\boldsymbol{\beta}}_1)$ will have a small value if $\hat{\boldsymbol{\beta}}_1$ is far from the modal value of the prior pdf for $\boldsymbol{\beta}_1$ and similar considerations relate to $p(\tilde{\boldsymbol{\beta}})$. Thus K_{0A}'s value depends on the extent to which sample estimates agree with prior information. Factor (iii) reflects relative design considerations. If $X'_1X_2 = 0$ and if $X'_1X_1 = I$ and $X'_2X_2 = I$, $|X'_1X_1|/|Z'Z| = 1$. Otherwise properties of the design matrices under H_0 and H_A generally affect the value of K_{0A}. Factor (iv) shows a partial dependence of K_{0A} on σ^2's value with the remaining dependence in the last factor of (6b). Since

$$\partial \ln K_{0A}/\partial\sigma^2 = [(\nu_1 s_1^2 - \nu_2 s_2^2)/\sigma^2 - k_2]/2\sigma^2\ ,$$

$$\partial \ln K_{0A}/\partial\sigma^2 > 0 \quad \text{if} \quad (\nu_1 s_1^2 - \nu_2 s_2^2)/k_2\sigma^2 > 1\ ,$$

and

$$\partial \ln K_{0A}/\partial\sigma^2 < 0 \quad \text{if} \quad (\nu_1 s_1^2 - \nu_2 s_2^2)/k_2\sigma^2 < 1.$$

Last factor (v), the likelihood ratio reflects goodness of fit considerations; note that $\nu_1 s_1^2 - \nu_2 s_2^2 \geqslant 0$. Thus if H_0 leads to a poor fit relative to H_A, the value of K_{0A} will diminish.

From (6a), we have

$$\begin{aligned}-2\ln K_{0A} \doteq\ & -2\ln[\Pi_0 p(\hat{\boldsymbol{\beta}}_1)/\Pi_A p(\tilde{\boldsymbol{\beta}})] + (\nu_1 s_1^2 - \nu_2 s_2^2)/\sigma^2 \\ & - k_2[\ln n/2\pi + \ln|\hat{V}'_{12}\hat{V}_{12}/n\sigma^2|^{1/k_2}]\ .\end{aligned} \tag{7}$$

The Akaike Information Criterion (AIC) for the present problem, with σ^2's value assumed known, is [see Sawa (1976, eq. (2.10))]

$$\begin{aligned}\text{AIC} &= -2[\ln p_0(y|\hat{\boldsymbol{\beta}}_1) - \ln p_A(y|\tilde{\boldsymbol{\beta}})] + 2(k_1 - k) \\ &= -2[-(\nu_1 s_1^2 - \nu_2 s_2^2)/2\sigma^2] - 2k_2 \\ &= (\nu_1 s_1^2 - \nu_2 s_2^2)/\sigma^2 - 2k_2\ ,\end{aligned} \tag{8}$$

where $p_0(y|\hat{\boldsymbol{\beta}}_1)$ and $p_A(y|\tilde{\boldsymbol{\beta}})$ are the likelihood functions in (4a) and (4b) evaluated at the ML estimates, $\boldsymbol{\beta}_1 = \hat{\boldsymbol{\beta}}_1$ and $\boldsymbol{\beta} = \tilde{\boldsymbol{\beta}}$. If AIC < 0, one chooses the simpler model, H_0 while if AIC > 0, one chooses the more complex model, H_A.

On comparing (7) and (8), it is seen that the last two terms in (7) are in the form of the AIC except that k_2 has a different coefficient. However the AIC in (8) does not include the first term on the right-hand side of (7). If $(\Pi_0/\Pi_A)[p(\hat{\boldsymbol{\beta}}_1)/p(\tilde{\boldsymbol{\beta}})] = 1$ and $\ln|\hat{V}_{21}'\hat{V}_{21}/n\sigma^2| + \ln n/2\pi = 2$ very special conditions, (7) and (8) become become identical.

As is well-known, the posterior odds ratio K_{0A} is a measure of relative strength of beliefs in the competing hypotheses H_0 and H_A. If we have a two-state–two-action loss structure that is symmetric, e.g.,

		State of world		Expected loss
		H_0 true	H_A true	
Actions	State H_0 true	0	L	$p_A L$
	State H_A true	L	0	$p_0 L$
	Posterior probabilities $K_{0A} = p_0/p_A$	p_0	p_A	

Then acting so as to minimize expected loss results in choosing the hypothesis with the higher posterior probability. That is if $K_{0A} = p_0/p_A > 1$, one chooses H_0 and if $K_{0A} < 1$, one chooses H_A. When $K_{0A} = 1$ one is indifferent in terms of expected loss. Applying these well-known considerations to (7), the posterior odds criterion (POC) under the loss structure shown above reduces to: if $-2\ln K_{0A} < 0$, choose H_0; if $-2\ln K_{0A} > 0$, choose H_A; and if $-2\ln K_{0A} = 0$, both hypotheses are equivalent in terms of expected loss since $p_0 = p_A$. Generalization to asymmetric loss structures is direct.

From (7) and (8), the relation of the posterior odds criterion (POC) and the AIC has been made explicit. It is seen that the AIC is a truncated version of the POC. The analysis above has been performed using an approximate POC and under the assumption that σ^2 has a known value. In practice, it is not necessary to employ the approximate POC as long as one has explicit prior pdfs, $p(\boldsymbol{\beta}_1)$ and $p(\boldsymbol{\beta})$ and can evaluate (3) analytically or numerically. Also, as shown in the appendix the assumption that σ^2 has a known value can be relaxed without much difficulty. Last, the POC can be computed for variants of the above problem and for many other problems.

Appendix

In the case that σ has an unknown value and we have just vague or diffuse prior information, we shall assume that σ and the regression coefficients are independently distributed and the prior pdf for σ under H_0 and H_A is in the usual diffuse form, $p(\sigma) \propto 1/\sigma, 0 < \sigma < \infty$. Then (3) becomes,

$$K_{0A} = \frac{\Pi_0}{\Pi_A} \frac{\int\int p(\boldsymbol{\beta}_1) l(\boldsymbol{\beta}_1, \sigma|y) \, \mathrm{d}\boldsymbol{\beta}_1 \, \mathrm{d}\sigma/\sigma}{\int\int\int p(\boldsymbol{\beta}_1, \boldsymbol{\beta}_2) l(\boldsymbol{\beta}_1, \boldsymbol{\beta}_2, \sigma|y) \, \mathrm{d}\boldsymbol{\beta}_1 \, \mathrm{d}\boldsymbol{\beta}_2 \, \mathrm{d}\sigma/\sigma} . \tag{9}$$

Using the likelihood functions in (4a) and (4b) and the linearizations of the prior pdfs for the regression coefficients, integration over the β's in (9) yields

$$K_{0A} \doteq \frac{\Pi_0}{\Pi_A} \frac{p(\hat{\boldsymbol{\beta}}_1)}{p(\tilde{\boldsymbol{\beta}})} \left(\frac{|X_1'X_1|}{|Z'Z|}\right)^{-1/2} \frac{1}{(\sqrt{2\pi})^{k_2}} \times \frac{\int_0^\infty \sigma^{-(k_1+1)} \exp\{-\nu_1 s_1^2/2\sigma^2\} \, \mathrm{d}\sigma}{\int_0^\infty \sigma^{-(k+1)} \exp\{-\nu_2 s_2^2/2\sigma^2\} \, \mathrm{d}\sigma} . \tag{10}$$

The integrals on the right-hand side of (10) can be evaluated using the results in Zellner (1971, p. 371) to yield

$$\begin{aligned} K_{0A} &\doteq \frac{\Pi_0}{\Pi_A} \frac{p(\hat{\boldsymbol{\beta}}_1)}{p(\tilde{\boldsymbol{\beta}})} \left(\frac{|X_1'X_1|}{|Z'Z|}\right)^{-1/2} \frac{1}{(\sqrt{2\pi})^{k_2}} \cdot \frac{\Gamma(\nu_1/2)(2/\nu_1 s_1^2)^{\nu_1/2}}{\Gamma(\nu_2/2)(2/\nu_2 s_2^2)^{\nu_2/2}} \\ &\doteq \frac{\Pi_0}{\Pi_A} \frac{p(\hat{\boldsymbol{\beta}}_1)}{p(\tilde{\boldsymbol{\beta}})} \left(\frac{n}{2\pi}\right)^{k_2/2} \left|\frac{\hat{V}_{12}'\hat{V}_{12}}{n s_2^2}\right|^{1/2} \left(\frac{\nu_2 s_2^2}{\nu_1 s_1^2}\right)^{\nu_1/2} \frac{\Gamma[(\nu_2+k_2)/2]}{\nu_2^{k_2/2}\Gamma(\nu_2/2)} , \end{aligned} \tag{11}$$

where $\Gamma(\cdot)$ is the gamma function. From (11), we have

$$\begin{aligned} -2 \ln K_{0A} \doteq & -2 \ln[\Pi_0 p(\hat{\boldsymbol{\beta}}_1)/\Pi_A p(\tilde{\boldsymbol{\beta}})] - \nu_1 \ln(\nu_2 s_2^2/\nu_1 s_1^2) \\ & - k_2 [\ln n/2\pi + |\hat{V}_{21}' \hat{V}_{21}/n s_2^2|^{1/k_2}] \\ & - 2 \ln\{\Gamma(\nu_1/2)/\nu_2^{k_2/2}\Gamma(\nu_2)\} , \end{aligned} \tag{12}$$

where $(\nu_2 + k_2)/2 = \nu_1/2$ has been employed. Note that the likelihood ratio for this problem with σ's value unknown is $(\nu_2 s_2^2/\nu_1 s_1^2)^{n/2}$, and thus the second term on the right-hand side of (12) is related to the likelihood ratio and reflects relative goodness of fit. The first and third terms on the right-hand side of (12) are identical in form to the corresponding terms in (7) in the text except that the sample quantity s_2^2, rather than σ^2, appears. The last term on the right-hand side of (12) is an additional term not appearing in (7).

References

Akaike, H., 1972, Information theory and the extension of the maximum likelihood principle, in: Proceedings of the Second International Symposium on Information Theory, 267–281.

Akaike, H., 1974, A new look at statistical model identification, IEEE Transactions on Automatic Control AC-19, 716–723.

Geisel, M.S., 1970, Comparing and choosing among parametric statistical models: A Bayesian analysis with macroeconomic applications, Unpublished doctoral dissertation (Graduate School of Business, University of Chicago, Chicago, IL).

Jeffreys, H., 1939, 1948, 1961 and 1967, Theory of probability (Oxford University Press, Oxford).

Lempers, F.B., 1971, Posterior probabilities of alternative linear models (University of Rotterdam Press, Rotterdam).

Sawa, T., 1976, Information criterion for the choice of regressions, Paper presented to Econometric Society Meeting, Sept. 1976, and to the 14th Meeting of the NBER-NSF Seminar on Bayesian Inference in Econometrics, June 1977.

Thornber, H., 1966, Applications of decision theory to econometrics, Unpublished doctoral dissertation (Department of Economics, University of Chicago, Chicago, IL).

Zellner, A., 1971, An introduction to Bayesian inference in econometrics (Wiley, New York).

[25]

BAYESIAN ANALYSIS, MODEL SELECTION AND PREDICTION

Arnold Zellner and Chung-ki Min
H.G.B. Alexander Research Foundation
Graduate School of Business, University of Chicago
Chicago, Illinois 60637

1. Introduction

In this paper, we shall consider (1) the derivation of prior distributions for use in Bayesian analysis, (2) Bayesian and non-Bayesian model selection procedures and (3) an application of Bayesian model selection procedures in the context of an applied forecasting problem. With respect to derivations of prior distributions, Jaynes (1980, p.618) has stated:

"It was from studying these flashes of intuition in Jeffreys that I became convinced that there must exist a general formal theory of determination of priors by logical analysis of the prior information–and that to develop it is today the top priority research problem of Bayesian theory."

As is well known, many including Jeffreys (1967), Lindley (1956), Savage (1961), Hartigan (1964), Novick and Hall (1965), Jaynes (1968), Box and Tiao (1973), Bernardo (1979), and Zellner (1971, 1977, 1990) have worked to solve the "top priority research problem of Bayesian theory," to use Jaynes' words. Herein we shall consider further the maximal data information prior (MDIP) distribution approach put forward and discussed in Zellner (1971, 1977, 1990). A MDIP is a solution to a well-defined optimization problem and its use provides maximal data information relative to the information in a prior density. Also, as shown below, a MDIP maximizes the expectation of the logarithm of the ratio of the likelihood function, $\ell(\theta|y)$, to the prior density, $\pi(\theta)$, i.e. $\ell n[\ell(\theta|y)/\pi(\theta)]$. Several explicit MDIPs will be presented including those for parameters of time series processes–see Phillips (1991) for a lively discussion of this topic.

With respect to model selection procedures, posterior odds and several non-Bayesian model selection criteria will be compared and discussed along with corrections for "data-mining"–see e.g. Jeffreys (1967, p. 254). After Jeffreys' seminal work many have employed posterior odds to evaluate alternative hypotheses and models. Herein, we shall show how posterior odds can be used in variable selection problems–see Geisel (1975), Zellner (1971, 1984), Zellner and Siow (1980), Leamer (1978), and the studies in Poirier (1991) for earlier discussions and applications of posterior odds in model selection and Geisel (1975), Diebold (1990), Min and Zellner (1990), and Palm and Zellner (1991) for their use in model and forecast combination. Finally, some of the techniques mentioned above will be applied in analyzing U.S. annual data to determine how to formulate a forecasting model for the growth rate of real gross national product. Posterior odds and forecasting results for 256 alternative models will be reviewed and compared.

The plan of the paper is as follows. In section 2, maximal data information prior densities will be discussed. Then in section 3, posterior odds used in model selection problems will be compared to several non-Bayesian model selection criteria. Section 4 is devoted to results of applying Bayesian and non-Bayesian techniques to select a forecasting model. A summary of results and concluding remarks are presented in section 5.

2. Maximal Data Information Prior (MDIP) Distributions

In Zellner (1971, 1977, 1990), MDIPs' rationale, derivation and explicit forms for many problems have been discussed. As pointed out earlier, the negative entropy of a joint density function $p(y,\theta)$ for observations and parameters is given by $-H = \int\int p(y,\theta)\ell n\ p(y,\theta)dyd\theta$, which on using $p(y,\theta) = \pi(\theta)f(y|\theta)$ can be expressed as

$$-H = \int I(\theta)\pi(\theta)d\theta + \int \pi(\theta)\ell n\ \pi(\theta)d\theta \tag{1}$$

with

$$I(\theta) = \int f(y|\theta)\ell n\ f(y|\theta)dy \tag{2}$$

the negative entropy associated with $f(y|\theta)$, the data density. In (1), $-H$, a measure of the information in $p(y,\theta)$, breaks up into two parts, $\int I(\theta)\pi(\theta)d\theta$, the average information in the data density, $f(y|\theta)$, and $\int \pi(\theta)\ell n\ \pi(\theta)d\theta$, the information in the prior density $\pi(\theta)$.

The criterion functional which is maximized with respect to choice of $\pi(\theta)$ subject to side conditions to yield a MDIP, denoted by $\pi_*(\theta)$ is:

$$G[\pi(\theta)] = \int I(\theta)\pi(\theta)d\theta - \int \pi(\theta)\ell n\ \pi(\theta)d\theta \tag{3}$$

which is just the difference of the two terms on the r.h.s. of (1). Thus $G[\pi(\theta)]$ is equal to the average information in the data density, $I(\theta)$ given in (2) minus the information in the prior density, $\pi(\theta)$. Maximizing G with respect to the choice of $\pi(\theta)$, subject to side conditions reflecting initial information, e.g. $\int \pi(\theta)d\theta = 1$, etc. provides a unique prior density which gives maximal data information relative to the information in the prior; that is the expression on the r.h.s. of (3) is maximized. Thus the information in the data density is featured by use of a MDIP.[1]

An alternative expression for $G[\pi(\theta)]$ in (3) is:

$$G[\pi(\theta)] = \int \left[\int g(\theta|y)\ell n\{\ell(\theta|y)/\pi(\theta)\}d\theta\right] h(y)dy \tag{4}$$

[1] In cases in which y is a vector $y' = (y_1, y_2, ... y_n)$ and the y_i's are independently and identically distributed, $f(y|\theta) = \prod_{i=1}^{n} f(y_i|\theta)$; then (2) is $I_n(\theta) = nI(\theta)$, where $I(\theta) = \int f(y_i|\theta)\ell n\ f(y_i|\theta)dy_i$, $i = 1, 2, ..., n$, and maximization of G in (3) yields the prior $\pi_*(\theta)$ that makes the average information in the data density $I_n(\theta)/n$ as large as possible relative to the information in the prior density.

where $p(y,\theta) = h(y)g(\theta|y)$, with $g(\theta|y)$ the posterior density of θ, $h(y)$ the marginal density of y and $\ell(\theta|y) \equiv f(y|\theta)$ is the well-known likelihood function. Thus (4) indicates that $G[\pi(\theta)]$ is an average of the logarithm of the likelihood-prior ratio, $\ell(\theta|y)/\pi(\theta)$, taken w.r.t. $g(\theta|y)$, the posterior density and $h(y)$, the marginal density of the observations. It follows that maximizing G w.r.t. choice of $\pi(\theta)$ **makes the likelihood function $\ell(\theta|y)$ as high as possible relative to the prior, $\pi(\theta)$.** This is indeed a satisfying feature of a MDIP.

As shown in Zellner (1971, 1977, 1990), on maximizing $G[\pi(\theta)]$ in (3) with respect to $\pi(\theta)$ subject to $\int \pi(\theta)d\theta = 1$, that is that $\pi(\theta)$ is a proper density, the solution is:

$$\pi_*(\theta) = c\exp\{I(\theta)\} \qquad a < \theta < b \tag{5}$$

with $I(\theta)$ given in (2) and $c^{-1} = \int_a^b e^{I(\theta)}d\theta$ is a normalizing constant. Thus if $I(\theta) =$ constant, $\pi_*(\theta)$ is a uniform density. Also for the class of densities $f[(y-\theta)/\sigma], \pi_*(\theta,\sigma) \propto 1/\sigma$ with $a_1 < \theta < b_1$ and $0 < a_2 < \sigma < b_2$. See Zellner (1977) for many other MDIPs.

Above, we have used just the side condition, $\int \pi(\theta)d\theta = 1$ in the maximization problem. If other side conditions are imposed reflecting given initial information, e.g. $\int \theta^i\pi(\theta)d\theta = \mu_i$, i = 1,2,...,m, then the MDIP is given by $\pi_*(\theta) \propto \exp\{I(\theta) + \lambda_1\theta + \lambda_2\theta^2 + \ldots + \lambda_m\theta^m\}$, where the λ_i are Lagrange multipliers which can be evaluated given values of the prior moments, μ_i, i = 1,2,...,m. If, e.g. m = 2 and $I(\theta) =$ const., $\pi_*(\theta)$ has a normal form. If m = 4 and $I(\theta) =$ const., $\pi_*(\theta)$ has an exponential quartic form. We see that side conditions reflecting initial information that may be available can readily be utilized in deriving MDIPs. Since all of these MDIPs are proper, they can be used in computing posterior odds. However, before discussing this topic, we shall consider some MDIPs for time series models.

For a **stationary** autoregressive model, $y_t - \mu = \rho(y_{t-1} - \mu) + \varepsilon_t$, where $\varepsilon_t \sim N(0,\sigma^2)$, $-1 < \rho < 1$, $Ey_t = \mu, \sigma^2_{y_t} = \sigma^2/(1-\rho^2)$, and $f(y_t|\mu,\rho,\sigma) \propto (1-\rho^2)^{1/2}\sigma^{-1}\exp\{-(1-\rho^2)(y_t-\mu)^2/2\sigma^2\}$. Then $\ell n\ f(y_t|\mu,\rho,\sigma) =$ const. $+\frac{1}{2}\ell n(1-\rho^2) - (1-\rho^2)(y_t-\mu)^2/2\sigma^2$ and $I(\theta) = E\ \ell n\ f(y_t|\mu,\rho,\sigma) =$ const. $+\frac{1}{2}\ell n(1-\rho^2) - \frac{1}{2}$ and thus $\pi_*(\mu,\rho,\sigma) \propto \exp\{I(\theta)\} = (1-\rho^2)^{\frac{1}{2}}/\sigma$ as pointed out in Zellner (1977). Thus μ, $\ell n\ \sigma$ and ρ are independently distributed with μ and $\ell n\ \sigma$ uniform and ρ having a density $\pi_*(\rho) \propto (1-\rho^2)^{\frac{1}{2}}$, $-1 < \rho < 1$, a form of a beta density symmetric about a mode at $\rho = 0$ and with a value equal to zero for $\rho = 1$ and $\rho = -1$.

Now for an AR(2) stationary process, $y_t - \mu = \rho_1(y_{t-1} - \mu) + \rho_2(y_{t-2} - \mu) + \varepsilon_t$, where $\varepsilon_t \sim N(0,\sigma^2), Ey_t = \mu$, $\sigma^2_{y_t} = \rho_1^2\sigma^2_{y_t} + \rho_2^2\sigma^2_{y_t} + 2\rho_1\rho_2\gamma_1 + \sigma^2$, with $E(y_{t-1}-\mu)(y_{t-2}-\mu) = \gamma_1$. Since $\gamma_1 = \rho_1\sigma^2_{y_t} + \rho_2\gamma_1$, $\gamma_1 = \sigma^2_{y_t}\rho_1/(1-\rho_2)$ and thus $\sigma^2_{y_2} = \sigma^2/[1-\rho_1^2-\rho_2^2-2\rho_1^2\rho_2/(1-\rho_2)]$. For this problem, the MDIP is,

$$\pi_*(\mu,\rho_1,\rho_2,\sigma) \propto [1 - \rho_1^2 - \rho_2^2 - 2\rho_1^2\rho_2/(1-\rho_2)]^{1/2}/\sigma \tag{6}$$

with ρ_1 and ρ_2 constrained to the region of stationarity in the ρ_1, ρ_2 plane, see Zellner (1971,p.196) for a graphical display of this region and Fig. 1 for a plot of the contours of this prior for ρ_1 and ρ_2. For higher order stationary autoregressive processes, similar procedures can be employed to obtain MDIP priors.

It is often the case that one may not be sure that an autoregressive process is stationary. For example in terms of an AR(1) process $y_t = \rho y_{t-1} + \varepsilon_t$, it may be that $\rho \geq 1$ which implies that the process is not stationary. If such a process begins at $t = 0$, $y_t = \rho^t y_0 + \varepsilon_t +$

$\rho\varepsilon_{t-1} + ... + \rho^{t-1}\varepsilon_1$ and if y_0 is fixed, $Ey_t = \rho^t y_0$, and $\sigma^2_{y_t} = [1+\rho^2+\rho^4+\rho^6+...+\rho^{2(t-1)}]\sigma^2$ under the assumption that the ε_t's are uncorrelated with zero means and common variance σ^2. If it is further assumed that the ε_t's are NID$(0,\sigma^2)$, then y_t has a normal density with mean $\rho^t y_0$ and variance $\sigma^2_{y_t}$, $f(y_t|y_0,\rho,\sigma) \propto \sigma^{-1}_{y_t} \exp\{-(y_t - \rho^t y_0)^2/2\sigma^2_{y_t}\}$ and the MDIP is $\pi_*(\sigma_{y_t}) \propto 1/\sigma_{y_t}$ or

$$\pi_*(\rho,\sigma) \propto 1/\sigma[1+\rho^2+\rho^4+...+\rho^{2(t-1)}]^{1/2} \tag{7}$$

In this case, the MDIP depends on t, as well as σ and ρ where σ and ρ are confined to finite, possibly very large, sub-regions of the parameter space, $0 < \sigma < \infty$ and $-\infty < \rho < \infty$. For $t = 1$, $\pi_*(\rho)$ is uniform while for $t > 1$, it has a mode at $\rho = 0$ and falls off symmetrically for $\rho \neq 0$. Given a MDIP relative to $f(y_t|\rho,\sigma,y_0)$, on observing y_t we can use the prior in (7) to obtain a posterior density for ρ and σ which can be used as a prior in the analysis of the next observation y_{t+1} and so on. If on the other hand we want the MDIP relative to the data density for $(y_t, y_{t+1}, ..., y_{t+n}) = (y_t, \boldsymbol{y}')$, then $f(y_t, \boldsymbol{y}|y_0,\rho,\sigma) = f(y_t|y_0,\rho,\sigma)f(\boldsymbol{y}|y_t,y_0,\rho,\sigma)$ and application of the MDIP approach to $f(\boldsymbol{y}|y_t,\rho,\sigma)$ alone yields $\pi_*(\rho,\sigma) \propto 1/\sigma$. On the other hand when the complete joint density is considered, the information in the $n+1$ observations $(y_t\boldsymbol{y})$ is $I_{n+1} = E[\ell n\ f(y_t|y_0,\rho,\sigma) + \ell n\ f(\boldsymbol{y}|y_t y_0,\rho,\sigma)] = -[\ell n\ \sigma_{y_t} + n\ \ell n\ \sigma]$, with $\sigma^2_{y_t} = \sigma^2[1+\rho^2+...+\rho^{2(t-1)}$. Then the information per observation, $I = I_{n+1}/(n+1)$ and $\pi_*(\rho,\sigma) \propto \exp\{I\} = 1/\sigma[1+\rho^2+\rho^4+...+\rho^{2(t-1)}]^{1/(n+1)}$. This prior density will tend to become uniform in ρ as the sample size n becomes large and/or t approaches 1.

The above analysis rests on the presumption that y_0 and t, the time the process has been operative, are both known. If one or the other of these quantities has an unknown value, then the density for y_t cannot be implemented. In such a case it is possible to formulate $f(y_{t+1}|y_t,\rho,\sigma)$ and use it to obtain the following MDIP, $\pi_*(\rho,\sigma) \propto 1/\sigma$, $0 < a_1 < \sigma < b_1$ and $-a_2 < \rho < b_2$. Thus, if it is thought that ρ might possibly be slightly larger than 1 but surely not less than -1, the density of ρ could be taken uniform with $-1 \leq \rho \leq 1.1$, as was done in Anton (1986) who used a beta prior for ρ defined over this interval.[2] Similarly, the MDIP associated with $f(y_{t+1}, y_{t+2}, ...y_{t+n}|y_t,\rho,\sigma)$ is $\pi_*(\rho,\sigma) \propto 1/\sigma$ with $0 < a_1 < \sigma < b_1$ and $a_2 < \rho < b_2$.

For higher order AR processes say an AR(3), $y_t = \alpha + \beta_1 y_{t-1} + \beta_2 y_{t-2} + \beta_3 y_{t-3} + \varepsilon_t$, with the ε_t's NID$(0,\sigma^2)$, if y_0, y_{-1} and y_{-2} have known values and if it is not known when the process started, we can consider the conditional density $f(y_1, y_2, ...y_n|y_0, y_{-1}, y_{-2}\boldsymbol{\theta})$, where $\boldsymbol{\theta}' = (\alpha_1, \beta_1, \beta_2, \beta_3)$ and the MDIP for the parameters is $\pi_*(\alpha, \beta_1, \beta_2, \beta_3, \sigma) \propto 1/\sigma$, defined over a finite region of the parameter space. See Hong (1989) for an analysis of this prior distribution's implications for the properties of the roots of the AR(3) process and for the amplitude and period of periodic components associated with complex conjugate roots. Of course if other side conditions are utilized in obtaining the MDIP, its form will not be that shown above. For example if we use side conditions for the first and second moments of α, β_1, β_2, and β_3, the MDIP for these parameters will be normally shaped and not uniform. While any of the above proper prior densities can be employed in computation of

[2] The interval $-1 \leq \rho \leq 1.1$ is equivalent to the interval $0 \leq \frac{1+\rho}{2.1} \leq 1$. Letting $\theta = (1+\rho)/2.1$, Anton (1986) used a proper beta density for θ such that the associated density for ρ had a modal value=1. In the MDIP approach, a beta density will be obtained as a solution if $I(\theta)$=const. and one uses the side conditions $\int_0^1 \pi(\theta)d(\theta) = 1$, $\int_0^1 \pi(\theta)\ell n\ \theta d\theta = c_1$ and $\int_0^1 \pi(\theta)\ell n\ (1-\theta)d\theta = c_2$.

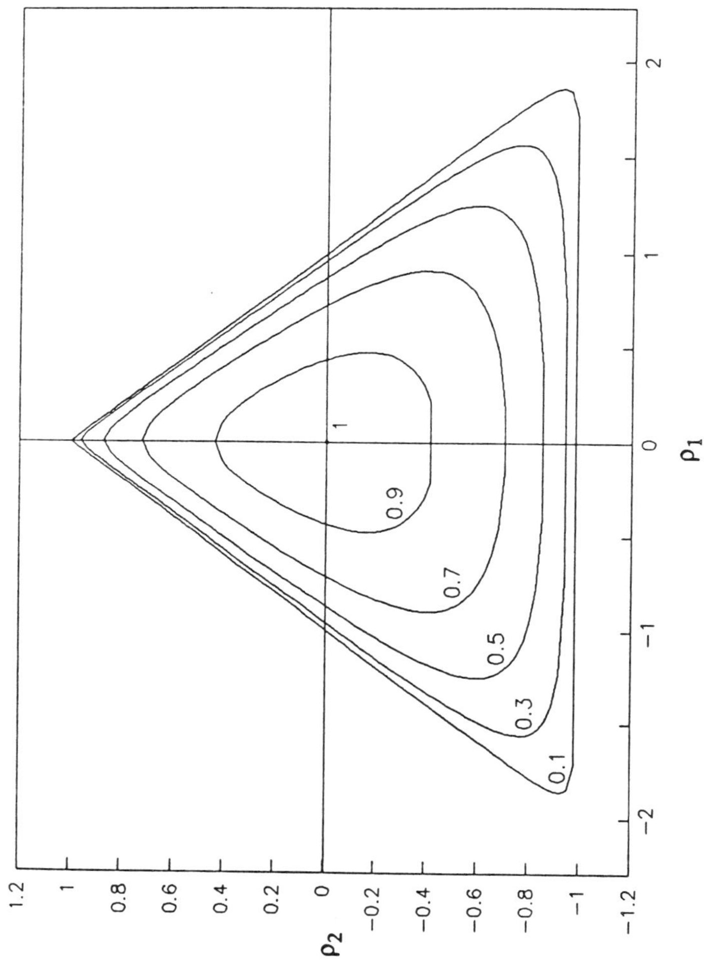

Fig. 1. Contours of the maximal data information prior density for parameters ρ_1 and ρ_2 of a stationary AR(2) model (The MDIP prior for ρ_1 and ρ_2, shown in (6) in the text is $\pi_*(\rho_1, \rho_2) \propto \left[1 - \rho_1^2 - \rho_2^2 - \frac{2\rho_1^2\rho_2}{1-\rho_2}\right]^{1/2}$ with $-2 < \rho_1 < 2$, $\rho_2 < \rho_1 + 1$ for $-2 < \rho_1 \leq 0$ and $\rho_2 < -\rho_1 + 1$ for $0 \leq \rho_1 < 2$.)

Bayes' factors, in the present paper we shall utilize the priors developed in Zellner and Siow (1980) for model selection–see Zellner (1984, Ch. 3.7), Connolly (1991), Moulton (1991), and Koop (1992) for applications of this approach which will also be employed in section 4 after a consideration of alternative models and model selection techniques in the next section.

3. Models and Model Selection Techniques

In considering models and model selection, it is relevant to consider the origins of models. First, in situations in which little is known, it may be wise to assume, as Jeffreys and others have suggested, that all variation is random unless shown otherwise. Use of simple "benchmark" models, e.g. random walk models may be a good starting point as was done in early analyses of stock prices and in other areas. In our empirical work to follow, we shall evaluate a random walk model along with others.

Second, some suggest starting with a broad model or a broad class of models and then testing to determine whether a member of the class is supported by the data. Of course this approach requires us to have enough information to specify a relevant class of models that is not too heavily parameterized or complicated.

Third, if one can enumerate all possible models under consideration, as will be done below, there may be a certain subclass, e.g. models with four input variables, that has more members than other subclasses and this fact, given no other relevant information may lead us to have an a priori preference for this particular subclass of models.

Fourth, many utilize intuition and background theoretical and empirical knowledge to formulate models. For example, in modeling the rate of growth of annual output denoted by y_t, it seemed important to us to have a model incorporating an oscillatory component to represent business cycles. Thus in Garcia-Ferrer et al. (1987), we considered an autoregression (AR) of order three, $y_t = \alpha + \beta_1 y_{t-1} + \beta_2 y_{t-2} + \beta_3 y_{t-3} + \varepsilon_t$. On finding that this model did not fit the data well in the vicinity of turning points, we added some leading indicator (LI) variables, to obtain the following AR(3)LI model:

$$\begin{aligned} y_t &= \beta_0 + \beta_1 y_{t-1} + \beta_2 y_{t-2} + \beta_3 y_{t-3} + \beta_4 S_{t-1} + \beta_5 S_{t-2} + \beta_6 M_{t-1} + \beta_7 W_{t-1} + u_t \\ &= \beta_0 \boldsymbol{x_t}'\beta + u_t \end{aligned} \tag{8}$$

where the subscript t denotes the value of a variable in the t'th year and the variables, y_t, S_t, M_t, and W_t are growth rates of real GNP, real stock prices, real money and real world stock prices, respectively and u_t is a non-autocorrelated zero-mean error term with variance σ^2 for all t. The relation in (8) has been derived from simple macroeconomic theories and shown to perform better in forecasting than random walk models and AR(3) models–see Garcia-Ferrer, et al. (1987) and Zellner and Hong (1989). Still we can ask, as Mittnik (1990) and Otter (1990) did, would models containing other combinations of the input variables be more strongly supported by the data? We shall present some results below bearing on this issue.

Finally, is it possible to produce a model such as that in (8) by a formal method, say by maximizing entropy subject to side conditions? The answer to this question is yes. That is, with $f(y_t|\beta_0 + \boldsymbol{x}_t'\beta, \sigma^2)$ the conditional density of y_t given $\beta_0 + \boldsymbol{x}_t'\beta$ and σ^2, consider maximizing $H = -\int f(y_t|\cdot)\, \ell n\, f(y_t|\cdot) dy_t$ subject to $\int f(y_t|\cdot) dy_t = 1$, $\int y_t f(y_t|\cdot) dy_t = \beta_0 + \boldsymbol{x}_t'\beta$, and $\int (y_t - \beta_0 - \boldsymbol{x}_t'\beta)^2 f(y_t|\cdot) dy_t = \sigma^2$. The solution to this problem is $f_*(y_t|\beta_0 +$

$x_t'\beta, \sigma^2$), a normal density function for y_t given $\beta_0 + x_t'\beta$ and σ^2 with mean $\beta_0 + x_t'\beta$ and variance σ^2. Thus maxent can be used to obtain the model in (8) and others as shown in Ryu (1991). However, the side conditions employed, for example, the variables to include in x_t can be varied and thus a model selection problem arises.

In approaching the problem of which variables to include in an equation like (8), we broadened the set to include the seven variables shown in (8) plus M_{t-2} to give a total of 8 input variables denoted by $(y_1, y_2, y_3, S_1, S_2, M_1, M_2, W_1)$ and the intercept β_0 which will be included in all 256 alternative models. That is, we have $N_r = \binom{n}{r} = n!/(n-r)!r!$ models containing r variables. Thus, $N_0 = 1$, $N_1 = 8$, $N_2 = 28$, $N_3 = 56$, $N_4 = 70$, $N_5 = 56$, $N_6 = 28$, $N_7 = 8$, and $N_8 = 1$, a total of 256 distinct models. It is to these 256 possible models that we shall apply model selection procedures.

For the purpose of deriving posterior odds on a particular model versus the broadest model, that containing all 8 input variables and an intercept, let us write

$$\begin{aligned} y_t = \beta_0 + \underset{1\times k_1}{x_{1t}'}\ \underset{k_1\times 1}{\beta_1} + \underset{1\times k_2}{x_{2t}'}\ \underset{k_2\times 1}{\beta_2} + u_t \end{aligned} \tag{9}$$

with $x_t' = (x_{1t}', x_{2t}')$, the vector of all $k_1 + k_2 = 8$ input variables. Then consider

$$H_A : \beta_2 = 0 \text{ versus } H_B : \beta_2 \neq 0 \tag{10}$$

H_A implies that a particular model with k_1 input variables, x_{1t}, is appropriate, while H_B implies that the model containing all 8 input variables is appropriate. Given a set of observations on y_t and x_t, the posterior odds relating to H_A vs. H_B, denoted by K_{AB}, is given by

$$K_{AB} = \frac{\pi_A}{\pi_B} \frac{\int p_A(y|\beta_1, \sigma)\pi_A(\beta_1, \sigma)d\beta_1 d\sigma}{\int p_B(y|\beta_1, \beta_2\sigma)\pi_B(\beta_1, \beta_2, \sigma)d\beta_2 d\beta_2 d\sigma} \tag{11}$$

where π_A/π_B is the prior odds on H_A vs. H_B and the second factor on the r.h.s. of (11) is the Bayes Factor. If we take $\pi_A/\pi_B = 1$, use normal likelihood functions and the prior densities employed in Zellner and Siow (1980, p.593), then as shown in this last reference

$$K_{AB} \doteq b(\nu_B/2)^{k_2/2}\{(1 - R_B^2)/(1 - R_A^2)\}^{(\nu_B - 1)/2} \tag{12}$$

where $b = \pi^{1/2}/\Gamma[(k_2 + 1)/2]$, R_A^2 and R_B^2 are the sample squared multiple correlation coefficients under H_A and H_B, respectively and $\nu_B = n - k_1 - k_2 - 1$. Alternatively, the posterior odds can be expressed as

$$K_{AB} \doteq b(\nu_B/2)^{k_2/2}(SSE_B/SSE_A)^{(\nu_B - 1)/2} \tag{13}$$

where SSE_B and SSE_A are the sums of squared least squares residuals under hypotheses B and A, respectively. We see that the better is the fit of the restricted model, M_A, as measured by R_A^2 or SSE_A, the higher the odds in favor of M_A vs. M_B, the broadest model. Also note that the larger is k_2, the number of input variables added under H_B, the greater the odds in favor of M_A vs. M_B, all else held constant. Of particular importance here is to note that K_{AB} in (13) is a monotonic function of SSE_A, all else held constant.

As will be seen, this is a characteristic of alternative model selection criteria considered below.

Some well known alternative model selection criteria are, with $p \equiv k_1 + 1$,

(i) Mean squared error: $MSE = SSE_A/(n-p)$
(ii) Schwarz's criterion: $SC = n\ \ell n(SSE_A/n) + p\ \ell n\ n$
(iii) Akaike's information criterion: $AIC = n\ \ell n\ SSE_A/n + 2p$
(iv) Mallows' $C(p)$: $C(p) = SSE_A/\hat{\sigma}^2 + 2p - n$
(v) Sawa's Bayesian information criterion: $BIC = n\ \ell n(SSE_A/n) + 2(p+2)q - 2q^2$

where $\hat{\sigma}^2 = SSE_B/(n - k_1 - k_2 - 1)$ and $q = \hat{\sigma}^2/(SSE_A/n)$. It is seen that the criteria (i)-(v) are all monotonic functions of SSE_A and **thus for given k_1 and n will give an identical ordering of models**. However, the "penalties" imposed for including additional input variables are different in the alternative model selection rules.

Now if we were to select the model, say M_A^*, with the highest posterior odds greater than one relative to the broadest model, this would be optimal given a zero-one loss function. However, it must be recognized that on screening 256 models, we may get one that appears to be a good one just by chance, a so-called "data-mining process." Jeffreys (1967, p.254) refers to this as a "selection problem". He suggests the following solution. If we put prior probability 1/2 on M_A and 1/2 on the $m = 255$ alternative models, each having prior probability q, then $(1-q)^m = 1/2$ is the probability that the m alternative models are false. Thus, Jeffreys obtains, $q = 1 - (2)^{-1/m} \doteq 1/m \log 2$ for large m. In our case, $m = 255$ and the prior odds $\pi_A/\pi_B = 1/2q \doteq m/2\log 2 = .7m = (.7)255 = 178.5$ rather than 1. Thus Jeffreys' correction for selection is quite large in our case.

Another way to think about the selection problem is to assign a probability, P_A to a null model, M_{A_0}, say the ARLI model in (8) and assign probability P_B to the remaining 255 models with $P_B/255$ assigned to each. Then the prior odds for M_{A_0} vs. any other model is equal to $P_A/(P_B/255) = 255P_A/P_B$. If, for example $P_A = 1/16$ and $P_B = 5/16$, then the prior odds for M_{A_0} versus any other model of the remaining 255 models is 51 to 1, again a large number. Note also that $P_A + P_B < 1$ so as to allow for the possibility that models not included in the 256 under consideration, for example nonlinear or time-varying parameter models, may be appropriate. As these considerations indicate, it is not always appropriate to use 1:1 prior odds in computing posterior odds. In what follows, we shall show the results of computing posterior odds based on 1:1 prior odds and then consider effects of adjustments for selection on results.

4. Model Selection and Prediction Results

Here we relate an annual output growth rate variable for the U.S., y_t, to y_{t-1}, y_{t-2}, y_{t-3}, S_{t-1}, S_{t-2}, M_{t-1}, M_{t-2}, and W_{t-1}, that is to 8 input variables which we denote by $cy_1y_2y_3S_1S_2M_1M_2W_1$, where c denotes a constant or intercept term. Thus there are 8 input variables in the broadest model. As indicated above, there is a total of 256 models, that is $\sum_{r=0}^{8}\binom{8}{r}$. Using the odds expression in (12), we computed the odds on each sub-model versus the broadest model, the model with 8 input variables and an intercept term for two periods, 1954-73 and 1954-87. Also for these periods and models we computed within sample mean squared errors, MSEs. The results for our autoregressive-leading indicator (AR(3)LI) model are shown in line 1 of Table 1. It is seen that for the periods 1954-73 and 1954-87, the posterior odds for the AR(3)LI model versus the broadest model, given in the last line of the table are 2.80 and 6.19, respectively. Note also that the within sample MSEs for the AR(3)LI model are 1.79 and 2.70 versus 1.81 and 2.80 for the broadest model. Finally, the root mean-squared error of forecast for 14 one year ahead forecasts, 1974-87 is 2.31 for the AR(3)LI model versus 2.47 for the broadest model. Thus in this comparison, the AR(3)LI model is preferred on all counts. The AR(3)LI model also dominates the AR(3) and random walk benchmark models in lines 8 and 9 of Table 1. The same can be said with respect to the models in lines 6 and 7 that were selected by Mittnik (1990) and Otter (1990) using Hankel matrix, state-space model identification techniques.

Table 1
Model Selection Using U.S. Annual Data, Posterior Odds, Mean Squared Error and Out-of-Sample Forecast RMSEs

	1954-73		1954-87		Out-of-Sample, 1974-87,
Model[a]	Odds[b]	MSE[c]	Odds[b]	MSE[c]	RMSE of Forecast[d]
1. AR(3)LI $cy_1y_2y_3S_1S_2M_1W_1$	2.80	1.79	6.19	2.70	2.31
2. $cy_3M_1M_2W_1$	17.42	1.57	2.74	3.46	2.74
3. $cy_3S_1M_1M_2W_1$	15.67	1.53	14.29	2.88	2.44
4. $cy_3S_1M_1W_1$	8.61	1.81	26.34	2.87	2.30
5. $cy_3S_1S_2M_1W_1$	12.88	1.59	59.92	2.56	2.22
6. $cS_1S_2M_1M_2W_1$[e]	2.18	2.27	1.68	3.44	2.59
7. $cy_1S_1M_1W_1$[f]	2.11	2.39	2.55	3.48	2.54
8. $AR(3), cy_1y_2y_3$	.04	5.41	.01	5.76	2.69
9. Random Walk for ℓn GNP with Drift, c	.06	4.89	.02	5.75	2.68
10. General Model, $cy_1y_2y_3S_1S_2M_1M_2W_1$	1.00	1.81	1.00	2.80	2.47

[a]The general model in line 10 is given by $y_t = c+\beta_1 y_{t-1}\beta_2 y_{t-2}+\beta_3 y_{t-3}+\beta_4 S_{t-1}+\beta_5 S_{t-2}+\beta_6 M_{t-1}+\beta_7 M_{t-2}+\beta_8 W_{t-1}+\epsilon_t$ and denoted by $cy_1y_2y_3S_1S_2M_1M_2W_1$ where y_t, S_t, M_t, and W_t denote the annual growth rates of real GNP, real stock prices, real money and real world stock prices, respectively. Other models are particular cases of the general model.

[b]Posterior odds are computed for each specific model versus the general model in line 10 with prior odds in each case taken 1:1. See equations (12) and (13) in the text and Zellner and Siow (1980), and Zellner (1984, Ch.3.7) for expressions for odds.

[c]Within sample mean squared error $= \sum_{t=1}^{N} \hat{\epsilon}_t^2/\nu$, where ν = degrees of freedom, or number of observations minus number of estimated parameters.

[d]One-step-ahead forecast errors, Δ_t were employed to compute RMSEs given by RMSE $= \sqrt{\sum_1^{14} \Delta_t^2/14}$.

[e]This is the model selected by Mittnik (1990), based on data 1953-73 for the U.S.

[f]This is the model selected by Otter (1990), based on data 1954-73 for the U.S.

As regards the models in lines 2-5, these were the ones most favored by the posterior odds and several other model selection criteria. While the model selection procedures favor these models, it has to be remembered that they were selected from 255 alternatives to the AR(3)LI model in line 1. There is a danger that the models so selected may fit the data by chance. If that is the case, the out-of-sample forecasting performance of these models with new data will usually not be as satisfactory as the within sample fits might imply. In the present instance, note that while the models in lines 2 to 5 appear to be favored relative to the AR(3)LI model in line 1 by posterior odds of from about 3 to 6 to 1 for the 1954-73 data, these models' out-of-sample RMSEs of forecast are either nearly the same as or higher than that of the AR(3)LI model. It would appear that use of 1:1 prior odds is not entirely satisfactory. If a Jeffreys-like correction for selection is introduced resulting in a prior odds favoring the AR(3)LI model, we would have some protection from effects of data-mining or of getting a model that happens to fit by chance. On the other hand, note that all the models in lines 2 to 5 have a special AR(3) structure–only y_{t-3} appears in these equations with an estimated coefficient that is negative. On examining the roots of $y_t = \beta_3 y_{t-3}$ with $\beta_3 < 0$, it is the case that there are two complex conjugate roots and a real root giving rise to an oscillatory component with a period of six years. Thus this particular formulation is a parsimonious way of obtaining a cycle.

These calculations indicate that the AR(3)LI model which we have employed in our past work–see e.g. Garcia-Ferrer et al. (1987) and Zellner and Hong (1989) is not grossly dominated by any of the 255 alternative models considered using data for the U.S. However, this is not to say that other types of models cannot dominate the AR(3)LI. Calculation of posterior odds for fixed versus time-varying parameter models for 18 countries in Min and Zellner (1990) shows that for many countries models with time-varying parameters drawn from hyper-distributions are favored and perform better in forecasting out-of-sample.

5. Summary and Concluding Remarks

In this paper, we have provided additional analysis of maximal data information prior densities. We have shown that they are the prior densities for which the posterior expectation of the logarithm of the ratio of a likelihood function to a prior is maximized. Thus the information in the likelihood function is emphasized in this sense by use of MDIPs. Also, we derived and discussed MDIPs for stationary and non-stationary autoregressive processes which are sensible and operational. Then we turned to a discussion of posterior odds and other model selection criteria. In analyzing alternative criteria, it was found that they all provided the same ordering of models with the same number of input variables since they are monotonic functions of the sum of squared least squares residuals. Further, attention was given to the problem of "selection effects" or "data-mining." In our application of posterior odds and other model selection techniques to the problem of evaluating 256 alternative models, it was found that a selection problem was probably present and that a Jeffreys-like correction to the prior odds was needed. Finally, it was found that the posterior odds approach is quite operational and yielded sensible results.

REFERENCES

Anton, L.A.M. (1986), "Empirical Regularities in Short-Run Exchange Rate Behavior," doctoral dissertation, Department of Economics, U. of Chicago.

Bernardo, J. (1979), "Reference Posterior Distributions for Bayesian Inference," J. of Royal Statistical Society, B, 41, 113-147.

Box, G.E.P. and G.C. Tiao (1973), Bayesian Inference in Statistical Analysis, Reading, MA: Addison-Wesley Publishing Co.

Connolly, R.A. (1991), "A Posterior Odds Analysis of the Weekend Effect," J. of Econometrics, 49, 51-104.

Diebold, F.X. (1990), "A Note on Bayesian Forecast Combination," ms., Department of Economics, U. of Pennsylvania, Philadelphia, PA.

Garcia-Ferrer, A., R.A. Highfield, F. Palm, and A. Zellner (1987), "Macroeconomic Forecasting Using Pooled International Data," J. of Business and Economic Statistics, 5, 53-67.

Geisel, M.S. (1975), "Bayesian Comparisons of Simple Macroeconomic Models," in S.E. Fienberg and A. Zellner, editors, Studies in Bayesian Econometrics and Statistics in Honor of Leonard J. Savage, Amsterdam, The Netherlands: North-Holland Publishing Co., 227-256.

Hartigan, J. (1964), "Invariant Prior Distributions," Annals of Mathematical Statistics, 35, 836-845.

Hong, C. (1989), "Forecasting Real Output Growth Rates and Cyclical Properties of Models: A Bayesian Approach," doctoral dissertation, Department of Economics, U. of Chicago.

Jaynes, E.T. (1968), "Prior Probabilities," IEEE Transactions on Systems Science and Cybernetics, SSC-4, 227-241.

______ (1980), "Discussion," in J.M. Bernardo, M.H. DeGroot, D.V. Lindley, and A.F.M. Smith, editors, Bayesian Statistics, Valencia, Spain: University Press, 618-629.

Jeffreys, H. (1967), Theory of Probability (3rd rev. ed.; 1st ed., 1939), London: Oxford U. Press.

Koop, G. (1992), " 'Objective' Unit Root Tests," J. Applied Econometrics, 7, 65-82.

Leamer, E. (1978), Specification Searches, New York: J. Wiley and Sons, Inc.

Lindley, D.V. (1956), "On a Measure of Information Provided by an Experiment," Annals of Mathematical Statistics, 27, 986-1005.

Min, C.K. and A. Zellner (1990), "Bayesian and Non-Bayesian Methods for Combining Models and Forecasts with Applications to Forecasting International Growth Rates," ms., H.G.B. Alexander Research Foundation, U. of Chicago, to appear in J. of Econometrics.

Mittnik, S. (1990), "Macroeconomic Forecasting Using Pooled International Data," J. of Business and Economic Statistics, 8, 205-208.

Moulton, B.R. (1991), "A Bayesian Approach to Regression Selection and Estimation with Application to a Price Index for Radio Services," J. of Econometrics, 49, 169-193.

Novick, M. and W. Hall (1965), "A Bayesian Indifference Procedure," J. of the American Statistical Association, 60, 1104-1117.

Otter, P.W. (1990), "Canonical Correlation in Multivariate Time Series Analysis with an Application to One-Year-Ahead and Multiyear-Ahead Macroeconomic Forecasting," J. of Business and Economic Statistics, 8, 453-457.

Palm, F.C. and A. Zellner (1991), "To Combine or Not to Combine? Issues of Combining Forecasts," ms., H.G.B. Alexander Research Foundation, U. of Chicago, to appear in the J. of Forecasting.

Phillips, P.C.B. (1991), "To Criticize the Critics: An Objective Bayesian Analysis of Stochastic Trends," J. of Applied Econometrics, 6, 333-364.

Poirier, D.J. (1991), Bayesian Empirical Studies in Economics and Finance, editor, Journal of Econometrics, 49, 1-304.

Ryu, H.K. (1991), "Orthogonal Basis and Maximum Entropy Estimation of Probability Density and Regression Functions," doctoral dissertation, Department of Economics, U. of Chicago.

Savage, L.J. (1961), The Subjective Basis of Statistical Practice, ms., Department of Statistics, U. of Michigan, Ann Arbor, MI.

Zellner, A. (1971), An Introduction to Bayesian Inference in Econometrics, New York: J. Wiley and Sons, Inc. (reprinted by Krieger Publishing Co., Malabar, FL, 1987).

______ (1977), "Maximal Data Information Prior Distributions," in A. Aykac and C. Brumat, eds., New Developments in the Applications of Bayesian Methods, Amsterdam: North-Holland Publishing Co., 211-232 and reprinted in A. Zellner (1984).

______ (1984), Basic Issues in Econometrics, Chicago: U. of Chicago Press.

Zellner, A. (1990), "Bayesian Methods and Entropy in Economics and Econometrics," in W.T. Grandy, Jr. and L.H. Schick (eds.), Maximum Entropy and Bayesian Methods, Dordrecht, Netherlands: Kluwer Academic Publishers, 17-31.

Zellner, A. and C. Hong (1989), "Forecasting International Growth Rates Using Bayesian Shrinkage and Other Procedures," J. of Econometrics, 40, 183-202.

______ and A. Siow (1980), "Posterior Odds Ratios for Selected Regression Hypotheses," in J.M. Bernardo, M.H. DeGroot, D.V. Lindley, and A.F.M. Smith, editors, Bayesian Statistics, Valencia, Spain: University Press, 585-603.7

C

Prediction and Control

[26]

PREDICTION AND DECISION PROBLEMS IN REGRESSION MODELS FROM THE BAYESIAN POINT OF VIEW*

ARNOLD ZELLNER AND V. KARUPPAN CHETTY
University of Chicago

In this paper we review the derivation of the predictive density function for the normal multiple regression model, state and prove a general theorem on optimal point prediction, and show how the predictive density can be employed in the analysis of an illustrative investment problem. Then we derive the predictive density function for the multivariate normal regression model and indicate how it can be used in the analysis of several problems.

1. INTRODUCTION

IN THIS paper we utilize a Bayesian approach to analyze several prediction and decision problems associated with normal regression models. In particular, for the multiple regression model, we review the derivation of the distribution of the next q observations and show how this distribution can be employed to make inferences about future values of the dependent variable. Then we state a rather general theorem which yields optimal points predictions for quadratic loss functions. Some of these results are utilized in an illustrative analysis of an investment decision. Here we obtain the distribution of the present value of a future income stream and show how it depends on the rate of discount and the length of the time horizon. Next, we generalize some of the results obtained for the multiple regression model to the multivariate normal regression model. We derive the distribution of a future vector observation. The properties of this distribution as well as its use in prediction and decision problems are discussed. We briefly indicate how this distribution can be employed in connection with economic policy, portfolio and farm management problems.

2. REVIEW OF PREDICTION IN THE MULTIPLE REGRESSION MODEL

2.1. *Specification of Model and Distribution of Next q Observations*

We assume that the $T\times 1$ observation vector y is generated by the usual normal multiple regression model

$$y = X\beta + u \tag{2.1}$$

where X is a $T\times K$ matrix with rank K of nonstochastic elements, β is a $K\times 1$ coefficient vector, and u is a $T\times 1$ error vector whose elements have zero means, common variance σ^2, and are normally and independently distributed. We shall further assume that our prior information is suitably represented by locally uniform and independent distributions for the elements of β and $\log\sigma$; cf. Jeffreys (1948), Savage (1961), and Box and Tiao (1962). That is, our prior distributions are:[1]

* This research was supported by the National Science Foundation under Grant GS-151. The authors wish to thank two unknown referees, Clifford Hildreth and John W. Pratt for suggestions which helped to improve the presentation. This paper appeared earlier as Systems Formulation and Methodology Workshop Paper 6403, University of Wisconsin, April, 1964.

[1] For use of non-diffused prior distributions in regression analysis see Raiffa and Schlaifer (1961) and Tiao and Zellner (1964a).

$$p(\beta) \propto \text{const.} \quad \text{and} \quad p(\sigma) \propto 1/\sigma \tag{2.2}$$

with $-\infty < \beta_\alpha < \infty$, $\alpha = 1, 2, \cdots, K$, and $0 < \sigma < \infty$. On combining the prior distributions in (2.2) with the likelihood function for the regression model by means of Bayes' Theorem, we obtain the following well-known posterior distribution of β and σ:

$$p(\beta, \sigma \mid y,) \propto \sigma^{-(T+1)} \exp\left\{-\frac{1}{2\sigma^2}(y - X\beta)'(y - X\beta)\right\}. \tag{2.3}$$

Now let the next q observations, $\tilde{y}' = (y_{T+1}, y_{T+2}, \cdots, y_{T+q})$ satisfy

$$\tilde{y} = \tilde{X}\beta + \tilde{u} \tag{2.4}$$

where $\tilde{X}$ is a $q \times K$ matrix of known nonstochastic elements, β is the $K \times 1$ coefficient vector appearing in (2.1), and $\tilde{u}$ is a $q \times 1$ vector of future errors, normally and independently distributed each with zero mean and common variance σ^2. We have then for the joint distribution of $\tilde{y}$, β, and σ:

$$p(\tilde{y}, \beta, \sigma \mid y) \propto \sigma^{-(T+q+1)} \exp\left\{-\frac{1}{2\sigma^2}(u'u + \tilde{u}'\tilde{u})\right\}.$$

Then on integrating with respect to σ and the elements of β, we obtain the following predictive probability density function for the q elements of $\tilde{y}$:

$$p(\tilde{y} \mid y) = c\{\nu + (\tilde{y} - \tilde{X}\hat{\beta})'H(\tilde{y} - \tilde{X}\hat{\beta})\}^{-(\nu+q/2)} \tag{2.5}$$

where

$$\nu = T - K,\ \hat{\beta} = (X'X)^{-1}X'y,\ H = (1/s^2)[I + \tilde{X}(X'X)^{-1}\tilde{X}']^{-1},\ \nu s^2 = (y - X\hat{\beta})'(y - X\hat{\beta})$$

and

$$c = \nu^{\nu/2}\left(\frac{\nu}{2} + \frac{q}{2} - 1\right)! \, |H|^{1/2}/\pi^{q/2}\left(\frac{\nu}{2} - 1\right)!.$$

The result in (2.5) is completely analogous to that derived in Raiffa and Schlaifer (1961, p. 345), which they call the unconditional distribution of $\tilde{y}$, except that here we work with the diffuse prior distributions in (2.2) rather than a non-diffuse natural conjugate prior distribution.[2] From (2.5), it is seen that the distribution of $\tilde{y}$ is in the multivariate Student t form—see, e.g., Dunnett and Sobel (1954), Cornish (1954), and Raiffa and Schlaifer (1961, p. 256 ff.). This implies that

$$E\tilde{y} = \tilde{X}\hat{\beta}, \tag{2.6}$$

$$E(\tilde{y} - E\tilde{y})(\tilde{y} - E\tilde{y})' = \frac{\hat{u}'\hat{u}}{T - K - 2}[I + \tilde{X}(X'X)^{-1}\tilde{X}'], \tag{2.7}$$

[2] See Theil (1963) and Tiao and Zellner (1964a) for considerations bearing on the use of non-diffuse natural conjugate prior distributions in regression analysis.

and the quantity

$$\frac{\tilde{y}(i) - \tilde{X}(i)\hat{\beta}}{\sqrt{h^{ii}}} \tag{2.8}$$

has a univariate Student t distribution with $T-K$ degrees of freedom where $\tilde{y}(i)$ is the ith element of $\tilde{y}$, $\tilde{X}(i)$ is the ith row of $\tilde{X}$, and h^{ii} is the $(i, i)^{\text{th}}$ element of H^{-1}. Further, from the properties of the multivariate t distribution, linear combinations of the elements of $\tilde{y}$ will be distributed in the multivariate t form —see Raiffa and Schaifer (1961, p. 258). In particular, with $d' = (d_1, d_2, \cdots, d_q)$, a nonstochastic vector, $V = d'\tilde{y}$ has a univariate t distribution with $T-K$ degrees of freedom; that is,

$$t_{T-K} = \frac{V - \hat{V}}{s\{d'[I + \tilde{X}(X'X)^{-1}\tilde{X}']^{-1}d\}^{1/2}} \tag{2.9}$$

where $\hat{V} = E(V) = d'\tilde{X}\hat{\beta}$. This result will be utilized in an illustrative analysis of an investment problem in Section 2.3.

2.2. *Theorem on Optimal Point Predictions*

Let B be any positive definite $q \times q$ matrix; if $\bar{y}$ is a point prediction of $\tilde{y}$, let $L = (\tilde{y} - \bar{y})'B(\tilde{y} - \bar{y})$ be our loss function. Then the point prediction that minimizes EL is $\bar{y} = E\tilde{y} = \tilde{X}\hat{\beta}$.

Proof:

$$\begin{aligned} EL &= E[(\tilde{y} - E\tilde{y}) - (\bar{y} - E\tilde{y})'B(\tilde{y} - E\tilde{y}) - (\bar{y} - E\tilde{y}) \\ &= E(\tilde{y} - E\tilde{y})'B(\tilde{y} - E\tilde{y}) + (\bar{y} - E\tilde{y})'B(\bar{y} - E\tilde{y}). \end{aligned}$$

Given that B is p.d., this last expression will be a minimum for $\bar{y} = E\tilde{y} = \tilde{X}\hat{\beta}$, q.e.d.

Thus, when it is appropriate to employ a quadratic loss function, our theorem tells us that the mean of the distribution of $\tilde{y}$ will be an optimal point predictor, "optimal" in the sense of minimizing our expected loss.[3]

2.3. *Illustrative Application to an Investment Decision*

We assume that the regression model in (2.1) "explains" net returns on an investment. At time T, with sample information covering periods $t = 1, 2, \cdots, T$, an investor wishes to make an inference about the present value of net returns for the next q future periods, $T+1, T+2, \cdots, T+q$. That is, he is concerned about

---

[3] It is apparent that the minimization procedure could also be carried through with inequality restrictions imposed on the elements of $\bar{y}$, in which case we would have a quadratic programming problem to solve.

$$V = \frac{\tilde{y}_{T+1}}{1+r} + \frac{\tilde{y}_{T+2}}{(1+r)^2} + \cdots + \frac{\tilde{y}_{T+q}}{(1+r)^q} \tag{2.10}$$

where $\tilde{y}_{T+i}$ is net return in the ith future period, t is the investor's rate of discount,[4] and V, a random variable, is the present value of the income stream for the next q periods.

Clearly V is a linear combination of the future y's and thus, as pointed out in Section 2.2, will have a univariate Student t distribution. That is, the quantity

$$t = (V - \hat{V})/s\{d'[I_q + \tilde{X}(X'X)^{-1}\tilde{X}']d\}^{1/2} \tag{2.11}$$

has a univariate Student t distribution with $T-K$ degrees of freedom where

$$d' = [(1+r)^{-1}, (1+r)^{-2}, \cdots, (1+r)^{-q}], \qquad \hat{y} = \tilde{X}\hat{\beta},$$
$$\hat{V} = d'\hat{y}, \quad \text{and} \quad s^2 = (T-K)^{-1}(y - X\hat{\beta})'(y - X\hat{\beta}).$$

If an investor has a utility function depending on V, say $U(V)$, its expectation can generally be evaluated given that we have the distribution of V. Also, if several investments are under consideration, the expected utility associated witn each can be computed and used to rank their relative attractiveness.

3. PREDICTION IN THE MULTIVARIATE REGRESSION MODEL

3.1. *Specification of Model and Distribution of Next Vector Observation*

In this Section we consider the traditional multivariate normal regression model

$$y_\alpha = X\beta_\alpha + u_\alpha \qquad \alpha = 1, 2, \cdots, m \tag{3.1}$$

where

y_α is a $T\times 1$ vector of observations on the αth "dependent" variable,
X is a $T\times K$ matrix of nonstochastic elements with rank K,
β_α is a $K\times 1$ regression coefficient vector in the αth equation, and
u_α is a $T\times 1$ vector of error terms.

We assume that the elements of the u_α vectors have zero means and a multivariate normal distribution with covariance matrix $\Sigma\otimes I_T$ where Σ is an $m\times m$ positive definite matrix with elements σ_{α}'. Further, it is assumed that the next vector observation[5] $y'_F = (y_{1F}, y_{2F}, \cdots, y_{mF})$ has elements generated by

$$y_{\alpha F} = X_F\beta_\alpha + u_{\alpha F} \qquad \alpha = 1, 2, \cdots, m \tag{3.2}$$

where X_F is a $1\times K$ matrix of known elements relating to period $T+1$, and the $u_{\alpha F}$ are future error terms, with zero means, distributed normally and independently of the u_α with covariance matrix Σ.

[4] It would be possible to use rates of discount specific to particular future periods if they were thought appropriate.

[5] Note that in this section the symbol y_F denotes a $m\times 1$ vector of observations on m variates for the period $T+1$.

In the present analysis, we follow the invariance theory of Jeffreys (1961, p. 179ff.) and use for our prior distribution of $\beta=(\beta_1, \beta_2, \cdots, \beta_m)$ and Σ

$$p(\beta, \Sigma) = p(\beta)p(\Sigma) \tag{3.3}$$

with

$$p(\beta) \propto \text{constant} \tag{3.4}$$

and

$$p(\Sigma) \propto |\Sigma|^{-(m+1)/2} \tag{3.5}$$

As pointed out in Tiao and Zellner (1964b), (3.5) implies

$$p(\Sigma^{-1}) \propto |\Sigma^{-1}|^{-(m+1)/2}, \tag{3.6}$$

a prior distribution derived and employed by Savage (1961) in the analysis of a set of means; see also Geisser and Cornfield (1963).

On combining these prior distributions with the likelihood function for the model in (3.1), namely,

$$l(\beta, \Sigma^{-1} | y) \propto |\Sigma^{-1}|^{T/2} \exp\{-\tfrac{1}{2} \operatorname{tr} \Sigma^{-1}A\}, \tag{3.7}$$

where $y=(y_1, y_2, \cdots, y_m)$ and

$$A = \{u_\alpha' u_l\}, \qquad \alpha, l = 1, 2, \cdots, m,$$

we obtain the following joint posterior distribution for β and Σ^{-1}:

$$p(\beta, \Sigma^{-1} | y) \propto |\Sigma^{-1}|^{T-(m+1)/2} \exp\{-\tfrac{1}{2} \operatorname{tr} \Sigma^{-1}A\}. \tag{3.8}$$

Further, we have that

$$\begin{aligned} p(y_F, \beta, \Sigma^{-1} | y) &= p(y_F | \beta, \Sigma^{-1}, y)p(\beta, \Sigma^{-1} | y) \\ &\propto |\Sigma^{-1}|^{(T-m)/2} \exp\{-\tfrac{1}{2} \operatorname{tr} \Sigma^{-1}(A + A_F)\} \end{aligned} \tag{3.9}$$

where the matrix A_F is given by,

$$A_F = \{u_{\alpha F} u_{lF}\}, \qquad \alpha, l = 1, 2, \cdots, m. \tag{3.10}$$

For integration over the distinct elements of Σ^{-1}, we utilize the properties of the Wishart distribution, see, e.g., Anderson (1958, p. 154), to obtain

$$\begin{aligned} p(y_F, \beta | y) &\propto |A + A_F|^{-(T+1)/2} \\ &\propto |S + B|^{-(T+1)/2} \end{aligned} \tag{3.11}$$

where $S=\{\bar{u}_\alpha \bar{u}_l\}$ and $B=\{(\beta_\alpha-\hat{\beta}_\alpha-\tilde{\beta}_\alpha)'Z'Z(\beta_l-\hat{\beta}_l-\tilde{\beta}_l)\}$ with $\bar{u}_\alpha=w_\alpha-Z\tilde{\beta}_\alpha$, $w_\alpha' = [(y_\alpha - X\hat{\beta}_\alpha) \vdots y_{\alpha F} - X_F\hat{\beta}_\alpha]$, $Z' = (X' \vdots X_F)$, $\tilde{\beta}_\alpha = (Z'Z)^{-1}Z'w_\alpha$ and $\hat{\beta}_\alpha=(X'X)^{-1}X'y_\alpha$. Then following the analysis in Tiao and Zellner (1964b), we find

$$\begin{aligned} p(y_F, \beta | y) \propto\ & |\bar{S}|^{-1/2} |\bar{S} + \bar{B}|^{-T/2} |Z'D_mZ|^{-1/2} \\ & \cdot\{c_m + (\beta_m - \eta_m)'Z'D_mZ(\beta_m - \eta_m)\}^{-(T+1)/2} \end{aligned} \tag{3.12}$$

where $\overline{S}+\overline{B}$ is the $(m-1)\times(m-1)$ matrix formed from $S+B$ by deleting the mth column and row, c_m is the reciprocal of the $(m, m)^{\text{th}}$ element of S^{-1},

$$D_m = I - \gamma(\overline{S} + \overline{B})^{-1}\gamma', \quad \gamma = (\gamma_1, \gamma_2, \cdots, \gamma_{m-1}), \quad \gamma_\alpha = Z\,(\beta_\alpha - \hat{\beta}_\alpha - \tilde{\beta}_\alpha),$$

$$\eta_m = \hat{\beta}_m + \tilde{\beta}_m + d\overline{S}^{-1}s, \quad d = (\beta_1 - \hat{\beta}_1 - \tilde{\beta}_1 \cdots \beta_{m-1}\,\hat{\beta}_{m-1} - \tilde{\beta}_{m-1})$$

and s is the mth column of S excluding the last element. Noting that (3.12) is in the multivariate t form, the integration over β_m can be performed. Then, by similar operations, the other elements of β can be integrated out to yield

$$p(y_F\,|y)\alpha \qquad \left|\{\hat{u}'_\alpha\,\hat{u}_e\} + q\{\hat{u}_{\alpha F}\,\hat{u}_{eF}\}\right|^{-(T-1-K)/2} \tag{3.13}$$

where $\hat{u}_{\alpha F}=y_{\alpha F}-X_F\hat{\beta}_\alpha$, $\hat{u}_\alpha=y_\alpha-X\hat{\beta}_\alpha$, and $q=[1+X_F(X'X)^{-1}X_F']^{-1}$. By some further algebra, the density function in (3.13) can be expressed as follows:[6]

$$\begin{aligned} p(y_F\mid y) \propto \{\nu + (y_{1F} - X_F\hat{\beta}_1 \cdots y_{mF} - X_F\hat{\beta}_m) \\ H(y_{1F} - X_F\hat{\beta}_1 \cdots y_{mF} - X_F\hat{\beta}_m)\}^{-(\nu+m)/2} \end{aligned} \tag{3.14}$$

where $\nu=T-K-(m-1)$ and $H=\nu q\{\hat{u}'_\alpha\hat{u}_l\}$. Then it is seen that the m elements of y_F are distributed in the multivariate t form which implies:

$$Ey_{\alpha F} = X_F\hat{\beta}_\alpha \qquad \alpha = 1, 2, \cdots, m \tag{3.15}$$

and

$$\begin{aligned} E(y_{\alpha F} - X_F\hat{\beta}_\alpha)(y_{lF} - X_F\hat{\beta}_l) &= \frac{\nu}{\nu-2}H^{-1} \\ &= \frac{1}{\nu-2}\{\hat{u}_\alpha'\hat{u}_e\}[1 + X_F(X'X)^{-1}X_F']. \end{aligned} \tag{3.16}$$

Also, as is well-known, the marginal distribution of a subset of the elements of y_F will be in the multivariate t form and that of a single element will be in the univariate t form—cf., e.g., Raiffa and Schlaiffer (1961, p. 258). In addition, as mentioned above, linear combinations of variables distributed in the multivariate t form will themselves be so distributed. In particular, for nonstochastic n_α, $\alpha=1, 2, \cdots, m$, the quantity W given by

$$W = n_1y_{1F} + n_2y_{2F} + \cdots + n_my_{mF} \tag{3.17}$$

is distributed in the univariate t form; that is, the quantity

$$t = \frac{W - \hat{W}}{[n'H^{-1}n]^{1/2}} \tag{3.18}$$

has a univariate t distribution with $\nu=T-K-(m-1)$ degrees of freedom where

$$\hat{W} = EW = \sum_{\alpha=1}^{m} n_\alpha X_F\hat{\beta}_\alpha \quad \text{and} \quad n' = (n_1n_2, \cdots, n_m).$$

[6] After this result was obtained and reported in the University of Wisconsin Systems Formulation and Methodology Workshop Paper 6403, April 20, 1964, alternative methods for deriving it have been suggested by a referee and friends. Also cf. Geisser (1964) for another derivation reported in a manuscript dated August, 1964.

4. DISCUSSION OF RESULTS AND APPLICATIONS

The distribution of y_F in (3.14) with mean vector and covariance matrix as given in (3.15) and (3.16), respectively, can be employed to make inferences about the "next" vector observation for multivariate normal regression models such as certain "unrestricted reduced form" systems or "reduced form" systems associated with "just identified" econometric models.[7] For these systems, this distribution will also play an important role in analyzing decision problems. That is, if $\bar{y}_F$ represents a set of known nonstochastic target values, as defined by Tinbergen (1956), for y_F, we clearly can determine the distribution of $y_F-\bar{y}_F$ from $p(y_F|y)$ in (3.19). Further, if we have a loss function which depends on $y_F-\bar{y}_F$, it seems possible to determine the X_F vector which minimizes the mathematical expectation of a loss function—cf. Theil (1961a), Fisher (1962), and the references cited in these works for some previous work on this problem. To illustrate how this problem can be approached utilizing our results, we consider a quadratic loss function $L=(y_F-\bar{y}_F)'G(y_F-\bar{y}_F)$, where $G=\{g_{\alpha l}\}$ is positive definite with known elements. Then,

$$
\begin{aligned}
EL &= \int (y_F-\bar{y}_F)'G(y_F-\bar{y}_F)p(y_F\mid y)dy_F \\
&= \int (y_F-Ey_F)'G(y_F-Ey_F)p(y_F\mid y)dy_F + (\bar{y}_F-Ey_F)'G(\bar{y}_F-Ey_F) \\
&= \int \Big[\sum_{\alpha,l} g_{\alpha l}(y_{\alpha F}-Ey_{\alpha F})(y_{lF}-Ey_{lF})\Big] p(y_F|y)dy_F \qquad (3.19) \\
&\quad + \sum_{\alpha,l} g_{\alpha l}(\bar{y}_{\alpha F}-Ey_{\alpha F})(\bar{y}_{lF}-Ey_{lF}) \\
&= \sum_{\alpha,l} g_{\alpha l}\{\bar{s}_{\alpha l}[1+X_F(X'X)^{-1}X_F'] + (\bar{y}_{\alpha F}-X_F\hat{\beta}_\alpha)(\bar{y}_{lF}-X_F\hat{\beta}_l)\}
\end{aligned}
$$

where (3.15) and (3.16) have been used and $\bar{s}_{\alpha l}\equiv\hat{u}_\alpha\hat{u}_l/(\nu-2)$. Then, on differentiating EL with respect to X_F we obtain

$$\frac{\partial EL}{\partial X_F} = \sum_{\alpha,l} g_{\alpha l}\{2\bar{s}_{\alpha l}(X'X)^{-1}+2\hat{\beta}_\alpha\hat{\beta}_l'\}X_F' - \sum_{\alpha,l} g_{\alpha l}(\hat{\beta}_\alpha\bar{y}_{lF}+\hat{\beta}_l\bar{y}_{\alpha F}).$$

The value of X_F which sets this derivative equal to zero, say $\overline{X}_F$, is given by

$$\overline{X}_F = \frac{1}{2}\Big[\sum_{\alpha,l} g_{\alpha l}\{\bar{s}_{\alpha l}(X'X)^{-1}+\hat{\beta}_\alpha\hat{\beta}_l'\Big]^{-1}\sum_{\alpha,l} g_{\alpha l}(\hat{\beta}_\alpha\bar{y}_{lF}+\hat{\beta}_l\bar{y}_{\alpha F}) \qquad (3.20)$$

which can be computed easily given that the $g_{\alpha l}$ are known.

Another result which seems useful in connection with applications is that given in equation (3.18) which gives the distribution of the linear combination of next period's y's shown in (3.17). If $y_{\alpha F}$ in (3.17) is interpreted as the net yield per dollars invested in the αth investment alternative and n_α as the number of dollars invested in the αth alternative, then W is the return of an investor's

[7] For results pertaining to the prediction problem for "fully recursive" models, see Zellner (1964).

portfolio in the future period $T+1$. Since the distribution of W is known for given $n_1, n_2, \cdots, n_m$, it is not difficult to compute the distribution of W for various sets of n's and thus give an investor a chance to observe the distributions of the future return of his portfolio as its composition is varied. Further, if the investor has fixed resources, say W_T, then the n's are constrained to satisfy

$$W_T = \sum_{\alpha=1}^{m} n_\alpha.$$

If one postulates a utility function for the investor, $U(W)$, it appears possible to find the n's which maximize $EU(W)$ subject to this constraint.

$$W_T = \sum_{\alpha=1}^{m} n_\alpha.$$

Another possible application of our results is suggested if we regard $y_{\alpha F}$ as the yield per acre of the αth crop in period $T+1$ and write

$$R = n_1 p_1 y_{1F} + n_2 p_2 y_{2F} + \cdots + n_m p_m y_{mF} \qquad (3.21)$$

where p_α is the net unit price for the αth crop, assumed known, and n_α is the number of acres devoted to the αth crop. Then R is the net return to a farmer and is a random variable since it depends on the random, as yet unobserved, $y_{\alpha F}$, $\alpha=1, 2, \cdots, m$. Given a choice of the n's, R is distributed in the univariate t form and thus the distribution of R can be studied as a function of the choice of the n's. If the farmer has a fixed number of acres, say N, then of course we must have

$$N = \sum_{\alpha=1}^{m} n_\alpha.$$

Further if one posits a utility function for the farmer, say U(R), it seems possible to find the n's which maximize E U(R) subject to the aforementioned constraint[8] and $n_\alpha \geq 0$, $\alpha=1, 2, \cdots, m$.

REFERENCES

Anderson, T. W. *An Introduction to Multivariate Statistical Analysis*, New York: J. Wiley and Sons, Inc., 1958.

Box, G. E. P. and Tiao, G. C. "A Further Look at Robustness via Bayes' Theorem," *Biometrika*, 49, 1962, 419–32.

Cornish, E. A. "The Multivariate t-Distribution Associated with a Set of Normal Sample Deviates," *Australian Journal of Physics*, 7, 1954, 531–542.

Dunnett, C. W. and Sobel, M. "A Bivariate Generalization of Student's t-Distribution with Tables for Special Cases," *Biometrika*, 41, 1954, 153–69.

Fisher, W. D. "Estimation in the Linear Decision Model," *International Economic Review*, 3, 1962, 1–29.

Geisser, S. and Cornfield, J. "Posterior Distribution for Multivariate Normal Parameters," *Journal of the Royal Statistical Society*, Series B, 25, 1963, 368–76.

[8] This problem has been formerly considered by Tintner (1955), Theil (1961b), and others from other points of view.

Geisser, S. "Bayesian Estimation in Multivariate Analysis," U. S. Army Mathematics Research Center, University of Wisconsin, Report No. 503, August, 1964.

Jeffreys, H. *Theory of Probability*, (2nd ed.), Oxford: Clarendon Press 1948.

——— *Theory of Probability*, (3rd ed.), Oxford: Clarendon Press 1961.

Raiffa, H. and Schlaifer, R. *Applied Statistical Decision Theory*, Harvard University, 1961.

Savage, L. J. *The Subjective Basis of Statistical Practice*, Manuscript, University of Michigan, 1961.

Theil, H. *Economic Forecasts and Policy*, (2nd ed.), Amsterdam: North-Holland Publishing Co., 1961.

——— "Some Reflections on Static Programming Under Uncertainty," *Weltwirtschafliches Archiv*, 87, 1961, 124–38.

——— "On the Use of Incomplete Prior Information in Regression Analysis," *Journal of the American Statistical Association*, 58, 1963, 401–14.

Tiao, G. C. and Zellner, A. "Bayes's Theorem and the Use of Prior Knowledge in Regression Analysis," *Biometrika*, 44, 1964a, 315–27.

Tiao, G. C. and Zellner, A. "On the Bayesian Estimation of Multivariate Regression," *Journal of the Royal Statistical Society*, Series B, 26, 1964b, 277–85.

Tinbergen, J. *Economic Policy: Principles and Design*, Amsterdam: North-Holland Publishing Co., 1956.

Tintner, G. "Stochastic Linear Programming with Application to Agricultural Economics," in *Proceedings of the Second Symposium on Linear Programming*, Washington, D.C., 1, 1955, 197–228.

Zellner, A. "On Prediction in 'Fully Recursive' Econometric Models," Systems Formulation and Methodology Workshop Paper 6402, University of Wisconsin 1964).

[27]

Reprinted from
THE FUTURE OF STATISTICS

SENSITIVITY OF CONTROL TO UNCERTAINTY AND FORM OF THE CRITERION FUNCTION*

Arnold Zellner and Martin S. Geisel
UNIVERSITY OF CHICAGO

1. Introduction

In a good deal of the literature on control problems in economics, control problems are solved under the assumptions that (1) the criterion function is quadratic and (2) the parameters of the process generating the data are known exactly.† Since there are many situations in which criterion functions are not quadratic and parameter values are unknown and have to be estimated, we became interested in exploring the sensitivity of solutions to control problems to the form of the criterion function and the state of our information about process parameters. The present paper contains the results of computations relating to these issues for a simple control problem.

2. Specification of Model and Problem

The process assumed to generate our data is a simple normal regression model, namely,

$$y_i = \beta x_i + u_i, \qquad i = 1, 2, \ldots, T \tag{1}$$

where y_i is the ith observation on the dependent variable, x_i is the ith value of a control variable in the sample, β is an unknown scalar parameter, and u_i is the ith unobserved random error term. We assume that the u_i's are normally and independently distributed, each with mean zero and common unknown variance σ^2. Further, we assume that we have little prior information about

* Research financed by the National Science Foundation under Grant GS-1350.
† See for example Simon (1956), Theil (1957), and Holt *et al.* (1960).

β and σ and represent this by assuming β and $\log \sigma$ are uniformly and independently distributed, that is ,†

$$p(\beta, \sigma) \propto 1/\sigma; \qquad -\infty < \beta < \infty, \quad 0 < \sigma < \infty. \tag{2}$$

The control problem which we consider is the following one. In the first future period, $i = T + 1$, we wish to set $w \equiv x_{T+1}$ so that $z \equiv y_{T+1}$ will be close to a given target value, denoted by a. The loss functions which we consider are given in (3) and (4):

$$L(z, a) = |z - a|^{\alpha}, \qquad \alpha = \tfrac{1}{2}, 1, 2, \text{and } 4 \tag{3}$$

and

$$L(z, a) = \begin{cases} k\,|z - a| & \text{for} \quad z \geq a \\ \ \ |z - a| & \text{for} \quad z < a \end{cases} \qquad \begin{matrix} k = 0.25, 0.50, 0.75, 1.0, \\ 1.5, 2.0, 3.0. \end{matrix} \tag{4}$$

In (3) we have a class of symmetric loss functions. The particular members which we use in our work include the square root of the absolute error, the absolute error, the squared error, and the error raised to the fourth power. In (4) we have a class of linear loss functions. For $z < a$, the loss is the absolute error while for $z \geq a$, the loss is given by $k|z - a|$ with k assigned the values shown in (4).

The solution to the control problem stated above will consist of finding a value of $w \equiv x_{T+1}$, such that the mathematical expectation of the loss function is minimal. The loss function $L(z, a)$ is random, since z is a random variable and its mathematical expectation is given by

$$EL(z, a) = \int_{-\infty}^{\infty} L(z, a) p(z \mid \underset{\sim}{y}, w)\, dz \tag{5}$$

where $p(z \mid \underset{\sim}{y}, w)$ is the predictive probability density function (pdf) for $z \equiv y_{T+1}$ given the sample observations $\underset{\sim}{y}$ and the setting of the control variable $w \equiv x_{T+1}$. For the present simple regression model with the prior assumptions in (2), the predictive pdf is given by‡

$$p(z \mid \underset{\sim}{y}, w) = \frac{\Gamma(m)}{\Gamma\left(\frac{1}{2}\right)\Gamma\left(\frac{2m-1}{2}\right)} \frac{g^{1/2}}{v^{1/2}} \left\{1 + \frac{(z - w\hat{\beta})^2 g}{v}\right\}^{-m} \tag{6}$$

† Diffuse improper prior pdfs like this one have been employed in many works, including Jeffreys (1961), Lindley (1965), Box and Tiao (1962), and Savage (1961). It would not be difficult to pursue the present analysis with "informative" prior pdfs in the "normal-gamma" form.

‡ See for example Raiffa and Schlaifer (1961), Geisser (1965), and Zellner and Chetty (1965).

where

$$\nu = T - 1, \qquad m = \frac{(\nu + 1)}{2}, \qquad \hat{\beta} = \frac{\sum_{i=1}^{T} x_i y_i}{\sum_{i=1}^{T} x_i^2},$$

$$g = \left\{ s^2 \left(1 + \frac{w^2}{\sum_{i=1}^{T} x_i^2} \right) \right\}^{-1}, \qquad \nu s^2 = \sum_{i=1}^{T} (y_i - \hat{\beta} x_i)^2.$$

As is well known, the predictive pdf in (6) is in the form of a student t pdf with mean $w\hat{\beta}$. Since g depends on w, the "spread" of the pdf, as well as its mean, depends on the value of the control variable w.

With $p(z \mid \underset{\sim}{y}, w)$ as shown in (6), it is the case that the expected loss, shown in (5), will depend on w. Then our problem is

$$\min_{w} EL(z, a). \tag{7}$$

The w solving this problem, say w^*, is the optimal control setting which can be inserted in (5) to compute the expected loss associated with the optimal setting. For the quadratic loss function $(z - a)^2$, the solution to this problem can easily be obtained analytically. It is†

$$w^* = \frac{a}{\hat{\beta}} \left(\frac{1}{1 + (1/t^2)} \right) \tag{8}$$

and

$$E(L \mid w = w^*) = \frac{\nu}{\nu - 2} s^2 + \frac{a^2}{1 + t^2}, \tag{9}$$

where‡

$$t^2 = \frac{\hat{\beta}^2 m_{xx}}{\bar{s}^2} \quad \text{with} \quad m_{xx} = \sum_{i=1}^{T} x_i^2, \quad \bar{s}^2 = \nu s^2/(\nu - 2).$$

It is interesting to observe that in (8) w^* is approximately equal to the target value, a, divided by the sample quantity $\hat{\beta}$, $a/\hat{\beta}$. This solution can be arrived at by substituting $w\hat{\beta}$ for z in $L = (z - a)^2$ and minimizing with respect to w, a procedure yielding the so-called "certainty equivalence" solution, say w_{ce}. Since $w^* \neq w_{ce}$, we should expect that our loss will be greater if we use w_{ce}

† The result in (8) is essentially the same as that reported by Fisher (1962), except that here we have assumed σ^2 unknown.

‡ Note that the quantity t is $[(\nu - 2)/\nu]^{1/2}$ times the usual t-statistic used to test the hypothesis that $\beta = 0$ in the sampling theory framework.

rather than w^*. In fact, on inserting $w = w_{ce}$ in (5) and evaluating the integral, it is found that

$$E(L \mid w = w_{ce}) = \frac{\nu}{\nu - 2} s^2 + \frac{a^2}{t^2} \tag{10}$$

which is indeed larger than the expected loss associated with the use of $w = w^*$, given in (9). The absolute gain in using w^* rather than w_{ce} is given by

$$E(L \mid w = w_{ce}) - E(L \mid w = w^*) = \frac{a^2}{t^2(1 + t^2)} \tag{11}$$

which is seen to depend on both a^2 and t^2.

3. Computations with Generated Data

While it is possible to obtain neat analytical results for quadratic loss functions without much difficulty, it is not very easy to get analytical solutions for the other loss functions shown above. Therefore, we decided to employ two sets of given data and to use numerical methods to get solutions to the control problem under various conditions.

The first set of data† was generated from the model shown in (1), with parameter values $T = 15$, $\beta = 2.0$, and $\sigma^2 = 9$. Further, the x's were drawn independently from a normal distribution with mean 0 and variance 0.64. A least squares fit yields

$$\underset{(.8259)}{\hat{y}_i = 1.5885 x_i} \tag{12}$$

where 1.5885 is the sample quantity $\hat{\beta} = \sum_{i=1}^{T} x_i y_i / \sum_{i=1}^{T} x_i^2$, 0.8259 is the square root of the sample quantity $s^2/\sum_{i=1}^{T} x_i^2$, and $s^2 = 5.7351$. With these sample quantities computed, the parameters of the predictive pdf $p(z \mid y, w)$ in (6) are known. Then, given a particular loss function, the integral giving expected loss in (5) was evaluated numerically for various values of w in a search for the minimizing or optimal value, say w^*.

The results of the above computations for the symmetric loss functions in (3), that is $L(z, a) = |z - a|^\alpha$, with $a = 4.0$, are shown in Table I(a)–(b) and graphically in Fig. 1. We see from Table I(a) that when our loss function is $|z - 4.0|^{1/2}$, the minimal expected loss, 1.4632, is obtained for $w = 1.9812$. As the exponent in the loss function assumes larger values, as might be anticipated, expected losses grow in absolute value and the optimal setting for w

† The data are given in the appendix.

TABLE I(a)

ABSOLUTE EXPECTED LOSS AS A FUNCTION OF THE CONTROL VARIABLE SETTING AND FORM OF THE LOSS FUNCTION

Value of control variable, w	Form of loss function			
	$\|z-4\|^{0.5}$	$\|z-4\|$	$\|z-4\|^2$	$\|z-4\|^4$
3.6780	1.7432	3.6031	20.853	1457.7
2.9895	1.5789	2.9680	14.370	731.98
2.5181[a]	1.4992	2.6783	11.743	496.01
2.1751	1.4681	2.5666	10.759	412.52
2.1035	1.4651	2.5554	10.652	402.44
2.0364	1.4636	2.5490	10.582	395.05
2.0045	1.4632	2.5473	—	—
1.9812	1.4632[c]	2.5467	—	—
1.9735	1.4632	2.5466	10.544	389.85
1.9584	1.4633	2.5465[c]	—	—
1.9144[b]	1.4638	2.5475	10.533[c]	386.42
1.8587	1.4653	2.5513	10.543	384.44
1.8061	1.4675	2.5576	10.572	383.67
1.7960	—	—	—	383.64[c]
1.7565	1.4704	2.5658	10.615	383.89
1.7095	1.4737	2.5757	10.672	384.93
1.5442	1.4906	2.6271	10.987	394.90
1.4080	1.5104	2.6887	11.383	410.36

[a] $w = 4/\hat{\beta}$, certainty equivalence solution for $|z-4|^2$

[b] $w = (4/\hat{\beta})[t^2/(1+t^2)]$, optimal solution for $|z-4|^2$.

[c] Minimal expected loss for given loss function.

gets smaller. For example, with the quartic loss function the minimum expected loss is about 383.6 and the associated optimal setting is given by $w = 1.7960$. Clearly as a given error becomes more costly, it pays to be more "conservative" in setting w.

In Table I(b), the results in Table I(a) are put on a relative basis. For each loss function, the expected losses associated with particular settings for w are expressed as a percentage of the minimal expected loss for that loss function. For example, with the quadratic loss function $|z-4|^2$, the minimal expected loss, labeled 100.0, is associated with the control variable setting $w = 1.9144$. On the other hand, the setting $w = 2.5181$, which is the certainty equivalence solution, leads to an expected loss for the quadratic loss function which is about 1.11 times the minimal expected loss associated with the optimal solution. From Table I(b), it is clear that use of the certainty equivalence setting for w can be costly in terms of expected loss, particularly if the loss function is quartic. On the other hand, the use of the optimal w for the

TABLE I(b)

RELATIVE EXPECTED LOSS AS A FUNCTION OF THE CONTROL VARIABLE SETTING AND FORM OF LOSS FUNCTION

Value of control variable, w	Form of loss function			
	$\|z-4\|^{0.5}$	$\|z-4\|$	$\|z-4\|^2$	$\|z-4\|^4$
3.6780	119.1	141.5	198.0	380.0
2.9895	107.9	116.6	136.4	190.8
2.5181[a]	102.5	105.2	111.5	129.3
2.1751	100.3	100.8	102.1	107.5
2.1035	100.1	100.3	101.1	104.9
2.0364	100.0	100.1	100.5	103.0
2.0045	100.0	100.0	—	—
1.9812	100.0[c]	100.0	—	—
1.9735	100.0	100.0	100.1	101.6
1.9584	100.0	100.0[c]	—	—
1.9144[b]	100.0	100.0	100.0[c]	100.7
1.8587	100.1	100.2	100.1	100.2
1.8061	100.3	100.4	100.4	100.0
1.7960	—	—	—	100.0[c]
1.7565	100.5	100.8	100.8	100.1
1.7095	100.7	101.1	101.3	100.3
1.5442	101.9	103.2	104.3	102.9
1.4080	103.2	105.6	108.1	107.0

[a] $w = 4/\hat{\beta}$, certainty equivalence solution for $|z-4|^2$.
[b] $w = (4)\hat{\beta}([t^2/(1+t^2)]$, optimal solution for $|z-4|^2$.
[c] Minimal expected loss, rescaled to equal 100.0, for each loss function.

quadratic case,

$$w = \frac{4}{\hat{\beta}} \frac{t^2}{1+t^2} = 1.9144,$$

is strikingly robust with respect to variation in the form of the loss function. If this setting is used when the loss function is quartic, there is just about a 0.7% greater loss than associated with the use of the optimal setting for the quartic case, namely, $w = 1.7960$. Of course, it should be realized that this finding occurs with respect to symmetric loss functions and, as shown below, will not in general be the case with asymmetric loss functions.

Above we have concentrated attention on how variation in the exponent of the loss function affects optimal control settings and expected losses. We now permit the target value a as well as the exponent to vary, and tabulate the ratio of expected loss associated with the use of the optimum w to expected

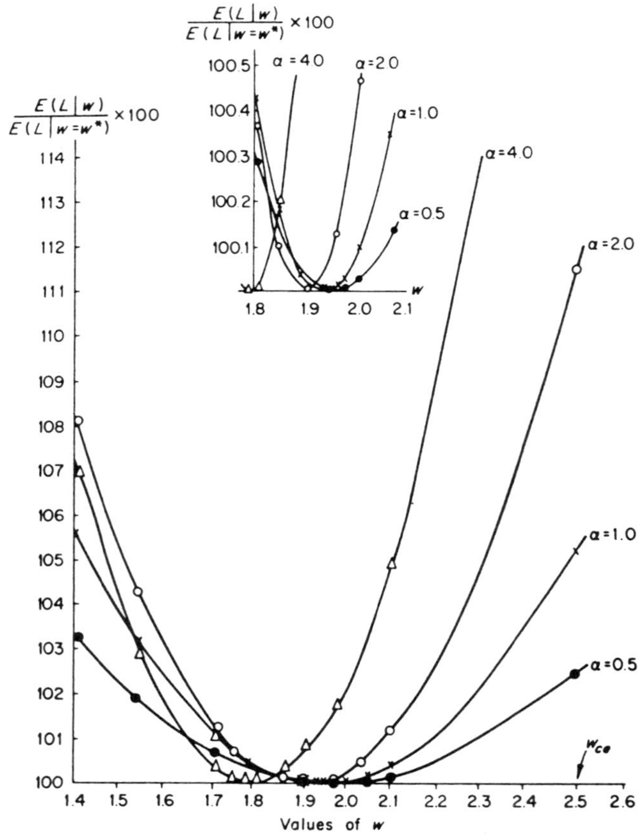

Fig. 1

loss associated with the use of the optimum w for the quadratic loss function and to that associated with the certainty equivalence solution. The results are shown in Tables II(a) and (b). Again it is seen that the optimal setting for the quadratic case is remarkably robust with respect to variation of a. However use of the certainty equivalence setting for w results in appreciable losses relative to those associated with the optimal settings. In particular, as the parameter a grows in size, the quality of the certainty equivalence solution deteriorates. For example, with $a = 5.0$ and with the loss function $|z - 5|^4$, the loss associated with the optimal solution is about 72% of the loss associated with the certainty equivalence solution. The fact that the certainty equivalence solution deteriorates as the parameter a grows larger is due to the

TABLE II(a)

RATIO OF EXPECTED LOSS ASSOCIATED WITH OPTIMAL SOLUTION TO EXPECTED LOSS ASSOCIATED WITH USE OF OPTIMAL SETTING OF w FOR THE QUADRATIC LOSS FUNCTION

Value of a	Form of loss function			
	$\|z-a\|^{0.5}$	$\|z-a\|$	$\|z-a\|^2$	$\|z-a\|^4$
3.0	0.9998	0.9998	1.0	0.9959
4.0	0.9996	0.9996	1.0	0.9928
5.0	0.9993	0.9994	1.0	0.9892

TABLE II(b)

RATIO OF EXPECTED LOSS ASSOCIATED WITH OPTIMAL SOLUTION TO EXPECTED LOSS ASSOCIATED WITH USE OF CERTAINTY EQUIVALENCE SETTING FOR w

Value of a	Form of loss function			
	$\|z-a\|^{0.5}$	$\|z-a\|$	$\|z-a\|^2$	$\|z-a\|^4$
3.0	0.9835	0.9661	0.9286	0.8410
4.0	0.9760	0.9508	0.8969	0.7735
5.0	0.9696	0.9379	0.8704	0.7185

fact that, with a large, the control variable w generally assumes values far from the origin and the spread of the predictive pdf is substantial relative to its spread for small a. Since the certainty equivalence approach does not take account of second and higher moments, it does not incorporate allowance for the greater spread of the predictive pdf associated with large values of w.

Above we have considered several symmetric loss functions. In some problems, there is no difficulty in selecting a specific loss function. However, in quite a few situations, it may be possible to argue that the loss function is symmetric but difficult to establish in its exact form. In such situations, one approach is to use a loss function of the form $L(z, a) = |z - a|^\alpha$ with a prior pdf for the parameter α, say $p(\alpha)$. Then the solution to the one period control problem is the solution to the following problem:

$$\min_w EL(z, a) = \min_w \iint |z - a|^\alpha p(\alpha) p(z \mid w, \underline{y})\, d\alpha\, dz. \tag{13}$$

As an example, suppose that a priori we find it reasonable to limit α to the range 0 to 6 and believe that the mean of α is 2. One pdf which has these properties is the following beta pdf:

$$p(\alpha) = \frac{\Gamma(p+q)}{\Gamma(p)\Gamma(q)}\left(\frac{\alpha}{6}\right)^{p-1}\left(1-\frac{\alpha}{6}\right)^{q-1} \tag{14}$$

with $p = 3$ and $q = 6$. Given that (14) is deemed a reasonable representation of a priori views regarding possible values of α, we can substitute from (14) into (13) and then determine the value of w that minimizes expected loss. For the data generated, the predictive pdf in (6), and with $a = 4.0$, we find the optimal control setting to be $w^* = 1.8061$ and the expected loss associated with this setting to be 44.78. Since our pdf for α permits values of α up to 6, the optimal control setting is pushed far in the direction of the optimal setting for the quartic loss function case. Further, on comparison with the expected loss associated with the quadratic loss function case, namely, 10.53, it is evident that uncertainty about the form of the loss function, which we have represented by (14), can be costly in terms of expected loss.

To consider this last point explicitly, write the expression for expected loss as follows:

$$E_{z,\alpha}L(z, \alpha, a) = \int\left[\int_0^c L(z, \alpha, a)p(\alpha)\, d\alpha\right]p(z \mid w, y)\, dz. \tag{15}$$

For our special case, the integral within brackets, say I, is given by

$$I = \int_0^c |z - a|^\alpha p(\alpha)\, d\alpha$$

$$= |z - a|^{\bar{\alpha}} \int_0^c |z - a|^{\alpha - \bar{\alpha}} p(\alpha)\, d\alpha,$$

when $\bar{\alpha} = \int_0^c \alpha p(\alpha)\, d\alpha$, the mean of α. Since for any given z, Jensen's inequality gives

$$\int_0^c |z - a|^\alpha\, p(\alpha)\, d\alpha \leq |z - a|^{\bar{\alpha}},$$

it follows that

$$E_{z,\alpha}L(z, \alpha, a) \geq E_z L(z, \bar{\alpha}, a). \tag{16}$$

Therefore, allowing α to be random in our loss function will lead to our expected loss being greater, or at best the same, as the loss experienced when we use a loss function with α set at its mean value. If we are indeed uncertain about the value of α and represent this state of affairs by choice of a pdf for $p(\alpha)$, there is of course no justification for setting α equal to its mean value and then solving our control problem with $\alpha = \bar{\alpha}$. As shown above, this leads to an incorrect setting for the control variable and an incorrect value for the associated expected loss.

We now turn from a consideration of symmetric loss functions to a consideration of the linear loss functions shown in (4). In Table III, we provide

TABLE III

COMPARISON OF OPTIMAL SOLUTIONS FOR ASYMMETRIC LINEAR LOSS FUNCTIONS WITH RESULTS OF EMPLOYING CERTAINTY EQUIVALENCE AND QUADRATIC CONTROL SETTINGS[a]

Slope parameter loss function[b]	Optimal setting $w = w^*$	Expected loss for $w = w^*$	Quadratic setting $w = w_q$	Expected loss for $w = w_q$	Certainty equivalence setting $w = w_{ce}$	Expected loss for $w = w_{ce}$
$k = 0.25$	3.4778	1.5518	1.9144	1.9519	2.5181	1.6739
$k = 0.50$	2.6214	2.0059	1.9144	2.1504	2.5181	2.0087
$k = 0.75$	2.2321	2.3131	1.9144	2.3489	2.5181	2.3435
$k = 1.0$	1.9584	2.5465	1.9144	2.5475	2.5181	2.6783
$k = 1.5$	1.6226	2.8937	1.9144	2.9446	2.5181	3.3478
$k = 2.0$	1.3928	3.1511	1.9144	3.3416	2.5181	4.0174
$k = 3.0$	1.0944	3.5272	1.9144	4.1358	2.5181	5.3565

[a] Computations based on generated data and predictive pdf in (6).

[b] The loss function employed is

$$L = \begin{cases} |z-4| & \text{for } -\infty < z \leq 4 \\ k|z-4| & \text{for } 4 < z < \infty. \end{cases}$$

the results of calculations giving the optimal control settings and the expected losses associated with them for a number of linear loss functions, using the data which we generated and the predictive pdf in (6). The target value in the loss function is 4.0. For $-\infty < z \leq 4$, the loss is $|z - 4.0|$, while for $4 < z < \infty$, the loss is $k\,|z - 4.0|$, where k is a slope parameter which has been assigned the values shown in the first column of Table III. Also shown in Table III are the control settings and losses for the certainty equivalence approach and for the solution of the control problem with a quadratic loss function, $L = |z - 4.0|^2$. These results give us an opportunity to assess how incorrect assumptions about the form of loss functions affect results.

With respect to optimal solutions for linear loss functions, the setting of w is quite sensitive to asymmetry of the loss function with optimal settings having the property of tending to keep costly outcomes or values of z from occurring. For example, with $k = 0.25$, values of z larger than 4 are much less costly than values less than 4, and the large value for $w = 3.4778$ reflects this fact. On the other hand, with $k = 3.0$, having z larger than 4 is quite costly, and the optimal value for $w = 1.0944$ is accordingly small.

As regards assessing the effects of the error of working with a symmetric quadratic loss function, when in fact the loss function is linear and asymmetric, the results in Table III show that this can indeed at times be a costly error. For example, let us consider the first line of Table III. If we employ a quadratic loss function, the optimal setting for w is 1.9144, whereas the optimal setting for the linear loss function with $k = 0.25$ is 3.4778. Using this linear loss function and setting $w = 1.9144$, the quadratic solution yields an expected loss equal to 1.9519. This constrats with the expected loss associated with the optimal w for this linear loss function, $w = 3.4778$, which yields an expected loss equal to 1.5518. Comparison of these expected losses, 1.95 and 1.55, indicates that it can be costly to use a symmetric quadratic loss function when in fact the loss function is linear and quite asymmetric.

It is also interesting to compare the relative performance of the approximate certainty equivalence and exact solutions for the quadratic case. For $k = 0.25$, the certainty equivalence solution is better in terms of expected loss than is the exact solution for the quadratic case. The reason for this is that the certainty equivalence control setting, w_{ce}, is larger than the control setting coming from the exact solution for the quadratic case, w_q, and closer to the optimal setting for $k = 0.25$. On the other hand, for $k = 3.0$, the situation is reversed, with w_q being closer to the optimal control setting than w_{ce}, and here the certainty equivalence solution w_{ce} yields much higher expected loss.

4. Control with a Simple Money Multiplier Model

To illustrate the practical relevance of some of the points made above, we turn to the analysis of a simple money multiplier model,

$$y_i = \beta x_i + u_i, \qquad i = 1, 2, \ldots, T, \tag{17}$$

where $y_i = Y_i - Y_{i-1}$ is the annual change in the Friedman-Meiselman (1963) income variable, $x_i = M_i - M_{i-1}$ the annual change in their money supply variable, β the "money multiplier," and u_i a disturbance term with $Eu_i = 0$ and $Eu_i^2 = \sigma^2$, $i = 1, 2, \ldots, T$. Further we assume that the u_i's are normally and independently distributed and that the x_i's are under control. Our problem is to implement (17) with data 1921–1929 for the U.S.† and then to determine the 1930 setting for $w \equiv x_{T+1}$, the change in the money supply, so as to have $z \equiv y_{T+1}$, the change in income, "close" to 10 billion dollars.

† The annual data were taken from Friedman and Meiselman (1963). Their money variable is currency in the hands of the public plus all commercial bank deposits while their income variable is consumption expenditures plus their autonomous quantity A.

With annual data 1921–1929, we have, for the least squares regression of y_i on x_i,

$$\hat{y}_i = \underset{(.8813)}{2.0676} x_i, \tag{18}$$

where the figure in parentheses is the conventional standard error, $\hat{\beta} = 2.0676$, and $s^2 = \sum_{i=1}^{T}(\hat{y}_i - \hat{\beta}x_i)^2/\nu = 0.2681 \times 10^8$ with $\nu = T - 1 = 8$.

Given these sample data and statistics, we first determine settings for w which minimize $E_z L(z, a) = E_z |z - 10|^\alpha$ for various values of α, namely, $\alpha = 0.5, 1.0, 2.0$, and 4.0. In Table IV expected loss as a function of the control

TABLE IV

ABSOLUTE EXPECTED LOSS AS A FUNCTION OF CONTROL VARIABLE SETTING AND FORM OF LOSS FUNCTION FOR THE MONEY MULTIPLIER MODEL

Value of control variable, w	Form of loss function			
	$\lvert z-10\rvert^{0.5}$	$\lvert z-10\rvert$	$\lvert z-10\rvert^2$	$\lvert z-10\rvert^4$
6.383	2.467	7.271	88.24	3.221×10^4
5.503	2.305	6.378	69.08	2.104×10^4
4.837[a]	2.222	5.931	60.03	1.619×10^4
4.314	2.189	5.751	56.24	1.402×10^4
4.201	2.186	5.734	55.80	1.370×10^4
4.093	2.185[c]	5.725	55.51	1.345×10^4
3.991	2.185	5.723[c]	55.34	1.326×10^4
2.893[b]	2.187	5.728	55.29[c]	1.310×10^4
3.801	2.190	5.738	55.34	1.299×10^4
3.712	2.193	5.753	55.46	1.292×10^4
3.628	2.198	5.772	55.66	1.287×10^4
3.548	2.203	5.795	55.92	1.284×10^4
3.470	2.209	5.821	56.24	1.284×10^4[c]
3.397	2.216	5.849	56.60	1.286×10^4
3.326	2.223	5.880	57.00	1.289×10^4
3.258	2.230	5.912	57.43	1.294×10^4
3.012	2.261	6.056	59.41	1.324×10^4

[a] $w = 10/\hat{\beta}$, certainty equivalence solution for $|z - 10|^2$.

[b] $w = (10/\hat{\beta})[t^2/(1 + t^2)]$, optimal solution for $|z - 10|^2$.

[c] Minimal expected loss for particular loss function.

setting w is tabulated for the four symmetric loss functions. It is seen that, as with the generated data, the optimal solution for the quadratic case shows remarkable robustness in that the expected losses associated with it are not far different from those associated with the optimal solutions for $\alpha = 0.5$, $\alpha = 1.0$, and $\alpha = 4.0$. On the other hand, the certainty equivalence solution, $w = 4.837$, is not too bad for $\alpha = 0.5$ and $\alpha = 1.0$, but does result in appreciably higher expected losses for $\alpha = 2.0$ and $\alpha = 4.0$. With $\alpha = 2.0$, the expected loss for the certainty equivalence solution is 60.029, whereas that associated with the optimal control setting is just 55.292. The difference for $\alpha = 4$ is even more striking. These findings are summarized in Table V.

TABLE V

Relative expected loss[a]	Form of loss function			
	$\lvert z-10\rvert^{0.5}$	$\lvert z-10\rvert$	$\lvert z-10\rvert^2$	$\lvert z-10\rvert^4$
$\dfrac{E(L \mid w = w^*)}{E(L \mid w = w_q)}$	0.999	0.999	1.000	0.980
$\dfrac{E(L \mid w = w^*)}{E(L \mid w = w_{ce})}$	0.983	0.965	0.921	0.793

[a] Below w^* is the optimal setting for each loss function, w_q is the optimal setting for the quadratic case, and w_{ce} is the certainty equivalence setting.

As regards linear loss functions for the present problem, we used the same set as employed with the generated data, namely,

$$L(z, 10) = \begin{cases} |z-10| & \text{for} \quad -\infty < z \leq 10 \\ k\,|z-10| & \text{for} \quad 10 < z < \infty, \end{cases} \tag{19}$$

with $k = 0.25, 0.50, 0.75, 1.0, 1.5, 2.0$, and 3.0. The results of our calculations are shown in Table VI.

TABLE VI

CONTROL SETTINGS AND ASSOCIATED EXPECTED LOSSES FOR VARIOUS LINEAR LOSS FUNCTIONS

Value of k[a]	Optimal setting, $w = w^*$	Expected loss for $w = w^*$	Optimal setting for quadratic, $w = w_q$	Expected loss for $w = w_q$	Certainty equivalence setting, $w = w_{ce}$	Expected loss for $w = w_{ce}$
0.25	6.5129	3.342	3.8935	4.311	4.8366	3.707
0.50	5.2327	4.422	3.8935	4.783	4.8366	4.448
0.75	4.4962	5.159	3.8935	5.255	4.8366	5.190
1.00	3.9907	5.723	3.8935	5.728	4.8366	5.931
1.50	3.3967	6.567	3.8935	6.672	4.8366	7.414
2.00	2.9842	7.196	3.8935	7.616	4.8366	8.897
3.00	2.4192	8.122	3.8935	9.505	4.8366	11.862

[a] k is defined in (19).

The picture which emerges from the figures in Table VI is much the same as that encountered with the generated data when linear loss functions were employed. First, both the optimal solution for the quadratic and the certainty equivalence solution can be far from the mark in certain situations, and neither seems in general to be better than the other under all conditions employed. Thus when loss functions are asymmetric, it is very important to use solutions appropriate for such functions rather than rely on results appropriate for symmetric quadratic loss functions.

5. Concluding Remarks

In conclusion, we would like to emphasize that the computation of optimal settings in the present examples was relatively simple because the model and loss functions employed were quite simple. More work has to be done to get simple procedures for more complicated problems. Second, in the examples above, we chose problems in which there was considerable uncertainty about the model's parameters. Of course if sample sizes are large and parameters are precisely estimated, some of the differences between optimal and approximate solutions would be much less pronounced. However, it is not only sample size that matters in this regard, but also the size of the target value. Third, and last, we have considered just one-period problems. Much work remains to obtain operational solutions to multiperiod problems.

Appendix

Data Generated under Conditions
Described in Section 3

i	y_i	x_t	u_i
1	−0.039	−1.026	2.013
2	2.730	1.542	−0.354
3	0.997	−0.532	2.061
4	−2.990	−0.253	−2.484
5	2.660	0.040	2.580
6	−0.624	−0.882	1.140
7	6.598	0.875	4.848
8	−0.669	−0.066	−0.537
9	0.348	−0.300	0.948
10	4.428	0.924	2.580
11	−0.723	0.699	−2.121
12	−0.632	0.182	−0.996
13	−2.434	1.078	−4.590
14	1.864	0.512	0.840
15	−1.620	−0.438	−0.744

REFERENCES

Box, G. E. P., and Tiao, G. C. (1962). A further look at robustness via Bayes' theorem. *Biometrika* **49**, 419–433.

Fisher, W. D. (1962). Estimation in the lin:ar decision model. *Intern. Economic Rev.* **3**, 1–29.

Friedman, M., and Meiselman, D. (1963). The relative stability of monetary velocity and the investment multiplier in the United States, 1897–1958. *In* "Commission on Money and Credit Research Study, *Stabilization Policies*." Prentice-Hall, Englewood Cliffs, New Jersey.

Geisser, S. (1965). Bayesian estimation in multivariate analysis. *Ann. Math. Stat.* **36**, 150–159.

Holt, C., Muth, J. F., Modigliani, F., and Simon, H. A. (1960). "Planning Production, Inventories and Work Force." Prentice-Hall, Englewood Cliffs, New Jersey.

Jeffreys, H. (1961). "Theory of Probability." Clarendon Press, Oxford.

Lindley, D. V. (1965). "Introduction to Probability and Statistics from a Bayesian Viewpoint: Part 2, Inference." Cambridge Univ. Press, London and New York.

Raiffa, H., and Schlaifer, R. (1961). "Applied Statistical Decision Theory." Harvard Univ. Press, Cambridge, Massachusetts.

Savage, L. J. (1961). The subjective basis of statistical practice. Unpublished.

Simon, H. A. (1956). Dynamic programming under uncertainty with a quadratic criterion function. *Econometrica* **24**, 74–81.

Theil, H. (1957). A Note on certainty equivalence in dynamic planning. *Econometrica* **25**, 346–349.

Zellner, A., and Chetty, V. K. (1965). Prediction and decision problems in regression models from the Bayesian point of view. *J. Am. Stat. Assoc.* **60**, 608–616.

DISCUSSION

J. C. Kiefer. I think the paper that we have just heard is a very important direction for research, both computational and theoretical, in statistics. It is an area that one might call "robustness under changes of the loss function," and there is very little meaningful work on this subject. It's part of the folklore of the subject that the loss function is not too important. We've seen today that that's not necessarily a correct assumption. The idea that the loss function isn't too important, I suspect, stems from the classical results that everyone knows, in the theory of normal estimation where the good old sample mean is optimum in almost anybody's sense for any reasonable symmetric loss function. Here (in Professor Zellner's model) we see even for quite sensible and nearby loss functions that that isn't the case. I think the use of the computer in this manner is quite good. This seems like an analytical problem that in general is fairly intractible. At the same time, I think it is important that one not get too lazy because of the availability of computers for problems like this. It is quite likely that analytically one cannot do a heck of a lot in finding measures of robustness that are any good for a large class of problems under changes of the loss function. But I think it is very tempting to try—as an analytical problem I'm sure Professor Zellner is familiar with—to get a priori bounds of some sort. Here one grinds into the bound formula the probability model and (say) two possible loss functions, and comes out with a very simple-minded bound on how badly off one could be if one had one loss function and used the procedure which is best in some sense for another loss function.

The problem that Professor Zellner mentioned of the more complicated model which one uses in dynamic programming should be very interesting, and I suspect he's right that the sensitivity may be even greater there. One problem of interest that comes to mind in relating the problem that he has discussed here with the problem where one sums over a large number of intervals, and which problem is also amenable to computer study, lies in the following question. A method of this sort gives you the best procedure if you look one time interval ahead, and in the dynamic programming problem the optimum procedure for a large number of time intervals is very hard to compute. It would be very nice to have, again, a priori theoretical bounds which would tell one the following: suppose that a procedure is good for one time interval ahead; how good is that particular procedure, which is easier to compute, for the problem where one really wants to sum the losses over a large number of intervals? There is not much work on this problem so far either, I think, and one can easily give trivial examples to show that the one-shot optimum procedure of this sort can be very very bad where one works for a large number of intervals. I'm sure that lots of you are familiar with that, but it seems very likely that for nice problems of this sort the one-shot optimum procedure should not be so bad.

I would like to talk a little bit about the subject of prior densities that has arisen at several points in the paper. Since in the discussion on statistical inference tomorrow I may not have a chance to say anything on my viewpoint on Bayesian Statistics, maybe some of it will creep in at this point, although that's not the entire purpose of these comments. Firstly, on the subject of the uniform a priori distribution which one assumes here for the location and scale parameter, there are now in the literature of statistics and related subjects, literally dozens of papers where one uses these a priori distributions. I would like to discuss the use of these on three grounds.

Firstly, if one uses such an a priori law, is it necessary to perform the actual computation of Bayes procedures? Well, the answer to that is often no, and the reason that the answer is often no goes back in the location-scale case to the early work of Pitman, in 1939 or so, on invariant estimators. Pitman discovered in that case that using the uniform a priori distribution on the group, when there is a group present as there is here, is roughly the same as computing the best invariant procedure. Or in the case of confidence or credible intervals, computing the Bayes procedure with respect to these particular improper a priori distributions is the same as computing the best invariant confidence intervals. This was demonstrated in a general setting by Charles Stein and finally published by him in the Bayes-Laplace volume that appeared a couple of years ago. So for many problems, one can dispense entirely with integrating with respect to this uniform a priori distribution on the group, when there is a group or a homogeneous space for the parameter, and instead compute directly the best invariant procedure; it amounts to the same thing. The literature is full of such unnecessary computations. Now the second question is, what happens when you use the resulting procedure? Can you get into any trouble? Again we're indebted to Charles Stein for the answer to this question. The answer is that for nice groups such as the location-scale group one does not lose a heck of a lot, but that for large groups (anything involving, for example, the full linear group, the 2×2 nonsingular matrices under multiplication, which arise in multivariate analysis), it is quite possible to have examples where this a priori distribution leads to very very bad statistical procedures. And one cannot know a priori in a problem involving the full linear group whether or not one is going to get into trouble by the use of the uniform a priori distribution on the group. That's not relevent to the present problem, I repeat, because here the location-scale group, the one-dimensional affine group, is what is involved. (One can still lose a little, in that invariant procedures here need not be admissible, but not as much as in certain multivariate examples where they are not even minimax.)

So one has to be a little bit careful about using that a priori distribution in statistics.

The third thing that I would like to say about the use of that a priori distribution, which is valid even in nice situations where one has, say, just the location parameter, concerns the rationale for using the uniform a priori distribution or locally uniform priors in the consequent stable estimation, as it is called in the Bayesian literature. It is really a large sample rationale. There is a theorem that goes back to von Mises around 1919 in the Bernoulli case and which was later proved in other cases by Kolmogorov, and in more subtle developments by LeCam and Wolfowitz, that says that, if the a priori distribution is sufficiently smooth, then for large sample sizes inference does not depend on what the particular a priori distribution was. Therefore one might as well, in such settings, use an a priori distribution for which the integration is easy to perform, namely, this constant density function in the case of the location parameter. But as far as I can tell, there are in this life an awful lot of small sample problems, and I really question, therefore, the whole rationale that is behind the use of locally uniform priors. I know I have several times argued with eminent Bayesians about where you get the prior from, and have received an answer which goes something like this: "Surely you wouldn't give more than a million-to-one odds that the parameter value is greater than a thousand as against less than a thousand." Well, I, perhaps, wouldn't bet at all on the problem one way or the other, and one might wonder whether a theory of inference should be based on insisting I do; but, agreeing for purposes of the present discussion that I would bet and that I wouldn't give a million to one odds, I am then put in the position of being trapped, because then in stable estimation the prior doesn't matter so much. If you put a million times as much density here as here, then for large sample sizes it doesn't matter. I think you all see the difficulty, which is, how large does the sample size have to be? So I think these locally uniform priors are to be dealt with with a little bit of care.

Finally, on the subject of this parameter alpha in equation (14), I'm sure that Professor Zellner has chosen this as an example of the kind of result one can get, where the conclusions are supposed to indicate something to the statistician faced with uncertainty. The use of examples of this sort has frequently been recommended. I think that the language that is often used calls this a "reference prior," and very often one says to the practical man, "Well, if your a priori distribution is such and such, then I can tell you that the following is true about various procedures." Now, again I will express some distrust about the use of a priori distributions of this kind, if they are to be used very much with the actual practical man rather than as an example which may show us the type of thing that goes on. I suspect, to start with, that the practical man when faced with a reference prior from an expert is simply snowed. Suppose some eminent statistician says, "Now look, if your reference prior is such and such then you should do such and such. Of course, I don't say that that is your prior, I say *if* this is your prior you should do such and such." I think again it is fairly clear to most of you what the typical customer will do. His alternative to submission is to be ornery and say "By golly, I don't think that that is my prior. Your reference prior is not what I want." In that case he is back out in the cold with the rest of us non-Bayesians. The other thing that worries me about reference priors is that they are sometimes chosen with the idea that a certain shape of the reference prior is a very reasonable one. For example, the present reference prior is one that goes up and comes down again like this. In the work of Box and Tiao on similar things, which is referred to in this paper, they discuss the use of reference priors which, in some parametrization of a certain parameter under interest, are taken to be any density function which is unimodal. And I'm very worried about the way in which the unsuspecting customer can get the wrong idea from some picture like this. It's very tempting to say, "Oh, gee, a unimodal a priori density of this index parameter, whatever it is, what can be more reasonable than that? Sure, I think that something like that is the case." But now you see that we have here in a subtle situation a classical difficulty with a priori distributions that has appeared in the literature for years. Everyone knows the example that says, if you tell me to use uniform a priori density to mean uncertainty, then all I have to do is to transform the parameter in some nonlinear way and uniform density becomes something else. For example, if this is the parameter β then the corresponding density of the parameter β^2 looks completely different, and the uniform density for β^2 is not a uniform density for β. Well, up here you have a more subtle example of the same sort of thing, because there is nothing whatsoever holy about the parametrization that is given in any one of these problems for a parameter of this kind. And what looks very natural, to say that we are going to consider just unimodal priors for this particular index parameter, when you think of it in this light, indicates some danger, because that particular parametrization is not at all holy; thus, in a different parametrization, the unimodal priors may be quite a different class, and may yield different conclusions from those obtained with the original class of priors.

Note added in proof: A brief statement on the remarks of Professor Zellner which follow, and which show me how unclear I must have been: I am not worried that flat priors will "always" be used, but rather that they will too often be used, with bad rationale; I am no more devoted to unbiased estimation than is Professor Zellner; the Jeffreys-Hartigan work he mentions is concerned with the invariant prior whose deviltry was mentioned above in my "second question on priors"; and I take little comfort from his reassurance that "the practical man will soon be quite familiar with [will be confident in his misuse of?—J.K.] prior distributions."

A. Zellner. I wish to thank Professor Kiefer for the thoughtful and interesting remarks which he has made with respect to our paper. As regards his observation that "robustness under changes of the loss function" is a relatively unexplored area, I'm completely in agreement and, in fact, it was curiosity about this issue that motivated me to undertake the present work. Next, on the question of getting analytical results, I am very sympathetic to the idea of having further work done. In the paper, we have some analytic results for the quadratic case. To extend this analysis to other loss functions would involve rather involved work with incomplete moments. As a first approach to the problem, we thought that use of readily available computer techniques would provide information about many aspects of the problem relatively easily.

On multiperiod stochastic dynamic programming problems, which Professor Kiefer mentioned, I have looked at two-period problems with quadratic loss functions and have found that the solution to the first-period problem is rather good relative to an approximate "full solution" obtained by backward induction—this for a simple regression model such as the one used in this paper. Thus what you conjectured appears to hold for the special cases I have studied. More work is in progress on this topic.

With regard to your points on prior densities, I doubt that I shall be able to cover all aspects of your remarks in the few minutes which I have available. After all, this discussion goes way back to the Reverend Bayes, Laplace, and others. First, with respect to flat or diffuse prior densities, it is not always the case that these will be used. There may very well be outside information which a policy maker or manager will want to use in his analysis. Being able to incorporate such information in his analysis by choice of a suitable prior pdf is indeed very useful.

Second, on the question about whether flat prior pdf's can lead to bad procedures, this is indeed true in certain multiparameter problems. In particular, in the classical errors-in-the-variable model, I have found that putting a flat prior pdf on the incidental parameters affects inferences vitally. Thus, in multiparameter problems, as in other problems, one has to be careful in choice of a prior pdf. As an aside, the method of maximum likelihood estimation has its well-known problems in problems with incidental parameters. Further, Stein has shown that maximum likelihood estimators for the cases of k means and for coefficients of the regression model are inadmissible, given not unusual loss functions, when the number of parameters is not small. In summary, multiparameter problems appear to pose problems which deserve attention no matter what system of inference is employed.

Third, on your point regarding small sample problems, if a uniform prior adequately represents one's information about the parameter, say θ, and if one has a loss function involving θ, say $L(\theta, \hat{\theta})$, where $\hat{\theta}$ is an estimate for θ, then in general one can find a $\hat{\theta}$ which minimizes expected loss, with the expectation being taken with respect to the posterior pdf for θ. This procedure is entirely consistent with the expected utility hypothesis and provides a point estimate which has a finite sample justification even though a flat or diffuse prior pdf is being used. Also, I would like to point out to sampling theorists in the audience that several Monte Carlo experiments have been performed to study the sampling properties of Bayesian estimators relative to other estimators with results uniformly favorable to the Bayesian estimators.

Fourth, with respect to reference priors and their use and misuse, I am sure that the possibility for misuse is not uniquely associated with reference priors. Also, many students in business and economics are being taught about prior distributions so that it is quite probable that *the practical man* will soon be quite familiar with prior distributions.

Last, on the parameterization problem and the choice of prior distribution, Professor Kiefer pointed out again that properties of the prior pdf will change depending on the parameterization. The same considerations apply to the finite sample properties of sampling theory procedures. For example, if $\hat{\theta}$ is an unbiased estimator for θ, $1/\hat{\theta}$ is not an unbiased estimator for $1/\theta$. Thus the parameterization problem has to be faced up to in both the Bayesian and sampling theory approaches. Jeffreys' and Hartigan's important work on invariant prior distributions represents an effort to get consistency in important respects in the choice of prior distributions and appears to me to be an important contribution to the problem which you raised. Finally, in many problems there appears to be a natural parameterization, and if there is, that eases life a great deal.

H. O. Hartley. I wonder whether you feel that if you had other data, the robustness issue would not alter? Do you feel that you should have investigated more data before looking at the robustness issue? Secondly, is it not the case that, in any *particular* economic problem, a *particular* loss function is relevant and no other?

A. Zellner. On the first question, the analytical results which we have pertain to the quadratic case and a very widely used approximate solution for the quadratic case. These are analytical results and are general. With respect to results for other loss functions, we used several sets of data in addition to those reported in the paper with results being fairly similar, that is analyses being quite sensitive to the form of the loss function and uncertainty about the parameters. Of course, in line with Professor Kiefer's suggestion, more analytical results would be welcome.

In response to you second question regarding loss functions or criterion functions in economic problems, in many important instances in econometrics there is considerable uncertainty about the form of the criterion function. For example, much work has been done on the problem of maximizing the mathematical expectation of a criterion function, usually called a social welfare function, which depends on variables of a macroeconometric model approximating the behavior of a national economy. Clearly there is much uncertainty and doubt about how to formulate a social welfare function for an economy. Should such a function reflect a planner's welfare judgments, a committee's welfare judgments, or some average of individuals' welfare judgments? These are just some of the problems involved in formulating criterion functions to be used in conjunction with models of economies, and the quadratic form, which is often employed, is nothing more than a rough first approximation in many instances. And even at the individual-firm level, there may be considerable uncertainty about the form of the profit function, since the form of this function depends on technological conditions of production, and supply and demand conditions, all or some of which may be imperfectly known. Thus, in actual cases, there will usually be some uncertainty about the form of the profit function. Of course, approximations are usually employed and the adequacy of these approximations is exactly what concerns us.

[28]

The Quality of Quantitative Economic Policy-making when Targets and Costs of Change are Mis-specified

BY ARNOLD ZELLNER*

1. Introduction

In a series of well-known works [3], [4], [5], [6], Tinbergen has made pioneering and outstanding contributions to the theory and application of quantitative economic policy-making (Q.E.P.). Tinbergen's approach to Q.E.P. involves the following elements: (*a*) a criterion or welfare function that depends on certain economic variables, (*b*) a classification of variables into categories, target and non-target endogenous variables and instrumental and non-instrumental exogenous variables, (*c*) an econometric model involving relationships for variables and (*d*) boundary conditions for selected variables. Within this framework, which resembles closely the framework of modern control theory, Tinbergen provided simple, operational procedures for solving for values of policy instrument variables at a time when modern control theory was in its infancy.

In addition to structuring and providing solutions to Q.E.P. problems, Tinbergen's work importantly involves consideration of the sensitivity of policy solutions to possible errors in

* Research financed in part by the National Science Foundation under grant GS-2347 and by income from the H. G. B. Alexander Fund, Graduate School of Business, University of Chicago. This paper was written when the author was Visiting Ford Research Professor in the Department of Economics, University of California at Berkeley. Some of the material in this paper was presented to the Econometrics Seminar, Stanford University. Several comments of T. W. Anderson were particularly helpful.

formulating welfare functions and economic models. He is quite aware of the many difficulties in formulating satisfactory welfare functions to be used in Q.E.P. With regard to the relation of welfare functions of individuals and of policy-makers, he writes, '. . . there may be a certain degree of similarity between individual welfare functions and that of the policy-maker. The more democratic is the community, the more will the citizens be able to further this similarity. . . .' [5, p. 14]. Also, he recognises possible inconsistencies between short-run and long-run targets, mentions a possible reconciliation by instrument differentiation, and then states, 'If, however, no differentiation is possible, the only possibility is a 'weighing of targets' or, in practical terms, a compromise' [5, p. 137]. Further, in a pragmatic fashion, he writes, 'For practical purposes it will often be desirable and sometimes possible to specify the mathematical form of the utility functions in such a way as to make them workable. For modest ranges of the relevant variables simple mathematical forms will always be acceptable as approximations. . . .' [3, p. 52]. He goes on to discuss the use of a quadratic approximation. As regards errors associated with econometric models, he explicitly considers errors in estimated parameter values and in functional forms of relationships [4, pp. 59–63]. Taken together, these considerations constitute a remarkable contribution to the problem of assessing the *robustness* of Q.E.P., a vital subject that has not received much attention in the literature.

In the present paper, attention is focused on several problems in which there is uncertainty about the values of parameters in a quadratic criterion function. That there may be uncertainty about the values of such parameters is emphasised by van Eijk and Sandee, who write:

> In principle, the coefficients of a welfare function can be estimated only by interviewing the policy-makers. They would have to answer a series of questions about the marginal rates of substitution for all target variables and in different situations. For the time being, however, a genuine interviewing of policy-makers is impossible. This means that interviews must be imaginary. . . . In short, the presumable outcome of a real interview must be forecast. [7, p. 4].

This quotation suggests that there may be considerable uncertainty about the values of parameters in policy-makers' welfare functions, even abstracting from aggregation problems and the relation of policy-makers' welfare functions to those of private citizens. Herein, for simple quadratic criterion functions, that Tinbergen has indicated may serve as useful approximations in certain instances,[1] the effects of errors in assigning target values and in assessing the cost of changing policy instruments are evaluated analytically.[2] Target values and cost of change are emphasised because it is believed that these are key considerations in Q.E.P.[3] In fact, Tinbergen points out that, 'the relevant starting point to any quantitative policy . . . is the set of target values. . . .' [5, p. 133]. Some of the issues analysed below are the following: (*a*) what are some well-defined conditions in which there may be uncertainty about the parameters of simple quadratic criterion functions? (*b*) can errors in assigning target values be offset by errors in assessing the cost of changing policy instrument variables? (*c*) do overly ambitious policy-makers who overstate target values and underestimate cost of change fare better in terms of expected utility than do more conservative policy-makers who understate target values and overestimate cost of change? These questions, and others, are considered below in connection with some problems in Q.E.P. involving control of regression processes.

2. Criterion Functions and the Control of a Simple Regression Process

In this section, an example illustrating uncertainties that can arise concerning values of a criterion function's parameters is presented. Then an analysis of the effects of such uncertainties on the solution to a one-period Q.E.P. problem is reported.

[1] See [9] for some results on the effects of departures from the quadratic form.

[2] In particular cases, sensitivity analysis can of course be performed numerically.

[3] In [8], the effects of departures from temporally independent preferences are analysed in a deterministic framework with extremely interesting findings.

Let the criterion function for individual i be the following quadratic loss function[4]

$$L_i = (y_i - a_i)^2 + c_i(x - x_0)^2 \qquad (i = 1, 2, \ldots, N) \qquad (1)$$

where y_i is an endogenous variable, a_i is the ith individual's target value for y_i, x is a policy instrument or control variable that a policy-maker controls, x_0 is the known initial value for x, and c_i is a positive parameter. If we attach a weight,[5] w_i, to individual i's loss, L_i, with $0 \leq w_i$ and $\sum_{i=1}^{N} w_i = 1$, and assume that it is appropriate to sum individuals' weighted losses to obtain total loss, L'_T, we obtain

$$L'_T = (y - \bar{a})^2 - 2(y - \bar{a})\sigma_{wa} + c(x - x_0)^2 + (1/N)\sum(\delta y_i - \delta a_i)^2 + \sum \delta w_i[(\delta y_i - \delta a_i)^2 + 2\,\delta y_i(y - \bar{a})] \qquad (2)$$

where summations extend from $i = 1$ to $i = N$, $y = \sum y_i/N$, $\bar{a} = \sum a_i/N$, $\delta y_i = y_i - y$, $\delta a_i = a_i - \bar{a}$, $\delta w_i = w_i - \bar{w}$, $\bar{w} = \sum w_i/N = 1/N$, $c = \sum w_i c_i$, and $\sigma_{wa} = \sum \delta w_i \delta a_i$. If we assume a simple regression connecting y_i and x, that is, $y_i = \beta x + u_i$ $(i = 1, 2, \ldots, N)$, then

$$y = \beta x + u \qquad (3)$$

where $u = \sum u_i/N$. Since $\delta y_i = y_i - y = u_i - u$ does not depend on x, we can express (2) as follows

$$L'_T = \left[y - \bar{a}\left(1 + \frac{\sigma_{wa}}{\bar{a}}\right)\right]^2 + c(x - x_0)^2 + \varepsilon \qquad (4)$$

where ε represents all other terms and is such that its mathematical expectation does not depend on the value of the control variable x. Thus we can write

$$L_T = L'_T - \varepsilon = (y - a)^2 + c(x - x_0)^2 \qquad (5)$$

where $a = \bar{a}(1 + \sigma_{wa}/\bar{a})$ and $c = \sum w_i c_i$.

[4] The ith individual's utility, U_i, can be regarded as given by $U_i = -L_i + K_i$, where K_i is an arbitrary constant.

[5] Tinbergen writes, 'Some of the controversies between communist policies and Western policies may be expressed in terms of the weight given to the interests of various social groups constituting society. . . . [6, p. 76].

In this problem, the policy-maker's criterion function is assumed to be (5) and (3) is his simple economic model for y. In Q.E.P., the policy-maker must assign numerical values to the parameters a and c of (5). The value of a depends on the mean, $\bar{a}$, of individuals' targets, and the covariance σ_{wa} between individuals' weights and target values. The parameter c in (5) is a weighted average of the individual c_is. Since the policy-maker generally does not know the values of the a_is and the c_is and since there is often considerable uncertainty about the appropriate w_is to employ, it is to be expected that the values assigned to a and to c in (5) will generally depart from 'true' values. To allow for such a departure, we write the loss function that the policy-maker actually uses as follows

$$L = (y - k_1 a)^2 + k_2 c(x - x_0)^2 \tag{6}$$

where the parameters k_1 and k_2 have been introduced to allow for errors in assigning values to a and c. Of course, if $k_1 = k_2 = 1$, there are no errors. In what follows, we shall assume k_1, $k_2 > 0$ and a, $c > 0$.

If a policy-maker minimises the mathematical expectation of L in (6) with respect to x, it is of interest to determine how his minimising value for x, say x^*, depends on the values of k_1 and k_2 and the extent to which it departs from the optimal value of x, say x^0, obtained with $k_1 = k_2 = 1$ in (6). Also, it is relevant to determine how $E(L_T \mid x = x^*)$, with L_T given in (5), depends on k_1 and k_2 and the extent to which it differs from minimal expected loss, namely $E(L_T \mid x = x^0)$. The analysis of these problems will be carried forward initially under the assumption that the value of β in (3) is known and positive. Then the case in which β's value is unknown will be analysed.

Analysis When Value of β is Known ($\beta > 0$)

In (3), we assume $Eu = 0$ and $Eu^2 = \sigma^2 < \infty$. Then the mathematical expectation of the loss function in (6) is

$$EL = (\beta x - k_1 a)^2 + k_2 c(x - x_0)^2 + \sigma^2 \tag{7}$$

The value of x, say x^*, that minimises expected loss in (7) is given by

$$x^* = (k_1 a/\beta + k_2 v x_0)/(1 + k_2 v) \tag{8}$$

where $v = c/\beta^2$. It is seen that x^* is a weighted average of k_1a/β and x_0 with weights 1 and k_2v, respectively. In addition, k_1a/β is the value of x that provides a minimal value to $(\beta x - k_1a)^2$, the first term on the right-hand side of (7), while x_0 is the value of x that gives a minimal value to the second term on the right-hand side of (7). Thus x^* is a weighted average of quantities, k_1a/β and x_0, that minimise individual components of expected loss.

For comparison with x^*, the optimal setting for x, say x^0, is the value of x that minimises (7) when $k_1 = k_2 = 1$. When $k_1 = k_2 = 1$, (7) is identical to (5), the true loss function. The optimal value for x is given by

$$\begin{aligned} x^0 &= (a/\beta + vx_0)/(1 + v) \\ &= x_0 - (\beta x_0 - a)/\beta(1 + v) \end{aligned} \tag{9}$$

From the first line of (9), it is seen that x^0 is a weighted average of a/β and x_0. Since x^* can differ from x^0, given $k_1, k_2 \neq 1$, use of $x = x^*$ can, of course, result in suboptimal Q.E.P., except in situations in which errors in assigning the target value and the value of the cost of change parameter happen to be offsetting (see below).

With respect to x^* in (8) and x^0 in (9), the following results are of interest.

(*a*) With $a/\beta > 0$, from (8) $\partial x^*/\partial k_1 > 0$; that is, with all other quantities held constant, x^* will be greater in value the larger is k_1.

(*b*) From (8), $\partial x^*/\partial k_2 = v(x_0 - k_1a/\beta)/(1 + k_2v)^2$. Then $\partial x^*/\partial k_2 > 0$ if $x_0 > k_1a/\beta$ and $\partial x^*/\partial k_2 < 0$ if $x_0 < k_1a/\beta$. The dependence of the algebraic sign of this derivative on x_0 and k_1a/β can be appreciated by remembering that x^* is a weighted average of x_0 and k_1a/β with weights k_2v and 1, respectively. Thus, when $x_0 > k_1a/\beta$, increasing k_2, an increase in the weight given to x_0, increases x^* while when $x_0 < k_1a/\beta$, increasing k_2 results in a decrease in x^*.

(*c*) With $k_1 = 1$, no error in the target value, overstatement of the cost of change parameter, i.e. $k_2 > 1$, results in x^* being closer to x_0 than is x^0, given $a/\beta \neq x_0$, while with having $k_2 < 1$ results in x^* being farther from x_0 than is x^0.

(*d*) With $k_2 = 1$, no error in assessing cost of change, $x^* > x^0$ if $k_1 > 1$ and $x^* < x^0$ if $k_1 < 1$.

(*e*) With $a, \beta > 0$, if $k_2/k_1 = 1$, i.e. equally proportionate over- or understatement of target value and cost of change, then $x^* > x^0$ if $k_2, k_1 > 1$ and $x^* < x^0$ if $k_1, k_2 < 1$.

(*f*) Values of k_1 and k_2, not necessarily equal to one, can be found such that $x^* = x^0$. Such values are obtained by equating (8) and (9) to yield

$$k_1 = -k_2 v(\beta x_0 - a)/a(1 + v) + \beta x^0/a \tag{10a}$$

or

$$\delta_1 = -\delta_2 \phi \gamma \tag{10b}$$

where

$$\delta_1 = k_1 - 1, \qquad \delta_2 = k_2 - 1,$$

$$\phi = v/(1 + v) \qquad \text{and} \qquad \gamma = (\beta x_0 - a)/a \tag{10c}$$

Since we assumed $k_1, k_2 > 0$ and $\delta_1, \delta_2 > -1$. The relations (10a) and (10b) indicate that errors in assessing the target value and cost of change can be offsetting. With $\gamma > 0$, *an overstatement of the cost of change, $\delta_2 > 0$, must be accompanied by an understatement of the target value, $\delta_1 < 0$, for x^* to equal x^0. With $\gamma < 0$, for minimal expected loss to be experienced, i.e. $x^* = x^0$, δ_1 and δ_2 must have the same algebraic signs which means joint over- or understatement of target value and cost of change.*[6]

Having investigated several features of the dependence of x^* on the values of k_1 and k_2, we now turn to a consideration of how the expectation of the 'true' loss function in (5) depends on k_1 and k_2 when $x = x^*$ with x^* shown in (8). After some straightforward algebra, we have

$$E(L_T \mid x = x^*) = \frac{a^2\gamma^2}{1 + v}\left[\left(\frac{w + z}{1 + z}\right)^2 + v\right] + \sigma^2 \tag{11a}$$

where $v = c/\beta^2$, $\gamma = (\beta x_0 - a)/a$, $w = \delta_1/\gamma$, $z = v\delta_2/(1 + v)$, $\delta_1 = k_1 - 1$ and $\delta_2 = k_2 - 1$. Since $w = z = 0$ when

[6] The dependence of these results on the algebraic sign of $\gamma = (\beta x_0 - a)/a$, can be appreciated if it is noted that x^0 is a weighted average of x_0 and a/β, with weights v and 1, respectively, while x^* is a weighted average of x_0 and $k_1 a/\beta$, with weights $k_2 v$ and 1, respectively. Thus, for example, if $a/\beta > x_0 (\gamma < 0)$ and $k_1 > 1$, k_2 must be less than 1 for x^0 to be equal to x^*. If $a/\beta < x_0 (\gamma > 0)$ and $k_1 > 1$, k_2 must be greater than 1 for x^0 to equal x^*.

$k_1 = k_2 = 1$ (or $\delta_1 = \delta_2 = 0$), minimal expected loss is

$$E(L_T \mid x = x^0) = a^2\gamma^2 v/(1 + v) + \sigma^2 \tag{11b}$$

A condition on w and z in (11a) sufficient to achieve minimal expected loss shown in (11b), is $w = -z$, which is the same as (10b). Further, expected loss in (11a) will be constant for values

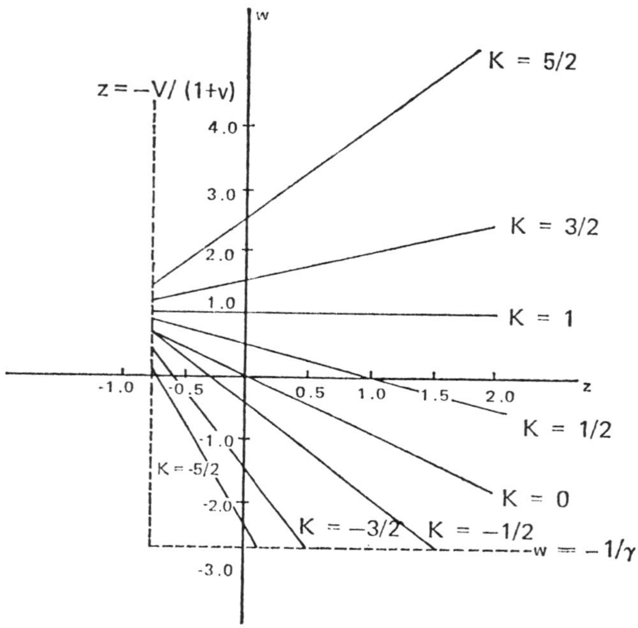

FIG. 7.1. *Contours Associated with* (11a)*

*In the figure it is assumed that $\gamma = (\beta x_o - a)/a > 0$.

of w and z satisfying $(w + z)/(1 + z) = K$, where K is a constant; that is

$$w = K + (K - 1)z \tag{12a}$$

or

$$\delta_1 = \gamma K + (K - 1)\gamma v \delta_2/(1 + v) \tag{12b}$$

with $w > -1/\gamma$ and $z > -v/(1 + v)$. The contours associated with (11a) are straight lines – see Figure 7.1. When $K = 0$, (12a) yields $w = -z$, the locus such that minimal expected loss is experienced. As K's absolute value increases, expected loss increases. Note that for $K < 1$, (12a) yields negatively sloped iso-loss lines while for $K > 1$ the slopes of these lines is positive. Thus the trade-off between z and w required to keep expected loss constant changes in nature as K goes through the value 1.

Next, we consider the behaviour of expected loss in (11a) for values of δ_1 and δ_2 such that $|\delta_1| = |\delta_2| = \delta$. Here, equal absolute proportionate errors in the target value and in the cost-of-change parameter's value are assumed. Under this condition we wish to compare expected losses for the following four cases.

Case 1 $\delta_1, \delta_2 > 0$; target and cost of change overstated, denoted (0, 0)

Case 2 $\delta_1 < 0$ and $\delta_2 > 0$; target understated and cost of change overstated, denoted (U, 0)

Case 3 $\delta_1, \delta_2 < 0$; target and cost of change understated, denoted by (U, U)

Case 4 $\delta_1 > 0$, $\delta_2 < 0$; target overstated and the cost of change understated, denoted (0, U)

For these four cases, with $|\delta_1| = |\delta_2| = \delta$, expected losses from (11a) were compared[7] with the following results for the six distinct pairwise comparisons where P $\equiv$ 'preferred to':

(1) Case 1 (0, 0) *v* Case 3 (U, U):	(0, 0)P(U, U)
(2) Case 2 (U, 0) *v* Case 4 (0, U):	(U, 0)P(0, U)
(3) Case 1 (0, 0) *v* Case 2 (U, 0):	(0, 0)P(U, 0) if $\gamma < 0$
	(U, 0)P(0, 0) if $\gamma > 0$
(4) Case 1 (0, 0) *v* Case 4 (0, U):	(0, 0)P(0, U) if $\gamma < \delta$
	(0, U)P(0, 0) if $\gamma > \delta$
(5) Case 2 (U, 0) *v* Case 3 (U, U):	(U, 0)P(U, U) if $\gamma > -\delta$
	(U, U)P(U, 0) if $\gamma < -\delta$
(6) Case 3 (U, U) *v* Case 4 (0, U):	(U, U)P(0, U) if $\gamma < 0$
	(0, U)P(U, U) if $\gamma > 0$

It is seen that in the first two comparisons, unqualified results are obtained. In comparison (1) equal proportionate overstatement of target and cost of change is preferred to equal

[7] This involves consideration of the quantity $(w + z)^2/(1 + z)^2$.

proportionate understatement of these two items. Heuristically, overstating the cost of change tends to prevent x from being varied, thereby saving on costs of change, and to prevent y from moving to an overstated target, thereby tending to keep closer to the true value. With equal proportionate understatement, x tends to be changed too much and y tends to be brought close to an understated target value. Excessive change in x and being away from the true target inflate expected loss.

In comparison (2), understating the target and overstating the cost of change, (U, 0), is preferred to overstating the target and understating the cost of change (0, U), when absolute proportionate errors are all equal in magnitude. Note that for (U, 0), understatement of the target and overstatement of cost of change tend to keep cost of change low. In the case of (0, U), an overstatement of the target and understatement of the cost of change, both errors tend to produce a large change in x, resulting in high costs of change, and to bring y close to a mistakenly high target, thereby raising the cost of being off target.

In four of the comparisons, (3)–(6), results depend on the initial conditions, as reflected in the value of $\gamma = (\beta x_0 - a)/a$, or the relation of the relative magnitudes of γ and δ, the absolute size of the proportionate errors. From the results in (1)–(6) above, we can make the following preference orderings.

Condition on $\gamma = (\beta x_0 - a)/a$	*Preference ordering*
$\gamma < -\delta$	(0, 0)P(U, U)P(U, 0)P(0, U)
$-\delta < \gamma < 0$	(0, 0)P(U, 0)P(U, U)P(0, U)
$0 < \gamma < \delta$	(U, 0)P(0, 0)P(0, U)P(U, U)
$\delta < \gamma$	(U, 0)P(0, U)P(0, 0)P(U, U)

We see that when $\gamma < 0$, that is $x_0 < a/\beta$, *the worst pair of errors to make is overstatement of the target and understatement of the cost of change*. When $\gamma > 0$, that is $x_0 > a/\beta$, *the worst pair of errors is understatement of the target and understatement of the cost of change*. In both these cases the error in assigning a target value involves a tendency to change the control variable too much (in the upward direction in the first instance and in the downward direction in the second). This tendency toward too much change is reinforced by the underassessment of the cost of change with the result that we end up, on average, off target and experience excessive costs of change.

For $\gamma < 0$, *the least damaging pair of errors is overstatement of the target accompanied by overstatement of the cost of change, errors that are offsetting. Similarly for* $\gamma > 0$, *that is* $x_0 > a/\beta$, *understatement of the target and overstatement of the cost of change tend to be offsetting errors and are least serious.* Note that with $\gamma > 0$, understatement of the target means adopting a target value that is farther from βx_0 than is appropriate; similarly when $\gamma < 0$ an overstated target value is farther from βx_0 than is appropriate. Both of these errors make for excessive movement of the control variable which is offset by the overstatement of the cost of change.

Analysis when Values of β and σ are Unknown

When the values of the parameters β and σ, involved in (2.3), are unknown and must be estimated, it is known that such uncertainty affects the optimal setting for the control variable x – see, for example, [1], [2], [9] and [10, ch. 11]. Assume that past sample data on y and x are available that have been generated by a simple normal regression model

$$y_t = \beta x_t + u_t \qquad (t = 1, 2, \ldots, T) \tag{13}$$

We assume that the u_ts are normally and independently distributed, each with zero mean and common standard deviation, σ. If we have little information about the values of β and σ and represent this state of ignorance by the following prior p.d.f.[8]

$$p(\beta, \sigma) \propto 1/\sigma \qquad (-\infty < \beta < \infty, 0 < \sigma < \infty) \tag{14}$$

where $\propto$ denotes proportionality, the predictive p.d.f. for $y \equiv y_{T+1}$ is well known to be in the following univariate Student t form – see, for example, [10, ch. 3]

$$p(y \mid \mathbf{y}) = \frac{\Gamma[(\nu + 1)/2]}{\Gamma(\frac{1}{2})\Gamma(\nu/2)} (g/\nu)^{1/2} \{1 + (g/\nu)(y - x\hat{\beta})^2\}^{-(\nu+1)/2} \tag{15}$$

[8] In (14) β and $\log \sigma$ are assumed independently and uniformly distributed, a representation of ignorance about the values of the parameters that has been employed extensively in the literature. The analysis presented below can be readily generalised to cases in which an informative prior p.d.f. in the 'normal-gamma' form is utilised.

where $\mathbf{y}' = (y_1, y_2, \ldots, y_T)$ denotes the given sample information, $x \equiv x_{T+1}$, the control variable's value in period $T + 1$, $\nu = T - 1$, $\hat{\beta} = \sum_{t=1}^{T} x_t y_t / \sum_{t=1}^{T} x_t^2$, $g^{-1} = s^2(1 + x^2/\sum_{t=1}^{T} x_t^2)$, $\nu s^2 = \sum_{t=1}^{T} (y_t - \hat{\beta}x_t)^2$, and Γ denotes the gamma function.

The mean and variance of the p.d.f. for y in (15) exist for $\nu > 2$ and are given by

$$E(y \mid \mathbf{y}, x) = x\hat{\beta}$$

and

$$\operatorname{var}(y \mid \mathbf{y}, x) = \nu s^2\left(1 + x^2 \Big/ \sum_{t=1}^{T} x_t^2\right) \Big/ (\nu - 2) \tag{16}$$

It is to be noted that both the mean and variance in (16) depend on $x \equiv x_{T+1}$.

With the above results set forth, we can now use (15) and (16) to obtain the mathematical expectation of the loss function in (6), namely

$$\begin{aligned} EL &= \operatorname{var}(y \mid \mathbf{y}, x) + (k_1 a - \hat{\beta}x)^2 + k_2 c(x - x_0)^2 \\ &= \bar{s}^2(1 + x^2/m) + (k_1 a - \hat{\beta}x)^2 + k_2 c(x - x_0)^2 \end{aligned} \tag{17}$$

where $\bar{s}^2 = \nu s^2/(\nu - 2)$, $m = \sum_{t=1}^{T} x_t^2$, and $x_0 \equiv x_T$. In the second line of (17), the first term will be minimised if $x = 0$, the second if $x = k_1 a/\hat{\beta}$ and the third if $x = x_0$. The value of x that minimises (17), say x^+, is a weighted average of these three quantities given by

$$\begin{aligned} x^+ &= [\hat{\beta}^2(k_1 a/\hat{\beta}) + k_2 c x_0]/(\hat{\beta}^2 + k_2 c + \bar{s}^2/m) \\ &= (k_1 a'/\hat{\beta} + k_2 \hat{v}' x_0)/(1 + k_2 \hat{v}') \end{aligned} \tag{18}$$

where $\hat{v}' = c/\hat{\beta}^2(1 + h^{-1})$, $a' = a/(1 + h^{-1})$ and $h^{-1} = \bar{s}^2/m\hat{\beta}^2$. $\hat{\beta}$ and $\bar{s}^2/m$ are the mean and variance, respectively, of the posterior p.d.f. for β. Thus h is the squared coefficient of variation associated with the posterior p.d.f. for β, a relative measure of the precision of estimation.[9]

On comparing (18), the setting for x and when β's and σ's values are unknown, with (8), the setting for x when β's value

[9] Note that h is of $O(T)$ and thus as T gets large $h^{-1} \to 0$.

is known, it is seen that $\hat{\beta}$ replaces β and an additional term, h^{-1}, appears in the denominator of (18). When T is large, (18) and (8) produce identical results. Also, the second line of (18) indicates that x^{+} can be viewed as a weighted average of $k_1 a'/\hat{\beta}$ and x_0 with weights 1 and $k_2\hat{v}'$, respectively.

If no errors are made in assigning the target value and the cost-of-change parameter's value ($k_1 = k_2 = 1$), the optimal setting for x is, with $\hat{v} = c/\hat{\beta}^2$,

$$\begin{aligned} x^{\otimes} &= (a/\hat{\beta} + \hat{v}x_0)/(1 + \hat{v} + h^{-1}) \\ &= (a'/\hat{\beta} + \hat{v}'x_0)/(1 + \hat{v}') \end{aligned} \tag{19}$$

which is (18) with $k_1 = k_2 = 1$. Since (18) and (19) are precisely similar in form to (8) and (9), respectively, the dependence of x^{+} on the values of k_1 and k_2 will be similar to that presented for x^{*} earlier. Also, as above, there are values of k_1 and k_2, not necessarily both equal to one, such that $x^{+} = x^{\otimes}$. These are given by

$$\delta_1 = -\delta_2\hat{\phi}'\gamma' \tag{20}$$

where $\delta_1 = k_1 - 1$, $\delta_2 = k_2 - 1$, $\hat{\phi}' = \hat{v}'/(1 + \hat{v}')$ and $\gamma' = (\hat{\beta}x_0 - a')/a'$. The relation in (20) is precisely similar in form to that in (10b). Since this is the case, the analysis following (10b) can be utilised in the present case as well and thus will not be repeated.

3. Control of a Multiple Regression Process with a Mis-specified Loss Function

A scalar endogenous policy variable y is assumed to be generated by the following multiple regression process

$$y = \mathbf{x}'\boldsymbol{\beta} + u \tag{21}$$

where $\mathbf{x}' = (x_1, x_2, \ldots, x_k)$. is a $1 \times k$ vector of control variables,[10] $\boldsymbol{\beta}' = (\beta_1, \beta_2, \ldots, \beta_k)$ is a $1 \times k$ vector of regression coefficients with known values, and u is a random disturbance term with $E(u) = 0$ and $E(u^2) = \sigma^2 < \infty$. Let our loss function be

$$L_T = (y - a)^2 + (\mathbf{x} - \mathbf{x}_0)'\mathbf{C}(\mathbf{x} - \mathbf{x}_0) \tag{22}$$

[10] Here we assume that all variables in $\mathbf{x}$ can be controlled. Generalisation to the case in which a subset of the variables is under control is direct and will not be presented – see, e.g. [10, Chap. 11].

where, as above, a is a given target value for y, $\mathbf{x}_0$ is a $k \times 1$ vector of given initial values for $\mathbf{x}$ and $\mathbf{C}$ is a positive definite symmetric matrix with elements having known values. Then the mathematical expectation of L_T is

$$E(L_T) = (\mathbf{x}'\boldsymbol{\beta} - a)^2 + (\mathbf{x} - \mathbf{x}_0)'\mathbf{C}(\mathbf{x} - \mathbf{x}_0) + \sigma^2 \tag{23}$$

On differentiating (23) with respect to the elements of $\mathbf{x}$, the value of $\mathbf{x}$ that minimises expected loss, say $\mathbf{x}^0$, is given by[11]

$$\begin{aligned} \mathbf{x}^0 &= (\mathbf{C} + \boldsymbol{\beta}\boldsymbol{\beta}')^{-1}(a\boldsymbol{\beta} + \mathbf{C}\mathbf{x}_0) \\ &= \mathbf{x}_0 - (\boldsymbol{\beta}'\mathbf{x}_0 - a)\mathbf{C}^{-1}\boldsymbol{\beta}/(1 + \boldsymbol{\beta}'\mathbf{C}^{-1}\boldsymbol{\beta}) \end{aligned} \tag{24}$$

It is interesting to observe that the computed value of y with $\mathbf{x} = \mathbf{x}^0$ can be expressed as

$$\boldsymbol{\beta}'\mathbf{x}^0 = \frac{1}{1 + \boldsymbol{\beta}'\mathbf{C}^{-1}\boldsymbol{\beta}}\boldsymbol{\beta}'\mathbf{x}_0 + \frac{\boldsymbol{\beta}'\mathbf{C}^{-1}\boldsymbol{\beta}}{1 + \boldsymbol{\beta}'\mathbf{C}^{-1}\boldsymbol{\beta}}a \tag{25}$$

a weighted average of $\boldsymbol{\beta}'\mathbf{x}_0$ and a. Further, expected loss given $\mathbf{x} = \mathbf{x}^0$ is:

$$E(L_T \mid \mathbf{x} = \mathbf{x}^0) = \sigma^2 + a^2\gamma^2/(1 + \boldsymbol{\beta}'\mathbf{C}^{-1}\boldsymbol{\beta}) \tag{26}$$

where $\gamma = (\boldsymbol{\beta}'\mathbf{x}_0 - a)/a$.

If, instead of (22) we employ the following loss function

$$L = (y - k_1 a)^2 + (\mathbf{x} - \mathbf{x}_0)'\mathbf{G}(\mathbf{x} - \mathbf{x}_0) \tag{27}$$

where, just as in Section 2, k_1 is a parameter introduced to allow for possible errors in specifying the target value. Also, $\mathbf{G}$ is assumed to be a positive definite symmetric matrix. If $k_1 = 1$ and $\mathbf{G} = \mathbf{C}$, (27) is identical to (22), the true loss function.

On computing the expected value of L in (27) and solving for the value of $\mathbf{x}$, say $\mathbf{x}^*$, that minimises $E(L)$ we obtain

$$\begin{aligned} \mathbf{x}^* &= (\mathbf{G} + \boldsymbol{\beta}\boldsymbol{\beta}')^{-1}(k_1 a\boldsymbol{\beta} + \mathbf{G}\mathbf{x}_0) \\ &= \mathbf{x}_0 - (\boldsymbol{\beta}'\mathbf{x}_0 - k_1 a)\mathbf{G}^{-1}\boldsymbol{\beta}/(1 + \boldsymbol{\beta}'\mathbf{G}^{-1}\boldsymbol{\beta}) \end{aligned} \tag{28}$$

[11] Note that

$$(\mathbf{C} + \boldsymbol{\beta}\boldsymbol{\beta}')^{-1} = \mathbf{C}^{-1} - \frac{\mathbf{C}^{-1}\boldsymbol{\beta}\boldsymbol{\beta}'\mathbf{C}^{-1}}{1 + \boldsymbol{\beta}'\mathbf{C}^{-1}\boldsymbol{\beta}}.$$

which is similar to (24) except that $k_1 a$ replaces a and $\mathbf{G}$ replaces $\mathbf{C}$. From (28), we have

$$\boldsymbol{\beta}'\mathbf{x}^* = \frac{1}{1 + \boldsymbol{\beta}'\mathbf{G}^{-1}\boldsymbol{\beta}}\boldsymbol{\beta}'\mathbf{x}_0 + \frac{\boldsymbol{\beta}'\mathbf{G}^{-1}\boldsymbol{\beta}}{1 + \boldsymbol{\beta}'\mathbf{G}^{-1}\boldsymbol{\beta}} k_1 a \tag{29}$$

a weighted average of $\boldsymbol{\beta}'\mathbf{x}_0$ and $k_1 a$. On comparing (25) with (29) it is seen that there is a difference in the weights and in (29) $k_1 a$ replaces a in (25).

From (24) and (28), we can relate $\mathbf{x}^*$ and $\mathbf{x}^0$ as follows:

$$\begin{aligned}\mathbf{x}^* = \mathbf{x}^0 &+ (\boldsymbol{\beta}'\mathbf{x}_0 - a) \\ &\times \left[\frac{1}{1 + \boldsymbol{\beta}'\mathbf{C}^{-1}\boldsymbol{\beta}}\mathbf{C}^{-1} - \frac{1}{1 + \boldsymbol{\beta}'\mathbf{G}^{-1}\boldsymbol{\beta}}\mathbf{G}^{-1}\right]\boldsymbol{\beta} \\ &+ \frac{\delta_1 a \mathbf{G}^{-1}\boldsymbol{\beta}}{1 + \boldsymbol{\beta}'\mathbf{G}^{-1}\boldsymbol{\beta}}\end{aligned} \tag{30}$$

where $\delta_1 = k_1 - 1$. We see from (30) that the elements of $\mathbf{x}^*$ depend directly on the value of δ_1, the error in specifying the target value. If in (30), $\delta_1 = 0$ and $\mathbf{G} = \mathbf{C}$, then $\mathbf{x}^* = \mathbf{x}^0$; however, these are not the only conditions under which $\mathbf{x}^* = \mathbf{x}^0$. Broader sufficient conditions for $\mathbf{x}^* = \mathbf{x}^0$ are obtained by equating the last two terms on the right-hand side of (30) to zero. This yields the following sufficient condition for $\mathbf{x}^*$ to be equal to $\mathbf{x}^0$:

$$\delta_1 = -\gamma\left(\frac{\boldsymbol{\beta}'\mathbf{G}^{-1}\boldsymbol{\beta} - \boldsymbol{\beta}'\mathbf{C}^{-1}\boldsymbol{\beta}}{\boldsymbol{\beta}'\mathbf{G}^{-1}\boldsymbol{\beta}}\right)\left(\frac{1}{1 + \boldsymbol{\beta}'\mathbf{C}^{-1}\boldsymbol{\beta}}\right) \tag{31}$$

where $\delta_1 = k_1 - 1$ and $\gamma = (\boldsymbol{\beta}'\mathbf{x}_0 - a)/a$. The condition in (31) parallels that given in (30) and indicates that errors in assigning values to the true target and to the cost-of-change matrix $\mathbf{C}$ can be offsetting.

The expected loss associated with $\mathbf{x} = \mathbf{x}^*$, derived from the true loss function in (22), is

$$E(L_T \mid \mathbf{x} = \mathbf{x}^*) = \sigma^2 + (\boldsymbol{\beta}'\mathbf{x}^* - a)^2 + (\mathbf{x}^* - \mathbf{x}_0)'\mathbf{C}(\mathbf{x}^* - \mathbf{x}_0) \tag{32}$$

Using $\boldsymbol{\beta}'\mathbf{x}^* - a = \boldsymbol{\beta}'\mathbf{x}^0 - a + \boldsymbol{\beta}'(\mathbf{x}^* - \mathbf{x}^0)$ and $\mathbf{x}^* - \mathbf{x}_0 = \mathbf{x}^0 - \mathbf{x}_0 + \mathbf{x}^* - \mathbf{x}^0$, we can write (32) as

$$\begin{aligned}E(L_T \mid \mathbf{x} = \mathbf{x}^*) = E(L_T \mid \mathbf{x} = \mathbf{x}^0) &+ (\mathbf{x}^* - \mathbf{x}^0)'(\mathbf{C} + \boldsymbol{\beta}\boldsymbol{\beta}')(\mathbf{x}^* - \mathbf{x}^0) \\ &+ 2(\mathbf{x}^* - \mathbf{x}^0)'\{\mathbf{C}(\mathbf{x}^0 - \mathbf{x}_0) + (\boldsymbol{\beta}'\mathbf{x}_0 - a)\boldsymbol{\beta}\}\end{aligned} \tag{33}$$

where $E(L_T \mid \mathbf{x} = \mathbf{x}^0)$, expected loss associated with the optimal setting of $\mathbf{x}$, is given in (26). The last two terms on the right-hand side of (33) represent the extent to which expected loss may be inflated if $\mathbf{x}^* \neq \mathbf{x}^0$. From (33) it is the case that

$$E(L_T \mid \mathbf{x} = \mathbf{x}^*) - E(L_T \mid \mathbf{x} = \mathbf{x}^0) \geq 0.$$

With regard to (21), if the elements of $\boldsymbol{\beta}$ and σ have unknown values and if we have past data on y and $\mathbf{x}$, we can proceed as in Section 2 to derive the predictive p.d.f. for y and use it to evaluate the mathematical expectations of the loss functions in (22) and (27). With diffuse prior assumptions regarding $\boldsymbol{\beta}$ and σ,[12] it is well known that the predictive p.d.f. for y in the standard linear normal regression model is in the univariate Student t form with mean $\mathbf{x}'\hat{\boldsymbol{\beta}}$ and variance $\bar{s}^2(1 + \mathbf{x}'\mathbf{M}^{-1}\mathbf{x})$ where $\hat{\boldsymbol{\beta}} = (\mathbf{X}'\mathbf{X})^{-1}\mathbf{X}'\mathbf{y}$, the sample least squares quantity, $\mathbf{M} = \mathbf{X}'\mathbf{X}$ and $\bar{s}^2 = \nu s^2/(\nu - 2)$ with $\nu s^2 = (\mathbf{y} - \mathbf{X}\hat{\boldsymbol{\beta}})'(\mathbf{y} - \mathbf{X}\hat{\boldsymbol{\beta}})$, the residual sum of squares and $\nu = T - k$, where T is the sample size.

Using the predictive p.d.f. for y, we obtain the following expression for the expectation of L in (27)

$$E(L) = \bar{s}^2(1 + \mathbf{x}'\mathbf{M}^{-1}\mathbf{x}) + (k_1 a - \mathbf{x}'\hat{\boldsymbol{\beta}})^2 + (\mathbf{x} - \mathbf{x}_0)'\mathbf{G}(\mathbf{x} - \mathbf{x}_0) \quad (34)$$

The value of $\mathbf{x}$ that minimises (34), say $\mathbf{x}^+$, is given by

$$\begin{aligned} \mathbf{x}^+ &= (\bar{s}^2\mathbf{M}^{-1} + \mathbf{G} + \hat{\boldsymbol{\beta}}\hat{\boldsymbol{\beta}}')^{-1}(k_1 a\hat{\boldsymbol{\beta}} + \mathbf{G}\mathbf{x}_0) \\ &= \hat{\mathbf{x}}_0 + (k_1 a - \hat{\boldsymbol{\beta}}'\hat{\mathbf{x}}_0)\bar{\mathbf{G}}^{-1}\hat{\boldsymbol{\beta}}/(1 + \hat{\boldsymbol{\beta}}'\bar{\mathbf{G}}^{-1}\hat{\boldsymbol{\beta}}) \end{aligned} \quad (35)$$

with $\hat{\mathbf{x}}_0 = (\mathbf{I} - \bar{s}^2\bar{\mathbf{G}}^{-1}\mathbf{M}^{-1})\mathbf{x}_0$ and $\bar{\mathbf{G}} = \mathbf{G} + \bar{s}^2\mathbf{M}^{-1}$. When $k_1 = 1$ and $\mathbf{G} = \mathbf{C}$, (35) yields the optimal setting for $\mathbf{x}$, say $\mathbf{x}^{\otimes}$, based on the true loss function in (22), namely

$$\mathbf{x}^{\otimes} = \tilde{\mathbf{x}}_0 + (a - \hat{\boldsymbol{\beta}}'\tilde{\mathbf{x}}_0)\bar{\mathbf{C}}^{-1}\hat{\boldsymbol{\beta}}/(1 + \hat{\boldsymbol{\beta}}'\bar{\mathbf{C}}^{-1}\hat{\boldsymbol{\beta}}) \quad (36)$$

with $\tilde{\mathbf{x}}_0 = (\mathbf{I} - \bar{s}^2\bar{\mathbf{C}}^{-1}\mathbf{M}^{-1})\mathbf{x}_0$ and $\bar{\mathbf{C}} = \mathbf{C} + \bar{s}^2\mathbf{M}^{-1}$. Since (35) and (36) are in the same forms as (24) and (28), the comparison of (35) and (36) is quite similar to that for (24) and (28) and thus will not be presented.

[12] We employ a widely used diffuse prior p.d.f. for $\boldsymbol{\beta}$ and σ, namely, $p(\boldsymbol{\beta}, \sigma) \propto 1/\sigma$ with $0 < \sigma < \infty$ and $-\infty < \beta_i < \infty$ $(i = 1, 2, \ldots, k)$. Here the elements of $\boldsymbol{\beta}$ and $\log \sigma$ are assumed uniformly and independently distributed.

4. Concluding Remarks

We have examined the consequence of employing mis-specified targets and costs of changing policy instrument variables within the context of one-period control of simple and multiple regression processes. As is to be expected, errors in formulating criterion functions will generally, but not always, lead to poorer, in terms of expected loss, policy decisions. In general, both for cases in which model parameters have known or unknown values, the iso-expected loss lines for mis-specified quadratic loss functions were found to be linear. Also, it was found that in general errors in assessing target values and cost of change can be offsetting. However, this of course does not imply that errors will always be offsetting in actual problems. In each practical problem, it appears important to assess the consequences of possible errors in assigning values to parameters of a criterion function as well as the consequences of other possible errors. This conclusion is far from novel and, in fact, is reflected in much of Tinbergen's work.

References

[1] FISHER, W., 'Estimation in the Linear Decision Model', *International Economic Review*, III (January 1962) 1–29.

[2] PRESCOTT, E. C., 'Adaptive Decision Rules for Macro Economic Planning', unpublished doctoral dissertation, Graduate School of Industrial Administration, Carnegie-Mellon University, 1967.

[3] TINBERGEN, J., *Centralization and Decentralization in Economic Policy*, (Amsterdam: North-Holland Publishing Co., 1954).

[4] TINBERGEN, J., *On the Theory of Economic Policy*, 2nd ed., (Amsterdam: North-Holland Publishing Co., 1955).

[5] TINBERGEN, J., *Economic Policy: Principles and Design*, (Amsterdam: North-Holland Publishing Co., 1956).

[6] TINBERGEN, J., *Central Planning*, (New Haven/London: Yale University Press, 1964).

[7] VAN EIJK, C. J. and SANDEE, J., 'Quantitative Determination of an Optimum Economic Policy', *Econometrica*, XXVII (January 1959) 1–13.

[8] WAN, H. Y., 'Optimal Saving Programs under Intertemporally Dependent Preferences', *International Economic Review*, XI (October 1970) 521–47.

[9] ZELLNER, A. and GEISEL, M. S., 'Sensitivity of Control to Uncertainty and Form of the Criterion Function', in *The Future of Statistics*, ed. D. G. Watts (New York: Academic Press, 1968, 269–89).

[10] ZELLNER, A., *An Introduction to Bayesian Inference in Econometrics*, (New York: John Wiley and Sons, Inc., 1971).

PART V

BAYESIAN FORECASTING

Journal of Econometrics 40 (1989) 183–202. North-Holland

FORECASTING INTERNATIONAL GROWTH RATES USING BAYESIAN SHRINKAGE AND OTHER PROCEDURES*

Arnold ZELLNER and Chansik HONG

University of Chicago, Chicago, IL 60637, USA

Forecasts of eighteen countries' annual output growth rates, 1974–1984, are computed using an autoregressive model containing leading indicator variables and various forecasting procedures. It is found that Bayesian shrinkage forecasting techniques produced improved forecasts in terms of an out-of-sample root-mean-square criterion relative to those provided by three naive models and by autoregressive models with and without leading indicator variables. The precision of our forecasts compares favorably with that of OECD forecasts derived from complicated econometric 'structural' macroeconometric models and subjected to judgmental adjustments. Work to rationalize our forecasting models in terms of macroeconomic theory is in progress.

1. Introduction

In our past work, Garcia-Ferrer, Highfield, Palm and Zellner (1987), we employed several methods to forecast growth rates of real output (GNP or GDP) for eight European Economic Community countries and the U.S. year by year for the period 1974–81. It was found that diffuse prior or least squares forecasts based on an autoregressive model of order 3 including leading indicator variables, denoted by AR(3)LI, were reasonably good in terms of forecast root mean squared error (RMSE) relative to those of three naive models and of AR(3) models without leading indicator variables. Also, it was found that certain shrinkage forecasting techniques produced improved forecasting results for many countries and that our simple mechanical forecasts compared favorably with OECD annual forecasts which were constructed using elaborate models and judgmental adjustments.

In the present paper our main objectives are to extend our earlier work by (1) providing further analysis of shrinkage forecasting techniques, (2) providing forecasting results for an extended time period, 1974–84, for our past sample of nine countries, (3) applying our forecasting techniques to data relating to nine additional countries, and (4) reporting results of forecasting experiments using a simple modification of our AR(3)LI model.

*Research financed by the National Science Foundation and by income from the H.G.B. Alexander Endowment Fund, Graduate School of Business, University of Chicago. Luis Mañas-Antón, Carla Inclan and Michael Zellner provided valuable research assistance.

0304-4076/89/$3.50

The importance of checking the forecasting performance of our techniques using new data is reflected in objectives (2) and (3) above. The modification of our AR(3)LI model, mentioned in (4), was motivated by macroeconomic considerations embedded in structural models currently being formulated, which yield reduced form equations similar in form to our AR(3)LI forecasting model and reduced form equations for other variables (e.g., the rate of inflation, employment growth, etc.) that will be subjected to forecasting tests in the future.

The plan of the paper is as follows. In section 2 our AR(3)LI model is explained and analysis yielding several shrinkage forecasts is presented. Also, an extended version of our AR(3)LI model is specified and it is shown how it can be employed to yield forecasts. Section 3 is devoted to a presentation of our data, and in section 4 previous forecasting results are compared with those relating to our broader data set and with those yielded by our extended AR(3)LI model. Finally, we provide a summary of results and some concluding remarks in section 5.

2. Model description and forecasting procedures

In this section, we shall describe the autoregressive-leading indicator (ARLI) model employed in our past work as well as some possible extensions of it. Then we shall consider various forecasting procedures of our ARLI models.

2.1. Model description

In Garcia-Ferrer et al. (1987), the following AR(3)LI model was employed to generate one-year-ahead forecasts of the growth rate of real output, y_{it}, for eight years, 1974–81, for nine countries:

$$y_{it} = \beta_{0i} + \beta_{1i}y_{it-1} + \beta_{2i}y_{it-2} + \beta_{3i}y_{it-3} + \beta_{4i}SR_{it-1} + \beta_{5i}SR_{it-2} + \beta_{6i}GM_{it-1} + \beta_{7i}WR_{t-1} + u_{it}, \qquad i=1,2,\dots,9, \quad t=1,2,\dots,T, \tag{1a}$$

or

$$y_i = X_i\beta_i + \boldsymbol{u}_i, \qquad i=1,2,\dots,9, \tag{1b}$$

where, with L being the lag operator and the subscript i, t denoting the value of a variable for the ith country in the tth time period,

y_{it} = rate of growth of output = $(1-L)\log O_{it}$ with O_{it} real output,
SR_{it} = real stock return = $(1-L)\log(SP_{it}/P_{it})$ with SP_{it} a stock price index and P_{it} a general price index,

GM_{it} = growth rate of real money supply = $(1-L)\log(M_{it}/P_{it})$ with M_{it} = nominal money supply,
WR_t = world return = median of countries' real stock returns, SR_{it} in period t,
β_{ji} = parameters for ith country, $j = 0, 1, \ldots, 7$,
u_{it} = disturbance term.

In (1b) the model for each country is expressed in matrix notation with y_{it} a typical element of y_i, $(1, y_{it-1}, y_{it-2}, y_{it-3}, SR_{it-1}, SR_{it-2}, GM_{it-1}, WR_{t-1})$ a typical row of X_i, and u_{it} a typical element of $\boldsymbol{u}_i$. Some comments regarding the model in (1) follow:

1) An autoregression of order 3 was chosen to permit the possibility of having two complex roots associated wth a cycle and a real root associated with a trend. Past calculations indicated that estimated roots had these properties for eight of nine countries. Also, use of just an AR(3) process without leading indicator variables did not perform well in actual forecasting. Use of leading indicator variables led to improved forecasts in most cases as measured by RMSEs of forecast.

2) The disturbance terms in (1) were found to be practically serially uncorrelated for most countries and not highly correlated across countries, results based on least squares analyses of (1) using initial annual data of 1951–73 for estimation. The introduction of the 'common effect' variable, WR_t, reduced contemporaneous disturbance terms' correlations considerably.

3) The leading indicator stock return variables and money growth rate variable apparently caught the effects of oil price shocks, policy changes, etc. in the period of fit 1951–73 and in our previous forecast period 1974–81. Here we are employing market variables to take rough account of expectational and other effects influencing countries' output growth rates.

4) Macroeconomic considerations suggest that a measure of world output growth and changes in countries' real exchange rates affect countries' exports and these should be included in our ARLI model. Since these variables are close to being white noise, they may be buried in the disturbance terms of (1). Below, we shall report some results using an ARLI model including a measure of world output growth.

5) In our past work, forecasts from (1), using least squares and some shrinkage forecast procedures, were reported. Also, some forecasts yielded by a time-varying-parameter version of (1) were reported. Some of these results will be presented below and compared with more recently obtained results.

In our forecasting experiments, we employ annual data, usually 1954–73, twenty observations with data from 1951–53 used for initial lagged values to fit our models.[1] Then the fitted models are employed to forecast outcomes for

[1]The 1954–73 period was used for all countries except Australia 1960–73, Canada 1959–73, Japan 1956–73, and Spain 1958–73.

1974 and subsequent years with the models re-estimated year by year. Multi-year-ahead forecasts have not as yet been calculated. For the forecast period 1974–81 (eight years), least squares forecasts using (1) have yielded forecast RMSEs ranging from 1.47 to 2.92 with a median of 2.23 percentage points for eight EEC countries and the U.S. – see table 2, line F of Garcia-Ferrer et al. (1987). Our 'η-shrinkage' forecasts described below, yielded forecast RMSEs ranging from 1.25 to 2.52 percentage points with a median of 1.78 percentage points – see table 4, line G3 of Garcia-Ferrer et al. (1987). Similar results for an extended time period and for nine additional countries are presented below.

2.2. Derivation and description of shrinkage forecasts

In this subsection, we provide derivations of several shrinkage forecasts, including the 'η-forecast' and the 'γ-forecast'. The performance of these forecasts will be compared with those of naive models and diffuse prior forecasts or least squares forecasts derived from the ARLI model, or variants of it, shown in (1).

The η-forecast involves averaging a forecast from (1), say a diffuse prior or least squares forecast for a particular country, $\hat{y}_{if}$, with the mean of all the N countries' forecasts, $\bar{\hat{y}}_f = \sum_{i=1}^{N} \hat{y}_{if}/N$, as follows:

$$\tilde{y}_{if}^{*} = \eta \bar{\hat{y}}_f + (1-\eta)\hat{y}_{if}$$

$$= \bar{\hat{y}}_f + (1-\eta)(\hat{y}_{if} - \bar{\hat{y}}_f). \tag{2}$$

From the second line of (2), it is seen that for $0 < \eta < 1$, a country's forecast, $\hat{y}_{if}$, is shrunk toward the average forecast $\bar{\hat{y}}_f$ for all countries.

One way to obtain an optimal forecast in the form of (2) is to employ the following predictive loss function:

$$L = (y_{if} - \tilde{y}_{if})^2 + c\left(\sum_{i=1}^{N} y_{if}/N - \tilde{y}_{if}\right)^2, \tag{3}$$

where $c > 0$ is a given constant, the y_{if}'s are the future unknown values, $i = 1, 2, \ldots, N$, and $\tilde{y}_{if}$ is some forecast. Note that the loss function in (3) incorporates an element of loss associated with being away from the mean outcome in its second term. Under the assumption that the y_{if}'s are independent [common influences have been represented by input variables in (1)] and have predictive probability density functions (pdfs) with mean m_i and variance v_i, $i = 1, 2, \ldots, N$. The predictive expectation of the loss function in (3) is

$$\mathrm{E}L = v_i + \left(m_i - \tilde{y}_{if}\right)^2 + c\left[\mathrm{E}\left(\bar{y}_f - \mathrm{E}\bar{y}_f\right)^2 + \left(\mathrm{E}\bar{y}_f - \tilde{y}_{if}\right)^2\right], \tag{4}$$

where $\bar{y}_f = \sum_{i=1}^N y_{if}/N$. On minimizing (4) with respect to $\tilde{y}_{if}$, the result is

$$\tilde{y}_{if}^* = \eta \sum_{i=1}^N m_i/N + (1-\eta)m_i, \tag{5}$$

where $\eta = c/(1+c)$. If diffuse priors for the $\boldsymbol{\beta}_i$'s in (1) are employed, the means of the predictive pdfs are $m_i = \boldsymbol{x}_{if}'\hat{\boldsymbol{\beta}}_i$, $i = 1, 2, \ldots, N$, where $\boldsymbol{x}_{if}'$ is a vector of observed inputs for the first future period and $\hat{\boldsymbol{\beta}}_i = (X_i'X_i)^{-1}X_i'y_i$, the least squares estimate for country i. Under these conditions (5) takes the form of (2) with $\hat{y}_{if} = \boldsymbol{x}_{if}'\hat{\boldsymbol{\beta}}_i$. This is the 'diffuse prior η-forecast'.

Another approach for obtaining relatively simple shrinkage forecasts is a slightly modified form of the Lindley–Smith (1972) procedure in which the coefficient vectors are assumed generated by

$$\boldsymbol{\beta}_i = \boldsymbol{\theta} + \boldsymbol{\delta}_i, \qquad i = 1, 2, \ldots, N, \tag{6}$$

with the $\boldsymbol{\delta}_i$'s assumed independently distributed, each having a $\mathrm{N}(\mathbf{0}, \phi^{-1}\sigma_u^2 I_k)$ distribution where $0 < \phi < \infty$, σ_u^2 is a common variance of u_{it} for all i and t, and $\boldsymbol{\theta}$ is a $k \times 1$ mean vector. If the u_{it}'s are assumed normally and independently distributed, each with zero mean and common variance σ_u^2, then a conditional point estimate for $\boldsymbol{\beta}' = (\boldsymbol{\beta}_1', \boldsymbol{\beta}_2', \ldots, \boldsymbol{\beta}_N')$, denoted by $\tilde{\boldsymbol{\beta}}_a$, an $Nk \times 1$ vector, is given by

$$\tilde{\boldsymbol{\beta}}_a = \left(Z'Z + \phi I_{Nk}\right)^{-1}\left(Z'Z\hat{\boldsymbol{\beta}} + \phi J\bar{\boldsymbol{\beta}}\right), \tag{7}$$

where Z is a block-diagonal matrix with $X_1, X_2, \ldots, X_N$ on the main diagonal, $\hat{\boldsymbol{\beta}} = (Z'Z)^{-1}Z'y$, where $y' = (y_1', y_2', \ldots, y_N')$, $\hat{\boldsymbol{\beta}}' = (\hat{\boldsymbol{\beta}}_1', \hat{\boldsymbol{\beta}}_2', \ldots, \hat{\boldsymbol{\beta}}_N')$ with $\hat{\boldsymbol{\beta}}_i = (X_i'X_i)^{-1}X_i'y_i$, $J' = (I_k, I_k, \ldots, I_k)$, and

$$\bar{\boldsymbol{\beta}} = \left(\sum_{i=1}^N X_i'X_i\right)^{-1} \sum_{i=1}^N X_i'X_i\hat{\boldsymbol{\beta}}_i, \tag{8}$$

a matrix-weighted average of the least squares estimates, the $\hat{\boldsymbol{\beta}}_i$'s, which replaces $\boldsymbol{\theta}$ in (7). Also, $\bar{\boldsymbol{\beta}}$ in (8) can be obtained by regressing $y' = (y_1', y_2', \ldots, y_N')$ on X, where $X' = (X_1', X_2', \ldots, X_N')$, that is, from one big regression in which it is assumed that the $\boldsymbol{\beta}_i$'s are equal. Point forecasts can be obtained using the coefficient estimate $\tilde{\boldsymbol{\beta}}_a$ for various selected values of ϕ. When ϕ is very large, (7) reduces approximately to (8). Forecasts based on the estimate in (8) have been reported earlier in Garcia-Ferrer et al. (1987). It should be recognized that, while the β_i's are probably not all the same, the bias introduced by assuming them to be may be more than offset in a MSE error sense by a reduction in variance.

As an alternative to the assumptions used in connection with (6), following the 'g-prior' approach of Zellner (1983, 1986), we assume that the δ_i's in (6) are independently distributed with normal distributions $\mathrm{N}[\mathbf{0}, (X_i'X_i)^{-1}\sigma_\delta^2]$. With this assumption and the earlier assumption made about the u_{it}'s, the joint pdf for $y' = (y_1', y_2', \ldots, y_N')$ and $\beta' = (\beta_1', \beta_2', \ldots, \beta_N')$ is proportional to

$$\exp\{-[(y - Z\beta)'(y - Z\beta) + \gamma(\beta - J\theta)'Z'Z(\beta - J\theta)]/2\sigma_u^2\}, \quad (9)$$

where $\gamma = \sigma_u^2/\sigma_\delta^2$ and other quantities have been defined in connection with (6)–(7). On completing the square on β in the exponential terms of (9), the mean of β given y, γ and θ is

$$\bar{\beta} = \left[(Z'Z)^{-1}Z'y + \gamma J\theta\right]/(1 + \gamma), \quad (10a)$$

with the ith subvector of $\bar{\beta}$ given by

$$\bar{\beta}_i = \left(\hat{\beta}_i + \gamma\theta\right)/(1 + \gamma), \quad (10b)$$

where $\hat{\beta}_i = (X_i'X_i)^{-1}X_i'y_i$, the least squares estimate for the ith country's data. Thus (10b) is a simple average of $\hat{\beta}_i$ and θ with $\gamma = \sigma_u^2/\sigma_\delta^2$ involved in the weights. When a diffuse prior pdf for θ, $p(\theta) \propto$ const., is employed, the posterior pdf for θ can be derived from (9) and employed to average the expression in (10) to obtain the marginal mean of β_i given γ and the data, namely,

$$\bar{\beta}_i^m = \left(\hat{\beta}_i + \gamma\tilde{\beta}\right)/(1 + \gamma), \quad (11)$$

with $\tilde{\beta}$ given in (8), the estimate resulting from a big regression in which the β_i's are assumed equal. As $\gamma = \sigma_u^2/\sigma_\delta^2$ grows in value, $\bar{\beta}_i^m \to \tilde{\beta}$ while as $\gamma = \sigma_u^2/\sigma_\sigma^2 \to 0$, $\bar{\beta}_i^m \to \hat{\beta}_i$, the ith country's least squares estimate.

If instead of assuming that δ_i has a $\mathrm{N}(\mathbf{0}, (X_i'X_i)^{-1}\sigma_\delta^2)$ distribution, we assume that the δ_i's are independently distributed, with a $\mathrm{N}[\mathbf{0}, (X_i'X_i)^{-1}\sigma_i^2]$ distribution, $i = 1, 2, \ldots, N$, then analysis similar to that presented in connection with (9) yields as the conditional mean of β_i,

$$\bar{\beta}_i^c = \left(\hat{\beta}_i + \gamma_i\theta\right)/(1 + \gamma_i), \qquad i = 1, 2, \ldots, N, \quad (12)$$

where $\gamma_i = \sigma_u^2/\sigma_i^2$. If we further condition on $\theta = \tilde{\beta}$, with $\tilde{\beta}$ given in (8), then

$$\bar{\beta}_i^c = \left(\hat{\beta}_i + \gamma_i\tilde{\beta}\right)/(1 + \gamma_i), \qquad i = 1, 2, \ldots, N, \quad (13)$$

which is similar to (11) except that the γ_i's are not all equal as is the case in (11).

Upon introducing prior pdfs for σ_u and σ_δ, or σ_u and the σ_i's, it is possible to compute the marginal distributions of the β_i's, a possibility to be explored in future work – see Miller and Fortney (1984) for interesting computations on a closely related problem. At present, we shall evaluate (13) for various values of γ_i and determine the quality of resulting forecasts. That is, the γ_i-forecast for country i is

$$\tilde{y}_{if} = \boldsymbol{x}'_{if}\bar{\boldsymbol{\beta}}_i^c, \tag{14}$$

with $\bar{\boldsymbol{\beta}}_i^c$ given in (13) and $\boldsymbol{x}'_{if}$ a given input vector.

In summary, we shall use the η-forecast in (2), the γ-forecast based on (11) and the γ_i-forecast in (14) in our forecasting experiments. Also, note that the η-forecasting approach can be applied to the γ-forecasts.

2.3. *Elaboration of the AR(3)LI model*

As mentioned previously, we think that it is advisable to add a variable reflecting world real income growth, denoted by w_t, to our AR(3)LI model in (1). Then our equation becomes

$$y_{it} = w_t\alpha_i + \boldsymbol{x}'_{it}\boldsymbol{\beta}_i + u_{it}, \qquad i = 1,2,\dots,N, \quad t = 1,2,\dots,T, \tag{15}$$

where $\boldsymbol{x}'_{it}\boldsymbol{\beta}_i$ represents the constant and other lagged variables in (1) and α_i is the ith country's coefficient of the world income growth rate variable, w_t. To forecast one period ahead using (15), it is clear that w_t must be forecasted. To do this we introduce the following equation for w_t which will be estimated and used to forecast w_t one period in the future:

$$\begin{aligned} w_t = {} & \pi_0 + \pi_1 w_{t-1} + \pi_2 w_{t-2} + \pi_3 w_{t-3} + \pi_4 MSR_{t-1} \\ & + \pi_5 MGM_{t-1} + v_t, \qquad t = 1,2,\dots,T, \end{aligned} \tag{16}$$

where MSR_t is the median of all countries' real stock returns, MGM_t the median of all countries' real money growth rates, v_t a disturbance term, and the π_i's are parameters. Thus (16) indicates that we are employing an AR(3)LI model for w_t, the rate of growth of world real income. As a proxy for w_t, we employ the median of all countries' real output growth rates.

Viewing (15)–(16), it is seen that we have a 'triangular' system. For the future period $f = T + 1$, we have

$$\mathrm{E}y_{if} = \mathrm{E}w_f\mathrm{E}\alpha_i + \boldsymbol{x}'_{if}\mathrm{E}\boldsymbol{\beta}_i, \qquad i = 1,2,\dots,N, \tag{15a}$$

$$\mathrm{E}w_f = z'_f\mathrm{E}\boldsymbol{\pi}, \tag{15b}$$

where Ey_{if} and Ew_f are means of the predictive pdfs for y_{if} and w_f, respectively, and $E\alpha_i$, $E\beta_i$ and $E\pi$ are posterior means of the parameters α_i, β_i and $\pi' = (\pi_0, \pi_1, \ldots, \pi_6)$, respectively. If the system in (15)–(16) is *fully recursive* and diffuse prior pdfs for all parameters are employed, $E\pi = \hat{\pi}$, $E\alpha_i = \hat{\alpha}_i$ and $E\beta_i = \hat{\beta}_i$, where $\hat{\pi}$, $\hat{\alpha}_i$ and $\hat{\beta}_i$ are least squares estimates – see Zellner (1971, ch. 8) and Bowman and Laporte (1975). Also, $Ew_f = z_f'\hat{\pi}$ and $Ey_{if} = z_f'\hat{\pi}\hat{\alpha}_i + x_{if}'\hat{\beta}_i$, $i = 1, 2, \ldots, N$. Thus forecasts under these assumptions are easily computed. If the system in (15)–(16) is not fully recursive, that is, the u_{it}'s and v_t are correlated, then the expectations in (15) have to reflect the non-recursive nature of the system. In the present work, we shall use a 'conditional' forecasting procedure which is equivalent to a 2SLS point forecast. That is, the parameters α_i and β_i in (15) are estimated by 2SLS, a conditional Bayesian estimate [see Zellner (1971, p. 266)], and these estimates along with a forecast of w_t from (16) are employed to obtain a forecast of y_{it} from (15). Such forecasts will be compared with those that assume that w_t's value in a forecast period is perfectly known, a 'perfect foresight' assumption. In current work, an unconditional Bayesian approach for analyzing (15)–(16) when the u_{it}'s and v_t are correlated is being developed.

3. Data

Annual data for eighteen countries employed in our work have been assembled in the main from the International Monetary Fund's International Financial Statistics data base and are available on a diskette from the authors for a nominal fee to cover costs. The output data include annual rates of growth of real output (GNP or GDP), of real stock prices and of real money for each country. In computing rates of growth of real stock prices, an index of nominal stock prices was deflated by an index of the price level for each country. Nominal money, M_1, was deflated by a general price index for each country to obtain a measure of real money.

Boxplots of output growth rates, real stock price growth rates and real money growth rates are shown in fig. 1. It is seen that the median growth rates exhibit a cyclical pattern with that for real stock prices having a considerably greater amplitude than those for output and real money growth rates. Also, as might be expected, the interquartile ranges of the real stock price growth rates are much larger than those of output and real money growth rates. Further, the interquartile ranges for growth rates of real stock prices appear to be slightly smaller in the vicinity of troughs than of peaks for the first half of the sample and the ranges for all three variables tend to be slightly larger in the vicinity of peaks than of troughs in many cases. Last, the plots of the real money and real stock price growth rates give some evidence of a slight lead relative to those of the output growth rates.

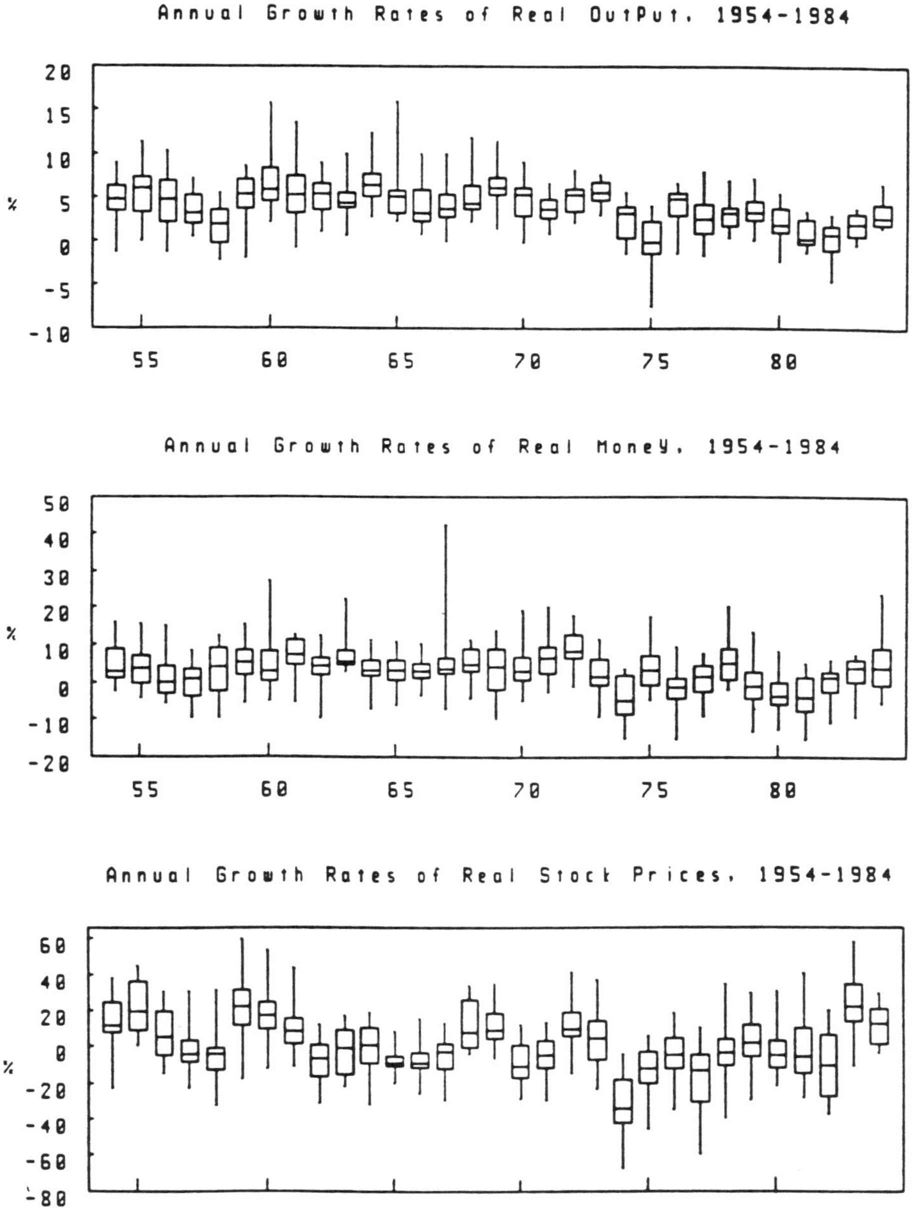

Fig. 1. Boxplots of data for eighteen countries, 1954–84.

It will be noted from the plots in fig. 1 that there are some apparently outlying points in the data with a number of them present in the data from the nine additional countries' data. These outlying data points are being subjected to close scrutiny and procedures for accommodating outlying data points are being considered for use in future work. In the present study, *all data were employed*, including outlying data without any special treatment given to them.

4. Forecasting results

In this section, we first compare RMSEs of one-year-ahead forecasts, 1974–81 (eight years), with those for the period 1974–84 for our original nine countries using various models and methods. Then for nine additional countries, forecasting results for the period 1974–84 are presented and compared with earlier results. Finally, the effects of introducing a world growth rate variable in our AR(3)LI model in (1a) on forecasting performance will be described.

4.1. Forecasting results for an expanded data set

Shown in table 1 are the RMSEs of forecast for nine countries for the periods 1974–81 and for 1974–84. Here and elsewhere, all models were re-estimated using data up to the forecast year. In the top panel of table 1, results for the eight one-year-ahead forecasts, 1974–81, are shown. It is seen that the median RMSE for the AR(3)LI model, 2.23 percentage points, is quite a bit lower than the median RMSEs for the naive models and for the AR(3) model. In addition, shrinkage or pooling techniques applied to the AR(3)LI model led to median RMSEs of 2.22 and 1.78, a very slight reduction in the former case and a somewhat larger reduction in the latter case, from 2.23 to 1.78 percentage points. Use of the η-shrinkage procedure, with $\eta = 0.5$, led to reduction of RMSEs for seven of the nine countries.

As regards the results for 1974–84, in the lower panel of table 1, the AR(3)LI model's median RMSE is 2.41 percentage points, a good deal lower than those associated with the naive models and the AR(3) model. Also, the two shrinkage or pooling procedures produced modest decreases in median RMSEs, from 2.41 to 2.26 and 2.31 and in six of nine cases for the η-shrinkage procedure. In these respects, the forecasting results parallel those obtained for the shorter period, 1974–81. However, note that there is an increase in the median RMSEs for the AR(3)LI model in going from the period 1974–81 to the longer period 1974–84 which also shows up in seven of the nine countries' RMSEs.

In table 2 RMSEs of forecast are shown for nine additional countries. Here the median RMSE for the AR(3)LI model, 3.33, is somewhat larger than that for the naive models and for the AR(3) model. However, the shrinkage or pooled forecasts based on the AR(3)LI model show substantial reductions in median RMSEs, from 3.33 to 2.01 and 2.32, which are similar in magnitude to

Table 1

Nine countries' RMSEs of one-year-ahead forecasts, 1974–81 and 1974–84.

Model	Belgium	Denmark	France	Germany	Ireland	Italy	Netherlands	U.K.	U.S.	Median
					(Percentage points)					
					(A) 1974–81					
A. NMI ($\hat{y}_t = 0$)	3.09	2.83	2.96	2.95	4.38	3.72	3.77	2.21	3.48	3.09
B. NMII ($\hat{y}_t = y_{t-1}$)	4.25	3.73	2.43	3.26	2.06	4.88	4.04	3.91	3.60	3.73
C. NMIII ($\hat{y}_t$ = past average)	3.23	3.48	3.05	3.87	1.88	3.90	3.74	2.95	2.81	3.23
D. AR(3)[a]	3.66	3.46	2.89	3.39	1.69	4.75	3.52	3.50	2.48	3.46
E. AR(3) LI[b]	1.56	2.92	2.43	1.47	1.83	2.57	2.63	2.23	1.82	2.23
1. Shrinkage(1)[c]	1.69	2.37	1.35	2.03	1.77	2.22	2.87	2.26	2.75	2.22
2. Shrinkage(2)[d]	1.68	2.21	1.61	1.25	1.52	2.01	2.52	2.46	1.78	1.78
					(B) 1974–84					
A. NMI ($\hat{y}_t = 0$)	2.59	2.78	2.65	2.69	4.02	3.27	3.32	2.30	3.79	2.78
B. NMII ($\hat{y}_t = y_{t-1}$)	3.53	3.56	2.20	2.91	2.85	4.26	3.57	3.69	3.89	3.56
C. NMIII ($\hat{y}_t$ = past average)	3.01	3.05	3.08	3.85	2.35	3.98	3.87	2.62	3.09	3.08
D. AR(3)[a]	2.98	2.96	2.47	3.10	2.29	4.34	3.35	3.21	3.01	3.01
E. AR(3)LI[b]	1.73	2.73	2.52	2.28	2.80	3.40	2.41	2.32	2.14	2.41
1. Shrinkage(1)[c]	1.96	2.26	1.66	2.00	2.14	2.45	2.53	2.39	2.79	2.26
2. Shrinkage(2)[d]	1.81	2.37	2.07	1.94	2.31	2.73	2.50	2.63	2.03	2.31

[a] Least squares forecasts from an AR(3) model for each country.
[b] Least squares forecasts from an AR(3) model with leading indicator variables shown in eq. (1a).
[c] Least squares forecasts with use of coefficient estimate in (8).
[d] Use of shrinkage equation in (2) with $\eta = 0.5$.

Table 2

Nine additional countries' RMSEs of one-year-ahead forecasts, 1974–84.

Model	Australia	Austria	Canada	Finland	Japan	Norway	Spain	Sweden[e]	Switzerland	Median[e]
	(Percentage points)									
A. NMI ($\hat{y}_t = 0$)	3.10	2.91	3.51	3.39	4.10	4.15	2.51	2.29	3.03	3.10
B. NMII ($\hat{y}_t = y_{t-1}$)	2.90	2.82	4.13	2.34	3.12	1.98	1.86	1.87	3.91	2.82
C. NMIII ($\hat{y}_t$ = past average)	2.75	3.11	3.31	2.88	4.98	1.76	3.56	2.43	4.48	3.11
D. AR(3)[a]	2.87	3.16	3.55	2.58	3.26	1.75	2.50	2.22	4.24	2.87
E. AR(3)LI[b]	3.34	2.71	3.68	3.37	3.33	1.62	2.06	2.32	3.45	3.33
1. Shrinkage(1)[c]	2.01	1.77	2.39	2.01	2.40	1.68	1.65	2.01	2.71	2.01
2. Shrinkage(2)[d]	2.32	2.12	2.92	2.42	2.51	1.52	1.81	2.29	3.25	2.32

[a] Least squares forecasts from an AR(3) model for each country.
[b] Least squares forecasts from an AR(3) model with leading indicator variables shown in eq. (1a).
[c] Least squares forecasts with use of coefficient estimate in (8).
[d] Use of shrinkage equation in (2) with $\eta = 0.5$.
[e] Based on ten forecasts, 1974–83.

Table 3

Summary statistics on forecasting results for eighteen countries, 1974–84.

Model	Median RMSE	Smallest RMSE	Largest RMSE
		(Percentage points)	
A. NMI ($\hat{y}_t = 0$)	3.07	2.29	4.15
B. NMII ($\hat{y}_t = y_{t-1}$)	3.02	1.87	4.26
C. NMIII ($\hat{y}_t$ = past average)	3.09	1.76	4.98
D. AR(3)[a]	3.00	1.75	4.34
E. AR(3)LI[b]	2.62	1.62	3.68
1. Shrinkage(1)[c]	2.14	1.65	2.79
2. Shrinkage(2)[d]	2.32	1.52	3.25

[a]Least squares forecasts from an AR(3) model for each country.
[b]Least squares forecasts from an AR(3) model with leading indicator variables shown in eq. (1a).
[c]Least squares forecasts with use of coefficient estimate in (8).
[d]Use of shrinkage equation in (2) with $\eta = 0.5$.

those reported for the original nine countries for the periods 1974–81 and 1974–84 in table 1. In table 2, on comparing the AR(3)LI country RMSEs with the corresponding shrinkage RMSEs, it is seen that there is a reduction of RMSEs in all but one case. Thus for the nine additional countries, shrinkage results in a somewhat greater improvement in forecasting results than in the case of the original nine countries.

In table 3 summary forecasting results for different models, applied to all eighteen countries' data to forecast year by year for the period 1974–84, are reported. It is seen that the AR(3)LI model's median RMSE is about 13 percent or more below those of the AR(3) and naive model's median RMSEs, 2.62 versus 3.00 or greater. Also, the shrinkage or pooled forecasts median RMSEs, 2.14 and 2.32, are 18 and 11 percent lower, respectively, than the AR(3) model's median RMSE. Further, the ranges of the AR(3)LI and shrinkage forecast RMSEs for the eighteen countries are much smaller than those for the AR(3) and naive models.

In summary, the shrinkage or pooling techniques used earlier in Garcia-Ferrer et al. (1987) for nine countries for the period 1974–81 are effective in reducing overall median RMSEs for an extended forecast period, 1974–84, and for nine additional countries.

The η-shrinkage forecast RMSEs, reported above, are based on the same value of $\eta = 0.5$ for all countries. It is of interest to see how sensitive forecasting results for individual countries are to variation in the value of η. In table 4 one-year-ahead forecast RMSEs are reported for selected values of η in (2) for each of the eighteen countries in our sample. When $\eta = 0$, the forecasts are least squares forecasts for each country based on our AR(3)LI model in (1a). When $\eta = 1.0$, the forecast for each country is the mean of the eighteen countries' least squares forecasts from the AR(3)LI model in (1a). On

Table 4

Sensitivity of η-forecast RMSEs to value of η for AR(3)LI model in (1a), 1974–84.[a]

Country	$\eta = 0$	$\eta = 0.25$	$\eta = 0.50$	$\eta = 0.75$	$\eta = 1.0$
	(Percentage points)				
Belgium	1.73	1.70	1.81	2.04	2.35
Denmark	2.73	2.49	2.37[b]	2.41	2.59
France	2.52	2.29	2.07	1.90	1.77[b]
Germany	2.28	2.04	1.94[b]	2.01	2.24
Ireland	2.80	2.51	2.31	2.24[b]	2.29
Italy	3.40	3.04	2.73	2.46	2.27[b]
Netherlands	2.41[b]	2.42	2.50	2.65	2.87
U.K.	2.32[b]	2.44	2.63	2.89	3.20
U.S.	2.14	1.97[b]	2.03	2.31	2.75
Australia	3.34	2.80	2.32	1.92	1.67[b]
Austria	2.71	2.40	2.12	1.87	1.67[b]
Canada	3.68	3.27	2.92	2.64	2.46[b]
Finland	3.37	2.86	2.42	2.10	1.95[b]
Japan	3.33	2.90	2.51	2.18	1.96[b]
Norway	1.62	1.53	1.52[b]	1.57	1.70
Spain	2.06	1.86	1.81[b]	1.91	2.14
Sweden[c]	2.32	2.28	2.29[b]	2.34	2.44
Switzerland	3.45	3.26	3.25[b]	3.42	3.75
Median RMSE	2.62	2.43	2.32	2.21	2.28
Range	1.62–3.68	1.53–3.27	1.52–3.25	1.57–3.42	1.67–3 75

[a]The η-forecast is $\tilde{y}_{if} = \eta\hat{y}_f + (1-\eta)\hat{y}_{if}$, where $\hat{y}_f$ is the mean of the eighteen countries' least squares forecasts and $\hat{y}_{if}$ is the ith country's least squares forecast.
[b]Minimum of entries in each row. The median of these RMSEs is 1.96.
[c]Based on ten forecasts, 1974–83.

viewing the median RMSEs at the bottom of table 4, it is seen that the median RMSEs are 2.43 for $\eta = 0.25$, 2.32 for $\eta = 0.5$, and 2.21 for $\eta = 0.75$, all below the median RMSE for $\eta = 0$, 2.62, that for individual countries' unshrunk least squares forecasts. For individual countries, the RMSEs change as the value of η changes. For example in the case of Belgium, the RMSEs vary from 1.73 for $\eta = 0$ to 2.35 for $\eta = 1.0$, with a minimum of 1.70 for $\eta = 0.25$. Two countries' minimal RMSEs occur for $\eta = 0$ and $\eta = 0.25$, six for $\eta = 0.50$, one of $\eta = 0.75$, and seven for $\eta = 1.0$. The median RMSE of these minimal values is 1.96, somewhat below that associated with the use of $\eta = 0.5$ for all countries, namely 2.32, or of $\eta = 0.75$, 2.21.

Just as it is of interest to determine the sensitivity of countries' η-forecasts to the values of η employed, it is of interest to determine how sensitive γ-forecasts for countries are to the value of γ employed. The coefficient estimate in (13) permits γ to be different for different countries. In table 5, RMSEs of forecast are reported for each country for selected values of γ. A zero value for γ yields AR(3)LI least squares forecasts while a very large value for γ results in an AR(3)LI forecast based on the pooled coefficient estimate

Table 5

Sensitivity of γ-forecast RMSEs to values of γ for AR(3)LI model in (1a), 1974–84.[a]

Country	$\gamma = 0$	$\gamma = 0.50$	$\gamma = 1.0$	$\gamma = 2.0$	$\gamma = 3.0$	$\gamma = 5.0$	$\gamma = 10^6$
	(Percentage points)						
Belgium	1.73	1.60[b]	1.62	1.69	1.74	1.81	1.96
Denmark	2.73	2.34	2.22	2.16[b]	2.16[b]	2.18	2.26
France	2.52	2.17	2.01	1.87	1.81	1.75	1.66[b]
Germany	2.28	1.91	1.81	1.80[b]	1.82	1.87	2.00
Ireland	2.80	2.47	2.34	2.24	2.20	2.17	2.14[b]
Italy	3.40	3.00	2.82	2.67	2.61	2.55	2.45[b]
Netherlands	2.41	2.34[b]	2.35	2.38	2.41	2.44	2.53
U.K.	2.32	2.19	2.18[b]	2.21	2.24	2.28	2.39
U.S.	2.14	1.98[b]	2.06	2.24	2.35	2.48	2.79
Australia	3.34	2.72	2.46	2.25	2.16	2.09	2.01[b]
Austria	2.71	2.35	2.19	2.03	1.96	1.89	1.77[b]
Canada	3.68	3.08	2.83	2.63	2.54	2.47	2.39[b]
Finland	3.37	2.70	2.42	2.20	2.12	2.06	2.01[b]
Japan	3.33	2.89	2.71	2.57	2.51	2.46	2.40[b]
Norway	1.62	1.54[b]	1.54[b]	1.56	1.58	1.61	1.68
Spain	2.06	1.85	1.76	1.70	1.68	1.66	1.65[b]
Sweden[c]	2.32	2.18	2.12	2.07	2.05	2.03	2.01[b]
Switzerland	3.45	2.98	2.81	2.71	2.68[b]	2.68[b]	2.71
Median RMSE	2.62	2.34	2.22	2.21	2.16	2.13[b]	2.14
Range	1.62–3.68	1.54–3.08	1.54–2.83	1.56–2.71	1.58–2.68	1.61–2.68	1.65–2.79

[a] The coefficient estimate in (13) was employed to compute forecasts. When $\gamma = 0$, the forecasts are least squares forecasts, and when $\gamma = 10^6$, they are produced using the coefficient estimate in (8).

[b] Minimum of entries in each row.

[c] Based on ten forecasts, 1974–83.

in (8). From the median RMSEs reported at the bottom of Table 5, it is seen that a common value of $\gamma = 5.0$ yields a median RMSE equal to 2.13, not far different from that associated with $\gamma = 10^6$, namely 2.14. However for individual countries, RMSEs show more substantial variation as γ assumes different values. For example in the case of Germany, the RMSE is 1.80 when $\gamma = 2.0$, quite a bit lower than the RMSE of 2.00 associated with $\gamma = 10^6$. Ten of the eighteen countries show minimal RMSEs for $\gamma = 10^6$, while the remaining eight countries have minima in the vicinity of $\gamma = 0.5$ to $\gamma = 1.0$ in five cases and of $\gamma = 2$ to $\gamma = 5$ for the remaining three. The median RMSE for these minimal values is 2.01, slightly lower than the median RMSE of 2.14 when $\gamma = 10^6$ is used for all countries. In future work it may be worthwhile to estimate γ for each country which will probably produce lower RMSEs for selected countries, e.g., Belgium, the Netherlands, the U.S., and Norway. Also combinations of γ-forecast and η-forecast procedures are under consideration as well as other shrinkage techniques.

Table 6

RMSEs and MAEs of one-year-ahead forecasts of the medians of eighteen countries' output growth rates, 1974–84.[a]

Model	RMSE	MAE
	(Percentage points)	
AR(3)	2.74	2.24
AR(3)LI[b]	1.48	1.24

[a]The initial estimation period is 1954–73 (20 years). Estimates are updated year by year in the forecast period.

[b]With w_t being median output growth rate in year t, the AR(3)LI model is $w_t = \pi_0 + \pi_1 w_{t-1} + \pi_2 w_{t-2} + \pi_3 w_{t-3} + \pi_4 x_{t-1} + \pi_5 z_{t-1} + \varepsilon_t$, where, for year t, x_t is the median of countries' growth rates of real stock prices and z_t the median of countries' growth rates of real money. This equation was employed to generate one-year-ahead least squares forecasts for each year, 1974–84

Table 7

RMSEs of one-year-ahead forecasts of annual real output growth rates employing an AR(3)LI model including the world growth rate, 1974–84.[a]

Country	AR(3)LI with w_t[b]	AR(3)LI with $\hat{w}_t(1)$[c]	AR(3)LI with $\hat{w}_t(2)$[d]	AR(3)LI without w_t[e]
	(Percentage points)			
Belgium	1.54	2.80	1.79	1.96
Denmark	1.77	2.74	2.04	2.26
France	1.03	1.90	1.36	1.66
Germany	0.90	2.39	1.35	2.00
Ireland	2.82	1.57	2.54	2.14
Italy	1.57	3.08	1.82	2.45
Netherlands	1.86	2.52	2.27	2.53
U.K.	1.87	2.67	2.21	2.39
U.S.	2.56	2.95	2.36	2.79
Australia	2.24	2.39	1.96	2.01
Austria	1.23	2.06	1.57	1.77
Canada	2.27	2.81	2.15	2.39
Finland	2.14	2.13	1.83	2.01
Japan	2.85	2.30	2.58	2.40
Norway	1.66	1.79	1.45	1.68
Spain	1.12	2.03	1.22	1.65
Sweden[f]	1.62	2.58	1.70	2.01
Switzerland	2.20	3.17	2.42	2.71
Median RMSE	1.82	2.46	1.90	2.14
Range	0.90–2.85	1.57–3.17	1.22–2.58	1.65–2.71

[a]The model employed is shown in (15) and estimated under the assumption that coefficients are the same for all countries.

[b]The value of w_t, the median output growth rate or world growth rate, is assumed known in the forecast period, a 'perfect foresight' assumption.

[c]$\hat{w}_t(1)$ is a forecast of w_t from an AR(3) model for w_t – see table 6.

[d]$\hat{w}_t(2)$ is a forecast of w_t from the AR(3)LI model described in footnote b of table 6.

[e]Least squares forecasts using AR(3)LI model in (1), with the use of coefficient estimate in (8).

[f]Based on ten forecasts, 1974–83.

4.2. Forecasting using a world output growth rate variable

As mentioned previously, countries' exports are influenced by world income or output. Thus a world output growth rate variable, denoted by w_t, was added to the AR(3)LI model as shown in eq. (15). The variable w_t is taken to be the median of the eighteen countries' output growth rates for the year t – see fig. 1 for a plot of w_t for the years in our sample.

To use (15) in forecasting, it is necessary to forecast w_t. The model for w_t in (16) was fitted by least squares, using data from 1954–73, and used to forecast the 1974 value and subsequent values with coefficient estimates updated year by year for the years 1974–84. The RMSEs of these one-year-ahead forecasts and of those yielded by an AR(3) model of w_t are reported, along with MAEs, in table 6. It is seen that the AR(3)LI model for w_t produced a RMSE of 1.48 and a MAE of 1.24, values much smaller than those associated with forecasts from an AR(3) model, 2.74 and 2.24, respectively. The forecasts from these two models for w_t, denoted by $\hat{w}_t(1)$ for the AR(3) forecasts and $\hat{w}_t(2)$ for the AR(3)LI model forecasts, were used to generate one-year-ahead forecasts from (15) for individual countries' output growth rates for 1974–84, using the γ-forecast with a very large value of γ. Also, for comparative purposes, a 'perfect foresight' model, one in which it is assumed that w_t is known exactly, was employed to generate forecasts with results given in the first column of table 7. With the w_t value assumed known in each forecast period, the median RMSE of the annual forecasts for the eighteen countries, 1974–84, is 1.82, a value much lower than those reported in table 3, and for the AR(3)LI model without the w_t variable it is 2.14, shown in the last column of table 7. In column 2 of table 7 are shown forecast RMSEs when w_t was forecasted using an AR(3) model. Since the AR(3) forecasts of w_t are not very good – see table 6 – the forecasts of country output growth rates based on them are in general not as good as those based on known values of the w_t variable. When the AR(3)LI model in (16) was used to produce forecasts of w_t, denoted by $\hat{w}_t(2)$, and these were used to forecast individual countries' output growth rates, the results, as shown in column 3 of table 7, were much better. The median RMSE associated with these forecasts is 1.90, not far different from the 'perfect foresight' median RMSE of 1.82. Also when $\hat{w}_t(2)$ was employed, sixteen of eighteen countries' RMSEs were reduced relative to the RMSEs for the AR(3)LI model without the world growth variable, shown in column 4 of table 7. Further, the median RMSE of 1.90, associated with the AR(3)LI world growth rate model, is smaller than all of those shown in table 3.

4.3. Comparisons with OECD forecast RMSEs

Smyth (1983) has presented a description of the forecasting procedures employed by the Organization for Economic Cooperation and Development

(OECD) to produce annual forecasts of seven countries' annual rates of growth of output, 1968–79. The OECD forecasts are derived from elaborate country econometric models and are subjected to judgmental adjustments by individuals not associated with the modeling process, that is by individuals from the OECD's various country desks – see Smyth (1983, p. 37). In Garcia-Ferrer et al. (1987, p. 61ff) comparisons of RMSEs of OECD forecasts for five countries with those provided by AR(3)LI models were presented and discussed. In table 8, OECD RMSEs of forecast for seven countries, 1968–79, are presented along with forecast RMSEs for the same countries, 1974-84, computed in the present study. While the forecast periods 1968–79 and 1974–84 are somewhat different and different methodologies were employed, it is still of interest to consider the relative forecasting performance of OECD and our forecasts.

From table 8, it is seen that the OECD forecast RMSEs have a median of 2.12 percentage points with a range of 1.45–4.40. The OECD's RMSE for Japan, 4.40 is quite large. The median RMSEs, in our study range from 2.52 for least squares forecasts from our AR(3)LI model in (1) to 2.15 for the forecasts obtained from our 'AR(3)LI world income' model in (15) using a forecasted value of w_t from (16) and a γ-forecast with $\gamma = 10^6$. The range of the forecast RMSEs in this latter case is 1.35–2.58. The RMSEs in line 3 of table 8 are smaller than the corresponding OECD RMSEs in five of seven cases. Large reductions were encountered for Germany, Italy and Japan, while smaller reductions appeared for France and the U.K. In the cases of the U.S. and Canada, the OECD RMSEs were smaller, much smaller for the U.S. These comparisons, however, must be qualified since the forecast period and data used by OECD are different from those employed in this study.

5. Summary and concluding remarks

We have presented the results of forecasting experiments for nine countries' annual output growth rates for the periods 1974–81 and 1974–84 and similar results for an additional sample of nine countries for the period 1974–84. In general, the forecasting experiments revealed that methods and models employed earlier in Garcia-Ferrer et al. (1987) worked reasonably well when applied to an extended sample of data and countries. The shrinkage forecasting procedures produced larger reductions in RMSEs of forecast for the nine additional countries than for the original sample of nine countries. The AR(3)LI model, incorporating a world income growth rate variable, forecasted the best and provided forecast RMSEs that compared favorably with the RMSEs of OECD forecasts for a subsample of seven countries that were produced using complex country models and judgmental adjustments.

In future work, we shall extend our forecasting experiments to include forecasts derived from time-varying parameter Bayesian state–space models –

Table 8

Comparison with OECD forecast RMSEs.

Forecasts	Canada	France	Germany	Italy	Japan	U.K.	U.S.	Median	Range
					(Percentage points)				
1. OECD, 1968–79	1.71	1.45	2.12	2.86	4.40	2.26	1.38	2.12	1.38–4.40
2. AR(3)LI, 1974–84									
a. Least squares	3.68	2.52	2.28	3.40	3.33	2.32	2.14	2.52	2.14–3.68
b. γ-forecasts[a]	2.39	1.66	2.00	2.45	2.40	2.39	2.79	2.39	1.66–2.79
c. η-forecasts[b]	2.92	2.07	1.94	2.73	2.51	2.63	2.03	2.51	1.94–2.92
3. AR(3)LI with forecasted w_t, 1974–84[c]	2.15	1.36	1.35	1.82	2.58	2.21	2.36	2.15	1.35–2.58

[a] $\gamma = 10^6$ in model for eighteen countries.
[b] $\eta = 0.5$ in model for eighteen countries.
[c] w_t, the world output growth rate was forecasted from (16) and used in (15) to produce γ-forecasts for eighteen countries with $\gamma = 10^6$.

see Garcia-Ferrer et al. (1987) and Highfield (1986) for macroeconomic forecasting results obtained with such models. In addition, as stated in section 1, macroeconomic structural equation systems are under consideration that have reduced form equations for the rate of growth of output that are similar in form to the forecasting equations used in our present and past work. These structural macroeconomic equation systems also yield reduced form equations for several additional variables that will be appraised in future forecasting experiments. With several forecasting equations per country, there will be an opportunity to experiment with a broader range of shrinkage forecasting techniques. Hopefully, this work will yield a set of forecasting equations for each country that yield reasonably good forecasts for a broad sample of countries and for a temporally expanded data set. Further, disaggregation will be studied by modeling components of GNP as fractions or log-fractions of relevant aggregates. In this way, we hope to 'iterate in' to satisfactory structural models for countries in the structural econometric modeling, time series analysis (SEMTSA) approach put forward in our past work – see Zellner and Palm (1974, 1975), Palm (1983) and Zellner (1979, 1984, 1987).

References

Bowman, H.W. and A.M. Laporte, 1975, Stochastic optimization in recursive equation systems with random parameters with an application to control of the money supply, in: S.E. Fienberg and A. Zellner, eds., Studies in Bayesian econometrics and statistics in honor of Leonard J. Savage (North Holland, Amsterdam) 441–462.

Garcia-Ferrer, A., R.A. Highfield, F. Palm and A. Zellner, 1987, Macro-economic forecasting using pooled international data, Journal of Business and Economic Statistics 5, 53–67.

Highfield, R.A., 1986, Forecasting with Bayesian state space models, Unpublished doctoral dissertation (Graduate School of Business, University of Chicago, Chicago, IL).

Lindley, D.V. and A.F.M. Smith, 1972, Bayes' estimates for the linear model, Journal of the Royal Statistical Society B 34, 1–41.

Miller, R.B. and W.G. Fortney, 1984, Industry-wide expense standards using random coefficient regression, Insurance: Mathematics and Economics 3, 19–33.

Palm, F.C., 1983, Structural econometric modeling and time series analysis: An integrated approach, in: A. Zellner, ed., Applied time series analysis of economic data, Proceedings of the conference on applied time series analysis of economic data, Oct. 13–15, 1981 (U.S. Government Printing Office, Washington, DC) 199–233.

Smyth, D.J., 1983, Short-run macroeconomic forecasting: The OECD performance, Journal of Forecasting 2, 37–49.

Zellner, A., 1971, An introduction to Bayesian inference in econometrics (Wiley, New York); reprinted in 1987 (Krieger, Melbourne, FL).

Zellner, A., 1979, Statistical analysis of econometric models, Journal of the American Statistical Association 74, 628–651.

Zellner, A., 1983, Applications of Bayesian analysis in econometrics, The Statistician 32, 23–34.

Zellner, A., 1984, Basic issues in econometrics (University of Chicago Press, Chicago, IL).

Zellner, A., 1986, On assessing prior distributions and Bayesian regression analysis with g-prior distributions, in: P.K. Goel and A. Zellner, eds., Bayesian inference and decision techniques: Essays in honor of Bruno de Finetti (North-Holland, Amsterdam) 233–243.

Zellner, A., 1987, Macroeconomics, econometrics and time series analysis, Revista Espanola de Economia 4, 3–9.

Zellner, A. and F.C. Palm, 1974, Time series analysis and simultaneous equation econometric models, Journal of Econometrics 2, 17–54.

Zellner, A. and F.C. Palm, 1975, Time series and structural analysis of monetary models of the U.S. economy, Sankhyā 37, 12–56.

Journal of Econometrics 49 (1991) 275–304. North-Holland

Forecasting turning points in international output growth rates using Bayesian exponentially weighted autoregression, time-varying parameter, and pooling techniques*

Arnold Zellner, Chansik Hong, and Chung-ki Min
University of Chicago, Chicago, IL 60637, USA

In this paper two variants of an autoregressive, leading indicator (ARLI) model are employed to forecast turning points in the growth rates of 18 countries' annual real output for the years 1974–1986. Bayesian predictive densities have been employed to compute probabilities of downturns and upturns, and they have been employed in a Bayesian decision-theoretic framework to obtain optimal turning point forecasts. Various versions of the ARLI model, with and without pooling of data across countries, produced about 70 to 80 percent correct forecasts for 158 turning points. Time-varying parameter models performed slightly better than others. The performance of the ARLI models in forecasting turning points is shown to be considerably better than that of four naive methods of forecasting turning points.

1. Introduction

In previous work, [Zellner, Hong, and Gulati (1990) and Zellner and Hong (1989)] the problem of forecasting turning points in economic time series was formulated and solved in a Bayesian decision-theoretic framework. The methodology was applied using a fixed parameter autoregressive, leading indicator (ARLI) model and unpooled data for 18 countries to forecast turning points over the period 1974–1985. In the present paper, we investigate the extent to which use of exponential weighting, time-varying parameter ARLI models, and pooling techniques leads to improved results in forecasting turning points for the same 18 countries over a slightly extended period, 1974–1986.

*Research financed in part by the National Science Foundation and by income from the H.G.B. Alexander Endowment Fund, Graduate School of Business, University of Chicago.

The methodology employed in this work has benefited from earlier work of Moore (1983), Moore and Zarnowitz (1982), Zarnowitz (1985), Wecker (1979), and Kling (1987). Just as Wecker and Kling have done, we employ a model for the observations and an explicit definition of a turning point, for example a downturn (DT) or an upturn (UT). Along with Kling, we allow for parameter uncertainty by adopting a Bayesian approach and computing probabilities of a DT or UT given past data from a model's predictive probability density function (pdf) for future observations. Having computed such probabilities from the data, we use them in a decision-theoretic framework with given loss structures to obtain optimal turning point forecasts which can readily be computed.

The plan of our paper is as follows. In section 2, we explain our models and methods. Section 3 is devoted to a description of our data. In section 4, the results of the present paper are compared to those computed using earlier methods and to turning point forecasting results yielded by naive models. Finally, section 5 presents a summary of results and some concluding remarks.

2. Models and methods

In this paper we use two alternative models for each country. The first is a fixed parameter, autoregressive model of order three with lagged leading indicator variables, denoted by FP/ARLI and given by

$$y_{it} = \boldsymbol{x}_{it}'\boldsymbol{\beta}_i + u_{it}, \qquad t = 1,2,\ldots,T, \quad i = 1,2,\ldots,N, \tag{1}$$

where the subscripts i and t denote the value of a variable for the ith country in the tth year. Further,

y_{it} = rate of growth of output, real GNP or GDP,
$\boldsymbol{\beta}_i$ = a $k \times 1$ parameter vector,
u_{it} = disturbance term, and
$\boldsymbol{x}_{it}' = (y_{it-1}, y_{it-2}, y_{it-3}, 1, SP_{it-1}, SP_{it-2}, GM_{it-1}, WSP_{t-1})$,

with

SP_{it} = rate of growth of real stock prices,
GM_{it} = rate of growth of real money, M_1, divided by a general price index, and
WSP_t = median of the SP_{it}'s for the tth year.

Eq. (1) incorporates an autoregression of order three to allow for the possibility of having two complex roots and a real root for the process. After using this assumption in Garcia-Ferrer et al. (1987) and in Zellner and Hong (1989), calculations reported in Hong (1989) show that with high posterior probabilities, there are two complex roots and one real root in the FP/ARLI model for each of the 18 countries in our sample. Further, lagged stock price change variables are introduced to capture expectational, policy change and other chaotic effects. The lagged growth rate of real money is employed to represent real balance effects and WSP_{t-1} to allow for common shocks hitting all countries. See Zellner, Hong, and Gulati (1990) for estimates of the parameters of (1) for each of the 18 countries in our sample and for evidence that the u_{it}'s are not very highly autocorrelated. In our present work, we shall assume that the u_{it}'s are independently distributed, each with a normal distribution with zero mean and variance σ_i^2.

We consider two variants of model (1), namely a fixed parameter variant as shown in (1), and a time-varying parameter (TVP) version, denoted by TVP/ARLI, as shown in (2),

$$y_{it} = \boldsymbol{x}_{it}'\boldsymbol{\beta}_{it} + u_{it}'. \tag{2}$$

Aggregation effects, policy changes, etc. may cause parameters to vary through time and hence the use of a time-varying parameter vector in (2). As in past work, Garcia-Ferrer et al. (1987), we assume that the $\boldsymbol{\beta}_{it}$'s are generated by a vector random walk process,

$$\boldsymbol{\beta}_{it} = \boldsymbol{\beta}_{it-1} + \boldsymbol{v}_{it}, \qquad i = 1,2,\dots,N, \quad t = 1,2,\dots,T, \tag{3}$$

where the $\boldsymbol{v}_{it}$'s are assumed independently and identically normally distributed with zero mean vector and covariance matrix $\phi\sigma_i^2 I_k$, with $0 < \phi < \infty$. When ϕ has a very small value, (2) behaves like a fixed parameter model.

The second model which we shall employ is that given in (1) with the addition of a world real income (WI) growth rate variable, w_t, namely,

$$y_{it} = w_t\alpha_i + \boldsymbol{x}_{it}'\boldsymbol{\beta}_i + \eta_{it}, \qquad i = 1,2,\dots,N, \quad t = 1,2,\dots,T, \tag{4}$$

where w_t is a measure of world income growth, which we represent by the median of the individual country growth rates, the y_{it}'s in the tth year and α_i is a scalar parameter. To use (4) to forecast, there is need for an equation to forecast the w_t's. It is assumed that the η_{it}'s are zero mean, independent normal error terms with variance $\sigma_{\eta i}^2$. In previous work we have found that an AR(3)LI model for w_t works well in forecasting – see Zellner and Hong (1989). This model is given by

$$w_t = \boldsymbol{z}_t'\boldsymbol{\pi} + \varepsilon_t, \qquad t = 1,2,\dots,T, \tag{5}$$

where

$\boldsymbol{\pi}$ = coefficient vector,
ε_t = white noise, zero mean normal disturbance term with variance σ_ε^2, and
$z_t' = (w_{t-1}, w_{t-2}, w_{t-3}, 1, WSP_{t-1}, WGM_{t-1})$,

with

WSP_t = median of countries' SP_{it}'s for year t, and
WGM_t = median of countries' GM_{it} for year t.

Thus the world output growth rate as represented by w_t is assumed generated by an AR(3)LI model.

As written in (4) and (5), the model is a fixed parameter model, denoted by FP/ARLI/WI. Just as above, it is possible to allow the coefficient vectors in (4) and (5) to be time-varying and generated by independent, vector random walk processes. This yields our time-varying parameter, autoregressive, leading indicator, world income model, denoted by TVP/ARLI/WI – see appendix for further details.

As regards pooling techniques – see Garcia-Ferrer et al. (1987) and Zellner and Hong (1989) for effects of pooling on the RMSEs of one-year-ahead point forecasts, we shall employ an extreme form of pooling in the present paper. That is we assume that the coefficient vectors in (1) and (2) are the same for all countries. Under this assumption, (1) and (2) become

$$y_{it} = \boldsymbol{x}_{it}'\boldsymbol{\beta} + u_{it}, \tag{1a}$$

and

$$y_{it} = \boldsymbol{x}_{it}\boldsymbol{\beta}_t + u_{it}'. \tag{2a}$$

Similarly, with respect to (4), we assume

$$y_{it} = w_t\alpha + \boldsymbol{x}_{it}'\boldsymbol{\beta} + \eta_{it}, \tag{4a}$$

and in the TVP version,

$$y_{it} = w_t\alpha_t + \boldsymbol{x}_{it}'\boldsymbol{\beta}_t + \eta_{it}', \tag{4b}$$

$$w_t = z_t'\boldsymbol{\pi}_t + \varepsilon_t'. \tag{5a}$$

While the assumptions embedded in (1a), (2a), (4a), and (4b) may seem extreme, it is the case that they have led to good point forecasts in previous work and, as will be seen, to good turning point forecasts in our present work.

Finally, we note that exponential weighting or 'discounting' [see, e.g., West et al. (1985) and Highfield (1986)] can be employed when parameters are thought to be changing in value through time. By putting heavier weight on recent observations and lighter weight on observations distant in time from the present, use of exponential weighting produces parameter estimates which can adapt to local changes in values of parameters. Thus exponential weighting is a procedure that permits adaptation to deterministic and/or random changes in parameters' values – see appendix for recursions for computing TVP estimates and exponentially weighted parameter estimates.

2.1. *Optimal turning point forecasts*

As mentioned above, if we have a model for the observations, a definition of a turning point, a predictive pdf, and a loss structure, we can compute an optimal turning point forecast. Above we have indicated several models which we shall use in our analyses. As regards definitions of turning points, we shall use the following definitions employed in our previous work.

Definition of a downturn (DT) in year $T+1$: If the annual growth rate observations for country i, y_{iT-2}, y_{iT-1}, y_{iT}, and y_{iT+1} satisfy

$$y_{iT-2}, y_{iT-1} < y_{iT} > y_{iT+1}, \tag{6a}$$

then a DT has occurred in period $T+1$ for country i, while if the growth rates satisfy

$$y_{iT-2}, y_{iT-1} < y_{iT} \leq y_{iT+1}, \tag{6b}$$

no DT (NDT) has occurred for country i in period $T+1$. Note that we condition the definition of a DT on two previous observations being below a third observation. Similarly, we define an upturn as follows:

Definition of an upturn (UT) in year $T+1$: If the growth rate observations satisfy

$$y_{i,T-2}, y_{iT-1} > y_{iT} < y_{iT+1}, \tag{7a}$$

an UT has occurred in period $T+1$, while if

$$y_{iT-2}, y_{iT-1} > y_{iT} \geq y_{iT+1}, \tag{7b}$$

no UT (NUT) has occurred for country i in period $T+1$.

While the definitions in (6) and (7) are not the only possible definitions, we shall use them because they are explicit and seem appropriate for data which may display a cycle with about a four-year period. Note that if there were a

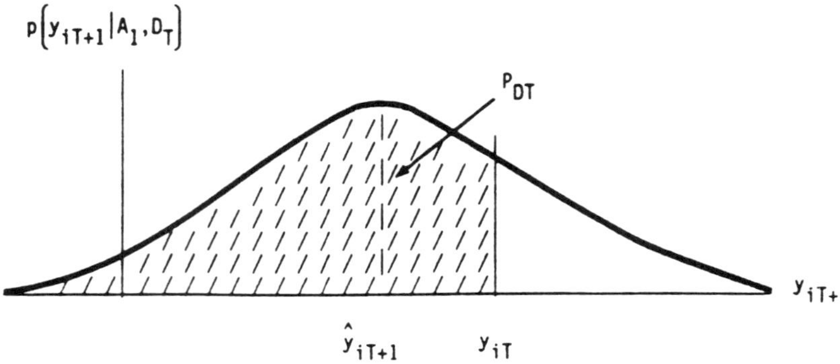

Fig. 1. Calculation of probability of a DT in period $T+1$.

deterministic cycle with an exact four-year period, the above definitions would be completely appropriate.

Having defined a DT in (6a), the problem of forecasting arises when we have observed y_{iT-2}, y_{iT-1}, and y_{iT}, and y_{iT+1} is a future, as yet unobserved, value. If a predictive pdf for y_{iT+1} given all past data, a model, and prior information through period T are available, then we can compute the probability of a DT, P_{DT}, as shown below.

As is well known, in general a predictive pdf for a future observation, e.g., y_{iT+1}, $-\infty < y_{iT+1} < \infty$, is given by

$$p(y_{iT+1}|D_T) = \int_\theta f(y_{iT+1}|\boldsymbol{\theta}, D_T)\pi(\boldsymbol{\theta}|D_T)\,\mathrm{d}\boldsymbol{\theta}, \tag{8}$$

where $f(y_{iT+1}|\boldsymbol{\theta}, D_T)$ is the pdf for y_{iT+1} given the parameter vector $\boldsymbol{\theta} \subset \Theta$, the parameter space, and D_T, the past sample and prior information as of time T, and $\pi(\boldsymbol{\theta}|D_T)$ is the posterior pdf for $\boldsymbol{\theta}$ yielded by Bayes' Theorem.

The predictive pdf in (8) is calculated for each of our models, as shown in the appendix, and used to evaluate probabilities of DT's and UT's. For example, using the definition in (6a) and the predictive pdf in (8), the probability of a DT for country i in the future period $T+1$, denoted by P_{DT}, is

$$P_{\mathrm{DT}} = \Pr(y_{iT+1} < y_{iT}|A_1, D_T) = \int_{-\infty}^{y_{iT}} p(y_{iT+1}|A_1, D_T)\,\mathrm{d}y_{iT+1}, \tag{9}$$

where A_1 denotes the condition: $y_{iT-2}, y_{iT-1} < y_{iT}$. In fig. 1, the calculation of (9) is illustrated. The shaded area in fig. 1 is just the probability of a DT. Note that it will be larger than $\frac{1}{2}$ if the given value $y_{iT} > \hat{y}_{iT+1}$, the modal value of the predictive pdf.

Once P_{DT} has been computed, we use its value in conjunction with the loss structure in table 1 to choose an optimal turning point forecast. In table 1 losses are scaled to be zero if forecasts are correct, while losses $c_1, c_2 > 0$ are incurred for incorrect forecasts. For example, if the forecast DT is selected and the outcome is NDT, then loss c_1 is incurred. Thus c_1 and c_2 represent

Table 1

Loss structure for forecasting downturn (DT) or no downturn (NDT).

	Outcomes		Expected losses
	DT	NDT	
Forecasts			
DT	0	c_1	$c_1(1-P_{\mathrm{DT}})$
NDT	c_2	0	$c_2 P_{\mathrm{DT}}$
Probabilities	P_{DT}	$1-P_{\mathrm{DT}}$	

costs associated with two possible errors, namely (1) forecast DT and the outcome is NDT and (2) forecast NDT and the outcome is DT. See Zellner et al. (1990) for analysis of expanded loss structures which take account of not only cost c_1 and c_2 but also costs associated with errors of point forecasts.

The expected losses, shown in table 1, associated with the choice of a DT or a NDT forecast, can be employed to choose the forecast which minimizes expected loss. For example, if $c_1(1-P_{\mathrm{DT}})<c_2P_{\mathrm{DT}}$, or equivalently,

$$1<(c_2/c_1)P_{\mathrm{DT}}/(1-P_{\mathrm{DT}}), \tag{10}$$

the optimal forecast is DT. Note that the decision rule in (10) involves a cost parameter ratio, c_2/c_1, and the odds on a DT occurring, $P_{\mathrm{DT}}/(1-P_{\mathrm{DT}})$. If $c_2/c_1=1$, the case of a symmetric loss structure, (10) will be satisfied if $P_{\mathrm{DT}}>0.5$. Thus for a symmetric loss structure, forecast DT if $P_{\mathrm{DT}}>\frac{1}{2}$ and NDT if P_{DT} does not satisfy this condition. On the other hand if $c_2/c_1=4$, (10) will be satisfied for $P_{\mathrm{DT}}>0.2$. Thus, if the computed value of P_{DT} is larger than 0.2 and $c_2/c_1=4$, the optimal forecast is DT. In this case, a less stringent condition on P_{DT} is employed to help avoid the relatively large cost c_2 associated with an incorrect NDT forecast. See Zellner and Hong (1988) for the results of forecasting turning points using various cost parameter ratios.

Since the methodology of forecasting UT's using the definition in (7), a predictive pdf for y_{iT+1}, and a loss structure similar to that in table 1 with positive costs c_3 and c_4, associated with incorrect UT and NUT forecasts, respectively, is similar to that explained above for forecasting DT's , it will not be presented. To minimize expected loss in choosing between UT and NUT forecasts, one chooses an UT forecast if $c_3(1-P_{\mathrm{UT}})<c_4P_{\mathrm{UT}}$, or equivalently,

$$1<(c_4/c_3)P_{\mathrm{UT}}/(1-P_{\mathrm{UT}}), \tag{11}$$

where P_{UT} is the probability of an upturn in period $T+1$ given by

$$P_{\mathrm{UT}}=\Pr(y_{iT+1}>y_{iT}|A_2,D_T)=\int_{y_{iT}}^{\infty}p(y_{i,T+1}|A_2,D_T)\,\mathrm{d}y_{iT+1}, \tag{12}$$

where A_2 is the condition $y_{iT-2}, y_{iT-1} > y_{iT}$. Thus when A_2 is satisfied, the probability of an UT, P_{UT}, can be computed from (12) and used to select between UT and NUT forecasts so as to minimize expected loss.

In our empirical work, from the annual output growth rate data for each country, we identify sequences of data satisfying condition A_1: $y_{it-2}, y_{it-1} < y_{it}$. Then we use the data through year t to compute the predictive pdf for y_{it+1} and use it to choose optimally between DT and NDT forecasts, as explained above. A similar procedure is employed with respect to making UT and NUT forecasts. These procedures were carried out for each of our two models using fixed parameters, time-varying parameters and exponential weighting, and the pooling techniques mentioned above. Similar calculations have been performed without employing pooling. The results will be presented after a brief discussion of our data.

3. Description of data

The data employed in our calculations have been taken from the IMF International Financial Statistics data base and are available on a diskette on request for a small fee to cover costs. The annual data 1954–1973 were used to estimate our models with data 1951–1953 serving as initial values. The models were used to compute probabilities of DT's and UT's for each country whenever a sequence of points satisfying the conditions for turning points, defined above was encountered, using all data prior to the year in which a turning point could occur.

The output data for each country, either annual real GNP or real GDP, were logged and first-differenced to yield annual growth rates. Similarly, nominal money, M_1, at the end of each year was deflated by a general price index, logged and first-differenced to yield the rate of change of real money. A general price index was used to deflate an annual stock price index, which was logged and first-differenced to yield a rate of growth of real stock prices for each country and year.

Shown in fig. 2 are boxplots for the three variables' data used in this study, rates of growth of real output, real money, and real stock prices. In each box, the horizontal line is the median of the 18 countries' values of a variable in a particular year, the height of each box is the interquartile range of the 18 countries' values of a variable in a given year while the points outside each box are observations in the tails of the distribution of values of a variable in a particular year. From fig. 2, it is seen that the countries' output, money, and real stock price growth rates tend to move up and down together with the variation of the growth rate of real stock prices considerably larger than that of the other two variables. It will be noted that there are several outlying points in the data. While we plan to deal with these outlying points in future work, in the present study *all data were employed with no adjustments for outliers or anything else.*

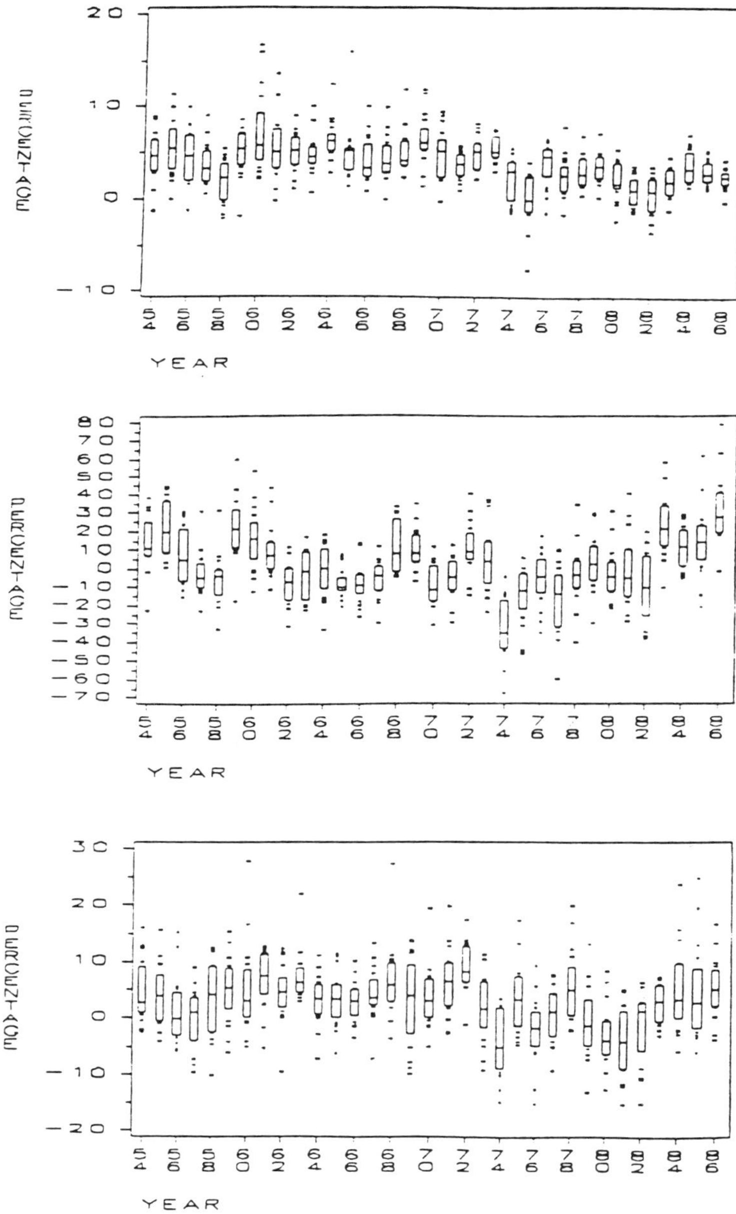

Fig. 2. Box plots of annual growth rates of real output (upper panel), real stock prices (middle panel), and real money (lower panel), 1954–86.

In fig. 2, the horizontal lines in the boxes are the median values of variables in each year used in our model for w_t in (5). It is indeed interesting to note that the annual median growth rates of output, real money, and real stock prices follow rather smooth paths through time which appear to exhibit fluctuations with about a four-year period as measured from peak to peak or trough to trough. See Zellner and Hong (1989) for evidence on the forecasting performance of the ARLI model for w_t in (5).

4. Results of forecasting turning points, 18 countries, 1974–86

For each country's data, 1974–86, we identified sequences of observations which satisfied our definitions of turning points given in (6) and (7). Then we used data from 1954 through year T, the year prior to the year in which a turning point did or did not occur, to compute the probabilities that a turning point would occur in year $T+1$ from the predictive pdf's of six alternative models. These probabilities were employed to make optimal turning point forecasts as explained above. Initially, we report results based on a symmetric loss structure for which a turning point is forecasted when its probability of occurrence is greater than 0.5. After reporting these results, others relating to asymmetric loss structures will be presented and discussed.

The six models used in making turning point forecasts are the following third-order ARLI models:

1. TVP/ARLI, a time-varying parameter ARLI model,
2. TVP/ARLI/WI, a time-varying parameter ARLI-world income model,
3. EW/ARLI, an exponentially weighted ARLI model,
4. EW/ARLI/WI, an exponentially weighted ARLI-world income model,
5. FP/ARLI, a fixed parameter ARLI model, and
6. FP/ARLI/WI, a fixed parameter ARLI-world income model.

Each model is implemented with no pooling of data across countries and with pooling using the methods discussed in section 2. Our calculations provide information relating to the following questions:

a. Do fixed parameter models perform as well as TVP and EW ARLI models?
b. Do the models containing the world income variable perform better than those that do not?
c. Does pooling of data across countries improve forecast performance?
d. Are DT and NDT forecasts better than UT and NUT forecasts?
e. Is the percentage of correct DT forecasts greater than the percentage of correct NDT forecasts?
f. Is the percentage of correct UT forecasts greater than the percentage of correct NUT forecasts?

The forecasting results presented below for 158 turning point forecasts will help to provide answers to these and other questions.

In table 2 and figs. 3 and 4 are shown results of using our six models to forecast 158 turning points for 18 countries, 1974–86. Using symmetric loss structures, we forecast a DT when $P_{DT} > \frac{1}{2}$ and an UT when $P_{UT} > \frac{1}{2}$. In the upper panel of table 2, the forecasting results using no data pooling are presented. The model that appears to have a slight edge over the others is the TVP/ARLI/WI model. Use of it led to 82 percent of 158 turning point forecasts being correct, that is 129 of 158 forecasts are correct. Of the 76 DT and NDT forecasts, use of this model produces 83 percent correct forecasts, while for 82 UT and NUT forecasts, 80 percent are correct. For the other models, of 158 turning point forecasts, the percentages correct vary from 72 to 77, somewhat lower than for the TVP/ARLI/WI model. Also for the 76 DT and NDT forecasts, use of these other models leads to percentages correct that range from 68 to 76, while for the 82 UT and NUT forecasts, the percentages correct vary from 74 to 77. Thus use of the TVP/ARLI/WI model produces approximately 80 percent correct forecasts overall and for DT/NDT and UT/NUT forecasts separately.

Note further, from panel A of table 2, that use of TVP models produces better results than fixed parameter models or models with exponential weighting. However, in most cases the differences in performance as measured by percentage of correct forecasts are not large. Further, from lines 1–4 of table 2, it is seen that addition of the world income variable leads to small improvements in forecasting performance relative to that of models not including it.

From panel B of table 2, which provides results based on data pooled across countries, use of the TVP/ARLI/WI and EW/ARLI/WI models results in 80 and 81 percent, respectively, of 158 turning point forecasts being correct. The percentages correct for the other models range from 74 to 79, with models incorporating the WI variable performing slightly better than those not including it. Also, the TVP and EW models tend to perform slightly better in forecasting turning points relative to the pooled fixed parameter models.[1] Similar conclusions emerge with respect to results for the 76 DT and NDT and 82 UT and NUT forecasts. In addition, with the use of pooling, the performance of the models for DT/NDT forecasting, 80 to 88 percent correct, is somewhat better than that for the UT/NUT forecasts, 67 to 76 percent correct.

As regards the comparison of the results in the upper and lower panels of table 2, it appears that pooling has led to improved DT and NDT forecasts but no improvement with respect to UT and NUT forecasts. It is the case

[1]The fixed parameter model denoted by FP/ARLI was employed in our past work without pooling – see Zellner, Hong, and Gulati (1990).

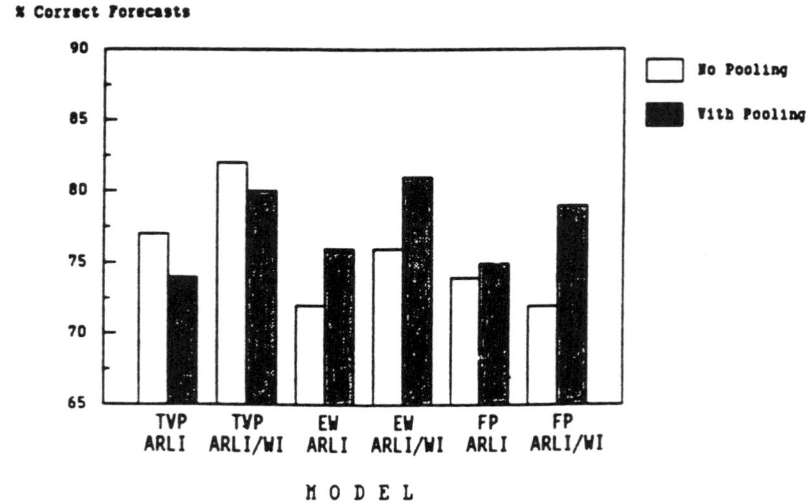

Fig. 3. Percentages of correct forecasts, 158 turning point forecasts, 1974–86.

that pooling leads to much better performance of the NDT forecasts, particularly for the FP models.

To summarize the results in table 2 and figs. 3 and 4, use of our methods and models leads to approximately 70 to 80 percent correct turning point forecasts based on 158 cases. While the differences in performance are not large in many instances, the results tend to favor the TVP/ARLI/WI model with or without pooling which produces 80 and 82 percent correct turning point forecasts, respectively.

To appraise the forecasting performance of various models and methods further, we have shown the numbers of incorrect forecasts in table 2. For example, the TVP/ARLI/WI model in line (2) produces 29 incorrect turning point forecasts, 13 associated with DT and NDT forecasts and 16 associated with UT and NUT forecasts. If the loss structure in table 1 is symmetric with $c_1 = c_2 = 1$ and if a similar one for UT/NUT forecasts is also symmetric with costs of incorrect forecasts equal to 1, then 29 is the realized loss associated with the use of the unpooled TVP/ARLI/WI model in forecasting turning points. From table 2, it is seen that realized losses range from 29 to 45 for panel A of the table and from 30 to 41 for panel B. These results can be compared to the results that are associated with an eternally pessimistic forecaster who always chooses DT and NUT forecasts – see table 3, line 2. Since there were 59 actual DT's out of 76 cases, the eternal pessimist has 17 of his DT forecasts wrong. Also, since there were 34 NUT's in 82 cases, 48 of his NUT forecasts are incorrect. Thus his total number of incorrect forecasts is $17 + 48 = 65$. As is evident, this eternally pessimistic

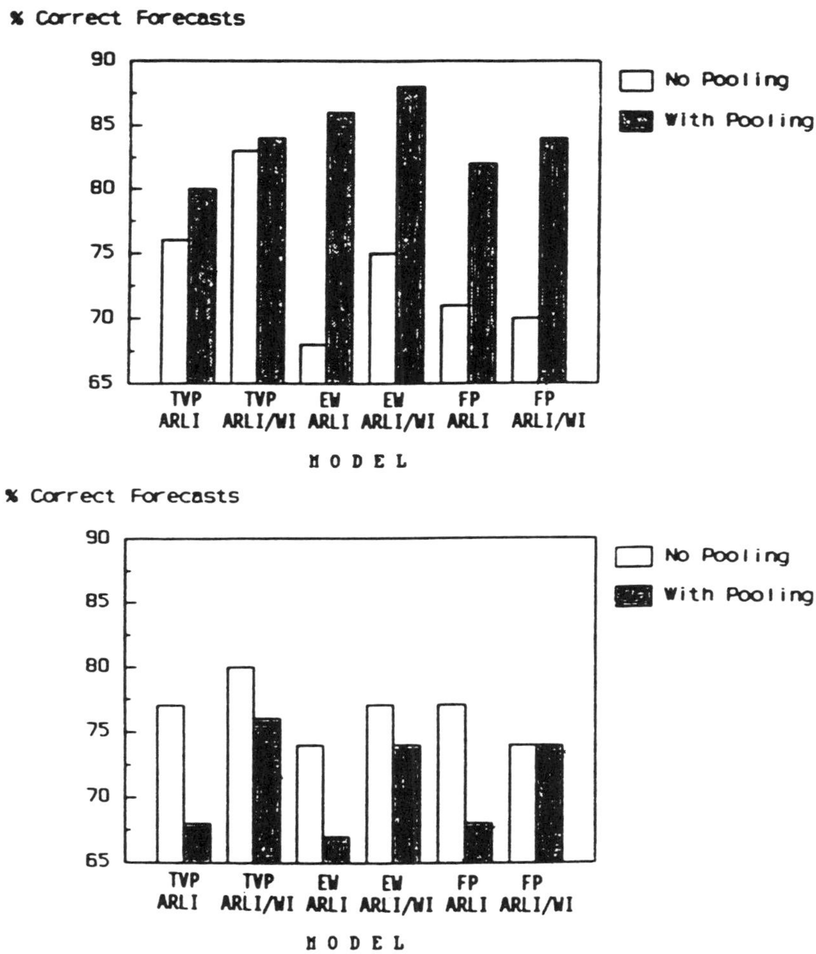

Fig. 4. Percentages of correct forecasts, 76 DT & NDT forecasts (upper panel) and 82 UT & NUT forecasts (lower panel), 1974–86.

forecaster experiences a realized loss under symmetric loss of 65 which is much higher than those associated with the models and methods in table 2. Thus, our turning point forecasts are much better than those of an 'eternal pessimist'.

Also shown in table 3 are the forecasting results of an 'eternal optimist' who always forecasts NDT and UT. The eternal optimist has 59 of his 76 NDT forecasts incorrect and 34 of his 82 UT forecasts incorrect. Thus his realized loss is $59 + 34 = 93$, much higher than that associated with the unpooled or pooled TVP/ARLI/WI model's realized losses, 29 and 32, respectively.

Table 2

Forecasting turning points in annual output growth rates, 18 countries, 1974–86, using various models.

Model	Percentage of correct forecasts			Number of incorrect forecasts		
	158 turning point forecasts	76 DT & NDT forecasts	82 UT & NUT forecasts	158 turning point forecasts	76 DT & NDT forecasts	82 UT & NUT forecasts
(A) *No pooling*						
1. TVP/ARLI	77	76	77	37	18	19
2. TVP/ARLI/WI	82	83	80	29	13	16
3. EW/ARLI	72	68	74	45	24	21
4. EW/ARLI/WI	76	75	77	38	19	19
5. FP/ARLI	74	71	77	41	22	19
6. FP/ARLI/WI	72	70	74	44	23	21
(B) *With pooling*						
7. TVP/ARLI	74	80	68	41	15	26
8. TVP/ARLI/WI	80	84	76	32	12	20
9. EW/ARLI	76	86	67	38	11	27
10. EW/ARLI/WI	81	88	74	30	9	21
11. FP/ARLI	75	82	68	40	14	26
12. FP/ARLI/WI	79	84	74	33	12	21

Table 3

Performance of various procedures for forecasting turning points in annual output growth rates for 18 countries, 1974–86.

	Number and percent of incorrect forecasts[a]		
Forecasting Procedures	76 DT/NDT forecasts	82 UT/NUT forecasts	158 turning point forecasts
1. Eternal[b] optimist	59 (78)	34 (41)	93 (59)
2. Eternal pessimist[c]	17 (22)	48 (59)	65 (41)
3. Deterministic four-year cycle[d]	17 (22)	34 (41)	51 (32)
4. TVP/ARLI/WI unpooled	13 (17)	16 (20)	29 (18)
TVP/ARLI/WI pooled	12 (16)	20 (24)	32 (20)

[a]Figures in parentheses are percentages of incorrect forecasts.
[b]An eternal optimist forecasts either NDT or UT.
[c]An eternal pessimist forecasts either DT or NUT.
[d]Forecasts either DT or UT.

The forecasting performance of a person who believes in a deterministic four-year cycle is shown in line 3 of table 3. Such a forecaster always forecasts DT and UT with the result that 17 of his 76 DT forecasts and 34 of his 82 UT forecasts are incorrect. Thus his total realized loss is 51, again higher than those produced using the unpooled or pooled TVP/ARLI/WI model, namely 29 and 32, respectively. Further, as can be seen from the figures in table 2, the realized losses for all of our models are much lower than those of the eternal optimist, the eternal pessimist, and the deterministic four-year cycle forecaster.

In summary, in terms of overall realized losses, the results in tables 2 and 3 indicate that all of our models performed better than the eternal optimist, the eternal pessimist, and the deterministic four-year cycle forecasters. The addition of the WI variable has led to improved forecasting as measured by overall realized loss.

We can also ask if our turning point forecasts are better than those yielding by coin-flipping, that is use of a probability of a DT $= \frac{1}{2}$ on all occasions or probability of an UT $= \frac{1}{2}$ on all occasions. Under these assumptions, the expected number of turning points $= \frac{1}{2} \times 158 = 79$. However, we observe 59 DT's and 48 UT's in our data using the definitions in (6) and (7). Thus there are 107 actual turning points observed. The number of TPT's observed, 107 of 158 cases, is much greater than the number predicted by a coin-flipping forecaster, 79. On the other hand, if we sum our computed

Table 4

Expected number of TPT's, DT's, and UT's.

Model	Unpooled	Pooled
1. TVP/ARLI	104	98
2. TVP/ARLI/WI	101	98
3. EW/ARLI	101	102
4. EW/ARLI/WI	98	99
5. FP/ARLI	100	98
6. FP/ARLI/WI	96	94
Actual number of TPT's	107	107
Total number of cases	158	158

probabilities of turning points for the 158 possible turning points, we obtain the results shown in table 4.

It is seen that the forecasting results of table 4 indicate that our methods and models yield an expected number of turning points much closer to the observed outcome, 107, than those of the pessimistic forecaster, the coin-flipping forecaster, and a deterministic four-year cycle forecaster who would forecast a DT or an UT on every occasion and thus would expect 158 turning points.

In tables 5 and 6 we display results for DT and NDT and UT and NUT forecasts separately for each model, with and without pooling. For the results without pooling, shown in panel A, we see that for all models the percentages of correct DT forecasts are much higher than those for NDT forecasts, with the former ranging from 89 to 95 percent and the latter from 39 to 58 percent. The poor performance of NDT forecasts was also encountered in Zellner, Hong, and Gulati (1990) using the unpooled FP/ARLI model. The percentages of the TVP/ARLI/WI model's DT and NDT forecasts correct, 94 and 58 respectively, are higher than the corresponding percentages for all other models except for the DT percentage correct for the FP/ARLI/WI model, 95. As regards the effects of pooling on forecasting performance, panel B of table 5 indicates that pooling has resulted in an increase in the percentage of correct NDT forecasts for all models and not much change in the percentages of correct DT forecasts.

Just as above, we can ask how many DT's a coin-flipper, an optimist, a pessimist, or a believer in a deterministic four-year cycle would predict. Since there are 76 cases meeting the conditions for forecasting either DT or NDT, a coin-flipper using $P_{\mathrm{DT}} = \frac{1}{2}$ in each case would expect 38 DT's. The optimist would expect zero DT's, while the other two forecasters would expect 76 DT's. The number of downturns actually observed is 59, far from zero, 38, or 76. If we sum the probabilities of DT's associated with the 76 cases in which a DT could occur according to our definition, the sums for our different

Table 5

Forecasting downturns (DT's) and no downturns (NDT's) in annual output growth rates, 18 countries, 1974–86.

Decision rule: Forecast DT if $P_{DT} > 0.5$.

Model	Forecast	Number: Correct	Number: Incorrect	Total	% correct	Realized loss
		(A) No pooling				
TVP/ARLI	DT	47	6	53	89	18
	NDT	11	12	23	48	
TVP/ARLI/WI	DT	49	3	52	94	13
	NDT	14	10	24	58	
EW/ARLI	DT	40	5	45	89	24
	NDT	12	19	31	39	
EW/ARLI/WI	DT	45	5	50	90	19
	NDT	12	14	26	46	
FP/ARLI	DT	40	3	43	93	22
	NDT	14	19	33	42	
FP/ARLI/WI	DT	38	2	40	95	23
	NDT	15	21	36	42	
		(B) With pooling				
TVP/ARLI	DT	50	6	56	89	15
	NDT	11	9	20	55	
TVP/ARLI/WI	DT	52	5	57	91	12
	NDT	12	7	19	63	
EW/ARLI	DT	52	4	56	93	11
	NDT	13	7	20	65	
EW/ARLI/WI	DT	53	3	56	95	9
	NDT	14	6	20	70	
FP/ARLI	DT	47	2	49	96	14
	NDT	15	12	27	56	
FP/ARLI/WI	DT	49	2	51	96	12
	NDT	15	10	25	60	

models range from 42 to 49 for our unpooled cases and from 44 to 51 for our pooled cases. Thus the expected number of DT's computed from our models are much closer to the actual number of observed DT's, 59, than are those of a coin-flipper, an optimist, a pessimist, and a believer in a deterministic four-year cycle.

In table 6, information is provided on the performance of our models and methods in forecasting UT's and NUT's. Without pooling, 71 to 81 percent of the UT forecasts are correct and 78 to 86 percent of the NUT forecasts are

Table 6

Forecasting upturns (UT's) and no upturns (NUT's) in annual output growth rates, 18 countries, 1974–86.

Decision rule: Forecast UT if $P_{UT} > 0.5$.

Model	Forecast	Number Correct	Number Incorrect	Total	% correct	Realized loss
		(A) *No pooling*				
TVP/ARLI	UT	44	15	59	75	19
	NUT	19	4	23	83	
TVP/ARLI/WI	UT	42	10	52	81	16
	NUT	24	6	30	80	
EW/ARLI	UT	45	18	63	71	21
	NUT	16	3	19	84	
EW/ARLI/WI	UT	42	13	55	76	19
	NUT	21	6	27	78	
FP/ARLI	UT	45	16	61	74	19
	NUT	18	3	21	86	
FP/ARLI/WI	UT	43	16	59	73	21
	NUT	18	5	23	78	
		(B) *With pooling*				
TVP/ARLI	UT	39	17	56	70	26
	NUT	17	9	26	65	
TVP/ARLI/WI	UT	40	12	52	77	20
	NUT	22	8	30	73	
EW/ARLI	UT	40	19	59	68	27
	NUT	15	8	23	65	
EW/ARLI/WI	UT	41	14	55	74	21
	NUT	20	7	27	74	
FP/ARLI	UT	41	19	60	68	26
	NUT	15	7	22	68	
FP/ARLI/WI	UT	39	12	51	76	21
	NUT	22	9	31	71	

correct. The numbers of incorrect forecasts range from 16 for the TVP/ARLI/WI model to 21 for the EW/ARLI and FP/ARLI/WI models. Again the TVP/ARLI/WI model performs slightly better than the other models. The same is true when pooling is used as shown in panel B of table 6, although the differences are rather small.

Further, there are 82 cases in which either an UT or a NUT forecast can be selected. A coin-flipper, $P_{UT} = \frac{1}{2}$, would expect 41 UT's. A pessimist who forecasts NUT in each case, equivalent to $P_{UT} = 0$, would expect zero UT's.

On the other hand, an optimist and a believer in a deterministic four-year cycle would expect 82 UT's. Actually, there were 48 UT's observed in our data, much greater than zero, much lower that 82, and somewhat larger than the coin-flipper's expected number, 41. On the basis of the sum of our computed probabilities, we found that our expected number of UT's ranged from 53 to 58 for our unpooled models and from 48 to 54 for our pooled models. The pooled TVP/ARLI/WI model's summed probabilities equals 47.6, very close indeed to the number of UT's observed, namely 48.

From the above turning point forecasting performance, based on symmetric loss structures or equivalently on the use of the decision rules, forecast a DT if $P_{\mathrm{DT}} > 0.5$, otherwise forecast a NDT, and forecast an UT if $P_{\mathrm{UT}} > 0.5$, otherwise forecast NUT, we conclude that the TVP/ARLI/WI model, with or without pooling, produces relatively good results. The margin of superiority of this model is generally small relative to other models which included the WI variable. These models perform quite a bit better than do four naive turning point forecasting procedures.

Shown in table 7 are probabilities of DT's by year and country computed from the pooled TVP/ARLI/WI model using the definition of a DT given in (6a). For example, the probability of a DT in 1974 for Belgium is 0.82 based on data through 1973. The asterisk affixed to this probability indicates that the Belgian output growth rate did indeed turn down in 1974. It is seen that eight countries have high DT probabilities in 1974 and that all experienced a DT in this year. In 1977, 13 countries have high probabilities of a DT and 12 experienced a downturn. Japan, with a probability of a DT equal to 0.83 did not experience a DT, as indicated by the dash affixed to its probability. However, one year later in 1978, the probability of a DT for Japan is 0.55, based on data through 1977, and it did experience a DT in 1978. In the majority of cases for which the probabilities of DT's are high, a DT was experienced. In addition to the exception noted for Japan in 1977, other exceptions are encountered for Switzerland in 1980, Norway in 1984, and Switzerland in 1985. In each of these cases, the probability of a DT in the next year is high and a DT occurred in that year. Thus for these cases, there is an early false DT forecast followed by a correct DT forecast in the next year. There are also several cases in which the probability of a DT is low and yet a DT occurred, for example for the UK in 1979 and for Austria, Denmark, and Switzerland in 1986. In general, however, when the probabilities of a DT are high, countries experienced DT's, while when they are low they did not experience DTs.

In table 8, the computed probabilities of UT's by country and year based on the pooled TVP/ARLI model are displayed. In general, when the computed probabilities are high, UT's occurred, while when they are low, UT's did not occur. For example in 1975, the probabilities of an UT are smaller than 0.50 for 10 countries and eight of them did not experience an

Table 7

Computed probabilities of downturns by year and country from the pooled TVP/ARLI/WI model[a].

	Year												
Country	1974	1975	1976	1977	1978	1979	1980	1981	1982	1983	1984	1985	1986
Australia				0.68*			0.83*					0.94*	
Austria				0.77*			0.95*				0.27*		0.36*
Belgium	0.82*			0.90*				0.89*				0.54*	
Canada	0.94*			0.91*							0.37–	0.88*	
Denmark				0.95*			0.80*			0.65*		0.33–	0.36*
Finland						0.13–	0.93*						0.69*
France				0.88*						0.44*			0.51–
Germany	0.87*			0.92*			0.85*				0.27–	0.64*	
Ireland					0.74*				0.92*			0.60*	
Italy	0.81*			0.87*			0.81*					0.66*	
Japan				0.83–	0.55*							0.67*	
Netherlands	0.88*			0.84*			0.70*				0.23–	0.58*	
Norway		0.98*		0.95*			0.78*				0.58–	0.66*	
Spain					0.51*						0.44–	0.31–	0.37–
Sweden	0.68*					0.12–	0.80*				0.37–	0.80*	
Switzerland					0.47*		0.74–	0.91*				0.57–	0.77*
United Kingdom	0.99*			0.85*		0.31*				0.32–	0.54*		0.44*
United States	0.86*			0.85*		0.52*					0.42–	0.96*	

[a]An asterisk (*) indicates that a downturn occurred while a dash (–) indicates that a downturn did not occur. See eq. (6a) in the text for the definition of a downturn.

Table 8

Computed probabilities of upturns by year and country from the pooled TVP/ARLI model.[a]

Country	1974	1975	1976	1977	1978	1979	1980	1981	1982	1983	1984	1985	1986
							Year						
Australia		0.11*			0.45*			0.77*		0.31*			
Austria	0.42–	0.12–	0.90*			0.92*			0.57*				
Belgium		0.04–	0.96*						0.74*				
Canada		0.05–	0.77*					0.56*		0.95*			
Denmark		0.25*				0.86*		0.80–	0.95*				
Finland		0.20–	0.89–	0.51*					0.71*				
France	0.39–	0.19–	0.89*					0.41–	0.55*		0.90*		
Germany		0.70–	0.98*					0.61–	0.64–	0.80*			
Ireland		0.04–	0.67–	0.73*			0.34*			0.15–	0.98*		
Italy			0.96*						0.53–	0.42*			
Japan		0.50*						0.12–	0.48–	0.31*			
Netherlands			0.97*			0.74*		0.63–	0.70–	0.89*			
Norway	0.82*				0.38*			0.12–	0.63	0.71*			
Spain		0.09–	0.70*			0.79–	0.81*		0.81*				
Sweden			0.54–	0.60–	0.82*			0.69–	0.82*				0.57–
Switzerland	0.30–	0.16–	0.99*						0.28	0.89*			
United Kingdom		0.55*						0.80*					
United States		0.52–	0.90*				0.57–	0.76*		0.91*			0.80*

[a]An asterisk (*) indicates that an upturn occurred while a dash (–) indicates that an upturn did not occur. See eq. (7a) in the text for the definition of an upturn.

UT. In this year four countries had probabilities of an UT equal to or greater than 0.50 and two experienced UT's. Germany with a probability of an UT of 0.70 did not experience an UT in 1975; however, for the next year, 1976, its probability of an UT is 0.98 and it did experience an upturn. Similarly for the U.S., its 1975 probability of an UT is 0.52 and it did not experience an UT, but its 1976 probability is 0.90 and its output growth rate did turn up in 1976. In several other cases, for example Denmark, Germany, and Sweden in 1981 and 1982, a similar pattern is encountered, namely high probabilities of UT's in 1981 and 1982 with UT's occurring in the latter but not in the former year. Also, in some cases, the probabilities of an UT are low and yet upturns occurred, for example for Australia and Denmark in 1975 and for Ireland in 1980. Overall, however, when the computed probabilities are high, UT's usually occurred, while when they are low, UT's did not occur.

We now turn to a presentation of some forecasting results for asymmetric loss structures. In table 9 are shown the results of DT/NDT forecasts for various values of the cost parameter ratio, c_2/c_1, where c_1 and c_2 are the cost parameters in the loss structure given in table 1. When $c_2/c_1 = 4$, it is optimal to forecast a DT when the probability of a DT is greater than 0.2. For this case, 74 DT forecasts were made with 59 or 80 percent correct. Only two NDT forecasts were made and both were correct. With $c_2 = 4c_1$, the cost of an erroneous NDT forecast is much greater than that of an erroneous DT forecast and thus fewer NDT forecasts are made. On the other hand, when $c_2/c_1 = \frac{1}{4}$, the optimal decision rule is to forecast a DT when the probability of a DT is greater than 0.8, a rule which leads to fewer DT forecasts being made. In this case, only 30 DT forecasts were made with 29 or 97 percent being correct. Of 46 NDT forecasts made in this case, 16 or 35 percent are correct. These results indicate that the turning point forecasting procedure adapts to variations in the cost parameter ratio, c_2/c_1. Similar results have been obtained for forecasting UT's and NUT's when the cost parameter ratio, or equivalently the 'cut-off' probability of an UT, is varied.

In table 10 the Brier scores for our turning point forecasts are presented. The Brier score, or quadratic probability score, denoted by B is defined as follows:

$$B = \sum_{i=1}^{N} 2(P_i - R_i)^2/N, \tag{13}$$

where N = number of forecasts, P_i = computed probability associated with the ith forecast, and R_i = realized outcome, equal to one if the forecasted event occurred and equal to zero if it did not. B ranges from 0 to 2, with $B = 0$ denoting perfect accuracy. B in (13) was computed separately for all our turning point forecasts and for DT/NDT and UT/NUT forecasts separately.

Table 9

Forecasting DT's and NDT's using asymmetric and symmetric loss structures and pooling.

Cost parameter ratio, c_2/c_1[a]	Decision rule	Forecasts	TVP/ARLI/WI pooled Number	TVP/ARLI/WI pooled % correct	EW/ARLI/WI pooled Number	EW/ARLI/WI pooled % correct	FP/ARLI/WI pooled Number	FP/ARLI/WI pooled % correct
4	$P_{DT} > 0.2$	DT	74	80	74	80	74	80
		NDT	2	100	2	100	2	100
7/3	$P_{DT} > 0.3$	DT	71	82	68	84	68	84
		NDT	5	80	8	75	8	75
3/2	$P_{DT} > 0.4$	DT	62	89	61	88	61	88
		NDT	14	71	15	67	15	67
1	$P_{DT} > 0.5$	DT	57	91	56	95	56	95
		NDT	19	63	20	70	20	70
2/3	$P_{DT} > 0.6$	DT	47	96	45	96	45	96
		NDT	29	52	31	48	31	48
3/7	$P_{DT} > 0.7$	DT	38	95	31	97	31	97
		NDT	38	40	45	36	45	36
1/4	$P_{DT} > 0.8$	DT	30	97	24	96	24	96
		NDT	46	35	52	31	52	31

[a] c_1 and c_2 are cost parameters in the loss structure in table 1.

Table 10

Brier scores for turning point forecasts, 18 countries, 1974–86.[a]

Model	All turning points	Downturns	Upturns
	(A) *No pooling*		
1. TVP/ARLI	0.32	0.30	0.35
2. TVP/ARLI/WI	0.31	0.30	0.33
3. EW/ARLI	0.37	0.35	0.38
4. EW/ARLI/WI	0.34	0.33	0.35
5. FP/ARLI	0.37	0.36	0.37
6. FP/ARLI/WI	0.35	0.36	0.35
	(B) *With pooling*		
7. TVP/ARLI	0.31	0.26	0.35
8. TVP/ARLI/WI	0.28	0.24	0.31
9. EW/ARLI	0.32	0.28	0.37
10. EW/ARLI/WI	0.29	0.26	0.32
11. FP/ARLI	0.34	0.30	0.37
12. FP/ARLI/WI	0.29	0.27	0.31

[a]The Brier score is defined in (13) in the text.

From the results in table 10, it is seen that the Brier scores for all turning points range from 0.31 to 0.37 for the six models when there is no pooling, and from 0.28 to 0.34 with pooling. With no pooling the TVP/ARLI/WI model has the lowest Brier score, namely 0.31. With pooling, this same model has the lowest score, 0.28. Thus according to the Brier score criterion, the TVP/ARLI/WI model performed the best with pooling resulting in about a 10 percent reduction in Brier score relative to the unpooled case. Further, the other models' performance also shows improvement with pooling in every case. The pooled EW/ARLI/WI model's Brier score, 0.29, is only slightly larger than that of the TVP/ARLI/WI model. It is also the case that inclusion of the WI variable improved performance in every case relative to that of corresponding models without this variable. Last, the FP/ARLI model performed the worst both in the case of no pooling and with pooling.

In the case of computed probabilities for DT's, the Brier scores range from 0.30 to 0.36 for models with no pooling and from 0.24 to 0.30 with pooling. In each case, the score is lower with pooling, indicating that pooling improves performance. The lowest and best score, 0.24, is encountered for the TVP/ARLI/WI model with pooling. A score of 0.26 is obtained for the pooled TVP/ARLI and EW/ARLI/WI models. That these three models, which allow for nonconstant parameters, exhibit the best performance is noteworthy.

With respect to computed probabilities for UT's, the models with no pooling have Brier scores of 0.33 to 0.38, while those with pooling range from

Table 11

Brier scores[a] for eternal optimist, eternal pessimist, coin-flipper, and deterministic four-year cycle forecasters.

Forecaster	76 DT/NDT forecasts	82 UT/NUT forecasts	158 turning point forecasts
Eternal optimist[b]	1.55	0.83	1.18
Eternal pessimist[c]	0.45	1.17	0.82
Coin-flipper[d]	0.50	0.50	0.50
Deterministic four-year cycle[e]	0.45	0.83	0.65

[a]The Brier score is defined in (13) in the text.

[b]The eternal optimist assumes that the probability of a DT = 0 and the probability of an UT = 1.0 on all occasions.

[c]The eternal pessimist assumes that the probability of a DT = 1.0 and the probability of an UT = 0 on all occasions.

[d]A coin-flipper assumes that the probability of a DT or of an UT = $\frac{1}{2}$ on all occasions.

[e]A deterministic four-year cycle forecaster assumes that the probability of an UT or a DT = 1.0 on all occasions.

0.31 to 0.37. In this case, pooling did not produce a uniform improvement in the scores. For the TVP/ARLI/WI model, its scores are 0.33 with no pooling and 0.31 with pooling, the latter being the lowest and best score. Also, the results in table 10 indicate that the Brier scores are somewhat better for probabilities associated with DT's than for those associated with UT's for all models except the FP/ARLI/WI model with no pooling where they were equal. Thus, using the Brier score measure, forecasting performance with respect to UT's is slightly poorer than that for DT's.

Finally, in table 11 are shown the Brier scores for the naive turning point forecasters, the eternal optimist, the eternal pessimist, the coin-flipper, and the believer in a deterministic four-year cycle. As can be seen, the Brier scores for all these naive forecasters are larger than all those in table 10 indicating that naive forecasters' performance is poorer than that associated with the models and methods reported in table 10.

5. Summary and concluding remarks

In this paper we have found that several ARLI models have produced rather good turning point forecasting performance. Using a Bayesian decision-theoretic approach to generate optimal turning point forecasts, it was found that about 80 percent of 158 turning point forecasts for 18 countries, 1974–86, are correct using the TVP/ARLI/WI and EW/ARLI/WI models with pooling – see table 2. For other models, with and without pooling, the percentage of correct turning point forecasts is 72 or higher. Similar good results are obtained for DT/NDT and UT/NUT forecasts separately. Also these forecasts are superior to those of several naive turning point forecasting procedures.

The weakest performance was encountered in the case of NDT forecasts where, with pooling, the percentage of correct forecasts ranges from 55 to 70. In the case of the pooled TVP/ARLI/WI model NDT forecasts, 7 of 19 NDT forecasts were incorrect. For these 7 cases, the computed probabilities of DT's have values from 0.27 to 0.47. Thus in each of these cases there is a substantial probability of a DT. If a more conservative rule for forecasting DT's were employed, say forecast a DT if $P_{\mathrm{DT}} > 0.2$, then, as shown in table 9, all NDT forecasts are correct.

Finally, the good performance of our models and methods in forecasting turning points in annual growth rates of real output for 18 countries, 1974–86, is surprising in view of the fact that this period is a difficult one from the point of view of forecasting. That is, the period is one in which two oil crises occurred, countries moved from fixed to flexible exchange rates, monetary policy in the U.S. changed dramatically in 1979, inflation and unemployment rates were very high in the late 1970s, etc. The fact that our models and methods performed so well without any ad hoc adjustments for the effects of these major events suggests that they may be useful to private and public decision-makers who are interested in computing probabilities of DT's and UT's and forecasting turning points.

Appendix: Calculation of posterior and predictive distributions

A.1. Fixed parameter models

Dropping the subscript i, the model in eq. (1) of the text is $y_t = \boldsymbol{x}_t'\boldsymbol{\beta} + u_t$, $t = 1, 2, \ldots, n$. The u_t's are assumed independently drawn from a zero-mean normal distribution with variance σ^2. A spread out normal-inverted gamma natural conjugate prior distribution for $\boldsymbol{\beta}$ and σ was employed, namely $p(\boldsymbol{\beta}|\sigma) \sim N(\mathbf{0}, \sigma^2 I \times 10^6)$ and $p(\sigma) \sim IG(\nu_0 s_0)$ with ν_0 and s_0 assigned very small values. Then at $t = 0$, with given initial values, the predictive probability density function (pdf) for y_1, denoted by $p(y_1|D_0)$, is in the univariate Student-t form. That is, $t_{\nu_0} = (y_1 - \boldsymbol{x}_i'\hat{\boldsymbol{\beta}}_0)/s_0 a_0$ has a univariate Student-t pdf with ν_0 degrees of freedom where $a_0^2 = 1 + \boldsymbol{x}_i'\boldsymbol{x}_i \cdot 10^6$ and $\hat{\boldsymbol{\beta}}_0 = \mathbf{0}$. By updating the posterior pdfs period by period by use of Bayes' Theorem, they can be employed to obtain predictive pdf's period by period. For example, the predictive pdf for y_{t+1} is in the univariate Student-t form and such that

$$t_{\nu_t} = \left(y_{t+1} - \boldsymbol{x}_{t+1}'\hat{\boldsymbol{\beta}}_t\right)/s_t a_t \qquad \text{(A.1)}$$

has a univariate Student-t pdf where

$$\hat{\boldsymbol{\beta}}_t = (Z_t'Z_t)^{-1}Z_t'y_t, \tag{A.2}$$

with $y_t' = (y_t, y_{t-1}, y_{t-2}, \ldots, y_1)$, $Z_t' = (\boldsymbol{x}_t, \boldsymbol{x}_{t-1}, \ldots, \boldsymbol{x}_1)$, and $a_t^2 = 1 + \boldsymbol{x}_{t+1}'(Z_t'Z_{t-1})^{-1}\boldsymbol{x}_{t+1}$. The following up-dating formulas were employed:

$$\hat{\boldsymbol{\beta}}_t = \hat{\boldsymbol{\beta}}_{t-1} + (Z_{t-1}'Z_{t-1})^{-1}\boldsymbol{x}_t(y_t - \boldsymbol{x}_t'\hat{\boldsymbol{\beta}}_{t-1})/\left[1 + \boldsymbol{x}_t'(Z_t'Z_t)^{-1}\boldsymbol{x}_t\right],$$

$$\nu_t = \nu_{t-1} + 1,$$

$$\nu_t s_t^2 = \nu_{t-1}s_{t-1} + (y_t - \boldsymbol{x}_t'\hat{\boldsymbol{\beta}}_t)^2 + (\hat{\boldsymbol{\beta}}_t - \hat{\boldsymbol{\beta}}_{t-1})'Z_{t-1}'Z_{t-1}(\hat{\boldsymbol{\beta}}_t - \hat{\boldsymbol{\beta}}_{t-1}).$$

$$(Z_t'Z_t)^{-1} = (Z_{t-1}'Z_{t-1})^{-1} - (Z_{t-1}'Z_{t-1})^{-1}\boldsymbol{x}_t\boldsymbol{x}_t'(Z_{t-1}'Z_{t-1})^{-1}$$
$$/\left[1 + \boldsymbol{x}_t'(Z_{t-1}'Z_{t-1})^{-1}\boldsymbol{x}_t\right].$$

To compute, for example, the probability that $y_{t+1} < y_t$, given information through t, $\Pr(y_{t+1} < y_t | D_t)$, use is made of (A.1) from which $y_{t+1} = s_t a_t t_{\nu_t} + \boldsymbol{x}_{t+1}'\hat{\boldsymbol{\beta}}_t$. Then $\Pr(y_{t+1} < y_t | D_t) = \Pr[t_{\nu_t} < (y_t - \boldsymbol{x}_{t+1}'\hat{\boldsymbol{\beta}}_t)/s_t a_t | D_t]$, which was evaluated using the Student-t distribution with ν_t degrees of freedom.

For the pooled ARLI model, all countries' parameters were assumed to be the same, that is, $\boldsymbol{\beta}_1 = \boldsymbol{\beta}_2 = \cdots = \boldsymbol{\beta}_{18}$ and $\sigma_1 = \sigma_2 = \cdots = \sigma_{18}$, and a spread-out natural conjugate prior pdf and the above methods were used to compute posterior and predictive pdf's period by period. The predictive pdf's were employed to compute probabilities of downturns and upturns, as explained above.

Similar procedures were employed to update posterior and predictive pdf's for the FP/ARLI/WI model in (4) and (5) of the text under the assumption that (4) and (5) constitute a fully recursive econometric model with constant variance, zero-mean independent normal disturbance terms. Using spread-out natural conjugate prior pdf's for the parameters, posterior and one-step-ahead predictive pdf's were computed using regression-like results in the literature [see Zellner (1971, ch. 8) and Bowman and Laporte (1975)]. The one-step-ahead predictive pdf for y_{t+1} from (4) was approximated by a univariate Student-t pdf conditioned on $w_{t+1} = \hat{w}_{t+1}$ is the mean of the predictive pdf for w_{t+1}. Such predictive pdf's were employed to compute probabilities of turning points. For the pooled FP/ARLI/WI model, countries' parameters were all assumed to be the same and the procedures described above were employed.

A.2. Time-varying parameter (TVP) models

For the TVP model (2)–(3), on dropping the subscript i for convenience, we have for $t = 1, 2, \ldots, n$

$$y_t = x_t'\beta_t + u_t, \qquad u_t\text{'s NID}(0, \sigma^2), \tag{A.3a}$$

$$\beta_t = \beta_{t-1} + v_t, \qquad v_t\text{'s NID}(0, \phi\sigma^2 I_k). \tag{A.3b}$$

The distribution on the initial coefficient vector, β_0, was taken to be a spread-out normal distribution for β_0 given σ with a zero-mean vector and an inverted gamma pdf for σ with small values for its parameters, ν_0 and s_0, $p(\sigma|\nu_0, s_0^2) \propto \sigma^{-(\nu_0+1)}\exp\{-\nu s_0^2/2\sigma^2\}$. Then by direct calculations, at time t, we have

$$\hat{\beta}_t = \hat{\beta}_{t-1} + V_{t-1}x_t(y_t - x_t'\hat{\beta}_{t-1})/(1 + x_t'V_{t-1}x_t),$$

$$\nu_t s_t^2 = \nu_{t-1}s_{t-1}^2 + (y_t - x_t'\hat{\beta}_t)^2 + (\hat{\beta}_t - \hat{\beta}_{t-1})'V_{t-1}^{-1}(\beta_t - \hat{\beta}_{t-1}),$$

$$\nu_t = \nu_{t-1} + 1,$$

$$V_t = V_{t-1} - V_{t-1}x_t x_t'V_{t-1}/(1 + x_t'V_{t-1}x_t) + \phi I_k,$$

and, for the predictive pdf,

$$p(y_{t+1}|D_t) \to t_{\nu_t} = (y_{t+1} - x_{t+1}'\hat{\beta}_t)/s_t a_t,$$

where

$$a_t^2 = 1 + x_{t+1}'V_t x_{t+1}.$$

The parameter ϕ was assigned a value equal to 0.30 in the above calculations.

For the pooled TVP/ARLI model, the equations in (A.3a)–(A.3b) are interpreted as relating to all countries, and recursions similar to those presented above, with $\phi = 0.01$, were employed to compute predictive pdf's and associated quantities. For the TVP/ARLI/WI model, it was assumed to be fully recursive, and recursions similar to those presented above were employed for each of its two equations country by country for the unpooled case and for the entire set of countries in the pooled case. The ϕ parameter was assigned a value of 0.20 for the unpooled case and 0.10 for the pooled case. The values of ϕ were selected to provide a minimum for the median of the 18 countries' RMSE's of one-year-ahead forecasts, 1974–86, for each model.

In the case of models employing exponential weighting, we have for the ith country (with the subscript i omitted for convenience):

$$W_t^{1/2} y_t = W_t^{1/2} X_t \boldsymbol{\beta} + \boldsymbol{\varepsilon}_t \tag{A.4a}$$

or

$$y_t^* = X_t^* \boldsymbol{\beta} + \boldsymbol{\varepsilon}_t, \tag{A.4b}$$

where

$$W_t = \begin{bmatrix} 1 & 0 & 0 & \cdots & 0 \\ 0 & \mu & 0 & \cdots & 0 \\ 0 & 0 & \mu^2 & \cdots & 0 \\ \vdots & \vdots & \vdots & \ddots & \vdots \\ 0 & 0 & 0 & \cdots & \mu^{t-1} \end{bmatrix},$$

$$y_t = \begin{pmatrix} y_t \\ y_{t-1} \\ \vdots \\ y_1 \end{pmatrix} \quad \text{and} \quad X_t = \begin{pmatrix} \boldsymbol{x}_t' \\ \boldsymbol{x}_{t-1}' \\ \vdots \\ \boldsymbol{x}_1' \end{pmatrix}.$$

We assume that $\boldsymbol{\varepsilon}_t \sim N(\mathbf{0}, \sigma^2 I_t)$. As is evident from (A.4a) or (A.4b), this can be considered to be a fixed parameter model with the current observations given heavier weights than those more removed from the present. With a diffuse prior for the parameters, the posterior mean of $\boldsymbol{\beta}$ with data through t is

$$\hat{\boldsymbol{\beta}}_t = (X_t' W_t X_t)^{-1} X_t' W_t y_t$$

and

$$s_t^2 = (y_t - X_t \hat{\boldsymbol{\beta}}_t)' (y_t - X_t \hat{\boldsymbol{\beta}}_t)/(t-k).$$

From the predictive pdf for y_{t+1},

$$t_{\nu_t} = (y_{t+1} - \boldsymbol{x}_{t+1}' \hat{\boldsymbol{\beta}}_t)/s_t a_t$$

has a univariate Student-t distribution with $\nu_t = t - k$ degrees of freedom, where $a_t^2 = 1 + \boldsymbol{x}_{t+1}'(X_t' W_t X_t)^{-1} \boldsymbol{x}_{t+1}$.

In the calculations presented in the text, the parameter μ was assigned a value equal to 0.95 for the unpooled ARLI and ARLI/WI models. In the case of pooling, wherein the value of $\boldsymbol{\beta}$ was assumed the same for all countries, μ was taken equal to 0.85 for the ARLI model and 0.80 for the

ARLI/WI model. These μ values provided a minimum value of the median RMSE's of one-year-ahead forecasts, 1974–86, for the 18 countries.

References

Bowman, H.W. and A.M. Laporte, 1975, Stochastic optimization in recursive equation systems with random parameters with an application to control of the money supply, in: S.E. Fienberg and A. Zellner, eds., Studies in Bayesian econometrics and statistics (North-Holland, Amsterdam) 441–462.

Garcia-Ferrer, A., R.A. Highfield, F. Palm, and A. Zellner, 1987, Macroeconomic forecasting using pooled international data, Journal of Business and Economic Statistics 5, 53–67.

Highfield, R.A., 1986, Forecasting with Bayesian state space models, Ph.D. dissertation (University of Chicago, Graduate School of Business, Chicago, IL).

Hong, C., 1989, Forecasting real output growth rates and cyclical properties of models: A Bayesian approach, Ph.D. dissertation (University of Chicago, Department of Economics, Chicago, IL).

Kling, J.L., 1987, Predicting the turning points of business and economic time series, Journal of Business 60, 201–238.

Moore, G.H., 1983, Business cycles, inflation and forecasting, 2nd ed. (Ballinger, Cambridge, MA).

Moore, G.H. and V. Zarnowitz, 1982, Sequential signals of recession and recovery, Journal of Business 55, 57–85.

Wecker, W.E., 1979, Predicting the turning points of a time series, Journal of Business 52, 35–50.

West, M., P.J. Harrison, and H.S. Migon, 1985, Dynamic generalized linear models and Bayesian forecasting, Journal of the American Statistical Association 80, 73–83.

Zarnowitz, V., 1985, Recent work on business cycles in historical perspective, Journal of Economic Literature 23, 523–580.

Zellner, A., 1971, An introduction to Bayesian inference in econometrics (Wiley, New York, NY).

Zellner, A. and C. Hong, 1988, Bayesian methods for forecasting turning points in economic time series: Sensitivity of forecasts to asymmetry of loss structures, to appear in: K. Lahiri and G. Moore, eds., Leading economic indicators: New approaches and forecasting records (Cambridge University Press, Cambridge).

Zellner, A. and C. Hong, 1989, Forecasting international growth rates using Bayesian shrinkage and other procedures, Journal of Econometrics 40, 183–202.

Zellner, A., C. Hong, and G.M. Gulati, 1990, Turning points in economic time series, loss structures and Bayesian forecasting, in: S. Geisser, J. Hodges, S.J. Press, and A. Zellner, eds., Bayesian and likelihood methods in statistics and econometrics: Essays in honor of George A. Barnard (North-Holland, Amsterdam) 371–393.

[31]

Journal of Forecasting, Vol. 11, 687–701 (1992)

To Combine or not to Combine? Issues of Combining Forecasts

FRANZ C. PALM
University of Limburg, The Netherlands

ARNOLD ZELLNER
University of Chicago, U.S.A.

ABSTRACT

This paper addresses issues such as: Does it always pay to combine individual forecasts of a variable? Should one combine an unbiased forecast with one that is heavily biased? Should one use optimal weights as suggested by Bates and Granger over twenty years ago? A simple model which accounts for the main features of individual forecasts is put forward. Bayesian analysis of the model using noninformative and informative prior probability densities is provided which extends and generalizes results obtained by Winkler (1981) and compared with non-Bayesian methods of combining forecasts relying explicitly on a statistical model for the individual forecasts. It is shown that in some instances it is sensible to use a simple average of individual forecasts instead of using Bates and Granger type weights. Finally, model uncertainty is considered and the issue of combining different models for individual forecasts is addressed.

KEY WORDS Combining forecasts Bayesian methods Model uncertainty

INTRODUCTION

In 1969 Bates and Granger published their path-breaking article. 'The combination of forecasts'. Since then, more than two hundred contributions on combining forecasts have appeared in the literature. On the occasion of the twentieth anniversary of the publication of Bates and Granger (1969), Granger (1989) reviewed the problem and Clemen (1989) surveyed the statistical literature on combining forecasts and produced an annotated bibliography of more than two hundred publications on this topic.

Both the *Journal of Forecasting* and the *International Journal of Forecasting* devoted, respectively, a complete issue and a section of an issue to the topic of combining forecasts. This shows that the topic is of importance, although the title of a contribution by J. S. Armstrong (1989): 'The end of the beginning—or the beginning of the end' indicates that issues of combining forecasts have not been completely resolved and that twenty years after

0277–6693/92/080687–15$12.50 *Received October 1990*
 Revised August 1991

the first publication, there was still some controversy about this topic. Among the issues that have not yet been completely resolved are questions such as: Does it always pay to combine? Should one combine an unbiased forecast with one that is heavily biased? Should one obtain the optimal weights as suggested by Bates and Granger (1969) or should one simply average the individual forecasts, a combining procedure that has worked well in practice (see also Clemen, 1989), and has been used on a large scale by, among others, the Blue Chip Company which publishes simple averages of economists' forecasts? The question of whether to combine forecasts or to combine information sets in the presence of model uncertainty has not yet been completely resolved either (see Diebold, 1989).

In this paper we address the above issues and provide guidelines for applied work on when and how to combine forecasts in a variety of contexts. In the next section a simple model which accounts for the main features of individual forecasts is put forward. A Bayesian analysis using noninformative and informative prior probability densities is provided which extends and generalizes results obtained by Winkler (1981). The third section is devoted to non-Bayesian methods for combining forecasts relying explicitly on a statistical model for the individual forecasts. Simple and weighted average forecasts (also called consensus forecasts) are presented. Mean squared error (MSE) and expected quadratic loss criteria are used to compare the performance of different ways to combine forecasts. In particular, we show that in some instances it is sensible to use a simple average of individual forecasts instead of using weights that have been derived by minimizing the mean squared forecast error of a linear combination of individual forecasts. In the fourth section we allow for model uncertainty, and address the issue of combining different models and of combining individual forecasts and different models for the forecasts. The final section contains a summary and some conclusions.

BAYESIAN METHODS FOR COMBINING FORECASTS

In this section we present operational Bayesian methods for combining individual forecasts which extend results originally put forward by Winkler (1981) by bringing in bias parameters explicitly and treating the unknown covariance case. The aim is to forecast the future value y^{f} of a variable y_t. Observations on y_t are available for the period $t = 1, 2, \dots T$. In addition, we assume that N forecasts, f_{it}, $i = 1, 2, \dots N$, are available for each observation y_t, $t = 1, 2, \dots T$ and for the future value y_{T+1} denoted by y^{f}.

The following simple model which incorporates important features of forecasts is assumed to hold for the individual forecasts:

$$f_{it} - y_t = \theta_i + \varepsilon_{it} + \eta_t \tag{1}$$

where θ_i denotes the bias of forecast i, ε_{it} is a forecast error specific to forecast i, and η_t denotes the overall uncertainty of the forecasts. Reasons for the presence of bias are: (1) forecasters do not use squared error loss (see e.g. Zellner, 1986, for asymmetric loss functions); (2) carelessness; (3) use of a poor or defective information set or incorrect model; and (4) errors of measurement. The error terms ε_{it} and η_t are assumed to be mutually and serially independent, and normally distributed with mean zero and constant variances σ_i^2 and σ_η^2. These assumptions could be relaxed by allowing for correlation between the ε_{it}'s, etc. The inclusion of a common component η_t in the forecast error reflects uncertainty which cannot be diversified by combining individual forecasts. It introduces positive contemporaneous correlation between individuals' forecast errors.

Winkler (1981) restricted himself to a model for unbiased forecasts. For $\theta_i = 0$,

$i = 1, 2, \ldots N$, he analyzed the model (1) under two alternative assumptions: (1) the covariance matrix of $u_{it} = \varepsilon_{it} + \eta_t$, Σ is known and equal to Σ_0, and (2) the covariance matrix is an unrestricted positive definite symmetric (PDS) matrix which has an inverted Wishart prior distribution $\text{IW}(\Sigma \mid \Sigma_0, \delta_0, N)$ with parameter matrix Σ_0 and $\delta_0 \geqslant N$. Winkler showed that the posterior consensus probability density function (pdf) of y^f given a vector of individual forecasts $f = (f_{1t+1}, \ldots f_{Nt+1})'$ is a normal pdf with mean $\mu^* = \iota'\Sigma_0^{-1}f/\iota'\Sigma_0^{-1}\iota$, where ι is an $N \times 1$ unit vector, and variance $\sigma^{*2} = \iota'\Sigma_0^{-1}\iota$ when Σ is known and equal to Σ_0. The mean of the predictive pdf of y^f is a linear combination of the individual forecasts with weights $\iota'\Sigma_0^{-1}/\iota'\Sigma_0^{-1}\iota$ equal to the weights which minimize the mean squared forecast error in a sampling theory framework when the covariance matrix of the forecast error is known.

When Σ is not known and has an inverted Wishart prior pdf, the posterior consensus pdf of y^f is a univariate Student-t pdf with $\delta_0 + N - 1$ degrees of freedom, mean $m^* = \iota'\Sigma_0^{-1}f/\iota'\Sigma_0^{-1}\iota$, and variance $(\delta_0 + N - 1)s^{*2}/(\delta_0 + N - 3)$, where $s^{*2} = [\delta_0 + (m^*\iota - f)'\Sigma_0^{-1}(m^*\iota - f)]/(\delta_0 + N - 1)\iota'\Sigma_0^{-1}\iota$.

Along the lines of Winkler's (1981) analysis, Lindley (1983) pays special attention to the case where experts' opinions are independent. Winkler's approach has been extended and applied by Agnew (1985) to combine unbiased forecasts of a vector of variables.

For model (1) with $\theta_i = 0$ and known matrix $\Sigma = \Sigma_\varepsilon + \sigma_\eta^2\iota\iota'$, where Σ_ε is a diagonal matrix with the i-th diagonal element being equal to σ_i^2, the results of Winkler (1981) specialize to yield the weights attached to f_i in the posterior mean of y^f. Applying the Binomial Inversion Theorem to Σ, we obtain

$$\Sigma^{-1} = \Sigma_\varepsilon^{-1} - \Sigma_\varepsilon^{-1}\iota(\sigma_\eta^{-2} + \iota'\Sigma_\varepsilon^{-1}\iota)^{-1}\iota'\Sigma_\varepsilon^{-1} \tag{2}$$

If we substitute this expression into Winkler's expression for μ^* and σ^{*2}, the result is

$$\mu^* = \iota'\Sigma_\varepsilon^{-1}f/\iota'\Sigma_\varepsilon^{-1}\iota = \Sigma_{i=1}^N(f_{iT+1}/\sigma_i^2)/\Sigma_{i=1}^N(1/\sigma_i^2) \tag{3}$$

and

$$\sigma^{*2} = \{\iota'\Sigma_\varepsilon^{-1}\iota[1 - \iota'\Sigma_\varepsilon^{-1}\iota/(\sigma_\eta^{-2} + \iota'\Sigma_\varepsilon^{-1}\iota)]\}^{-1} \tag{4}$$

Notice that μ^* in equation (3) does not depend on σ_η^2.

The results of Winkler (1981) can be generalized in a straightforward way when $\theta = (\theta_1, \ldots, \theta_N)'$ is not known and Σ is either known or is an unknown PDS matrix.

Anandalingam and Chen (1989) provide a Bayesian model for combining forecasts from experts who might be biased and correlated with each other. The combination procedure involves debiasing and then combining unbiased forecasts. With the exception of the bias, the parameters in their model are assumed to be known (and time dependent). We consider the case where the bias and the covariance matrix Σ are unknown and have a prior probability distribution, and we derive the posterior predictive distribution, the mean of which is a 'debiased' combined forecast.

Assume that the forecast errors have been observed for T periods and are generated by

$$f_{it} - y_t = \theta_i + u_{it}, \qquad u_t \sim \text{IN}(0, \Sigma) \tag{5}$$

where $u_t = (u_{1t}, \ldots u_{Nt})'$. Then the model specializes to become a multivariate regression model which in matrix notation is represented as follows:

$$\underset{T \times N}{Y} = \underset{T \times 1\;1 \times N}{\iota\theta'} + \underset{T \times N}{U} \tag{6}$$

where typical elements of Y', U', and θ are, respectively, $Y_{it} = f_{it} - y_t$, u_{it}, and θ_i. Results given in Zellner (1971, Chapter 8) can be used to derive the Bayesian predictive pdf of y_{T+1} given f_{iT+1}, $i = 1, 2, \ldots N$.

Along with Zellner (1971), we take the following prior pdfs on Σ^{-1} and θ, assumed to be independently distributed:

$$P(\Sigma^{-1}) \propto |\Sigma^{-1}|^{-(N+1)/2} \tag{7}$$

$$\text{and } P(\theta) \propto \text{constant.} \tag{8}$$

The likelihood function can be expressed as

$$L(\theta, \Sigma^{-1} | Y) \propto |\Sigma|^{-T/2} \exp[-\tfrac{1}{2} \operatorname{tr} S\Sigma^{-1} - \tfrac{1}{2} \operatorname{tr} (\theta - \hat{\theta})\iota'\iota(\theta - \hat{\theta})'\Sigma^{-1}] \tag{9}$$

with $S = (Y - \iota\hat{\theta}')'(Y - \iota\hat{\theta}')$, and $\hat{\theta}' = (\iota'\iota)^{-1}\iota'Y = \iota'Y/T$.

We can combine the prior pdfs in equations (7) and (8) with the likelihood function to obtain the posterior pdf for the parameters θ and Σ^{-1}. The predictive pdf of y_{T+1} given f_{iT+1}, $i = 1, 2, \ldots N$, can be obtained from expression (8.57) on p. 235 in Zellner (1971), which in the present case specializes to become

$$P(y_{T+1} | f_{T+1}, Y) \propto |S|^{-T/2}\{1 + [T/(T+1)]\,(y_{T+1}\iota - \mu)'S^{-1}(y_{T+1}\iota - \mu)\}^{-T/2} \tag{10}$$

where $\mu = f_{T+1} - \hat{\theta}$. Completing the square, introducing $\mu^*\iota$ with $\mu^* = \iota'S^{-1}\mu/\iota'S^{-1}\iota$, expression (10) becomes

$$P(y_{T+1} | f_{T+1}, Y) \propto [1 + (y_{T+1} - \mu^*)^2/(T-1)s^{*2}]^{-T/2} \tag{11}$$

where $(T-1)s^{*2} = [T + 1 + T(\mu^*\iota - \mu)'S^{-1}(\mu^*\iota - \mu)]/T(\iota'S^{-1}\iota)$.

The predictive pdf of y_{T+1} given f_{iT+1}, $i = 1, 2, \ldots N$, and TN observations on past forecast errors Y_{it} when the bias vector and the covariance matrix are not known is a Student-t pdf with mean $\mu^* = \iota'S^{-1}(f_{T+1} - \hat{\theta})/\iota'S^{-1}\iota$ and variance $s^{*2}(T-1)/(T-3)$ and $T-1$ degrees of freedom. The mean of the predictive pdf is a linear combination of the individual forecasts corrected for an estimate of the bias.

When the forecast error is assumed to have the structure specified in model (1) with σ_i^2, $i = 1, \ldots N$, and σ_η^2 being unknown parameters, a multivariate regression model with unobserved error-components results, for which Bayesian solutions have been given by Hill (1965, 1976), among others. As shown by Hill (1965), conditionally on the variance ratio $\sigma_\varepsilon^2/\sigma_\eta^2$, assuming it to be constant across forecasts, the analysis is very simple. Bayesian solutions of the general variance-component model however are numerically complex. A flexible and tractable approach which is expected to yield a solution that is close to that of model (1) consists in taking model (5) as a starting point, leaving Σ unrestricted and using an informative prior that is centered at a parameter matrix of the form of $\Sigma_\varepsilon + \sigma_\eta^2\iota\iota'$.

In line with the above, we assume that the prior pdf of Σ^{-1} is a Wishart pdf:

$$P(\Sigma^{-1}) \propto |\Sigma^{-1}|^{(\nu - N - 1)/2} |\Sigma_0^{-1}|^{-\nu/2} \exp\{-\tfrac{1}{2} \operatorname{tr} \Sigma_0 \Sigma^{-1}\} \tag{12}$$

with ν degrees of freedom and Σ_0 being an $N \times N$ PDS matrix of the form $\Sigma_0 = \Sigma_{\varepsilon 0} + \sigma_{\eta 0}^2\iota\iota'$, where $\Sigma_{\varepsilon 0}$ is a known diagonal matrix and $\sigma_{\eta 0}^2$ is a known variance parameter. The posterior pdf for the parameters θ and Σ^{-1} is proportional to the product of the prior pdfs in expressions (12) and (8) with the likelihood function (9).

Along the lines of the derivation in Zellner (1971, pp. 234–5), we can derive the predictive pdf of y_{T+1} given forecasts f_{iT+1}, $i = 1, \ldots N$,

$$P(y_{T+1} | f_{T+1}, Y) \propto [1 + (y_{T+1} - \mu^{**})^2/(T-1)s^{**2}]^{-(T+\nu)/2} \tag{13}$$

$$\text{where } \mu^{**} = \iota'\bar{S}^{-1}\mu/\iota'\bar{S}^{-1}\iota \tag{14}$$

with $\bar{S} = S + \Sigma_0$ and $(T-1)s^{**2} = [T + 1 + T(\mu^{**}\iota - \mu)'\bar{S}^{-1}(\mu^{**}\iota - \mu)]/[T(\iota'\bar{S}^{-1}\iota)]$.

When an informative Wishart pdf is used for Σ^{-1}, the predictive pdf of y_{T+1} given f_{iT+1}, $i = 1, 2, \ldots N$, and TN observations on past forecasts errors is a Student-t pdf with mean μ^{**} and $(T + \nu - 1)$ degrees of freedom. Again, the mean of the posterior pdf is a linear combination of the individual forecasts corrected for an estimate of the bias with weights that depend on Σ_0 and on the sum of squares of the residuals S. When the variance-component structure in model (1) is indeed appropriate, in large sample, the sum of squares of the residuals divided by the sample size will also have that structure and so will $\bar{S}$. In that case, for large T, μ^{**} will depend on the estimate of Σ_ε. Obviously, with the complete predictive pdf given, there is no need to use the posterior mean as a combined point forecast in all situations. For instance, for an asymmetric loss function the optimal combined point forecast can also be obtained analytically in many cases and it differs from the posterior mean.

Finally, when little or no sample evidence on the past performance of the individual forecasts is available, one will have to use an informative prior for the bias vector θ and for the covariance matrix Σ. To conclude, this section generalizes Winkler's analysis by bringing in bias parameters explicitly and treating the unknown covariance case and yields procedures that are quite operational.

NON-BAYESIAN METHODS FOR COMBINING FORECASTS

In this section we present the non-Bayesian approach to combining forecasts and compare it with the Bayesian solution given in the previous section. Consider again the model given in equation (1). When no or little evidence is available on the past performance of the individual forecasts (e.g. if T is small or equal to zero) the past record of the N individual forecasters cannot be used to estimate the bias terms and the variances σ_i^2 and σ_η^2. If no prior information about individual forecasters' performance and the values of the parameters in equation (1) is available, a natural solution would be to give equal weights to the individual forecasts reflecting an equal degree of confidence in the quality of individual forecasts. This is, in fact, the solution adopted by the Blue Chip Index forecasters, who use a simple average of individual forecasts. Averaging equation (1) over i yields

$$\bar{f}_t - y_t = \bar{\theta} + \bar{\varepsilon}_t + \eta_t \tag{15}$$

where $\bar{f}_t$, $\bar{\theta}$, and $\bar{\varepsilon}_t$ denote simple averages over $i = 1, 2, \ldots N$, e.g. $\bar{f}_t = \Sigma_i f_{it}/N$. When the biases of the individual forecasts average to zero, that is, when $\bar{\theta} = 0$, the simple average $\bar{f}_t$ is an unbiased forecast of y_t. The simple average $\bar{f}_t$ also has a Bayesian interpretation. The weight attached to an individual forecast reflects the degree of confidence in the forecast. In absence of information about the quality of the forecasts, equal variances will be attached to the individual forecast errors. With no sample evidence available, $\bar{S}$ in equation (14) $\bar{S} = \Sigma_{\varepsilon 0} + \sigma_{\eta 0}^2 \iota\iota'$ with $\Sigma_{\varepsilon 0} = \sigma_\varepsilon^2 I$. Using the Binomial Inversion Theorem, the predictive mean μ^{**} can be expressed as $\mu^{**} = \iota' f/N$, which is a simple average of individual forecasts.

Under quadratic loss, the point forecast which minimizes expected loss is the mean of the combined predictive density. In equation (1) the mean of y_t equals $f_{it} - \theta_i$, so that the optimal point forecast is

$$\hat{y}_t^{\text{f}} = \Sigma_{i=1}^N r(f_{it} - \theta_i) = \bar{f}_t \tag{16}$$

with $r = 1/N$ and $\bar{\theta} = 0$. For the simple average to minimize expected quadratic loss it is necessary that the biases of the individual forecasts average to zero. A sufficient condition for this to be true is that the individual forecasts are equal to the expectation of y^{f}.

The use of the simple average forecast in equation (15) instead of individual forecasts has the advantage of averaging the individual biases and the individual components of the forecast error. These biases may average to zero and the variance of the average of the individual forecast errors $\Sigma_i \sigma_i^2/N^2$ will go to zero as $N \rightarrow \infty$ and thus the mean square error (MSE) associated with the simple average $\bar{f}_t$ approaches σ_η^2, given that $\bar{\theta} = 0$.

Bates and Granger (1969) proposed the use of optimal weights, that is, weights which minimize the MSE associated with a weighted average of unbiased forecasts. To implement this approach, one has to eliminate the bias terms θ_i and then get the weights of the forecasts expressed in terms of the variances of the forecast errors. This requires that the bias and variance parameters θ_i, σ_i^2 are given or can be estimated with some precision from forecast evidence. More formally, their approach consists of minimizing the variance of the combined forecast error:

$$\min_\alpha \alpha'\Sigma\alpha \text{ subject to } \alpha'\iota = 1 \tag{17}$$

where α is the vector of weights, Σ is the covariance matrix of the individuals forecast errors, and ι is a vector with all elements equal to one. The optimal weights are given by

$$\alpha^* = (\iota'\Sigma^{-1}\iota)^{-1}\Sigma^{-1}\iota \tag{18}$$

and the value of the criterion function is

$$S(\alpha^*) = \alpha^{*\prime}\Sigma^{-1}\alpha^* = (\iota'\Sigma^{-1}\iota)^{-1} \tag{19}$$

When the individual forecasts are possibly biased, the weights which minimize the MSE

$$\text{MSE}(\alpha = \alpha'\Sigma + \theta\theta')\alpha \text{ subject to } \alpha'\iota = 1 \tag{20}$$

are

$$\tilde{\alpha} = [\iota'(\Sigma + \theta\theta')^{-1}\iota]^{-1}[\Sigma + \theta\theta']^{-1}\iota, \tag{21}$$

where θ is the bias vector. The minimum of the MSE is

$$\text{MSE}(\tilde{\alpha}) = [\iota'(\Sigma + \theta\theta')^{-1}\iota]^{-1} \tag{22}$$

It is obvious that $\tilde{\alpha}$ specializes to become equal to α^* when $\theta = 0$. Also notice that the value of equation (22) exceeds that of equation (19). This follows directly from the fact that $(\Sigma + \theta\theta') - \Sigma$ is a positive semi-definite matrix.

When equal weights are used, i.e. $\alpha = \iota/N$, the MSE becomes

$$\text{MSE}(\iota/N) = \iota'(\Sigma + \theta\theta')\iota/N^2 \tag{23}$$

which is larger than or equal to $\text{MSE}(\tilde{\alpha})$ even if $\theta = 0$.

When Σ and/or θ are not known and an estimate or an approximate value $\bar{\Sigma}$ and/or $\bar{\theta}$ has to be used which is fixed, the MSE associated with the equal weights combined forecast (23) may be smaller than that associated with weights based on $\bar{\Sigma}$ after correcting for bias using $\bar{\theta}$:

$$\text{MSE}(\hat{\alpha}^*) = [\iota'\bar{\Sigma}^{-1}(\bar{\Sigma} + (\theta - \bar{\theta})(\theta - \bar{\theta})')\bar{\Sigma}^{-1}\iota]/(\iota'\bar{\Sigma}^{-1}\iota)^2 \tag{24}$$

The MSE in equation (23) may even be smaller than that in equation (24) if Σ is known, i.e. $\bar{\Sigma} = \Sigma$.

It should be noted that the problem of estimating θ and Σ is similar to that of estimating the expected return and the covariance matrix of the vector of returns needed to compute an optimal portfolio of assets. This problem has been analyzed by Brown (1978) and Jorion

(1985), among others. Jorion (1985) argues that estimation risk due to uncertain mean returns has a considerable impact on optimal portfolio selection, whereas the covariance matrix is estimated with relative precision. He finds that the out-of-sample performance of optimal portfolios with estimated parameters in practice was not superior to that of a hold-the-market strategy for several countries.

When the forecast errors are generated by model (1), the matrix Σ has a special structure $\Sigma = \Sigma_\varepsilon + \sigma_\eta^2 \iota\iota'$, where Σ_ε is a diagonal matrix with σ_i^2 being a typical element of the main diagonal. The weights α will not depend on σ_η^2 and the MSE(α) in equation (20) becomes

$$\mathrm{MSE}(\alpha) = \alpha'(\Sigma_\varepsilon + \theta\theta')\alpha + \sigma_\eta^2 \tag{25}$$

Since the optimal weights α^* do not depend on σ_η^2, there is no way to reduce or eliminate the uncertainty due to the common component η_t. Model (1) yields a possible explanation of a finding by Newbold and Granger (1974) and Winkler and Makridakis (1983) that the best combinations were those that ignored correlations among forecast errors. Also, to the extent that a redundancy in the forecast errors (i.e. a positive correlation) is due to a common component η_t, the redundancy does not affect the weighting scheme (see e.g. Maines, 1990, on redundancy).

To illustrate the above, we assume that model (1) holds with $N = 2$. The results are summarized in Table I, where α_1 denotes the weight associated with $f_{1t}(\alpha_2 = 1 - \alpha_1)$.

Table I. Weights and MSEs associated with combined forecasts based on model (1) for $N = 2$

Weight		Minimum MSE		Minimum MSE when $\sigma_1^2 = \sigma_2^2$	
$\alpha_i = 1$		$\sigma_i^2 + \theta_i^2 + \sigma_\eta^2$	(26)		
$\alpha_1^* = \sigma_2^2/(\sigma_1^2 + \sigma_2^2)$ $(\theta_1 = \theta_2 = 0)$	(27)	$(\sigma_1^2\sigma_2^4 + \sigma_1^4\sigma_2^2)/(\sigma_1^2 + \sigma_2^2)^2 + \sigma_\eta^2$	(28)	$\sigma_1^2/2 + \sigma_\eta^2$	(29)
$\tilde{\alpha}_1 = \dfrac{\sigma_2^2 + \theta_2(\theta_2 - \theta_1)}{\sigma_1^2 + \sigma_2^2 + (\theta_1 - \theta_2)^2}$	(30)	$\dfrac{\sigma_1^2\sigma_2^2 + \sigma_1^2\theta_2^2 + \sigma_2^2\theta_1^2}{\sigma_1^2 + \sigma_2^2 + (\theta_1 - \theta_2)^2} + \sigma_\eta^2$	(31)	$\sigma_1^2\left[\dfrac{\sigma_1^2 + \theta_1^2 + \theta_2^2}{2\sigma_1^2 + (\theta_1 - \theta_2)^2}\right] + \sigma_\eta^2$	(32)
$\tilde{\alpha}_1(\theta_1 = 0) = \dfrac{\sigma_2^2 + \theta_2^2}{\sigma_1^2 + \sigma_2^2 + \theta_2^2}$	(33)	$\dfrac{\sigma_1^2(\sigma_2^2 + \theta_2^2)}{\sigma_1^2 + \sigma_2^2 + \theta_2^2} + \sigma_\eta^2$	(34)	$\sigma_1^2\left[\dfrac{\sigma_1^2 + \theta_2^2}{2\sigma_1^2 + \theta_2^2}\right] + \sigma_\eta^2$	(35)
$\alpha_i = \frac{1}{2}$	(36)	$\frac{1}{4}(\sigma_1^2 + \sigma_2^2 + (\theta_1 + \theta_2)^2) + \sigma_\eta^2$	(37)	$\sigma_1^2/2 + (\theta_1 + \theta_2)^2 + \sigma_\eta^2$	(38)

Obviously, expression (28) in Table I is smaller than equation (26) even if θ_i in (26) is zero, indicating that combining unbiased forecasts leads to a reduction in MSE. As shown above, correcting for bias, assumed known, first and then combining the forecasts yields a smaller MSE than combining the biased forecasts (compare expressions (31) and (34) to (28) and (32) and (35) to (29)).

Comparing equation (32) to (26) in Table I for $i = 1$, it follows that the MSE of the combined forecast is smaller than that of f_{1t} iff

$$\sigma_1^2[\sigma_1^2 - 2\theta_1\theta_2] + \theta_1^2[2\sigma_1^2 + (\theta_1 - \theta_2)^2] \geqslant 0 \tag{39}$$

a condition that is always satisfied if one of the forecasts f_{1t} and f_{2t} is unbiased. When $\theta_1 = 0$, the MSE of the combined forecast (34) is always smaller than that of f_{1t}: $\sigma_1^2 + \sigma_\eta^2$. Here, we see that the reduction of the variance of the forecast error more than offsets the increase of the MSE due to the bias in f_{2t}. There is a trade-off between variance reduction as a result of combining and the introduction of some bias through the inclusion of f_{2t} in the forecast.

In practice the variances σ_i^2 and the biases θ_i are not known. In some instances they can be estimated from evidence on past forecast performance. When θ_i is constant through time, θ_i can be estimated as the sample average of past forecast errors $f_{it} - y_t$, $t = 1, 2 \dots T$. Similarly, the variance of past forecast errors yields an estimate of $\sigma_i^2 + \sigma_\eta^2$. From the covariance between the errors of two different forecasts, the parameter σ_η^2 can be estimated. The mean can also be estimated using Stein's estimator provided the forecast error covariance matrix is known or an outside estimate is available.

Asymptotically efficient estimates of the parameters θ_i can be obtained by estimating the set of equations

$$(f_{it} - y_t) = \theta_i + \varepsilon_{it} + \eta_t \tag{40}$$

as a system of seemingly unrelated regressions with all the contemporaneous disturbance covariances equal to σ_η^2.

An approach suggested by Granger and Ramanathan (1984) which consists of regressing the actual data on the forecasts

$$y_t = \gamma + \Sigma_i \beta_i f_{it} + u_t \qquad t = 1, 2, \dots T \tag{41}$$

possibly restricting γ to zero and the sum of the β_i's to one will not be feasible if T is smaller than the number of parameters to be estimated. The use of prior information on β in the form of a g-prior pdf advocated by Diebold and Pauly (1990) also requires some evidence on past forecast performance. The regression function (41) will be well defined with constant parameters when e.g. y_t and f_{it}, $i = 1, 2, \dots N$, are jointly normally distributed $t = 1, 2, \dots T$ but not under very general conditions. The regression approach applied to equation (40) will be more efficient than that in (41). The number of degrees of freedom will be, respectively, $TN - (2N + 1)$ and $T - (N + 1)$. Also if the bias θ_i in equation (40) is time-varying (say, θ_{it}), one can specify a process for θ_{it}, estimate its parameters, obtain a forecast of the bias, and correct the forecast of y_t^f by the forecast of θ_{it}. Diebold and Pauly (1987) use weighted least squares to create time-varying parameter combination models.

The fact that in order to correct for bias and compute the optimal weights, the bias and the variance parameters have to be adequately modeled and estimated will often lead to specification and estimation errors. When sampling errors are taken into account, a simple average of the individual forecasts such as used in the Blue Chip Index may outperform the combined forecast using estimates of the optimal weights and of biases (see also Einhorn and Hogarth, 1975). Obviously, in the population the forecasts using the weights (27) and (30), given in Table I, dominate the simple average of forecasts which for $N = 2$ has a MSE given in equation (37). It outperforms the single forecast f_{1t} if expression (26) is larger than (37). This is even the case when f_{1t} is unbiased (i.e. $\theta_1 = 0$) and $\sigma_1^2 = \sigma_2^2$ as long as $\sigma_2^2 \leqslant 2\sigma_1^2$. For empirical evidence on mean forecasts which outperform individual forecasts, the reader is referred to e.g. Ashton (1982) and O'Brien (1988).

The simple average of forecasts will also perform quite well when the specification for the forecast error in model (1) is generalized to account for a multiplicative bias γ_i in addition to an additive possibly time-varying bias θ_{it}:

$$y_t = \gamma_i f_{it} - \theta_{it} - \varepsilon_{it} - \eta_t \tag{42}$$

Averaging expression (42) over i yields

$$y_t = \bar{\gamma}\bar{f}_t + (1/N)\Sigma_i(\gamma_i - \bar{\gamma})(f_{it} - \bar{f}_t) - \bar{\theta}_t - \bar{\varepsilon}_t - \eta_t \tag{43}$$

where $\bar{\gamma} = 1/N\ \Sigma_i\ \gamma_i$. The simple average of individual forecasts will be an unbiased forecast with MSE equal to $1/N^2\ \Sigma_i\ \sigma_i^2 + \sigma_\eta^2$ when the following conditions are satisfied:

$$1/N\ \Sigma_i(\gamma_i - \bar{\gamma})(f_{it} - \bar{f}_t) - \bar{\theta}_t = 0 \tag{44a}$$

$$\bar{\gamma} = 1 \tag{44b}$$

These conditions may be satisfied or nearly satisfied in many applications. For instance, when the θ_{it}'s average to zero, the multiplicative biases average to one and the covariance between the multiplicative bias and the individual forecasts is zero, conditions (44) are met. When sufficient observations on y_t and $\bar{f}_t$ are available, the conditions in (44) can be tested as restrictions on the coefficients of a nonhomogeneous regression of y_t on $\bar{f}_t$.

To summarize: the results of this section show that, in general, combining individual forecasts into a simple or a weighted average can lead to a reduction in MSE and that there can be valid reasons for using a simple average of the individual forecasts instead of using 'optimal' weights. A simple average will be appropriate when we have insufficient information on the past performance of individual forecasters. Specification and sampling errors involved in estimating the bias and variance parameters needed to obtain the weights of an optimal combination of forecasts can affect the MSE in such a way that the resulting forecast is outperformed by a simple average. Also, if the parameters are time-varying it may be difficult (if not impossible) to estimate appropriate weights.

In other words, equal weight averages may be more robust with respect to specification uncertainty, time-variation of parameters, and estimation errors than weighted averages with weights that have to be estimated. These conclusions are not inconsistent with empirical findings such as those of Ashton and Ashton (1985) that, in some instances, differential weighting schemes with weights based on *ex-post* performance are more accurate than equal weighting.

Finally, when a quadratic loss function is used, the optimal Bayesian combined forecast is the posterior mean which uses the same weights as in the non-Bayesian approach when the covariance matrix of the forecast errors Σ is known. When Σ has to be estimated from observed forecast performance, in large samples both approaches will rely on the same weights provided the model for the forecast errors with constant Σ and θ is appropriate.

MODEL UNCERTAINTY AND COMBINING FORECASTS

In this section we consider issues that arise when individual forecasts are generated by a set of alternative models which is not necessarily exhaustive, i.e. with probabilities that add to less than one. We use the following notation: $h_i(y^f \mid D)$ is the predictive pdf of model M_i given the data D, with mean $E_i(y^f \mid D) = m_i$ and variance v_i, p_i is the posterior probability of M_i, and $\Delta_{ij} = (m_i - m_j)^2$. We assume that the loss function is squared predictive error. For earlier contributions to Bayesian model-combining techniques, the reader is referred to Bunn (1977) and Geisel (1975).

It is well known (see also Zellner, 1989) that when the models or the predictive pdfs are mutually exclusive and exhaustive, the optimal point forecast (which minimizes expected quadratic loss) is the mean of the combined predictive pdf:

$$m_c = E_c(y^f \mid D) = \Sigma_{j=1}^{N} p_j m_j \tag{45}$$

For $N = 2$, we have $m_c = p_1 m_1 + (1 - p_1) m_2$ and expected loss associated with m_c is equal to the variance of the predictive pdf:

$$v_c = p_1 v_1 + (1 - p_1) v_2 + p_1(1 - p_1)\,\Delta_{12} \tag{46}$$

Expected loss associated with the point forecast $f = m_1$ is

$$E_c\{(y^f - f)^2 \mid f = m_1\} = p_1 v_1 + (1 - p_1) v_2 + (1 - p_1)\,\Delta_{12} \tag{47}$$

The difference between equations (47) and (46) equals $\Delta_{12}(1 - p_1)^2 \geqslant 0$, so that the combined forecast is preferred to the forecast based on M_1 (or M_2) only. The difference between equations (46) and (47) will be obviously close to zero if p_1 is close to one or m_1 is close to m_2.

In general, when the set of models considered is exhaustive and forecasting can be done conditionally on any model, the combined forecast is preferred to the individual one. However, as shown by Min and Zellner (1990), if the models are not exhaustive or if not all the models are potentially used in forecasting, combining forecasts is not always optimal.

To discuss the impact of nonexhaustivity of the models, we consider a simplified version of the loss structure put forward in Min and Zellner (1990). Their loss structure is based on mean squared predictive error.

The combined forecast associated with M_c is equal to $(1 - w)m_1 + wm_2$ and $w = p_2/(p_1 + p_2)$. To obtain the complete quadratic loss structure, one has to add v_j to the elements of the jth column of Table II. When this is done, for each of the forecasting models expected loss is increased by the same amount $p_1 v_1 + p_2 v_2 + p_3 v_3$, so that the difference in expected loss across models is not affected. Therefore, this sum has not been included in the loss structure. We assume that the models M_1 and M_2 are specified and that M_3 is representative of all the models that could be entertained with some confidence for the problem at hand.

The prior probabilities of M_j, $j = 1, 2, 3$, can be specified and add up to one. However, as the models in class M_3 are not specified, sample evidence cannot be brought to bear on M_3. The posterior odds of M_1 versus M_2, p_1/p_2, and the value of $w = p_2/(p_1 + p_2)$ can be obtained but the posterior probability associated with M_3 cannot be computed. There is an argument for using the information that we have, that is, the prior probability attached to M_3 as it reflects our degree of confidence in the class of models M_3 even though as long as the model M_3 (or class of models) is not specified, sample evidence cannot be used to learn about the probability of M_3. The use of prior probability of M_3, p_3 throughout the analysis allows us to obtain the values of p_1 and p_2 given that $p_3 = 1 - p_1 - p_2$.

Table II.

Forecasting conditional on	State M_1	M_2	M_3	Expected loss
M_1	0	Δ_{12}	Δ_{13}	$\Delta_{12} p_2 + p_3 \Delta_{13}$
M_2	Δ_{12}	0	Δ_{23}	$\Delta_{12} p_1 + p_3 \Delta_{23}$
M_c	$w^2 \Delta_{12}$	$(1 - w)^2 \Delta_{12}$	Δ_{c3}	$\Delta_{12}[w^2 p_1 + (1 - w)^2 p_2] + p_3 \Delta_{c3}$
Probability	p_1	p_2	p_3	

The choice among M_1, M_2, and M_c outlined in Table II can only be made if we have some knowledge about the sign and size of m_3 or, equivalently, about Δ_{13} or Δ_{23}, as m_1 and m_2 are known. When m_3 is exactly known, one can include it in the combined forecast to obtain

$$f_c = \Sigma_{j=1}^{3} p_j m_j \tag{48}$$

If we have some knowledge about the sign and size of m_3 we may be able to implement the Bayesian forecast selection rule given in Table II. Model M_1 will be preferred to M_2 iff $E_1(L) \leqslant E_2(L)$, where L denotes the loss function. This is equivalent to

$$\Delta_{12}(p_1 - p_2) \geqslant p_3(m_1^2 - m_2^2) + 2p_3(m_2 - m_1)m_3 \tag{49}$$

which is a linear inequality in m_3. Similarly, M_c yields lower expected loss than M_1 iff $E_1(L) - E_c(L) \geqslant 0$, which is equivalent to the condition

$$\Delta_{12}[p_2(2w - w^2) - w^2 p_1] + p_3[(1 - w^2)m_1^2 - (1 - w)^2 m_2^2 - 2w(1 - w)m_1 m_2] - 2(1 - w)p_3(m_1 - m_2)m_3 \geqslant 0 \tag{50}$$

Finally, M_c is preferred to M_2 iff $E_2(L) - E_c(L) \geqslant 0$ or, equivalently,

$$\Delta_{12}[(1 - w^2)p_1 - (1 - w)^2 p_2] + p_3[(1 - w^2)m_1^2 - (1 - w)^2 m_2^2 - 2w(1 - w)m_1 m_2] - 2m_3 p_3 w(m_2 - m_1) \geqslant 0 \tag{51}$$

To obtain more insight into the meaning of these conditions, consider the following example with $p_1 = p_2 = \frac{1}{4}$, ($p_3 = \frac{1}{2}$), $m_1 = 5$, $m_2 = 10$. For these values, expected loss as a function of m_3 is

$$\begin{aligned} E_1(L \mid m_3) &= 18.75 - 5m_3 + \tfrac{1}{2}m_3^2 \\ E_2(L \mid m_3) &= 56.25 - 10m_3 + \tfrac{1}{2}m_3^2 \\ E_c(L \mid m_3) &= 31.25 - 7.5m_3 + \tfrac{1}{2}m_3^2 \end{aligned} \tag{52}$$

Since the difference between any pair of these expressions does not depend on m_3^2, we can limit ourselves to the linear part, i.e. to $E_i^*(L \mid m_3) = E_i(L \mid m_3) - \frac{1}{2}m_3^2$, $i = 1, 2, c$, for which the graphs are given in Figure 1. As shown in Figure 1, for values of $m_3 \leqslant 5$, the line

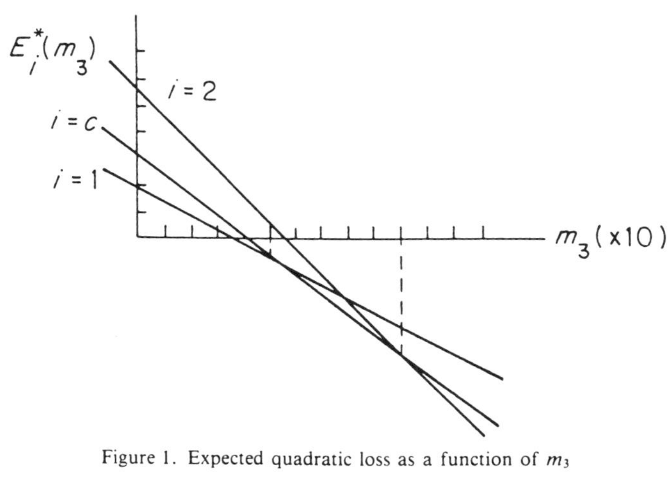

Figure 1. Expected quadratic loss as a function of m_3

corresponding to $E_1^*(m_3)$ is the lowest one. When $5 < m_3 \leqslant 10$, the lowest line is that representing $E_c^*(m_3)$. When $m_3 > 10$, $E_2^*(m_3)$ yields lowest expected loss. The following choice minimizes expected quadratic loss. Choose M_1 if one believes that $m_3 \leqslant 5$, choose M_c if $5 < m_3 \leqslant 10$, choose M_2 if $10 < m_3$.

Some comments are in order.

1. The optimal rule can be interpreted as follows. Always choose the point forecast which is closest to the value of m_3. For example, when $m_3 \leqslant 5$, choose the individual forecast m_1. For extreme values of m_3, individual forecasts will be preferred to the combined forecast. The fact that we entertain M_3 means that with probability p_3 the mean of y^f equals m_3. Choosing a forecast that is close to m_3 in the metric of the expected loss function is equivalent to eliminating part of the bias resulting from the fact that M_3 is not included in the choice set.
2. As the example shows, when the set of models considered is not exhaustive it is not possible to choose an optimal point forecast without having some *a priori* information about m_3. In other words, when one believes *a priori* that forecasts based on M_1 and M_2 only are biased, one has to bring in some information on m_3 to de-bias the forecast partly or completely.
3. In pairwise comparisons of the expected losses of M_1 or M_2 with that of the combined forecast, it is sufficient that $\Delta_{c3} - \Delta_{13} < 0$ to prefer the combined forecast to that obtained from M_1 as $w^2 p_1 + (1-w)^2 p_2 \leqslant p_1, p_2$. Similarly, in order to choose the combined forecast instead of that obtained from M_2, it is sufficient that $\Delta_{c3} - \Delta_{23} < 0$.
4. It is obvious that when the value of m_3 is exactly known, the combined forecast (48) can be computed. This achieves minimum expected loss for a quadratic loss function and is therefore preferred to m_1, m_2, and m_c based on m_1 and m_2.
5. The comments made above also apply if M_3 in Table II is a benchmark model as in Min and Zellner (1990). In particular, as already pointed out, the combined forecast m_c is not always preferred to the individual forecasts m_1 and m_2.
6. When we do not have prior information on p_3 and m_3, which implies that the prior odds of M_1 versus M_2, not the individual prior probabilities of M_1 and M_2, are given, an additional principle has to be brought in, as shown by Zellner (1989) and Min and Zellner (1990). The density combining M_1 and M_2 is chosen by minimizing a distance metric expressing the distance between densities based on M_1 and M_2 and a combining density. If one uses a quadratic distance metric, Zellner (1989) and Min and Zellner (1990) show that the combined density combines $h_1(y^f \mid D)$ and $h_2(y^f \mid D)$ with weights equal to w and $1-w$, respectively. As notice above, the value of w can be directly obtained from the posterior odds ratio p_1/p_2. This choice of the combining pdf basically means that one uses the combined pdf conditional on the event that y^f being generated by M_1 or M_2: $h_c(y^f \mid D, M_1 \cup M_2)$. To the extent that we are confident that M_1 and M_2 represent the states of the world with a high probability, an analysis based on M_1 and M_2 only will probably suffice. When this is not the case, it may be safer to proceed as outlined in this section instead of making statements conditional on M_1 or M_2.
7. Bayes' theorem can be applied to combine forecasts of experts allowing for model uncertainty. For each model j for the forecasts, the mean of the predictive pdf of y_{T+1} given f_{iT+1}, $i = 1, 2, \ldots N$, can be obtained along the lines of the approach presented in the second section of this paper. Under quadratic loss, the forecast which minimizes expected loss is then given by

$$E(y^f \mid D) = \Sigma_{j=1}^{J} p_j \mu_j^* \qquad (53)$$

where p_j is the posterior probability that the forecast error is generated by model j and μ_j^* denotes the mean of the predictive pdf conditionally on model j (see e.g. expressions (11) or (14)).

8. Winkler's (1981) approach and its extensions presented in the second section rely on the assumption that forecasters use the same model. It may, however, be wise to assume that forecasters use different models and that the models are not known to an investigator. One forecaster may have an asymmetric loss function which makes it perfectly rational for him to produce a biased forecast while another may give forecasts that have serially correlated forecast errors. When there is much empirical evidence on the forecasters' performance in the past, one can choose a model on the basis of this evidence. If all forecasters are found to use the same model (e.g. model (5) with normally distributed forecast errors) the approach put forward in the second section naturally applies. If there is no or little evidence on the performance of forecasters, it will be more appropriate to put a probability on each forecaster reflecting the confidence in his statements, entertain several models for each forecaster with a probability attached to each model, and combine the forecasts. For a quadratic loss function, the optimal combined point forecast will then be

$$E(y^{\mathrm{f}} \mid D) = \Sigma_{i=1}^{N} p_i \, \Sigma_{j=1}^{J} p_{ij} m_{ij} \tag{54}$$

where p_i is the posterior probability attached to the statements of forecaster i, p_{ij} is the posterior probability that model j is appropriate for forecaster i, and m_{ij} is the expected value of y^{f} conditional on model j for forecast i. As evidence on forecast errors becomes available, the uncertainty about a forecaster's reliability and the appropriateness of the different models will decrease.

Hopefully, this analysis will lead to selecting the same model for the different forecasts. This discussion again emphasizes the importance of analyzing data on forecast performance for the derivation of combined forecasts.

CONCLUSIONS

In this paper we have considered several issues related to combining forecasts. The main point is that in order to determine when and how to combine forecasts it is sensible to rely explicitly on a model for the individual forecasts. If there is uncertainty about the model, alternative models which represent our best state of knowledge can be specified and entertained.

Model for the individual forecasts can be used to assess the properties of different combined forecasts such as a simple average or weighted averages of individual forecasts. The advantages of a simple average of forecasts compared with a weighted average are:

1. Its weights are known and do not have to be estimated, an important advantage if there is little evidence on the performance of individual forecasts or if the parameters of the model generating the forecasts are time-varying;
2. In many situations a simple average of forecasts will achieve a substantial reduction in variance and bias through averaging out individual bias;
3. It will often dominate, in terms of MSE, forecasts based on optimal weighting if proper account is taken of the effect of sampling errors and model uncertainty on the estimates of the weights.

The Bayesian approach to combining forecasts can be adopted to obtain optimal weights even if there is little information available on the performance of the individual forecasts. It can also account for model uncertainty when the set of models considered is both exhaustive and non-exhaustive. When the models are not exhaustive, an additional principle has to be introduced to determine the weights for the models under consideration. Also, information about the models or class of models not included in the analysis can be useful to determine whether inference conditional on the models entertained yields sensible results. Finally, the approach put forward above can be extended to apply to the problem of forecasting a vector of variables.

ACKNOWLEDGEMENTS

This research has been done when the first author was a visiting professor at the Graduate School of Business, University of Chicago. He gratefully acknowledges the hospitality of the Graduate School of Business, University of Chicago. The research was financed by the National Science Foundation and by income from the H. G. B. Alexander Endowment Fund, Graduate School of Business, University of Chicago.
The authors thank Derek Bunn and two referees for their useful comments.

REFERENCES

Agnew, C. E., 'Bayesian consensus forecasts of macroeconomic variables', *Journal of Forecasting*, **4** (1985), 363–76.

Anandalingam, G. and Chen, L., 'Linear combinations of forecasts: a general Bayesian model', *Journal of Forecasting*, **8** (1989), 199–214.

Armstrong, J. S., 'Combining forecasts: the end of the beginning or the beginning of the end', *International Journal of Forecasting*, **5** (1989), 584–88.

Ashton, A. H., 'An empirical study of budget-related predictions of corporate executives', *Journal of Accounting Research*, **20** (1982), 440–49.

Ashton, A. H. and Ashton, R. H., 'Aggregating subjective forecasts: some empirical evidence', *Management Science*, **31** (1985), 1499–1508.

Bates, J. M. and Granger, C. W. J., 'The combination of forecasts', *Operations Research Quarterly*, **20** (1969), 319–25.

Brown, S. J., 'The portfolio choice problem: comparison of certainty equivalence and optimal Bayes portfolios', *Communications in Statistics* B, **7** (1978), 321–334.

Bunn, D. W., 'A comparative evaluation of the outperformance and minimum variance procedures for the linear synthesis of forecasts', *Operations Research Quarterly*, **28** (1977), 653–62.

Clemen, R. T., 'Combining forecasts: a review and annotated bibliography', *International Journal of Forecasting*, **5** (1989) 559–84.

Diebold, F. X., 'Forecast combination and encompassing, reconciling two divergent literatures', *International Journal of Forecasting*, **5** (1989), 589–92.

Diebold, F. X. and Pauly, P., 'Structural change and the combination of forecasts', *Journal of Forecasting*, **6** (1987), 21–40.

Diebold, F. X. and Pauly, P., 'The use of prior information in forecast combination', *International Journal of Forecasting*, **6** (1990), 503–8.

Einhorn, H. J. and Hogarth, R. M., 'Unit weighting schemes for decision making', *Organizational Behavior and Human Performance*, **13** (1975), 171–92.

Geisel, M. S., 'Bayesian comparison of simple macroeconomic models', reprinted from the *Journal of Banking, Money and Credit* in Fienberg, S. E. and Zellner, A. (eds), *Studies in Bayesian Econometrics and Statistics*, Amsterdam: North-Holland, 1975.

Granger, C. W. J., 'Invited review: combining forecasts—twenty years later', *Journal of Forecasting*, **8** (1989), 167–73.
Granger, C. W. J. and Ramanathan, R., 'Improved methods of forecasting', *Journal of Forecasting*, **3** (1984), 197–204.
Hill, B. M., 'Inference about variance components in the one way model', *Journal of the American Statistical Association*, **60** (1965), 806–25.
Hill, B. M., 'Exact and approximate Bayesian solutions for inference about variance components and multivariate inadmissibility', Technical Report, University of Michigan, Ann Arbor, 1976.
Jorion, P., 'International portfolio diversification with estimation risk', *Journal of Business*, **58** (1985), 259–78.
Lindley, D., 'Reconciliation of probability distributions', *Operations Research*, **31** (1983), 866–80.
Maines, L. A., *Data Redundancy and Consensus Forecasting Judgements*, University of Chicago, Graduate School of Business, Unpublished doctoral dissertation, 1990.
Min, Chung-Ki and Zellner, A., 'Bayesian and non-Bayesian methods for combining models and forecasts with applications', Technical Report, H. G. B. Alexander Research Foundation, Graduate School of Business, University of Chicago, to appear in *Journal of Econometrics* (1990).
Newbold, P. and Granger, C. W. J., 'Experience with forecasting univariate time series and the combination of forecasts' (with discussion), *Journal of the Royal Statistical Society*, A, **137** (1974), 131–49.
O'Brien, P., 'Analysts' forecasts as earnings expectations', *Journal of Accounting and Economics*, **10** (1988), 53–83.
Winkler, R. L., 'Combining probability distributions from dependent information sources', *Management Science*, **27** (1981), 479–88.
Winkler, R. L., 'Combining forecasts: a philosophical basis and some current issues', *International Journal of Forecasting*, **5** (1989), 605–9.
Winkler, R. L. and Makridakis, S., 'The combination of forecasts', *Journal of the Royal Statistical Society*, A, **146** (1983), 150–57.
Zellner, A., *An Introduction to Bayesian Inference in Econometrics*, New York: Wiley, 1971.
Zellner, A., 'Biased predictors, rationality and the evaluation of forecasts', *Economics Letters*, **21** (1986), 45–8.
Zellner, A., 'Bayesian and non-Bayesian methods for combining models and forecasts', Manuscript, H. G. B. Alexander Research Foundation, Graduate School of Business, University of Chicago, 1989.

Authors' biographies:
Franz Palm is Professor of Econometrics at the Faculty of Economics and Business Administration, University of Limburg. He has a PhD in economics from the University of Louvain. He was a research associate at CORE and a Professor of Economics at the Free University, Amsterdam before joining the University of Limburg. His research has been concentrated on theoretical and applied econometrics with emphasis on time series analysis.

Arnold Zellner is the H. G. B. Alexander Distinguished Service Professor in the Graduate School of Business, University of Chicago. He has an AB in physics from Harvard, a PhD in economics from the University of California at Berkeley, and an honorary doctorate from the Autonomous University of Madrid. Before accepting an appointment at the University of Chicago in 1966, he held appointments in economics at the University of Washington in Seattle and the University of Wisconsin in Madison. His current and past research has been concentrated on theoretical and applied econometrics with emphasis on Bayesian inference and decision techniques, time series analysis, and forecasting.

Authors' addresses:
Franz C. Palm, Faculty of Economics and Business Administration, University of Limburg, PO Box 616, 6200 MD, Maastricht, The Netherlands.

Arnold Zellner, Graduate School of Business, University of Chicago, 1101 East 58th Street, Chicago, IL 60637, USA.

Journal of Econometrics 56 (1993) 89–118. North-Holland

Bayesian and non-Bayesian methods for combining models and forecasts with applications to forecasting international growth rates*

Chung-ki Min and Arnold Zellner

University of Chicago, Chicago, IL 60637, USA

Bayesian methods for combining models and their forecasts are developed and applied using various time series models which yield forecasts of output growth rates for eighteen countries, 1974–87. Optimal Bayesian combining procedures involve use of posterior odds for alternative models which are derived and computed for several fixed parameter vs. time-varying parameter models. These odds shed light on the parameter constancy issue and are also used in predictive tests to decide whether or not to combine forecasts of alternative models since, as indicated in the paper, it is not always optimal to combine forecasts. Finally, various Bayesian pooling techniques and non-Bayesian forecast combining techniques are described and their performance in forecasting one-year-ahead output growth rates of eighteen countries, 1974–87, is reported.

1. Introduction

In past work, Garcia-Ferrer et al. (1987) and Zellner and Hong (1989), variants of a relatively simple autoregressive model of order three containing lagged leading indicator variables, called an ARLI model, provided good one-year-ahead forecasts of annual output growth rates for eighteen industrial countries, 1974–84. In Zellner, Hong, and Gulati (1990) and Zellner, Hong, and Min (1991), this ARLI model and variants of it produced good turning point forecasts, about 70 to 80 percent of 158 turning points correctly forecasted. In Hong (1989), the ARLI model's cyclical properties were analyzed and its forecasting performance was shown to be slightly superior to that of a version of Barro's 'money surprise' model. Lesage (1989) and Lesage and Magura (1990) have used ARLI models to forecast employment growth rates and turning

Correspondence to: Arnold Zellner, Graduate School of Business, University of Chicago, 1101 East 58th Street, Chicago, IL 60637, USA.

* Research financed by the National Science Foundation and by income from the H.G.B. Alexander Endowment Fund, Graduate School of Business, University of Chicago. Cheng Hsiao and two referees provided comments which were helpful in revising an earlier version of this paper.

points in them for eight metropolitan labor markets with satisfactory results. Blattberg and George (1991) used similar techniques in successfully forecasting sales of different brands of a product.

Some of our past work has involved use of fixed parameter models (FPMs) and time-varying parameter models (TVPMs). In the present paper, we derive and compute posterior odds relating to our FPMs and TVPMs using data for eighteen countries, 1973–87. While there are many reasons, Lucas effects, aggregation effects, wars, etc., for believing that parameters may be time-varying, economic theorists' models are generally fixed parameter models. Our calculated posterior odds will shed some light on the parameter constancy issue and are used to choose between FPMs' and TVPMs' forecasts year-by-year. As an alternative approach, we consider procedures for combining models and/or their forecasts; see Clemen (1989), an extensive review paper discussing theoretical and empirical work on various methods of combining forecasts. While a good portion of the literature suggests that combining forecasts will produce better forecasting performance, we show that this argument is not true in general. Clearly, combining a good forecast with a bad forecast will not usually produce a combined forecast that performs better than the good forecast, a point also made in Diebold (1990a). To cope with this problem, we develop a Bayesian forecast selection rule based on a predictive loss criterion which indicates which of several forecasts, including a combined forecast, is optimal. This rule is applied in forecasting experiments and results are compared to those of other procedures. Thus, our paper provides analytical and empirical results on the questions: (1) Fixed or time-varying parameters? (2) To combine or not to combine? (3) If so, how? and (4) How do different procedures perform in actual forecasting?

The plan of the paper is as follows. Section 2 deals with the issue of whether or not to combine forecasts. In section 3, it is indicated how posterior odds can be used to choose between or among models in such a way as to minimize expected loss. Then, building on the work of Geisel (1975) and Zellner (1989), posterior odds are employed to combine forecasts of two or more exhaustive or nonexhaustive models. Further, a Bayesian forecast selection rule is derived using a predictive loss structure which can be employed to decide when to use a combined forecast and when to use some other forecast. Section 4 contains a brief review of our data and models and presentation of the results of our forecasting experiments. A summary of results and some concluding remarks are presented in section 5.

2. To combine or not to combine forecasts?

In previous analyses of whether to combine or average different forecasts, it has been customary to assume that all forecasts being combined are unbiased; see, e.g., Reid (1968, 1969) and Bates and Granger (1969) which, according to

Clemen (1989), '... are considered by most forecasters to be the seminal works in the area of combining forecasts'. While some works [e.g., Clemen (1989), Diebold and Pauly (1987), and Palm and Zellner (1990)] discuss the possibility of forecasts being biased, the earlier analytical results, based on the assumption of no bias [see also Winkler (1981) for a Bayesian analysis involving the assumption of no bias], are still influential. Therefore we review briefly the analysis leading to the conclusion that combining forecasts leads to improved performance in terms of a mean squared error reduction. Let f_1 and f_2 be two unbiased forecasts, that is, $f_1 = y + e_1$ and $f_2 = y + e_2$, where y is the actual, random outcome and e_1 and e_2 are zero mean errors with variances σ_1^2 and σ_2^2, respectively, and covariance σ_{12}. Bates and Granger (1969) showed that with $f_c = wf_1 + (1-w)f_2$, a combined forecast, then $\mathrm{E}(f_c - y)^2$ will be minimized if $w = w^* = (\sigma_2^2 - \sigma_{12})/(\sigma_1^2 + \sigma_2^2 - 2\sigma_{12})$ and $\mathrm{E}(f_c - y)^2 < \mathrm{E}(f_i - y)^2$, $i = 1, 2$. Thus using a value of w different from 0 or 1 is considered to be optimal. However, as is almost obvious, this conclusion depends critically on the unbiasedness assumption, as shown below. Also, it should be noted that only linear combinations of the two forecasts are being considered. It may be that limiting the analysis to linear combinations is unduly restrictive. Further, it has been noted that the optimal weight, w^*, depends on parameters with unknown values which must be estimated. How estimation of the weights affects the optimality conclusion requires analysis. If past observations are available on $y - f_1$ and $y - f_2$ and it is assumed that these pairs are iid bivariate normal with a zero mean vector, then the predictive density can be derived and used to evaluate $\mathrm{E}(y - f_c)^2$ and to obtain an optimal value of w. Again it will be optimal always to combine forecasts. In this case the weights are functions of sample quantities and prior parameters. However, the analysis is based on the assumption that the e's have zero means and constant covariance matrix for all sample points, very strong assumptions which clearly will not hold in all situations.

We now consider what happens when the condition of zero bias is relaxed. Let f_1 and f_2 be two forecasts and y the random variable being forecasted. Assume $f_1 - y = e_1$ and $f_2 - y = \theta + e_2$, where e_1 and e_2 have zero means, variance σ^2, and covariance σ_{12}, and θ is the unknown bias associated with f_2. Then we have $MSE_1 = \mathrm{E}(y - f_1)^2 = \sigma^2$ and $MSE_2 = \mathrm{E}(y - f_2)^2 = \theta^2 + \sigma^2$. If we consider a combined forecast, $f_c = wf_1 + (1-w)f_2$, where w is a weight, $0 < w < 1$, then $f_c - y = w(f_1 - y) + (1-w)(f_2 - y)$ and $MSE_c = \mathrm{E}(f_c - y)^2 = w^2\sigma^2 + (1-w)^2(\sigma^2 + \theta^2) + 2w(1-w)\sigma_{12}$. It follows that

$$MSE_c - MSE_1 = (1-w)\sigma^2[(1-w)(\theta/\sigma)^2 - 2w(1-\rho_{12})], \tag{1}$$

where $\rho_{12} = \sigma_{12}/\sigma^2$. From (1), $MSE_c - MSE_1 > 0$ if $(1-w)(\theta/\sigma)^2 - 2w(1-\rho_{12}) > 0$ or

$$(\theta/\sigma)^2 > (2w/(1-w))(1-\rho_{12}). \tag{2}$$

As $\rho_{12} \to 1$, this condition approaches $(\theta/\sigma)^2 > 0$ which will be satisfied for $|\theta| > 0$. As $\rho_{12} \to -1$, the condition approaches $(\theta/\sigma)^2 > 4w/(1-w)$. For example, if $w = \frac{1}{2}$, the value of w^* above when $\sigma_1^2 = \sigma_2^2 = \sigma^2$, (2) yields $(\theta/\sigma)^2 > 2(1-\rho_{12})$ or $|\theta|/\sigma > 1.414(1-\rho_{12})^{1/2}$ or $|\theta|/\sigma > 1.414$ if $\rho_{12} = 0$. If $w = \frac{1}{2}$, and $\rho_{12} = 0$, $MSE_c - MSE_1 = (\sigma^2/2)[(\theta/\sigma)^2/2 - 1]$ which can be large if $|\theta|/\sigma$ is large. Thus use of weights appropriate for unbiased forecasts can be quite suboptimal when forecasts are biased.

In summary, it is seen that the presence of bias in forecasts can result in combined forecasts being worse in terms of *MSE* than individual forecasts. See also Diebold (1990a, b) for additional analysis of these issues. Forecasts may be biased for many reasons including use of shrinkage techniques or asymmetric loss functions and presence of specification, measurement and other errors. Correcting for such time-varying biases and defining 'optimal weights' is difficult. Thus it is important to appraise models and their forecasts carefully before combining them. In the next section, we show how posterior odds can be utilized to compare models and choose between or among them and their forecasts.

3. Posterior odds and choosing models and their forecasts

3.1. Posterior odds and choosing models

Many, including Jeffreys (1967), Geisel (1975), Rossi (1983, 1985), Leamer (1978), Schwarz (1978), and Zellner (1971, 1984), have indicated how posterior odds can be calculated and used to discriminate between or among different hypotheses and models. In the present context, we compute posterior odds for a fixed parameter model (FPM) vs. a time-varying parameter model (TVPM), denoted by $K = P_1/P_2$, where P_1 and P_2 are the posterior probabilities associated with the FPM and the TVPM, respectively – see appendix A for a derivation of the explicit expression for K used in our calculations. It has been recognized in many works that parameters of aggregate time series models may not be constant because of (1) aggregation effects, (2) effects of policy changes, that is Lucas effects, (3) effects of major events such as strikes, wars, oil crises, etc., (4) adaptive optimization on the part of economic agents, (5) changes in factor prices that induce changes in choice of production techniques, and (6) changes in tastes and preferences, perhaps due to generational, advertising, and educational effects. Given so many reasons for anticipating parameters to be time-varying, it is indeed of interest to compute odds on FPMs vs. TVPMs country-by-country to gain information relating to this important issue.

Further, if we have a standard two-state, two-action loss structure, shown in table 1, we can make the choice of model which yields lower expected loss. If the loss structure is symmetric, $c_1 = c_2$, then we choose FPM if $K > 1$ and TVPM if $K < 1$. As has been explained in the literature, posterior odds reflect goodness of

Table 1

A loss structure[a] for model selection.

		State of world		
		FPM[b]	TVPM[c]	Expected loss
Choices	FPM	0	c_1	$c_1 P_2$
	TVPM	c_2	0	$c_2 P_1$
Posterior probability		P_1	P_2	

[a] c_1 and c_2 are positive losses associated with incorrect choices.
[b] FPM denotes a fixed parameter model.
[c] TVPM denotes a time-varying parameter model.

fit, extent to which prior information is in accord with sample information, relative precision of sample, and prior information and sample size – see the references cited above for discussions of these points.

For each year in our forecast period, we compute the posterior odds $K = P_1/P_2$ based on data up to the forecast year. If $K > 1$, we choose the FPM and use the mean of its predictive density for the forecast year's output growth rate as our point forecast. If $K < 1$, we choose the TVPM and use its one-year-ahead predictive mean as our point forecast. These calculations are performed year-by-year, with K being updated, beginning with forecast year 1974 and ending with forecast year 1987.

Since it is possible that neither the FPM nor the TVPM may perform well relative to a benchmark model (BMM), we expand the 2×2 loss structure in table 1 to a 3×3 loss structure relating to choices of the FPM, the TVPM, and the BMM using posterior odds computed for pairs of these models year-by-year. Using these posterior odds and assuming that the 3×3 loss structure is symmetric, we choose among the three models so as to minimize expected loss and use the mean of the predictive density of the model so chosen as our point forecast. The BMM used in these calculations is an autoregressive model of order three.

3.2. *Combining and choosing among individual and combined forecasts*

Given that posterior odds are available relating to two mutually exclusive models, the issue arises as to how they can be used to combine the models and/or their forecasts – see Diebold (1990b) for further discussion. If the models are considered to be exhaustive, then their probabilities add to one and the expectation of a future observation, y, is just

$$\begin{aligned} \mathrm{E}(y) &= P_1 \mathrm{E}(y \mid \mathrm{FPM}) + (1 - P_1)\mathrm{E}(y \mid \mathrm{TVPM}) \\ &= (K/(1+K))\mathrm{E}(y \mid \mathrm{FPM}) + (1/(1+K))\mathrm{E}(y \mid \mathrm{TVPM}). \end{aligned} \tag{3}$$

This last expression is the predictive mean of a future observation given that we average over the two models. Clearly if K is large, favoring the FPM, its mean will get a large weight and that of the TVPM will receive a small weight. Since the large value of K reflects good *relative* performance of the FPM in the past, it seems appropriate to give its forecast heavier weight in the combined forecast. The predictive mean in (3) will be used as a point forecast in our forecasting experiments.

It should be appreciated that the posterior odds is just a relative measure which does not assure that the absolute performance of either model is satisfactory. For example, if $K = 19$, $P_1 = 0.95$, a high probability associated with the FPM. However, this high probability is attained under the assumption that the two models are exhaustive. If they are not, then it is clear that with $K = 19$, P_1 might have any value and the mean in (3) cannot be computed. Since there is another rationalization of the combining formula in (3) other than the exhaustive assumption, see below, we shall use the predictive mean in (3), the Bayesian combined forecast (BCF) in our forecasting experiments.

When the models considered, say a FPM and a TVPM, do not constitute an exhaustive set, as is usually the case since there are many additional models which can be considered, the posterior odds $K = P_1/P_2$ can be computed, but it is not possible to evaluate P_1 and P_2 separately and thus the first line of eq. (3) cannot be implemented. For this problem, Zellner (1989) introduced the concept of a combining predictive density which is closest in a distance metric sense to the predictive densities of the nonexhaustive models and suggested use of the mean of the combined predictive density as a point forecast. With a weighted quadratic distance measure, it was shown that for two nonexhaustive models, e.g., FPM and TVPM, the mean of the combined predictive density is given by the second line of (3). Note that to compute the second line of (3) only the value of $K = P_1/P_2$ and the means of the two models' predictive densities are needed. The values of P_1 and P_2 individually are not needed.

In (3) it has been assumed that the available forecasts will be averaged. If we consider the predictive squared error loss structure in table 2, wherein possible forecast choices and different states of the world are shown, it is optimal to choose the combined forecast when only three alternatives are available, namely choose the FPM's forecast, the TVPM's forecast, or the combined forecast, shown in (3). This can be seen from consideration of table 2. The entries in table 2 are expected squared error losses given that model i's forecast, $\bar{y}_i$, the predictive mean, is used and model j is appropriate. In this situation, expected loss, computed using M_j's predictive density, is $\mathrm{E}_j(y-\bar{y}_i)^2 = \mathrm{E}_j[y-\bar{y}_j-(\bar{y}_i-\bar{y}_j)]^2 = \mathrm{E}_j(y-\bar{y}_j)^2 + (\bar{y}_i-\bar{y}_j)^2 \equiv v_j + \Delta_{ij}$, where $\bar{y}_j$ is the mean of M_j's predictive density, v_j is the variance of M_j's predictive density, and $\Delta_{ij} \equiv (\bar{y}_i - \bar{y}_j)^2$. On comparing expected losses shown in table 2, which can be done without assuming the models (M_1, M_2, and M_c) are exhaustive, we have $\mathrm{E}(L \mid M_c)/\mathrm{E}(L \mid M_1) = P_1/(P_1+P_2) < 1$ and $\mathrm{E}(L \mid M_c)/\mathrm{E}(L \mid M_2) = P_2/(P_1+P_2) < 1$. Thus, choice

Table 2

Predictive squared error loss structure.[a]

Model Choice set	State M_1 (FPM)	M_2 (TVPM)	M_c	Expected loss $(P_1 v_1 + P_2 v_2 + P_c \bar{v}) +$
M_1 (FPM)	v_1	$v_2 + \Delta_{12}$	$\bar{v} + w_2 \Delta_{12}$	$\Delta_{12} w_2 (P_1 + P_2 + P_c)$
M_2 (TVPM)	$v_1 + \Delta_{12}$	v_2	$\bar{v} + w_1 \Delta_{12}$	$\Delta_{12} w_1 (P_1 + P_2 + P_c)$
M_c	$v_1 + w_2^2 \Delta_{12}$	$v_2 + w_1^2 \Delta_{12}$	$\bar{v} + w_1 w_2 \Delta_{12}$	$\Delta_{12} w_1 w_2 (P_1 + P_2 + P_c)$
Posterior probability	P_1	P_2	P_c	

[a] In the table, $v_i = E(y - \bar{y}_i)^2$, the variance of M_i's predictive density, $\bar{v} = w_1 v_1 + w_2 v_2$, with $w_i = P_i/(P_1 + P_2)$ and $\Delta_{12} = (\bar{y}_1 - \bar{y}_2)^2$, where $\bar{y}_1$ and $\bar{y}_2$ are predictive means of M_1's and M_2's predictive densities, respectively. Each nondiagonal entry in the table is given by $E_j(y - \bar{y}_i)^2$, where E_j indicates that M_j's predictive density is employed.

Table 3

Simplified extended predictive squared error loss structure.[a]

Model Choice set	State M_1 (FPM)	M_2 (TVPM)	M_c	M_4 (BMM)	Expected loss[b]
M_1 (FPM)	0	Δ_{12}	$w_2 \Delta_{12}$	Δ_{14}	$\Delta_{12} w_2 (P_1 + P_2 + P_c) + P_4 \Delta_{14}$
M_2 (TVPM)	Δ_{12}	0	$w_1 \Delta_{12}$	Δ_{24}	$\Delta_{12} w_1 (P_1 + P_2 + P_c) + P_4 \Delta_{24}$
M_c	$w_2^2 \Delta_{12}$	$w_1^2 \Delta_{12}$	$w_1 w_2 \Delta_{12}$	Δ_{c4}	$\Delta_{12} w_1 w_2 (P_1 + P_2 + P_c) + P_4 \Delta_{c4}$
Posterior probability	P_1	P_2	P_c	P_4	

[a] See footnote of table 2. Also $\Delta_{i4} = (\bar{y}_i - \bar{y}_4)^2$, where $\bar{y}_i$ and $\bar{y}_4$ are predictive means of M_i and M_4 respectively. $\bar{y}_c = w_1 \bar{y}_1 + w_2 \bar{y}_2$, where $w_1 = P_1/(P_1 + P_2) = K/(1 + K)$ and $w_2 = P_2/(P_1 + P_2) = 1/(1 + K)$ where $K = P_1/P_2$.

[b] Common elements in the columns of the loss structure, e.g., v_1, v_2, $\bar{v}$, etc. – see table 2 – have been omitted.

of the combined forecast (3) leads to minimal expected loss for the predictive squared error loss structure in table 2.

To allow for other models not explicitly considered in table 2, we consider the 3×4 loss structure in table 3 which includes M_4, e.g., a benchmark model. On comparing expected losses given in the last column of table 3, it is seen that the combined forecast is not always preferred because of additional terms involving $\Delta_{14} = (\bar{y}_1 - \bar{y}_4)^2$, $\Delta_{24} = (\bar{y}_2 - \bar{y}_4)^2$, and $\Delta_{c4} = (\bar{y}_c - \bar{y}_4)^2$, where $\bar{y}_4$ is the predictive mean of M_4 and $\bar{y}_c$ is the predictive mean of M_c.

Given that M_4 is specified, say an AR(3) benchmark model, posterior odds can be calculated and used in connection with table 3 to choose the forecast which leads to minimal expected loss.

3.3. Regression combining techniques

Having discussed general Bayesian procedures for combining models and their forecasts, we now briefly take up regression procedures which have been used extensively in the literature [see, e.g., Nelson (1972), Granger and Ramanathan (1984), Diebold and Pauly (1987), and the discussion and references in Clemen (1989)]. In this approach past actual values of a variable are related to forecasts of them provided by two or more forecasting methods; that is, the following regression relation is considered:

$$A_t = \alpha + \beta_1 f_{1t} + \beta_2 f_{2t} + u_t, \qquad t = 1, \ldots, n. \tag{4}$$

In regressing A_t on f_{1t} and f_{2t}, two forecasts, some impose the constraint that the intercept is zero and the regression coefficients add to one. Others allow for a nonzero intercept to allow for possible bias and may or may not impose the condition that the regression coefficients sum to one. Further, others have used ridge regression techniques, time-varying parameter techniques, etc. in analyzing the relation between A_t and f_{1t} and f_{2t}. In our opinion, the basic problem is one of identifying or determining the form of this relationship. If it is not determined very well, it's not at all clear that the regression approach will yield satisfactory results. Note that in comparison to the Bayesian combining relationship in (3), there the weights or coefficients on the individual forecasts, the predictive means, are not least squares weights. Further, if f_{1t} and f_{2t} are derived from imperfect models, say models with important variables omitted, then they will have biases that usually vary through time in a complicated way. Modelling such time-varying biases seems difficult. Be that as it may, below we implement variants of the regression approach to produce combined forecasts and compare measures of their forecasting performance to those of other approaches described above and in section 4.

4. Models, methods, and their performance in forecasting

In this section, we first describe the models which we shall employ to forecast annual output growth rates for eighteen countries year-by-year, 1974–87, using data 1951–73 to 'start up' the forecasting experiments. Second, we enumerate the forecasting techniques which will be employed. Third, we briefly describe the data employed in our calculations and then present the empirical forecasting results for uncombined, combined, unpooled, and pooled forecasts derived from use of fixed and time-varying parameter models.

4.1. Models and forecasts

The two models that we employ, used in our previous work, are (i) a univariate autoregression of order three containing lagged leading indicator variables, denoted by ARLI, and (ii) the ARLI model expanded to include a measure of current world income growth, denoted by ARLI/WI. As explained in our past work, an autoregression of order three is employed to allow for the possibility of there being two complex conjugate roots, giving rise to an oscillatory component, and one real root associated with a local trend. In Hong (1989) it is shown that ARLI models for eighteen countries have two complex roots and one real root with high posterior probability in each case. Further, Hong established that the oscillatory components are damped in each case and have periods in the vicinity of four to six years. Also, each estimated real root has amplitude less than one. As regards the leading indicator variables, lagged growth rates of real stock prices and of real money, they are introduced to take account of informational, expectational, and policy effects as well as a real balance effect exerted by money supply changes – see Zellner, Huang, and Chau (1965) for discussion and estimation of a real balance effect. The world income variable, introduced in Zellner and Hong (1989), is a variable affecting countries' exports. Further discussion of the economic rationale of the ARLI models is provided by Hong (1989) who shows that they can be produced as the reduced form equations of particular IS-LM models. Also, Zellner and Manas-Anton (1986) link up the ARLI models to reduced form equations obtained from particular aggregate supply and demand models and Min (1990) shows their relation to generalized real business cycle models.

In what follows, we utilize fixed parameter (FP) and time-varying parameter (TVP) versions of the ARLI and ARLI/WI models. The FP versions of these models, denoted by FP/ARLI and FP/ARLI/WI, are utilized with and without pooling. The unpooled FP/ARLI model is

$$y_{it} = \boldsymbol{x}'_{it}\boldsymbol{\beta}_i + u_{it}, \qquad i = 1, 2, \ldots, N, \quad t = 1, 2, \ldots, T, \tag{5}$$

where y_{it} = annual output growth rate for ith country in the tth year, $\boldsymbol{\beta}_i = k \times 1$ parameter vector, $\boldsymbol{x}'_{it} = (1, y_{it-1}, y_{it-2}, y_{it-3}, SR_{it-1}, SR_{it-2}, GM_{it-1}, MSR_{t-1})$, where SR_{it} = rate of growth of real stock prices, GM_{it} = rate of growth of real money, M_1, MSR_t = median of SR_{it}'s in year t, and u_{it} is an error term. We assume that the u_{it}'s are NID$(0, \sigma_i^2)$. Using a diffuse prior density for $\boldsymbol{\beta}_i$ and σ_i, it is well-known that given initial values 1951–53 and data 1954–73, the predictive density for 1974's output growth rate is in the univariate Student-t form with mean equal to the least squares forecast of the 1974 growth rate – see, e.g., Zellner (1971). This is our 1974 forecast for the FP/ARLI model without pooling. To forecast 1975's output growth rate, we update our predictive density to incorporate the 1974 data and use its mean, the least squares forecast of 1975's growth rate, as our forecast. Proceeding in this way year-by-year, we obtain a sequence of unpooled FP/ARLI forecasts, recursively calculated, for the years 1974–87.

The FP/ARLI/WI model, introduced and used in Zellner and Hong (1989), involves adding a scalar world income growth rate variable, w_t, to (5) as follows:

$$y_{it} = w_t \alpha_i + \boldsymbol{x}'_{it} \boldsymbol{\beta}_i + u_{it}, \tag{6}$$

where α_i is a scalar parameter and w_t is the median of the y_{it}'s in the tth year. The term $\boldsymbol{x}'_{it}\boldsymbol{\beta}_i$ is exactly the same as that in (5). To use (6) to forecast, it is necessary to forecast w_t. To accomplish this, we use the following ARLI model for w_t, an autoregression of order three including lagged leading indicator variables:

$$w_t = \boldsymbol{z}'_t \boldsymbol{\pi} + \varepsilon_t, \tag{7}$$

where $\boldsymbol{z}'_t = (1, w_{t-1}, w_{t-2}, w_{t-3}, MSR_{t-1}, MGM_{t-1})$, with MSR_t = median growth rate of real stock prices for year t and MGM_t = median growth rate of real money for year t with the median taken relative to the eighteen countries for year t. The ε_t's in (7) are assumed independently drawn from a zero mean normal distribution with variance σ^2_w. For country i, (6) and (7) constitute a triangular system which we assume to be fully recursive, that is u_{it} and ε_t are assumed to be independently distributed for all t. Using the recursive assumption and a diffuse prior density for the parameters, we implement the model using data 1951–53 for initial values and 1954–73, twenty years, for estimation and obtain an approximate predictive density for the output growth rate in 1974. The mean of this predictive density is our forecast for 1974. On updating the posterior and predictive densities year-by-year, a sequence of forecasts for the remaining years, 1975–87, is obtained. These are our unpooled FP/ARLI/WI forecasts.

Our unpooled, time-varying parameter models are simple elaborations of the above fixed parameter models. Our unpooled TVP/ARLI model is (5) with $\boldsymbol{\beta}_i$ made time-varying, that is,

$$y_{it} = \boldsymbol{x}'_{it} \boldsymbol{\beta}_{it} + u_{it} \tag{8}$$

and

$$\boldsymbol{\beta}_{it} = \boldsymbol{\beta}_{it-1} + \boldsymbol{v}_{it}. \tag{9}$$

Thus in our unpooled TVP/ARLI model, used in Garcia-Ferrer et al. (1987), the random coefficient vector is assumed to be generated by a vector random walk process. Given start-up data, 1951–73, and an initial state distribution, recursive calculations provide a predictive density for the 1974 output growth rate, the mean of which is our point forecast. On recursively updating the predictive densities year-by-year, their means are our unpooled TVP/ARLI model forecasts, 1974–87.

Our unpooled TVP/ARLI/WI model is that shown in (6) and (7) with the parameters made random as shown below:

$$y_{it} = w_t \alpha_{it} + x'_{it} \beta_{it} + u_{it}, \tag{10}$$

$$w_t = z'_t \pi_t + \varepsilon_t, \tag{11}$$

$$\begin{pmatrix} \alpha_{it} \\ \beta_{it} \\ \pi_t \end{pmatrix} = \begin{pmatrix} \alpha_{it-1} \\ \beta_{it-1} \\ \pi_{t-1} \end{pmatrix} + \begin{pmatrix} \eta_{it} \\ v_{it} \\ \delta_t \end{pmatrix} \tag{12}$$

Recursive algorithms and needed assumptions for obtaining and updating predictive densities for this system are given in appendix B. The means of these successively updated predictive densities, 1974–87, are our unpooled TVP/ARLI/WI forecasts.

We have also computed separate forecasts and combined forecasts for alternative models under various pooling assumptions. In pooling the FP/ARLI model in (5), we assumed that the coefficient vectors, the β_i's, satisfy $\beta_{it} = \theta + \varepsilon_{it}$ with $\varepsilon_{it} \sim N(0, \sigma^2 \Sigma)$, an assumption implying that the β_{it}'s have a common mean vector θ, a model similar to those used by, among others, Swamy (1971) and Lindley and Smith (1972). In connection with the FP/ARLI/WI model in (6) and (7), we make a similar assumption, namely that $\alpha_{it} = \alpha + v_{it}$ and $\beta_{it} = \theta + \varepsilon_{it}$, where $\varepsilon^*_{it} = (v_{it} \varepsilon'_{it})'$ is assumed $N(0, \sigma^2 \Sigma^*)$. Thus we allow the β's to be time-varying with a fixed hyperparameter vector θ and call these models fixed hyperparameter ARLI models, denoted by FHP/ARLI. Using these pooling assumptions, individual and combined forecasts were computed year-by-year, 1974–87 – see appendix C for details.

The pooled TVHP/ARLI model is given by

$$y_t = X_t \beta_t + u_t, \qquad u_t \sim N(0, \sigma^2 I_m),$$

with $y'_t = (y_{1t}, y_{2t}, \dots, y_{mt})$, $\beta'_t = (\beta'_{1t}, \beta'_{2t}, \dots, \beta'_{mt})$, $u'_t = (u_{1t}, u_{2t}, \dots, u_{mt})$, and

$$X_t = \begin{pmatrix} x'_{1t} & 0' & \cdots & 0' \\ 0' & x'_{2t} & \cdots & 0' \\ \vdots & \vdots & \ddots & \vdots \\ 0' & 0' & \cdots & x'_{mt} \end{pmatrix}.$$

Further it is assumed that $\beta_t = B\theta_t + \varepsilon_t$ and $\theta_t = \theta_{t-1} + \eta_t$, where ε_t is $N(0, \sigma^2 \Omega)$ and is independent of η_t which is $N(0, \phi\sigma^2 I_k)$ with $\Omega = I_n \otimes \Sigma$ and $B' = [I_k I_k \dots I_k]$.

It is seen that with these pooling assumptions the $\boldsymbol{\beta}_{it}$ given $\boldsymbol{\theta}_t$ have a common mean vector, $\boldsymbol{\theta}_t$, which is assumed generated by a random walk. See appendix B for further description and analysis of this pooling model and a similar one for the TVP/ARLI/WI model.

4.2. *Description of data*

The data used in our study have been taken from the University of Chicago IMF International Financial Statistics data base and are available on a diskette on request for a small fee to cover costs. Annual data 1954–73 were used to estimate our models with data 1951–53 serving as initial values. Then the one-year-ahead forecasts described in the preceding sections were computed year-by-year, 1974–87, with models being updated each year.

The output data for each country, either annual real GNP or real GDP, were logged and first-differenced to yield annual growth rates. For each country nominal money, M_1, at the end of each year was deflated by a general price index, logged and first-differenced to yield the rate of change of real money. Similarly, for each country a general price index was used to deflate an annual stock price index, which was logged and first-differenced to yield a rate of growth of real stock prices. Symbolically the variables are given by: $y_{it} = (1 - L)\ln(O_{it}/P_{it})$, $SR_{it} = (1 - L)\ln(SP_{it}/P_{it})$, and $GM_{it} = (1 - L)\ln(M_{it}/P_{it})$, where L is the lag operator, $L^n x_t = x_{t-n}$, and for the ith country in the tth year, O_{it} = output level, P_{it} = general price index, SP_{it} = stock price index, and M_{it} = nominal M_1. Data plots are given in Garcia-Ferrer et al. (1987) and Zellner and Hong (1989).

4.3. *Empirical results*

In tables 4a and 4b, we present posterior odds for FP vs. TVP models, based on data 1973–87. For the unpooled ARLI and ARLI/WI models, twelve and eight countries, respectively, have odds of less than 0.50, that is greater than 2:1 in favor of the TVP models. For the ARLI model, only three countries have odds greater than or equal to 2:1 in favor of the FP model. Thus without the WI variable, there is a general preference for TVP for twelve countries. However, with the WI variable included, the odds favor TVPs for eight countries and FPs for seven. For the pooled ARLI model, TVHPs are favored for twelve countries. However, for the pooled ARLI/WI model, all countries' odds are between 0.50 and 1.99, somewhat inconclusive with respect to whether the hyperparameter vector is fixed or time-varying.

Tables 5, 6a, and 6b provide *RMSE*'s of forecasts for unpooled and pooled FP and TVP models as well as those for two decision-theoretic model selection procedures. For the unpooled ARLI model, it is seen from table 6a that the

Table 4a

Posterior odds for fixed vs. time-varying parameter models computed from annual data, 1973–87.[a]

	Models			
	Unpooled		Pooled	
Country	ARLI[b]	ARLI/WI[c]	ARLI[d]	ARLI/WI[e]
Australia	0.16	0.34	1.01	0.94
Austria	0.01	0.07	0.21	1.02
Belgium	1.64	0.45	0.40	1.57
Canada	0.41	2.21	0.57	1.72
Denmark	0.01	0.39	0.05	0.99
Finland	0.78	1.68	0.36	0.76
France	0.01	1.30	0.47	1.42
Germany	0.00	0.01	0.16	1.43
Ireland	6.18	3.04	1.87	1.50
Italy	0.01	0.06	0.32	1.21
Japan	0.45	0.48	6.81	0.91
Netherlands	0.41	2.32	0.10	1.92
Norway	6.60	4.95	1.36	1.09
Spain	0.27	0.75	0.57	1.35
Sweden	0.25	0.34	0.04	0.58
Switzerland	0.71	4.25	0.21	0.91
UK	0.22	8.67	0.13	0.82
USA	9.18	17.31	0.24	0.81

[a] Posterior distributions for models parameters, computed using annual data 1954–72, were used to form Bayes factors for the period 1973–87. Prior odds were set 1:1 in all cases. The entries in the table are odds in favor of fixed parameter (FP) models.
[b] Unpooled FP/ARLI vs. unpooled TVP/ARLI.
[c] Unpooled FP/ARLI/WI vs. unpooled TVP/ARLI/WI.
[d] Pooled FHP/ARLI vs. pooled TVHP/ARLI.
[e] Pooled FHP/ARLI/WI vs. pooled TVHP/ARLI/WI.

Table 4b

Number of countries by values of posterior odds, fixed vs. time-varying parameter models, annual data, 1973–87.[a]

	Models			
Posterior odds,	Unpooled		Pooled	
FP vs. TVP[b]	ARLI	ARLI/WI	ARLI	ARLI/WI
0–0.49	12	8	12	0
0.50–0.99	2	1	2	8
1.00–1.99	1	2	3	10
$\geq$ 2.00	3	7	1	0

[a] Tabulation of data from table 4a. See footnotes of table 4a and appendix for procedures employed in computing odds.
[b] Posterior odds for fixed parameter (FP) vs. time-varying parameter (TVP) models.

Table 5

Root mean squared errors (*RMSE*'s) of the one-year-ahead forecasts of annual real output growth rates for 1974–87 for several variants of fixed parameter autoregressive leading indicator (FP/ARLI), world income (FP/ARLI/WI) models, and an AR(3) benchmark model.[a]

	Models				
	AR(3)	Unpooled ARLI	Unpooled ARLI/WI	Pooled ARLI[b]	Pooled ARLI/WI[b]
Country	(percentage points)				
Australia	2.72	3.21	3.45	2.10	2.27
Austria	2.58	2.75	2.84	1.97	1.80
Belgium	2.80	1.71	1.54	1.87	1.94
Canada	2.79	2.70	2.52	2.08	1.74
Denmark	2.89	3.40	3.21	2.77	2.39
Finland	2.45	3.22	3.32	1.85	1.69
France	2.16	2.51	2.27	1.58	1.62
Germany	2.90	2.82	2.82	1.97	1.56
Ireland	2.38	2.56	2.68	2.19	2.48
Italy	3.88	3.19	3.68	2.35	2.14
Japan	2.93	3.02	3.46	2.13	2.23
Netherlands	2.86	1.92	1.67	1.92	1.60
Norway	1.76	1.72	1.84	1.71	1.44
Spain	2.35	2.17	2.28	1.68	1.21
Sweden	1.98	2.11	2.15	2.30	1.76
Switzerland	3.77	3.24	3.15	2.56	2.43
UK	2.94	2.13	1.50	2.20	1.92
USA	2.69	2.31	1.99	2.53	2.08
Median	2.76	2.63	2.60	2.09	1.86
Minimum	1.76	1.71	1.50	1.58	1.21
Maximum	3.88	3.40	3.68	2.77	2.48

[a] See text for discussion of models.
[b] In these cases the hyperparameters are fixed.

lowest median *RMSE*'s are those for the TVP and LS_2 cases, where LS_2 denotes a 2×2 symmetric loss structure, table 1 with $c_1 = c_2$, relating to the choice of FP and TVP models' forecasts. For the pooled ARLI model, the lowest median *RMSE* is that for the TVHP version, 1.95, with 1.97 being the *RMSE* for the LS_2 forecasts. Note that pooling has resulted in about a 15% reduction in median *RMSE*'s. Further, in the case of the pooled ARLI forecasts, the minimal and maximal *RMSE*'s are lower in every case. For the four methods reported in table 6a, it appears that for both the pooled cases the TVP and LS_2 forecasts are overall the best and are not very different. For each of the forecasting methods, pooling models performed better than corresponding unpooled models according to the median *RMSE* criterion and generated improved *RMSE*'s in many individual cases.

Table 6a

Root mean squared errors (*RMSE*'s) of various one-year-ahead forecasts of annual real output growth rates, 1974–87, using an autoregressive leading indicator (ARLI) model.

Country	Unpooled ARLI				Pooled ARLI			
	FP	TVP	LS_2[a]	LS_3[b]	FHP	TVHP	LS_2[a]	LS_3[b]
	(percentage points)							
Australia	3.21	3.01	3.06	2.72[c]	2.10[c]	2.16	2.36	2.46
Austria	2.75	1.99	1.92[c]	2.33	1.97	1.67[c]	1.82	2.32
Belgium	1.71	1.70[c]	1.71	1.71	1.87	1.70[c]	1.93	1.93
Canada	2.70[c]	2.96	2.96	3.28	2.08	2.05[c]	2.07	2.07
Denmark	3.40	2.67[c]	2.67[c]	2.67[c]	2.77	2.48	2.45	2.36[c]
Finland	3.22	2.97	3.22	2.45[c]	1.85	1.53[c]	1.54	1.69
France	2.51	2.27[c]	2.27[c]	2.61	1.58	1.23[c]	1.37	1.49
Germany	2.82	1.91[c]	1.91[c]	2.81	1.97	1.54[c]	1.55	1.55
Ireland	2.56	2.90	2.56	2.38[c]	2.19[c]	2.26	2.28	2.33
Italy	3.19	2.32[c]	2.32[c]	3.39	2.35	2.18[c]	2.22	2.22
Japan	3.02[c]	3.16	3.16	3.16	2.13[c]	2.46	2.13[c]	2.13[c]
Netherlands	1.92[c]	2.05	2.04	2.29	1.92	1.42[c]	1.59	1.59
Norway	1.72[c]	2.31	1.96	1.76	1.71[c]	1.95	1.94	1.76
Spain	2.17	1.87[c]	1.93	2.08	1.68	1.51[c]	1.52	1.52
Sweden	2.11	1.93[c]	1.97	2.03	2.30	1.90[c]	1.99	2.03
Switzerland	3.24	3.18[c]	3.25	3.25	2.56	2.11[c]	2.35	2.52
UK	2.13	1.90[c]	1.98	1.98	2.20	1.96[c]	1.96[c]	1.96[c]
USA	2.31[c]	2.46	2.43	2.37	2.53	2.46[c]	2.49	3.09
Median	2.63	2.31	2.29[c]	2.41	2.09	1.95[c]	1.97	2.05
Minimum	1.71	1.70	1.71	1.71	1.58	1.23	1.37	1.49
Maximum	3.40	3.18	3.25	3.39	2.77	2.48	2.49	3.09

[a] Year-by-year model selection based on the 2×2 symmetric loss structure (LS_2) in table 1. The forecast of the model favored by posterior odds is employed.

[b] Year-by-year model selection based on a 3×3 symmetric loss structure (LS_3), involving FP, TVP, and an AR(3) benchmark model. The forecast of the model chosen to minimize expected loss was used.

[c] Minimal value in row.

Table 6b

Root mean squared errors (*RMSE*'s) of various one-year-ahead forecasts of annual real output growth rates, 1974–87, using an autoregressive leading indicator–world income (ARLI/WI) model.[a]

Country	Unpooled ARLI/WI				Pooled ARLI/WI			
	FP	TVP	LS_2[a]	LS_3[b] (percentage points)	FHP	TVHP	LS_2[a]	LS_3[b]
Australia	3.45	3.23	3.23	3.10[c]	2.27[c]	2.27[c]	2.29	2.29
Austria	2.84	2.13[c]	2.14	2.42	1.80	1.70[c]	1.70[c]	1.70[c]
Belgium	1.54[c]	1.78	1.77	1.77	1.94	1.84[c]	1.94	1.94
Canada	2.52[c]	2.78	2.64	3.01	1.74[c]	1.78	1.74[c]	1.74[c]
Denmark	3.21	2.68[c]	2.95	2.95	2.39	2.14[c]	2.26	2.26
Finland	3.32	3.16	3.32	2.91[c]	1.69	1.58[c]	1.58[c]	1.58[c]
France	2.27	1.74[c]	2.29	2.33	1.62	1.43[c]	1.60	1.60
Germany	2.82	1.98[c]	1.98[c]	2.84	1.56	1.23[c]	1.42	1.42
Ireland	2.68	3.15	3.35	2.38[c]	2.48[c]	2.53	2.49	2.49
Italy	3.68	2.51[c]	2.58	3.52	2.14	1.98[c]	2.18	2.18
Japan	3.46	3.32[c]	3.32[c]	3.32[c]	2.23[c]	2.30	2.28	2.28
Netherlands	1.67[c]	1.74	1.86	2.28	1.60	1.39[c]	1.64	1.64
Norway	1.84	2.56	2.19	1.76[c]	1.44[c]	1.55	1.56	1.56
Spain	2.28	1.85[c]	2.15	2.37	1.21	1.17[c]	1.21	1.21
Sweden	2.15	1.90[c]	1.97	2.09	1.76	1.50[c]	1.51	1.51
Switzerland	3.15	3.04[c]	3.17	3.17	2.43	2.23[c]	2.38	2.38
UK	1.50	1.39[c]	1.50	1.50	1.92	1.69[c]	1.87	1.87
USA	1.99[c]	2.22	2.19	2.19	2.08	1.94[c]	2.02	2.02
Median	2.60	2.37	2.24[c]	2.40	1.86	1.74[c]	1.80	1.80
Minimum	1.50	1.39	1.50	1.50	1.21	1.17	1.21	1.21
Maximum	3.68	3.32	3.35	3.52	2.48	2.53	2.49	2.49

[a,b,c] See footnotes of table 6a.

Table 6c

RMSE's for pooled and unpooled ARLI/WI models' forecasts by country, 1974–87.

RMSE (%)	Countries									Freq.	Prop.
					(A) Pooled TVHPM						
1.00–1.49	FRN	GER	NET	SPN						4	0.22
1.50–1.99	AUR	BEL	CAN	FIN	ITY	NOR	SWD	UKM	USA	9	0.50
2.00–2.49	AUL	DEN	JAP	SWZ						4	0.22
2.50–2.99	IRE									1	0.06
3.00–3.49										0	0.00
		Median = 1.74			Minimum = 1.17			Maximum = 2.53		18	1.00
					(B) Unpooled TVPM						
1.00–1.49	UKM									1	0.06
1.50–1.99	BEL	FRN	GER	NET	SPN	SWD				6	0.33
2.00–2.49	AUR	USA								2	0.11
2.50–2.99	CAN	DEN	ITY	NOR						4	0.22
3.00–3.49	AUL	FIN	IRE	JAP	SWZ					5	0.28
		Median = 2.37			Minimum = 1.39			Maximum = 3.32		18	1.00
					(C) Pooled FHPM						
1.00–1.49	NOR	SPN								2	0.11
1.50–1.99	AUR	BEL	CAN	FIN	FRN	GER	NET	SWD	UKM	9	0.50
2.00–2.49	AUL	DEN	IRE	ITY	JAP	SWZ	USA			7	0.39
2.50–2.99										0	0.00
3.00–3.49										0	0.00
		Median = 1.86			Minimum = 1.21			Maximum = 2.48		18	1.00
					(D) Unpooled FPM						
1.00–1.49										0	0.00
1.50–1.99	BEL	NET	NOR	UKM	USA					5	0.28
2.00–2.49	FRN	SPN	SWD							3	0.17
2.50–2.99	AUR	CAN	GER	IRE						4	0.22
3.00–3.49	AUL	DEN	FIN	JAP	SWZ					5	0.28
3.50–3.99	ITY									1	0.06
		Median = 2.60			Minimum = 1.50			Maximum = 3.68		18	1.01

Table 6b shows that pooled ARLI/WI models' forecasts and model selection procedures perform better than their unpooled counterparts. For example, the median *RMSE*'s for the pooled models and procedures range from 1.74 to 1.86, while for the unpooled models and procedures they range from 2.24 to 2.60. For the unpooled case, the LS_2 forecast had a median *RMSE* of 2.24, slightly below those of the forecasts of the FP, TVP, and LS_3 procedures. For the pooled ARLI/WI model all procedures worked about equally well in terms of median *RMSE*.

To show more clearly the effects of using pooled and unpooled models, table 6c presents *RMSE*'s for unpooled and pooled ARLI/WI forecasts by *RMSE* interval and by country. On comparing panels A and B, it is seen that the pooled TVHPM forecast *RMSE*'s are much more highly concentrated in the vicinity of 1.5 to 2.5 than is the case of *RMSE*'s for forecasts of the unpooled TVPM. The same conclusion holds for the comparison of the distribution of countries by *RMSE* for the pooled and unpooled FPMs. Finally, on comparing panels A and C, it is seen that the results for the pooled TVHPMs and the pooled FHPMs are similar.

In table 7, we form 2×2 tables relating models preferred by the posterior odds criterion and relative size of *RMSE*'s of FPMs and TVPMs. In general there appears to be a positive association present. For example, for the unpooled ARLI model, when the posterior odds favor the FPM, three of four *RMSE*'s

Table 7

Number of countries by posterior odds and root mean squared error (*RMSE*) of forecast by type of model, annual output growth rates, 1974–87.

(A) Unpooled models

		ARLI Lower RMSE for:			ARLI/WI Lower RMSE for:		
		FPM	TVPM	Total	FPM	TVPM	Total
Posterior odds favor:	FPM	3	1	4	5	4	9
	TVPM	3	11	14	1	8	9
	Total	6	12	18	6	12	18

(B) Pooled models

		ARLI Lower RMSE for:			ARLI/WI Lower RMSE for:		
		FHPM	TVHPM	Total	FHPM	TVHPM	Total
Posterior odds favor:	FHPM	4	0	4	3	7	10
	TVHPM	0	14	14	2	6	8
	Total	4	14	18	5	13	18

Table 8a

Root mean squared errors of one-year-ahead forecasts of annual real output growth rates, 1974–87, using model combining approaches based on posterior odds and a predictive model selection rule.

	Unpooled ARLI				Pooled ARLI			
Country	FP	TVP	COMB$_1$[a]	COMB$_2$[b]	FHP	TVHP	COMB$_1$[a]	COMB$_2$[b]
Australia	3.21	3.01	3.07	2.93[c]	2.10	2.16	2.11	2.06[c]
Austria	2.75	1.99[c]	2.05	2.11	1.97	1.67[c]	1.77	1.81
Belgium	1.71	1.70[c]	1.72	1.72	1.87	1.70[c]	1.75	1.80
Canada	2.70	2.96	2.97	2.65[c]	2.08	2.05[c]	2.05[c]	2.06
Denmark	3.40	2.67[c]	2.79	2.69	2.77	2.48[c]	2.51	2.48[c]
Finland	3.22	2.97[c]	3.08	3.03	1.85	1.53[c]	1.63	1.62
France	2.51	2.27[c]	2.32	2.43	1.58	1.23[c]	1.34	1.41
Germany	2.82	1.91[c]	1.96	1.96	1.97	1.54[c]	1.66	1.67
Ireland	2.56[c]	2.90	2.69	2.66	2.19	2.26	2.20	2.08[c]
Italy	3.19	2.32[c]	2.38	2.76	2.35	2.18[c]	2.23	2.33
Japan	3.02[c]	3.16	3.07	3.06	2.13[c]	2.46	2.23	2.24
Netherlands	1.92[c]	2.05	1.98	1.92[c]	1.92	1.42[c]	1.55	1.64
Norway	1.72[c]	2.31	1.99	1.74	1.71	1.95	1.79	1.58[c]
Spain	2.17	1.87[c]	1.93	1.87[c]	1.68	1.51[c]	1.56	1.60
Sweden	2.11	1.93[c]	1.96	1.95	2.30	1.90[c]	1.97	1.96
Switzerland	3.24	3.18[c]	3.25	3.21	2.56	2.11[c]	2.29	2.40
UK	2.13	1.90[c]	1.97	1.97	2.20	1.96[c]	2.00	2.00
USA	2.31[c]	2.46	2.40	2.35	2.53	2.46[c]	2.48	2.49
Median	2.63	2.31[c]	2.35	2.39	2.09	1.95[c]	1.99	1.98
Minimum	1.71	1.70	1.72	1.72	1.58	1.23	1.34	1.41
Maximum	3.40	3.18	3.25	3.21	2.77	2.48	2.51	2.49

[a] A Bayesian model combining procedure in which two models, such as a FP and a TVP mode., are combined with weights based on their posterior odds ratio – see eq. (3) in text.

[b] A Bayesian predictive model selection rule in which the model with the minimum expected loss is used to forecast at each time period. Choice is made among the forecasts of three models, a FPM, a TVPM, and a combined model, using the predictive loss structure in table 3.

[c] Minimal value in row.

were smaller for the FPM model, while when they favor the TVPM, eleven of fourteen favored TVPMs had lower *RMSE*'s. In the case of the pooled ARLI models, the association is perfect, while for the pooled ARLI/WI model the association is somewhat weaker.

In tables 8a and 8b, we present forecasting results relating to two combining methods. One, COMB$_1$, is an average of the predictive means of FP and TVP models with posterior probabilities or posterior odds involved in the weights – see (3) above. The other, COMB$_2$, is the forecast provided by the Bayesian predictive loss model selection rule described in connection with the predictive loss structures in table 3 of section 3. We see from tables 8a and 8b that, while pooling resulted in improved performance of the combining procedures, the combining procedures did not in general produce large or even moderately

Table 8b

Root mean squared errors of one-year-ahead forecasts of annual real output growth rates, 1974–87, using model combining approaches based on posterior odds and a predictive model selection rule.

	Unpooled ARLI/WI				Pooled ARLI/WI			
Country	FP	TVP	COMB$_1$[a]	COMB$_2$[b]	FHP	TVHP	COMB$_1$[a]	COMB$_2$[b]
Australia	3.45	3.23	3.29	3.13[c]	2.27	2.27	2.26[c]	2.27
Austria	2.84	2.13[c]	2.22	2.27	1.80	1.70[c]	1.73	1.75
Belgium	1.54[c]	1.78	1.74	1.74	1.94	1.84[c]	1.89	1.91
Canada	2.52[c]	2.78	2.66	2.61	1.74	1.78	1.74	1.73[c]
Denmark	3.21	2.68[c]	2.86	2.89	2.39	2.14[c]	2.25	2.22
Finland	3.32	3.16[c]	3.22	3.25	1.69	1.58[c]	1.62	1.58[c]
France	2.27	1.74[c]	2.00	2.03	1.62	1.43[c]	1.52	1.55
Germany	2.82	1.98[c]	2.01	1.99	1.56	1.23[c]	1.39	1.43
Ireland	2.68	3.15	3.03	2.55[c]	2.48	2.53	2.50	2.45[c]
Italy	3.68	2.51[c]	2.75	3.03	2.14	1.98[c]	2.05	2.12
Japan	3.46	3.32[c]	3.40	3.36	2.23[c]	2.30	2.26	2.30
Netherlands	1.67	1.74	1.71	1.66[c]	1.60	1.39[c]	1.51	1.53
Norway	1.84[c]	2.56	2.22	1.87	1.44	1.55	1.48	1.40[c]
Spain	2.28	1.85[c]	1.99	1.98	1.21	1.17	1.16[c]	1.19
Sweden	2.15	1.90[c]	1.92	2.00	1.76	1.50[c]	1.59	1.61
Switzerland	3.15	3.04[c]	3.15	3.17	2.43	2.23[c]	2.32	2.40
UK	1.50	1.39[c]	1.47	1.48	1.92	1.69[c]	1.79	1.79
USA	1.99[c]	2.22	2.10	2.12	2.08	1.94[c]	2.00	2.01
Median	2.60	2.37	2.22	2.20[c]	1.86	1.74[c]	1.77	1.77
Minimum	1.50	1.39	1.47	1.48	1.21	1.17	1.16	1.19
Maximum	3.68	3.32	3.40	3.36	2.48	2.53	2.50	2.45

[a, b, c] See footnotes of table 8a.

large decreases in median *RMSE*'s. Further there was not much difference in the performance of the two combining procedures. The pooled TVHP forecasts tended to perform best and also better in terms of median *RMSE* than equally-weighted averages of FP and TVP forecasts (results available on request).

Tables 9a and 9b present the results of employing three regression combining techniques. It is seen that the regression combining technique A, involving a zero intercept and free regression coefficients, produces the lowest median *RMSE*'s in all but one case, with values of 2.24 and 2.21 for the unpooled cases (see table 9a) and 2.25 and 1.89 for the pooled cases (see table 9b). These values are not far different from corresponding *RMSE*'s for the Bayesian COMB$_1$ method, namely 2.35, 2.22, 1.99, and 1.77. However, the regression combining methods A, B, and C produced rather different results. Further none of the regression combining techniques produced a median *RMSE* lower than that associated with the pooled FHP and TVHP ARLI/WI models, namely 1.86 and 1.74.

Table 9a

RMSE's of combined annual output growth rate forecasts of FP and TVP models using regression methods, 1974–87.[a]

	Combined unpooled ARLI forecasts[b]			Combined unpooled ARLI WI forecasts[b]		
Country	A	B	C	A	B	C
Australia	2.40[c]	3.43	2.49	2.61	3.26	2.55[c]
Austria	1.86[c]	2.12	2.31	1.70[c]	2.03	2.17
Belgium	1.71[c]	1.79	3.08	1.85[c]	1.85[c]	2.48
Canada	2.54[c]	3.46	2.73	3.22	3.22	2.90[c]
Denmark	2.02[c]	2.70	2.44	2.35[c]	2.70	2.67
Finland	2.43	3.07	2.39[c]	2.46	3.46	2.40[c]
France	3.97[c]	4.19	4.25	1.95[c]	1.97	2.65
Germany	1.59[c]	1.95	1.62	1.59[c]	2.04	1.60
Ireland	3.03	2.83	2.80[c]	3.45	3.09	3.07[c]
Italy	2.23[c]	3.20	2.75	2.84[c]	3.51	2.86
Japan	3.51[c]	3.55	4.06	4.20	4.55	3.88[c]
Netherlands	1.54[c]	2.06	1.94	1.76[c]	2.51	2.15
Norway	2.40	2.92	1.96[c]	2.59	3.02	1.94[c]
Spain	1.41[c]	2.42	2.65	1.48[c]	2.99	2.87
Sweden	1.75[c]	1.95	1.77	1.57[c]	1.96	1.85
Switzerland	2.96[c]	3.90	3.49	2.95[c]	3.66	3.37
UK	2.10[c]	2.25	2.38	1.41[c]	1.51	2.11
USA	2.24[c]	2.50	2.34	2.08[c]	2.30	2.18
Median	2.24[c]	2.77	2.46	2.21[c]	2.84	2.51
Minimum	1.41	1.79	1.62	1.41	1.51	1.60
Maximum	3.97	4.19	4.25	4.20	4.55	3.88

[a] Regression weights for the 1974 regression combined forecast were obtained by regressing actual outcomes on FP and TVP models' forecasts for the years 1965–73. Weights for subsequent years were obtained by updating the 1974 weights year-by-year.

[b] Methods A, B, and C involve regression of actual outcomes on FP and TVP models' forecasts under the following conditions. For regression A, the intercept is zero and the regression coefficients are unrestricted. For regression B, the intercept is zero and the regression coefficients are constrained to add to one. For regression C, the intercept and the coefficients are unrestricted. See Granger and Ramanthan (1984) for further discussion of these methods.

[c] Minimal value in row.

5. Summary and concluding remarks

From our empirical results, we have reached the following conclusions:

(1) While there is some evidence in terms of posterior odds values for time-varying parameter models in the case of unpooled models and pooled ARLI models, for the pooled ARLI/WI models the evidence is not as decisive with respect to models with time-varying hyperparameters. However, use of TVP models led to reductions in *RMSE*'s of forecast in many cases.

Table 9b

RMSE's of combined annual output growth rate forecasts of FHP and TVHP models using regression methods, 1974–87.[a]

	Combined pooled ARLI forecasts[b]			Combined pooled ARLI/WI forecasts[b]		
Country	A	B	C	A	B	C
Australia	2.69	2.54	2.49[c]	2.64	2.61	2.40[c]
Austria	1.86[c]	1.96	2.00	1.68[c]	1.77	1.91
Belgium	2.25	2.06[c]	2.62	1.90[c]	1.92	2.61
Canada	2.36	2.30	2.21[c]	1.97[c]	2.16	2.25
Denmark	2.26[c]	2.98	2.51	2.15[c]	2.39	2.18
Finland	1.82	1.70[c]	1.79	1.73	1.74	1.72[c]
France	2.62	2.25	1.93[c]	1.46[c]	1.48	2.15
Germany	1.69	1.88	1.63[c]	1.08[c]	1.10	1.16
Ireland	2.66	2.36[c]	2.78	2.67	2.66	2.37[c]
Italy	2.55[c]	2.67	2.66	2.04	2.03[c]	2.31
Japan	2.86	2.56[c]	4.19	3.47[c]	3.48	3.82
Netherlands	1.50	1.47[c]	1.59	1.89	1.84[c]	2.21
Norway	1.90	1.88[c]	2.01	1.60	1.57[c]	1.95
Spain	1.69[c]	1.69[c]	2.54	1.19	1.18[c]	1.63
Sweden	1.94	1.83[c]	2.00	1.63	1.56[c]	1.78
Switzerland	2.46[c]	2.58	2.56	2.47	2.44[c]	2.66
UK	1.90[c]	2.08	2.11	1.84[c]	2.08	1.95
USA	2.70	2.65[c]	2.83	2.25[c]	2.27	2.41
Median	2.25	2.17[c]	2.35	1.89[c]	1.98	2.20
Minimum	1.50	1.47	1.59	1.08	1.10	1.16
Maximum	2.86	2.98	4.19	3.47	3.48	3.82

[a,b,c] See footnotes of table 9a.

(2) The use of various Bayesian and non-Bayesian forecast combining methods did not produce much, if any, reduction in *RMSE*'s vis-à-vis those for uncombined forecasts. Different Bayesian combining methods did not produce very different results, whereas different regression combining techniques did and were not as good as certain Bayesian procedures.

(3) Several new procedures for selecting forecasts and for combining forecasts were developed. Further investigation of their theoretical properties and more experience in using them would be desirable. Also, posterior odds for fixed versus time-varying parameter models have been derived, computed and shown to be useful in evaluating alternative models.

(4) As shown graphically in table 6c, the pooling techniques that we utilized in connection with fixed and time-varying parameter models led to substantial reductions in forecast *RMSE*'s – see also tables 5, 6a, and 6b.

Finally, our forecasting methods and models have been applied using data for a period, 1951–87, which includes the Korean War, the Vietnamese War, price and credit controls, strikes, oil crises, exchange rate policy changes, major changes in U.S. monetary policy, etc. That our models and methods worked to produce good forecasting results without any special adjustments for these and other major events is indeed satisfying. It appears that our leading indicator stock market and money variables reflect information about major events before output adjustments take place. Thus the output growth rate responds with a lag to changes in lagged indicator variables and such variables reflect anticipated systematic and nonsystematic shocks, which produce oscillatory behavior of output growth rates, a topic to be explored further in future research.

Appendix A: Posterior odds for fixed vs. time-varying parameter models

Denote the hypothesis of fixed parameters by H_0 and that of time-varying parameters by H_1. Then, as is well-known, the posterior odds, denoted by K, is given by

$$K = \Pr(H_0 | y, Z, I_0)/\Pr(H_1 | y, Z, I_1)$$

$$= \frac{\Pr(H_0 | I_0)}{\Pr(H_1 | I_1)} \frac{f(y | H_0, Z, I_0)}{f(y | H_1, Z, I_1)}, \tag{A.1}$$

where $\Pr(H_0 | I_0)/\Pr(H_1 | I_1)$ is the prior odds, taken equal to one in our calculations, and the second factor on the right-hand side of (A.1) is the Bayes factor. Here y = data, Z = data on the exogenous or input variables, and I_0 and I_1 denote prior information under hypotheses H_0 and H_1, respectively.

In our calculations, we employed the data 1951–72 to obtain posterior densities for the parameters (see appendix B) which were employed as prior densities in computing the posterior odds incorporating data for 1973. The Bayes factors were computed readily by exploiting the fact that the marginal density of the observation vector $y' = (y_{t+1}, y_{t+2}, \dots, y_T)$ can be expressed as follows:

$$f(y | H_i, Z, I_i)$$

$$= p(y_{t+1}, y_{t+2}, \dots, y_T | Z, H_i, I_i)$$

$$= p(y_{t+1} | Z_{t+1}, H_i, I_i) p(y_{t+2} | y_{t+1} Z_{t+1}, Z_{t+2}, H_i, I_i)$$

$$\times \dots \times p(y_T | y_{t+1}, y_{t+2}, \dots, y_{T-1} Z_{t+1}, \dots, Z_T, H_i, I_i), \tag{A.2}$$

for $i = 0, 1$, with each factor on the right-hand side of (A.2) in the univariate Student-t form. Thus the computation of the Bayes factor in (A.1), period-by-period, is relatively straightforward. Explicitly,

$$P(y_\tau \mid y_{t+1}, \ldots, y_{\tau-1}, Z_{t+1}, \ldots, Z_\tau, H_i, I_i)$$

$$= a_{i\tau}\{1 + h_{i\tau}(y_\tau - \hat{y}_{i\tau\mid\tau-1})^2/v_{\tau-1}\}^{-(v_{\tau-1}+1)/2}.$$

Then,

$$f(y \mid H_i, Z, I_i) = \prod_{\tau=t+1}^{T} a_{i\tau}\{1 + h_{i\tau}(y_\tau - \hat{y}_{i\tau\mid\tau-1})^2/v_{\tau-1}\}^{-(v_{\tau-1}+1)/2}.$$

Appendix B: Updating procedures for predictive density of time-varying parameter models

B.1. Unpooled case

The model for the ith country, where for convenience we drop the subscript i, is

$$y_t = x_t'\beta_t + u_t, \qquad u_t\text{'s NID}(0, \sigma^2),$$

$$\beta_t = \beta_{t-1} + \varepsilon_t, \qquad \varepsilon_t\text{'s NID}(0, \phi\sigma^2 I),$$

for $t = 1, 2, \ldots$. Our distributions on the initial state, with I_0 denoting initial information, are given by

$$p(\beta_0 \mid \sigma, I_0) \sim \text{N}(0, 10^6 \times \sigma^2 I),$$

$$p(\sigma \mid I_0) \sim \text{IG}(v_0, s_0^2),$$

where v_0 and s_0^2 have been assigned small values in the inverted gamma density for σ.

Then the recursive algorithm for updating parameters is

Posterior mean:

$$\hat{\beta}_t = \hat{\beta}_{t-1} + V_{t-1}x_t(y_t - x_t'\hat{\beta}_{t-1})/(1 + x_t'V_{t-1}x_t),$$

Residual SS:

$$\nu_t s_t^2 = \nu_{t-1} s_{t-1}^2 + (y_t - \boldsymbol{x}_t'\hat{\boldsymbol{\beta}}_t)^2 + (\hat{\boldsymbol{\beta}}_t - \hat{\boldsymbol{\beta}}_{t-1})' V_{t-1}^{-1} (\hat{\boldsymbol{\beta}}_t - \hat{\boldsymbol{\beta}}_{t-1}),$$

Degrees of freedom:

$$\nu_t = \nu_{t-1} + 1,$$

where

$$V_t \equiv \sigma^{-2} \operatorname{cov}(\boldsymbol{\beta}_{t+1} \mid \sigma, D_t) = C_t + \phi I,$$

$$C_t \equiv \sigma^{-2} \operatorname{cov}(\boldsymbol{\beta}_t \mid \sigma, D_t) = V_{t-1} - V_{t-1}\boldsymbol{x}_t\boldsymbol{x}_t' V_{t-1}/(1 + \boldsymbol{x}_t' V_{t-1} \boldsymbol{x}_t),$$

$$D_t = \text{information available at } t.$$

At time t, the predictive density for y_{t+1} is

$$p(y_{t+1} \mid \boldsymbol{x}_{t+1}, D_t, I_0) = \iint p(y_{t+1} \mid \boldsymbol{x}_{t+1}, \boldsymbol{\beta}_{t+1}, \sigma)$$

$$\times\, p(\boldsymbol{\beta}_{t+1}, \sigma \mid D_t, I_0)\,\mathrm{d}\boldsymbol{\beta}_{t+1}\,\mathrm{d}\sigma,$$

$$\propto \left\{ \nu_t + \left(\frac{y_{t+1} - \boldsymbol{x}_{t+1}'\hat{\boldsymbol{\beta}}_t}{s_t a_t} \right)^2 \right\}^{-(\nu_t+1)/2},$$

and thus

$$t_{\nu_t} = (y_{t+1} - \boldsymbol{x}_{t+1}'\hat{\boldsymbol{\beta}}_t)/s_t a_t,$$

has a univariate Student-t density with ν_t degrees of freedom where $a_t^2 = 1 + \boldsymbol{x}_{t+1}' V_t \boldsymbol{x}_{t+1}$.

B.2. A pooled case

Using the notation introduced in section 4, the model for the observation vector $\boldsymbol{y}_t' = (y_{1t}, y_{2t}, \dots, y_{mt})$ is

$$\boldsymbol{y}_t = X_t \boldsymbol{\beta}_t + \boldsymbol{u}_t, \qquad \boldsymbol{u}_t \sim \mathrm{NID}(0, \sigma^2 I_m), \tag{B.1a}$$

$$\boldsymbol{\beta}_t = B\theta_t + \varepsilon_t, \qquad \varepsilon_t \sim \mathrm{NID}(0, \sigma^2 \Omega), \tag{B.1b}$$

$$\theta_t = \theta_{t-1} + \boldsymbol{\eta}_t, \qquad \boldsymbol{\eta}_t \sim \mathrm{NID}(0, \phi\sigma^2 I_k), \tag{B.1c}$$

with $\boldsymbol{u}_t$, ε_t, and $\boldsymbol{\eta}_t$ assumed independently distributed. Then the following densities are relevant: The joint density for (y_t, β_t, θ_t) given $(\sigma, \phi, \Omega, \theta_{t-1})$ is

$$p(y_t, \beta_t, \theta_t \mid \theta_{t-1}, \sigma, X_t, \phi, \Omega)$$

$$= p(y_t \mid \beta_t, \sigma, X_t) p(\beta_t \mid \theta_t, \sigma, \Omega) p(\theta_t \mid \theta_{t-1}, \sigma, \phi)$$

$$\propto \sigma^{-m} \exp\{-(y_t - X_t\beta_t)'(y_t - X_t\beta_t)/2\sigma^2\}$$

$$\times \sigma^{-mk} |\Omega|^{-1/2} \exp\{-(\beta_t - B\theta_t)'\Omega^{-1}(\beta_t - B\theta_t)/2\sigma^2\}$$

$$\times (\phi\sigma^2)^{-k/2} \exp\{-(\theta_t - \theta_{t-1})'(\theta_t - \theta_{t-1})/2\phi\sigma^2\}. \quad \text{(B.2)}$$

The posterior joint density for β_t, θ_t, σ given y_t, X_t, D_{t-1}, ϕ, Ω, where $D_{t-1} \equiv$ information available at time $t-1$ is

$$p(\beta_t, \theta_t, \sigma \mid y_t, X_t, D_{t-1}, \phi, \Omega) \propto p(y_t \mid \beta_t, \theta_t, \sigma, X_t, D_{t-1}, \phi, \Omega)$$

$$\times p(\beta_t, \theta_t, \sigma \mid X_t, D_{t-1}, \phi, \Omega). \quad \text{(B.3)}$$

On substituting from (B.1b) in (B.1a), we have

$$y_t = X_t(B\theta_t + \varepsilon_t) + \boldsymbol{u}_t = X_t B\theta_t + X_t\varepsilon_t + \boldsymbol{u}_t. \quad \text{(B.4)}$$

Then, using (B.4) and (B.1c), with a normal–IG density for (θ_1, σ), namely $p(\theta_1, \sigma \mid D_0, \phi) = p(\theta_1 \mid \sigma, D_0, \phi) p(\sigma \mid D_0)$ with the first factor $\mathrm{N}(\theta_0, \sigma^2 V_0)$ and the second $\mathrm{IG}(\nu_0, s_0^2)$, we obtain the following joint posterior density for (θ_t, σ):

$$p(\theta_t, \sigma \mid D_t, \phi, \Omega) \sim \mathrm{N}(\hat{\theta}_{t|t}, \sigma^2 C_t) \times \mathrm{IG}(\nu_t, s_t^2) \quad \text{(B.5)}$$

and

$$p(\theta_t \mid D_t, \phi, \Omega) = c_t |C_t|^{-1/2} \{\nu_t s_t^2 + (\theta_t - \hat{\theta}_{t|t})'$$

$$\times C_t^{-1}(\theta_t - \hat{\theta}_{t|t})\}^{-(k+\nu_t)/2}, \quad \text{(B.6)}$$

where

$$c_t = \Gamma\left(\frac{\nu_t + k}{2}\right)(\nu_t s_t^2)^{\nu_t/2} / \pi^{k/2} \Gamma(\nu_t/2),$$

$$\hat{\theta}_{t|t} = \{V_{t-1}^{-1} + B' X_t' M_t X_t B\}^{-1} \{V_{t-1}^{-1} \hat{\theta}_{t-1|t-1} + B' X_t' M_t y_t\},$$

$$M_t \equiv \{I_m + X_t'\Omega X_t\}^{-1},$$

$$C_t \equiv \sigma^{-2}\operatorname{var}(\theta_t | D_t) = \{V_{t-1}^{-1} + B'X_t'M_tX_tB\}^{-1},$$

$$V_t = C_t + \phi I,$$

$$\nu_t = \nu_{t-1} + m_t,$$

$$\nu_t s_t^2 = \nu_{t-1} s_{t-1}^2 + (y_t - X_tB\hat{\theta}_{t|t})' M_t (y_t - X_tB\hat{\theta}_{t|t})$$
$$+ (\hat{\theta}_{t|t} - \hat{\theta}_{t-1|t-1})' V_{t-1}^{-1} (\theta_{t|t} - \hat{\theta}_{t-1|t-1}).$$

The marginal density for $\boldsymbol{\beta}_t$ given D_t, σ, and Ω is obtained from (B.5) in conjunction with (B.1b), namely,

$$p(\boldsymbol{\beta}_t, \sigma | D_{t-1}, \phi, \Omega) = \mathrm{N}[B\hat{\theta}_{t-1|t-1}, \sigma^2(BV_{t-1}B' + \Omega)]$$
$$\times \mathrm{IG}(\nu_{t-1}, s_{t-1}^2). \quad \text{(B.7)}$$

On combining (B.7) with the likelihood function for (B.1a) via Bayes' Theorem and integrating out σ, the result is

$$p(\boldsymbol{\beta}_t | D_t, \phi, \Omega) = \frac{\Gamma[(\nu_t + mk)/2]}{\pi^{mk/2}\Gamma(\nu_t/2)} (\nu_t s_t^2)^{\nu_t/2} |R_t|^{-1/2}$$
$$\times \{\nu_t s_t^2 + (\boldsymbol{\beta}_t - \hat{\boldsymbol{\beta}}_{t|t})' R_t^{-1} (\boldsymbol{\beta}_t - \hat{\boldsymbol{\beta}}_{t|t})\}^{-(mk+\nu_t)/2}, \quad \text{(B.8)}$$

where, in addition to quantities defined above,

$$\hat{\boldsymbol{\beta}}_{t|t} = \{X_t'X_t + A_t\}^{-1}\{X_t'y_t + A_tB\hat{\theta}_{t-1|t-1}\}, \quad \text{(B.9)}$$

$$A_t \equiv (BV_{t-1}B' + \Omega)^{-1}, \quad R_t \equiv \sigma^{-2}\operatorname{var}(\boldsymbol{\beta}_t | D_t) = (X_t'X_t + A_t)^{-1}.$$

Finally, the predictive density for y_{t+1} is obtained by integrating the joint density for y_{t+1} and σ given X_{t+1}, D_t with respect to σ. We have

$$p(y_{t+1}, \sigma | X_{t+1}, D_t) = \int p(y_{t+1} | X_{t+1}, \boldsymbol{\beta}_{t+1}, \sigma)$$
$$\times p(\boldsymbol{\beta}_{t+1}, \sigma | D_t) \mathrm{d}\boldsymbol{\beta}_{t+1}, \quad \text{(B.10)}$$

and on integrating (B.10) over σ, the result is

$$p(y_{t+1} \mid X_{t+1}, D_t) = a_t \mid P_t \mid^{-1/2} \{\nu_t s_t^2 + (y_{t+1} - \hat{y}_{t+1})' \times P_t^{-1}(y_{t+1} - \hat{y}_{t+1})\}^{-(m+\nu_t)/2}, \tag{B.11}$$

where

$$a_t = \Gamma\left(\frac{\nu_t + m}{2}\right)(\nu_t s_t^2)^{\nu_t/2} / \pi^{m/2}\,\Gamma(\nu_t/2),$$

$$\hat{y}_{t+1} = X_{t+1} B \hat{\theta}_{t|t},$$

$$P_t \equiv \sigma^{-2}\,\mathrm{var}(y_{t+1} \mid D_t) = X_{t+1} A_{t+1}^{-1} X'_{t+1} + I_m.$$

It is seen that (B.11) is in the form of a multivariate Student-t density, and thus the marginal density for an element of y_{t+1}, say $y_{i,t+1}$, has a univariate Student-t density [see, e.g., Zellner (1971, app. B.2)].

B.3. Values for hyperparameters, (ϕ, Σ)

The values of (ϕ, Σ), with $\Omega = I_n \otimes \Sigma$, have been selected to minimize *RMSE*'s of one-year-ahead forecasts for the pre-forecast period 1969–73. (The values of (ϕ, Σ) presented below are for the models with variables in percentage points.) Subsequent calculations for 1974 and following years have been conditioned on these values.

(1) Unpooled TVP models:

$$\begin{cases} \phi = 0.3 & \text{for ARLI,} \\ \phi = 0.2 & \text{for ARLI/WI;} \end{cases}$$

(2) Pooled TVP models:

$$\begin{cases} \phi = 0.1 \quad \text{for ARLI and ARLI/WI,} \\ \Sigma = \begin{bmatrix} 0.1 & & & 0 \\ & 0.002 & & \\ & & \ddots & \\ 0 & & & 0.002 \end{bmatrix}; \end{cases}$$

(3) Pooled FP models:

$$\left\{ \Sigma = \begin{bmatrix} 0.1 & & & 0 \\ & 0.002 & & \\ & & \ddots & \\ 0 & & & 0.002 \end{bmatrix}; \right.$$

(4) The equation for world real income growth rate: $\phi = 0.5$.

Appendix C: Updating procedures for predictive density of a pooled FHP model

Using the notation introduced in section 4, the model for the observation vector $y_t' = (y_{1t}, y_{2t}, \ldots, y_{mt})$ is

$$y_t = X_t \beta_t + u_t, \qquad u_t \sim \text{NID}(0, \sigma^2 I_m),$$

$$\beta_t = B\theta + \varepsilon_t, \qquad \varepsilon_t \sim \text{NID}(0, \sigma^2 \Omega).$$

These will be equivalent to model (B.1a)–(B.1c) when ϕ is set to zero. Therefore the same updating procedures can be used with ϕ set to zero.

References

Bates, J.M. and C.W.J. Granger, 1969, The combination of forecasts, Operations Research Quarterly 20, 451–468.

Blattberg, R.C. and E.I. George, 1991, Shrinkage estimation of price and promotional elasticities: Seemingly unrelated equations, Journal of the American Statistical Association 86, 304–315.

Clemen, R.T., 1989, Combining forecasts: A review and annotated bibliography, International Journal of Forecasting 5, 559–583.

Diebold, F.X., 1990a, Forecast combination and encompassing: Reconciliation of two divergent literatures, International Journal of Forecasting 5, 589–592.

Diebold, F.X., 1990b, A note on Bayesian forecast combination procedures, in: A. Westlund and P. Haski, eds., Economic structural change: Analysis and forecasting (Springer-Verlag, New York, NY) 1–8.

Diebold, F.X. and P. Pauly, 1987, Structural change and the combination of forecasts, Journal of Forecasting 6, 21–40.

Garcia-Ferrer, A., R.A. Highfield, F.C. Palm, and A. Zellner, 1987, Macroeconomic forecasting using pooled international data, Journal of Business and Economic Statistics 5, 53–67.

Geisel, M.S., 1975., Bayesian comparisons of simple macroeconomic models, in: S.E. Fienberg and A. Zellner, eds., Studies in Bayesian econometrics and statistics in honor of Leonard J. Savage (North-Holland, Amsterdam) 227–256.

Granger, C.W.J. and R. Ramanathan, 1984, Improved methods of combining forecasts, Journal of Forecasting 3, 197–204.

Hong, C., 1989, Forecasting real output growth rates and cyclical properties of models: A Bayesian approach, Ph.D. dissertation (University of Chicago, Chicago, IL).
Jeffreys, H., 1967, Theory of probability (Oxford University Press, London).
Leamer, E.E., 1978, Specification searches (Wiley, New York, NY).
Lesage, J.P., 1989, Forecasting turning points in metropolitan employment growth rates using Bayesian exponentially weighted autoregression, time-varying parameter and pooling techniques, Journal of Business and Economic Statistics, forthcoming.
Lesage, J.P. and M. Magura, 1990, Using Bayesian techniques for data pooling in regional payroll forecasting, Journal of Business and Economic Statistics 8, 127–136.
Lindley, D. and A.F.M. Smith, 1972, Bayes estimates for the linear model, Journal of the Royal Statistical Society B 34, 1–41.
Min, C., 1990, Forecasting models with time-varying parameter, pooling and Bayesian model-combining approaches: Economic theory and application, Thesis proposal paper (University of Chicago, Chicago, IL).
Nelson, C.R., 1972, The predictive performance of the FRB–MIT–Penn model of the U.S. economy, American Economic Review 62, 902–917.
Palm, F.C. and A. Zellner, 1990, To combine or not to combine forecasts?, Paper in progress (H.G.B. Alexander Research Foundation, University of Chicago, IL).
Reid, D.J., 1968, Combining three estimates of gross domestic product, Economica 35, 431–444.
Reid, D.J., 1969, A comparative study of time series prediction techniques on economic data, Ph.D. dissertation (University of Nottingham, Nottingham).
Rossi, P.E., 1983, Specification and analysis of econometric production models, Ph.D. dissertation (University of Chicago, Chicago, IL).
Rossi, P.E., 1985, Comparison of alternative functional forms in production, Journal of Econometrics 30, 345–361.
Schwarz, G., 1978, Estimating the dimension of a model, Annals of Statistics 6, 441–464.
Swamy, P.A.V.B., 1971, Statistical inference in random coefficient models (Springer-Verlag, Berlin).
Winkler, R.L., 1981, Combining probability distributions from dependent information sources, Management Science, 27, 479–488.
Zellner, A., 1971, An introduction to Bayesian inference in econometrics (Wiley, New York, NY).
Zellner, A., 1984, Basic issues in econometrics (University of Chicago Press, Chicago, IL).
Zellner, A., 1989, Bayesian and non-Bayesian methods for combining models and forecasts, Manuscript (H.G.B. Alexander Research Foundation, University of Chicago, IL).
Zellner, A. and C. Hong, 1989, Forecasting international growth rates using Bayesian shrinkage and other procedures, Journal of Econometrics 40, 183–202.
Zellner, A. and L.A. Manas-Anton, 1986, Macroeconomic theory and international macroeconomic forecasting, Manuscript (H.G.B. Alexander Research Foundation, University of Chicago, Chicago, IL).
Zellner, A., C. Hong, and G.M. Gulati, 1990, Turning points in economic time series, loss structures and Bayesian forecasting, in: S. Geisser, J.S. Hodges, S.J. Press, and A. Zellner, eds., Bayesian and likelihood methods in statistics and econometrics (North-Holland, Amsterdam) 371–393.
Zellner, A., C. Hong, and C. Min, 1991, Forecasting turning points in international output growth rates using Bayesian exponentially weighted autoregression, time varying parameter and pooling techniques, Journal of Econometrics, 49, 275–304.
Zellner, A., D.S. Huang, and L.C. Chau, 1965, Further analysis of the short-run consumption function with emphasis on the role of liquid assets, Econometrica 33, 571–581.

Name index

Economists of the Twentieth Century

Monetarism and Macroeconomic Policy
Thomas Mayer

Studies in Fiscal Federalism
Wallace E. Oates

The World Economy in Perspective
Essays in International Trade and European Integration
Herbert Giersch

Towards a New Economics
Critical Essays on Ecology, Distribution and Other Themes
Kenneth E. Boulding

Studies in Positive and Normative Economics
Martin J. Bailey

The Collected Essays of Richard E. Quandt (2 volumes)
Richard E. Quandt

International Trade Theory and Policy
Selected Essays of W. Max Corden
W. Max Corden

Organization and Technology in Capitalist Development
William Lazonick

Studies in Human Capital
Collected Essays of Jacob Mincer, Volume 1
Jacob Mincer

Studies in Labor Supply
Collected Essays of Jacob Mincer, Volume 2
Jacob Mincer

Macroeconomics and Economic Policy
The Selected Essays of Assar Lindbeck
Volume I
Assar Lindbeck

The Welfare State
The Selected Essays of Assar Lindbeck
Volume II
Assar Lindbeck

Classical Economics, Public Expenditure and Growth
Walter Eltis

Money, Interest Rates and Inflation
Frederic S. Mishkin

The Public Choice Approach to Politics
Dennis C. Mueller

The Liberal Economic Order
Volume I Essays on International Economics
Volume II Money, Cycles and Related Themes
Gottfried Haberler
Edited by Anthony Y.C. Koo

Economic Growth and Business Cycles
Prices and the Process of Cyclical Development
Paolo Sylos Labini

International Adjustment, Money and Trade
Theory and Measurement for Economic Policy
Volume I
Herbert G. Grubel

International Capital and Service Flows
Theory and Measurement for Economic Policy
Volume II
Herbert G. Grubel

Unintended Effects of Government Policies
Theory and Measurement for Economic Policy
Volume III
Herbert G. Grubel

The Economics of Competitive Enterprise
Selected Essays of P.W.S. Andrews
Edited by Frederic S. Lee and Peter E. Earl

The Repressed Economy
Causes, Consequences, Reform
Deepak Lal

Economic Theory and Market Socialism
Selected Essays of Oskar Lange
Edited by Tadeusz Kowalik

Trade, Development and Political Economy
Selected Essays of Ronald Findlay
Ronald Findlay

General Equilibrium Theory
The Collected Essays of Takashi Negishi
Volume I
Takashi Negishi

The History of Economics
The Collected Essays of Takashi Negishi
Volume II
Takashi Negishi

Studies in Econometric Theory
The Collected Essays of Takeshi Amemiya
Takeshi Amemiya

Exchange Rates and the Monetary System
Selected Essays of Peter B. Kenen
Peter B. Kenen

Econometric Methods and Applications (2 volumes)
G.S. Maddala

National Accounting and Economic Theory
The Collected Papers of Dan Usher, Volume I
Dan Usher

Welfare Economics and Public Finance
The Collected Papers of Dan Usher, Volume II
Dan Usher

Economic Theory and Capitalist Society
The Selected Essays of Shigeto Tsuru, Volume I
Shigeto Tsuru

Methodology, Money and the Firm
The Collected Essays of D.P. O'Brien
(2 volumes)
D.P. O'Brien

Economic Theory and Financial Policy
The Selected Essays of Jacques J. Polak
(2 volumes)
Jacques J. Polak

Sturdy Econometrics
Edward E. Leamer

The Emergence of Economic Ideas
Essays in the History of Economics
Nathan Rosenberg

Productivity Change, Public Goods and Transaction Costs
Essays at the Boundaries of Microeconomics
Yoram Barzel

Reflections on Economic Development
The Selected Essays of Michael P. Todaro
Michael P. Todaro

The Economic Development of Modern Japan
The Selected Essays of Shigeto Tsuru
Volume II
Shigeto Tsuru

Money, Credit and Policy
Allan H. Meltzer

Macroeconomics and Monetary Theory
The Selected Essays of Meghnad Desai
Volume I
Meghnad Desai

Poverty, Famine and Economic Development
The Selected Essays of Meghnad Desai
Volume II
Meghnad Desai

Explaining the Economic Performance of Nations
Essays in Time and Space
Angus Maddison

Economic Doctrine and Method
Selected Papers of R.W. Clower
Robert W. Clower

Economic Theory and Reality
Selected Essays on their Disparities and Reconciliation
Tibor Scitovsky

Doing Economic Research
Essays on the Applied Methodology of Economics
Thomas Mayer

Institutions and Development Strategies
The Selected Essays of Irma Adelman
Volume I
Irma Adelman

Dynamics and Income Distribution
The Selected Essays of Irma Adelman
Volume II
Irma Adelman

The Economics of Growth and Development
Selected Essays of A.P. Thirlwall
A.P. Thirlwall

Theoretical and Applied Econometrics
The Selected Papers of Phoebus J. Dhrymes
Phoebus J. Dhrymes

Innovation, Technology and the Economy
The Selected Essays of Edwin Mansfield
(2 volumes)
Edwin Mansfield

Economic Theory and Policy in Context
The Selected Essays of R.D. Collison Black
R.D. Collison Black

Location Economics
Theoretical Underpinnings and Applications
Melvin L. Greenhut

Spatial Microeconomics
Theoretical Underpinnings and Applications
Melvin L. Greenhut

Capitalism, Socialism and Post-Keynesianism
Selected Essays of G.C. Harcourt
G.C. Harcourt

Time Series Analysis and Macroeconometric Modelling
The Collected Papers of Kenneth F. Wallis
Kenneth F. Wallis

Foundations of Modern Econometrics
The Selected Essays of Ragnar Frisch
(2 volumes)
Edited by Olav Bjerkholt

Growth, the Environment and the Distribution of Incomes
Essays by a Sceptical Optimist
Wilfred Beckerman

The Economics of Environmental Regulation
Wallace E. Oates

Econometrics, Macroeconomics and Economic Policy
Selected Papers of Carl F. Christ
Carl F. Christ

Strategic Approaches to the International Economy
Selected Essays of Koichi Hamada
Koichi Hamada

Economic Analysis and Political Ideology
The Selected Essays of Karl Brunner
Volume One
Edited by Thomas Lys

Growth Theory and Technical Change
The Selected Essays of Ryuzo Sato
Volume One
Ryuzo Sato

Industrialization, Inequality and Economic Growth
Jeffrey G. Williamson

Economic Theory and Public Decisions
Selected Essays of Robert Dorfman
Robert Dorfman

The Logic of Action One
Method, Money and the Austrian School
Murray N. Rothbard

The Logic of Action Two
Applications and Criticism from the Austrian School
Murray N. Rothbard

Bayesian Analysis in Econometrics and Statistics
The Zellner View and Papers
Arnold Zellner

On the Foundations of Monopolistic Competition and Economic Geography
The Selected Essays of B. Curtis Eaton and Richard G. Lipsey
B. Curtis Eaton and Richard G. Lipsey

Microeconomics, Growth and Political Economy
The Selected Essays of Richard G. Lipsey
Volume One
Richard G. Lipsey

Macroeconomic Theory and Policy
The Selected Essays of Richard G. Lipsey
Volume Two
Richard G. Lipsey

Employment, Labor Unions and Wages
The Collected Essays of Orley Ashenfelter
Volume One
Edited by Kevin F. Hallock

Education, Training and Discrimination
The Collected Essays of Orley Ashenfelter
Volume Two
Edited by Kevin F. Hallock

Economic Institutions and the Demand and Supply of Labour
The Collected Essays of Orley Ashenfelter
Volume Three
Edited by Kevin F. Hallock

Monetary Theory and Monetary Policy
The Selected Essays of Karl Brunner
Volume Two
Edited by Thomas Lys

Macroeconomic Issues from a Keynesian Perspective
Selected Essays of A.P. Thirlwall
Volume Two
A.P. Thirlwall

Economics Against the Grain Volume One
Microeconomics, Industrial Organization and Related Themes
Julian L. Simon

Economics Against the Grain Volume Two
Population Economics, Natural Resources and Related Themes
Julian L. Simon

Money and Macroeconomics
The Selected Essays of David Laidler
David Laidler